Experimental Methods in Polymer Chemistry

Physical Principles and Applications

Experimental Methods in Polymer Chemistry

Physical Principles and Applications

JAN F. RABEK
Department of Polymer Technology,
The Royal Institute of Technology, Stockholm,
Sweden

A Wiley–Interscience Publication

JOHN WILEY & SONS
Chichester . New York . Brisbane . Toronto

British Library Cataloguing in Publication Data:

Rabek, Jan F.
    Experimental methods in polymer chemistry.
    1. Polymers and polymerization—Analysis
    I. Title
547'.84   QD139.P6   79-40511

ISBN 0 471 27604 9

*Typeset by John Wright and Sons Ltd at The Stonebridge Press, Bristol BS4 5NU
and printed at The Pitman Press, Bath, Avon*

To the memory of my father, Professor
T. I. Rabek, founder of the Polymer
Technology Institute (1950–65) at Wrocław
Technical University (Poland), an out-
standing scientist and teacher who intro-
duced me to the study of polymer chemistry.

Preface

Investigations in the field of macromolecular chemistry require an adequate knowledge of physicochemical methods in this branch of science. Also, a basic understanding of instrumentation is essential for the successful interpretation of experimental results. A modern polymer chemist must be acquainted with the principles of these methods and their practical applications, i.e. the physical principles involved in the phenomena under study, the characteristic procedures, including the description of measuring operations, apparatus, and methods of sampling, and the application range of a particular method with the necessary calculations and representation of results. I have found that, in their day-to-day work, many polymer chemists need practical assistance in the form of explicit suggestions without complicated mathematical and theoretical deliberations.

An attempt is made in this book to present information in a simple, lucid, and ordered manner without going into the details of physicochemical theory. It is not the purpose of this book to cover the topic of instrumentation in great specificity or to supply information on the selection of suitable instruments for specific needs. Only the features contributing in important ways to modern experimental techniques are described. Experimental and measurement techniques are briefly discussed, but where practical information is required it is given with sufficient details to facilitate the procedure, e.g. preparation of samples or cleaning of measuring cells. Such information may in some cases be most useful. For the same reason the UV, IR, and NMR spectra of most common solvents used in spectral analysis are included, as well as tables for conversion of the various units used in spectroscopy.

In response to the recognized need for uniformity, IUPAC recommends the use of Système International (SI) units, which are essentially standardized MKS units. In the United States, however, the majority of chemists still prefer to use CGS units together with such common non-metric units as the calorie and the atmosphere. SI units appear in increasing frequency in textbooks and especially in journals; is it therefore important to be familiar with both the CGS and the SI systems of units. Until SI units are sufficiently well established it will often be necessary to convert from one system into another. For that reason both CGS and SI systems are used side by side in the present volume. Chapter 40 is devoted to CGS–SI interconversions and to the elucidation the most frequently used constants, such as the universal gas constant, the Avogadro number, and the Boltzmann constant.

Literature which refers to the entire topic under discussion (general bibliography is placed at the beginning of each chapter. Bibliographies

(B: 1–1482) and references (R: 2001–7349) dealing with particular problems are found at the end of each section. The 1482 bibliographies and 5349 references indicated do not constitute a complete review of the literature dealing with the extensive range of problems examined in this book, but they may nevertheless be valuable for further study.

This book is intended for graduate students and practising polymer chemists requiring a clear and concise introduction to the experimental chemistry of polymers together with practical information about experimental and measurement techniques and the interpretation and representation of results.

The text of the book consists of relatively brief sections presenting a concise and systematic description of the physical and experimental chemistry of polymers. According to the teaching experience of the author, such representation is most suitable for students, graduate students, and technologists among whom the degree of knowledge and practical experience may vary enormously. This applies in particular to students from undeveloped countries, where industries are only beginning to be introduced and where subject matter must be presented in a simple and clear manner.

The book should also be of assistance for the study of advanced monographs and original papers.

I wish to record my indebtedness to Drs W. J. Barnes, W. K. R. Barnikol, W. Bartensen, D. C. Bassett, G. Challa, H. D. Chanzy, S. Claesson, F. R. Damont, M. W. Darlington, J. F. Dawkins, B. C. Edwards, A. Gałęski, J. Grebowicz, C. C. Gryte, T. Hasimoto, D. Heikens, M. Iguchi, F. K. Keith, G. Kosztersznitz, M. Kryszewski, M. W. Ladd, P. J. Lemstra, A. J. Lovinger, Lau Chi-Ming, J. H. Magill, I. Murase, T. Nagasawa, H. D. Noether, F. J. Padden Jr, T. Pakuła, A. J. Pennings, P. J. Phillips, F. C. Price, E. J. Roche, R. Salovey, G. V. Schultz, Y. Shimomura, G. R. Smith, P. Smith, T. Stone, M. Takayanagi, R. E. Sievers, R. G. Vadimsky, K. Watanabe, W. Whitney, B. Wunderlich, G. Yadon, and A. Zwijnenburg, all of whom have supplied excellent photographs which are included in this book.

I am also grateful for the permissions granted by the many journal and book editors and publishers to reproduce material in this volume.

Finally, I must acknowledge my gratitude for their patience to the members of my family who, during the time in which this book was being prepared, have had to forego my company during innumerable evenings, weekends, and holidays.

. . . to know something about all
and all about something.

Contents

Chapter 1

Overview of the Structures of Polymers

General bibliography: 124, 127, 402, 424, 456, 457, 503, 637, 680, 737, 747, 804, 871, 909, 943, 1040, 1078, 1081, 1083, 1120, 1134, 1139, 1178, 1244, 1256, 1289, 1298, 1299, 1302, 1336, 1349, 1350, 1396, 1429.

1.1 GENERAL DEFINITIONS

A very large molecule is called a MACROMOLECULE or a MACROMOLE-CULAR COMPOUND or a HIGH POLYMER MOLECULE. When a polymer contains only a small number of structural units it usually is called an OLIGOMER.

A HIGH POLYMER is a MACROMOLECULAR SUBSTANCE and is described by the term POLYMER, by the name (e.g. poly(vinyl chloride)), and by the formula (e.g. $\text{+ CH}_2\text{—CHCl +}_n$), where n is the number of structural units (base units) called MERS (MONOMERIC UNITS).

A MACROMOLECULAR COMPOUND always consists of a mixture of HOMOLOGOUS POLYMERIC COMPOUNDS. The slight differences in composition are due to the presence of end groups, occasional branching, variations in orientation of mers, irregularity in the sequence of different types of mers (in copolymers), and sometimes other irregularities.

A REGULAR POLYMER is a polymer which is built from base units of the same type.

The BASE UNIT is the smallest possible repeating unit.

Structure nomenclatures for regular linear polymers are given in publications in the Bibliography (B: 1453, 1454).

ISOMERIC POLYMERS are polymers which have essentially the same percentage composition, but differ with regard to their molecular structures.

A HOMOPOLYMER is a polymer consisting of molecules containing (apart from end groups, branch junctions, and other minor irregularities) either a single type, or two or more chemically different types, in a regular sequence.

A COPOLYMER is a polymer consisting of molecules containing large numbers of units of two or more chemically different types in an irregular sequence.

A product of addition copolymerization is considered to be a copolymer, even if it is believed to consist of only two types of mer alternating regularly in the polymer chain. The regularity in such ALTERNATING COPOLYMERS is probably never rigorous.

A TERPOLYMER is a copolymer in which the locations (relative locations within the polymer chain in a linear copolymer) of two (or three or four or more) chemically different types of unit form an irregular pattern.

A LINEAR POLYMER (COPOLYMER) is a polymer in which the units in each molecule are linked together in a chain-like structure (Fig. 1.1).

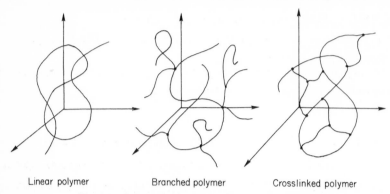

Linear polymer   Branched polymer   Crosslinked polymer

Fig. 1.1   Schematic drawings of backbone structures

A BRANCHED POLYMER is a polymer composed of molecules having a branched structure: chain-like between the junctions and between the branch chains and the branch junctions (Fig. 1.1).

A REPEATING UNIT of a linear polymer is that part of the molecular chain of such composition that the complete macromolecule (neglecting end groups, which are always chemically different from other structural units in the chain, branch junctions, etc.) might hypothetically be produced by bonding together a large number of these units.

The MONOMERIC UNIT or MER of a linear homopolymer is that unit of the molecule which contains the same kind and number of atoms as the real or hypothetical monomer.

The base unit and the mer may be the same or different. In poly(ethylene), for example, the mer ($CH_2{=}CH_2$) is larger than the base unit (CH_2).

In polymers prepared by addition polymerization the molecular weight of the structural units (e.g. $CH_2{-}CHCl$) is equal to the molecular weight of the monomer (e.g. $CH_2{=}CHCl$).

In polymers prepared by condensation reactions the molecular weight of structural units (e.g. $NH{-}(CH_2)_n{-}CO$) always differs from the molecular weight of the monomer (e.g. $NH_2{-}(CH_2)_n{-}COOH$) because small molecules (e.g. H_2O) are eliminated during the reaction.

The DEGREE OF POLYMERIZATION may be described as the (average) number of base units (including that at the beginning and that at the end of the macromolecule chain) in a molecule.

The DEGREE OF POLYMERIZATION (DP) is related to the molecular weight of the polymer as follows:

$$DP = M/m, \qquad (1.1)$$

where
   M is the molecular weight of the polymer,
   m is the molecular weight of the structural unit.
   A CROSSLINKING POLYMER is a polymer composed of macromolecules containing a three-dimensional network (Fig. 1.1).

POLYMER SHAPE should be considered as having two aspects:

   (i) the POLYMERIC CONFIGURATION, which is the polymer shape formed by primary valency bonding;
   (ii) the POLYMERIC CONFORMATION, which is the polymer shape due to rotation around primary valency bonds.

1.2 STEREOCHEMISTRY OF MACROMOLECULES

STRUCTURAL REGULARITY may be considered as having two aspects:

   (i) RECURRENCE REGULARITY (STRUCTURAL ISOMERISM)—the regularity with which the repeat recurs along the polymer chain.
   (ii) STEREOREGULARITY (TACTICITY) (STERIC ISOMERISM)—the regularity of the spatial order of the repeat unit.

1.2.1 Structural Isomerism

The vinyl polymerization of such monomers as $CH_2{=}CRR'$ (where $R \neq R'$, $R' \neq H$) usually occurs by HEAD-TO-TAIL ADDITION reactions, and yields STRUCTURALLY REGULAR POLYMERS with sequences of the type $\{CH_2{-}CRR'{-}CH_2{-}CRR'\}$.

The CRR' group is called the HEAD and the CH_2 group is called the TAIL. Polymers contain the HEAD-TO-HEAD

$$(-CH_2{-}CRR'{-}CRR'{-}CH_2{-})$$

and the TAIL-TO-TAIL $(-CRR'{-}CH_2{-}CH_2{-}CRR'{-})$ STRUCTURES (sometimes called SYNCEPHALIC SEQUENCES).

Monomers such as

$$CH_2{=}CR{-}CH{=}CH_2$$
$$1 \quad\; 2 \quad\; 3 \quad\; 4$$

$(R \neq H)$ can undergo 1,4-, 1,2-, or 3,4-polymerization. Two last polymerizations yield structurally regular sequences (structures **(I)** and **(II)**, respectively).

$$
\begin{array}{cc}
\begin{array}{l}
-CH_2{-}CR{-}\\
\quad\;\; |\\
\quad\;\; CH\\
\quad\;\; \|\\
\quad\;\; CH_2
\end{array}
&
\begin{array}{l}
-CH_2{-}CH{-}\\
\quad\;\; |\\
\quad\;\; CR\\
\quad\;\; \|\\
\quad\;\; CH_2
\end{array}
\\[2em]
\textbf{(I)} & \textbf{(II)}
\end{array}
$$

1.2.2 Steric Isomerism

The presence of stereoisomeric centres is responsible for different configurations of a regular polymer. The STEREOISOMERIC CENTRES present in macromolecular chains are

- (i) double bonds responsible for the *cis–trans* configurations (*CIS–TRANS ISOMERISM*);
- (ii) carbon atoms bonded to two different groups (R, R') (TETRAHEDRAL STEREOISOMERIC CENTRES).

1.2.3 Cis–trans Isomerism

The 1,4-polymerization of such monomers as CH_2=CR—CH=CH_2 (R = H) yields *CIS* (**III**) or *TRANS* (**IV**) ISOMERISM in the main chain. Structurally regular polymers such as (**III**) and (**IV**) are called *CIS*-TACTIC and *TRANS*-TACTIC, respectively.

(**III**)   (**IV**)

1.2.4 Tetrahedral Stereoisomerism

An ATACTIC POLYMER is a polymer in which the stereoisomeric configurations at all main chain sites are completely random.

An ISOTACTIC POLYMER is a tactic polymer in which the stereoisomeric configurations around the main chain are the same.

A SYNDIOTACTIC POLYMER is a tactic polymer in which the stereoisomeric configurations around the main chain sites alternate regularly.

In Fig. 1.2 are shown

- (i) Fisher projections for tactic polymers of general formula
  (—CH_2—CHR—)$_n$—;
- (ii) the hypothetically extended zigzag chains of two isotactic and two syndiotactic polymers, having one and two atoms per chain unit, respectively.

Two bonds of the chain adjacent to the carbon atom constituting THE STEREOISOMERIC CENTRE are indicated by the (+) and (−) signs (Fig. 1.3). Two monomeric units are identical from the configurational viewpoint when corresponding bonds are characterized by the same set of (+) and (−) signs.

In isotactic polymers of general structure $(CH_2$—CHR$)_n$ two monomeric units are configurationally ENANTIOMORPHOUS when corresponding bonds are characterized by a set of opposite signs:

$$-CH_2\!\!\underset{(-)}{\overset{(+)}{\rule{0pt}{0pt}}}\!\!CH\!\!\underset{\mathclap{|\,}}{\overset{(-)}{\rule{0pt}{0pt}}}\!\!\underset{R}{}\!\!CH_2\!\!\underset{(-)}{\overset{(+)}{\rule{0pt}{0pt}}}\!\!CH\!\!\underset{\mathclap{|\,}}{\overset{(-)}{\rule{0pt}{0pt}}}\!\!\underset{R}{}\!\!CH_2\!\!\underset{(-)}{\overset{(+)}{\rule{0pt}{0pt}}}\!\!CH\!\!-$$

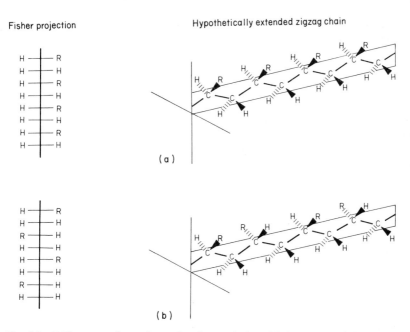

Fig. 1.2 Different conformations of polymer chains. (a) Isotactic poly(propylene); (b) syndiotactic poly(propylene)

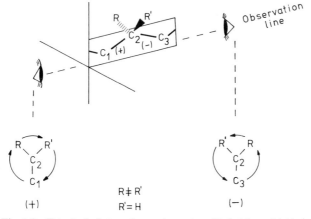

Fig. 1.3 Tetrahedral stereoisomeric centres. Definition of (+) and (−) bonds

In syndiotactic polymers of general structure $(-CH_2-CHR)_n-$ the bonds are characterized by a set of the same signs:

$$-CH_2 \overset{(+)}{-} CH \overset{(-)}{-} CH_2 \overset{(-)}{-} CH \overset{(+)}{-} CH_2 \overset{(+)}{-} CH-$$
$$\begin{array}{ccc} | & | & | \\ R & R & R \end{array}$$

When characterizing the steric isomerism of polymers of general formula $+CH_2-CRR'+_n$ it is convenient to refer to the structures of localized segments of the chains.

A sequence of two successive monomer units in a structurally regular polymer chain such as $+CH_2-CRR'+_n$ is called a TACTIC PLACEMENT or a TACTIC DYAD.

An ISOTACTIC PLACEMENT or ISOTACTIC DYAD and a SYNDIOTACTIC PLACEMENT or SYNDIOTACTIC DYAD in the hypothetically extended (zigzag), conformation are shown in Fig. 1.4.

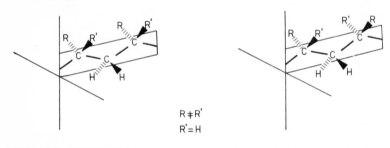

$$R \neq R'$$
$$R' = H$$

ISOTACTIC PLACEMENT
or ISOTACTIC DYAD

SYNDIOTACTIC PLACEMENT
or SYNDIOTACTIC DYAD

MESO METHYLENE GROUP
with HETEROSTERIC
SYMMETRY

RACEMIC METHYLENE GROUP
with HOMOSTERIC
SYMMETRY

Fig. 1.4 Different types of tactic placements or tactic dyads. (a) Isotactic placement or isotactic dyad (*meso* methylene group with heterosteric symmetry; (b) syndiotactic dyad (racemic methylene group with homosteric symmetry)

The central methylene group of an isotactic placement is referred as a *MESO* methylene group or an *m* UNIT.

The central methylene group of a syndiotactic placement is referred as a RACEMIC methylene group or an *r* UNIT.

A sequence of three successive monomer units in a structurally regular polymer chain such as $+CH_2-CRR'+_n$ is called a TACTIC TRIAD.

All three main chain sites of steric isomerism have the same configuration in an ISOTACTIC TRIAD. The central unit in an isotactic triad is called an *i* UNIT and is equivalent to an *mm* SEQUENCE.

The main chain sites of steric isomerism alternate in a SYNDIOTACTIC TRIAD (Fig. 1.5). The central unit in a syndiotactic triad is called an *s* UNIT and is equivalent to an *rr* SEQUENCE.

In a HETEROTACTIC TRIAD (Fig. 1.5) two adjacent main chain sites of steric isomerism have the same configuration and the third site has the opposite configuration. The central unit of a heterotactic triad is called an *h* UNIT and is equivalent to an *mr* or *rm* SEQUENCE.

Six tactic tetrad sequences of monomer units in a polymer of the type $+CH_2\!-\!CRR'\!+_n$ are: *mmm, mmr $\equiv$ rmm, rmr, mrr $\equiv$ rrm, mrm,* and *rrr.*

The symmetry of a methylene group in which the protons are differentiated by symmetry is called HETEROSTERIC, and that of a methylene group in which the protons are equivalent has been termed HOMOSTERIC.

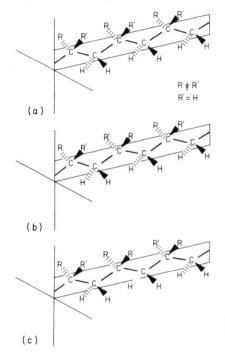

(a)

$R \neq R'$
$R' = H$

(b)

(c)

Fig. 1.5 Different conformations of tactic triads. (a) Isotactic triad, *i* unit or *mm* sequence; (b) syndiotactic triad, *s* unit or *rr* sequence; (c) heterotactic triad, *h*, unit or *rm* sequence

Tactic polymers of the type $+CH_2\!-\!CRR'\!+_n$ are MONOTACTIC in the sense that the base unit or structural repeat unit contains only one main chain site of isomerism.

A DI-ISOTACTIC POLYMER is a polymer the base unit of which contains as components of the main chain two carbon atoms each having two different lateral substituents, and the steric orientation of which is such that it makes the molecule isotactic with respect to the configuration around corresponding chain atoms of either type considered separately.

There are two types of di-isotactic polymer, differing with regard to the steric configurations around these two types of carbon atoms:

(i) an ERYTHRO-DI-ISOTACTIC polymer is a di-isotactic polymer in which the configurations at the two main-chain sites of steric isomerism in the base unit are alike.

8

(a) In a hypothetically extended (zigzag conformation) molecule of an ERYTHRO-di-isotactic polymer of general formula $+CHRCHR'+_n$ the substituents of one kind (R) are all on one side of the plane containing the chain atoms and the substituents of the other (R') are all on the other side.

(b) In a Fischer projection, all R and R' substituents are on the same side of the line representing the main chain.

(c) In a Newman representation of an eclipsed formation of two consecutive chain atoms and their attached atoms or groups, in which the next succeeding chain atom is superimposed over the next preceding chain atom, R is superimposed over R' and H over H.

(ii) a THREO-DI-ISOTACTIC polymer is a di-isotactic polymer in which the configurations at the two main-chain sites of steric isomerism in each base unit are opposite.

(a) In hypothetically extended (zigzag conformation) molecule of a THREO-di-isotactic polymer of general formula $+CHRCHR'+_n$, the substituents of both kinds, R and R', are all on the same side of the plane containing the chain atoms.

(b) In a Fischer projection, R and R' are on opposite sides of the line representing the main chain.

(c) In a Newman representation of an eclipsed formation of two consecutive chain atoms and their attached atoms and groups, in

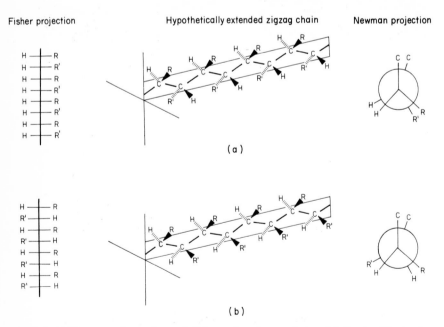

Fig. 1.6 Different conformations of di-isotactic polymer chains. (a) *Erytho*-di-isotactic configuration; (b) *threo*-di-isotactic configuration

which the next succeeding chain atom is superimposed over the next preceding chain atom, R is superimposed over H and H and R'.

Bibliography: 648.

In di-isotactic polymers of general formula $+CHR—CHR'+_n$ two tetrahedral stereoisomeric centres give two configurations defined as previously:

(i) an *erythro*-di-isotactic configuration characterized by the pairs of signs $(-)(+)$ or $(+)(-)$ (Fig. 1.7),

$$—CH\underline{^{(+)(-)}}CH\underline{^{(+)(-)}}CH\underline{^{(+)(-)}}CH\underline{^{(+)(-)}}$$

R	R'	R	R'

(ii) a *threo*-di-isotactic configuration characterized by the pairs of signs $(-)(-)$ or $(+)(+)$ (Fig. 1.7),

$$—CH\underline{^{(+)(+)}}CH\underline{^{(-)(-)}}CH\underline{^{(+)(+)}}CH\underline{^{(-)(-)}}$$

R	R'	R	R'

Bibliography: 127, 135, 175, 177, 327, 336, 385, 402, 428, 457, 637, 648, 680, 737, 966–968, 1078, 1091, 1120, 1136, 1178, 1289, 1299, 1336, 1349, 1396, 1416, 1452.
References: 2513, 2520, 2923, 3520, 5283, 5284, 5286, 5458, 5471, 5974.

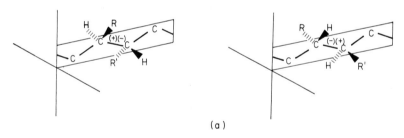

(a)

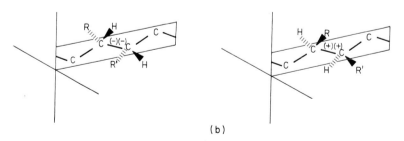

(b)

Fig. 1.7   (a) *Erythro* and (b) *threo* relative di-isotactic configurations. The pairs $(-, +)$ or $(+, -)$ define a relative configuration ERYTHRO, whereas the pairs $(-, -)$ or $(+, +)$ define a relative configuration THREO.

The definitions of DI-SYNDIOTACTIC polymers (*erythro* or *threo* configuration) are analogous to those of the corresponding di-isotactic polymers.

1.2.5 Sequence Statistics

The distribution of configurational sequences in vinyl polymer chains is of considerable theoretical interest, and in particular for the study of stereospecific addition polymerization.

The sequence statistics most commonly considered are:

(i) The BERNOULLI TRIAL. In a Bernoullian statistical process the result of an event is characterized by a single probability which is independent of the results of all previous events.

(ii) The FIRST-ORDER MARKOV (MARKOFFIAN) TRIAL. In the Markoffian method the distribution of tactic placements is dependent only on the nature of the preceding sequence of tactic placements.

The more detailed characteristics of some of these distributions and of the generating mechanism have been discussed elsewhere.

Bibliography: 175, 578, 843, 872, 1416.
References: 2924, 5968.

Computer programs are available for calculating copolymer and terpolymer structures using Markoffian statistics, see B: 579 and R: 3924.

1.3 COPOLYMERS

General bibliography: 10, 27, 225, 226, 270–272, 933, 936, 978, 1046, 1244.

1.3.1 Types of Copolymer

Copolymers can be divided into various groups according to their structures (Fig. 1.8):

(i) RANDOM COPOLYMERS are produced in bulk, solution, aqueous suspension, or emulsion using free-radical initiators of the peroxide type or redox systems. Initiation through irradiation is also possible.

(ii) RANDOM COPOLYMERS WITH ORDERED SEQUENCE DISTRIBUTION are produced by controlled feeding of the monomers during the copolymerization reaction.

(iii) ALTERNATING COPOLYMERS are produced by special copolymerization processes. The reactivity of polar monomers can be enhanced by complexing them with a metal halide or organo-aluminium halide. These complexed monomers participate in a one-electron transfer reaction with either an uncomplexed monomer or another electron-donor monomer.

(iv) GRAFT COPOLYMERS (cf. Section 1.3.2).

(v) BLOCK COPOLYMERS (cf. Section 1.3.3).

(vi) CROSSLINKED COPOLYMERS are produced by generating free-radical sites on a preformed polymer in the presence of a monomer. Grafting occurs on this site and combination termination results in cross-linking.

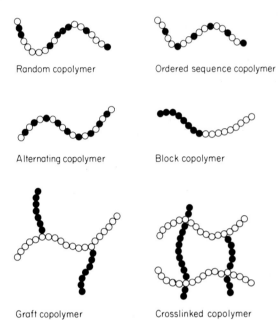

Random copolymer   Ordered sequence copolymer

Alternating copolymer   Block copolymer

Graft copolymer   Crosslinked copolymer

Fig. 1.8   Schematic representation of different types of copolymer

1.3.2 Graft Copolymers

All GRAFT COPOLYMERS are built from a polymer backbone (A) to which a number of sequences (B) are grafted (Fig. 1.9).

Graft copolymers are generally prepared by free-radical, anionic, or cationic addition or ring-opening polymerization of a monomer in the presence of a preformed reactive polymer.

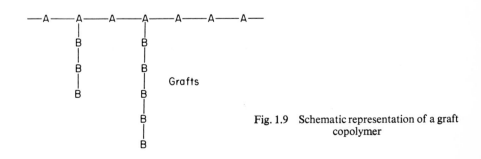

Fig. 1.9   Schematic representation of a graft copolymer

12

The grafting reactions and resulting grafted samples are characterized according to the following parameters:

(i) TOTAL CONVERSION (%)
$$= \frac{\text{Total weight of polymer formed}}{\text{Weight of monomer charged}} \times 100;$$

(ii) GRAFTING RATIO (%) $= \dfrac{\text{Weight of polymer in grafts}}{\text{Weight of substrate}} \times 100;$

(iii) GRAFTING EFFICIENCY (%)
$$= \frac{\text{Weight of polymer in grafts}}{\text{Total weight of polymer formed}} \times 100;$$

(iv) ADD-ON (%) $= \dfrac{\text{Weight of polymer in grafts}}{\text{Total weight of grafted polymer}} \times 100;$
(substrate + grafts)

(v) FREQUENCY OF GRAFTS = Average number of grafted sequences per grafted chain.

It is difficult to characterize the structures of graft copolymers with accuracy. Also, the length of graft segments and their polydispersity are difficult to measure. An additional complication is the unanswered question of the spacing of graft junction points along the backbone. To date, these problems remain largely unsolved.

Bibliography: 225, 226, 270–272, 933, 1046, 1244.

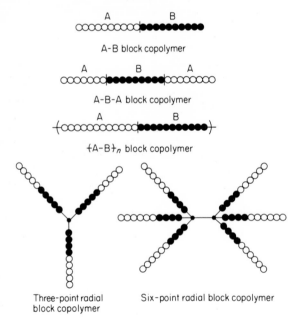

A-B block copolymer

A-B-A block copolymer

+A-B+$_n$ block copolymer

Three-point radial block copolymer

Six-point radial block copolymer

Fig. 1.10 Schematic representation of different types of block copolymer

1.3.3 Block Copolymers

BLOCK COPOLYMERS are built of chemically dissimilar, terminally connected segments (Fig. 1.10).

Block copolymers are generally prepared by sequential anionic addition or ring-opening polymerization or step-growth polymerization.

Block copolymers are more difficult to characterize than homopolymers and uniform statistical copolymers. The length of blocks and the block copolymer structure (e.g. A—B versus A—B—A) is difficult to determine. The number of segments in a block copolymer can sometimes be deduced from the synthetic technique employed. In A—B and in A—B—A block copolymers there are, by definition, one and two junction points, respectively. In $\text{-}(\text{A—B})_n$ block copolymers, the distance between intersegment linkages can be deduced from a knowledge of the block molecular weight.

Bibliography: 10, 11, 27, 156, 225, 226, 270–272, 416, 933, 935, 936, 978, 1244, 1281, 1431.

1.3.4 Sequence Distribution and Tacticity in Copolymers

Table 1.1 Recognizable features of copolymer structure with different distribution of A and B monomer units (taken from H. J. Harwood, Chapter 11, in *Characterization of Materials and Research: Ceramics and Polymers*, edited by J. J. Burke and V. Weiss. Syracuse: Syracuse University Press, 1975, p. 316)

Monomers	A			B
Dyads	AA	AB	BA	BB
Triads	AAA			BBB
	BAA			ABB
	AAB			BBA
	BAB			ABA
Tetrads	AAAA	AABA	ABAA	ABBA
	BAAA	AABB	BBAA	BBBA
	AAAB	BABA	ABAB	ABBB
	BAAB	BABB	BBAB	BBBB
Pentads	AAAAA	AAABA	ABAAA	ABABA
	BAAAA	BBAAA	BBAAA	BBABA
	AAAAB	BAABA	ABAAB	ABABB
	BAAAB	BAABB	BBAAB	BBABB
	ABBBA	AABBA	ABBAA	AABAA
	BBBBA	AABBB	BBBAA	BABAA
	ABBBB	BABBA	ABBAB	AABAB
	ABBBB	BABBB	BBBAB	BABAB

Three different copolymers

Random copolymer:	AAABAABBABBB
Alternating copolymer:	ABABABABABAB
Block copolymer:	AAABBBAAABBB

have different distributions of A and B units along the main chain. The properties of polymers having such varied structures can be quite different, even when
they have the same composition. Most properties of copolymers are dependent
on the relative amounts of various dyads, triads, tetrads, pentads, or higher
sequences present (cf. Table 1.1). Measurements of different chemical and
physical properties provide useful information about the structure of copolymers.

The chemical reactivity of functional groups on copolymers depends on the
steric effects, neighbouring group participation, and ionic interaction, e.g. the
reactivity of A units in an AB copolymer may vary depending on whether the
units are centred in AAA, AAB (BAA), or BAB triads. Configurational effects
may also cause A units in the various types of triad to have different reactivities.

The physical properties of copolymers are often dependent on the distribution
of monomer sequences, but most of them are observed as averages of the
contribution of the various types of structural units present.

Bibliography: 11, 272, 576–578, 944.
References: 3129, 3704, 5743, 6210, 6496.

1.3.5 Determination of the Intermolecular Compositional Heterogenity in Copolymers

INTERMOLECULAR COMPOSITIONAL HETEROGENITY IN CO-
POLYMERS may arise from one of two sources:

   (i) polymerization conditions;
   (ii) deliberate blending.

Methods for characterization of compositional heterogeneity include

   (i) gel permeation chromatography (GPC) with simultaneous differential
      refractive index and ultraviolet and/or infrared detection;
   (ii) thin-layer chromatography (TLC);
   (iii) equilibrium density gradient ultracentrifugation (DGU).

These methods provide useful information about the reactions occurring during
synthesis and/or the nature of commercial products.

References: 4686, 5294, 6126, 6509.

1.3.6 Study of Sequence Distribution and Tacticity

Methods used to characterize the arrangement of monomer units and the configurational structures of homopolymers and copolymers are

(i) the physical methods:
    (a) the degree of crystallization;
    (b) the glass transition temperature;
    (c) the melting point;
    (d) studies on heats of polymerization (copolymerization);
    (e) infrared spectroscopy;
    (f) Raman spectroscopy;
    (g) neutron spectroscopy;
    (h) NMR spectroscopy;
    (i) dipole moment measurements.
(ii) the chemical methods:
    (a) selective degradation followed by analysis of the degradation fragments;
    (b) pyrolysis–gas chromatography;
    (c) pyrolysis–mass spectroscopy;
    (d) intra- and intersequence cyclization;
    (e) studies of polymer reactivity.

1.3.7 Intra- and Intersequence Cyclization

CYCLIZATION REACTIONS occur when neighbouring substituents on a polymer or copolymer chain can react to form cyclic structures containing five- or six-membered rings. Cyclization reactions can be used very effectively in studies of copolymer and terpolymer structures.

Bibliography: 578.
References: 3811, 4330–4332, 6351, 6352.

1.4 CONFORMATION OF MACROMOLECULES

The CONFORMATION OF A MOLECULE is defined as one of the distinct spatial positions taken up by its atoms or groups of atoms through rotation around single bonds or as a change in the relative positions of a given unit joined by several single bonds.

Molecules with non-superimposable conformations are called CONFORMERS, ROTAMERS, or CONFORMATIONAL ISOMERS.

Detailed information on the existence and stability of conformers can be obtained from energy calculations.

Bibliography: 106, 135, 175, 402, 457, 471, 630, 637, 1083, 1134, 1349.
References: 2006, 2031, 2287, 2288, 3134, 3417, 3359, 3350, 4899, 5089, 5908, 6681.

1.4.1 Internal Rotation Angles in Polymers

A useful way of describing the conformation of the main polymer chain is to specify the sequence of internal rotation angles (Fig. 1.11).

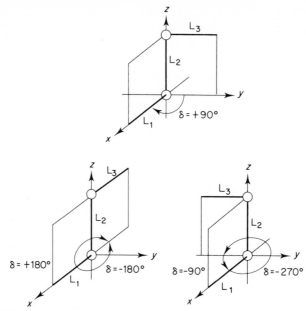

Fig. 1.11 Conventions for specification of internal rotation angles (δ)

The INTERNAL ROTATION ANGLE (δ) is the angle defined by the two planes L_1L_2 and L_2L_3, which is seen to express the conformation of the chain around the bond L_2:

(i) δ is positive for clockwise rotation of less than 180°;
(ii) δ is negative for counterclockwise rotations of less than 180°.

The orientation of a given bond with respect to the two preceding bonds in the chain is designated by BUNN'S (A, B, C) NOTATION or TADOKORO'S (T, G, $\bar{G}$) NOTATION (Fig. 1.12):

(i) A (T), TRANS ($\delta = 180°$);
(ii) B (G), LEFT GAUCHE ($\delta = +60°$);
(iii) C ($\bar{G}$), RIGHT GAUCHE ($\delta = -60°$).

These three basic bond sequences in single-bonded carbon chains (Fig. 1.13) are the only ones that are permitted by the principle of staggered bonding. When the chain motif, or geometrical repeating unit, comprises more than one bond sequence of the same kind, it is denoted by the appropriate numerical subscript (e.g. T_2, $(TG)_3$, etc.). Mixed modes such as $TGT\bar{G}$ or $(TG)_2(T\bar{G})_2$ may also be specified.

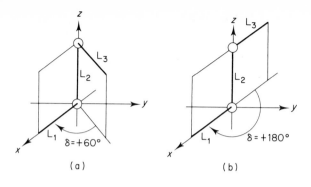

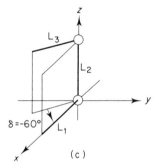

Fig. 1.12 Bunn's (A, B, C) convention and Tadokoro's *T*, *G*, *Ḡ*) convention for the specification of carbon-chain conformations. (a) Left *gauche*, B (*G*) conformation; (b) *trans*, A (T) conformation; (c) right *gauche*, C (*Ḡ*) conformation

Fig. 1.13 The three bond sequences in single-bonded carbon chains that conform to the principle of staggered bonds: (a) A (*T*) *TRANS* conformation; (b) B (*G*) LEFT *GAUCHE* conformation; (c) C (*Ḡ*) RIGHT *GAUCHE* conformation

Different chain conformations based on Tadokoro's notation (*T*, *G*, *Ḡ*) and characteristic identity period are shown in Table 1.2, and some of them in Fig. 1.14.

The preference for various conformations is expressed in terms of the CHARACTERISTIC RATIO (*Z*), given by

$$Z = \bar{r}_{0f}^2/nl^2, \tag{1.2}$$

where

$\bar{r}_{0f}^2$ is the mean-square end-to-end distance of the freely rotating chain (cf. eq. (2.30));

n is the number of bonds;

l is the length of the bond.

The characteristic ratio (*Z*) can be predicted by statical mechanics from a knowledge of the energetics of the conformations of a relatively short section of chain

18

Table 1.2 Characteristic identity periods in different chain conformations

Mode	Identity period (Å)
T_2	2.5
T_4	5.0
$TGT\bar{G}$	4.4
G_4	3.6
$(TG)_3$	6.2
$(T\bar{G})_3$	6.2
$T_3GT_3\bar{G}$	8.8
$(TG)_2(T\bar{G})_2$	8.5

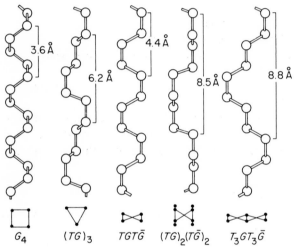

Fig. 1.14   Five chain conformations based on the T, G, and $\bar{G}$ staggered bond orientations. (From R: 2642)

or of model compounds. It can also be determined experimentally (cf. Section 2.11).

Bibliography: 16, 457, 1427.
References: 2002, 2004, 2005, 2125, 2151, 2286, 2422, 2423, 2444, 2503, 2525–2527, 2640–2642, 2645–2647, 2902, 2977, 3410, 3417, 3422, 3550, 3620, 3721, 3854, 3969, 4618, 4764, 4943, 4985, 4986, 5087, 5138, 5189, 5315, 5447, 5449, 5451, 5452, 5453, 5455, 6113, 6202, 6306, 6367, 6567, 6624–6626, 6628, 6634, 6668, 6713, 6743, 6811, 7102, 7270, 7277, 7278.

1.4.2 Helical Conformational Nomenclature

Many natural and synthetic polymers have a HELICAL STRUCTURE, which can be considered as a SCREW STRUCTURE. This type of structure can be described by a special HELICAL CONFORMATIONAL NOMENCLATURE.

There are two such systems of nomenclature (Fig. 1.15):

(i) the HELICAL POINT NET SYSTEM (H),

$$\Delta\Phi_H = 2\pi\, p/P = 2\pi\, t/u, \tag{1.3}$$

$$\Delta z_H = p = c/u; \tag{1.4}$$

(ii) the SCREW AXIS SYSTEM (S),

$$\Delta\Phi_S = 2\pi/u, \tag{1.5}$$

$$\Delta z_S = \gamma\,(c/u). \tag{1.6}$$

The two systems of notation are related by the expression

$$\gamma t = \varepsilon u + 1, \tag{1.7}$$

where
$\Delta\Phi$ is the angle of rotation,
Δz is a translation along the helix axis Z,
P is a translation equivalent to one turn of the helix (pitch of the helix) along the helix axis Z,
p is a projection of the distance between consecutive equivalent points on the helix along the helix axis Z,
t is the number of turns of the helix in the identity period c (an integer),
c is the crystallographic identity period parallel to the helix axis Z,
u is the number of points, or motifs, on the helix corresponding to period c (an integer),
γ is a positive integer > 0,
ε is a positive integer $\geqslant 0$.
The x and y axes of the component atoms with respect to some arbitrary origin of helical structure are not specified.

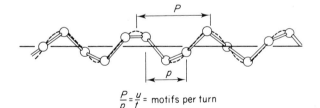

$$\frac{P}{p} = \frac{u}{t} = \text{motifs per turn}$$

Fig. 1.15   Helical nomenclature

It is important to underline that, instead of monomer unit or mer, both systems use the term MOTIF PER TURN (P/p or u/t) represented by one point on the helix.

Bibliography: 16, 135, 273, 402, 457, 928, 1048, 1134, 1427.
References: 2051, 2502, 2503, 2902, 2903, 2917, 3677, 4171, 4613, 4775, 4943, 5183, 5189, 5285, 5287, 5290, 5315, 5407, 5448, 5454, 5456, 5459, 5460, 5467, 5954, 6362, 6363, 6611, 6669, 6689, 6696, 6764, 7018, 7032, 7217, 7268.

Helical point net system crystallographic data are tabulated in publications in the Bibliography (B: 16, 958).

1.5 BRANCHED POLYMERS

A BRANCHED POLYMER is a polymer composed of molecules having branched structures. Depending on reaction conditions, growing polymer chains may undergo (Fig. 1.16):

(i) SHORT-CHAIN BRANCHING;
(ii) LONG-CHAIN BRANCHING.

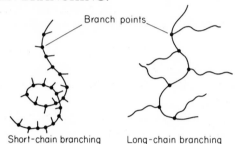

Short-chain branching   Long-chain branching

Fig. 1.16 Schematic representation of short- and long-chain branching

In long- and short-chain branching, each of the branches has the same chemical constitution as the main chain. Polymer properties are affected by the amount and distribution of short and long branches.

Most branched polymers can be divided into three groups (Fig. 1.17):

(i) regular stars;
(ii) regular combs;
(iii) random trees.

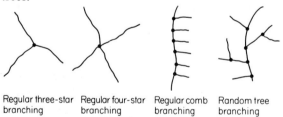

Regular three-star Regular four-star Regular comb Random tree
branching   branching   branching   branching

Fig. 1.17 Schematic representation of different types of branching

Branching in polymer molecules has been described mathematically as a function of the number of branch points it contains. This mathematical function is called THE NUMBER OF BRANCH POINTS and is denoted by g (cf. Section 25.15). The determination of g by intrinsic viscosity and GPC measurements is discussed in Sections 9.1.4 and 25.15.

Bibliography: 532, 538, 630, 1226.
References: 2099, 3818–3821, 5137, 5139, 6432, 6582.

1.6 CROSSLINKED POLYMERS

CROSSLINKED POLYMERS are polymers which have three-dimensional network structures, and for that reason are insoluble.

The CROSSLINK can be of the same or of a different structure as that which occurs in the main chain.

The crosslinked product which is swollen by solvent is called a GEL. Gels of very small size (300–1000 μm) are called MICROGELS.

Microgels, because of their high branch densities, behave as tightly packed spheres and are suspendable in solvents.

Gel formation occurs after a particular conversion, called the GEL POINT. Crosslinked polymers are characterized by

(i) the NETWORK CHAIN LENGTH, which is the number of chain links between two branch points in the network;
(ii) the BRANCH POINT, which is the point from which more than two chains radiate;
(iii) the NUMBER-AVERAGE MOLECULAR WEIGHT OF A NETWORK CHAIN, which is defined as

$$(\overline{M}_n)_c = \overline{M}_m/\zeta_c, \qquad (1.8)$$

where

$(\overline{M}_n)_c$ is the number-average molecular weight of a network chain,
$\overline{M}_m$ is the average molecular weight of the monomeric unit,
ζ_c is the degree of branching, defined as

$$\zeta_c = \frac{\text{Moles of crosslinked monomeric units}}{\text{Total moles of monomeric unit present}}; \qquad (1.9)$$

(iv) the CROSSLINK DENSITY (or the DEGREE OF CROSSLINKING) (Γ) is the number of crosslinked monomeric units per primary chain, and is given by

$$\Gamma = (\overline{M}_n)_0/(\overline{M}_n)_c, \qquad (1.10)$$

where

$(\overline{M}_n)_0$ is the number-average molecular weight of the primary chain (a primary chain is the linear molecule before crosslinking),
$(\overline{M}_n)_c$ is the number-average molecular weight of a network chain;
(v) the MOLAR CONCENTRATION OF EFFECTIVE NETWORK CHAINS, which is defined by

$$[M_c]_{\text{eff}} = [M_c]\,(1 - 2\,(\overline{M}_n)_c/(\overline{M}_n)_0) \qquad (1.11)$$

(in moles per gram (CGS) or moles per kilogram (SI)), where
$[M_c]_{\text{eff}}$ is the molar concentration of effective network chains,
$[M_c]$ is the molar concentration of all the chains present for $(\overline{M}_n) > (\overline{M}_n)_c$,
$(\overline{M}_n)_0$ and $(\overline{M}_n)_c$ are as defined above.

Equation (1.11) cannot be applied for very extensive crosslinking because the number of free ends is too high. The number of free chain ends increases with a decrease in the number-average molecular weight of the primary chain.

Bibliography: 286, 402, 482, 630, 1107, 1245, 1246.
References: 2290, 3250, 3259, 3718, 3735, 3736, 3751, 4086, 4412, 4555, 5271, 5356, 6391.

Crosslinked polymers contain several NETWORK DEFECTS such as (Fig. 1.18)

(i) unreacted functionalities or chain ends;

(ii) closed loops;

(iii) entanglements.

These defects influence the elastic properties of polymer networks.

Bibliography: 455, 630, 1203, 1246.
References: 3248, 3249, 3992, 4702, 5736, 5909, 6485, 6809, 6815.

Unreacted functionalities or chain ends

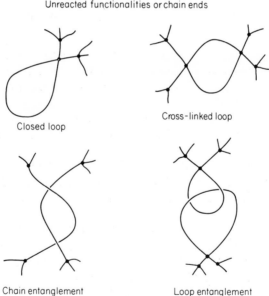

Closed loop

Cross-linked loop

Chain entanglement

Loop entanglement

Fig. 1.18   Schematic representation of network defects

1.7 POLYMER BLENDS

General bibliography: 153, 777, 868, 1046, 1128, 1244.

A POLYMER BLEND (POLYBLEND) is a physical mixture of two or more different polymers and/or copolymers which are not linked covalently.

Polyblends can be prepared by the following methods:

(i) mechanical mixing on rubber mills or extruders;
(ii) polymerization of one monomer in the presence of another polymer;
(iii) evaporation or precipitation from mixture of polymer solutions;
(iv) coagulation of a mixture of polymer lattices.

COMPATIBILITY in polyblends cannot be defined unequivocally. Compatibility has its origin in the mixing of two liquids. When liquids are mixed to form a homogeneous and single-phase mixture, they are said to be COMPATIBLE.

From thermodynamic, kinetic, and mechanical considerations, a homogeneous single-phase polyblend is improbable.

COMPATIBILITY IN POLYBLENDS is a representation of how close a polyblend can approach a single-phase mixture and is a relative measure of the degree of heterogeneity of the polyblend.

References: 4051, 4343, 4481, 5334, 5730, 5799, 6295.

1.7.1 Determination of Compatibility of Polyblends

There are several methods which can be applied for the determination of compatibility of polyblends:

(i) Solution in common solvents. If phase separation occurs, the components are incompatible. Since phase separation is affected by polymer concentration and by temperature, this test is quite arbitrary and gives only relative results.

(ii) Film casting. Films are cast from a homogeneous dilute solution of the polyblend. An opaque and crumbly film indicates incompatibility. This test is arbitrary and crude.

(iii) The appearance of fused products. The fused polyblend is pressed into a flat sheet. Transparency of the sheet signifies compatibility, whereas an opaque appearance means incompatibility. This test is arbitrary and crude.

(iv) The glass transition temperature. If the polyblend shows two distinct glass transition temperatures corresponding to parent polymers, it is incompatible. If the polyblend shows one transition temperature only, the system is compatible (cf. Section 32.5.8).

(v) Dynamic-mechanical measurement. This is the most sensitive method and has been used intensively. When the damping curves from a torsion pendulum test are obtained for the parent components and for the polyblend and the results are compared, a compatible polyblend will show a damping maximum between those of the parent polymers, whereas the incompatible polyblend gives two damping maxima at temperatures corresponding to those of the parent components. Dynamic-mechanical measurement can also give information on the shear modulus or tensile modulus. If the modulus–temperature curve shows multiple transitions, the polyblend is incompatible.

(vi) Microscopy. A phase-contrast light microscopy can detect heterogeneity at the 0.2–10 μm level. In many cases electron microscopy can show heterogeneity to a very fine scale, down to 0.01 μm (Fig. 1.19). Heterogeneity revealed by microscopy is a relative property.

(vii) Small-angle X-ray scattering (SAXS).

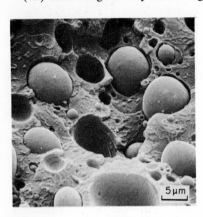

Fig. 1.19 Scanning electron micrograph of a fracture surface of a polyblend (poly(styrene)–poly(ethylene 75/25 by weight). (From R: 3982; reproduced by courtesy of D. Heikens and W. Barentsen and with the permission of IPC Science and Technology Press)

Describing compatibility it is desirable to give information on

(i) the composition of the polyblend and the characterization of its components;
(ii) the sample's history as to the method of preparation;
(iii) the method and instrument used to determine compatibility;
(iv) the experimental results and conclusions.

References: 2199, 2982, 3881, 3982, 4938, 5031, 6379, 6383, 6587, 7046, 7052.

Extensive data on compatible polymers are collected in the Polymer Handbook (B: 154).

Chapter 2

Study of Interactions Between Polymers and Solvents

General bibliography: 94, 127, 247, 386, 402, 456, 457, 622, 623, 747, 943, 1067, 1178, 1289, 1302, 1307, 1336, 1429.

Interaction between macromolecules and various solvents may cause localized disruption of intermolecular cohesive forces resulting in structural rearrangement.

AN IDEAL SOLVENT is characterized by its ability to dissolve any amount of polymer in the temperature range bounded by the crystallization temperature of the solution or a low-temperature demixing and the temperature at which the vapour pressure of the solution exceeds 1 torr (1 torr $= \frac{1}{760}$ atm $= 133.322$ N m^{-2} (SI)). In practice such ideal solvents for polymers do not exist.

A NON-SOLVENT is characterized by its inability to dissolve any amount of polymer at any temperature under atmospheric pressure.

The list of solvents and non-solvents for common polymers is given in Table 2.1. More detailed data are collected in the Polymer Handbook (B: 487).

Some polymers are well dissolved by mixed solvents. The miscibility of different solvents is shown in Fig. 2.1.

Extensive data of physical constants of the most common solvents are collected in the Polymer Handbook (B: 452).

Several solvents may cause a certain degree of SOLVATION of the polymer (the initial interaction with a polymer). SOLUBILITY and SOLUBILIZATION may be a second stage of solvation.

The RESIDUAL SOLVENT CONTENT in the polymer sample can be determined by heating the cast films under vacuum until a constant weight is obtained. The weight change is assumed to be due to release of residual solvent.

Bibliography: 141, 160, 161, 262, 574, 766, 777, 872.
References: 2162–2165, 2428, 2429, 3193, 3203, 4044, 4120, 4449, 5704, 5705, 5993, 6308, 6876–6878.

2.1 VOLUME CHANGE OF MIXING

The VOLUME CHANGE OF MIXING can be observed when solvent and polymeric solute are mixed. The measured volume (V) differs from the combined volumes (V_0) of the two components before mixing.

Table 2.1 Solvents and non-solvents for polymers (data collected from B: 487; reproduced by permission of John Wiley & Sons)

Solvents and non-solvents†

Polymer	Acetone	Alcohols (methanol, ethanol)	Aliphatic hydrocarbons (hexane, heptane)	Aromatic hydrocarbons (benzene, toluene)	Carbon disulphide	Carbon tetrachloride	Chloroform	Cyclohexane	Cyclohexanone	Dimethyl formamide	Dioxane	Esters	Ethers	Nitrobenzene	Methylene chloride	Tetrahydrofuran	Water
Poly(acrylic acid), atactic	−	(+)	−	−						+−		−	−	−		−	(+)+
Poly(acrylamide)	−	−	−	−						−		−	−			+	+
Poly(dienes):																	
Poly(butadiene)	−	−	+	++	−	++	−	++	−	−	−	−	−			++	−
Poly(isoprene)	−	−	+	+	−	+	−	+	−	−	−	−	−			+	−
Poly(ethylene)																	
High density	−	−	+ >133 °C	−	−	−	−	−	−	−	−	−	−	+ >200 °C		−	(ⓛ)
Low density	−	−	+ >80 °C	−	−	−	−	−	−	−	−	−	−			−	(ⓛ)
Poly(isobutene)	−	−	+	+++	+	+++	++	+	−	−	++	−	−			+++	−
Poly(methyl acrylate)	++	−	−	++	−	++	+	−	−	++	++	++	−			++	−
Poly(methyl methacrylate)	++	−	−	++	−	++	+	−	−	++	−	++	−			++	−
Poly(propylene), atactic	−	−	+ >35 °C	−	−	−	−	+ >35 °C	−	−	−	−	−			−	−
Poly(silanes), general			++	++		+	+	+	+		+		−			+	
Poly(styrene)	+	−	−	++	+	+	++	+	+	−	++	−	−			+	(ⓛ)
Poly(vinyl acetal)	−	+	−	+	−	+	+	−	+	+++	−	−	−			++	−
Poly(vinyl acetate)	−	+	−	+	−	−	+	−	+	+++	+	−	−			++	−
Poly(vinyl alcohol)	−	+	−	−	−	−	−	−	+	+	−	−	−			−	+
Poly(vinyl chloride)																	
High molecular weight	−	−	−	−	−	−	−	−	+	++	−	−		+		+	−
Low molecular weight	−	−	−	−	−	−	+	−	+	++	−	−		++	+	+	−
Poly(vinyl ethers), general	−	−	−	+	+	−	+	−	−	++	+				++	++	−
Poly(vinylidene chloride)	−	−	−	−	−	−	−	−	−	+	−					+ Hot	−

† + indicates that the liquid acts as a solvent; − indicates that it acts as a non-solvent.

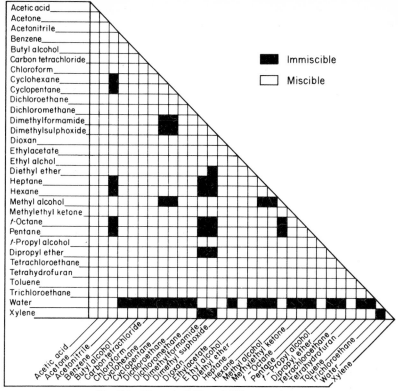

Fig. 2.1 The miscibility of different solvents

The EXCESS COEFFICIENT OF THERMAL EXPANSION (α) is determined by differentiating the analytical expression for the observed volume changes (V) as a function of temperature (T):

$$\alpha = \frac{1}{V}\left(\frac{\partial V}{\partial T}\right)_{p=\text{const.}} \tag{2.1}$$

References: 3407, 3408, 3280, 5613.

2.2 FREE VOLUME CONCEPT IN SOLUTIONS

The FREE VOLUME (V_f) in solution at a temperature T is defined as the volume generated by thermal motion of molecules and is defined as

$$V_f = V_T - V_0, \tag{2.2}$$

where

V_T is the total volume of the solution at temperature T (in kelvin),

V_0 is the theoretical molar volume of the most dense packing of the solvent molecules at 0 K.

28

The FREE VOLUME FRACTION (v_f) is defined as the ratio of the free volume (V_f) over the entire volume (V_T):

$$v_f = \frac{V_f}{V_T} = \frac{V_T - V_0}{V_T} \qquad (2.3)$$

(cf. also Section 32.2).

Bibliography: 585.
References: 3095–3097, 3211, 3477, 4055, 4886, 5601, 6186.

2.3 PRECIPITATION CURVES AND THE THETA TEMPERATURE

The PRECIPITATION TEMPERATURE is that temperature at which detectable turbidity can first be observed visually. The precipitation temperature (T_p) is determined upon slow cooling (1–2 °C per 10 min). This temperature should agree, within 0.2 °C, with the temperature at which the turbidity disappears upon warming.

The precipitation temperature of a given polymer in a given solvent as the function of the volume fraction of the solute can be presented in the form of the PHASE DIAGRAM (Fig. 2.2).

The CRITICAL PRECIPITATION TEMPERATURE (T_c) is determined from the phase diagram as the maximum point of the plot of the precipitation temperature versus the volume fraction of solute (Fig. 2.2). The curve shown in Fig. 2.2 is called the CLOUD-POINT CURVE or the PRECIPITATION CURVE.

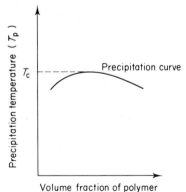

Fig. 2.2   Phase diagram for a polymer–solvent diagram

The precipitation curves plotted as T_c (in degrees Celsius) versus $\bar{v}M^{1/2}$ (where $\bar{v}$ is the specific volume of polymer and M is the molecular weight of polymer) yield straight lines (Fig. 2.3).

The THETA TEMPERATURE (the FLORY TEMPERATURE) (θ) is defined as the temperature at which polymer–solvent interactions are zero, i.e. linear chain molecules have unperturbed dimensions.

At the theta temperature, the second virial coefficient $A_2 = 0$ (cf. Section 2.10). Some examples of theta temperatures for common polymer–solvent systems are given in Table 2.2. More detailed data are collected in the Polymer Handbook (B: 404).

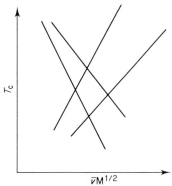

Fig. 2.3   Precipitation curves plotted
as T_c versus $\bar{v}M^{1/2}$

Table 2.2   Theta solvents for polymers. (From B:404; reproduced by permission of John Wiley & Sons.)

Polymer	Theta solvent Name	Composition	θ (°C)
Poly(1,4-butadiene)			
97% *cis*	*n*-Heptane		−1
90% *cis*	Hexane/heptane	50/50	5
94.6% *cis*	3-Pentanone		10.3
90% *cis*	5-Methyl-2-hexanone		12.6
90% *cis*	Hexane/heptane	75/25	20
90% *cis*	Isobutyl acetate		20.5
90% *cis*	5-Methyl-2-hexanone/2-pentanone	3 : 1	22.3
94.6% *cis*	3-Pentanone/2-pentanone	3 : 2	30.0
94.6% *cis*	Butanone		30
94.6% *cis*	5-Methyl-2-hexanone/2-pentanone	1 : 1	32.7
97% *cis*	*n*-Propyl acetate		35.5
96.4% *cis*	5-Methyl-2-hexanone/2-pentanone	1 : 3	46.2
90% *cis*	2-Pentanone		59.7
Poly(chloroprene)	Butanone		25
	Cyclohexane		45.5
	Cyclopentane		56.3
Poly(1,4-isoprene)			
cis (natural rubber)	2-Pentanone		14.5
	Hexane/isopropanol	1 : 1	21
	Butanone		25
96% *cis*	*n*-Heptane/*n*-propanol	69.5/30.5 (w/w)	25
trans (gutta percha)	*n*-Propyl acetate		60

Table 2.2 (*cont.*)

Poly(1-butene), atactic	Toluene		−46
	Isoamyl acetate		23
	Phenetole		61
	Anisole		83
			86.2
Poly(1-butene), isotactic	Cyclohexane/*n*-propanol	69 /31	35
	Anisole		89.1
	Ethylbenzene		−24.0
Poly(isobutene)	Toluene		−13.0
	Chlorobenzene/*n*-propanol	79.7/20.3	14.0
	Chloroform/*n*-propanol	79.5/20.5	14.0
	Ethyl *n*-caprylate		22
	Benzene		24
Poly(ethylene)	*n*-Pentane		∼ 85
	Diphenylene oxide		∼118
	Biphenyl		125
			125
			127.5
	n-Hexane		133
	1-Dodecanol		137.3
			138
	Diphenylmethane		142.2
	Bis(2-ethylhexyl) adipate		145
			145
			170
	Bis(2-ethylhexyl) sebacate		150
	n-Decanol		153.3
	Anisole		153.5
	Diphenyl ether		161.4
			163.9
			∼165
	p-Nonyl phenol		162.4
	p-Octyl phenol		174.5
	n-Octanol		180.1
	Benzyl phenyl ether		191.5
	p-t-Amyl alcohol		199.2
	Nitrobenzene		>200
	Dibutyl phthalate		>200
Poly(propylene), atactic	Carbon tetrachloride/*n*-propanol	74/26	25.0
	Carbon tetrachloride/*n*-butanol	67/33	25.0
	n-Hexane/*n*-butanol	68/32	25.0
	n-Hexane/*n*-propanol	78/22	25.0
	Methylcyclohexane/*n*-propanol	69/31	25.0
	Methylcyclohexane/*n*-butanol	66/34	25.0
	i-Amyl acetate		34
	n-Amyl acetate		36.6
	n-Butyl acetate		58.5
	i-Butyl acetate		65.5
	1-Chloronaphthalene		68
			74
	n-Propyl acetate		85.5
	Cyclohexanone		92
	Diphenyl ether		153
			153.3
	i-Amyl acetate		70
	Biphenyl		125.1

	Diphenyl ether		142.8
			145
			146.2
	Dibenzyl ether		183.2
Poly(acrylic acid) atactic	Dioxane		30
Poly(methacrylic acid), atactic	0.002 M HCl in water		30
	0.5 N NaCl in water		43
	0.05 N NaCl in water		68
Poly(methyl methacrylate), atactic	Chloroform		-273 ± 50
	Dichloroethane		-233 ± 50
	Benzene		-223 ± 50
	Methyl methacrylate		-163 ± 50
	Butanone		~ -98
	Ethyl acetate		~ -98
	Toluene		-65 ± 10
	Acetone		-55 ± 10
	Benzene/n-hexane	70/30	20
	Benzene/isopropanol	38/62	20
	Dioxane/n-hexane	59/41	20
	3-Methyl-2-butanone/n-hexane	83/17	20
	Dioxane/cyclohexane	53/47	20
Poly(styrene), atactic	n-Butyl formate		-9
	1-Chloro-n-decane		$+6.6$
	Hexyl-m-xylene		12.5
	Cyclohexane/toluene	86.9/13.1	15
	trans-Decalin		18.2
	trans-/cis-Decalin	79.6/23.1	19.3
	trans-Decalin		23.8
	Benzene/n-hexane	39/61	20
	Benzene/isopropanol	66/34	20
	Dioxane/n-hexane	38/62	20
	Dioxane/isopropanol	55/45	20
	Butanone/isopropanol	85.7/14.3	23
	Benzene/cyclohexanol	38.4/61.6	25
	Benzene/n-hexane	34.7/65.3	24
	Benzene/methanol	77.8/22.2	25
	Benzene/isopropanol	64.2/35.8	25
	Butanone/methanol	88.7/11.3	25
	Cyclohexane		34
Poly(vinyl acetate), atactic	Methanol		6
			6
	Ethanol/methanol	80/20	17
		60/40	26.5
		50/50	34
		40/60	36
	Butanone/isopropanol	73.2/26.8	25
	3-Methyl-butanone/n-heptane	73.2/26.8	25
	3-Heptanone		29
	3-Methyl-butanone/n-heptane	72.7/27.3	30
	Acetone/isopropanol	23/77	30
	6-Methyl-3-heptanone		66
Poly(vinyl alcohol)	Water		~ 97
Poly(vinyl chloride), atactic	Tetrahydrofuran/water	100/11.9	30
		100/ 9.5	30

32

The theta temperature for branched polymers is always lower than that for linear homologues in the same solvent. It depends on the length and on the number of branches, i.e. on the molecular structure of the polymer. When the degree of branching is low, there is coincidence between the values of the theta temperature for linear and branched polymers.

The precipitation curves plotted with the coordinates $(T-\theta)M^{\frac{1}{2}}$ versus $\bar{v}M^{\frac{1}{2}}$ give MOLECULAR WEIGHT-INDEPENDENT PRECIPITATION CURVES, i.e. curves obtained for different homologues of the same polymer superposed one on top of the other (Fig. 2.4).

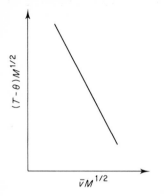

Fig. 2.4 The molecular weight-independent precipitation curve

The theta temperature for a linear polymer is a characteristic of a given polymer–solvent system, and is independent of the molecular weight of the polymer.

Bibliography: 125, 262, 367, 403, 456, 766, 1336.
References: 2026, 2500, 2693, 2818, 2831, 2985, 3055, 3109, 3110, 3121, 3176, 3276, 3402, 3719, 3964, 4299, 4483, 4579, 4657, 4660, 4662, 4664, 4667, 4717, 4789, 4790, 5421, 5422, 5495, 5578, 5583, 5586–5588, 5976, 6009, 6066, 6148, 6250, 6380–6382, 6474, 6421, 6475, 6477, 6478, 6510, 6595, 6639, 6921, 7147, 7151.

2.3.1 Methods to Determine Theta Solvents

There are several methods available for the determination of the theta temperature (θ):

(i) The PHASE EQUILIBRIUM method. In this method the temperature for phase separation (the critical precipitation temperature, T_c) is determined for a number of different concentrations of a polymer of known number-average molecular weight ($\overline{M}_n$) and the maximum critical precipitation temperature T_c is noted. The THETA TEMPERATURE (θ) can be calculated from the plot of T_c^{-1} versus $M^{-\frac{1}{2}}$ on Fig. 2.5, using the FLORY EQUATION,

$$\frac{1}{T_c} = \frac{1}{\theta} + \frac{V_1}{\bar{v}\theta\Psi M^{1/2}}, \tag{2.4}$$

where

 T_c is the critical precipitation temperature,
 θ is the theta temperature,
 V_1 is the molar volume of the solvent,
 $\bar{v}$ is the specific volume of the polymer,
 Ψ is the entropy of mixing,
 M is the molecular weight of the polymer.

The plot yields $1/\theta$ as the intercept at the ordinate and the slope is $\tan \alpha = V_1/\bar{v}\theta\Psi$. If they are used for mixed solvents, the data must also be extrapolated to infinite dilution. The method can be used for separation in two liquid phases only.

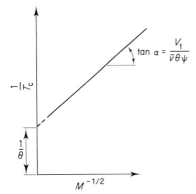

Fig. 2.5 The plot of $1/T_c$ versus $M^{-1/2}$

(ii) The SECOND VIRIAL COEFFICIENT method. This method is based on the fact that, at the theta temperature, the second virial coefficients $A_2 = 0$ (cf. Section 2.10). The slope of (π/c) (cf. Fig. 5.2) is zero if the solvent is a theta solvent. All methods in which the second virial coefficient can be determined, such as ebulliscopy (cf. Section 6.1.1), cryoscopy (cf. Section 6.2.1), sedimentation equilibrium (cf. Section 8.3.1), or light scattering (cf. Section 13.1.6) may be used.

(iii) The CLOUD POINT TITRATION method. In this method polymer solutions of different concentrations are titrated with a non-solvent until the first sign of cloudiness (cf. Section 4.2). The logarithms of the non-solvent concentration at the cloud point is then plotted against the logarithm of the polymer concentration at the cloud point and extrapolated to 100% polymer. The solvent/non-solvent mixture at this point corresponds to the theta mixture. Knowledge of the molecular weight is not necessary in this method.

(iv) The VISCOSITY–MOLECULAR WEIGHT RELATIONSHIP method. This method makes use of the fact that the exponent (a) in the Mark–Houwink–Sakurada equation (cf. Section 9.1.2) is equal to 0.5 for a random coil in a theta solvent.

Bibliography: 404, 456.
References: See references in Section 2.3.

2.4 UPPER AND LOWER CRITICAL SOLUTION TEMPERATURES

Cloud-point curves or precipitation curves for the different polymer–solvent systems have different shapes (Figs. 2.6, 2.7, and 2.8). The maxima and minima

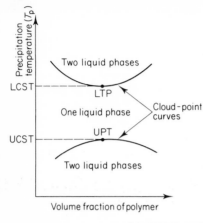

Fig. 2.6 Phase diagram for a polymer–solvent system where UCST<LCST (e.g. polystyrene–cyclohexanone)

on these curves are termed UPPER PRECIPITATION THRESHOLD (UPT) and LOWER PRECIPITATION THRESHOLD (LTP). These curves indicate the UPPER CRITICAL SOLUTION TEMPERATURE (UCST) and LOWER CRITICAL SOLUTION TEMPERATURE (LCST). (Note: The terms LCST and UCST have nothing to do with the actual magnitude of the temperature at which demixing occurs.)

The PHASE DIAGRAM of a polymer solution shows two regions of limited miscibility:

(i) below the upper critical solution temperature (UCST) associated with the theta temperature;
(ii) above the lower critical solution temperature (LCST).

The different polymer–solvent systems may have completely different phase diagrams. For some systems, UCST < LCST (Fig. 2.6), but for others, e.g. occurring in some highly polar systems (e.g. poly(oxyethylene)–water), UCST > LCST and a closed solubility loop is found (Fig. 2.7).

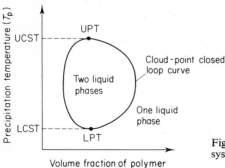

Fig. 2.7 Phase diagram for a polymer-solvent system where UCST>LCST (e.g. poly(oxyethy-lene)–water or poly(vinyl alcohol)–water)

In some cases precipitation curves have much more complicated shapes (e.g. poly(styrene)–acetone) (Fig. 2.8).

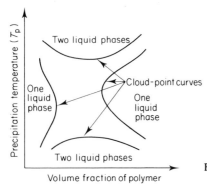

Fig. 2.8 Phase diagram for the poly-(styrene)–acetone system

Bibliography: 21, 125, 262, 456, 767, 1302.
References: 2056, 2224, 2225, 2243, 2295, 2374, 2819, 2996, 2999, 3000, 3001, 3064, 3122, 3123, 3133, 3281, 3469, 3504, 3719, 3861, 4667, 4791, 4792, 4794, 4866, 4923, 5042, 5202, 5203, 5495, 5497, 5546, 5702, 5704, 5706, 5707, 6145–6149, 6309, 6380, 6414, 6429, 6686, 7149.

2.5 SOLUBILITY OF POLYMERS IN MIXED SOLVENTS

There are three possible combinations of low molecular weight liquids in mixed solvents:

(i) solvent–nonsolvent;
(ii) non-solvent–non-solvent (a combination of two non-solvents producing a good mixed solvent for a polymer is termed COSOLVENCY);
(iii) solvent–solvent.

The CLOUD-POINT CURVES for a monotone change of solvent power with composition of the mixed solvent is shown schematically in Fig. 2.9. Each

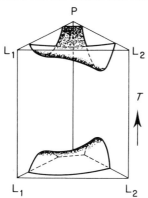

Fig. 2.9 Binodal surface for a ternary system and constant pressure showing a monotone decrease in solubility of polymer (P) when the composition of the mixed solvent is changed from pure L_1 to pure L_2. (From R: 7146; reproduced by permission of John Wiley & Sons)

point on the shadowed surfaces represents a CLOUD POINT (i.e. turbidity which can be observed visually). At temperatures between these surfaces the components are completely miscible, whereas solutions represented by points on the other side of the surfaces separate into two phases. For a narrow molecular

36

weight distribution of the polymer, the cloud-point surfaces are identical with the binodal surfaces, i.e. the end points of all tielines are situated on the cloud-point surfaces.

Bibliography: 125, 262, 764.
References: 2441, 2998, 3100, 3172, 4146, 4955, 5882.

2.6 INCOMPATIBILITY OF POLYMER SOLUTIONS

INCOMPATIBILITY of polymer solutions is the phenomenon in which quite dilute solutions of different polymers in the same solvent often do not mix, but separate into two phases. Upon decreasing the polymer concentration or increasing the temperature sufficiently, the two-phase system transforms into a one-phase system.

The practical way to investigate the incompatibility of polymer systems is to determine the SPINODAL—the curve that separates the unstable and meta-stable regions in the phase diagram. The spinodal (Fig. 2.10) can be found from the set of extrapolated points where the light scattering of the still homogeneous phase goes to infinity (i.e. from a Zimm plot, from the intersection of the extrapolated line $\theta = 0$ with the horizontal axis, Fig. 2.11).

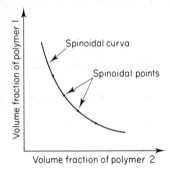

Fig. 2.10 Schematic diagram of a spinodal at constant temperature. The spinodal points are found by extrapolation of $R_{\theta \to 0}^{-1}$ to zero at a fixed volume fraction of one polymer and varying the volume fraction of the other. (From R: 6913; reproduced by permission of John Wiley & Sons)

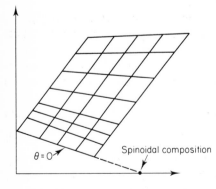

Fig. 2.11 Zimm plot (cf. Section 13.1.6) for the excess scattering of the system polymer 1 + polymer 2 in a given solvent. The spinoidal composition is found from the intercept of the $\theta = 0$ line with the horizontal axis. (From R: 6913; reproduced by permission of John Wiley & Sons)

Bibliography: 262, 777.
References: 2497, 2747, 3133, 3177, 3719, 4656, 4665, 4713, 4794, 5202, 6065, 6250, 6307, 6913–6915, 6978.

2.7 THERMODYNAMIC DESCRIPTION OF SOLUBILITY

A thermodynamic description of polymer–solvent mixing (dissolving) is given by the following equations:

(i) for an amorphous polymer,

$$\Delta G_m = \Delta H_m - T \Delta S_m \tag{2.5}$$

(in calories per mole (CGS) or joules per mole (SI));
(ii) for a crystalline polymer,

$$\Delta G_m = \Delta H_m - T \Delta S_m + \Delta H_f \tag{2.6}$$

(in calories per mole (CGS) or joules per mole (SI));
where
ΔG_m is the change in free energy of the system upon mixing (in calories per mole (CGS) or joules per mole (SI)),
ΔH_m is the change in enthalpy of the system upon mixing (in calories per mole (CGS) or joules per mole (SI)),
ΔS_m is the change in entropy of the system upon mixing (in calories per mole per degree Celsius (CGS) or joules per mole per kelvin (SI)),
ΔH_f is the heat of fusion of repeat units of polymer (in calories per mole (CGS) or joules per mole (SI)),
T is the thermodynamic temperature (in kelvins).
As yet there is no theoretical solution for the solubility of semicrystalline polymers. The problem of predicing crystalline solubility is a difficult one (B: 622 and R: 2642).
The FREE ENERGY (ΔG_m) for a polymer-solvent system is given by

$$\Delta G_m = RT[n_1 \ln v_1 + n_2 \ln v_2 + \chi v_1 v_2 (n_1 + mn_2)], \tag{2.7}$$

where
R is the universal gas constant (cf. Section 40.2),
T is the thermodynamic temperature (in kelvins),
n_1 and n_2 are the number of moles of solvent and polymer, respectively,
$m = V_2/V_1$ is the ratio of the molar volume of polymer and solvent, respectively,
v_1 and v_2 are the volume fraction of solvent and polymer, respectively (note: in many publications instead of v_1 and v_2 the symbols φ_1 and φ_2 are used) and are defined as

$$v_1 = n_1/(n_1 + mn_2), \tag{2.8}$$

$$v_2 = mn_2/(n_1 + mn_2), \tag{2.9}$$

χ is the Flory–Huggins interaction parameter (dimensionless) (cf. Section 2.9).
In eq. (2.7) the first two terms indicate the configurational entropy of mixing and are always negative. The third term must either be a very small positive number or negative in order to obtain a negative change in free energy (ΔG_m). Any process will occur when its change in free energy (ΔG_m) is negative (i.e. a

large increase in entropy followed by a small increase in enthalpy or a decrease in enthalpy).

The ENTROPY OF MIXING (ΔS_m) for a polymer–solvent system is always positive and increases upon mixing. In a solid polymer macromolecules are tangled, and molecular motion is limited to segmental Brownian motion. In solution polymer macromolecules are untangled and chains may translate about each other, may separate one from another, and may possess different configurations. The number of degrees of freedom increases, drastically increasing entropy.

The ENTHALPY OF MIXING (ΔH_m) (also called the HEAT OF MIXING) for a polymer–solvent system is given by the HILDEBRAND–SCATCHARD EQUATION:

$$\Delta H_m = V_m \left[\left(\frac{\Delta E_1}{V_1} \right)^{1/2} - \left(\frac{\Delta E_2}{V_2} \right)^{1/2} \right]^2 v_1 v_2 \qquad (2.10)$$

(in calories per mole (CGS) or joules per mole (SI)), where:

V_m is the total volume of the mixture,

ΔE_1 and ΔE_2 are the energies of vaporization of the solvent and the polymer, respectively (in calories per mole (CGS) or joules per mole (SI)).

The term $\Delta E/V$ is called the COHESIVE ENERGY DENSITY and is the energy of vaporization per mole, i.e. a measure of the amount of energy needed to overcome all intermolecular forces in 1 mol of the liquid.

The square root of the cohesive energy density is called the SOLUBILITY PARAMETER and has been given a special symbol (δ):

$$\delta = \left(\frac{\Delta E}{V} \right)^{1/2} \qquad (2.11)$$

(in $(\text{cal cm}^{-3})^{1/2}$ (CGS) or $(\text{J m}^{-3})^{1/2}$ (SI)).

Equation (2.10) for the enthalpy of mixing (ΔH_m) can have another form:

$$\Delta H_m = V_m (\delta_1 - \delta_2)^2 v_1 v_2 \qquad (2.12)$$

(in calories per mole (CGS) or joules per mole (SI)), where δ_1 and δ_2 are the solubility parameters of solvent and polymer, respectively.

A general rule is that

(i) the best solvent for a given polymer is one whose solubility parameter is equal or close to the solubility parameter of a polymer, i.e. when $\delta_1 \approx \delta_2$ ($\delta_1 - \delta_2 < 1.7$–$2.0$), then $\Delta H \approx 0$ and solubility between solvent and polymer occurs;

(ii) when $\delta_1 - \delta_2 > 2.0$, then $\Delta H \gg 0$ and solubility between solvent and polymer does not occur.

The solubility parameter of a solvent (δ_1) can be obtained from its enthalpy of vaporization (ΔH_1):

$$\delta_1 = [(\Delta H_1 - RT)/V_1]^{1/2} \qquad (2.13)$$

The SOLUBILITY PARAMETER OF THE SOLVENT MIXTURE (δ_{mix}) is given by

$$\delta_{mix} = \frac{x_1 V_1 \delta_1 + x_2 V_2 \delta_2}{x_1 V_1 + x_2 V_2}, \tag{2.14}$$

where

x_1 and x_2 are the molar fractions of components 1 and 2,
V_1 and V_2 are the molar volumes of components 1 and 2,
δ_1 and δ_2 are the solubility parameters of components 1 and 2.

Equation (2.14) is valid as long as there is no volume change on mixing.

A thermodynamic description of polymer–solvent mixing (dissolving) can also be expressed in the form of chemical potentials (μ).

The difference between the chemical potential of solvent in solution (μ_1) and in pure solvent (μ_1°) is given by

$$\Delta\mu = \mu_1 - \mu_1^\circ = RT\,[\ln(1-v_2) + (1 - 1/x)\,v_2 + \chi v_2^2], \tag{2.15}$$

where x is the average number of segments per polymer chain.

Solubility between solvent and polymer occurs when $\Delta\mu$ is negative.

Bibliography: 21, 24, 94, 109, 228, 229, 262, 456, 458, 494, 622, 623, 1067, 1146.
References: 2008, 2062, 2063, 2187, 2188, 2359, 2418, 2440, 2812, 3003, 3400, 3413, 3872, 3990, 4122, 4140, 4162, 4165–4168, 4359–4365, 4447, 4659, 4664, 5072, 5093, 5392, 5423, 5703, 5704, 5893, 6025, 6180, 6276, 6278, 6286–6289, 6398, 6399, 6476, 6690, 6692, 6706, 7144.

Note: The Polymer Handbook collects

(i) extensive data on heat, entropy, and volume changes for polymer–liquid systems (B: 164);
(ii) heats of solution of some common polymers (B: 323);
(iii) extensive data on polymer–solvent interaction and solubility parameters (B: 230, 1413).

Heats of mixing (ΔH_m) of dilute polymer solutions are determined by micro-calorimetric measurements.

Bibliography: 243.
References: 2096, 3137, 3570, 4152, 4355, 4363, 4901, 5880, 6588, 6806.

2.8 THREE-DIMENSIONAL SOLUBILITY ANALYSIS

The THREE-DIMENSIONAL SOLUBILITY PARAMETER (δ) can be calculated from

$$\delta = \sqrt{(\delta_d^2 + \delta_h^2 + \delta_p^2)}, \tag{2.16}$$

where

δ_d is the dispersion force,
δ_h is the hydrogen-bonding capacity (which is proportional to the strength of the hydrogen bonding),
δ_p is the polarity and induction interaction.

Solubility parameters for different solvents are given in Table 2.3.

Table 2.3 Solubility parameters for solvents (from B: 93; reproduced with permission of John Wiley & Sons)

Solvent	δ_d	δ_p	δ_h
Methanol	7.42	6.0	10.9
Ethanol	7.73	4.3	9.5
1-Propanol	7.75	3.3	8.5
2-Propanol	7.70	3.0	8.0
1-Butanol	7.81	2.8	7.7
Isobutanol	7.4	2.8	7.8
1-Pentanol	7.81	2.2	6.8
2-Ethyl butanol	7.70	2.1	6.6
2-Ethyl hexanol	7.78	1.6	5.8
Methyl isobutyl carbinol	7.47	1.6	6.0
Cyclohexanol	8.50	2.0	6.6
Ethylene glycol	8.25	5.4	12.7
Propylene glycol	8.24	4.6	11.4
1,3-Butanediol	8.10	4.9	10.5
Glycerol	8.46	5.9	14.3
m-Cresol	8.82	2.5	6.3
Diethylene glycol	7.86	7.2	10.0
Dipropylene glycol†	7.77	9.9	9.0
Ethyl lactate	7.80	3.7	6.1
n-Butyl lactate	7.65	3.2	5.0
Diacetone alcohol	7.65	4.0	5.3
Acetic acid†	7.1	3.9	6.6
Formic acid†	7.0	5.8	8.1
Butyric acid†	7.3	2.0	5.2
Acetic anhydride	7.5	5.4	4.7
Water†	6.0	15.3	16.7
Water–urea (satd)†	8.8	13.0	15.2
2-Methoxyethanol	7.9	4.5	8.0
2-Ethoxyethanol	7.85	4.5	7.0
2-Butoxyethanol	7.76	3.1	5.9
2-Ethoxyethyl acetate	7.78	2.3	5.2
2-(2-Methoxyethanoxy)ethanol	7.90	3.8	6.2
2-(2-Butoxyethoxy)ethanol	7.80	3.4	5.2
Dimethyl diethylene glycol	7.70	3.0	4.5
Ethyl acetate	7.44	2.6	4.5
n-Butyl acetate	7.67	1.8	3.1
Isobutyl acetate	7.35	1.8	3.7
Isoamyl acetate	7.45	1.5	3.4
Isobutyl isobutyrate	7.38	1.4	2.9
γ-Butyrolactone	9.26	8.1	3.6
Propylene carbonate	9.83	8.8	2.0
Ethylene carbonate	9.50	10.6	2.5
Diethyl ether	7.05	1.4	2.5
Furan	8.70	0.9	2.6
Tetrahydrofuran	8.22	2.8	3.9
Dioxane†	9.30	0.9	3.6
Methylal†	7.35	0.9	4.2
Anisole†	8.7	2.0	3.3

Table 2.3 (*cont.*)

Solvent	δ_d	δ_p	δ_h
Carbon tetrachloride	8.65	0.0	0.0
Chlorofrom	8.65	1.5	2.8
Methylene chloride	8.91	3.1	3.0
Ethylene chloride	9.20	2.6	2.0
1,1,1-Trichloroethane	8.25	2.1	1.0
1-Chlorobutane	7.95	2.7	1.0
Trichloroethylene	8.78	1.5	2.6
2,2-Dichlorodiethyl ether†	9.20	4.4	1.5
Chlorobenzene	9.28	2.1	1.0
o-Dichlorobenzene	9.35	3.1	1.6
Cyclohexyl chloride	8.50	2.7	1.0
Chloropropanol	8.58	2.8	7.2
Epichlorohydrin	9.30	5.0	1.8
1-Bromonaphthalene	9.94	1.5	2.0
Acetone	7.58	5.1	3.4
Methyl ethyl ketone	7.77	4.4	2.5
Methyl isobutyl ketone	7.49	3.0	2.0
Methyl isoamyl ketone	7.80	2.8	2.0
Diisobutyl ketone	7.77	1.8	2.0
Isophorone	8.10	4.0	3.6
Acetophenone	8.55	4.2	1.8
Cyclohexanone	8.65	4.1	2.5
Mesityloxide	7.97	3.5	3.0
Benzaldehyde	9.15	4.2	2.6
Acetonitrile	7.50	8.8	3.0
Butyronitrile	7.50	6.1	2.5
Nitromethane	7.70	9.2	2.5
Nitroethane	7.80	7.6	2.2
2-Nitropropane	7.90	5.9	2.0
Nitrobenzene	8.60	6.0	2.0
Formamide	8.40	12.8	9.3
Dimethyl formamide	8.52	6.7	5.5
Aniline	9.53	2.5	5.0
Ethanolamine	8.35	7.6	10.4
Dipropylamine	7.50	0.7	2.0
Diethylamone	7.30	1.1	3.0
Cyclohexylamine	8.45	1.5	3.2
Pyridine	9.25	4.3	2.9
Morpholine	9.20	2.4	4.5
Diethylenetriamine	8.15	6.5	7.0
Benzonitrile	8.50	6.5	2.5
n-Methyl-2-pyrrolidone	8.75	6.0	3.5
2-Pyrrolidone	9.5	8.5	5.5
Dimethyl acetamide	8.2	5.6	5.0
Tetramethyl urea	8.2	4.0	5.4
Ethylene cyanohydrin	8.4	9.2	8.6
Hexamethyl phosphoramide	9.0	4.2	5.5
Trimethyl phosphate	8.2	7.8	5.0
Triethyl phosphate	8.2	5.6	4.5

Table 2.3 (*cont.*)

Solvent	δ_d	δ_p	δ_h
Diethyl sulphide	8.25	1.5	1.0
Carbon disulphide	9.97	0.0	0.0
Dimethyl sulphoxide	9.00	8.0	5.0
Dimethyl sulphone	9.3	9.5	6.0
Benzene	8.95	0.5	1.0
Toluene	8.82	0.7	1.0
Xylene	8.65	0.5	1.5
Ethylbenzene	8.70	0.3	0.7
Styrene	9.07	0.5	2.0
Metralin	9.35	1.0	1.4
Hexane	7.24	0.0	0.0
Cyclohexane	8.18	0.0	0.0

† Values uncertain.

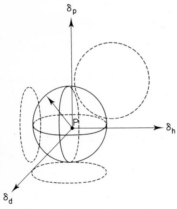

Fig. 2.12 Sketch of a typical volume of interaction for Hansen's three-dimensional solubility parameter concept

A solubility plot for a given polymer can be represented by a point (P) around which a sphere can be constructed (Fig. 2.12). This sphere defines the zone of solubility for a given polymer.

The general rule is that

(i) any liquid lying within the sphere is a true solvent for the polymer;
(ii) any liquid lying outside the sphere is a non-solvent;
(iii) any liquid lying close to the boundary of the sphere may be partially miscible with the polymer.

Three-dimensional solubility analysis is useful in solving several engineering problems, especially in the paint and coating industries.

Bibliography: 93.
References: 2484, 3024, 3025, 3523, 3888, 3889, 3891, 4379, 4616, 6761.

2.9 THE FLORY–HUGGINS INTERACTION PARAMETER

The FLORY INTERACTION PARAMETER (often called the FLORY–HUGGINS INTERACTION PARAMETER) (χ) is a measure of the interaction between any given solvent and a given polymer. It is a free-energy parameter and contains both entropy and enthalpy terms:

(i) for non-polar or slightly polar polymer–solvent systems, χ is given by

$$\chi = \chi_S + \chi_H \qquad (2.17)$$

(dimensionless), where

χ_S is the entropy term, equal to the inverse of the lattice coordination number (z) (i.e. equal to the inverse of the number of nearest neighbours of a molecule or segment in solution) (χ_S is a temperature-dependent quantity),

χ_H is the enthalpy term given by

$$\chi_H = (V_1/RT)(\delta_2 - \delta_1)^2, \qquad (2.18)$$

V_1 is the molar volume of the solvent,

R is the universal gas constant (cf. Section 40.2),

T is the thermodynamic temperature (in kelvins),

δ_1 and δ_2 are the solubility parameters of solvent and polymer, respectively;

(ii) for polar polymer–solvent systems, the solubility parameter of a polymer (δ_2) in eq. (2.18) should be replaced by a three-dimensional solubility parameter (cf. Section 2.8).

The Flory interaction parameter (χ) is a temperature-dependent quantity and is also concentration dependent over a wide range of polymer concentrations. The parameter (χ) must be measured for every polymer–solvent pair.

Table 2.4   Different methods for the determination of the Flory–Huggins interaction parameter (χ)

Method	References (R)
1. Equilibrium swelling	2493, 2494, 2535, 2559, 2578, 3223, 3424, 3403, 4313, 5066, 5383, 5763, 6310, 6691
2. Osmotic pressure	4727, 5579, 6610
3. Vapour pressure	2057, 2300, 2302, 2303, 2493, 2494, 2848, 3608, 5424, 5468, 6690
4. Sedimentation—equilibrium ultracentrifugation	2841
5. Viscosity	2535, 2577, 3003, 3416, 3762, 5066, 6310, 6433, 6614
6. Optical studies	2107, 2968, 3002, 3421
7. Inverse gas chromatography	2581, 3136, 3138, 3165, 3168, 3578, 3580, 3871, 4891–4894, 5470, 5585, 5708, 6260, 6606, 6607, 6622, 6690, 6766

The Flory–Huggins interaction parameter (χ) can be calculated from measurements of the second virial coefficient (A_2) (cf. Section 2.10) and also by other different methods (Table 2.4).

Bibliography: 141, 254, 262, 456, 872, 943, 1302.
References: 2440, 2663, 2693, 3124, 3281, 3407, 3414, 3419, 3423, 3425, 3591, 4075, 4077, 4162, 4472, 5612, 5703, 5705, 6162, 6163, 6700.

2.10 THE SECOND VIRIAL COEFFICIENT

The SECOND VIRIAL COEFFICIENT (A_2) is a measure of the interaction between a solvent and a polymer and is expressed according to the complex Flory–Huggins–Krigbaum theory as

$$A_2 = (\tfrac{1}{2} - \chi)\left(\frac{\bar{v}^2}{V_1}\right) J(X) = \Psi\left(1 - \frac{\theta}{T}\right)\left(\frac{\bar{v}^2}{V_1}\right) J(X), \qquad (2.19)$$

where
   χ is the Flory interaction parameter (cf. Section 2.9),
   $\bar{v}$ is the specific volume of the polymer,
   V_1 is the molar volume of the solvent,
   Ψ is the entropy of mixing,
   θ is the theta temperature,
   T is the temperature at which interaction between solvent and polymer occurs,
   $J(X)$ is a quantity which depends on molecular weight, the nature of the solvent, and the temperature and becomes equal to unity at the theta temperature, where $\chi = \tfrac{1}{2}$.
The second virial coefficient can be calculated by different methods and is then expressed as follows:

   (i) Osmometry (cf. Section 5.2):

$$A_2 = \frac{\rho_s}{\rho_p^2 M_s}(\tfrac{1}{2} - \chi). \qquad (2.20)$$

   (ii) Ebulliometry (cf. Section 6.1.1):

$$A_2 = \frac{RT^2}{\rho_p^2 \Delta H_s}(\tfrac{1}{2} - \chi). \qquad (2.21)$$

   (iii) Cryoscopy (cf. Section 6.2.1):

$$A_2 = \frac{RT^2}{\rho_p^2 \Delta H_s}(\tfrac{1}{2} - \chi). \qquad (2.22)$$

   (iv) Light scattering (cf. Section 13.1.6):

$$A_2 = \frac{\rho_s}{\rho_p^2 M_s}(\tfrac{1}{2} - \chi). \qquad (2.23)$$

Descriptions of the symbols are given in the relevant sections. The units of A_2 depend upon the way in which it is expressed.

A general rule is that

(i) for 'good solvents', A_2 is high;
(ii) for 'poor solvents', A_2 is low;
(iii) at $A_2 = 0$ the polymer precipitates from solution.

The second virial coefficient (A_2) is dependent on several factors:

(i) the temperature (A_2 decreases with increasing temperature);
(ii) the polymer–solvent system;
(iii) the molecular weight and molecular weight distribution (MWD) of the polymer;
(iv) the tacticity of the polymer;
(v) the molecular shape of macroions.

Bibliography: 1429.
References: 2337, 2740, 2750, 2751, 2741, 2752, 3003, 3053, 3206, 3302, 3420, 3571, 3822, 4085, 4152, 4295, 4463, 4746, 4751, 4761, 4770, 4850, 5514, 5608, 5640, 5690, 5832, 5879, 6094, 6281, 6284, 6287, 6417, 6578, 6595, 6923, 7143, 7145, 7331.

Extensive data of second virial coefficients of polymers in solutions are collected in The Polymer Handbook (B: 410).

2.11 UNPERTURBED DIMENSIONS OF LINEAR CHAIN MACROMOLECULES

The SIZE OF MACROMOLECULES (DIMENSION) in solution cannot strictly be defined because a single molecular coil alters in shape with time and these conformational changes are due to Brownian motion. The molecular size varies from molecule to molecule of identical mass and structure, and it can only be described in terms of average properties (Fig. 2.13):

(i) the MEAN-SQUARE AVERAGE END-TO-END DISTANCE ($\bar{r}^2$) of a linear chain molecule, given by

$$\bar{r}^2 = \bar{r}_0^2 \alpha^2 \qquad (2.24)$$

(in square centimetres (CGS) or square metres (SI)), where
$\bar{r}_0^2$ is the UNPERTURBATED MEAN-SQUARE AVERAGE-END-TO-END DISTANCE (in square centimetres (CGS) or square metres (SI)),
α is the expansion factor;

(ii) the SQUARE AVERAGE RADIUS OF GYRATION ($\bar{s}^2$), given by

$$\bar{s}^2 = \bar{s}_0^2 \alpha^2 \qquad (2.25)$$

in square centimetres (CGS) or square metres (SI), where
$\bar{s}_0$ is the UNPERTURBED SQUARE AVERAGE RADIUS OF GYRATION,
α is the expansion factor.

For linear polymers,

$$\bar{r}^2 = 6\bar{s}^2. \tag{2.26}$$

PERTURBED DIMENSIONS are the chain dimensions in a given solvent, where intermolecular interactions between macromolecules and solvent occur.

UNPERTURBED DIMENSIONS are the chain dimensions in a given solvent, determined solely by bond lengths and angles.

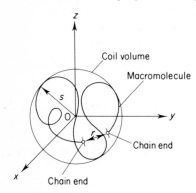

Fig. 2.13 Schematic representation of a single molecular coil. The centre of gravity of the macromolecule is at O, r is the chain end-to-end distance, and s is the radius of gyration about the centre of gravity of the macromolecule

The THETA TEMPERATURE (the FLORY TEMPERATURE) is that temperature of a given solvent $T = \theta$ where linear chain molecules have unperturbed dimensions.

Extensive data on theta solvents are collected in The Polymer Handbook (B: 404).

The unperturbed dimensions of a linear polymer chain ($\bar{r}_0^2$ or $\bar{s}_0^2$) vary with molecular weight ($\overline{M}_n$). Values of the type $\bar{r}_0^2/\overline{M}_n$ or $\bar{s}_0^2/\overline{M}_n$ depend only on chain structure and not upon a solvent or a temperature.

The dimensions of polymer chains in diluted solutions are determined by

(i) SHORT-RANGE INTERACTIONS such as bond angle restrictions and steric restrictions to internal rotation. These are determined by the unperturbed mean-square end-to-end distance $(\bar{r}_0^2)$ or by the unperturbed square average radius of gyration $(\bar{s}_0^2)$. A measure of the effect of short-range interactions is given by the quantity C_∞, defined as

$$C_\infty = \lim_{n \to \infty} \bar{r}_0^2 \Big/ \sum_i n_i l_i, \tag{2.27}$$

where n_i is the number of the ith kind of bond of length l_i.

(ii) LONG-RANGE INTERACTIONS, which depend upon the polymer-solvent interactions and are determined by the expansion factor (α) and the excluded volume parameter (z), which are corellated by

$$\lim_{z \to \infty} \alpha^x = \text{constant} \times z, \tag{2.28}$$

where x is a constant independent of z. Several values between 1 and 6.67 for x have been predicted, from different theories.

The EXPANSION FACTOR (α) is the ratio of the chain dimension in a given solvent (perturbed dimensions) ($\bar{r}^2$ or $\bar{s}^2$) and in a theta solvent (unperturbed dimensions) ($\bar{r}_0^2$ or $\bar{s}_0^2$) at a specific temperature:

$$\alpha = \frac{(\bar{r}^2)^{1/2}}{(\bar{r}_0^2)^{1/2}} = \frac{(\bar{s}^2)^{1/2}}{(\bar{s}_0^2)^{1/2}} \tag{2.29}$$

(dimensionless).

If the polymer is in a good solvent, the expansion factor (α) is greater than unity; this means that the actual perturbed dimensions ($\bar{r}^2$ or $\bar{s}^2$) exceed the unperturbed dimensions ($\bar{r}_0^2$ or $\bar{s}_0^2$). The expansion factor (α) also increases with molecular weight ($\overline{M}_n$), because the number of uncompensated interactions between chain segments increases with the number of segments when $T > \theta$. The expansion factor (α) is also related to the solvent power.

The EXCLUDED VOLUME is the volume from which a given polymer molecule effectively excludes all others. The excluded volume is a result of the virtual repulsion between polymer molecules arising from their spatial requirements. Each molecule in a very dilute solution in a good solvent will tend to exclude all others from the volume which it occupies.

The FREELY ROTATING STATE is a hypothetical state of the chain in which the bond angle restrictions are retained but the steric hindrances to internal rotation are relaxed.

The MEAN-SQUARE END-TO-END DISTANCE OF THE FREELY ROTATING CHAIN ($\bar{r}_{0f}^2$) can be calculated from the given basic structure of the chain using

$$\bar{r}_{0f}^2 = nl^2 \frac{(1+\cos\theta)}{(1-\cos\theta)}, \tag{2.30}$$

where

n is the number of bonds,
l is the length of bond,
θ is the valence bond angle.

The EFFECT OF STERIC HINDRANCE ON THE AVERAGE CHAIN DIMENSION (σ) (also called the POLYMER CHAIN STIFFNESS PARAMETER) is given by

$$\sigma = \frac{r_0}{r_{0f}} = \frac{(\bar{r}_0^2)^{1/2}}{(\bar{r}_{0f}^2)^{1/2}} \tag{2.31}$$

(dimensionless). The quantity σ is independent of the number of bonds (n).

The unperturbed dimensions of linear polymer chains can be determined by the following methods:

(i) limiting viscosity number measurements (cf. Section 2.13);
(ii) light scattering, e.g.
    (a) dissymmetry method in a theta solvent (cf. Section 13.1.5),
    (b) Zimm's plot in a theta solvent (cf. Section 13.1.6);

(iii) small-angle X-ray scattering;

(iv) stress–temperature dependence of undiluted or swollen samples as a function of r_0.

Bibliography: 402, 456, 457, 943, 1307, 1429, 1430.
References: 2031, 2040, 2052–2054, 2147, 2148, 2217, 2339, 2358, 2419, 2444, 2499, 2533, 2542, 2579, 2651, 2666, 2825, 2831, 2864, 3114, 3135, 3147, 3148, 3179, 3187, 3188, 3197, 3198, 3204, 3402, 3416, 3459–3461, 3499, 3547, 3805, 3840, 3841, 4483, 4695, 4724, 4767, 4771, 4792, 4805, 4844, 4851, 4877, 4878, 4940, 5086, 5088, 5118, 5141, 5142, 5152, 5182, 5347, 5387, 5405, 5524, 5547, 5699, 5836, 5852, 5881, 6094, 6105, 6114, 6154, 6191, 6338, 6341, 6451, 6723, 6836, 6867, 6899, 6922, 6923, 6994, 7042, 7145, 7326.

Unperturbed dimensions of linear chain molecules are collected in The Polymer Handbook (B: 786).

2.12 UNPERTURBED DIMENSIONS OF BRANCHED MACROMOLECULES

Branched macromolecules have a higher average segment density than linear chain unbranched macromolecules of the same molecular weight, and have a lower coil volume.

Branched macromolecules in solution are determined by

(i) The NUMBER OF BRANCH POINTS (g), which is the ratio of the chain dimension of the branched ($\bar{s}_{0B}^2$) and linear ($\bar{s}_{0L}^2$) macromolecules of the same molecular weight in a theta solvent,

$$g = \frac{\bar{s}_{0B}^2}{\bar{s}_{0L}^2} \qquad (2.32)$$

(dimensionless), where $\bar{s}^2$ is the radius of gyration. The NUMBER OF BRANCH POINTS (g) determines the number of branch points for a randomly branched polymer having trifunctional and tetrafunctional branch points for both mono- and polydisperse systems.

(ii) The EXPANSION FACTOR (α_B) is the ratio of the chain dimension of the same branched polymer in a good solvent ($\bar{r}_B^2$ or $\bar{s}_B^2$) and in a theta solvent ($\bar{r}_{0B}^2$ or $\bar{s}_{0B}^2$),

$$\alpha_B = \frac{(\bar{r}_B^2)^{1/2}}{(\bar{r}_{0B}^2)^{1/2}} = \frac{(\bar{s}_B^2)^{1/2}}{(\bar{s}_{0B}^2)^{1/2}} \qquad (2.33)$$

(dimensionless).

The parameters g and α_B can be obtained for any branched polymer; however, they require characterization in a theta solvent.

Bibliography: 456.
References: 2693, 2744, 3843, 4087, 4746, 4877, 5140, 5153, 5513, 6969, 6979, 7338.

2.13 MEASUREMENTS OF THE UNPERTURBED CHAIN DIMENSIONS FROM THE VISCOSITY OF DILUTE POLYMER SOLUTIONS

The EXPANSION OF THE COIL in a dilute solution is given by the FLORY–KRIGBAUM EQUATION:

$$\alpha^5 - \alpha^3 = 2C_M \, \Psi (1 - \theta/T) \, M^{1/2}, \tag{2.34}$$

where

Ψ is the entropy of mixing parameter,
θ is the theta temperature,
T is the temperature at which expansion of the coil occurs,
M is the molecular weight of the polymer,
C_M is a quantity defined by

$$C_M = \frac{9}{2^{5/2} \, \pi^{3/2}} \left(\frac{\bar{v}^2}{N_A V_1} \right) \left(\frac{M}{\bar{r}_0^2} \right)^{3/2}, \tag{2.35}$$

$\bar{v}$ is the specific volume of a polymer,
N_A is the Avogadro number (cf. Section 40.3),
V_1 is the molar volume of solvent,
$\bar{r}_0$ is the unperturbed mean-square end-to-end distance.

The expansion of the coil in a dilute solution depends upon its molecular weight and upon the thermodynamic interactions with the solvent. At temperatures above the theta temperature the expansion of the coil will be increased. As the temperature is lowered the expansion of the coil will be decreased, reaching zero at the theta temperature. At this temperature the molecular coil has its unperturbed dimensions.

The UNPERTURBED CHAIN DIMENSIONS of linear polymers can be determined from the following equations:

(i) The FLORY–FOX EQUATION,

$$[\eta] \, \overline{M} = \Phi V_h = \Phi (\bar{r}^2)^{3/2}, \tag{2.36}$$

where

$[\eta]$ is the limiting viscosity number (cf. Section 9.1),
$\overline{M}$ is the average molecular weight of the polymer,
Φ is a function related to the hydrodynamic behaviour of macromolecules in solution—for a narrow fraction and theta condition, $\Phi = \Phi_0 = 2.8 \times 10^{21} \; \text{mol}^{-1}$ (Φ_0 is also called the FLORY CONSTANT),
V_h is the hydrodynamic volume (i.e. the volume occupied by the macromolecule) (in cubic centimetres (CGS) or cubic metres (SI),
$\bar{r}^2$ is the mean-square end-to-end distance.

Replacing $\bar{r}^2$ in eq. (2.36) by $\bar{r}_0^2 \alpha^2$ and isolating the term $\bar{r}_0^2 \overline{M}^{-1}$, we obtain an equation which allows us to calculate the unperturbed chain dimensions ($\bar{r}_0^2$):

$$[\eta] = \Phi_0 (\bar{r}_0^2 \, \overline{M}^{-1})^{3/2} \, \overline{M}^{1/2} \alpha^3 = K_\theta \, \overline{M}^{1/2} \alpha^3, \tag{2.37}$$

where

α is the expansion factor, given by

$$\alpha^3 = [\eta]/[\eta]_\theta, \tag{2.38}$$

$[\eta]_\theta$ is the value of $[\eta]$ at the theta temperature,

K_θ is a constant for a given polymer, independent of solvent, temperature, and molecular weight (except in the range of low molecular weight) and given by

$$K_\theta = \Phi_0(\bar{r}_0^2 \, \overline{M}^{-1})^{3/2}, \tag{2.39}$$

K_θ can be obtained from measurement of the limiting viscosity number at the theta temperature,

$$[\eta]_\theta = K_\theta \, \overline{M}^{1/2} \tag{2.40}$$

(ii) The KURATA–STOCKMAYER–FIXMAN EQUATION (Fig. 2.14) (R: 4771, 6583),

$$[\eta] \, \overline{M}^{-1/2} = K_\theta + 0.51 B \Phi_0 \, \overline{M}^{1/2} \tag{2.41}$$

where

$$B = 2\bar{v}^2(\tfrac{1}{2} - \chi)/N_A V_1, \tag{2.42}$$

$\bar{v}$ is the specific volume of a polymer,

V_1 is the molar volume of a solvent,

χ is the Flory interaction parameter (cf. Section 2.9),

N_A is the Avogadro number (cf. Section 40.3).

A plot of $[\eta]/\overline{M}^{1/2}$ versus $\overline{M}^{1/2}$ is a straight line with intercept at the ordinate equal to K_θ and slope $\tan\alpha = 0.51\Phi_0 B$ (Fig. 2.14).

(iii) The FOX–FLORY–SCHAEFGEN EQUATION (Fig. 2.15) (R: 3416),

$$[\eta]^{2/3} \, \overline{M}^{-1/3} = K^{2/3} + K_\theta^{2/3} \, C\overline{M}[\eta]^{-1}, \tag{2.43}$$

where

C is a constant for the polymer–solvent system.

(iv) The BERRY EQUATION (Fig. 2.16) (R: 2376),

$$[\eta]^{1/2} \, \overline{M}^{-1/4} = K_\theta^{1/2} + 0.42 K_\theta^{1/2} \, \Phi_0 \, B\overline{M}[\eta]^{-1}. \tag{2.44}$$

(v) The INAGAKI–SUZUKI–KURATA EQUATION (Fig. 2.17) (R: 4220),

$$[\eta]^{4/5} \, \overline{M}^{-2/5} = 0.786 K_\theta^{4/5} + 0.454 K_\theta^{2/5} \, \Phi_0^{2/3} B^{2/3} \, \overline{M}^{1/3}. \tag{2.45}$$

(vi) The KAMIDE–MOORE EQUATION (Fig. 2.17) (R: 4387),

$$-\ln K + \ln\{2[(a - 1/2)^{-1} - 2]^{-1} + 1\} = (a - 1/2)\ln M_0 - \ln K_\theta, \tag{2.46}$$

where

K and a are constants of the Mark–Houwink–Sakurada equation,

M_0 is the geometric mean of the range of molecular weights of the samples used.

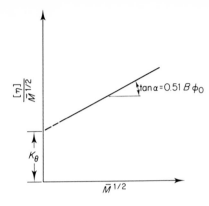

Fig. 2.14 A typical Kurata–Stockmayer –Fixman plot for a linear polymer in a given solvent at constant temperature

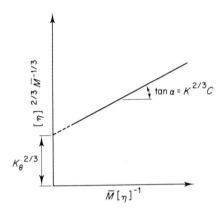

Fig. 2.15 A typical Fox–Flory-Schaefgen plot for a linear polymer in a given solvent at constant temperature

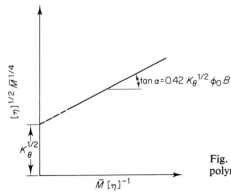

Fig. 2.16 A typical Berry plot for a linear polymer in a given solvent at constant temperature

52

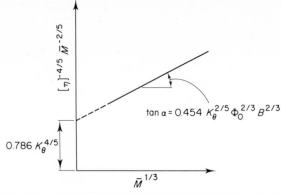

Fig. 2.17   A typical Inagaki–Suzuki–Kurata plot for a linear
polymer in a given solvent at constant temperature

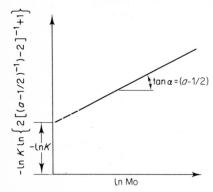

Fig. 2.18   A typical Kamide–Moore plot
for a linear polymer in a given solvent at
constant temperature

Bibliography: 255, 456, 457, 943, 1429.
References: 2231, 2237, 2298, 2393, 2450, 2452, 2650, 2994, 3179, 3194–3198, 3202, 3205, 3206, 3402, 3408, 3415, 3416, 3452, 3805, 3843, 4445, 4454, 4463, 4479, 4483, 4572, 4575, 4771–4773, 5085, 5189, 5388, 5491, 5516, 5770, 5771, 5925, 5961, 6104, 6154, 6303, 6506, 6583, 6638, 6769, 6985, 7145, 7218, 7221, 7252, 7253.

2.14 FORMATION OF LIQUID-CRYSTALLINE SOLUTIONS

Linear flexible-chain polymers in solution are random coiled (Fig. 2.19(a)). If the chains are made of stiff units and they are linked so as to extend the chain in one direction, then they are ROD-LIKE and can be represented in an idealized way by straight lines in a random array (Fig. 2.19(b)). Association with the solvent may contribute to rigidity and to the volume occupied by each polymer molecule.

As the concentration of rod-like molecules is increased, additional polymer may be dissolved by forming regions in which the polymer chains with associated solvent approach a parallel arrangement (Fig. 2.19(c)). These ordered regions are a MESOMORPHIC or LIQUID-CRYSTALLINE STATE and form a phase incompatible with the ISOTROPIC PHASE.

If the rod-like chains are in an approximate parallel array but not organized endwise, the ordered phase is called NEMATIC (Fig. 2.19(d)).

In melts of low molecular weight, liquid-crystalline compounds in which the molecules are of equal length and terminate with groups which provide polar association, layered states are called SMETIC.

The ordered phase is organized in microscopic domains of varying size and direction of orientation. Liquid-crystalline solutions are optically anisotropic, i.e. depolarize plane-polarized light. Solutions of randomly arranged polymer molecules are optically isotropic.

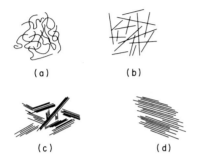

(a)   (b)

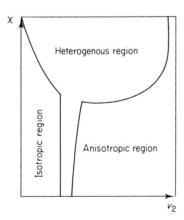

(c)   (d)

Fig. 2.19 Schematic representation of polymer states in solution. (a) Random coils; (b) random rods; (c) rods in liquid-crystalline arrays; (d) nematic state for one array

The behaviour of rigid rod-like polymers in solution is shown in a diagram which plots the solvent–polymer interaction parameter (the Flory–Huggins parameter) (χ) versus the volume fraction (v_2) of rod-like polymers (Fig. 2.20). At a constant χ value, as the polymer volume fraction (v_2) is increased, there is first a single phase with random polymer arrangement, then a narrow heterogeneous region of the random phase mixed with a second ordered or TACTOIDAL PHASE, and finally the system is wholly TACTOIDAL.

χ

Heterogenous region

Isotropic region

Anisotropic region

v_2

Fig. 2.20 Diagram relating the solvent–polymer interaction parameter (χ) and the volume fraction (v_2) of rod-like polymer molecules with the formation of a 'tactoidal' solution phase

Many factors affect formation of liquid-crystalline solutions:

(i) the polymer structure (extended chains);
(ii) the molecular weight (must exceed a minimum value);
(iii) the solubility (must be efficient to exceed the critical concentration);
(iv) the temperature (affects solubility and changes the liquid-crystalline range).

There are several simple effects which are characteristic of liquid-crystalline solutions:

(i) Opalescence. The liquid-crystalline solutions appear cloudy or turbid even though they contain no undissolved material. This is due to diffraction of light passing through liquid-crystalline domains which have differing directions of alignment and size. When such a system is stirred, even gently, a pearly or opalescent appearance is readily seen. This fades rapidly when stirring ceases.

(ii) Depolarization of plane-polarized light. A thin layer of solution, placed between crossed polarizers in a microscope, exhibit coloured domains (blue, green, red, yellow) which vary with the system and its history. The sample shown retains some directional character from shear. Others may be finely mottled or have far less colour.

(iii) The CRITICAL CONCENTRATION POINT. Solutions of extended chain polymers exhibit a sudden change in bulk viscosity with increasing polymer concentration at the point where liquid-crystalline order begins (Fig. 2.21). On the left of Fig. 2.21, solution viscosity rises in the normal

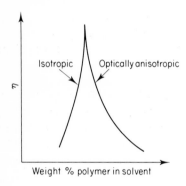

Fig. 2.21 Effect of polymer concentration and the formation of a liquid-crystalline solution on the bulk viscosity

way for isotropic solutions and then, at a CRITICAL CONCENTRATION POINT, the system will only accommodate more polymer if an ordered, liquid-crystalline phase can form. This occurs and is followed by decreasing viscosity and increasing amounts of ordered phase. When such mixed phases are not too viscous, they may separate upon standing or be separated in a centrifuge. The ordered, anisotropic phase, being more concentrated in polymer than the isotropic phase, appears as the lower layer when the solvent density is less than that of the polymer.

References: 3404, 3405, 5351.

2.15 SWELLING OF CROSSLINKED POLYMERS IN SOLVENTS

When a crosslinked polymer is placed in a liquid which is a solvent for the uncrosslinked polymer, it swells to an extent which is dependent upon the interaction between the polymer and the solvent, defined by

(i) a swelling potential;
(ii) the elastic potential (which is determined by the crosslink density).

Equilibrium is reached when these two potentials, which act in opposite directions, are equal (the increase of elastic energy of the chains forming the network balances the decrease in the free energy consequent to mixing of polymer segments with solvent molecules). This leads to the well-known FLORY SWELLING EQUATION:

$$\frac{1}{(\overline{M}_n)_c} = \frac{2}{(\overline{M}_n)_0} - \frac{(\overline{v}/V_1)[\ln(1-v_s)+v_s+\chi v_s^2]}{[v_s^{1/3}-\frac{1}{2}v_s]}, \tag{2.47}$$

where

$(\overline{M}_n)_c$ is the number-average molecular weight of a network chain,
$(\overline{M}_n)_0$ is the number-average molecular weight of the primary chain (a primary
　　chain is the linear molecule before crosslinking),
$\overline{v}$ is the specific volume of polymer,
V_1 is the molar volume of the solvent,
v_s is the final swollen equilibrium polymer volume fraction,
χ is the Flory polymer–solvent interaction parameter (cf. Section 2.9).

The value of the swelling equilibrium is related to the nature of the polymer–solvent system and provides information concerning the nature of the cross-linking and reinforcement.

In order to determine the FINAL SWOLLEN EQUILIBRIUM POLYMER VOLUME FRACTION (in per cent) it is necessary to place a sample of known density in the chosen solvent until weight measurements indicate the saturation of the polymer by liquid. Assuming that no extractables are present, and that all the absorbed solvent causes swelling, the final swollen equilibrium polymer volume fraction (v_s) is given by

$$v_s = \frac{(W_1 - W_0)\rho_p}{W_0 \rho_s} \times 100 \tag{2.48}$$

where

W_0 is the sample weight before swelling,
W_1 is the sample weight after swelling,
ρ_p is the density of polymer sample,
ρ_s is the density of the solvent.

In order to obtain the CURVE OF SWELLING RATE (Fig. 2.22), it is necessary to make a series of gravimetric measurements and determine the percentage volume swelling or weight increase as a function of time. Accurate determination of the swelling–time relationship requires a large number of experimental measurements, especially in the early stages of swelling. The gravimetric method usually requires an interruption of the swelling process, and this involves experimental difficulties. The accuracy of data obtained by this method is low.

56

Sol–gel analysis is carried out by soaking polymer samples in a given solvent at a given temperature (e.g. 20 °C) for a given time (from minutes up to hundreds of hours). After the soaking, the gel is obtained by drying the residual sample *in vacuo* at room temperature.

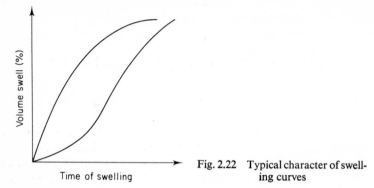

Fig. 2.22 Typical character of swelling curves

The SOL is determined as the difference in weight before and after the swelling experiments.

The SOL FRACTION (in per cent) is expressed by

$$\text{Sol fraction} = \frac{\text{Sol}}{\text{Sol} + \text{Gel}} \times 100. \qquad (2.49)$$

Bibliography: 456.
References: 2589, 2617, 2618, 2826, 3433, 3434, 3686, 3845, 4043, 4227, 4376, 4379, 4818, 5008, 5009, 5198, 5237, 5479, 5764, 6057, 6063, 6319, 6320, 6486, 6799, 6829, 7296, 7297.

Chapter 3

Molecular Weight Averages

General bibliography: 23, 125, 127, 159, 402, 424, 456, 681, 871, 1022, 1082, 1120, 1336, 1350, 1396.

The use of dimensionless molecular weights can lead to problems in attempting to devise a SI unit system for calculation purposes. For that reason it is sometimes recommended that the MOLECULAR MASS be used instead of molecular weight; MOLECULAR MASS can be alternatively described as the MOLAR MASS, defined as the mass of 1 mole of the substance (in kilograms per mole (SI)).

In the SI system, the MOLE is a fundamental quantity defined as the mass of a number of particles equal to the number of atoms present in exactly 0.012 kg of carbon-12. This number, equal to 6.023×10^{23} mol^{-1} is the Avogadro number (cf. Section 40.3).

The following molecular weight (and molecular mass) terminology is used because different techniques of molecular weight (molecular mass) determination measure completely independent polymer properties, and the data obtained vary widely.

The NUMBER-AVERAGE MOLECULAR WEIGHT (the NUMBER-AVERAGE MOLAR MASS) ($\overline{M}_n$) is defined as the total weight of all solute species divided by the total number of moles present:

$$\overline{M}_n = \sum_i N_i M_i \Big/ \sum_i N_i, \tag{3.1}$$

where

N_i is the number of moles of solute of species i,
M_i is the molecular weight of species i,
$N_i M_i$ is the actual weight of species i.

The number-average molecular weight ($\overline{M}_n$) is highly sensitive to the presence of a small number fraction of low-molecular weight macromolecules.

The number-average molecular weight ($\overline{M}_n$) can be obtained from membrane osmometry, ebulliometry, cryoscopy, and end-group analysis, which are all dependent on the number of molecules present.

The WEIGHT-AVERAGE MOLECULAR WEIGHT (the WEIGHT-AVERAGE MOLAR MASS ($\overline{M}_w$) is defined as

$$\overline{M}_w = \sum_i N_i M_i^2 \Big/ \sum_i N_i M_i. \tag{3.2}$$

The weight-average molecular weight ($\overline{M}_w$) is highly sensitive to the presence of small amounts by weight of high-molecular weight macromolecules.

The weight-average molecular weight ($\overline{M}_w$) can be obtained from light scattering and/or equilibrium ultracentrifugation, which are dependent on the weight of solute present.

The Z and $Z+1$ AVERAGE MOLECULAR WEIGHTS (the Z and $Z+1$ AVERAGE MOLAR MASSES) ($\overline{M}_Z$) and ($\overline{M}_{Z+1}$) are defined as

$$\overline{M}_Z = \sum_i N_i M_i^3 \bigg/ \sum_i N_i M_i^2, \tag{3.3}$$

$$\overline{M}_{Z+1} = \sum_i N_i M_i^4 \bigg/ \sum_i N_i M_i^3. \tag{3.4}$$

It is possible to calculate even higher averages by increasing the exponents of M_i.

The VISCOSITY-AVERAGE MOLECULAR WEIGHT (the VISCOSITY-AVERAGE MOLAR MASS) ($\overline{M}_v$) is defined as

$$\overline{M}_v = \left[\sum_i N_i M_i^{a+1} \bigg/ \sum_i N_i M_i \right]^{1/a}. \tag{3.5}$$

If $a = 1$, then $\overline{M}_v = \overline{M}_w$.

The viscosity-average molecular weight ($\overline{M}_v$) can be obtained only from viscosity measurements (cf. Section 9.1.2).

Between the different molecular weight averages we observe the following relationship (Fig. 3.2):

$$\overline{M}_n < \overline{M}_v \leqslant \overline{M}_w < \overline{M}_Z < \overline{M}_{Z+1} \tag{3.6}$$

Note: in this book all equations presented are defined in terms of molecular weight averages instead of molar mass averages.

Bibliography: 23, 116, 124, 125, 127, 159, 248, 250, 309, 630, 943, 1289, 1366.
References: 2222, 4023, 5023, 5310.

3.1 METHODS OF EXPRESSING MOLECULAR WEIGHT DISTRIBUTION: GRAPHICAL PRESENTATION

Three types of graph of molecular weight distribution (MWD) are used:

(i) The INTEGRAL MOLECULAR WEIGHT DISTRIBUTION CURVE is obtained by plotting the cumulative weight fraction versus molecular weight (Fig. 3.1). The CUMULATIVE WEIGHT FRACTION corresponding to a molecular weight, say M_1, is the weight fraction of all molecular species having molecular weight smaller than or equal to M_1.

The function representing the relation between the cumulative weight fraction and the molecular weight is called the INTEGRAL DISTRIBUTION FUNCTION, $I(M)$.

(ii) The DIFFERENTIAL MOLECULAR WEIGHT DISTRIBUTION CURVE is obtained by plotting the weight fraction versus molecular weight (Fig. 3.2). The total area bounded by the differential molecular weight distribution curve and the abscissa is unity.

The function representing the relation between the weight fraction and the molecular weight is called the DIFFERENTIAL WEIGHT DISTRI-BUTION FUNCTION, $W(M)$.

The differential molecular weight distribution curve can be obtained by graphically differentiating the integral molecular weight distribution curve.

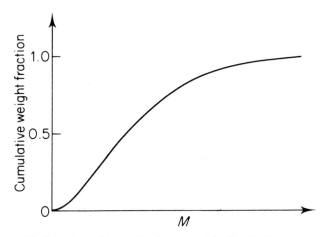

Fig. 3.1   Typical integral molecular weight distribution curve

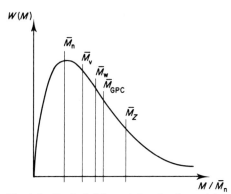

Fig. 3.2   Typical differential molecular weight
distribution curve

(iii) The NUMBER DISTRIBUTION CURVE is obtained by plotting the mole fraction versus molecular weight.

The function representing the relation between the mole fraction and molecular weight is called the NUMBER DISTRIBUTION FUNC-TION, $N(M)$.

60

The following equations describe the relations between the various functions discussed above:

$$dI(M)/dM = W(M),\tag{3.7}$$

$$I(M) = \int_0^M W(M)\,dM,\tag{3.8}$$

$$N(M) = \frac{W(M)}{M}\Bigg/\int_0^\infty \frac{W(M)}{M}\,dM,\tag{3.9}$$

$$\int_0^\infty W(M)\,dM = 1,\tag{3.10}$$

$$I(\infty) = 1,\tag{3.11}$$

$$\int_0^\infty N(M)\,dM = 1.\tag{3.12}$$

Bibliography: 35, 125, 309, 630, 689, 1022, 1309.
References: 2306, 2491, 3076, 3927, 4666, 5398, 6861.

3.2 DISTRIBUTION FUNCTIONS

The molecular weight distribution of polymers can sometimes be described by analytical functions of two or more parameters. Many molecular weight distribution functions have been proposed. The most common are

(i) the SCHULTZ DISTRIBUTION FUNCTION;
(ii) the TUNG DISTRIBUTION FUNCTION;
(iii) the LOG-NORMAL DISTRIBUTION FUNCTION.

More details can be found in the original publications.

Bibliography: 72, 527, 630, 1314.
References: 2822, 3519, 3714, 4137, 4817, 6272, 6860, 7053.

3.3 MOLECULAR WEIGHT DISTRIBUTION INDICES

A single number or index has also been used to describe the molecular weight distribution of a polymer. Several indices for molecular weight distribution have been proposed:

(i) the POLYDISPERSITY INDEX, defined as $\overline{M}_w/\overline{M}_n$;
(ii) the INHOMOGENITY FACTOR, U, defined as

$$U = (\overline{M}_w/\overline{M}_n) - 1;\tag{3.13}$$

(iii) the SQUARE ROOT OF THE INHOMOGENITY FACTOR, Δ, defined as

$$\Delta = U^{1/2} = [(\overline{M}_w/\overline{M}_n) - 1]^{1/2};\tag{3.14}$$

(iv) the POLYDISPERSITY INDEX FOR THE HIGHER MOLECULAR
WEIGHT ENDS, g, defined as

$$g = [(\overline{M}_Z/\overline{M}_w) - 1]^{1/2} \qquad (3.15)$$

The most commonly used index is the polydispersity index, defined as the
ratio of weight-average molecular weight to number-average molecular weight
$\overline{M}_w/\overline{M}_n$. This ratio is unity for a polymer of uniform molecular weight and
becomes larger as the distribution becomes broader. It is dependent on the
synthetic conditions under which the polymer is produced (Table 3.1).

Table 3.1   Typical ranges of $\overline{M}_w/\overline{M}_n$ in synthetic polymers (from:
127, 309; reproduced by permission of John Wiley & Sons)

Polymer	$\overline{M}_w/\overline{M}_n$
Hypothetical monodisperse polymer	1.000
Actual 'monodisperse' 'living' polymers	1.01–1.05
Addition polymer, termination by coupling	1.5
Addition polymer, termination by disproportionation, or condensation polymer	2.0
High conversion vinyl polymers	2–5
Polymers made with autoacceleration	5–10
Coordination polymers	8–30
Branched polymers	20–50

Bibliography: 1219, 1309.
References: 4132, 4978, 6275, 7332.

Polymolecularity correction factors are collected in The Polymer Handbook
(B: 72).

Chapter 4

Fractionation of Polymers

General bibliography: 23, 124, 125, 127, 247, 309, 402, 456, 681, 1082, 1112, 1314, 1350.

The FRACTIONATION of a polymer means the separation of that substance into its different molecular species in order to obtain homogeneous fractions.

The differences between macromolecules may be described by reference to three main properties:

   (i) molecular weight;
   (ii) chemical composition;
   (iii) molecular configuration and structure.

The common name FRACTIONATION of polymers refers to fractionation according to molecular weight.

Fractionation methods can be divided into the following groups:

   (i) fractionation by solubility;
   (ii) fractionation by chromatography;
   (iii) fractionation by sedimentation;
   (iv) fractionation by diffusion;
   (v) fractionation by ultrafiltration through porous membranes;
   (vi) fractionation by zone melting.

Fractionation data are usually arranged according to Table 4.1. The fractions are arranged in the order of ascending molecular weight. The weight fractions $w_1, w_2, ..., w_i$ are the weights of each fraction divided by the total weight of the sample. The average molecular weights of fractions are usually determined by viscosity measurement, or some other method.

The fractions obtained are not uniform in molecular weight distribution (Fig. 4.1). The distributions of the fractions overlap quite extensively. This overlapping may occur to a greater or lesser degree, depending on the efficiency of the fractionation, but is unavoidable in any practical method of separation.

Bibliography: 98, 247, 275, 309, 319, 331, 370, 486, 560, 649, 689, 707, 765, 766, 937, 1137, 1163, 1164, 1186, 1231, 1302, 1311, 1352.
References: 3015, 3016, 3832.

Table 4.1   Typical fractionation data

Fraction	Weight fraction	Average molecular weight
1	w_1	M_1
2	w_2	M_2
3	w_3	M_3
.	.	.
.	.	.
.	.	.
i	w_i	M_i

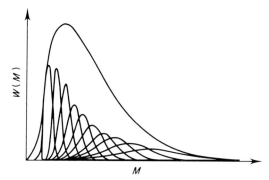

Fig. 4.1   Different distribution of fractions

4.1 FRACTIONATION BY SOLUBILITY

4.1.1 Fractional Precipitation

FRACTIONAL PRECIPITATION is the successive precipitation of polymer fractions from a solution by one of the following methods:

(i)   addition of non-solvent or precipitant;
(ii) elimination of solvent by evaporation;
(iii) lowering the temperature;
(iv) pressure variation at the lower critical solution temperature.

In methods (i)–(iii) the highest molecular weight fractions precipitate first, whereas in method (iv) the lower molecular weight fractions precipitate first.

Bibliography: 319, 707, 765, 770, 937, 1112, 1311.
References: 4384, 4663.

4.1.2 The Non-solvent Addition Method

The NON-SOLVENT ADDITION METHOD is a method whereby successive precipitation of polymer species from solution occurs by addition of a miscible non-solvent.

A polymer sample dissolved in a suitable solvent (0.1–1.0 wt %) is placed in a three-necked fractionation flask (Fig. 4.2(a)). A non-solvent is gradually added

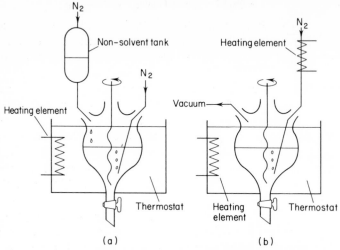

Fig. 4.2 (a) Fractionation flask used in the non-solvent addition method.
(b) Fractionation flask used in the solvent evaporation method

during vigorous agitation at constant temperature under an inert gas atmosphere. After a certain amount of non-solvent has been added, the addition of one more drop of non-solvent causes a turbidity that is not removed by agitation. The solution is then allowed to ripen for several hours. After this procedure a further amount of non-solvent is added until the solution turns milky. The solution is then warmed through several degrees to make it transparent again, and is further cooled gradually to the original temperature with vigorous agitation. After several hours the precipitated polymer fraction may be separated through a tap at the bottom of the fractionation flask, or by decanting, siphoning, or drawing it off by means of a syringe.

The remaining liquid is treated with a further volume of non-solvent, using the same procedure.

Polymer fractions are separated and dried. The final fraction is obtained by evaporating the liquid.

The great disadvantage of this fractional precipitation method is that every fraction contains an appreciable amount of the lower molecular weight components, and this effect is called the TAIL EFFECT.

In order to decrease this effect it is necessary to refractionate each fraction. In order to avoid the expansion of the liquid volume it is recommended that one combines fractional precipitation with an evaporation process. The disadvantage of this method is that fractional precipitation with 4–6 refractionations may take up to one month to carry out.

Bibliography: 560, 770, 1256.
References: 2492, 2529, 2701, 3401, 3902, 4621, 4622, 5075, 5227, 5240, 5246, 5303, 5391, 5457, 5551, 5575, 6296, 6783, 6879, 6882.

The solvent–non-solvent systems for different polymers are collected in Table 2.1. More detailed data are collected in the Polymer Handbook (B: 487).

4.1.3 The Solvent Evaporation Method

The SOLVENT EVAPORATION METHOD is a method whereby successive precipitation of polymer species from a solution of the polymer in a solvent–non-solvent mixture occurs by controlled evaporation of the more volatile solvent.

A polymer sample dissolved in a suitable solvent (0.1–1.0 wt %) is placed into a three-necked fractionation flask (Fig. 4.2(b)). A non-solvent is gradually added during vigorous agitation at a constant temperature under an inert gas atmosphere until the cloud point is reached. Evaporation of the solvent–non-solvent mixture is carried out by a warm, dry gas stream which passes through the flask, from which the atmosphere is removed by a water pump. The degree of precipitation may be controlled by a turbidimeter. When the solution reaches some fixed degree of turbidity, the air stream is slowed down and then stopped. The solution is then warmed through several degrees to make it transparent again, and is further cooled gradually to the original temperature with vigorous agitation. After several hours the precipitated polymer fraction may be separated through a tap at the bottom of the fractionation flask, or by decanting, siphoning, or drawing it off by means of a syringe. The advantages of this method are that

(i) the volume of the system decreases as fractionation proceeds;
(ii) continuous change of the solvent–non-solvent composition is possible;
(iii) local concentration of the non-solvent can be avoided;
(iv) it is possible to double the experimental scale using the same equipment.

Bibliography: 560, 770.

4.1.4 The Temperature Lowering Method

The TEMPERATURE LOWERING METHOD is a method whereby successive precipitation of polymer species from a solution occurs by controlled cooling.

A polymer sample is dissolved in a suitable warm solvent. Care must be taken that degradation of the polymer does not occur when high temperatures are involved. The cooling process is accompanied by vigorous agitation and should be sufficiently slow to approach quasi-equilibrium; usually it takes several hours. The size of fraction is determined by the temperature difference. Addition of a small amount of non-solvent may improve the fractionation process. The procedure for this method is simple and does not require any special equipment; a typical fractionation flask (Fig. 4.2) may be used.

Bibliography: 560, 770.
References: 2313, 2828–2830, 3421, 4257, 4471, 4687, 5443.

4.1.5 The Pressure Variation Method

The PRESSURE VARIATION METHOD is a method whereby successive precipitation is carried out at a lower critical solution temperature under isothermal conditions by changing the pressure of the system. The lowest molecular weight fraction precipitates first. The efficiency of the method is poor in the range of low molecular weights.

4.2 TURBIDIMETRIC TITRATION

TURBIDIMETRIC COMPOSITION TITRATION is an analytical method for measuring molecular weight distributions by continuous precipitation of polymer molecules from a very dilute solution by progressive addition of non-solvent. The higher molecular weight molecules precipitate first and increase the TURBIDITY (i.e. there is an increase in the optical density of the solution) (τ) (cf. Section 13.1.2). As more non-solvent is added, additional lower molecular weight molecules precipitate and the turbidity is continuously increased.

By calibrating with fractions of known molecular weight, the relationship between the amount of non-solvent added and the molecular weight may be established. The magnitude of the turbidity is a measure of the amount of polymer.

TURBIDIMETRIC TEMPERATURE TITRATION is an analytical method for measuring molecular weight distributions by continuous precipitation of polymer molecules from a very dilute solution by progressive change of temperature.

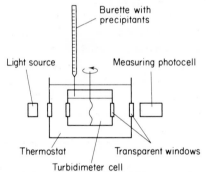

Fig. 4.3 Apparatus for turbidimetric titration

Turbidimetric titration equipment (Fig. 4.3) is composed of the following parts:

(i) a monochromatic light source;
(ii) a turbidimeter cell (25–100 ml) placed in a thermostat, in which the temperature should be varied within wide limits 20–200 °C±0.01 °C;
(iii) a stirring system (the rate of stirring in the turbidimeter cell must be as slow as possible in order to avoid the formation of unstable solutions but simultaneously to make the solution homogeneous);
(iv) an injection device used to add non-solvent—it must have a constant reproducible flow rate, which should be recorded (the rate of non-solvent addition lies between 0.05 and 2.0 ml min^{-1});

(v) a photocell with amplification and recording units (the scattered-light intensity is measured at a 90° angle).

Depending on the average molecular weight and the solvent–non-solvent system used, the initial polymer sample concentration will range from about 0.5 to 50 mg per 100 ml. At a higher concentration coagulation and flocculation may occur.

The turbidimetric titration which measures only the first detectable turbidity (CLOUD-POINT TITRATION) should be carried out

(i) with different polymer concentrations, and the results are then extrapolated to 100 wt% of polymer concentration;
(ii) with one concentration, but with a number of samples having different molecular weights.

Turbidimetric titration, which measures increased turbidity, is usually carried out with only one concentration of one polymer sample.

There are many sources of error in turbidimetric titration. The most important corrections are as follows:

(i) Corrections for the change in the volume of liquid. Concentration of non-solvent in liquid is not linearly related to the volume of non-solvent injected into the turbidimeter cell.
(ii) Corrections for the change of polymer concentration in liquid. Concentration of polymer does not change linearly with the volume of non-solvent injected into the turbidimeter cell.
(iii) Corrections for the concentration of precipitated polymer.
(iv) Corrections for the change in relative refractive index of the precipitated particles and a liquid. For homopolymers the polymer refractive index does not change. For copolymers there is a difference in refractive index between the comonomers.
(v) Particle size corrections (cf. theory of light scattering, Section 13.1.3).

Turbidimetric titration is a useful method for the study of:

(i) the molecular weight distribution of polymer samples.
(ii) Polymerization processes. Discontinuous or continuous processing, polymerization in bulk, in solution, or in emulsion, variation of the temperature, the catalyst, or the rate of monomer addition.
(iii) Degradation processes in bulk or in a solution.
(iv) Copolymerization processes (particular block and grafting processes.) Considerable information concerning the extent of copolymer formation and the amount of homopolymer may be obtained.

Bibliography: 275, 309, 325, 403, 512, 560, 607, 689, 774, 775, 1017, 1111, 1164, 1171, 1311, 1325, 1366.
References: 2023, 2083, 2084, 2310, 2420, 2509, 2685, 2863, 2912, 2973, 3013, 3014, 3301, 3303, 3304, 3584, 3650, 3705, 3715, 3785, 3796, 3798, 3842, 3913, 3920, 3941, 4084, 4090, 4138, 4211, 4697, 4747, 4748, 4793, 5110, 5226, 5328, 5329, 5350, 5435, 5624, 5684, 5697, 5762, 5733, 5734, 5940, 6135, 6245, 6303, 6350, 6504, 6525, 6563, 6615, 6720, 6760, 6872, 6896.

4.3 SUMMATIVE FRACTIONATION

SUMMATIVE FRACTIONATION is a method which consists of dissolving the polymer in a solvent and slowly adding a relatively large volume (one third) of a non-solvent. The mixture is centrifuged and the precipitated polymer sample is separated from the liquid. This procedure is repeated using a stronger non-solvent to remove more polymer from solution. The polymer from each solution is characterized as to amount and molecular weight to permit construction of molecular weight distribution curves.

Bibliography: 84, 689.
References: 2306, 2406, 2529, 2582–2584, 2967, 3699, 5084, 5966, 6495.

4.4 FRACTIONAL SOLUTION

FRACTIONAL SOLUTION is a method which involves preparing the polymer in an appropriate physical state and then extracting fractions of increasing molecular weight by use of a series of eluants of increasing solvent power. In this method the lowest molecular weight fraction is extracted first, whereas the highest molecular weight fraction is the last.

There are several methods of fractional solution:

(i) DIRECT EXTRACTION is a direct and successive extraction of polymer sample with a liquid of increasing solvent power in an extractor (a Soxhlet) at an appropriate temperature.

Bibliography: 331, 370, 406, 560.
References: 3142, 4100, 4517, 4729, 4995, 5420, 6928, 7214.

(ii) FILM EXTRACTION. This can be carried out by two methods:

(a) Batch operation. A 5–10 μm film of the polymer is deposited on an aluminium foil by dipping the foil into the polymer solution. The film is then dried and cut up into pieces approximately 1 cm × 3 cm in size. For 500–1000 mg of polymer the surface area should be 600–1200 cm^2. The pieces of coated foil are then placed in an Erlenmayer flask and extracted with 100 ml of a suitable solvent–non-solvent mixture by gentle swirling. After equilibrium has been reached, the eluant is removed, and a second portion of somewhat greater solvent added. This procedure is repeated until the fractionation is completed.

(b) Continuous operation. The polymer solution is applied as a thin coating on both sides of a slowly moving belt. On passing through a drying region the solvent evaporates. The thin film on the belt is extracted in a series of tubes containing solvent–non-solvent mixtures of increasing solvent power kept at constant temperature.

Bibliography: 406, 560, 707.
References: 2357, 3534–3536, 3833.

(iii) COLUMN EXTRACTION is a method whereby the polymer is deposited on an inactive support (e.g. sand or glass beads) packed in a column, and successive elution is carried out with a liquid of increasing solvent power. The liquid gradient ranged from 100% non-solvent to 100% solvent with a logarithmic change in solvent composition with time. A common way to produce such a gradient is to have a mixing vessel with a good stirrer initially filled with a non-solvent and to replace the volume removed from the mixing chamber into the column by addition of solvent at an equal rate.

Polymer molecules move up the column, being eluted by the liquid mixture which passes through the column under nitrogen pressure (Fig. 4.4). The column can be heated and extraction can be made by the boiling solvent–non-solvent mixture. If the column has a temperature gradient the column extraction becomes precipitation chromatography (cf. Section 4.7.2).

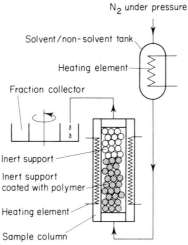

Fig. 4.4   Apparatus for fractionation by
the column elution method

Molecular weight distribution can be determined by collecting eluted fractions, recovering the polymer, measuring the amount and molecular weight of polymer in each fraction, and constructing integral and differential distribution curves from the data.

Bibliography: 80, 199, 406, 707, 1057, 1186.
References: 2193, 2248, 2392, 2505, 2702, 2709, 3050, 3052, 3342, 3446, 3474, 3590, 3625, 3812, 3911, 3912, 4022, 4066, 4134, 4531, 4728, 4816, 4976, 4977, 5380, 5551, 5553, 5589, 5686, 5858, 5876, 6243, 6279, 6314, 6386, 6742, 6781, 6826, 7212, 7213.

(iv) EXTRACTION OF A COACERVATE. In this method a successive extraction of polymer is carried out from a coacervate. The COACERVATE is the polymer-rich liquid phase which separates upon addition of a non-solvent to a polymer solution.

In this method non-solvent is added to the solution until essentially all of the polymer is in the coacervate. The dilute polymer solution, containing the lowest molecular weight fraction, is removed and the polymer in it isolated. The coacervate is then extracted with a mixture having slightly greater solvent power. This procedure is repeated until the fractionation is completed.

Bibliography: 406, 707, 1111.
References: 2391, 2657, 3076, 3081, 3254, 3763, 3905, 4383, 4385, 4386, 4388, 4390–4392, 4395–4399, 4976, 4977, 5443, 5576, 5595, 5996, 6805, 6807, 6860.

4.5 FRACTIONATION BY DISTRIBUTION BETWEEN IMMISCIBLE SOLVENTS

DISTRIBUTION BETWEEN IMMISCIBLE SOLVENTS is a method whereby polymer molecules are distributed according to molecular size between two immiscible liquids of different solvent power. The separation is carried out by a common countercurrent extraction.

References: 2587, 2754, 2755, 3254, 4448, 6879, 6752.

If the solvents are miscible at high temperatures but show phase separation at a lower temperature, it is also possible to carry out a quantitative separation of chemically different polymers dissolved in a suitable homogeneous solvent mixture.

Reference: 4761.

4.6 FRACTIONAL CRYSTALLIZATION

FRACTIONAL CRYSTALLIZATION is a method whereby successive separation of macromolecules from polymer solutions is carried out by crystallization at different temperatures. The crystallization can be induced by fast stirring. The high molecular weight macromolecules crystallize first, settling on the stirrer as long thin fibrillar crystals. The method is poorly reproducible.

References: 2058–2060, 2395, 2495, 2496, 2498, 4378, 4473, 4477, 4522, 4661, 5750, 5752, 5643, 5884, 5903, 7227, 7228, 7229.

4.7 FRACTIONATION BY CHROMATOGRAPHY

4.7.1 Adsorption Chromatography

ADSORPTION CHROMATOGRAPHY (chromatography on an active support) is a method in which the adsorption of polymer species on an active support depends on molecular weight.

There are two general methods of adsorption chromatography:

(i) FRONTAL ANALYSIS. In this method the active support is packed in a column. A solution of polymer passes down the column and is collected when leaving it (Fig. 4.5). The concentration in the effluent changes with the volume collected, and presents successive fronts due to the different adsorption of molecular species on the active support (cf. also Chapter 23).

(ii) ELUTION ANALYSIS can be carried out by two methods:

(a) COLUMN ELUTION. In this method a small quantity of polymer is adsorbed on the upper portion of the support packed in the column. The elution of polymer species is carried out with a suitable solvent. Different polymer species move down the column with different rates and they are collected in the effluent from the column. In the GRADIENT COLUMN ELUTION METHOD the eluant is a liquid of increasing solvent power (cf. also Section 23.13).

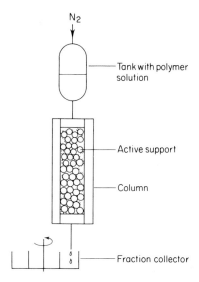

Fig. 4.5 Apparatus for chromatographic fractionation by the frontal analysis method

(b) THIN LAYER CHROMATOGRAPHY (TLC). In this method a small quantity of polymer is adsorbed on a glass plate covered by silica gel (Section 23.15). The glass plate is then placed vertically in a tank containing a suitable solvent. As the solvent moves upwards it carried each component with it, but at different rates. This method allows separation of polymer species according to chemical composition, molecular weight, and stereoregularity. The crucial factor in determining molecular weight distribution is the accurate analysis of the polymer concentration as a function of distance along its elution path. The most accurate method is a direct densitometric scanning of the chromatogram by spectrodensitometer.

References: 4215, 4403, 4404, 5311, 5625–5627, 5630 (cf. also references in Section 23.18).

4.7.2 Precipitation Chromatography

PRECIPITATION CHROMATOGRAPHY (also called the BAKER–WILLIAMS METHOD) is a method whereby a small quantity of polymer is deposited on an inactive support, e.g. glass beads (40–70 μm diameter), packed in a column, and successive elution is carried out with a liquid of increasing solvent power. The liquid gradient ranges from 100% of non-solvent to 100% of solvent with a logarithmic change in solvent composition with time.

A temperature gradient is set along the column. The upper part of the column has a higher temperature (100–200 °C) than the bottom (room temperature) (Fig. 4.6).

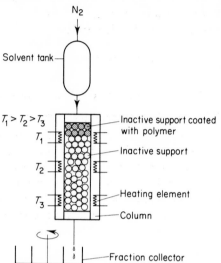

Fig. 4.6 Apparatus for chromatographic fractionation by the Baker–Williams method

Polymer species move down the column, being in a continuous exchange between a precipitated phase and a saturated solution. The distribution depends on molecular size. Molecular weight distribution can be determined by collecting eluted fractions, recovering the polymer, measuring the amount and molecular weight of polymer in each fraction, and constructing integral and differential weight distribution curves from the data.

Precipitation chromatography can also be carried out in the absence of a temperature gradient—it then becomes a continuous fractionation by fractional solution (column extraction method) (cf. Section 4.4).

Bibliography: 80, 152, 689, 853, 1056, 1057, 1163, 1172, 1186.
References: 2206, 2470, 2637, 2702, 2703, 2705, 2713, 2797, 2964, 2965, 3052, 3091, 3428, 3625, 3672, 3784, 3812, 3813, 3839, 3890, 4005, 4177, 4325, 4349, 4375, 4531, 4532, 4600, 4601, 4620, 4642, 4728, 5231, 5257, 5327, 5341, 5554, 5589, 5679, 5766, 5858, 5865, 6058, 6239, 6240, 6241, 6242, 6279, 6285, 6455, 6457, 6487, 6885, 7026, 7215.

4.7.3 Gel Permeation Chromatography

GEL PERMEATION CHROMATOGRAPHY (GPC) is a method in which the separation of polymer molecules is based on the different volumes inside the porous gel particles that are accessible to solute molecules of different size.

A column is packed with small gel particles which have pores of variable size. Solvent fills the interstitial space as well as all the pores of the gel. The sample is dissolved, introduced into the column, and eluted, all with the same solvent. As a result of the restriction imposed by the pore size on the larger molecules, there is a greater pore volume available to the small solute molecules which diffuse freely into and through the gel pores. The end effect is that the small molecules have a longer path length and they, therefore, are more strongly retarded during elution. The molecule separation depends only on the chain length and appears to be insensitive to structure (cf. Section 25.3).

Bibliography: 30, 32, 689 (cf. also references in Section 25.3).

4.7.4 Phase Distribution Chromatography

PHASE DISTRIBUTION CHROMATOGRAPHY (PDC) is a method in which the separation of polymers is achieved by partitioning the sample between the solvent and a polymer phase of very high molecular weight which is coated as a thin layer on an inactive support, e.g. glass beads (0.1 mm diameter). Separation increases with a decrease in the temperature of the solvent, which must be held below the theta-temperature of the sample (Fig. 4.7).

The separation is carried out in the apparatus shown in Fig. 4.8.

Reference: 2758.

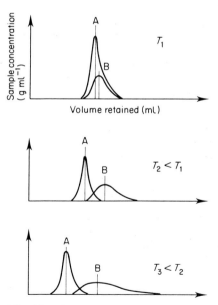

Fig. 4.7 The separation efficiency of phase distribution chromatography (PDC) of two samples A and B as a function of temperature (T)

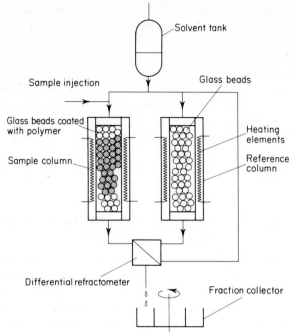

Fig. 4.8   Apparatus for phase distribution chromatography

4.8 FRACTIONATION BY SEDIMENTATION

Ultracentrifugation is a classical method for determining the molecular weights of high polymers, and their molecular weight distribution curves. There are three general methods applied in sedimentation analysis:

(i) the sedimentation velocity method (cf. Section 8.1);
(ii) the sedimentation equilibrium method (cf. Section 8.2);
(iii) the density gradient method (cf. Section 8.4).

Bibliography: 689, 890 (cf. also references in Chapter 8).

4.9 FRACTIONATION BY DIFFUSION

Fractionation of polymers according to their diffusion is another analytical technique. There are two general methods available:

(i) ISOTHERMAL DIFFUSION (at constant temperature). In this method polymer molecules diffuse at different rates from a solution into a boundary, depending on their molecular weight. The diffusion constants can be determined by optical methods and related to the polydispersity of the sample (cf. Section 8.1.5).

Bibliography: 224, 689.
References: 2696, 2697, 3061, 3062, 3508, 3658, 4273, 4274, 5112, 5279, 5321, 5322, 6929.

(ii) THERMAL DIFFUSION. In this method the temperature gradient induces thermal circulation of the molecules in a polymer solution, and separates them. The polymer solution is injected into the cascade column. Each cascade is a reservoir which has a temperature gradient between the upper and lower surfaces. Under this temperature gradient the molecules migrate towards the lower surface. The content of the bottom reservoir of column 1 (cascade 1) becomes the feed to column 2 (cascade 2), the bottom of column 2 becomes the feed to column 3, and so on for eight cascades, in which only the bottom product is passed on.

Bibliography: 408, 689.
References: 3636–3639, 3641, 3642, 3648, 3834, 4083, 4136, 4673, 4676, 4677, 4719, 4774, 4812–4815, 5400, 6758, 6775, 6776.

4.10 FRACTIONATION BY ULTRAFILTRATION THROUGH POROUS MEMBRANES

ULTRAFILTRATION THROUGH POROUS MEMBRANES is a method for the diffusion of molecules through a series of membranes of different porosity. The rate of diffusion depends on the molecular size and the degree of permeability of the membranes. These highly porous membranes are made of pure biologically inert cellulose nitrate, cellulose acetate, regenerated cellulose, and other polymers, and are classified as SURFACE FILTERS. In contrast to DEPTH FILTERS which are made of fibrous materials, it is their exceptionally homogeneous porous structure and uniform pore diameter which gives them such a precise retention efficiency. The greater proportion of substance filtered out of a solution is retained on the surface (SIEVE EFFECT).

DEPTH FILTERS function both mechanically and by adsorption through the thickness of the filter matrix. They consist of irregularly placed fibres or very thick sheets of compressed granular material. Differences in fibre or grain size, as well as in packing down and in thickness, mean that a certain amount of variation can occur in the channel size of a depth filter. Depth filters are commonly used as efficient prefilters. The filtering is not affected by changes of pressure within the filtering system or by the pH value of the solution being filtered, and is only slightly impaired by substances with an active interfacial tension. Filtering effect and flow rate can only be reduced by the clogging of the filter surface.

MEMBRANE-LAMINATES—these are produced by a coating process in which a uniformly thin and porous cellulose nitrate film is bonded to the surface of a cardboard support made of highly refined cellulose. This film has the sponge-like structure typical of surface filters with well-defined pore sizes. The wet strength of the support gives the membrane-laminates the strength they need to stand up to the handling received in filtering process using press. The membrane-laminates remain operationally stable even when used to filter fluids with pulsating flow currents. The film layer determines the output, chemical stability, and retention efficiency of the filter. It functions like a micro-fine sieve.

76

With a suitable strong support, porous membrane filters can withstand pressures up to 800 atm cm^{-2}. The maximum pressure is limited by the construction of the filtering apparatus. In practice, the working pressure in filtering equipment in general does not need to exceed 5–6 atm above atmospheric. Most of membranes commercially produced have a temperature resistance of up to 80–125 °C.

Bibliography: 249.
References: 2208, 2571, 3713, 3763, 3999, 4000, 4088, 5258, 6095, 6523, 6949, 7088.

4.11 FRACTIONATION BY ZONE MELTING

FRACTIONATION BY ZONE MELTING is a method based on the presence of different solute concentrations in a solid- and a liquid-solution phase at equilibrium. A poor solid solvent is packed in a column and a small amount of polymer is placed on top of the solid solvent and is further dissolved by heating a narrow zone (Fig. 4.9).

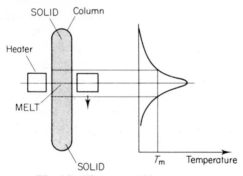

Fig. 4.9   The zone melting apparatus

After solidification, a lower zone is heated, melted, and resolidified. All the column is treated in this way, step by step. Polymer species move down the column at different rates depending on their molecular size during the molten state. At the end the polymer is distributed throughout the entire column and recovered by cutting the solid into several portions and removing the solvent.

Bibliography: 249.
References: 3288, 4959, 4976, 4977, 5734, 6130.

4.12 FIELD-FLOW FRACTIONATION

FIELD-FLOW FRACTIONATION (FFF) is an elution method whereby an external force field or gradient is applied, at right angles to the flow, across the entire face of the channel. The field interacts with the solute, forcing it towards the far wall and thus forming a thin steady-state layer against the wall.

The thickness of the layer is different for each polymer and depends on the strength of the coupling between the field and the macromolecules, and on the diffusion coefficient peculiar to them. The layer which flows along the channel (Fig. 4.10) has a parabolic profile and the fluid velocity is greatest near the centre of the channel. Larger macromolecules are forced more towards the wall than are smaller ones. In this elution method small particles are eluted first and large particles last.

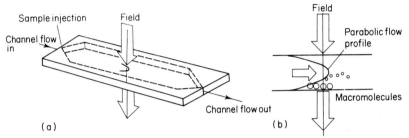

Fig. 4.10   Schematic representation of field-flow fractionation. (a) Flow channel; (b) parabolic flow profile in the channel

The external force fields are applied by using centrifugation or gravitational forces, electrical potential or thermal differences, and other fields.

Apart from the actual column fractionation, all other aspects of this technique, e.g. sample injection, separation, detection, data handling, and the possibility of sample collection, are the same as in the common fractional chromatographic method. Field-flow fractionation can be successfully applied for separating macromolecules that may lie anywhere in the broad molecular weight range from 10^3 to 10^{12}, corresponding roughly to particle sizes from 10^{-3} to 1 µm.

Bibliography: 510, 511.
References: 3636, 3644, 3646, 3647, 6775, 7230, 7231.

Chapter 5

Membrane Osmometry

General bibliography: 23, 114, 125, 127, 309, 681, 1040, 1082, 1301, 1307, 1350.

When the pure solvent is separated by a semipermeable membrane from the solvent in the solution, their chemical potentials are unequal. This causes a flow from the solvent, through the membrane, to the solution. The pressure difference across the membrane at equilibrium is called the OSMOTIC PRESSURE. This pressure difference is measured as the difference of the liquid heights between the solution and solvent capillaries (Δh) (Fig. 5.1).

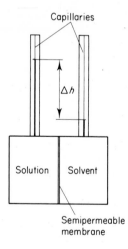

Fig. 5.1 Basic principles of osmotic pressure measurement

The practical range of molecular weights measured by membrane osmometry is from 5×10^3 to 5×10^5 and even 10^6. The lower limit is determined by the permeability of the membrane (diffusion of low molecular weight species), whereas the upper limit is determined by the smallest osmotic pressure which can be measured.

Bibliography: 77, 306, 635, 779, 1003, 1323, 1365.
References: 2869, 4085, 4804, 6092, 6397, 7142.

5.1 DETERMINATION OF THE NUMBER-AVERAGE MOLECULAR WEIGHT ($\overline{M}_n$) FROM OSMOTIC MEASUREMENTS

For a polydisperse linear polymer, the number-average molecular weight ($\overline{M}_n$) can be calculated from osmotic data using

$$\frac{\pi}{c} = \frac{RT}{\overline{M}_n} + Bc + Cc^2 + ..., \tag{5.1}$$

$$\frac{\pi}{c} = RT\left(\frac{1}{\overline{M}_n} + A_2 c + A_3 c^2 + ...\right), \tag{5.2}$$

$$\frac{\pi}{c} = \frac{RT}{\overline{M}_n}(1 + \Gamma_2 c + \Gamma_3 c^2 + ...) \tag{5.3}$$

(in dyne l g^{-1} cm^{-1} (CGS) or J kg^{-1} (SI)), where
π is the OSMOTIC PRESSURE, given by

$$\pi = \rho g \, \Delta h, \tag{5.4}$$

ρ is the solvent density,
g is the acceleration due to gravity (0.981 m s^{-2}),
Δh is the pressure expressed in centimetres of solvent,
c is the solution concentration,
π/c is called the REDUCED OSMOTIC PRESSURE,
R is the universal gas constant (cf. Section 40.2),
T is the thermodynamic temperature (in kelvins),
B, A, Γ_2 and C, A_3, Γ_3 are virial coefficients,
$\overline{M}_n$ is the number-average molecular weight.
In order to obtain $\overline{M}_n$ in the correct units it is necessary to express

π/c in dyne l g^{-1} cm^{-1} and $R = 0.082\ 07$ l atm mol^{-1} K^{-1} (CGS),

π/c in J kg^{-1} and $R = 8.314$ J K^{-1} mol^{-1} (SI).

The third term ($A_3 c^2$) is negligibly small, and eq. (5.2) will have the form

$$\pi/c = RT(1/\overline{M}_n + A_2 c). \tag{5.5}$$

Thus, a plot of π/c versus c is a straight line with intercept equal to $RT/\overline{M}_n$ and slope A_2 (Fig. 5.2). If measurements are made in the θ condition, π/c is independent of c and is equal to $RT/\overline{M}_n$:

$$\pi/c = RT/\overline{M}_n. \tag{5.6}$$

In practice the osmotic pressure is measured as the difference of the liquid heights between solution and solvent capillaries (Δh) (in centimetres). An extrapolation to infinite dilution requires measurements of the osmotic pressure at several concentrations (e.g. 2.0, 4.0, 6.0, and 8.0 g l^{-1}). The number-average molecular weight $(\overline{M}_n)$ can be calculated from the plot on Fig. 5.2, or from

$$\overline{M}_n = \frac{RT}{(\pi/c)_{c=0}}. \tag{5.7}$$

In the osmotic data treatment it is very important to express experimental data in the proper units.

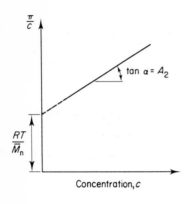

Fig. 5.2 Plot of the reduced osmotic pressure (π/c) versus concentration (c)

5.2 DETERMINATION OF A SECOND VIRIAL COEFFICIENT (A_2) FROM THE OSMOTIC MEASUREMENTS

The second virial coefficient (A_2) is expressed in osmometry as

$$A_2 = \frac{\rho_s}{\rho_p^2 M_s}(\tfrac{1}{2}-\chi), \tag{5.8}$$

where
ρ_s is the density of the solvent,
ρ_p is the density of the polymer,
M_s is the molecular weight of the solvent,
χ is the Flory interaction parameter (cf. Section 2.9).

The units of the second virial coefficient depend upon the way in which it is expressed (Table 5.1).

The second virial coefficient (A_2) can be calculated experimentally from values of (π/c) measured at two concentrations c_1 and c_2 using

$$A_2 = \frac{(\pi/c)_2-(\pi/c)_1}{RT(c_2-c_1)} \tag{5.9}$$

(in l mol g^{-2} (CGS) or m^3 mol kg^{-2} (SI)).

The second virial coefficient (A_2) can also be measured from the slope $\tan \alpha$ in Fig. 5.2.

Table 5.1 Different ways of expressing the reduced osmotic pressure
(π/c) and the second virial coefficient

Equation	Second virial coefficient, B, A_2, Γ_2	
	CGS units	SI units
$\dfrac{\pi}{c} = \dfrac{RT}{\overline{M}_n} + Bc + Cc^2$	dyne l g^{-2}	J m^3 kg^{-2}
$\dfrac{\pi}{c} = RT\left(\dfrac{1}{\overline{M}_n} + A_2\,c + A_3\,c^2\right)$	l mol g^{-2}	m^3 mol kg^{-2}
$\dfrac{\pi}{c} = \dfrac{RT}{\overline{M}_n}\left((1 + \Gamma_2\,c + \Gamma_3\,c^2)\right)$	l g^{-1}	m^3 kg^{-1}

5.3 OSMOTIC PRESSURE OF HIGHLY CONCENTRATED MACROMOLECULAR SOLUTIONS

The reduced osmotic pressure (π/c) for highly concentrated macromolecular solutions is given by

$$\frac{\pi}{c} = \frac{RT}{\overline{M}_n}(1 + \Gamma_2\,c + \Gamma_3\,c^2 + ...), \qquad (5.10)$$

where the factor $(1 + \Gamma_2\,c + \Gamma_3\,c^2 + ...)$ gives the relative deviation from the ideal solution. The virial coefficients Γ_2 and Γ_3 are functions of the solute/solvent system, the temperature, the solute molecular weight, and the molecular weight distribution.

Rearrangement of eq. (5.10) gives

$$\left(\frac{\pi/c}{A_1} - 1\right)\frac{1}{c} = \Gamma_2 + \Gamma_3\,c, \qquad (5.11)$$

where $A_1 = \dfrac{RT}{\overline{M}_n}$.

A plot of $[(\pi c/A_1) - 1](1/c)$ versus c is linear and yields Γ_2 as the intercept at the ordinate and the slope $\tan\alpha = \Gamma_3$ (Fig. 5.3).

Bibliography: 456.
References: 2367, 6947, 6948.

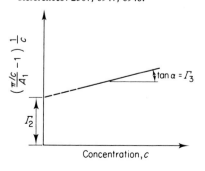

Fig. 5.3 Plot of $\left(\dfrac{\pi/c}{A_1} - 1\right)\dfrac{1}{c}$ versus concentration (c)

5.4 MEMBRANE OSMOMETERS

Membrane osmometers can be divided into two general groups: static osmometers and dynamic osmometers.

STATIC OSMOMETERS are built in two types:

(i) vertical (Helfritz-type) osmometers (Fig. 5.4);
(ii) horizontal (Schultz-type) osmometers (Fig. 5.5).

A static osmometer consists of two cells:

(i) a solution cells, shown in Figs 5.4 and 5.5;
(ii) a solvent cell, which is simply a glass container where the solution cell is placed.

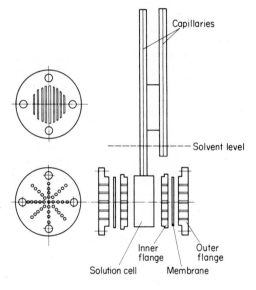

FRONT VIEW   SIDE VIEW
of outer flange   of exploded osmometer

Fig. 5.4   Cell assembly in the vertical (Helfritz-type) osmometer

One or two semipermeable membranes separate the two cells. After filling one cell with pure solvent and the other with a solution of known concentration, the osmometer is placed in a thermostat (± 0.01 °C). After attainment of the isochronogenous pressure, the difference of the liquid heights in capillaries attached to each cell is measured. The ISOCHRONOGENOUS PRESSURE is that pressure at which total volume flow is zero. The principal disadvantage of the static method is the length of time (from several hours to several days) required for attainment of the equilibrium osmotic pressure. This time is largely determined by the time required for the solvent to flow from the solvent cell to the solution cell across the semipermeable membrane.

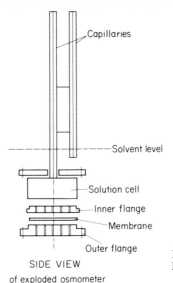

Capillaries

Solvent level

Solution cell

Inner flange

Membrane

Outer flange

SIDE VIEW

of exploded osmometer

Fig. 5.5 Cell assembly in the horizontal (Schultz-type) osmo-meter

Bibliography: 635, 1040, 1323.
References: 2609, 2869, 3316, 3317, 3505, 3553, 3651, 3998, 5107, 5247, 6182, 6271, 6273, 6507, 7335.

DYNAMIC OSMOMETERS are built as automatic devices which measure the flow rate through the membrane and adjust the pressure such that the net flow is zero. The principal advantage of the dynamic method is that the iso-chronogenous pressure is measured after short times (5–10 min) and there is less solute permeation.

The most common dynamic osmometers are

(i) The MECHROLAB OSMOMETER (Fig. 5.6). In this osmometer a horizontal membrane separates two cells, one for the solvent (lower) and a second for the solution (upper). A glass capillary is attached to the cell which contains the solvent. One end of the capillary is connected with a solvent tank through a flexible tube. A bubble of air is introduced into the capillary. The bubble interrupts a lamp beam illuminating a photo-cell and this is used to balance the osmotic flow by a servo system. The difference between the liquid heights at the solution level in the solution cell and at the solvent level in the solvent tank is the measure of the osmotic pressure.

(ii) The SHELL OSMOMETER (Fig. 5.7). In this osmometer a horizontal membrane separates two cells, one for the solvent (upper) and a second for the solution (lower). The solution cell has a flexible metal diaphragm as one wall that acts as one plate of a condenser. This condenser is part of a servomechanism that adjusts the pressure in the solvent cell. Transfer of solvent through the membrane deflects this diaphragm; the resulting change in the frequency of the circuit generates a signal which operates

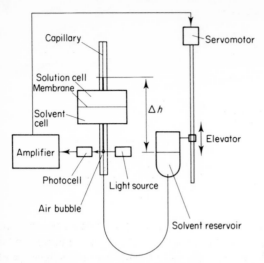

Fig. 5.6   Mechrolab osmometer

a servomechanism. Equilibrium is observed on a pen-recorder trace actuated by this servomechanism, and the slope of the trace indicates any degree of solute permeation through the membrane. A sample volume of 8–10 ml is used, and after 15–30 min equilibrium osmotic pressure can be read with an accuracy of ±0.01 cm. The practical molecular weight range is from about 10 000 to about 300 000 and a temperature range of 20–60 °C can be used (up to 130 °C with modifications).

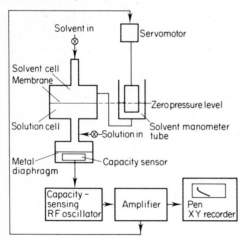

Fig. 5.7   Shell's osmometer

(iii) The MELABS OSMOMETER (Fig. 5.8). In this type of osmometer there is no servomechanism. The osmotic pressure is measured directly by a strain gauge attached to a flexible diaphragm in the solvent cell.

Bibliography: 45, 125, 309, 400, 723, 1323.
References: 2610, 2932, 4095, 5659, 6017, 6088.

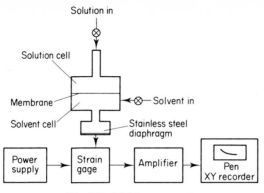

Fig. 5.8   Melabs osmometer

The data from automatic osmometers can be computerized. The input data include details of the solvent, details of weighings, details of dilutions, and the measured head of solvent pressure. The output data are π/c values, $(\pi/c)_0$, $\overline{M}_n$, the second virial coefficient, and the standard deviations.

Bibliography: 311.

5.4.1 Membranes

Membranes used in osmometry should be made from a material which is not dissolved by the solvent used. Gel cellophane membranes in a water-soaked condition are the most widely used type. They are produced by precipitating cellulose xanthate and are never dried out after their manufacture. Other commercially available membranes are made from cellulose hydrate, cellulose acetate, cellulose nitrate, polyurethanes, poly(chlorotrifluoroethylene) (Kel-F type), and even Vycor glass with a 40 Å mesh.

The water-soaked membranes must be CONDITIONED before use in an organic solvent. In this procedure water is replaced by the solvent. If the solvent is not miscible with water, the conditioning must first be carried out with ethanol or acetone and subsequently with the solvent which is to be used in the experiment. Instructions for this procedure are usually included with commercially available membranes.

Bibliography: 45, 282, 307, 398, 635.
References: 2093, 2301, 2738, 2911, 2931, 3300, 3401, 3553, 3651, 4114, 4213, 4766, 5106, 5167, 5179, 5248–5250, 5696, 5816, 6060, 6416, 6524, 6882, 6932, 6933, 6946, 7166, 7233.

5.4.2 Experimental Problems in Membrane Osmometry

Experimental problems in membrane osmometry include the following:

(i) Membrane leakage is difficult to find in static osmometry, whereas in dynamic automatic osmometry it can easily be observed by instability of the baseline.

(ii) Membrane asymmetry can be observed when both cells in the osmometer are filled with identical solvents, at which time a pressure difference will arise between the two cells. This effect is due to leakage, membrane compression, solute contamination, and temperature gradients.

(iii) Ballooning is caused by pressure differentials and can be detected by measuring the pressure change as the solvent is added to or removed from the solvent cell. This effect is due to the viscoelastic nature of the membrane used.

(iv) Rapid membrane degradation.

(v) Presence of dissolved air. Solvent and solution must be degassed before insertion into the cells.

Bibliography: 1323.
References: 3305, 4762, 5815, 6290, 6521, 6522, 6524, 6944, 6945, 6997.

Chapter 6

Colligative Property Methods

General bibliography: 23, 125, 127, 128, 309, 402, 681, 1082, 1350.

COLLIGATIVE PROPERTIES OF A SOLUTION are properties which are dependent on the number of molecules present.
Colligative property methods include

(i) ebulliometry;
(ii) cryoscopy;
(iii) isothermal distillation—the isopiestic method;
(iv) vapour pressure osmometry.

The measured effects are proportional to the molar concentration of the solute. The number-average molecular weight $(\overline{M}_n)$ can easily be calculated when the weight concentration is known. Each method requires measurements at different concentrations and an extrapolation to infinite dilutions. The magnitude of the various effects for different methods is compared and shown in Table 6.1.

Table 6.1 Colligative properties of polymers in solution

Effect	Number–average molecular weight of polymer		
	$\overline{M}_n = 1000$	$\overline{M}_n = 5000$	$\overline{M}_n = 20\,000$
Boiling point elevation (°C)	$2.5{-}3.0 \times 10^{-2}$	5×10^{-3}	1.5×10^{-3}
Freezing point depression (°C)	5×10^{-2}	1×10^{-2}	2.5×10^{-3}
Vapour pressure lowering (torr)	8×10^{-2}	1.5×10^{-2}	4×10^{-3}

6.1 EBULLIOMETRY

EBULLIOMETRY is based on the difference between the boiling temperature of a solution and the boiling temperature of the pure solvent.
For polydisperse linear polymers, the number-average molecular weight $(\overline{M}_n)$ can be calculated from ebulliometric data using

$$\frac{\Delta T}{c} = \frac{K_e}{\overline{M}_n} + Bc + Cc^2 + \ldots, \tag{6.1}$$

$$\frac{\Delta T}{c} = K_e \left(\frac{1}{\overline{M}_n} + A_2 c + A_3 c^2 + \ldots \right), \tag{6.2}$$

$$\frac{\Delta T}{c} = \frac{K_e}{\overline{M}_n}(1+\Gamma_2 c+\Gamma_3 c^2+...),\tag{6.3}$$

(in K l g^{-1} (CGS) or K m^3 kg^{-1} (SI)), where
ΔT is the BOILING-POINT ELEVATION (in kelvins),
c is the concentration of the polymer,
$\Delta T/c$ is the REDUCED BOILING-POINT ELEVATION,
K_e is the ebullioscopic constant (a measure of the elevation of the boiling point caused by 1 mol of solute) given by

$$K_e = RT_s^2 M_s/\rho_s \Delta H_s,\tag{6.4}$$

where
R is the universal gas constant (cf. Section 40.2),
T_s is the solvent boiling point (in kelvins),
M_s is the molecular weight of solvent,
ρ_s is the solvent density,
ΔH_s is the heat of solvent evaporation,
B, A_2, Γ_2 and C, A_3, Γ_3 are virial coefficients.
(The units of K_e depend upon the way in which one expresses all the other symbols: R, T_s, M_s, ρ_s and ΔH_s.)

The ebullioscopic constants for a number of solvents are listed in Table 6.2. The pressure correction dK_e/dp in Table 6.2 has to be added to K_e for each

Table 6.2   Ebullioscopic constants for various solvents (CGS system)

Solvent	Boiling point (°C)	$10^{-3} \times K_e$ (°C mol^{-1})	dK_e/dp (torr^{-1})
Acetic acid	118.0	3.07	0.8
Acetone	56.2	1.71	0.4
Benzene	80.1	2.54	0.7
Bromobenzene	156.1	6.12	1.6
Carbon disulphide	40.3	2.34	0.6
Carbon tetrachloride	76.7	5.03	1.3
Chloroform	61.2	3.64	0.9
Cyclohexane	80.4	2.79	0.7
Ethanol	78.3	1.19	0.3
Ethyl acetate	77.1	2.77	0.7
Ethyl ether	34.6	2.10	0.5
n-Hexane	68.7	2.80	0.7
Iodobenzene	188.4	8.87	2.1
Methanol	64.7	0.84	0.2
Methyl acetate	57.1	2.15	0.5
Methyl ethyl ketone	79.6	2.28	0.4
Nitrobenzene	210.8	5.24	1.3
n-Octane	125.6	4.25	1.0
Phenol	181.8	3.56	0.9
Toluene	110.7	3.33	0.8
Water	100.0	0.51	0.1

millimetre of pressure above 760 torr (1 torr $= \frac{1}{760}$ atm $= 133.322$ N m^{-2}) for each millimetre below this figure.

The number-average molecular weight ($\overline{M}_n$) can be calculated from the plot on Fig. 6.1.

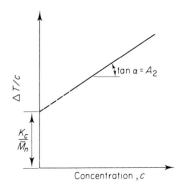

Fig. 6.1   Plot of the reduced boiling point elevation ($\Delta T/c$) versus concentration (c)

Ebulliometry can be applied to the determination of molecular weights of polymers up to 50 000 and even to 100 000 and 170 000.

Bibliography: 124, 423, 424, 518, 810, 1133, 1446.
References: 3675.

6.1.1 Determination of a Second Virial Coefficient (A_2) from Ebulliometric Measurements

The second virial coefficient (A_2) is expressed in ebulliometry as

$$A_2 = \frac{RT^2}{\rho_p^2 \, \Delta H_s}(\tfrac{1}{2} - \chi), \qquad (6.5)$$

where

R is the universal gas constant (cf. Section 40.2),
T is the thermodynamic temperature (in kelvins),
ρ_p is the density of polymer,
ΔH_s is the heat of solvent evaporation,
χ is the Flory interaction parameter (cf. Section 2.9).

The second virial coefficient (A_2) can be measured from the plot of $\Delta T/c$ versus concentration (c) as a curve of slope $\tan \alpha = A_2$ (Fig. 6.1).

6.1.2 Ebulliometer

The main part of an ebulliometer (Fig. 6.2) consists of a small boiler heated internally by a platinum heater. Just above the heater is the bell of a Cottrel pump in which bubbles of vapour lift the boiling solution to the outlet of the pump and over the thermocouple pocket. The end of the thermocouple pocket contains a thermopile measuring junction. The reference junction is placed at a point which is surrounded by a double jacket of vapour. After condensation the vapour returns to the boiler via the dripper. The ebulliometer is placed in a Dewar.

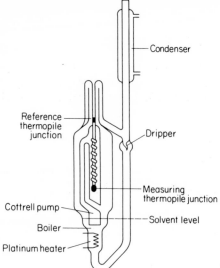

Fig. 6.2 Ebulliometer

The sensitivity with which temperatures can be measured is the limiting factor in the application of ebulliometry to polymer molecular weight determinations. Thermistors and thermopiles are the most commonly used sensors $(1 \times 10^{-5} - 2 \times 10^{-6} \, ^\circ C)$.

Bibliography: 517, 518, 1133.
References: 2143, 2432–2435, 3051, 3163, 3338, 3432, 3673, 3674, 3676, 3907, 4057, 4585, 4871, 5983, 6083, 6253, 6443, 7317.

One of the newest developments is the rotating ebulliometer. The apparatus is equipped with rotating vessels to establish complete equilibrium between liquid and vapour (Fig. 6.3). The boiling temperature can be chosen by reducing the pressure by means of a pressure regulator in the range 100–760 torr

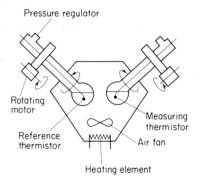

Fig. 6.3 Rotating ebulliometer

(1 torr $= \frac{1}{760}$ atm $= 133.322$ N m^{-2}). The elevation of the boiling point is measured by two thermistors in a Wheatstone bridge circuit.

Reference: 6483.

6.2 CRYOSCOPY

CRYOSCOPY is based on the difference between the freezing temperature of a solution and the freezing temperature of the pure solvent.

For a polydisperse linear polymer, the number-average molecular weight ($\overline{M}_n$) can be calculated from cryoscopic data using

$$\frac{\Delta T}{c} = \frac{K_c}{\overline{M}_n} + Bc + Cc^2 + ..., \tag{6.6}$$

$$\frac{\Delta T}{c} = K_c\left(\frac{1}{\overline{M}_n} + A_2 c + A_3 c^2 + ...\right), \tag{6.7}$$

$$\frac{\Delta T}{c} = \frac{K_c}{\overline{M}_n}(1 + \Gamma_2 c + \Gamma_3 c^2 + ...) \tag{6.8}$$

(in K l g^{-1} (CGS) or K m^3 kg^{-1} (SI)), where
ΔT is the DEPRESSION OF THE MELTING POINT (in kelvins),
c is the concentration of the polymer,
$\Delta T/c$ is the REDUCED DEPRESSION OF THE MELTING POINT,
K_c is the cryoscopic constant (a measure of the depression of the melting-point caused by 1 mol of solute), given by

$$K_c = -RT_s M_s/\rho_s \Delta H_s, \tag{6.9}$$

where
R is the universal gas constant (cf. Section 40.2),
T_s is the solvent freezing-point (in kelvins),
M_s is the molecular weight of a solvent,
ρ_s is the solvent density,
ΔH_s is the heat of solvent fusion,
B, A_2, Γ_2 and C, A_3, Γ_3 are virial coefficients.
(the units of K_c depend upon the way in which one expresses all other symbols: $R, T_s, M_s, \rho_s,$ and ΔH_s.)

The cryoscopic constants for a number of solvents are listed in Table 6.3.

The number-average molecular weight ($\overline{M}_n$) can be calculated from the plot on Fig. 6.4.

Cryoscopy can be applied to the determination of molecular weights of polymers up to 50 000.

Bibliography: 125, 515, 518, 1133.
References: 2155, 4787.

92

Table 6.3 Cryoscopic constants for various solvents
(CGS system)

Solvent	Melting point (°C)	$10^{-3} \times K_c$ (°C mol^{-1})
Acetic acid	16.6	3.9
Benzene	5.5	5.1
Camphor	178.0	40.0
Dimethylsulphoxide	18.4	4.8
Diphenyl	70.0	8.0
Diphenylamine	53.0	86.0
Ethylene dibromide	10.0	11.8
Naphthalene	80.2	6.8
Nitrobenzene	5.7	7.0
Phenol	41.0	7.4
Succinonitrile	58.8	20.3
Triphenylmethane	92.5	12.4
Urethane	50.0	5.1
Water	0.0	1.8

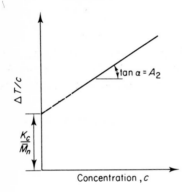

Fig. 6.4 Plot of the reduced depression of the melting point ($\Delta T/c$) versus concentration

6.2.1 Determination of a Second Virial Coefficient (A_2) from the Cryoscopic Measurements

The second virial coefficient (A_2) is expressed in cryometry as

$$A_2 = \frac{RT^2}{\rho_p \Delta H_s}(\tfrac{1}{2} - \chi),$$ (6.10)

where

R is the universal gas constant (cf. Section 40.2),
T is the thermodynamic temperature (in kelvins),
ρ_p is the density of polymer,
ΔH_s is the heat of solvent fusion,
χ is the Flory interaction parameter (cf. Section 2.9).

The second virial coefficient (A_2) can be measured from the plot of $\Delta T/c$ versus concentration (c) as a curve of slope $\tan \alpha = A_2$ (Fig. 6.4).

6.2.2 Cryoscopic Cells

The main part of a cryoscopic cell (Fig. 6.5) consists of a Pyrex glass tube with an air jacket. The lower part of the cryoscopic cell is immersed in a thermostatic bath maintained at about 1.5 K below the melting point of the solvent. The

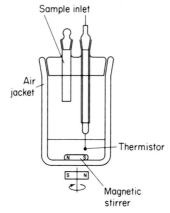

Sample inlet

Air jacket

Thermistor

Magnetic stirrer

Fig. 6.5   Cryoscopic cell

liquid in the cryoscopic cell is stirred by a magnetic stirrer. A thermistor is used as a sensor (1×10^{-4} °C). The sample is introduced by the tube. About 5 g of the purified solvent are placed into the weighed cryoscopic cell. The cell is

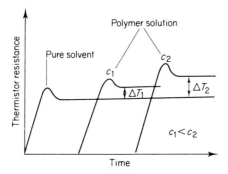

Fig. 6.6   Typical freezing curves

then put in the thermostatic bath and a freezing curve recorded for the pure solvent (Fig. 6.6). The cell is then rewarmed, a known amount of the sample is introduced through the tube, and a second freezing curve is recorded.

Bibliography: 515.
References: 2155, 2378, 3432, 3668, 3669, 4485, 5466, 6389, 6569, 6884, 6958, 7138.

6.3 ISOTHERMAL DISTILLATION—THE ISOPIESTIC METHOD

The ISOPIESTIC METHOD is based on the isothermal transfer of solvent due to concentration gradients. Distillation will occur from one solution to the other until the vapour pressures are equal (ISOPIESTIC CONDITIONS).

The number-average molecular weight ($\overline{M}_n$) can be calculated from

$$\frac{1}{\overline{M}_n} = \frac{1}{M_s} \lim_{c \to 0} \frac{c_s}{c_p}, \tag{6.11}$$

where

M_s is the molecular weight of the standard substance,
c_s is the concentration of the standard solution,
c_p is the concentration of the polymer solution.

The practical limit of molecular weights measured by the isopiestic method is up to 20 000 (an accuracy $\pm 5\%$).

Bibliography: 125, 1133, 1446.

6.3.1 Isopiestic Distillation Apparatus

The isopiestic distillation apparatus is built in the form of an H-tube (Fig. 6.7). The two tubes, A (reference solution of known weight concentration and known molar concentration) and B (polymer solution of known weight concentration

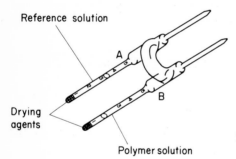

Reference solution

A

B

Drying agents

Polymer solution

Fig. 6.7   H-tube for molecular weight determination by the isopiestic method

and unknown molar concentration), contain the two solutions (c. 5 ml) sealed under vacuum. During equilibrium the H-tube is immersed in a thermostat ($\pm 0.001\ ^\circ\mathrm{C}$) and is continually rocked, so that solutions surge from one end of the tube to the other several times per minute over a period of a few days up to several weeks. Longer periods are required for more dilute solutions. Commonly used solvents include acetone at 25 $^\circ$C and benzene at 45 $^\circ$C. Liquids with high vapour pressures give more rapid equilibration. As a reference standard azobenzene (mol. wt $= 182.2$) and methyl stearate (mol. wt $= 298.5$) are often used. Other standards should have the highest purity and a low volatility. Small amounts of calcium sulphate are added in order to remove any trace of moisture. The concentration of the isopiestic solutions (when equilibrium is reached) are usually determined from measurement of the final volumes of the liquids in the two arms of the H-tube. The limiting concentration ratio ($\lim_{c \to 0} (c_s/c_p)$) at infinite dilution ($c = 0$) can be found by extrapolation of the equilibrium results obtained from a minimum of four different H-tubes filled with solutions at different concentrations.

Bibliography: 125, 1133, 1446.
References: 2862, 3609, 6129, 6396.

6.4 VAPOUR PRESSURE OSMOMETRY

Vapour pressure osmometry is based on RAOULT'S LAW, which states that in an ideal solution the partial vapour pressure of each component is proportional to its mole fraction. The vapour pressure lowering (Δp) is given by

$$\Delta p = p_1 - p_2 = p_1 n_2 = p_1 \frac{cV_1}{M_2},$$ (6.12)

where

Δp is a measure of the vapour pressure lowering,
p_1 is the vapour pressure of pure solvent,
p_2 is the vapour pressure of the solution,
n_2 is the mole fraction of the solution,
c is the solution concentration,
$V_1 = M_1/\rho_1$ is the molar volume of the solvent,
M_1 is the molecular weight of the solvent,
ρ_1 is the solvent density,
M_2 is the molecular weight of the solute.
The limiting vapour pressure lowering, $(\Delta p/c)_{c=0}$, is given by

$$\left(\frac{\Delta p}{c}\right)_{c=0} = p_1 V_1 \frac{1}{M_2}.$$ (6.13)

If the solute is a polymer sample, the value M_2 can be substituted by the number-average molecular weight $(\overline{M}_n)$:

$$\left(\frac{\Delta p}{c}\right)_{c=0} = p_1 V_1 \frac{1}{\overline{M}_n}.$$ (6.14)

Vapour pressure osmometry can be applied to any soluble polymer whose molecular weight does not exceed 20 000.

Bibliography: 125, 309, 866.
References: 2382, 3181, 4393, 4394, 4397, 6983.

6.4.1 Vapour Pressure Osmometers

Vapour pressure osmometers are built of two parts: a thermally insulated measurement chamber (Fig. 6.8) and an electronic data recording device. Two thermistor beads, which form two arms of a bridge circuit, are placed in a thermostatted chamber saturated with solvent vapour. A drop of solvent is placed on one thermistor (the reference bead) and a drop of solution on the

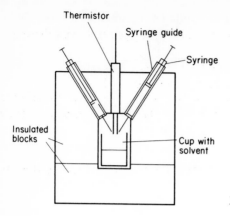

Fig. 6.8   Vapour pressure osmometer

other (the measuring bead) by the use of guided syringes. Solvent condenses on the measuring bead because of the lower vapour pressure of the solution, thereby warming the measuring bead, and the difference in temperature between the two beads results in a bridge imbalance.

At steady state, the temperature difference $(\Delta T)_s$ is given by

$$(\Delta T)_s = K_s(c/\overline{M}_n + A_2 c^2 + A_3 c^3 + ...),\qquad(6.15)$$

where

K_s is a calibration constant at the steady state,

c is the solution concentration,

$\overline{M}_n$ is the number-average molecular weight of the polymer,

A_2 and A_3 are the second and third virial coefficients of the polymer, respectively.

In practice, the apparatus is calibrated for a given solvent, temperature, and thermistor probe by measurements with a substance of known molecular weight

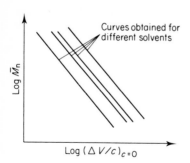

Fig. 6.9   Plot of log $\overline{M}_n$ versus log$(\Delta V/c)_{c=0}$

(e.g. naphthalene, benzil) at a number of solution concentrations. The results are then calculated from

$$\overline{M}_n = \frac{k}{(\Delta V/c)_{c=0}},\qquad(6.16)$$

where

k is an apparatus constant for a given solvent, temperature, and thermistor assembly,

ΔV is the bridge imbalance for a concentration, c.

The plot of $\log \overline{M}_n$ versus $\log (\Delta V/c)_{c=0}$ is linear (Fig. 6.9).

Several additional factors such as a drop size, solute volatility, and diffusion-controlled processes have an effect on vapour pressure osmometry measurements.

Bibliography: 127, 518, 925, 1333.
References: 2022, 2382, 2383, 2614, 2704, 3181, 4394, 5216, 5364, 5381, 5474, 5693, 6409, 6410, 6759, 6803, 6911, 6983, 7029.

Chapter 7

End-group Analysis

General bibliography: 23, 125, 127, 357, 362, 456, 681, 1082.

END-GROUP ANALYSIS is usually limited to linear polymers and estimates the number-average molecular weight $(\overline{M}_n)$:

(i) the total number of end groups is twice the number of polymer molecules;
(ii) if each polymer molecule contains one end group of a characteristic kind, then the number of end groups of this kind is equal to the number of polymer molecules.

The number-average molecular weight $(\overline{M}_n)$ can be calculated from

$$\overline{M}_n = n \times 10^6/m \tag{7.1}$$

where
$n = 1$–2 is the number of groups which can be determined per macromolecule,
m is the end-group concentration (in microequivalents per gram (CGS) or equivalents per kilogram (SI)).

The chemical reactivity of functional groups is independent of the molecular weight of the polymer molecules to which they are attached after the molecular weight has exceeded a few hundred grams per mole. The practical upper limit of molecular weights measured by end-group analysis is 50 000.

The nature of the end groups should be known for a routine control analysis (Table 7.1).

The common methods for end-group analysis are non-aqueous potentiometric titration techniques, whereas IR, NMR, and mass spectrometry have only limited application. End groups can also be determined by specific reactions with radioactive reagents, chromophoric reagents (dyes), or reagents containing elements not present in the bulk of the polymer.

In some cases end-group analysis is applied to determine the number-average molecular weight $(\overline{M}_n)$ of branched polymers obtained from multifunctional monomers.

The number-average molecular weight $(\overline{M}_n)$ in this case can be calculated from

$$\overline{M}_n = \frac{2x \times 10^6}{x - y}, \tag{7.2}$$

where

x is the total concentration of end groups (in microequivalents per gram (CGS) or equivalents per kilogram (SI)),

y is the number of branch points (in microequivalents per gram (CGS) or equivalents per kilogram (SI)).

Bibliography: 456, 495, 596, 1006, 1065.
References: 2019, 6346.

Table 7.1 End-group analysis

Polymer	Type of end group	Published in
Aliphatic poly(amides)	Amine end groups	B: 495, 1065; R: 2588, 2915, 3594, 5352, 5672, 6366, 6999, 7103, 7307
	Carboxyl end groups	B: 495, 1065; R: 2284, 3166, 3594, 5147, 5721, 6244, 6999, 7153
Aromatic poly(amides)	Amine end groups	B: 495; R: 3175, 4721, 5897
	Carboxyl end groups	B: 495; R: 3175, 4721, 5897
Poly(esters)	Hydroxyl end groups	B: 1212; R: 2291, 2664, 2767, 2770, 2944, 3770, 4538, 4641, 4674, 5036, 5376, 5709, 5854
	Carboxyl end groups	B: 357, 495; R: 3374, 4537 4674, 5036, 5148, 5854, 5979, 6926, 7004, 7006
Poly(nitryles)	Nitryle end groups	R: 4411
Poly(oxyalkylene glycols) and poly(urethanes)	Hydroxyl end groups	B: 357; R: 2439, 2620, 2662, 2664, 3026, 3885, 4011, 5049, 5658, 5997, 6572, 7096
	Isocyanide end groups	B: 357; R: 3058, 3768, 4756, 5395
	Terminal unsaturation	R: 5833, 7322
Vinyl polymers	Different end groups contained elements such as initiator fragments or chain transfer fragments	B: 59; R: 2085, 2137, 2233, 2336, 2386, 2683, 2827, 2963, 3406, 3394, 3631, 3632, 4405, 4406, 4536, 4569, 5191, 5379, 5461, 5666-5671, 5765, 6098, 6804, 6916, 7211

Chapter 8

Ultracentrifugation

General bibliography: 6, 68, 125, 140, 181, 281, 335, 399, 489, 491, 888, 1148, 1167, 1344, 1355, 1397–1400.

ULTRACENTRIFUGATION (SEDIMENTATION ANALYSIS) is a technique based on measurement of the composition of dissolved particles in solution placed in a fast rotating rotor.

The following types of sedimentation analysis are used:

(i) sedimentation velocity analysis (Section 8.1);
(ii) sedimentation-diffusion equilibrium analysis (Section 8.2);
(iii) sedimentation 'approach-to-equilibrium' analysis (Section 8.3);
(iv) sedimentation equilibrium in a density gradient analysis (Section 8.4).

8.1 SEDIMENTATION VELOCITY ANALYSIS

SEDIMENTATION VELOCITY ANALYSIS is a technique which measures the velocity with which particles in a polymer solution, or a colloidal suspension, are displaced under the influence of a strong centrifugal force.

During ultracentrifugation of a polymer solution all dissolved macromolecules move towards the bottom of the solution. This process is called SEDIMENTATION TRANSPORT. After a time interval (t) there is a layer of pure solvent at the meniscus and macromolecules settle in a narrow layer near the bottom of the cell (Fig. 8.1).

The velocity with which the macromolecules move towards the bottom of the cell is called the SEDIMENTATION VELOCITY.

The sedimentation velocity depends on

(i) the centrifugal force;
(ii) the buoyancy;
(iii) the molecular weight;
(iv) the friction factor of the macromolecules.

The boundary layer between solvent and sedimenting polymer solution moves towards the bottom of the cell (Fig. 8.1) at a velocity defined by the sedimentation velocity. This boundary is not sharp and it broadens with time because, in a polydisperse polymer solution, the various macromolecules sediment at different sedimentation velocities. The back-diffusion which starts when differences in concentration arise also has some effect on the shape of the boundary.

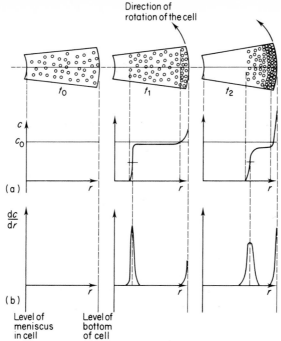

Direction of
rotation of the cell

Fig. 8.1 Sedimentation transport in the ultracentrifugation
cell during rotation. (a) Change of concentration throughout the cell; (b) the concentration gradient obtained in the
cell. c is concentration; r is the distance from the centre of
rotation

Two types of graph represent sedimentation transport:

(i) the concentration (c) throughout the cell during sedimentation transport
is expressed as a function of the distance from the rotational centre (r),
$c = f(r)$ (Fig. 8.1(a));

(ii) the concentration gradient ($\mathrm{d}c/\mathrm{d}r$) is obtained by differentiating the concentration curve (Fig. 8.1(b))—the rate of displacement of the concen-
tration curve is a measure of the sedimentation velocity.

The change in concentration of particles in a polymer solution ($\partial c/\partial t$) during
the equilibrium established between sedimentation and diffusion is given by the
LAMM DIFFERENTIAL ULTRACENTRIFUGATION EQUATION:

$$\left(\frac{\partial c}{\partial t}\right)_r = -\frac{1}{r}\frac{\partial}{\partial r}\left(S\omega^2 r^2 c - Dr\frac{\partial c}{\partial r}\right), \tag{8.1}$$

where
r is the distance from the centre of rotation to a considered point (volume) of
polymer solution (in centimetres (CGS) or metres (SI)),
c is the solution concentration before ultracentrifugation (in grams per cubic
centimetre (CGS) or kilograms per cubic metre (SI)),

ω is the angular velocity (the rotor speed) (in degrees per second (CGS) or radians per second (SI)),

S is the sedimentation coefficient (in seconds),

D is the diffusion coefficient (in square centimetres per second (CGS) or square metres per second (SI)).

The SEDIMENTATION COEFFICIENT (S) is the sedimentation velocity (v_S) per unit field ($\omega^2 r$) (also called the CENTRIFUGAL ACCELERATION):

$$S = v_S/\omega^2 r = (1 - \bar{v}\rho) M/N_A f. \tag{8.2}$$

The sedimentation velocity is determined by

$$v_S = \omega^2 r(1 - \bar{v}\rho) M/N_A f, \tag{8.3}$$

where

$\bar{v}$ is the specific volume of the solute (in cubic centimetres per gram (CGS) or cubic metres per kilogram (SI)),

ρ is the density of the solution (in grams per cubic centimetre (CGS) or kilograms per cubic metre (SI)),

$1 - \bar{v}\rho$ is called the BUOYANCY FACTOR—for dilute solutions,

$$1 - \bar{v}\rho = \frac{\mathrm{d}\rho}{\mathrm{d}c}, \tag{8.4}$$

M is the polymer molecular weight,

N_A is the Avogadro number (cf. Section 40.3),

f is the frictional coefficient.

A sedimentation coefficient (S) is measured in the unit called the svedberg (S)

$$1 \text{ SVEDBERG} = 1 \times 10^{-13} \text{ s.}$$

The DIFFUSION COEFFICIENT (D) is given by

$$D = \frac{RT}{N_A f}(1 + MA_2 c + MA_3 c^2 + ...), \tag{8.5}$$

where

R is the universal gas constant (cf. Section 40.2),

T is the thermodynamic temperature (in kelvins),

A_2 and A_3 are the virial coefficients.

The FRICTIONAL COEFFICIENT (f) is proportional to the average linear dimension of the macromolecules and to the solvent viscosity:

$$f = P(\bar{r}^2)^{1/2} \eta, \tag{8.6}$$

where

$P = 6\pi f_A$,

f_A is the asymmetry factor—the relationship between the radius of gyration and the radius most suitable in describing the molecule,

$\bar{r}^2$ is the square average end-to-end distance of the macromolecules,

η is the solvent viscosity coefficient.

Extensive data on sedimentation coefficients, diffusion coefficients, and frictional coefficients are collected in The Polymer Handbook (B: 410).

The sedimentation velocity is very dependent on:

(i) the solution concentration, because the viscosity has an effect on the frictional coefficient of the macromolecules (f) and hence on the sedimentation coefficient (S);

(ii) the hydrostatic pressure, which is built up to a few hundred atmospheres in the cell during rotation—this has an effect on

    (a) the partial specific volume of the solute ($\bar{v}$),

    (b) the density of the solution (ρ),

    (c) the frictional coefficient of the macromolecules (f).

Measurement of the sedimentation velocity gives information about

(i) the shape of the polymer particles;

(ii) the polymer molecular weight (M), defined as the sedimentation diffusion-average molecular weight;

(iii) molecular weight distribution (MWD);

(iv) the degree of branching in branched polymers (in combination with other methods);

(v) the parameters of reaction equilibria between polymers.

Bibliography: 1167, 1278.
References: 2138, 2210, 2309, 2399, 2436, 2876, 3541, 3542, 3691, 3726, 3843, 4103, 4315, 4571, 4584, 4658, 4685, 4811, 5076, 5513, 5515, 5537, 5727, 6094, 6632, 6641, 6991, 6992, 7036, 7107–7110.

8.1.1 Determination of the Shapes of Particles by Sedimentation Velocity Analysis

Combination of eq. (8.6) with the Flory–Fox equation (2.36) (cf. Section 2.13) gives a CONSTANT SHAPE FACTOR (β):

$$\beta = \Phi^{1/3} P^{-1} = \eta[\eta]^{1/3} M^{1/3}/f, \qquad (8.7)$$

where

Φ is a function related to the hydrodynamic behaviour of macromolecules,

$P = 6\pi f_{A}$,

f_{A} is the asymmetry factor—the relationship between the radius of gyration and the radius most suitable in describing the molecule,

η is the solvent viscosity coefficient,

$[\eta]$ is the limiting viscosity number,

M is the molecular weight of the polymer,

f is the frictional coefficient.

For a flexible macromolecule, $\beta = 2.5 \times 10^6$. For other particles (e.g. spheres, thin discs, thin rods, circular cylinders, random coils, polydisperse random coils, and stiff coils) the β value differs from the value mentioned above.

References: 5057, 5060.

8.1.2 Determination of Polymer Molecular Weights by Sedimentation Velocity Analysis

In sedimentation velocity analysis, the polymer molecular weight (M) is generally not equal to the weight-average molecular weight $(\overline{M}_w)$. It is called the SEDIMENTATION DIFFUSION-AVERAGE MOLECULAR WEIGHT.

The polymer molecular weight (M) can be obtained from eqs (8.2) and (8.5):

$$\frac{S}{D} = \frac{M+(1-\bar{v}\rho)}{RT(1+MA_2c+...)}. \tag{8.8}$$

The apparent polymer molecular weight (M_{app}) is then given by the SVEDBERG EQUATION:

$$M_{app} = \frac{RT}{(1-\bar{v})}\frac{S}{D}, \tag{8.9}$$

where

S is the sedimentation coefficient,
D is the diffusion coefficient,
$\bar{v}$ is the partial specific volume of the solute,
ρ is the density of the solution,
R is the universal gas constant (cf. Section 40.2),
T is the thermodynamic temperature (in kelvins),
A_2 is the second virial coefficient,
c is the solution concentration.

An extrapolation to infinite dilution requires measurements of M_{app} at several concentrations (c), e.g. 2.0, 4.0, 6.0, and 8.0 g l^{-1}. Then the polymer molecular

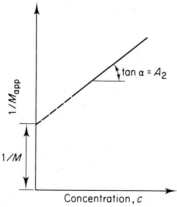

Fig. 8.2   Plot of $1/M_{app}$ versus concentration (c)

weight (M) can be calculated from the plot on Fig. 8.2 using

$$\frac{1}{M_{app}} = \frac{1}{M} + A_2 c + \dots \qquad (8.10)$$

The plot yields $1/M$ as the intercept at the ordinate and the slope is $\tan \alpha = A_2$ (the second virial coefficient).

Another method uses the extrapolation of S and D to zero concentration; the polymer molecular weight can then be calculated directly from

$$M = \frac{RT}{(1 - \bar{v}\rho)}\left(\frac{(S)_{c=0}}{(D)_{c=0}}\right), \qquad (8.11)$$

where

$(S)_{c=0}$ is the extrapolated sedimentation coefficient at zero concentration,

$(D)_{c=0}$ is the extrapolated diffusion coefficient at zero concentration.

8.1.3 Determination of the Molecular Weight Distribution (MWD) by Sedimentation Velocity Analysis

During ultracentrifugation, macromolecules of high molecular weight sediment towards the bottom with a specific velocity higher than macromolecules of low molecular weight.

The distribution of the sedimentation velocities is given by the concentration gradient which is established within the boundary region and its change with respect to time.

References: 6864, 6866.

8.1.4 Determination of the Sedimentation Constant

The SEDIMENTATION COEFFICIENT (S) is obtained from

$$S = v_S/\omega^2 r \qquad (8.12)$$

in svedbergs (S), where

v_S is the sedimentation velocity,

$\omega^2 r$ is the centrifugal acceleration,

ω is the angular velocity (the rotor speed),

r is the distance from the centre of rotation to the point (volume) of the polymer solution under consideration.

The dilute solution of the polymer (e.g. 2.0 g l^{-1}) is placed into the rotor of the centrifuge in a quartz cell. During rotation of the centrifuge, the macromolecules slowly sediment towards the bottom with constant velocity under the influence of centrifugal force (1–2 mm h^{-1}). The boundary between the solution and the solvent (more or less sharp) moves towards the bottom of the cell. The velocity with which this boundary moves is the SEDIMENTATION VELOCITY (v_S).

The sedimentation velocity (v_S) can be determined by the following methods (cf. Section 8.5):

(i) absorption techniques;
(ii) the Rayleigh interferometric optical system;
(iii) the Schlieren optical system.

The velocity with which the distance (r) between the rotor axis and the maximum of the concentration gradient increases with time (t) is the SEDIMENTATION VELOCITY (v_S).

The SEDIMENTATION COEFFICIENT (S) is then given by

$$S = \frac{v_S}{\omega^2 r} = \left(\frac{r_2 - r_1}{t_2 - t_1}\right)\frac{2}{\omega^2(r_1 + r_2)}, \qquad (8.12)$$

where r_1 and r_2 are the distances of the boundary between solution and solvent (maximum of the concentration gradient) from the centre of rotation at times t_1 and t_2.

The measurements of the sedimentation velocity (v_S) must be carried out at different concentrations (e.g. 2.0, 4.0, 6.0, 8.0 g l^{-1}). Then the sedimentation coefficient $(S)_{c=0}$ can be calculated from the plot on Fig. 8.3, using

$$\frac{1}{S} = \frac{1}{(S)_{c=0}} + k_S c. \qquad (8.13)$$

The plot yields $1/(S)_{c=0}$ as the intercept at the ordinate, and the slope is $\tan\alpha = k_S$. For many polymer solutions in good solvents, $k_S \approx 1.6[\eta]$, where $[\eta]$ is the limiting viscosity number (or intrinsic viscosity).

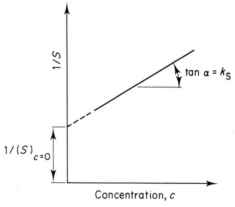

Fig. 8.3   Plot of $1/S$ versus concentration (c)

The constant k_S is not identical with the second virial coefficient (A_2), but $A_2 = k_S + k_D$. (For the constant k_D, cf. Section 8.1.5.)

8.1.5 Determination of the Diffusion Coefficient

The DIFFUSION COEFFICIENT (D) is given by

$$D = \frac{RT}{N_A f}(1 + MA_2 c + MA_3 c^2 + ...)$$ (8.14)

(in square centimetres per second (CGS) or square metres per second (SI)), where

R is the universal gas constant (cf. Section 40.2),
T is the thermodynamic temperature (in kelvins),
N_A is the Avogadro number (cf. Section 40.3),
f is the frictional coefficient,
M is the polymer molecular weight,
A_2 and A_3 are the virial coefficients,
c is the solution concentration.

The diffusion coefficient (D) can be determined by measuring the concentration gradient at different times (t), i.e. the rate of spreading of an initially sharp boundary formed between solvent and solution. Figure 8.4 shows a

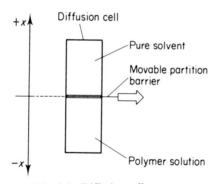

Fig. 8.4 Diffusion cell

schematic diagram of the experimental arrangement. In the lower part of the diffusion cell is placed the polymer solution, which is separated by a barrier from the upper part of the cell, which contains pure solvent. After removal of the barrier, the macromolecules from the lower solution layer begin to diffuse into the pure solvent; thus the concentration above the barrier increases in the same measure as the concentration below the barrier decreases. The concentration gradient remains greatest close to the boundary and decreases with increasing distance upwards and downwards from the boundary in the form of a Gaussian curve (Fig. 8.5). The concentration gradient curves become flattened as the time of measurement (t) increases (Fig. 8.6). In most cases the diffusion curves differ from Gaussian curves more or less strongly, and it is necessary to use the areas under the diffusion curves for the determination of the diffusion coefficient (D). The values of the diffusion constants obtained by this method

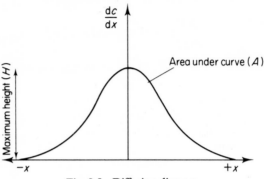

Fig. 8.5   Diffusion diagram

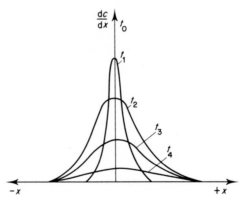

Fig. 8.6   Change of the concentration gradient (dc/dx) with time (t)

are called the HEIGHT–AREA DIFFUSION COEFFICIENTS (D_A). The average value of D_A can be calculated from the plot on Fig. 8.7 as $\tan \alpha = D_A$.

The measurement of the change of the concentration gradient (dc/dx) must be carried out at different concentrations (e.g. 2.0, 4.0, 6.0, and 8.0 g l^{-1}). Then the diffusion coefficient ($(D)_{c=0}$) can be calculated from the plot on Fig. 8.8 using

$$\frac{1}{D} = \frac{1}{(D)_{c=0}} + k_D c. \tag{8.15}$$

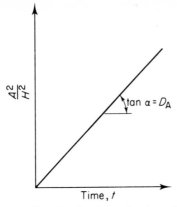

Fig. 8.7 Plot for the determination of the diffusion coefficient (D_A). A is the area under the diffusion curve and H is the maximum height of the diffusion curve (cf. Fig. 8.5)

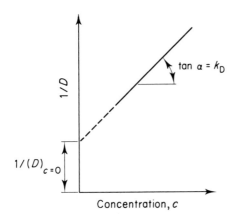

Fig. 8.8 Plot of $1/D$ versus concentration (c)

The plot yields $1/(D)_{c=0}$ as the intercept at the ordinate, and the slope is $\tan \alpha = k_D$. The constant k_D is not identical with the second virial coefficient (A_2), but $A_2 = k_D + k_S$. (For the constant k_S, cf. Section 8.1.4.)

Many different methods are used for the formation of the sharp boundary between solvent and solution:

(i) shearing or pulling away a barrier (membrane) separating the solution from the solvent (Fig. 8.4);

(ii) siphoning through a fixed capillary (Fig. 8.9);

(iii) siphoning through a moving capillary (Fig. 8.10);

(iv) siphoning through a slit in the cell wall (flow junction cell).

For measuring the change of the concentration gradient the Raleigh interferometer is commonly used with a helium–neon laser light source (Fig. 8.11).

Bibliography: 529, 1276.
References: 2613, 2861, 2867, 2875, 2878, 2988, 3018, 4340, 4367, 5868, 6629, 6646.

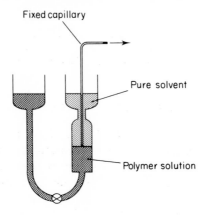

Fig. 8.9 Boundary formation by siphoning through a fixed capillary. The
diffusion cell must be open at top and
bottom to allow symmetrical flow of
solvent and solution. (From R: 2875;
reproduced by permission of the Royal
Swedish Academy of Sciences)

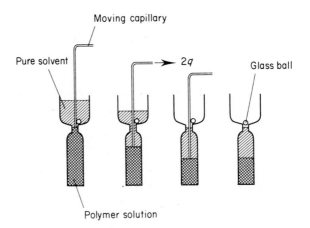

Fig. 8.10 Boundary formation by siphoning through a
capillary moving downwards in the cell together with the
boundary. The cell is closed at the bottom. Symmetrical flow
at the boundary is obtained when the rate of pumping is
adjusted to be twice the capillary sweep rate (q). (From
R:2875; reproduced by permission of the Royal Swedish
Academy of Sciences)

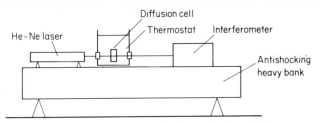

Fig. 8.11 Measuring device for the determination of the concentration gradients in the diffusion cell

8.2 SEDIMENTATION-DIFFUSION EQUILIBRIUM ANALYSIS

SEDIMENTATION-DIFFUSION EQUILIBRIUM ANALYSIS is a technique which measures the change in concentration of particles in a polymer solution, or a colloidal suspension, during the complete equilibrium established between sedimentation and diffusion under the influence of a weak centrifugal field.

Measurements of the sedimentation equilibrium give information about:

(i) the weight-average molecular weight ($\overline{M}_w$) and the Z-average molecular weight ($\overline{M}_Z$) (only if the density increments and refractive index increments are equal for all polymer components);

(ii) the molecular weight distribution of the polymer;

(iii) the structure of branched polymers.

Bibliography: 1167.
References: 2842, 3543, 3544, 3546, 5390, 5623, 6055, 6093, 6096, 6097, 6246, 6990, 7076.

8.2.1 Determination of the Weight-average Molecular Weight ($\overline{M}_w$) by Sedimentation-Diffusion Equilibrium Analysis

At complete sedimentation-diffusion equilibrium, the concentration at every point in the solution column is invariant with time, $(dc/dt)_r = 0$ (i.e. the net transport at every point in the solution column equals zero). The Lamm equation (eq. (8.1)) therefore becomes

$$\frac{S}{D} = \frac{1}{\omega^2 cr}\left(\frac{dc}{dr}\right). \tag{8.16}$$

Replacing S/D in the Svedberg equation (eq. (8.8)) by eq. (8.16) gives the following equation for the weight molecular weight (M_w):

$$M_w = \frac{RT}{\omega^2(1-\bar{v}\rho)}\frac{1}{cr}\left(\frac{dc}{dr}\right), \tag{8.17}$$

where

R is the universal gas constant (cf. Section 40.2),
T is the thermodynamic temperature (in kelvins),
ω is the angular velocity (the rotor speed),
$\bar{v}$ is the partial specific volume of the solute,
ρ is the density of the solution,
c is the solution concentration before centrifugation,
r is the distance from the centre of rotation to the point of the polymer solution column under consideration.

Integration of eq. (8.16) gives

$$M_w = \frac{2RT}{\omega^2(1-\bar{v}\rho)}\left(\frac{c_2-c_1}{c_0(r_2^2-r_1^2)}\right), \qquad (8.18)$$

where index 1 refers to the meniscus and index 2 refers to the bottom of the cell.

The concentration (c) in the ultracentrifugation cell can be determined by the following methods (cf. also Section 8.5):

(i) Absorption measurements at a number of points in the solution column in the cell.

(ii) Interferometric data, which directly yield the absolute concentrations throughout the solution column.

(iii) Measurements with the Schlieren optical technique. The concentration (c) in this method is determined by integration of dc/dr. (Note: measurement of the concentration gradient (dc/dr) requires knowledge of the specific refractive increment (dn/dc):

$$\frac{dc}{dr} = \frac{dn}{dr}\bigg/\frac{dn}{dc}. \qquad (8.19)$$

Equation (8.18) can be written in the form

$$M_w = \frac{2RT}{\omega^2(1-\bar{v}\rho)}\frac{d(\ln c)}{d(r^2)} \qquad (8.20)$$

and the value of $d(\ln c)/d(r^2)$ can be calculated from the plot of $\ln c$ versus r^2 (Fig. 8.12). A straight line in Fig. 8.12 is obtained only for monodisperse and homogeneous polymer samples. A polydisperse or heterogeneous polymer sample gives a curved line.

Sedimentation of polydisperse polymers gives not only differences in concentration, but also in redistribution of the polymer component (the large macromolecules sedimenting at a faster rate than the smaller ones). Molecules of the highest molecular weight will be at the point where concentration is highest. The weight molecular weight (M_w) will differ from one place to another through the solution column (Fig. 8.13). In determining the weight-average

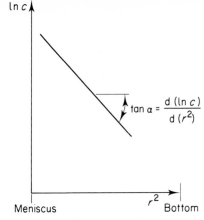

Fig. 8.12 Plot of ln c versus r^2 at a given rotor speed

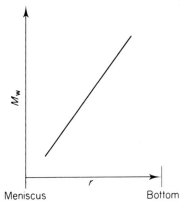

Fig. 8.13 Variation of M_w through the solution column

molecular weight ($\overline{M}_w$) of polydisperse polymers, appropriate integration must be carried out.

The time required for equilibrium to be reached in 2–3 mm layers in a polymer solution varies from a few hours to many days.

8.2.2 Determination of the Z-average Molecular Weight ($\overline{M}_Z$) by Sedimentation-diffusion Equilibrium Analysis

The Z-average molecular weight ($\overline{M}_Z$) can be calculated from the equation

$$\overline{M}_Z = \frac{RT}{\omega^2(1-\bar{v}\rho)}\left[\frac{1}{r_2}\left(\frac{dc}{dr}\right)_2 - \frac{1}{r_1}\left(\frac{dc}{dr}\right)_1 \middle/ (c_2-c_1)\right], \qquad (8.21)$$

where

R is the universal gas constant (cf. Section 40.2)

T is the thermodynamic temperature (in kelvins),

ω is the angular velocity (the rotor speed),

$\bar{v}$ is the partial specific volume of the solute (given in the literature),

ρ is the density of the solution,

c is the solution concentration in the equilibrium,

r is the distance from the centre of rotation to the point in the polymer solution under consideration, index 1 refers to the meniscus and index 2 refers to the bottom in the cell.

8.2.3 Determination of the Molecular Weight Distribution (MWD) by Sedimentation-diffusion Equilibrium Analysis

The molecular weight distribution (MWD) is determined from a change of concentration (c) throughout the solution column. If the column height is small, it is almost impossible to determine the complete molecular weight distribution. Better results can be obtained from measurements made at various rotor speeds.

Bibliography: 61.
References: 3210, 5912, 6630, 7074, 7075.

The data from ultracentrifugation can be computerized and molecular weights (M_w, M_Z, M_{Z+1}) and the molecular weight distribution (MWD) can then be calculated from information relating to a series of equilibria.

References: 4685, 6247, 6248, 6999.

8.3 SEDIMENTATION 'APPROACH-TO-EQUILIBRIUM' ANALYSIS

SEDIMENTATION APPROACH-TO-EQUILIBRIUM ANALYSIS (the ARCHIBALD METHOD) measures the change in local concentration in a polymer solution or a colloidal suspension, during the quasi-equilibrium (approach-to-equilibrium) between sedimentation and diffusion under the influence of a weak centrifugal field.

Measurements of the sedimentation approach-to-equilibrium give information about the weight-average molecular weight ($\overline{M}_w$).

Bibliography: 1167.
References: 2139, 6640, 6827, 7293.

8.3.1 Determination of the Weight-average Molecular Weight ($\overline{M}_w$) by the Archibald Method

In the sedimentation-diffusion quasi-equilibrium (approach-to-equilibrium), the concentrations at the bottom and at the meniscus are invariant with respect to time, $dc/dt = 0$ (i.e. the net transport at either end of the solution column equals zero).

The apparent weight molecular weight $(M_w)_{app}$ is given by

$$(M_w)_{app} = \frac{RT}{\omega^2(1-\bar{v}\rho)\,cr}\frac{1}{cr}\left(\frac{dc}{dr}\right),\qquad(8.22)$$

where

R is the universal gas constant (cf. Section 40.2),

T is the temperature (in kelvins),

ω is the angular velocity (the rotor speed),

$\bar{v}$ is the partial specific volume of the solute,

ρ is the density of the solution,

c is the solution concentration before centrifugation,

r is the distance from the centre of rotation to the point in the polymer solution column under consideration.

The concentration (c) at the meniscus or at the bottom (measured using Schlieren optics) (cf. Section 8.5) is determined by integration of dc/dr and is given by

$$c = c_0 \exp(-2S\omega^2 t),\qquad(8.23)$$

where

S is the sedimentation constant,

t is the time of centrifugation.

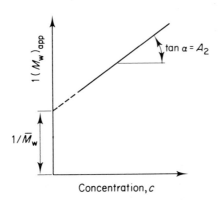

Fig. 8.14   Plot of $1/(M_w)_{app}$ versus concentration (c)

An extrapolation to infinite dilution requires measurements of $(M_w)_{app}$ at several concentrations (c), e.g. 2.0, 4.0, 6.0, and 8.0 g l^{-1}. Then the weight-average molecular weight $(\bar{M}_w)$ can be calculated from the plot on Fig. 8.14 using

$$\frac{1}{(M_w)_{app}} = \frac{1}{\bar{M}_w} + A_2 c + \dots.\qquad(8.24)$$

The plot yields $1/\overline{M}_w$ as the intercept at the ordinate, and the slope is $\tan \alpha = A_2$ (the second virial coefficient).

The Archibald method is less accurate for polydisperse polymers than sedimentation-diffusion equilibrium analysis.

8.4 SEDIMENTATION EQUILIBRIUM IN A DENSITY GRADIENT ANALYSIS

SEDIMENTATION EQUILIBRIUM IN A DENSITY GRADIENT ANALYSIS is a technique which measures the variation in density gradient in a polymer solution produced across the liquid column of a mixture of light and heavy solvents during the equilibrium established between sedimentation and diffusion under the influence of a weak centrifugal field.

Measurements of sedimentation equilibrium in a density gradient give information about

(i) the weight-average molecular weight $(\overline{M}_w)$;
(ii) the chemical inhomogenity of macromolecules.

Bibliography: 613, 1167.
References: 3098, 3313–3315, 3967, 4027, 4028, 4197, 4884, 4992, 5241, 5913, 5950, 6509.

8.4.1 Determination of the Weight-average Molecular Weight $(\overline{M}_w)$ by Density Gradient Analysis

A density gradient is produced in a mixture of a light and a heavy solvent (e.g. benzene and CBr_4) during a rotation at a given speed in an ultracentrifuge. The macromolecules will sediment from the meniscus towards the bottom of the cell and float from the bottom towards the meniscus (Fig. 8.15). At equilibrium, the macromolecules will collect at a point at which the density of mixed solvents exactly corresponds to the density of the macromolecule in solution. This position is at a distance r_0 from the centre of rotation.

The weight molecular weight (M_w) is given by

$$M_w = -\frac{2RT}{\omega^2 r_0 (\mathrm{d}\rho/\mathrm{d}r) \bar{v}_s \bar{j}_s c} \left(\frac{\mathrm{d}c}{\mathrm{d}[(r-r_0)^2]}\right), \qquad (8.25)$$

where

R is the universal gas constant (cf. Section 40.2),

T is the thermodynamic temperature (in kelvins),

ω is the angular velocity (the rotor speed),

r_0 is the distance from the centre of rotation to the point where the density of the macromolecule in solution corresponds exactly to the density of the mixed solvents,

r is the distance from the centre of rotation to the point under consideration in the polymer solution column,

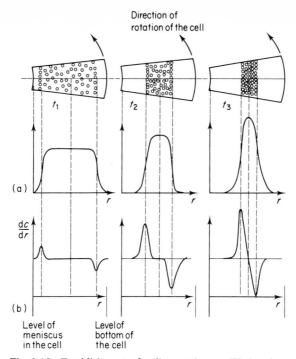

Fig. 8.15   Establishment of sedimentation equilibrium in a density gradient analysis. (a) Change of concentration throughout the cell; (b) the concentration gradient obtained in the cell. c is the concentration; r is the distance from the centre of rotation

ρ is the density of the mixed solvents at the point under consideration in the polymer solution column,

$\bar{v}_s$ is the partial specific volume of the solvated polymer,

f_s is the solvation factor,

c is the solution concentration.

Integration of eq. (8.25) gives the concentration c as a function of $(r-r_0)$:

$$c = c_0 \exp \left[-(r-r_0)^2 / 2\sigma^2 \right], \tag{8.26}$$

where

c_0 is the concentration at the centre of the band,

$$\sigma^2 = \frac{RT}{\omega^2 \, r_0 (\mathrm{d}\rho/\mathrm{d}r) \, \bar{v}_s f_s} \frac{1}{M_w}.$$

Equation (8.26) shows that at equilibrium the macromolecules collect in a band, with the concentration showing a Gaussian distribution about a distance r_0 and with a standard deviation σ.

The weight-average molecular weight ($\overline{M}_w$) can be determined by integration of M_w over the full width of the band. It is necessary to perform measurements at several concentrations and extrapolate results to zero concentration (infinite dilution).

8.4.2 Determination of the Chemical Inhomogeneity of Macromolecules by Density Gradient Analysis

Macromolecules of different densities (chemical inhomogeneity) give separate bands in density gradient analysis. The analysis of results is very complicated and requires additional data from other measurements such as molecular weight distribution and heterogeneity in density as a function of the molecular weight.

8.5 ANALYTICAL ULTRACENTRIFUGES

Analytical ultracentrifuges are built of the following parts (Fig. 8.16):

(i) Mechanical parts
   (a) A rotor which is an ellipsoidal duraluminium or titanium block with 2, 4, or 6 holes in which standard cells can be placed.
   (b) An electric motor which drives the rotor via a gear system and a flexible spindle. The angular velocity of the electric motor can be varied from 800 to 80 000 rev min^{-1}. The rotor speed is continuously compared with a reference speed of a synchronized motor with differential gear drive. In this way the rotor speed is kept constant. In the newest ultracentrifuges the rotor is suspended in a magnetic field and has no direct contact with the driving motor once it reaches operating speed. For protection against accidental damage, the rotor

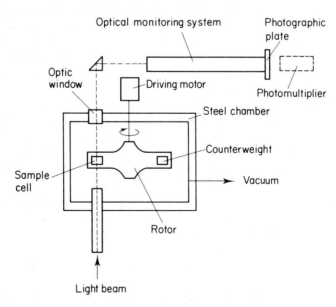

Fig. 8.16   Analytical ultracentrifuge

is driven in a steel chamber. In order to keep frictional heating to a minimum, the steel chamber is evacuated to 10^{-6} torr (1 torr $\frac{1}{760}$ atm $=133.322$ M m^{-2}) by a rotational oil-diffusion vacuum pump. With the aid of a cooling–heating system the temperature of the rotor is regulated from $-10\,°C$ to $+40\,°C \mp 0.1\,°C$. Application of high temperatures (above 100 °C) involves a number of practical problems (e.g. the lenses of the optical system are very liable to get fogged with condensing oil vapour).

Bibliography: 281, 1167, 1400.
References: 2303, 4685, 6054.

(ii) An optical system. Changes occurring in the cells can be monitored by an optical system. Light from a source passes through the cell and is further reflected by the mirror in the direction of the monitoring system—which includes a phase plate and a swingout mirror—and is registered directly on photographic plates.

For measuring the concentration, or concentration gradient, as a function of the height in the liquid column, three optical monitoring systems are available:

(a) The Schlieren optical system. In this system, the change in concentration (c) with distance from the centre of rotation (r) is optically differentiated with the aid of a special optical set-up, so that the concentration gradient (dc/dr) is observed as a function of distance (r). It is possible to observe the boundary as a maximum of the concentration gradient directly on a photographic plate. The photograph obtained is a two-dimensional plot, showing the refractive index index increment of a solution (dn/dc) as a function of the distance in the liquid column (Fig. 8.17(a)). The concentration gradient (dc/dr) is proportional to the refractive index increment of a solution (dn/dc) (cf. Section 8.2.1).

(b) The interference optical system. In this system the number or displacement of interference lines is measured (Fig. 8.17(b)). The number of interference lines is proportional to the difference in refractive indices and, thereby, the difference in concentrations. The interferometric method does not measure the concentration gradient, but this can be found by differentiation from the interferometric data.

(c) The absorption optical system (Fig. 8.17(c)). In this system the absorption of ultraviolet or visible light is measured by a photosensitive electrical device, e.g. a photomultiplier.

Bibliography: 529, 1167, 1276, 1400.
References: 2821, 3427, 4222, 4963, 4964, 5010, 6035, 6629, 6631, 6647.

(iii) Cells. Many different cells are commercially available. The choice of cell depends on experimental conditions:

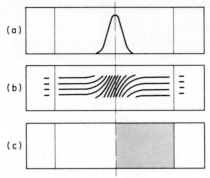

(a)

(b)

(c)

Fig. 8.17 Image of a diffusion boundary between two layers of different concentration obtained by three different optical techniques. (a) Schlieren optics; (b) interference optics; (c) absorption optics

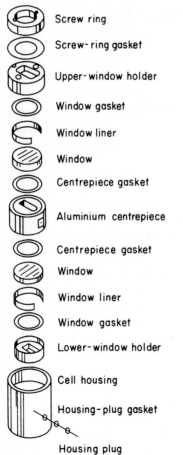

Screw ring

Screw-ring gasket

Upper-window holder

Window gasket

Window liner

Window

Centrepiece gasket

Aluminium centrepiece

Centrepiece gasket

Window

Window liner

Window gasket

Lower-window holder

Cell housing

Housing-plug gasket

Housing plug

Fig. 8.18 Dismantled ultracentrifuge cell. (From B: 1400; reproduced by permission of John Wiley & Sons)

(a) the type of sedimentation analysis (sedimentation velocity analysis or sedimentation equilibrium analysis);

(b) the optical system to be used (Schlieren, interference, or absorption optical system);

(c) the temperature, solvent, and concentration.

The design of one commercial cell is shown in Fig. 8.18. Two optically flat windows made from quartz or sapphire are placed on a centrepiece and are held in position by an outer casing of aluminium alloy, threaded at one end for a retainer ring. The solution is introduced into the sector-shaped channel in the centrepiece through a small hole. The walls of the channel delimit a sector of a cylinder defined by the axis of rotation, so that solute molecules will always sediment (and diffuse) in a radial direction in order to avoid convection currents. Tight seals are obtained between windows and centrepiece and also at the filling hole by the use of gaskets. Double-sector centrepieces, with two sector-shaped channels side by side, are more useful than a one-sector centrepiece. These permit simultaneous observation of a solution and its solvent (necessary with the interference optical system) and make a separate experiment to measure redistribution of components in or compression of the solvent unnecessary. The most common lengths for the optical path are 1.5, 3, 6, 12, 18, and 30 mm. In order to achieve proper stability of the rotor it is necessary to keep the difference in weight between two loaded cells (or between a loaded cell and its counterbalance) within 0.5 g.

Bibliography: 1167, 1400.
References: 7294, 7295.

8.6 PREPARATIVE ULTRACENTRIFUGATION

Preparative ultracentrifugation is used for the separation of substances of differing molecular weights and is especially suitable for the separation of compact macro molecules (e.g. biopolymers). There are three kinds of preparative ultracentrifugation method:

(i) Normal ultracentrifugation. The particles sediment in pure solvent or in relatively dilute salt solutions. The density of the solvent or the salt solution is practically constant over the whole cell (Fig. 8.19(a)). The more rapidly moving particles collect at the bottom, but they are always contaminated with the more slowly sedimenting particles. The slower moving molecules cannot be separated quantitatively.

(ii) Band ultracentrifugation (or zone ultracentrifugation). The particles sediment in a mixture of solvents which does not differ much with density. In the first stage, a density gradient is produced by ultracentrifugation of the solvent mixture alone. In the second stage, a very thin film of polymer solution is added to the top of a solvent layer. During further ultracentrifugation, the faster molecules in this film separate from the

122

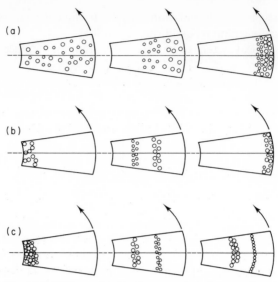

Fig. 8.19 Different kinds of preparative ultracentrifugation method. (a) Normal ultracentrifugation; (b) band
ultracentrifugation; (c) density gradient ultracentrifugation

slower ones, and form one or more bands (Fig. 8.19(b)). The isolated
macromolecules are of high purity.

(iii) Density gradient ultracentrifugation. The particles sediment in a mixture
of a light and a heavy solvent (e.g. benzene and CBr_4). At equilibrium
conditions, under the influence of the centrifugal force, the particles will
collect at a point at which the density of the mixed solvents exactly
corresponds to the density of the molecules in solution (Fig. 8.19(c)) (cf.
Section 8.4).

Chapter 9

Viscosimetric Methods

General bibliography: 23, 125, 127, 301, 309, 621, 681, 917, 922, 943, 1082, 1120, 1350, 1396.

DEFORMATION OF A POLYMER SAMPLE occurs when an applied force changes the sample's shape and size. These deformations can be

(i) REVERSIBLE (ELASTIC). When a force is applied to a perfectly elastic polymer sample, a finite deformation occurs. After removal of the force, the elastic sample returns to its original size and shape. The energy applied for the reversible deformation is completely recovered.

(ii) IRREVERSIBLE. When a force is applied to a perfectly viscous sample, the deformation changes with time and the sample flows. The energy applied for the irreversible deformation is not recovered.

Many polymers exhibit an elastic (reversible) deformation for short periods of time and an irreversible deformation if the force is applied for a longer period of time.

The SHEAR STRESS (τ) in a reversible deformation is given by

$$\tau = F/A \tag{9.1}$$

(in dynes per square centimetre (CGS) or newtons per square metre (SI)), where
F is the applied force (Fig. 9.1) (in dynes (CGS) or newtons (SI)),
A is the area (in square centimetres (CGS) or square metres (SI)).

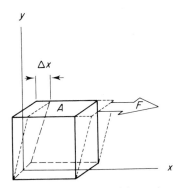

Fig. 9.1 Reversible deformation.
A is the area and F is the applied
force

The SHEAR STRAIN (γ) is the amount of deformation given by

$$\gamma = \Delta x / \Delta y. \tag{9.2}$$

For an ideal elastic material, the shear strain is directly proportional to the shear stress.

HOOKE'S LAW for reversible deformation gives the relation between shear stress (τ) and shear strain (γ):

$$\tau = F/A = G\gamma, \tag{9.3}$$

where G is the SHEAR MODULUS.

A fluid polymer sample constrained between two plates can be considered as a series of parallel layers (Fig. 9.2). The upper plate moves with a velocity (v)

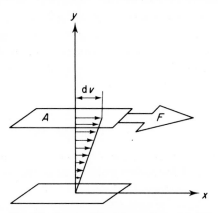

Fig. 9.2   Irreversible deformation. A is the
area and F is the applied force

relative to the lower plate under a shearing force (F) per area (A). Particular parallel layers of liquid move with velocities which are proportional to the distance of the fluid layer from the lower plate (laminar flow).

For ideal viscous fluids, the shear stress is directly proportional to the rate of change of velocity with distance (dv/dy).

HOOKE'S LAW for irreversible deformation gives the relation between shear stress (τ) and shear strain (γ):

$$\tau = \frac{F}{A} = \eta\left(\frac{dv}{dy}\right). \tag{9.4}$$

Since

$$\frac{dv}{dy} = \frac{d(dx/dt)}{dy} = \frac{d(dx/dy)}{dt}, \tag{9.5}$$

eq. (9.4) can be rewritten as

$$\tau = \eta\dot{\gamma} \tag{9.6}$$

where
η is the viscosity,
$\dot{\gamma} = \mathrm{d}\gamma/\mathrm{d}t$ is called the SHEAR RATE (VELOCITY GRADIENT) (in s^{-1}).
The VISCOSITY (η) is defined for a given shear stress (τ) and velocity gradient $(\mathrm{d}(\mathrm{d}x/\mathrm{d}y)/\mathrm{d}t)$ by the equation

$$\eta = \frac{\tau}{\dot{\gamma}} \qquad (9.7)$$

(in dyne s cm^{-2} (CGS) or N s m^{-2} (SI)). The unit (dyne s cm^{-2}) is called the POISE. A CENTIPOISE is equal to 10^{-2} poise. Pure liquids have a viscosity of c. 10^{-2} poise and for undiluted (and molten) polymers we find viscosities of 10^{12} poise and greater.
If the viscosity (η) is independent of the velocity gradient, the fluid is called a NEWTONIAN FLUID.
If the viscosity (η) is dependent on the velocity gradient, the fluid is called a NON-NEWTONIAN FLUID.
The VISCOSITY can also be defined as

$$\eta = \frac{\zeta}{\dot{\gamma}^2}, \qquad (9.8)$$

where ζ is the work dissipated in overcoming viscous resistance to flow in a unit volume of solution per unit time.
Measurement of the viscosities of polymer systems gives information about:

(i) the viscosity-average molecular weight ($\overline{M}_v$) (cf. Section 9.1.2);
(ii) the molecular weight distribution;
(iii) unperturbed chain dimensions (cf. Section 2.13);
(iv) viscosity characteristics for flow behaviour of concentrated solutions and, in particular, melted polymer samples, i.e. in relation to selecting parameters for polymer processing techniques such as extrusion, calendering, and moulding.

The LIMITING VISCOSITY NUMBER (the INTRINSIC VISCOSITY) ($[\eta]$) is defined as the fractional increase in the viscosity of one unit of solvent due to the addition of 1 g (CGS) or 1 kg (SI) of non-interacting polymer molecules.
In order to eliminate the influence of intermolecular interactions, the limiting viscosity number (the intrinsic viscosity) is always calculated by extrapolation of the viscosity number (the reduced specific viscosity) or the logarithmic viscosity number (the inherent viscosity) to zero concentration.
All the above quantities are collected together in Table 9.1.

Bibliography: 255, 787, 992, 1466.
References: 2012, 2025, 2400, 3191, 5268, 5388, 5832, 4424, 6708.

Table 9.1 Definitions of viscosity quantities

Official names (used in this book)	Common names (should be avoided)	Quantity
Viscosity coefficient	Viscosity	η
Viscosity ratio	Relative viscosity	$\eta_{rel} = \eta/\eta_0$
	Specific viscosity	$\eta_{sp} = \eta_{rel} - 1$
Viscosity number	Reduced specific viscosity	$\eta_{red} = (\eta_{rel} - 1)/c$
Logarithmic viscosity number	Inherent viscosity	$\eta_{inh} = \ln \eta_{rel}/c$
Limiting viscosity number	Instrinsic viscosity	$[\eta] = \lim_{c \to 0} (\eta_{sp}/c)$
Limiting viscosity number	Instrinsic viscosity	$[\eta] = \lim_{c \to 0} (\eta_{rel}/c)$

9.1 VISCOSITY MEASUREMENTS FOR DILUTED POLYMERS

In dilute solutions we find the application of the following viscosity-related quantities:

(i) the VISCOSITY RATIO (the RELATIVE VISCOSITY) (η_{rel}), which is given by the ratio of the outflow time for the solution (t) to the outflow time for the pure solvent (t_0),

$$\eta_{rel} = t/t_0 \qquad (9.9)$$

(dimensionless);

(ii) the SPECIFIC VISCOSITY (η_{sp}), which is the relative increment in viscosity of the solution over the viscosity of the solvent,

$$\eta_{sp} = (\eta - \eta_0)/\eta_0 = \eta_{rel} - 1 \qquad (9.10)$$

(dimensionless);

(iii) the VISCOSITY NUMBER (the REDUCED SPECIFIC VISCOSITY (η_{red}) is the specifiic viscosity taken per unit concentration (c),

$$\eta_{red} = \eta_{sp}/c = (\eta_{rel} - 1)/c \qquad (9.11)$$

(in decilitres per gram (CGS) or cubic metres per kilogram (SI)), where c is the concentration of polymer (in grams per decilitre (CGS) or kilograms per cubic metre (SI));

(iv) the LOGARITHMIC VISCOSITY NUMBER (the INHERENT VISCOSITY) (η_{inh}) is defined as

$$\eta_{inh} = \ln \eta_{rel}/c \qquad (9.12)$$

(in decilitres per gram (CGS) or cubic metres per kilogram (SI));

(v) the LIMITING VISCOSITY NUMBER (the INTRINSIC VISCOSITY) ($[\eta]$) is the viscosity number (the reduced specific viscosity) or the logarithmic viscosity number (the inherent viscosity) extrapolated to $c = 0$,

$$[\eta] = \lim_{c \to 0}\left(\frac{\eta_{sp}}{c}\right) = \lim_{c \to 0}\left(\frac{\eta_{rel}}{c}\right). \tag{9.13}$$

An extrapolation to infinite dilution requires measurements of the viscosity at several concentration (at least four concentrations, e.g. 0.05, 0.10, 0.15, 0.20 g per 100 ml). The sample concentration should not be too large because additional effects may then arise from intermolecular forces and entanglements between chains (for very large molecular weights).

There are several empirical equations for calculation of the limiting viscosity number (intrinsic viscosity):

(i) the HUGGINS EQUATION (R: 4163),

$$\eta_{sp}/c = [\eta] + k'[\eta]^2 c; \tag{9.14}$$

(ii) the KRAEMER EQUATION (R: 4701),

$$\eta_{rel}/c = [\eta] + k''[\eta]^2 c; \tag{9.15}$$

(iii) the SHULZ–BLASCHKE EQUATION (R: 6280),

$$\eta_{sp}/c = [\eta] + k'''[\eta]\,\eta_{sp}; \tag{9.16}$$

where k', k'', and k''' are constants for a given polymer at a given temperature in a given solvent. k' and k'' are related by the equation

$$k' - k'' \approx \tfrac{1}{2}. \tag{9.17}$$

The value of k' is usually in the range: $0.3 < k' < 0.4$ and increases as solvent power decreases. For a given polymer–solvent system, k' is not usually sensitive

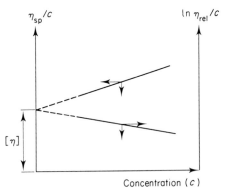

Fig. 9.3 Plot of η_{sp}/c or $\ln \eta_{rel}/c$ versus concentration

to molecular weight. The constants k', k'', and k''' can be determined from conventional measurements at a series of concentrations of a given polymer in a given solvent at a given temperature.

All three equations (eqs (9.14)–(9.16)) yield linear plots with the intercept equal to $[\eta]$ at $c = 0$ (Fig. 9.3).

Bibliography: 125, 621, 787, 942.
References: 2400, 2760, 3007, 3500, 3793, 4194, 4219, 4731, 4949, 5399, 5607, 5744, 5802, 6121, 6159, 6160, 7216.

Extensive data of Huggins and Schulz–Blaschke coefficients are collected in The Polymer Handbook (B: 1277).

9.1.1 One-point Method

This method involves measurement of the viscosity at a single concentration and then calculation of the limiting viscosity number (intrinsic viscosity) $[\eta]$ from one of eqs (9.14)–(9.16) if k' or k'' or k''' is known.

By combination of eqs (9.14), (9.15), and (9.17) it is possible to eliminate $k' - k''$ and obtain the SOLOMON–CIUTA EQUATION

$$[\eta] = 2(\eta_{sp} - \ln \eta_{rel})^{1/2}/c \tag{9.18}$$

The one-point method is accurate only when there is a linear relation between c and η_{sp}/c and/or η_{rel}.

References: 4788, 5403, 6376, 6480.

9.1.2 The Viscosity-average Molecular Weight

For polydisperse linear polymers the VISCOSITY-AVERAGE MOLECULAR WEIGHT ($\overline{M}_v$) is given by the MARK–HOUWINK–SAKURADA EQUATION:

$$[\eta] = K\overline{M}_v^a. \tag{9.19}$$

Extensive tables of constants K and a are available in the literature (B: 786, R: 2025).

The VISCOSITY-AVERAGE MOLECULAR WEIGHT ($\overline{M}_v$) is defined as

$$\overline{M}_v = [\Sigma N_i M_i^{1+a}/\Sigma N_i M_i]^{1/a}, \tag{9.20}$$

where K and a are constant for a given polymer at a given temperature in a given solvent.

The viscosity-average molecular weight ($\overline{M}_v$) is not available from experiments other than viscosity measurements.

A typical plot of log $[\eta]$ versus log $\overline{M}_v$ for a given polymer in a given solvent at a given temperature is shown in Fig. 9.4. In this type of plot one observes deviation from linearity for low molecular weight. It is possible to avoid this deviation from linearity when results are plotted as $1/[\eta]$ versus $1/\overline{M}_v^{1/2}$ (Fig. 9.5).

In the latter plot linearity is observed even for molecular weights as low as 5000.

The viscosity-average molecular weight $(\overline{M}_v)$ lies between the number-average $(\overline{M}_n)$ and weight-average $(\overline{M}_w)$ molecular weights (Fig. 3.2):

$$\overline{M}_n < \overline{M}_v < \overline{M}_w.$$

$\overline{M}_v$ lies closer to $\overline{M}_w$ than to $\overline{M}_n$.

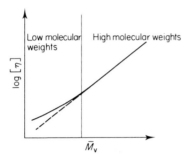

Fig. 9.4 Mark–Houwink–Sakurada relation for a given polymer in solution

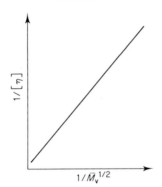

Fig. 9.5 Plot of $1/[\eta]$ versus $1/\overline{M}_v^{1/2}$

The constant a usually varies within the range $0.5 < a < 0.8$. Higher values are sometimes obtained for stiff and/or short molecules.

The constant a is related to the solvent power and to the expansion factor α. It depends on the thermodynamic interactions between polymer segments and solvent molecules (cf. Section 2.11).

Constants K and a are determined by plotting $\log [\eta]$ versus $\log \overline{M}_n$ or $\log \overline{M}_w$ (Fig. 9.6):

$$\log [\eta] = \log K + a \log M. \tag{9.21}$$

In order to determine values of K and a some absolute methods for the determination of molecular weight ($\overline{M}_n$ or $\overline{M}_w$) must be applied, e.g. osmotic pressure, light scattering, or equilibrium sedimentation measurements.

General rules are that

(i) if a polymer sample used for measuring constants K and a has a broad molecular weight distribution, the values determined for K and a may have serious errors;

(ii) if a polymer sample has the same $\overline{M}_w/\overline{M}_n$ ratio, only the constant a will be correct, whereas the constant K will not;

(iii) if log $[\eta]$ is plotted versus log $\overline{M}_n$, the constant K will be too high;

(iv) if log $[\eta]$ is plotted versus log $\overline{M}_w$, the constant K will be too low.

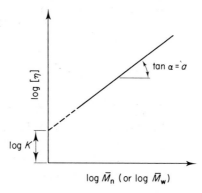

Fig. 9.6   Plot of log$[\eta]$ versus log $\overline{M}_n$ or
log $\overline{M}_w$ for a given polymer

Monodisperse polymer samples are seldom available and so carefully selected fractions of the given polymer are normally used for the determination of K and a.

The application of viscosity measurements to obtain the viscosity-average molecular weight $(\overline{M}_v)$ is additionally complicated by the following factors:

(i) the influence of molecular weight distribution;

(ii) the occurrence of branching;

(iii) the existence of a compositional and sequential composition of segments in the case of stereospecific polymers and copolymers;

(iv) the existence of agglomerates;

(v) the solvation of macromolecules;

(vi) entanglements between chains;

(vii) the drag effect;

(viii) physical factors such as

(a) adsorption of polymer molecules on wall capillaries,

(b) cleavage of chains by shearing,

(c) local heating due to viscous energy disipation.

References: 2025, 2091, 2102, 2170, 2304, 2379, 2504, 2649, 2694, 2860, 3179, 3180, 3192, 3200, 3201, 3306, 3318, 3519, 4040, 4069, 4187, 4382, 4551, 4599, 4771, 4802, 4917, 4949, 4990, 5076, 5083, 5118, 5217, 5251, 5293, 5389, 5442, 5524, 5770, 5812, 6336, 6488, 6635, 6787–6789, 6893, 6985, 7043, 7219, 7288, 7306.

9.1.3 Viscosity Measurements for Diluted Copolymers

For linear heterogeneous copolymers there exists no one single equation relating the limiting viscosity number (the intrinsic viscosity) $[\eta]$ to the viscosity-average molecular weight $(\overline{M}_v)$. For example, the three different copolymer types

Random copolymer:	AAABAABBABBB
Alternating copolymer:	ABABABABABAB
Block copolymer:	AAABBBAAABBB

have different distributions of A and B units along the main chain. All of these different copolymers may have the same molecular weight but they will have different limiting viscosity numbers (intrinsic viscosities). The Mark–Houwink–Sakurada equation (eq. (9.19)) is not valid for linear hetereogeneous copolymers.

Bibliography: 100, 255, 1214.
References: 3120, 3199, 4210, 4344, 4345, 5761, 5948, 5962, 6585, 6903.

9.1.4 Viscosity Measurements for Diluted Branched Polymers

For branched polymers the Mark–Houwink–Sakurada equation (eq. (9.19)) is again not valid. The viscosity-average molecular weight $(\overline{M}_v)$ calculated from this equation will be lower for branched polymers than for equivalent linear polymers (especially in the higher molecular weight region). For correct calculations knowledge of

(i) the molecular weight distribution function;
(ii) the branch point distribution function;
(iii) the type of branch structure present

is required.

Bibliography: 255, 534.
References: 2377, 2468, 2757, 3204, 3741, 3843, 4087, 4696, 4747, 4769, 5048, 5345, 5430, 5464, 5513, 5616, 5632, 5677, 6902, 6903, 6339, 7337.

9.1.5 Capillary Viscometers for Viscosity Measurements on Dilute Solutions

The capillary viscometers used for dilute solution measurements are made of glass. They are operated by filling with a suitable volume of liquid, drawing the liquid level to a point above the upper mark above the bulb, and measuring the time required for the liquid meniscus to fall from the upper mark to the lower mark. The flow time is related to the viscosity of the liquid and is determined by the driving pressure, using an equation known as POISEUILLE'S LAW:

$$\eta = \pi R^4 P/8lQ = \pi R^4 Pt/8lV, \qquad (9.22)$$

where

R is the radius of the capillary,
P is the pressure driving the fluid through the capillary,
l is the length of the capillary,

$Q = V/t$ is the volumetric flow rate,
V is the volume of liquid,
t is the time of flow.

Commonly, the driving pressure is a result of the difference in height of the two levels of the liquid in the viscometer (h). Higher driving pressures can be applied to the top level of the liquid by a monostat located externally to the viscometer.

The shear rate ($\dot{\gamma}$) in the capillary is given by

$$\dot{\gamma} = -\frac{dv}{dr} = \frac{Pr}{2l\eta},$$ (9.23)

where
  r is the distance from the centre of the capillary,
  v is the velocity of the fluid.
For laminar flow the velocity at a distance r from the centre of the capillary is given by

$$v(r) = P(R^2 - r^2)/4\eta l$$ (9.24)

The velocity varies parabolically from zero at the wall of the capillary ($r = R$) to a maximum value of

$$v_{max} = PR^2/4\eta l.$$ (9.25)

The shear stress (τ) in the capillary is given by

$$\tau = F/A = PR/2l,$$ (9.26)

where
  $F = \pi r^2 P$ is the force at a radial distance r from the centre of the capillary ($r_{max} = R$ and then $F = \pi R^2 P$),
  $A = 2\pi R l$ is the surface area of the capillary.
There are many sources of error in capillary viscometry, depending upon the viscosity level, the nature of the fluid, and the geometric design of the capillary viscometer. The corrections can be grouped into two categories:

  (i) corrections to the shear rate, i.e.
    (a) correction for the departure of the fluid from Newtonian behaviour (the Rabinowitz correction),
    (b) fluid compressibility,
    (c) dimensional changes of the capillary at different temperatures;
  (ii) corrections to the shear stress, i.e.
    (a) kinetic energy correction (the Hagenbach correction),
    (b) Coutte or entrance correction (required because the liquid velocity is not parallel to the axis of the capillary tube),
    (c) elastic energy correction.

These corrections are important for dilute solutions of polymers, whereas some of them can be ignored for undiluted polymers.

The Poisseuille equation (eq. (9.22)), after a kinematic energy correction and an entrance correction, has the form

$$\eta = A\rho t - B\rho/t = A\rho t(1 - B/At^2), \qquad (9.27)$$

where

η is the viscosity of the liquid,

ρ is the liquid density,

A and B are constants for the particular viscometer used.

A and B can be obtained graphically by plotting $(\eta/\rho)/t$ versus $1/t^2$. The intercept of the line through the data points gives A and the slope gives $(-B/A)$.

9.1.6 Relative Viscosity Measurements

The RELATIVE VISCOSITY is given by definition as

$$\eta_{rel} = \frac{\text{Viscosity of the solution}}{\text{Viscosity of solvent}} = \frac{\eta}{\eta_0}, \qquad (9.28)$$

and thus

$$\eta_{rel} = \frac{\eta}{\eta_0} = \frac{\rho t(1 - B/At^2)}{\rho_0 t_0(1 - B/At_0^2)}, \qquad (9.29)$$

where the subscript 0 refers to the pure solvent.

In dilute solutions the ratio ρ/ρ_0 is usually close to unity, so that

$$\eta_{rel} = \frac{t(1 - B/At^2)}{t_0(1 - B/At_0^2)}. \qquad (9.30)$$

If the viscometer has an outflow time greater than 100 s for the pure solvent, the kinetic energy corrections B/At^2 are negligible compared to unity, and then

$$\eta_{rel} = t/t_0. \qquad (9.31)$$

The main types of capillary viscometer (Fig. 9.7) are

(i) The OSTWALD VISCOMETER. The driving pressure in this visco-meter is proportional to the difference in the heights of the two levels in the U-tube (h). For reproducible measurements it is necessary to take the same volume of liquid in each measurement.

(ii) The CANNON–FANSKE VISCOMETER. The inclination of the two arms of the U-tube to the vertical has the effect of placing the centres of the two liquid surfaces along an axis which is vertical even if there is a small degree of error in the vertical mounting of the viscometer.

(iii) The UBBELOHDE VISCOMETER. The driving pressure in this visco-meter is determined by the distance from the level of the liquid in the bulb to the level which is the bottom of the capillary (h). The require-ment for the same volume of liquid in each measurement is eliminated in this viscometer. Some of these viscometers have a large reservoir from

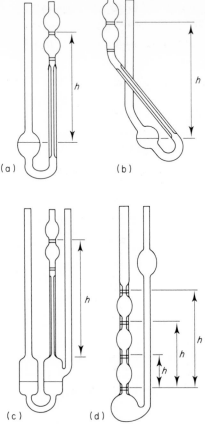

Fig. 9.7 Different types of viscometer. (a)
Ostwald viscometer; (b) Cannon–Fanske
viscometer; (c) Ubbelohde viscometer; (d)
variable shear stress viscometer

which it is possible to add solvent to the solution and in this way lower
the concentration without the necessity of emptying the viscometer.

(iv) The VARIABLE SHEAR STRESS VISCOMETER. The shear stress (τ)
in a capillary viscometer is proportional to the difference in heights of
the levels (h). In the Ubbelhode viscometer the shear stress decreases
with decreasing height. By changing the design of a viscometer, the shear
stress can differ for each of the bulbs. This type of viscometer is called a
variable shear stress viscometer.

All of these types of viscometer are available commercially in 10 ml or 1 ml
(semimicro) sizes with a variety of capillary diameters, giving a selection of
efflux times.

Viscometers have to be mounted vertically because otherwise an additional
error is introduced. Prior to measuring the viscosity of solution, filtration of

dust is desirable in order to avoid irreproducibility of readings. Viscometers should be thermostatted to ±0.01 °C in order to measure the limiting viscosity number (intrinsic viscosity) with an accuracy of 1 %.

Bibliography: 917, 992.
References: 2220, 2873, 3700, 3915, 4468, 4490, 4726, 5190, 6298, 6874.

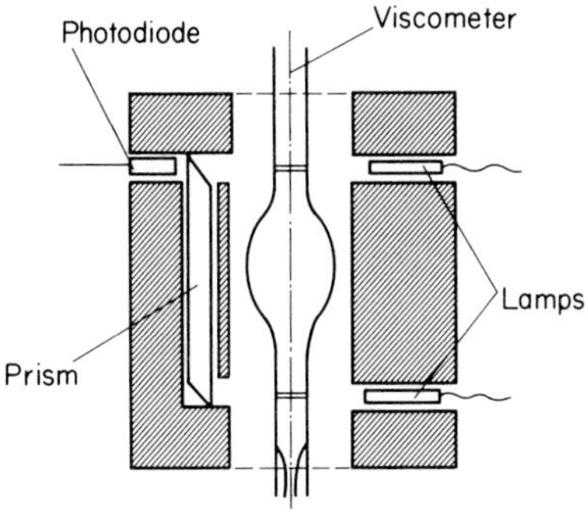

Fig. 9.8 Flow detector in an automatic viscometer. (From R: 3742; reproduced by permission of John Wiley & Sons)

In the newest developments, Ostwald- or Ubbelohde-type viscometers are surrounded by a glass thermostatic jacket fed by liquid from a separate thermostatic bath. Liquid flow in the viscometer is monitored by a single photocell, which is activated by one lamp directly and by another lamp via a prism (Fig. 9.8). The photocell signals are used to activate:

(i) an electronic timer system—an overall accuracy of ±0.01 s in timing is obtained by using a quartz crystal oscillator time base and digital printout of efflux times;
(ii) an air pump for automatic flow control;
(iii) an automatic burette for automated sample loading.

The block diagram in Fig. 9.9 shows the typical logic used. The system can be programmed to measure the efflux time of a polymer solution of concentration c, and then, by automatic dilution, the concentration can be diluted to, say, $2c/3$, $c/2$, $c/3$, $c/6$. Calculation of the limiting viscosity number (intrinsic

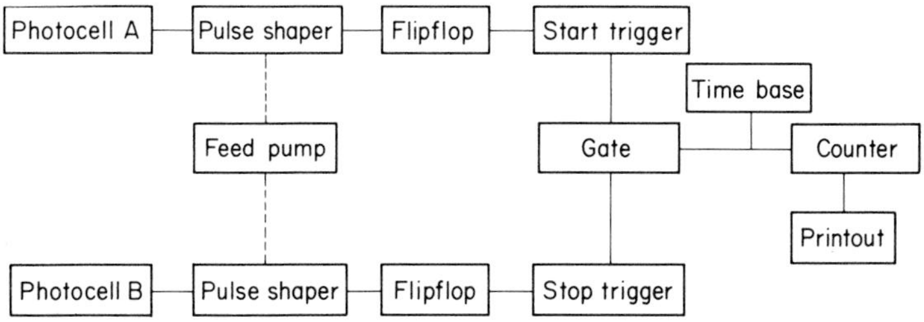

Fig. 9.9 Logic for automatic viscometry. (From B: 723, reproduced by permission of John Wiley & Sons)

viscosity) ($[\eta]$) from the usual plot of $(t-t_0)/t_0 c$ is carried out with the aid of a desk-top computer.

Many developments of autoviscometers are described in detail in original publications.

Bibliography: 723.
References: 2873, 2946, 3132, 3686, 3742, 4170, 4552, 4820, 4839, 4887, 5253, 5254, 6411, 6446, 6526.

The data from automatic viscometers can be computerized. The input data include data efflux times and solute concentrations (including $c = 0$ for pure solvent). The output data are the limiting viscosity number (intrinsic viscosity) ($[\eta]$), the Huggins constant, a plot of η_{sp}/c versus c, the slope of this plot, and the standard deviation.

Bibliography: 312.
References: 2681, 2946.

9.2 VISCOSITY MEASUREMENTS FOR UNDILUTED POLYMERS

The following methods are applied for the measurement of the viscosity of undiluted (melt) polymers:

(i) capillary or slit die viscometry, where flow of fluids occurs under applied pressure;

(ii) rotational viscometry, which uses viscometers with concentric cylinder geometry or with cone and plate geometry;

(iii) tensile creep and stress relaxation methods, which are useful in the very high viscosity range (not discussed in this book).

Bibliography: 110, 217, 255, 259, 301, 353, 395, 436, 533, 922, 1177, 1185, 1283, 1339, 1345, 1429.
References: 2380, 2807, 3113, 3231, 3657, 3738–3740, 3880, 4255, 4256, 5221, 5759, 5760, 5780, 6119, 6143, 6158, 6299, 6658, 6886, 6904, 7000.

9.2.1 Capillary Viscometers for Viscosity Measurements on Undiluted Polymers

There are two systems of operation for capillary rheometers:

(i) the extrusion pressure is constant (controlled by a pressure regulator) whereas the volumetric flow rate (or shear rate) is measured as the dependent variable;

(ii) the volumetric flow rate (or shear rate) is constant (controlled by a mechanically driven piston) whereas the extrusion pressure is measured as the dependent variable.

Capillary rheometers are also classified depending upon the method involved in obtaining shear rate–shear stress data:

(i) shear rate as the independent variable with shear stress as the dependent variable;

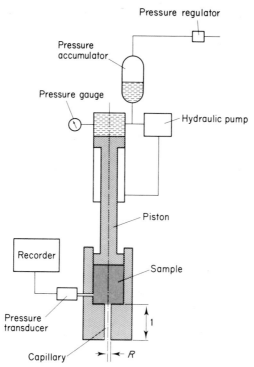

Fig. 9.10   Constant pressure extrusion rheometer

(ii) shear stress as the independent variable and shear rate as the dependent variable.

The capillary viscometers used for more viscous polymer liquids (capillary rheometers) are made from steel. A typical extrusion-type capillary rheometer consists of a container (barrel) with a fine-bore tube (capillary) attached to the bottom (Fig. 9.10). The high pressure forces the polymer liquid to pass from the container through the capillary tube. The fluid is forced through the capillary by the action of a plunger which is driven at selected speeds. The volumetric flow rate is fixed by the plunger speed and by the dimensions of the barrel.

The viscosity of the polymer liquid (η) is given by

$$\eta = \tau/\dot{\gamma} \qquad (9.7)$$

where
τ is the shear stress, and is given by

$$\tau = F/A = PR/2l, \qquad (9.26)$$

$F = \pi r^2 P$ is the force at a radial distance (r) from the centre of the capillary ($r_{max} = R$ and then $F = \pi R^2 P$),
$A = 2\pi Rl$ is the surface area of the capillary,
P is the pressure driving the fluid through the capillary,
$\dot{\gamma}$ is the shear rate.

138

For Newtonian fluids only,

$$\dot{\gamma} = 4Q/\pi R^3. \tag{9.32}$$

For the general case, however,

$$\dot{\gamma} = \frac{4Q}{\pi R^3}\left[\frac{3}{4} + \frac{1}{4}\left(\frac{d\,[\ln(4Q/\pi R^3)]}{d\,(\ln(PR/2l))}\right)\right], \tag{9.33}$$

where
$Q = V/t$ is the volumetric flow rate,
V is the volume of liquid,
t is the time of flow,
R is the radius of the capillary,
l is the length of the capillary.
The pressure (P) is measured by the pressure transducer, is recorded on a strip chart, and is used to calculate the shear stress (τ) by using eq. (9.32). The range of shear rates presently available to capillary rheometers is restricted on the low side to about $\dot{\gamma} = 1\text{--}0.1$ s^{-1}.

Automated read-out, computer data processing, and computer plotting permit easy acquisition and treatment of many data.

Bibliography: 255, 533, 754, 1339, 1357.
References: 2190, 2218, 3737, 3876, 4482, 5244, 6118, 6740.

9.2.2 Slit Die Rheometer

A SLIT DIE RHEOMETER is a rheometer with a rectangular cross-section which can be obtained by the system of two parallel plates (Fig. 9.11). The slit

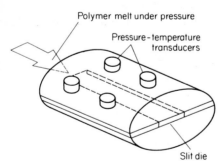

Polymer melt under pressure

Pressure-temperature transducers

Slit die   Fig. 9.11   Extrusion slit die

and capillary rheometers produce essentially the same information on the viscous and elastic properties of polymeric melts.

References: 3874, 3875, 3877, 3878, 3882, 4847.

9.2.3 The Concentric Cylinder Rotational Viscometer

The cylinder rotational viscometer is applied for measuring the viscosity of diluted polymer solutions. In Fig. 9.12 a schematic diagram of a concentric cylinder viscometer is shown.

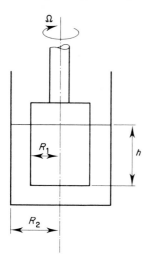

Fig. 9.12 Rotational viscometer with concentric
cylinder

In this type of viscometer the polymer sample is placed between the inner and outer cylinders. The sample is sheared by rotating the inner or the outer cylinder at a constant angular velocity (Ω) (in degrees per second (CGS) or radians per second (SI)).

The shear stress is given by

$$\tau = M/2\pi r^2 h, \qquad (9.34)$$

where

M is the torque (in dynes per centimetre (CGS) or newtons per metre (SI)),
r is the radial distance from the axis of rotation ($R_1 \leqslant r \leqslant R_2$ where R_1 and R_2 are the radii of the inner and outer cylinders),
h is the depth of immersion of the inner cylinder in the fluid being sheared.

The shear rate is given by

$$\dot{\gamma} = -r\frac{d\omega}{dr}, \qquad (9.35)$$

where ω is the angular velocity.

The viscosity of the polymer sample (η) is given by

$$\eta = \frac{\tau}{\dot{\gamma}} = -\frac{M}{2\pi h\, d\omega}\left(\frac{dr}{r^3}\right). \qquad (9.36)$$

After integration of eq. (9.36) between $r = R_1$, where $\omega = 0$, and $r = R_2$, where $\omega = \Omega$, the MARGULES EQUATION for the viscosity of a polymer sample is obtained:

$$\eta = \frac{M}{4\pi h\Omega}\left(\frac{1}{R_1^2} - \frac{1}{R_2^2}\right). \qquad (9.37)$$

The Margules equation is valid only for Newtonian fluids, when the flow is laminar, when there is no slippage at the instrument walls, and where the system is isothermal.

140

It is important to notice that both shear stress (τ) and shear rate ($\dot{\gamma}$) vary inversely as the square of the radial distance (r) from the axis of rotation. They are normally evaluated at either the surface of the inner ($r = R_1$) or the outer ($r = R_2$) cylinder. Shear stress (τ) and shear rate ($\dot{\gamma}$) should both be evaluated at the same position in the viscometer.

Rotational viscometers can be applied to obtain continuous measurements.

Bibliography: 1357.
References: 4504, 4931, 7070, 7333.

9.2.4 The Plate–Cone Rheometer

The plate–cone rheometer is applied for measuring the viscosity of more viscous polymer liquids, e.g. polymer melts. In Fig. 9.13 a schematic diagram of

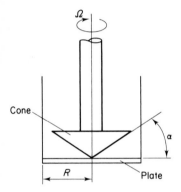

Cone

α

R

Plate

Fig. 9.13 Rotational viscometer with cone and plate

a plane–cone rheometer geometry is shown. The polymer melt is placed within the gap between the cone and the flat circular plate. The cone angle (α) is defined as the angle between the conical surface and the surface of the horizontal plate. The angle α is usually small (1–5 °C). The cone is rotated at a constant angular velocity (Ω).

The shear stress is given by

$$\tau = 3M/2\pi R^3, \tag{9.38}$$

where

M is the torque (in dynes per centimetre (CGS) or newtons per metre (SI)),
R is the radius of the cone.

The shear rate is given by

$$\dot{\gamma} = \Omega/\alpha, \tag{9.39}$$

where

Ω is the constant angular velocity (in degrees per second (CGS) or radians per second (SI)),
α is the cone angle (in degrees (CGS) or radians (SI)).

The viscosity of the polymer melt (η) is given by

$$\eta = \tau/\dot{\gamma} = 3\alpha M/2\pi R^3 \, \Omega = kM/\Omega \qquad (9.40)$$

where $k = 3\alpha/2\pi R^3$ is a constant calculated from the geometry of the instrument. The range of shear rates presently available for plate–cone rheometers is restricted on the low side to about $\dot{\gamma} = 10^{-3} \text{ s}^{-1}$.

Bibliography: 255, 533, 1339, 1357, 2381, 3739, 5760.

Chapter 10

Optical Methods in Polymer Research

General bibliography: 58, 87, 114, 139, 166, 167, 188, 207, 240, 241, 468, 647, 744, 782, 813, 849, 873, 874, 981, 990, 991, 1086, 1206, 1220, 1230, 1243, 1257, 1415.

10.1 PROPERTIES OF ELECTROMAGNETIC RADIATION

ELECTROMAGNETIC RADIATION is a form of energy, travelling through space unaccompanied by any matter. The behaviour of electromagnetic radiation can be attributed either to its wave-like character or to its corpuscular character. In Fig. 10.1 is shown a plane-polarized electromagnetic wave of a single

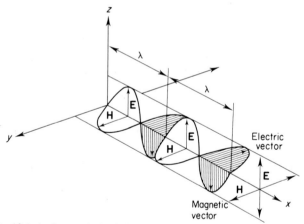

Fig. 10.1 A plane-polarized electromagnetic wave of a single frequency.
E is the electric vector and H is the magnetic vector

frequency, i.e. a MONOCHROMATIC BEAM. The PLANE-POLARIZED electromagnetic field means that the electric vector E vibrates in a single plane and the magnetic field vector H vibrates in another plane perpendicular to the electric field. In practice most electromagnetic radiation is UNPOLARIZED, i.e. has electric and magnetic vectors at all orientations perpendicular to the direction of propagation.

Different kinds of electromagnetic radiation are usually characterized by either the wavelength λ or the frequency ν.

WAVELENGTH (λ) is defined as the length of the cycle, or the distance between successive maxima or minima (Fig. 10.1).

The following units of wavelength are commonly used (CGS and SI):

$$1\ \mu m = 10^{-6}\ m = 10^{-4}\ cm \quad (\mu m = \text{micrometre}),$$
$$1\ nm = 10^{-9}\ m = 10^{-7}\ cm \quad (nm = \text{nanometre}),$$
$$1\ \text{Å} = 10^{-10}\ m = 10^{-8}\ cm \quad (\text{Å} = \text{ångström}).$$

1 ÅNGSTRÖM is a unit of length equal to 1/6438.4696 of the wavelength of the red line of Cd. (Note: do not use the old unit nomenclature, e.g. micron and millimicron.)

Most workers in the field of optical instrument design, photochemistry, and optics express wavelengths in nanometers, whereas spectroscopists commonly specify wavelengths in ångströms.

FREQUENCY (v) is the number of cycles per unit time (in s^{-1} or hertz). (Note: 1 Hz = 1 cycle s^{-1}.)

The wavelength and the frequency are related by

$$\lambda = v/\nu = c/\nu, \tag{10.1}$$

where v is the velocity of propagation.

All electromagnetic radiation travels through a vacuum with the same velocity (c) which has the value $2.9979 \times 10^{10}\ cm\ s^{-1}$ (CGS) or $2.9979 \times 10^{8}\ m\ s^{-1}$ (SI).

The frequency is the only true characteristic of a particular radiation; both the velocity (v) and the wavelength (λ) depend on the nature of the medium in which the electromagnetic wave travels.

The WAVENUMBER ($\bar{\nu}$) is the number of waves per unit length:

$$\bar{\nu} = 1/\lambda = \nu/c \tag{10.2}$$

(in cm^{-1} (CGS) or m^{-1} (SI)).

The units of wavenumber ($\bar{\nu}$) and wavelength (λ) are related as follows:

$$cm^{-1} = \frac{1}{\mu m} \times 10^4 \quad \text{(CGS)}, \qquad m^{-1} = \frac{1}{\mu m} \times 10^6 \quad \text{(SI)}. \tag{10.3}$$

Conversion tables of wavenumber (cm^{-1}) to wavelength (μm) are given in Appendix B (see pp. 648–649).

The electromagnetic spectrum is shown in Fig. 10.2.

According to the corpuscular theory of electromagnetic radiation, the energy takes the form of small bundles of energy or particles, called photons.

PHOTONS are quantized electromagnetic waves.

PHONONS are quantized vibrational waves.

In each case the energy of the molecule is given by the PLANCK EQUATION:

$$E = h\nu \tag{10.4}$$

where

h is the Planck constant, which has the value $6.626 \times 10^{-27}\ erg\ s$ (CGS), or $6.626 \times 10^{-34}\ J\ s$ (SI),

v is the frequency of light in the case of photons or the frequency of the vibration in the case of phonons (in s^{-1}).

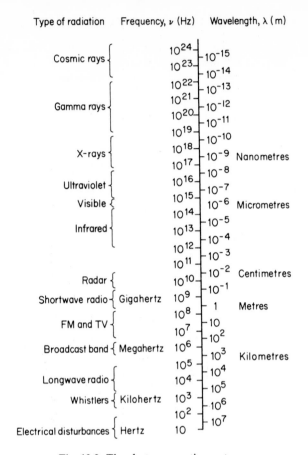

Fig. 10.2 The electromagnetic spectrum

10.2 SPECTROSCOPIC METHODS IN POLYMER RESEARCH

General bibliography: 53, 63, 65, 71, 81, 102, 190, 192, 207, 212, 240, 274, 337, 364, 389, 390, 405, 476, 612, 693, 730, 795, 878, 961, 962, 991, 1012, 1037, 1105, 1158, 1179, 1192, 1213, 1292, 1305, 1379, 1393.

SPECTROSCOPIC METHODS are based on the measurement of the intensity and the wavelength of the radiative energy and the spectra caused by transitions between characteristic energy states.

In Fig. 10.3 are presented the various regions of the electromagnetic spectrum, and names are given for some of the spectroscopic methods used in each region. All spectra can be divided into three fundamental types (Table 10.1):

(i) absorption spectra (cf. Chapters 14 and 15);
(ii) emission spectra (cf. Chapter 16);
(iii) scattering spectra (Raman spectra) (cf. Chapter 17).

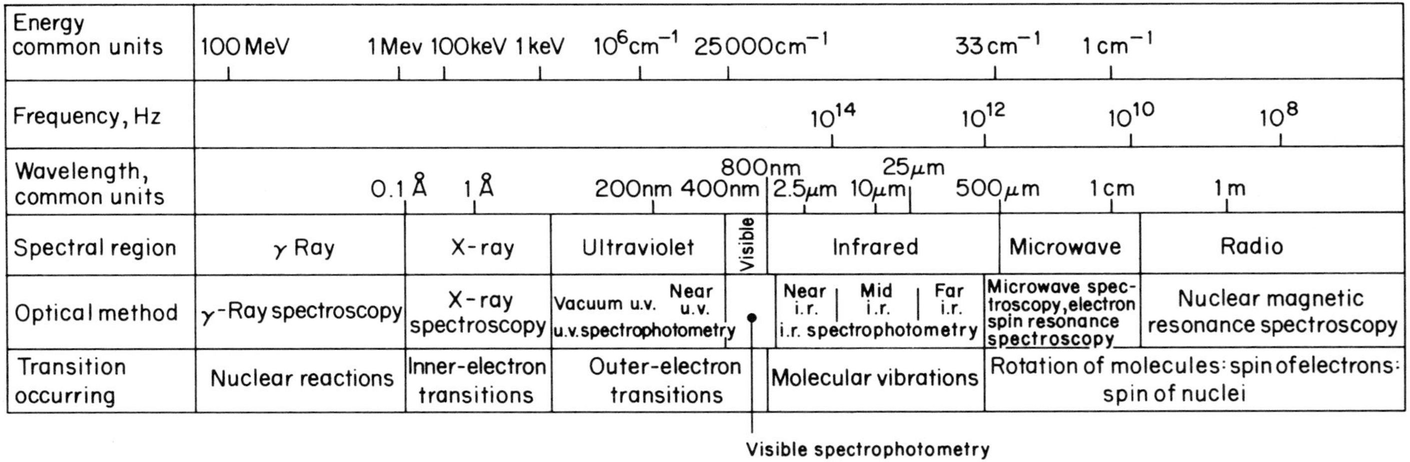

Energy common units	100 MeV	1 MeV	100 keV	1 keV		10^6 cm⁻¹	25000 cm⁻¹			33 cm⁻¹	1 cm⁻¹
Frequency, Hz					10^{14}			10^{12}		10^{10}	10^8
Wavelength, common units	0.1 Å	1 Å		200 nm 400 nm	800 nm 2.5 μm	10 μm 25 μm	500 μm		1 cm		1 m
Spectral region	γ Ray	X-ray		Ultraviolet	Visible	Infrared			Microwave		Radio
Optical method	γ-Ray spectroscopy	X-ray spectroscopy		Vacuum u.v. Near u.v. spectrophotometry		Near i.r. | Mid i.r. | Far i.r. i.r. spectrophotometry			Microwave spectroscopy, electron spin resonance spectroscopy		Nuclear magnetic resonance spectroscopy
Transition occurring	Nuclear reactions	Inner-electron transitions		Outer-electron transitions		Molecular vibrations			Rotation of molecules: spin of electrons: spin of nuclei		

Visible spectrophotometry

Fig. 10.3 The electromagnetic spectrum and optical methods for its study (reproduced by permission of McGraw-Hill)

Table 10.1 Classification of spectroscopic methods according to the types of spectra they exhibit (reproduced by permission of McGraw-Hill)

Absorption spectra	Emission spectra	Raman spectra
Ultraviolet and visible spectrophotometry Infrared spectrophotometry Microwave spectroscopy Circular dichroism spectrometry X-ray absorption spectroscopy Nuclear magnetic resonance (NMR) spectroscopy Electron spin resonance (ESR) spectroscopy	Emission spectroscopy Spectrofluorometry Spectrophosphorimetry	Raman spectroscopy

10.2.1 Absorption Spectroscopy

The total energy of a molecule consists of:

(i) electronic energy (E_{elec});
(ii) vibrational energy (E_{vib});
(iii) rotational energy (E_{rot});
(iv) translational energy (E_{trans}).

The change ΔE is the energy of a molecule upon absorption of electromagnetic radiation is given by

$$\Delta E = \Delta E_{\text{elec}} + \Delta E_{\text{vib}} + \Delta E_{\text{rot}} + \Delta E_{\text{trans}}, \tag{10.5}$$

where

ΔE_{elec} is the spacing between allowed electronic energy levels,

ΔE_{vib} is the spacing between allowed vibrational energy levels (ΔE_{vib} values are generally about 10 times less than ΔE_{elec} values),

ΔE_{rot} is the spacing between allowed rotational energy levels (ΔE_{rot} values are about 10 or 100 times less than ΔE_{vib} values),

ΔE_{trans} is the spacing between allowed translational energy levels (ΔE_{trans} values are extremely small and are unimportant in absorption spectroscopy).

When a beam of monochromatic radiation is transmitted through a polymer sample it may, depending on its wavelength (λ), be partially absorbed by specific molecular groups present in the sample. An experimental plot of transmitted intensity versus the wavelength of the radiation will exhibit minima at wavelengths of maximum absorption. Such a plot is called the ABSORPTION SPECTRUM OF THE SAMPLE.

In Fig. 10.4 an absorption spectrum at different wavelengths is shown schematically, corresponding to the

ultraviolet (UV) region ($\lambda = 0.1$–0.4 μm),
visible (VI) region ($\lambda = 0.4$–0.8 μm),
infrared (IR) region ($\lambda = 1$–50 μm).

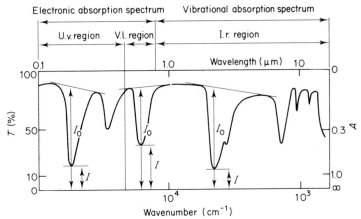

Fig. 10.4 Schematic absorption spectrum

A given absorption peak (band) is characterized by

(i) the position of an absorption peak (band);
(ii) the intensity of peak (band) absorbance.

The positions of UV and VI absorption peaks are specified in the same units as their wavelengths (λ) (i.e. ångströms, micrometres, or nanometres).

The positions of IR absorption bands can be specified in the same units as their wavelengths (λ) (i.e. ångströms, micrometres, or nanometres), frequencies (v) (i.e. s^{-1} or hertz), or wavenumbers ($\bar{v}$) (i.e. cm^{-1}).

10.3 NON-SPECTROSCOPIC OPTICAL METHODS IN POLYMER RESEARCH

NON-SPECTROSCOPIC OPTICAL METHODS do not measure spectra, do not involve transitions between characteristic energy states, and instead are based on interactions between electromagnetic radiation and matter which result in a change in direction or a change in the physical properties of the electromagnetic radiation. The specific mechanisms of interaction involved in non-spectroscopic methods are as follows:

(i) REFRACTION, which is the bending of radiation as it passes from one material into another and can be attributed to a difference in velocity in the two media (cf. Chapter 11).
(ii) REFLECTION, which occurs whenever radiation is incident upon a boundary between two materials across which there is a change in refractive index.
(iii) SCATTERING, which is the interaction of electromagnetic radiation with particles in its path by inducing oscillations in the electric charges of the matter. The dipoles that are induced radiate secondary waves in all directions (cf. Chapter 13).

(iv) INTERFERENCE, which can be defined as the modification of the intensity of radiation waves by combining or superimposing two or more waves:
 (a) DESTRUCTIVE INTERFERENCE occurs when intensity is decreased;
 (b) CONSTRUCTIVE INTERFERENCE occurs when intensity is increased. This phenomenon is applied in optical interference filters which can be used in the near-UV, visible, and near-IF regions (cf. Section 10.4).

(v) DIFFRACTION, which occurs when waves of electromagnetic radiation pass through a slit or travel past the edge of any opaque obstacle; they always spread to some extent into the region which is not directly exposed to the incoming waves.

(vi) DISPERSION, which can be defined as the separation of a mixture of wavelengths into its component wavelengths.

(vii) POLARIZATION, which occurs if the amplitudes of the vectors become unsymmetrical about the direction of travel; the elimination of all but one plane results in PLANE POLARIZATION (cf. Section 10.3.1).

(viii) DICHROISM, which is a selective absorption of one vibrational plane of radiation (cf. Section 35.5).

10.3.1 Polarized Radiation

Most sources produce electromagnetic radiation in which the vibrations of the electric and magnetic vectors occur with equal amplitude at all orientations perpendicular to the direction of propagation (Fig. 10.5(a)). Such radiation is named UNPOLARIZED RADIATION.

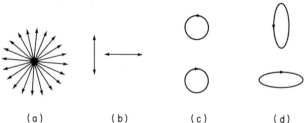

| (a) | (b) | (c) | (d) |

Fig. 10.5   The path that the electric amplitude vector traces out for different states of polarization. (a) Unpolarized light; (b) linear polarization; (c) circular polarization; (d) elliptical polarization

If any material medium affects the amplitude of unpolarized radiation differently, the electromagnetic radiation will be POLARIZED. When one vibration is completely eliminated the radiation is LINEARLY POLARIZED (or PLANE POLARIZED) (Fig. 10.5(b)).

The electrical vibration may rotate continuously in a clockwise sense, i.e. the beam may be right- or left-CIRCULARLY POLARIZED (Fig. 10.5(c)). The amplitude of a circularly polarized wave rotates in a clockwise manner, tracing out a helix in space (Fig. 10.6), and when viewed from behind the helix looks like an ellipse (Fig. 10.5(d)). The beam in such a projection is called ELLIPTICALLY POLARIZED.

The relative intensities of the two orthogonal vibrational components in a beam may be determined by using pairs of polarizers: the POLARIZER and the ANALYSER. They resolve light into orthogonal vibrations and selectively eliminate one set of such vibrations. If linearly polarized light is incident, the intensity transmitted by the analyser is

(i) 100% if the direction of vibration of incident radiation is parallel to the characteristic vibrational direction of the analyser (Fig. 10.5(a));
(ii) between 100% and zero if resolved into a favoured and a rejected component;
(iii) Zero if the direction of vibration of incident light is perpendicular to the favoured direction of the analyser—in this case the incident radiation generates no component in the vibrational direction passed (Fig. 10.5(b)).

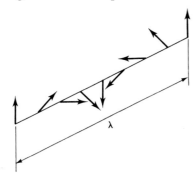

Fig. 10.6 The amplitude of a circularly polarized wave rotates in a clockwise manner tracing out a helix in space. Only one wavelength is illustrated

The DEGREE OF POLARIZATION (*P*) (or POLARIZANCE) is given by

$$P = \frac{I' - I''}{I' + I''}, \qquad (10.6)$$

where I' and I'' are the maximum and minimum intensities observed through an analyser rotated through 360°.

The effectiveness of a polarizer is determined by

(i) efficiency;
(ii) spectral range of operation;
(iii) angular aperture.

Polarizers can be divided into the following categories:

(i) DICHROIC POLARIZERS selectively absorb one of the orthogonal vibration components (Fig. 10.7). The following materials are applied as dichroic polarizers:

(a) Polaroid sheets of types H and K, which are frequently used in the visible and near-IR regions. They are prepared from poly(vinyl

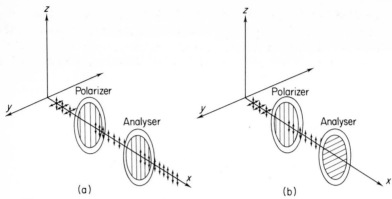

Fig. 10.7 Detection of linearly polarized radiation using Polaroid discs.
(a) Uncrossed polarizers; (b) crossed polarizers

alcohol) films. The H-sheets are first stretched to align the molecules in the direction of extension and then impregnated with iodine. The I_2 molecules complex with the polymer and are thereby aligned. The component of incident light with vibrations parallel to the bond axis of I_2 molecules is absorbed, whereas the component at right angles is transmitted. In the type K film selective absorption is achieved in a film of poly(vinyl alcohol) by catalytic dehydration followed by parallel alignment of double bonds by stretching.

(b) Films of pyrolytic graphite, used for polarization in the infrared from 2500 cm^{-1} to the microwave region.

(ii) REFLECTION POLARIZERS partially polarize light by reflection from a dielectric (Fig. 10.8). The reflectances of the two orthoganally polarized components are given by FRESNEL'S EQUATIONS:

$$R_\perp = \frac{1}{2}\frac{\sin^2(\Phi - \Phi')}{\sin^2(\Phi + \Phi')}, \tag{10.7}$$

$$R_\| = \frac{1}{2}\frac{\tan^2(\Phi - \Phi')}{\tan^2(\Phi + \Phi')}, \tag{10.8}$$

where
Φ is the angle of incidence (and reflection),
Φ' is the angle of refraction.

Fig. 10.8 Polarization of light by a reflection polarizer with plates mounted at the polarizing angle

The degree of polarization can be increased by adding more plates, since each new one will eliminate more of the vertical component. Reflection polarizers are widely used in the IR range (between four and six silver chloride, selenium, or germanium plates are employed).

(iii) DOUBLE-REFRACTION POLARIZERS are constructed from sections of calcite (CaCO₃) crystals. The crystal resolves the incident radiation into beams that are transmitted at different angles because of their different indices of refraction (Fig. 10.9). Double-refraction polarizers

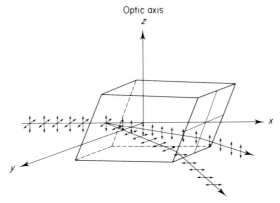

Fig. 10.9 Polarization of light by a double-refraction polarizer made of a calcite crystal

produce two complete spectral sets of linearly polarized light. Double-refraction polarized light produced by using CaCO₃ is used for polarization in the range from 220 to 1800 nm (2200 Å to 1.8 μm): whereas polarized light produced by using uniaxial crystals of calcium dihydrogen phosphate is useful in the UV range up to about 185 nm.

NICOL PRISMS (or NICOL POLARIZERS) are made from calcite crystals with a minimum of alternation of the natural crystal faces (Fig. 10.10). An incident ray is resolved into an ordinary and an extraordinary ray. By splitting the crystal at a suitable angle and cementing the halves with Canada balsam, an isotropic substance of $n_D = 1.55$,

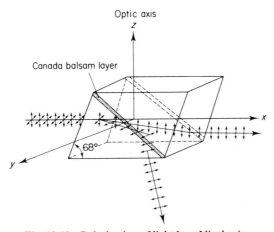

Fig. 10.10   Polarization of light by a Nicol prism

152

the ordinary ray is completely reflected at the interface. Only the extra-ordinary ray travels in the original direction. It appears as a beam parallel to the entering beam, but displaced laterally from it. Nicol prisms can only be used for polarization in the visible range, since Canada balsam is opaque to the UV region.

GLAN–THOMPSON PRISMS are made from calcite crystals which are trimmed so that the optic axis is perpendicular at the front and rear faces. Polarizers with an air gap between the prisms are useful for the polarization of light in the UV region, whereas prisms cemented with Canada balsam can be used only for polarization in the visible range.

ROCHON–WOLLASTON PRISMS are made from calcite crystals or quartz pieces which are cut and then cemented together with glycerine or castor oil (transparent down to 230 nm) (Fig. 10.11).

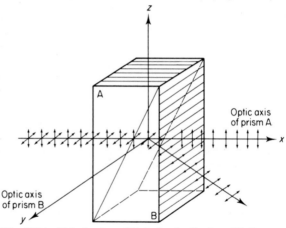

Fig. 10.11   Polarization of light by the Rochon–Wollaston prism

Bibliography: 468, 981, 1205, 1206, 1243, 1257, 1275, 1353, 1415.

10.4 OPTICAL MATERIALS

In photometric instruments the range of transparency of substances suitable for absorption cells, lenses, prisms, and windows is a critical factor.

The UV transmission of many materials is given in Table 10.2. Some crystals have several absorption bands. Table 10.3 lists a number of crystals and the short-wavelength edge of their absorption bands. Knowledge of the transmission properties of materials is particularly important in selecting window materials for IF or UV applications.

FIBRE OPTIC BUNDLES are flexible, transparent fibres produced from quartz (for the UV region) and synthetic polymers (for the VI region) and are applied to the 'transportation of light' into hidden parts of instruments or experimental devices. Transmission depends on total internal reflection at the walls (Fig. 10.12).

Table 10.2 Approximate wavelength limits for transmission of various optical materials and water near room temperature (from B: 242; reproduced by permission of John Wiley & Sons)

Material	Thickness (mm)	50%	30%	10%
		Approximate λ (Å) for percentage transmission indicated		
Window glass (standard)	1	3160	3120	3070
	3	3300	3230	3140
	10	3520	3420	3300
Optical (white crown) glass	1.8	3270	3200	3090
Pyrex (Corning 774)	1	3060	2970	2800
	2	3170	3090	2970
	4	3300	3190	3100
Corex D (Pyrex, Corning 9-53 (9700))	1	2780	2670	2500
	2	2880	2800	2670
	4	3040	2920	2810
Corex A	2.9	2480	2430	2400
Vycor 790	2			>2540
Vycor 791	1	2150	2130	2120
	2	2230	2170	2130
	4	2360	2250	2170
Quartz, crystal	5	1850		
	10	1930	1920	1860
Quartz, clear fused (General Electric Co.)	10	1940	1810	1720
Suprasil (Englehard Industries, Inc.)	10	1700	1680	1660
Sapphire, synthetic (Linde Air Products)	0.5			1425
Fluorite (CaF$_2$), natural	5	1350		
	10	1570	1450	1380
CaF$_2$, synthetic (Linde Air Products)	3			1220
Lithium fluoride (synthetic)	5	1070		
	10	1420	1270	1150
Plexiglas (polymethylmethacrylate)	2.5	3220	3100	2970
	5.0	3380	3250	3110
	10.0	3500	3420	3260
Water (distilled)	20	1880	1860	1850
	40	1920	1880	1860
	80	2020	1940	1880

Table 10.3 Some crystals and their absorption band edges

Material	Band edge(nm)	Material	Band edge (nm)
Quartz (SiO$_2$)	120	TiO$_2$	414
Sapphire (Al$_2$O$_3$)	140	Agl	444
Calcite (CaCO$_3$)	200	ZnSc	480
NaCl	210	CdS	510
Diamond	232	GaP	550
ZnS	345	Cu$_2$O	590
ZnO	390	GaAs	885
AgCl	390	PbS	3550

154

Fig. 10.12   Internal reflection in a bundle of optic
fibres

FILTERS are devices that absorb, reflect, or deviate all frequencies from an optical path except for one band or region. There are different types of filters:

(i) ABSORPTION GLASS FILTERS selectively remove part of the spectrum of a source. In Fig. 10.13 the transmission curves for a number of these filters are shown.

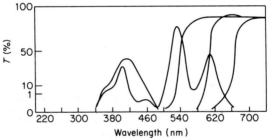

Fig. 10.13   Transmission curves for some absorption
filters

(ii) INTERFERENCE FILTERS are optical filters based on interference between multiple reflected rays (Fig. 10.14). They are made from extremely thin layers of dielectric sandwiched between semireflecting metallic films, usually of silver. Interference filters can be made for the region

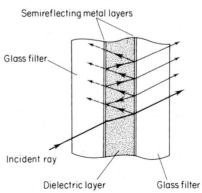

Fig. 10.14   Cross-section of an interference filter

from about 2000 cm^{-1} (5 μm) through into the near-UV. Transmission band characteristics of interference filters are shown in Fig. 10.15.

(iii) LIQUID TRANSMISSION FILTERS are made by dissolution of inorganic salts (NaNO$_2$, K$_2$Cr$_2$O$_7$) or organic food-colouring dyes in water and they can be applied in the region of 4000–7200 Å.

(iv) LIQUID BAND-PASS FILTERS are made for the separation of narrow spectral ranges of passing light. Composition data and characteristic selected liquid band-pass filters are shown in Table 10.4. For the 2537 Å selected wavelength a special filter composition is used, which is given in Table 10.5.

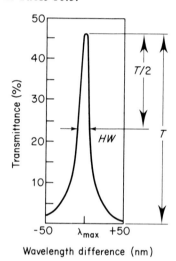

Fig. 10.15 Transmission band characteristics of interference filters. λ_{max} is the wavelength of the peak, T is the transmittance (%); HW is the half-width

Table 10.4 Composition and characteristic of narrow-band pass-filters (from R: 7140; reproduced by permission of Pergamon Press)

No.	Transmission Å	Filter composition
1	2600–3900	200 g COCl$_2$.6H$_2$O per litre + 100 g NiCl$_2$.6H$_2$O per litre in a mixture of 65% ethanol and 35% water containing 1 mol HCl per litre†
2	2900–3900	200 g CoCl$_2$.6H$_2$O per litre + 100 g NiCl$_2$.6H$_2$O per litre in a mixture of 55% dimethylformamide and 45% water containing 1 mol HCl per litre
3	3100–3900	200 g COCl$_2$.6H$_2$O per litre + 100 g NiCl$_2$.6H$_2$O per litre in a mixture of 35% ethanol, 25% acetone, and 40% water containing 1 mol HCl per litre
4	2600–4600	200 g COCl$_2$.6H$_2$O per litre in a mixture of 70% ethanol and 20% water containing 1 mol HCl per litre
5	2900–4500	200 g COCl$_2$.6H$_2$O per litre in a mixture of 65% dimethylformamide and 35% water containing 1 mol HCl per litre
6	3100–4600	200 g CoCl$_2$.6H$_2$O per litre in a mixture of 75% acetone and 25% water containing 1 mol HCl per litre

† The mixed solvent was prepared as follows: first 65 ml of 99.5% ethanol is brought into a 100 ml volumetric flask, 20 ml 5 M-HCl solution is added, and the flask is filled to the mark with water. Same procedure for the other solvents.

Table 10.5   Composition data of filter solutions for isolation of the 2537 Å wavelength from the medium-pressure mercury lamp (from B: 242; reproduced by permission of John Wiley & Sons)

Path length (cm)	Components
5	27.6 g $NiSO_4.6H_2O$ per 100 cm³ aqueous solution
5	8.4 g $CoSO_4.7H_2O$ per 100 cm³ aqueous solution
1	0.108 g I_2 + 0.155 g KI per 1000 cm³ water
5	Cl_2 gas at 1 atm and 25 °C

Design applications of physical and geometrical optics are given in several publications.

Bibliography: 242, 647, 744, 709, 782, 813, 873, 874, 981, 1086, 1230.
References: 4223, 7140.

10.5 LIGHT SOURCES

For spectroscopic measurements a continuous source of relatively low intensity is desirable. The intensity of the source (number of photons emitted per second) should be stable with time.

Many manufacturers do not supply emission spectra with the lamp they provide. In order to determine the emission spectra, photometers (radiometers) are used. There are two types of photometer:

(i) photodiode radiometers (photometers), which read the signal current from the photodiode—this signal current is proportional to the radiation input;

(ii) photomultiplier radiometers (photometers), which read the signal current from the photomultiplier—this signal current is proportional to the radiation input.

These instruments can be used for a wide variety of relative radiation (light) measurements.

The spectrum of the lamp can be obtained using a monochromator and a radiometer (Fig. 10.16). The lamp is placed in a fixed position at the entrance

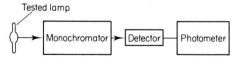

Fig. 10.16   Block diagram of instrument used for measuring the emission spectra of light sources

slit of the monochromator. The intensity of the light incident upon the probe is measured at intervals from 2 to 50 nm over the wavelength range 250–800 nm. The intensity readings are then plotted versus wavelength after correction for the change in dispersion of the prism over the range investigated. The prism dispersion data can be obtained from information published by the mono-chromator manufacturer.

For photochemical study a light source which emits a few intense frequencies is desirable.

Bibliography: 451, 1227.

10.5.1 Deuterium Light Sources

A deuterium arc is concentrated on a 1 mm diameter source which produces a line-free continuum from below 180 nm to 400 nm (Fig. 10.17). The total VI and

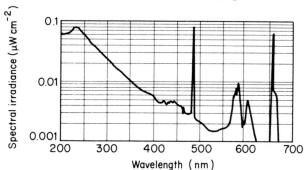

Fig. 10.17   Spectral output of the deuterium source. (Reproduced by permission of Oriel from Oriel Catalogue)

IR output is very low, thus producing low stray light levels. The deuterium light source is applied in many commercial UV spectrometers.

10.5.2 Rare-gas Lamps

Medium-pressure, rare-gas lamps provide a useful source of closely spaced frequencies in the UV and short-wavelength VI regions (Table 10.6).

High pressure xenon arc lamps have a continuous spectrum from 190 to 750 nm and are excellent for UV–VI spectroscopic applications.

Table 10.6   Wavelength range available from medium-pressure rare-gas lamps

Rare gas	Spectral range (nm)
Argon	390–440
Helium	380–450
Krypton	430–450
Neon	340–360
Xenon	460–480

10.5.3 Mercury Arc Lamps

Mercury arcs are classified according to the pressure within the arc envelope: low, medium, or high pressure. Each type emits a characteristic spectrum and intensity and has specific applications in spectroscopy and photochemical work.

Low-pressure mercury arcs in which the vapour pressure of mercury is about 10^{-3} torr (1 torr $= \frac{1}{760}$ atm $= 133.322$ N m^{-2}) emit radiation primarily at two so-called resonance frequencies, 1849 and 2537 Å. The 2537 Å line is the most intense, whereas the intensity of the 1849 Å line depends on the conditions of excitation. One of the features of low-pressure arcs is their long life (2000–10 000 h or more) and dependability.

Medium-pressure mercury arcs use a pressure of mercury vapour from 10^{-3} torr to 1 atm. The light intensity of the arc increases with increasing pressure, and the number of different frequencies emitted also increased.

The spectra of low- and medium-pressure mercury arcs are compared in Table 10.7.

Table 10.7 Energy distribution in low- and medium-pressure mercury arcs (from B: 242; reproduced by permission of John Wiley & Sons)

| Wavelength (Å) | Relative energy | |
	Low-pressure mercury arc	Medium-pressure mercury arc
13 673	—	15.3
11 287	—	12.6
10 140	—	40.6
5770–5790	10.14	76.5
5461	0.88	93.0
4358	1.00	77.5
4045–4078	0.39	42.2
3650–3663	0.54	100.0
3341	0.03	9.3
3126–3132	0.60	49.9
3022–3028	0.06	23.9
2967	0.20	16.6
2894	0.04	6.0
2804	0.02	9.3
2753	0.03	2.7
2700	—	4.0
2652–2655	0.05	15.3
2571	—	6.0
2537	100.00	16.6
2482	0.01	8.6
2400	—	7.3
2380	—	8.6
2360	—	6.0
2320	—	8.0
2224	—	14.0

High-pressure mercury arcs operate at internal pressures of several hundred atmospheres. The emission spectrum is characterized by broad bands instead of the discrete line spectrum of low- and medium-pressure mercury arcs. The high-pressure lamps are used mainly in photochemical work. The average life of such a lamp ranges from 10 to 80 h, depending on the cooling method and the number of starts. These lamps must be cooled with forced air or water to prevent envelope failure. Water must be circulated around the arc at very high velocities to prevent formation of steam bubbles and also should be of high purity for maximum transmittance and minimum electrical conductivity.

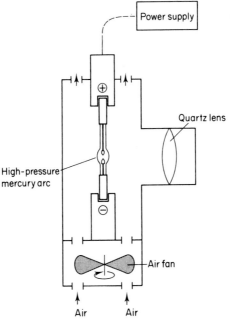

Fig. 10.18 View of the housing for a high-pressure mercury arc cooled with air

Typical housings for high-pressure mercury arcs cooled with air and/or water are shown in Figs 10.18 and 10.19, respectively.

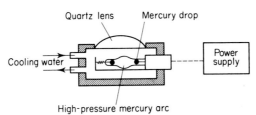

Fig. 10.19 View of the housing for a high-pressure mercury arc cooled with water

Bibliography: 242, 451, 1086.

10.6 LASERS

The word LASER is an acronym for 'light amplification by stimulated emission of radiation'.

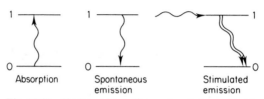

Fig. 10.20   Various absorption and emission processes

In Fig. 10.20 the basic processes which occur in lasers are presented:

(i) ABSORPTION occurs when a photon is absorbed by an atom. The energy of the photon is converted into internal energy of the atom, i.e. an electron is raised to a higher energy level.

$$\text{Rate of absorption} = IBN_0, \qquad (10.9)$$

where

I is the intensity of the radiation,
B is the EINSTEIN COEFFICIENT OF INDUCED ABSORPTION,
N_0 is the number of atoms in the ground state.

(ii) SPONTANEOUS EMISSION occurs when the excited atom returns to the ground state. The typical lifetime of an excited state is 10^{-9}–10^{-4} s.

$$\text{Rate of spontaneous emission} = AN_1, \qquad (10.10)$$

where

A is the EINSTEIN COEFFICIENT FOR SPONTANEOUS EMISSION,
N_1 is the number of atoms in the excited state.

(iii) STIMULATED EMISSION occurs when an additional photon having precisely the energy of the absorbed photon strikes an atom in the excited state. The incoming photon is now joined by a second photon from the excited state atom, resulting in a gain or an amplification of the photons. The second photon wave is precisely in phase with the photon wave that triggered its release, giving a perfectly COHERENT in-phase beam of radiation.

$$\text{Rate of stimulated emission} = ICN_1, \qquad (10.11)$$

where C is the EINSTEIN COEFFICIENT FOR STIMULATED EMISSION.

Considering these three processes, (i), (ii), and (iii), it can be seen that: the rate of increase of the population of the level 1 is given by

$$\text{Rate}_{0\rightarrow 1} = IBN_0;$$ (10.12)

the rate of the depopulation of the level 2 is given by

$$\text{Rate}_{1\rightarrow 0} = AN_1 + ICN_1;$$ (10.13)

the equilibrium is therefore

$$IBN_0 = AN_1 + ICN_1.$$ (10.14)

The minimum requirement for laser action is that there must be a POPU-LATION INVERSION, i.e.

$$ICN_1 > IBN_0 \qquad \text{or} \qquad N_1 > N_0,$$

otherwise absorption will dominate emission. In order to cause a population inversion it is necessary to achieve suitable conditions for laser action. The build-up of a population of atoms in an excited state is sometimes called OPTICAL PUMPING.

In order to amplify the stimulated emission the laser action is staged in the OPTICAL RESONANCE CAVITY which is a long tube closed off by two mirrors, one of which is semitransparent (Fig. 10.21). The spacing between the mirrors is an integral multiple of the desired wavelength. In the optical reso-nance cavity the build-up of energy then occurs at the desired wavelength (reso-nant frequency) and loss of energy occurs at other wavelengths. Photons travel-ling along the axis of the tube are reflected by the end mirrors, whereas photons travelling in other direction are lost. As photons pass through the tube, stimulated emission causes a rapid growth in the photon intensity, and this is

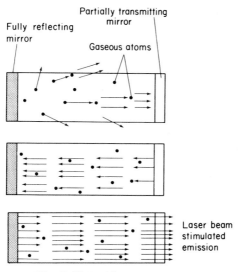

Fig. 10.21 Build-up of laser action

augmented by the repeated reflections. One of the mirrors is semi-transparent, and a portion of the travelling beam is able to leave the tube through this mirror.

The most important properties of lasers are as follows:

(i) Laser radiation is highly monochromatic (narrow line width).

(ii) Laser beams are very intense. All the laser energy is concentrated in a very narrow, parallel beam having a relatively constant cross-sectional area over great distances.

(iii) Laser beams, being highly focused, permit samples of extremely small size to be analysed or irradiated.

(iv) Laser beams, being very well collimated, simplify the optical geometry applied in instruments used in spectroscopy.

(v) Laser beams have well-defined polarization and for that reason are especially useful for accurate depolarization measurements.

Precaution. Do not look into a laser beam even if wearing safety glasses. Care should be taken that the beam is not reflected into the eyes when mirrors are being adjusted.

Bibliography: 209, 279, 365, 393, 485, 497, 523, 695, 742, 811, 939, 1038, 1064, 1152, 1156, 1210, 1319, 1358, 1376.

10.6.1 The Helium–Neon Laser

The helium–neon laser has the orange-red colour emission of 6329 Å with a power output of the order of milliwatts. The laser transmission occurs between

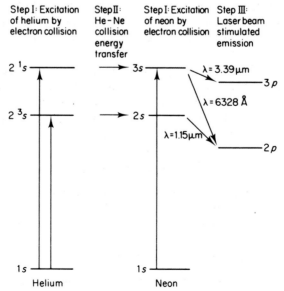

Fig. 10.22 Schematic diagram of energy levels involved in a helium-neon gas laser

energy levels in the neon, and the helium is used to pump the neon and obtain a population inversion. When an electric current passes through helium, the atoms are raised to excited states by collision with free electrons, and they then cascade down the energy levels. Those that arrive in the 2^3s and 2^1s levels remain there for a long period. Atoms gradually accumulate in these levels, which thus acquire large populations. When an excited helium atom collides with an unexcited neon atom, the excitation is transferred to the neon atom. Two other emissions occur at 3.39 µm and 1.15 µm (Fig. 10.22). A typical He–Ne laser is shown in Fig. 10.23. It consists of a Pyrex tube (35 cm long and 2 mm in diameter) with electrodes on the side and fused silica windows set at Brewster's angle on the ends of the tube. In this manner the

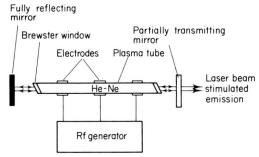

Fig. 10.23   Schematic diagram of a helium–neon gas
laser

laser output is polarized. The laser tube is first evacuated and then filled with helium at about 2.5 torr and neon at 0.5 torr (1 torr $= \frac{1}{760}$ atm $= 133.322$ N m^{-2}). The electrodes in the tube are connected to a high voltage of about 4000 V DC. Mirrors are placed outside the tube and aligned perpendicular to the axis to within a few seconds of arc. One mirror usually has the maximum reflectivity of 99.99% and the other is the output mirror with reflectivity of 99.0%. The mirrors must be made from multiple-layer dielectric films deposited on optically polished material. Metallic mirrors cannot be used because of high (4%) absorption. Early lasers used plane mirrors, but most gas lasers now have spherical mirrors because they are easier to align.

10.6.2 The Argon-ion Laser

The argon-ion laser has the blue-green colour emission of 4880, 4965, and 5145 Å with power outputs of the order of 500 mW to 1 W or more. Unfortunately, the lifetime of the argon-ion laser is short, making it too costly for general use at present.

The excitation in an argon-ion laser involves two steps:

(i) An inelastic collision with electrons takes argon atoms (with a $3s^2\,3p^6$ outer-electron configuration) to the Ar$^+$ state (with a $3s^2\,3p^5$ outer-electron configuration).

(ii) Another electron collision excites outer electrons of the argon ion (Ar^+) to one of several 4p states, and a subsequent transition to a 4s state results in an emission line. Thus, transitions between various levels in the 4p and 4s states result in emission of lines at 4880, 4965, and 5145 Å.

The 4880 Å line is the most intense and is the only emission if the amount of argon is kept small.

10.6.3 The Ruby Laser

The ruby laser belongs to the category known as solid-state lasers. Such lasers generally have higher power outputs than do gas lasers. Population inversion is obtained in ruby lasers by optical pumping.

The ruby is a single crystal of aluminium oxide within which a small proportion of the aluminium ions are replaced by Cr^{3+} ions. The depth of the red colour depends on the concentration of Cr^{3+} ions:

(i) light-pink ruby has a Cr^{3+} concentration of 0.05% or less;
(ii) dark-red ruby has a Cr^{3+} concentration of up to 1%.

When white light is incident upon a ruby crystal, blue and green are absorbed while the red is reflected. In addition to these absorption bands, which account for the colour, there are two narrow fluorescent lines at 6943 Å (R_1 level) and 6928 Å (R_2 level). The laser transition at 6943 Å generally takes place between the R_1 level and the ground state. Using special mirrors having high reflectivity at 6928 Å and low reflectivity at 6943 Å, the laser can be made to operate on the R_2 line (Fig. 10.24).

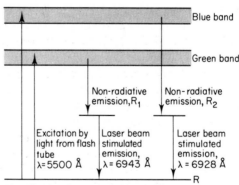

Fig. 10.24. Schematic diagram of energy levels involved in a ruby laser

The ruby laser is optically pumped using flash-lamps filled with xenon. When this light is absorbed by the Cr^{3+} ions, they are excited to the band levels. The ions quickly decay to the R levels by creating photons. The R levels are metastable and R_1 has a lifetime of 3.2×10^{-2} s. As the pumping continues, the ions accumulate in the R levels, and a large population inversion develops and laser action follows.

A typical ruby laser is shown in Fig. 10.25. A xenon flash-lamp, which is linear, is located along one focus of a cylindrical elliptical cavity. The ruby rod of 3–20 mm diameter and 20–250 mm in length is placed along the other focus. The elliptical cavity focuses almost all of the light from the flash-lamp into the ruby rod. Most xenon lamps cannot operate continuously at the high temperature needed to pump the ruby, and therefore are pulsed. One end of the ruby rod is cut into a rooftop shape and polished. The half-angle of the rooftop is

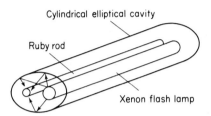

Fig. 10.25 Schematic representation of the optical cavity containing the ruby rod and the xenon flash-lamp

greater than the critical angle. Light propagating along the axis of the rod and incident on the rooftop is reflected through 180° and back down the axis. The other end of the rod is cut at Brewster's angle, so that light of one polarization experiences no reflection at this surface. The output mirror is mounted externally.

The efficiency of the original ruby laser is very small and does not usually exceed the value of 10^{-3}. For example, when the flash pulses have an input energy of 1000 W s, the output energy of the ruby laser has an energy less than 1 W s. The output power can be significantly increased in specially modified ruby lasers called GIANT PULSE LASERS, by Q-switching.

The Q-SWITCHING technique is so named because the Q-FACTOR (energy stored/energy lost per second) of the cavity is switched from a low value to a high value. Larger power outputs can be attained if the output mirror is closed off while the pumping is taking place. Then the population inversion will build up past the threshold value that corresponds to the condition for an unobstructed mirror. Now, when the mirror is opened to the cavity, the excess inversion gives a very high initial gain, and a large pulse is produced as the inversion is reduced to the steady-state condition.

There are two ways to switch the Q-factor of the cavity:

(i) By mounting the mirror on a rotating shaft. Each time the mirror is perpendicular to the axis of the rod, a giant pulse appears in the output. Pulse widths of the order of 100 ns and peak powers of the order of 100 MW can be produced in this manner.

(ii) By placing a saturable absorber in the cavity. Initially, the absorption coefficient is high. Light from the ruby rod then bleaches the absorber (e.g. vanadyl phthalocyanine), making it transparent. The light is then reflected by the high-reflectivity mirror, and its inversion is well above the threshold. The excess inversion is quickly dumped into a pulse, giving it a high peak power which may last less than 20 ns.

10.6.4 Organic Dye Lasers

The dye laser belongs to the category known as liquid-state lasers. The operation of the dye laser is based upon fluorescent transitions in large organic molecules in solution, e.g. in water or alcohol. They have power outputs of the

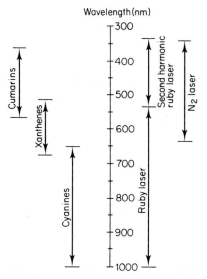

Fig. 10.26 Schematic diagram showing the wavelength range associated with the different dyes and pump lasers

order of 10^7–10^8 W and pulse durations of 10^{-9} s. Hundreds of commercially produced dyes are able to produce laser emissions in the wavelength range from 3400 Å to 1.2 μm (Fig. 10.26). Dyes which may serve as laser-active media must contain a chromophoric structure (indicated below by heavy lines), e.g.

7-Hydroxycumarin (blue)

Rhodamine 6G (orange-red)

Merocyanine (red)

The wavelength of the fluorescence is normally related to the length of the chromophore, with longer chromophores giving longer wavelengths, i.e. more red emission. The properties of any particular dye laser are determined by the spectral and molecular characteristics of the dye which is used.

Operation of the dye laser depends on the electronic transitions to the excited singlet and further triplet states. The laser process begins with the absorption of light from an excitation source (usually from a giant pulse laser or specially constructed flash-lamps) which raises the dye molecules from the ground state (S_0) to higher-lying excited singlet states ($^1S, {}^2S$) (Fig. 10.27). The laser emission

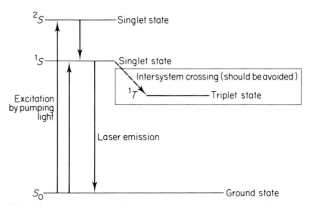

Fig. 10.27   Schematic diagram of energy levels in a dye laser

occurs from the excited singlet state (^1S) to the ground state (S_0). A fraction of the dye molecules will undergo intersystem crossing to the triplet state (T_1). Molecules which reach the triplet state are trapped there and are removed from the laser process. It is most important for the laser process that the intersystem crossing of molecules to the triplet state is avoided. This may be ensured by providing an excitation source which reaches laser threshold before the stimulated emission. For that light sources which provide several hundred kilowatts in a fraction of a microsecond are required (e.g. giant pulse lasers or special flash-lamps). Recently techniques for controlling the triplet state concentration by chemical quenchers and by mechanical flow have been developed.

Dye lasers can be divided into three groups:

(i) laser-pumped;
(ii) flashlamp-excited;
(iii) continuous-wave (CW).

A LASER-PUMPED DYE LASER is shown in Fig. 10.28. It can be arranged in one of two geometrical positions:

(i) tranverse excitation geometry;
(ii) longitudinal excitation geometry.

168

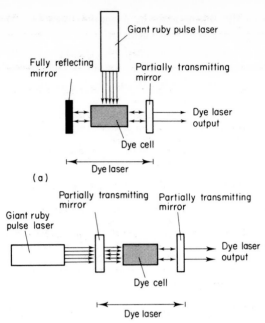

Fig. 10.28   Geometries utilized in the organic dye laser.
(a) Transverse excitation geometry; (b) longitudinal
excitation energy

Using giant ruby or Nd-glass lasers it is possible to produce dye-laser emission between wavelengths from 3400 Å to 1.2 µm (Fig. 10.26).

Bibliography: 545, 1152, 1235.

10.7 OPTICAL DETECTORS

Optical detectors are applied in any photometric method and they can be divided into

(i) photoemission detectors;
(ii) photoconductive detectors;
(iii) photovoltaic detectors;
(iv) thermal detectors;
(v) photographic detectors.

Spectral characteristics of some optical detectors are given in Table 10.8.

10.7.1 Photoemission Detectors

Photoemission detectors employ the PHOTOELECTRIC EFFECT, i.e. the ejection of electrons from alkali metals such as caesium, sodium, and potassium upon which radiant energy impinges. Such a photoelectric detector is called a PHOTOCELL (Fig. 10.29). The cathode is coated with one of the alkali metals.

Table 10.8 Spectral characteristics of some radiation detectors

Type	Sensitive element	Best frequency range (μm)
Photon detectors		
Photomultiplier tube	Group I and V metal oxide on cathode	0.16–0.7
Phototube	Alkali metal oxide	0.2–1
Photovoltaic cell	Semiconductor on metal	0.4–0.8
Photographic film or plate	AgX grains in emulsion	0.2–1.2
Photoconductive cell	I.Pbs, PbSe, InSb	0.7–4.5
	II.Ge: Cu, Au, or Zn activated (operated at ~5 K)	2–15 or 2–100
Thermal detectors		
Thermocouple (thermopile)	Junction of dissimilar metals with blackened strip as absorber	0.8–40
Bolometer (bridge detector)	Resistance wire or thermistor chip with blackened strip	0.9–40
Pneumatic cell (Golay cell)	Blackened membrane of gas chamber	0.8–1000
Pyroelectric	Ferroelectric crystal under permanent electric polarization	0.8–1000

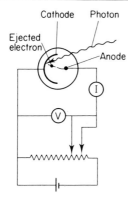

Fig. 10.29 A photocell, where the voltage giving a null current is proportional to the frequency of the incident light

A photon with energy $E = h\nu$ strikes the cathode and causes an electron to be ejected. If the electrostatic potential difference between the cathode and the anode is positive, electrons are attracted to the anode and a current is detected in the ammeter.

A photocell is characterized by a spectral response curve (Fig. 10.30), where the QUANTUM EFFICIENCY OF A PHOTOCELL (Q) is defined as the number of electrons released per incident quantum. Specific details on optical and electrical characteristics can generally be obtained from manufacturers' literature.

PHOTOMULTIPLIERS are devices which multiply a single photoelectron by five or more orders of magnitude. In addition to a photoemissive cathode they have several DYNODES which are maintained at successively higher potentials (Fig. 10.31). Generally the potential difference is increased in about 100 V steps from one dynode to the next. The potential difference accelerates the electrons so that when they strike the surface of the dynode each one causes three or four secondary electrons to be ejected. These secondary electrons are

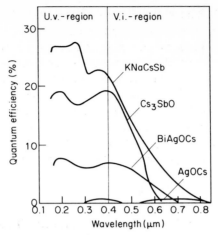

Fig. 10.30 Spectral sensitivity for several standard photocathode surfaces

then accelerated toward the next dynode and the process repeated. In this manner, a single electron at the cathode can be multiplied to one million electrons at the anode after ten stages. The last dynode, maintained at a high positive voltage with respect to the cathode, is grounded so that the output will be at low voltage.

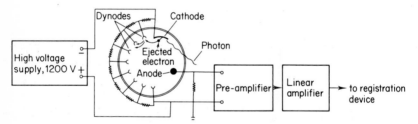

Fig. 10.31 Schematic diagram for a photomultiplier measuring system

As a result of their large internal amplification, photomultiplier tubes can be used only at low light intensities. When the power is on, exposure of a tube to even moderate levels of illumination can cause irreversible changes in electrode surfaces. For that reason photomultipliers are always mounted in light-tight housings with a shutter to control the entrance port.

A high-voltage supply capable of providing up to about 1200 V is necessary for the operation of the photomultiplier. Although most photomultipliers are operated at between 900 and 1000 V, the high-voltage supply should be adjustable over the range from 500 to 1200 V. The amplification of the photomultiplier is extremely dependent on the high voltage. For this reason, the high-voltage supply should also be very stable.

The preamplifier should be located as close to the photomultiplier as possible. Its function is to convert the weak (high-impedance) electrical pulse produced by the photomultiplier into strong (low-impedance) pulses capable of being conducted through a cable to the linear amplifier.

Measuring very low light intensities is usually carried out by the technique called PHOTON COUNTING. At low light intensities, a photomultiplier tube output consists of individual current pulses that can be counted. These photon-induced pulses must be distinguished from dark-current pulses that originate within the tube from other causes. Their amplitude is substantially less. A pulse amplitude discriminator, which can be an operational amplifier with differential input, can be set to pass only pulses whose height exceeds a minimum figure. The number of pulses that pass is counted as the read-out of the device.

Bibliography: 87, 139, 470, 702, 837, 1275.

10.7.2 Photoconductive and Photovoltaic Detectors

PHOTOCONDUCTIVE DETECTORS are solid detectors known as SEMI-CONDUCTORS whose electrical resistance decreases substantially under illumination.

PHOTOVOLTAIC DETECTORS are a special type of semiconductor which may generate a potential by optical excitation.

Bibliography: 9, 25, 980, 1334, 1414.

10.7.3 Thermal Detectors

THERMAL DETECTORS are solid detectors which can indicate temperature increases when light is absorbed. They are applied in the near-, middle-, and far-IR.

A THERMOCOUPLE consist of a junction of dissimilar metals. Thermo-couples used as radiation detectors are coated with lampblack and absorb almost all the incident radiation.

BOLOMETERS are detectors in which the resistance of a metal or a semi-conductor depends on its ambient temperature.

A THERMOPILE is a large number of thermocouples in series.

A GOLAY CELL is a detector which involves the expansion of heated gas (this device is equally sensitive at all wavelengths).

Bibliography: 647, 782, 874, 980, 1215, 1229, 1391.

10.7.4 Photographic Film

A photographic emulsion is a sensitive detector for the detection of radiation. An emulsion consists of silver halide crystals dispersed in a transparent water-expandable medium such as gelatin and is available on film or on thin glass plates. On exposure, the silver halide crystals receiving radiation build a latent image. Subsequent chemical development produces a black deposit of silver at

172

the site of the latent image. Emulsions give high spatial resolution since their silver halide crystals are small. The useful range is mainly in the UV and VI regions, but with special sensitizing it can be extended into the near-IR (230–880 nm). Different emulsions are produced for different speeds (low, medium, and high) and different contrasts (medium, high, and very high). Speed is identified by an ASA rating, e.g. an ASA of 400 represents a very fast emulsion and one of ASA 0.004 a very slow one. Contrast is identified by a gamma value (1–5). The response of an emulsion is plotted in Fig. 10.32. The DENSITY (D)

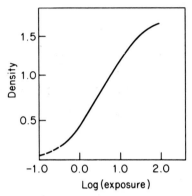

Fig. 10.32   Plot of image density versus the logrithm of exposure for a photographic emulsion

of the deposit is related to the logarithm of the EXPOSURE, which is the product of radiation intensity and time. The contrast of an emulsion is proportional to the slope of the curve and the speed is related to the range over which the curve is linear. A fast, low-contrast emulsion yields a curve of low slope whose linear range occurs at very low exposure.

If a very low intensity is to be recorded and a long exposure is not feasible, an adequate response cannot be obtained even with a fast emulsion of adequate contrast. In this case a prefogging technique can be applied by flashing an open film or plate by lamp several times at some distance. It is then used to record the desired low-intensity signal. If prefogging is done correctly, even very low-intensity radiation will bring the overall response into the linear range.

If a very high intensity is to be recorded, neutral-density filters or a stepped rotating sector whose transmission varies along its radius can be placed in the optical path. The radiation is then attenuated by a precise, known amount.

The density of the deposit is ordinarily determined by photoelectric scanning using a spectrodensitomer. The background density is subtracted before the ratio to the internal standard is taken.

Bibliography: 772, 839, 1293, 1300.

Chapter 11

Refractive Index

General bibliography: 125, 167, 263, 757, 912, 991.

The REFRACTIVE INDEX (INDEX OF REFRACTION) (n) of a material is the ratio of the velocity of light in a vacuum to its velocity in the material under study, and is defined by SNELL'S LAW (Fig. 11.1):

$$n = \frac{\sin \alpha}{\sin \beta} = \frac{\sin \alpha'}{\sin \beta'}, \qquad (11.1)$$

where
  α is the angle of incidence,
  β is the angle of refraction.
The numerical value of the refractive index (n) depends not only on the material but also on the wavelength of the light (λ) (Fig. 11.2).

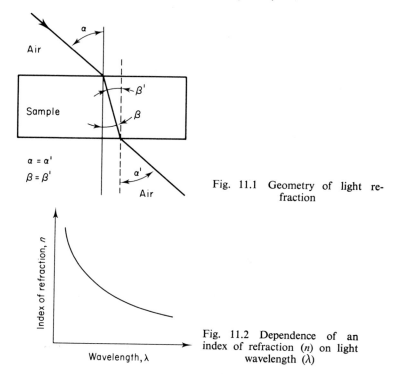

Fig. 11.1 Geometry of light refraction

Fig. 11.2 Dependence of an index of refraction (n) on light wavelength (λ)

The refractive index (n) of a material can be shown to depend on its molecular structure, and this relation is given by the LORENTZ–LORENTZ EQUATION:

$$\frac{n^2-1}{n^2+2} = \tfrac{4}{3}\pi P = \frac{\rho R_\lambda}{M}, \tag{11.2}$$

where
P is the polarizability of the molecules,

$$P = Na, \tag{11.3}$$

N is the number of molecules per unit volume,
a is the polarizability of one molecule, i.e. a measure of the mobility of the electrons in the molecule—the value of a is large when a molecule contains a large number of electrons,
ρ is the density of the material,
R_λ is the molar refractivity (or molecular refractivity) at constant wavelength,
M is the molecular weight of material.

Table 11.1 Refractive indexes of some common polymers
(From B: 757; reproduced by permission of John Wiley
& Sons)

Polymer	Refractive index
Poly(tetrafluoroethylene)	1.35–1.38
Poly(chlorotrifluoroethylene)	1.39–1.43
Cellulose acetate	1.46–1.50
Poly(vinyl acetate)	1.47–1.49
Poly(methyl methacrylate)	1.485–1.49
Poly(propylene)	1.49
Poly(vinyl alcohol)	1.49–1.53
Phenol-formaldehyde resin	1.5–1.7
Poly(isobutylene)	1.505–1.51
Poly(ethylene)	1.512–1.519 (25 °C)
Poly(isoprene) (natural rubber)	1.519 (25 °C)
Poly (butadiene)	1.52
Poly(isoprene) (synthetic rubber)	1.5219 (20 °C)
Poly(amide)	1.54
Poly(vinyl chloride)	1.54–1.56
Poly(styrene)	1.59–1.60
Poly (vinylidene chloride)	1.60–1.63

The polarizability (P) of one carbon–hydrogen polymer will not be significantly different from that of any other. Thus, by eq. (11.3) neither will there be any great difference between their refractive indices (a common value for n is ~1.5 for carbon–hydrogen polymers). Introduction of halogen atoms instead of hydrogen in polymer molecules will change the polarizability and therefore the refractive index (cf. Table 11.1).

The REFRACTIVE INDEX DIFFERENCE (dn) of a polymer in solution is given by

$$dn = n - n_1, \qquad (11.4)$$

$$n_1 + n_2 = n, \qquad (11.5)$$

where
n is the refractive index of the solution,
n_1 is the refractive index of the solvent,
n_2 is the refractive index of a polymer.
The refractive index of a polymer in solution depends on
 (i) the temperature,
 (ii) the wavelength of the light,
 (iii) the nature of the solvent,
 (iv) the weight concentration,
 (v) the molecular weight of the solute.
The dependence upon the molecular weight of the solute is particularly strong for low molecular weight polymers, i.e. oligomers. When the degree of polymerization (DP) is greater than 10 monomer units, the refractive index is a measure of the weight of polymer per unit volume and is practically independent of the molecular weight of the polymer being measured.

The SPECIFIC REFRACTIVE INDEX INCREMENT OF A SOLUTION (dn/dc) is defined as

$$\frac{dn}{dc} = \frac{n - n_1}{c}, \qquad (11.6)$$

where c is the concentration of the solute.
The specific refractive index increment of a solution (dn/dc) for $c \to 0$ can be calculated from the LORENTZ–LORENTZ EQUATION:

$$\frac{dn}{dc} = \frac{(n_1^2 + 2)^2}{6 n_1 \rho_2} \left(\frac{\rho_2 R_2}{M_2} - \frac{\rho_1 R_1}{M_1} \right), \qquad (11.7)$$

where
ρ_1 and ρ_2 are the densities of solvent and polymer, respectively,
n_1 is the refractive index of the solvent,
R_1 and R_2 are the molecular refractions of the solvent and of the repeating units of the polymer, respectively,
M_1 and M_2 are the molecular weights of the solvent and of the repeating units of the polymer, respectively.
In those cases where higher concentrations and different temperatures are used, it is convenient to express the concentration as the weight fraction (w).
The SPECIFIC REFRACTIVE INDEX INCREMENT OF A SOLUTION (dn/dc) can be expressed in terms of dn/dw:

$$\frac{dn}{dw} = \left[\rho + w \left(\frac{d\rho}{dw} \right) \right] \frac{dn}{dc}, \qquad (11.8)$$

176

where

$c = \rho w$ is the concentration of the solute,
ρ is the density of the solution,
w is the weight fraction of the solute.

At very low concentrations, $w(d\rho/dw)$ can be neglected and eq. (11.8) becomes

$$\frac{dn}{dw} = \rho \frac{dn}{dc}. \tag{11.9}$$

The main applications of the refractive index (n), refractive index difference (dn), and specific refractive index increment (dn/dc) are in

(i) ultracentrifugation (cf. Chapter 8);
(ii) light scattering (cf. Chapter 13);
(iii) chromatography (cf. Chapter 25);
(iv) birefringence (cf. Section 35.3).

Bibliography: 125, 167, 753, 912.
References: 2034, 2114, 2131, 2257, 2328, 2329, 2691, 2692, 3476, 3477, 3767, 3997, 4041, 4042, 4173, 4174, 4224, 4440, 4819, 4835, 4967, 4968, 5082, 5493, 5998, 6020, 6031, 6091, 6249, 6277, 6873, 6986, 7316.

Extensive tables of specific refractive index increments for many polymer-solvent systems are available in the literature (B: 155, 283, 401, 651).

11.1 DETERMINATION OF THE REFRACTIVE INDEX INCREMENT

Differences in refractive indices are measured directly in special differential refractometers as the deviation of a light monochromatic beam passing through a refractive index boundary in the divided cell (Fig. 11.3). The solution is placed

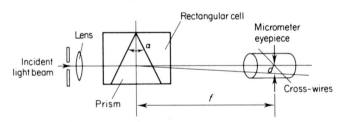

Fig. 11.3   Divided cell differential refractometer

in the hollow 60° prism and is surrounded by pure solvent in a rectangular cell. The optical system focuses the image of the fine slit on the cross-wires of the micrometer eyepiece. The refractive index difference (dn) can be calculated from

$$dn = n - n_1 = \frac{d}{2f \tan(\alpha/2)}, \tag{11.10}$$

where

n is the refractive index of the solution,

n_1 is the refractive index of the solvent,

d is the deflection,

f is the fixed focal length of the lens,

α is the apex angle of the cell.

The refractive index increment (dn/dc) can be obtained from the slope of the line given by plotting n versus c.

The differential cell is filled and is emptied by using a syringe. The cell must be tightly sealed in order to avoid the escape of vapour, which causes temperature gradients in the cell, and hence refractive index changes. The cell must be thermostatted to $\pm 0.01°$ C. The cell has two fixed positions which are $180°$ apart, oriented such that the front and the rear windows of the cell are perpendicular to the light beam. This arrangement permits the determination of the difference in the deflection resulting when the beam passes first the solvent and then the solution (position 1) or when it passes the solution before the solvent (position 2).

In the first stage of the measurement, both cell compartments are filled with solvent. The deflection for the solvent–solvent is given by

$$d_1 = d_{1\text{ (in position 2)}} - d_{1\text{ (in position 1)}}. \qquad (11.11)$$

In the second stage of the measurement, the solvent is removed from one compartment and solution is added. The deflection for the solution–solvent is given by

$$d_2 = d_{2\text{ (in position 2)}} - d_{2\text{ (in position 1)}}. \qquad (11.12)$$

The deflection corrected for solvent zero (d) is given by

$$d = d_2 - d_1. \qquad (11.13)$$

The refractive index difference (dn) can be calculated from

$$dn = kd, \qquad (11.14)$$

where k is the cell constant supplied by the manufacturer.

Bibliography: 263.
References: 2038, 2574, 2712, 6282.

11.2 AUTOMATIC DIFFERENTIAL REFRACTOMETER

In some types of measurement (e.g. GPC chromatography) an automatic differential refractometer is applied (Fig. 11.4). A beam of light from the incident lamp passes through the slit and the lens which collimates the light sources. The parallel beam passes through the cell containing the sample and reference liquids to the mirror. The mirror reflects the beam back through the sample and reference cell to the lens, which focuses it upon the detector. The location of the focused beam (rather than its intensity) is determined by

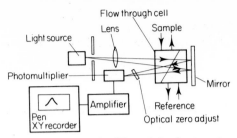

Fig. 11.4 Automatic differential refractometer

the angle of deflection resulting from the difference in refractive indices between the two parts of the cell. As the beam moves to the detector, an output signal is generated. This signal is amplified and registered on a recorder. The optical zero glass deflects the beam from side to side to adjust for the zero output signal. Differential refractometers are very sensitive to temperature. The noise factor increases with increasing temperature of operation.

References: 2257, 2404, 4322, 5336, 5523.

11.3 DETERMINATION OF THE SPECIFIC VOLUME OF THE SOLUTE FROM THE REFRACTIVE INDEX OF THE SOLUTION

The relation between the specific volume of the solute ($\bar{v}$) and the density increment ($d\rho/dw$) is

$$1 - \bar{v}\rho = \left(\frac{1-w}{\rho}\right)\left(\frac{d\rho}{dw}\right), \tag{11.15}$$

where
$\bar{v}$ is the specific volume of the solute,
ρ is the density of the solution,
w is the weight fraction of the solute.

The DENSITY INCREMENT is defined as $d\rho/dw$ and is related to $d\rho/dc$ as follows:

$$\frac{d\rho}{dw} = \left[\rho + w\left(\frac{d\rho}{dw}\right)\right]\frac{d\rho}{dc} \tag{11.16}$$

or

$$\frac{d\rho}{dw} = \rho\left(\frac{d\rho}{dc}\right)\left[1 - w\left(\frac{d\rho}{dc}\right)\right]. \tag{11.17}$$

The general relation between the specific refractive index increment and the density increment can be calculated from the LORENTZ–LORENTZ EQUATION:

$$\left(\frac{n^2-1}{n^2+2}\right)\frac{1}{\rho} = (1-w)R_1 + wR_2, \tag{11.18}$$

where

n is the refractive index,

R_1 is the specific refraction of the pure solvent and is given by

$$R_1 = \left(\frac{n_1^2-1}{n_1^2+2}\right)\rho,$$

R_2 is the specific refraction of the pure solute (polymer) and is given by

$$R_2 = \left(\frac{n_2^2-1}{n_2^2+2}\right)\rho,$$

subscript 1 relates to the pure solvent and subscript 2 to the pure solute (polymer).

The specific refraction (R) is a constant for a given substance at a given wavelength.

Equation (11.18) can be written as

$$\rho = \frac{n^2-1}{n^2+2}\left(\frac{1}{R_1+(R_2-R_1)w}\right). \tag{11.19}$$

Differentiation of (11.19) with respect to w gives the relation between the density increment $(d\rho/dw)$ and the specific refractive index increment (dn/dw):

$$\frac{d\rho}{dw} = \frac{1}{[R_1+(R_2-R_1)w]}\left(\frac{6n}{(n^2+2)^2}\right)\frac{dn}{dw}$$

$$-\frac{R_2-R_1}{[R_1+(R_2-R_1)w]^2}\left(\frac{n^2-1}{n^2+2}\right). \tag{11.20}$$

This method can be considerably simplified by the use of graphical extrapolation.

References: 2257, 2508, 3995, 3996, 4172, 4173, 5523, 5595, 5640, 6053, 6249, 6251.

Chapter 12

Optical Activity in Macromolecular Systems

General bibliography: 3, 115, 175, 232, 241, 274, 289, 329, 330, 374, 467, 682, 693, 844, 849, 945, 991, 1159, 1295, 1343, 1418, 1439.

Polymers possessing an asymmetric carbon in the main chain (e.g. poly(α-amino acids), $-\!(\!NH\!-\!CHR\!-\!CO\!)_n\!-$) or in a side group (e.g. poly(3-methyl-pent-1-ene), $-\!(\!CH_2\!-\!CHR\!)_n\!-$, where $R = -CH(CH_3)(C_2H_5)$) are optically active.

OPTICAL ACTIVITY consists of the rotation of linearly polarized light plane around the axis of propagation of the beam. This phenomenon occurs when light passes through some polymers in the solid or the liquid state, and may be explained as the interaction between matter and circularly polarized light of the D (right) and L (left) type.

The fundamental conditions for optical activity in polymers are as follows:

(i) the macromolecule should not possess mirror symmetry
(ii) the number of structural elements having asymmetric structure of a given D or L type must be different, otherwise, although the structural element may be optically active, the total effect would be zero since the rotation caused by the D elements would be compensated by an equal and opposite rotation caused by the L elements.

The presence of ASYMMETRIC CENTRES can be studied by

(i) OPTICAL ROTATORY DISPERSION (CIRCULAR BIREFRIN-GENCE) when different refractive indices are involved ($n_D \neq n_L$);
(ii) MOLECULAR CIRCULAR DICHROISM when different absorption coefficients are involved ($\varepsilon_D = \varepsilon_L$).

Extensive data on optically active polymers are collected in The Polymer Handbook (B: 526).

12.1 OPTICAL ROTATORY DISPERSION

OPTICAL ROTATORY DISPERSION (ORD) is the rotation of plane polar-ized light by an optically active polymer as a function of the wavelength of the incident light.

The ROTATION OF POLARIZATION (α) is given by

$$\alpha = \frac{\pi d}{\lambda}(n_L - n_D),\qquad(12.1)$$

where

α is expressed in degrees (CGS) or radians (SI),

d is the distance which the light travelled through an optical active sample,

λ is the wavelength of the light,

n_L and n_D are different refractive indices for circularly left- and right-polarized light.

The SPECIFIC ROTATION or the OPTICAL ACTIVITY ($[\alpha]_\lambda^T$) is defined by

$$[\alpha]_\lambda^T = \frac{100\alpha}{cl},$$

(12.2)

where

α is the angle (in degrees (CGS) or radians (SI)) measured in a polarimeter at a temperature T using monochromatic light with wavelength λ,

c is the concentration of the optically active polymer sample (in grams per cubic centimetre (CGS) or kilograms per cubic metre (SI)),

l is the path length (the thickness of the sample) (in decimetres (CGS) or metres (SI)).

The MOLAR ROTATION (or MOLECULAR ROTATION) ($[\varphi]_\lambda^T$) is defined by

$$[\varphi]_\lambda^T = 10\alpha/c_1\, l = M[\alpha]_\lambda^T/1000$$

(12.3)

(in deg. cm^2 dmol^{-1} (CGS) or rad m^2 mol^{-1} (SI)), where

M is the molecular weight of the optically examined compounds (in the case of polymers M is the molecular weight of an exactly determined structural unit (monomeric unit)),

c_1 is the concentration of the optically active sample in decimoles per litre (CGS) or moles per cubic metre (SI),

l is the path length (the thickness of the sample) (in decimetres (CGS) or metres (SI)).

In modern spectropolarimeters it is possible to measure the value of the angle α when it is of the order of millidegrees (CGS) or milliradians (SI) and to use path lengths of the order of 1 mm, i.e. less than the decimetre path in visual polarimeters.

In order to compare observed rotations in different solvents, it is necessary to apply a correction for the refractive index of the solvent. The molar rotation corrected in this way is termed the REDUCED MOLAR ROTATION ($(\Phi_{red})_\lambda^T$) and is given by

$$(\Phi_{red})_\lambda^T = \frac{3}{(n^2+2)} \frac{M[\alpha]_\lambda^T}{100}$$

(12.4)

(in deg. cm^2 dmol^{-1} (CGS) or rad m^2 mol^{-1} (SI)), where

n is the refractive index of the solvent.

12.2 CIRCULAR DICHROISM

CIRCULAR DICHROISM (CD) is the unequal absorption of plane polarized light by an optically active polymer as a function of the wavelength of the incident light.

For an optically active polymer sample the MOLECULAR CIRCULAR DICHROISM ($[\Delta\varepsilon]_\lambda^T$) is given by

$$[\Delta\varepsilon]_\lambda^T = \varepsilon_L - \varepsilon_D = \frac{1}{cl}\left(\log\frac{I_0}{I_L} - \log\frac{I_0}{I_D}\right) = \frac{\log(I_D/I_L)}{cl} \qquad (12.5)$$

(in deg. cm^2 dmol^{-1} (CGS) or rad m^2 mol^{-1} (SI)), where

ε_L and ε_D are the different molar absorptivities for left- and right-circularly polarized light at wavelength λ,

I_0 is the intensity of the incident light at wavelength λ,

I_L and I_D are the different intensities of the left- and right-circularly polarized light at wavelength λ,

l is the path length (in centimetres (CGS) or metres (SI)),

c is the concentration of the optically active sample (in moles per litre (CGS) or moles per cubic metre (SI)).

Sometimes the term MOLECULAR ELLIPTICITY ($[\theta]_\lambda^T$) is used. This is given by

$$[\theta]_\lambda^T = 3300[\Delta\varepsilon]_\lambda^T. \qquad (12.6)$$

12.3 THE COTTON EFFECT

The dependence of optical dispersion (ORD) and circular dichroism (CD) on wavelength is termed the COTTON EFFECT, and can be presented in the form of three different dispersion curves:

(i) The ANOMALOUS ROTATORY DISPERSION CURVE (Fig. 12.1). The fundamental property of a dispersion band is that the intensity of optical activity at the band centre is zero, but that it extends with measurable intensity far out in both directions away from the centre. Curves are called positive and negative depending on whether the peak or the trough (sometimes called the positive or negative extremums, respectively) occurs at longer wavelength.

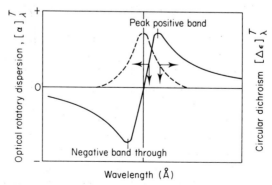

Fig. 12.1 Types of anomalous dispersion curve: ———, optical rotatory dispersion (ORD); — — —, circular dichroism (CD)

(ii) The PLAIN DISPERSION CURVE (Fig. 12.2). Plain curves are characterized by increasing rotations with decreasing wavelength, and are termed positive or negative depending upon whether they curve upward or downward as they approach shorter wavelengths.

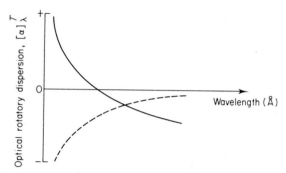

Fig. 12.2 Types of plane dispersion curve: ———, plain positive curve; — — —, plain negative curve

(iii) The MULTIPLE COTTON EFFECT CURVE. These curves exhibit several peaks and troughs. They are found in wavelength regions associated with multiple absorption bands.

The Cotton effect can also be divided into

(i) the INTRINSIC COTTON EFFECT, which arises from asymmetric chromophores already present in the macromolecule;

(ii) the EXTRINSIC COTTON EFFECT, which arises from induced asymmetric chromophores.

12.4 SPECTROPOLARIMETERS

For recording optical rotatory dispersion (ORD) and circular dichroism (CD) one uses commercially available spectropolarimeters with circular dichroism attachments.

The basic components of a spectropolarimeter are shown in Fig. 12.3. Light from a monochromatic source passes through a polarizer and becomes plane polarized, i.e. its vibrations are restricted to one plane, the orientation of which is determined by the orientation of the polarizer. After traversing the sample, the light reaches the analyser, a prism similar to polarizer. With a simple polarimeter, the detector is the human eye or a phototube.

The intensity of the light reaching the detector varies as the analyser is rotated and is at a minimum when the plane of transmission of the analyser is at 90° to the plane of polarization of the light incident upon it. In principle, the zero setting of the polarimeter is first determined in the absence of an optically active sample, using only solvent in a cell. Next the sample cell is filled with a solution of the sample, and the angle through which the analyser

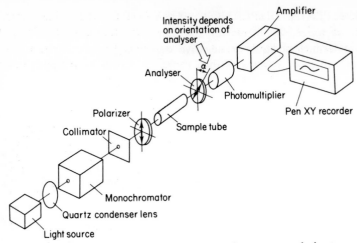

Fig. 12.3   Optical system and components in a spectropolarimeter

must be rotated to extinguish the light again gives the optical rotation of the solution.

Automatic recording spectropolarimeters apply the null reading system in one of two ways:

(i) by using a mechanical null balance device in which the analyser or polarizer is turned by a servometer (Fig. 12.4);

(ii) by using an electrical Faraday effect null balance in which the analyser and polarizer positions are fixed and the rotation of the sample is compensated for electrically by using the Faraday effect (the optical activity disappears in the magnetic field).

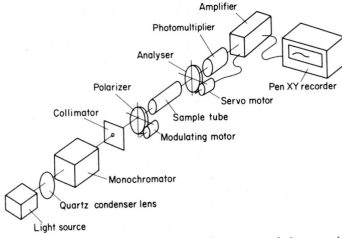

Fig. 12.4   Optical system and components of a spectropolarimeter using a mechanical null-balance system

A FARADAY CELL is a cylindrical cell (usually containing water) placed inside a coil which is connected to a DC source (Fig. 12.5). The plane of polarization of polarized light passing through the cell is rotated. If the coil is connected to an alternating source, the direction of the plane of polarization oscillates at the same frequency as the alternating current.

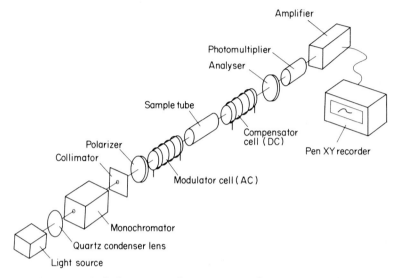

Fig. 12.5 Optical system and components of a spectropolarimeter using an electrical Faraday effect null-balance system

The measurement of the circular dichroism of an optically active polymer involves measuring the difference between the absorption coefficients of the sample in right- and left-circularly polarized light. To obtain circularly polarized light the beam of radiation must first be plane-polarized, after which the polarized beam must be passed through a device which will resolve it into right- and left-circularly polarized components. This is done by retarding one component relative to the other by exactly one quarter of a wavelength. There are two types of commercially produced devices for resolving plane-polarized light into its circularly polarized components.

(i) Prism modulators such as the FRESNEL RHOMB which are placed in the reference beam. By placing the sample in both beams of the spectrophotometer the difference in absorbance due to right- and left-circularly polarized light is automatically measured.

(ii) Electro-optic modulators such as the POCKELS CELL which consist of a z-cut plate of ammonium dihydrogen phosphate, a tetragonal crystal which has only one optical axis in the absence of an electric field (the z axis, along which the light beam passes), and two optical axes when an electric field is applied, i.e. it exhibits the POCKELS EFFECT. By applying a high alternating voltage (of the order of kilovolts) across the

186

z axis, the electro-optically induced axes will alternately be $+45°$ and $-45°$ to the original plane of polarization. The net effect of this is to transform the plane-polarized light first into right- and then into left-circularly polarized light in the course of one cycle of the applied alternating voltage. If the sample is circularly dichroic, i.e. if it absorbs right- and left-circularly polarized radiation differently, then the net light beam monitored by the detector will be elliptically polarized. Thus electro-optic modulators permit a direct measurement of circular dichroism.

The cells used for ORD and CD measurements can be of two types:

(i) completely fused cells;
(ii) screw-cap type cells.

It is important to avoid any strains in the cell. Strains may be detected by the deviations from linearity of the baseline when the cell is filled with water. The cells should always be placed in the instrument with the same end toward the light beam.

The instrument should be calibrated with camphorsulphonic acid ($1 \, \text{mg} \, \text{ml}^{-1}$) in water in a cell with a path length of 1 cm which has a circular dichroism peak at 290 nm with an amplitude of $+0.009$ absorbance unit. In the optical rotatory dispersion spectrum such a cell has an amplitude of $0.4°$ with a peak at 305–270 nm.

12.5 APPLICATION OF ORD AND CD IN POLYMER RESEARCH

Studies of optical rotatory dispersion (ORD) and circular dichroism (CD) are especially important in the study of optically active polymers (e.g. vinyl polymers with optically activity in the polymer backbone, polyesters, polyaldehydes, polyethers) and biopolymers, e.g. polypeptides and proteins in the α-helical configuration.

Bibliography: 175, 427, 682, 693, 931, 1041, 1042, 1173, 1418, 1439.
References: 2003, 2110, 2111, 2543, 2648, 2716, 2809, 2857, 2942, 2943, 2980, 3125, 3130, 3146, 3708–3713, 4175, 4654, 5076, 5119, 5535, 5644, 5645, 5838–5840, 6059, 6292, 6293, 6816, 6838, 6839, 6892, 7222, 7291, 7299.

Chapter 13

Light Scattering

General bibliography: 99, 125–127, 166, 236, 263, 264, 309, 320, 381, 382, 425, 500, 650, 733–735, 776, 797, 846, 899, 943, 994, 995, 1016, 1250, 1289, 1308, 1331.

Light is not scattered by a perfectly homogeneous medium, whereas if light passes through an inhomogeneous medium it is scattered in all directions.

Solutions of macromolecules can always be considered as inhomogeneous media. The random diffusive motion of macromolecules causes the formation of areas of differing concentration. These areas of different concentration have different dielectric constants and hence different refractive indices to the bulk of the fluid, and thus act as scattering centres.

We can consider two types of light scattering:

(i) ELASTIC LIGHT SCATTERING (called RAYLEIGH SCATTERING), where the scattered light has the same frequency as the incident light (called the RAYLEIGH LINE). The intensity of the scattered light (the intensity of the Rayleigh line) is measured by ordinary wide-angle (cf. Section 13.1) and narrow-angle (cf. Section 13.3) spectroscopy.

When the scattering centres move, they develop a low-frequency Doppler shift from the incident frequency, and the scattered light is no longer purely monochromatic (QUASIELASTIC SCATTERING). Instead of just the one Rayleigh line, we observe a RAYLEIGH PEAK. The shape of the Rayleigh peak can be measured by the use of LIGHT-BEATING SPECTROSCOPY (also called RAYLEIGH LINEWIDTH SPECTROSCOPY) (cf. Section 13.4).

(ii) INELASTIC SCATTERING (sometimes called RAYLEIGH–BRILLOUIN SCATTERING), where the scattered light has a different frequency to the incident light. Inelastic light scattering is measured by RAYLEIGH–BRILLOUIN SPECTROSCOPY (cf. Section 13.5).

The light scattered from large particles can be seen by the naked eye and is called the TYNDALL EFFECT.

The intensity of the scattered light in dilute polymer solutions is only about 10^{-4} of the incident light intensity and can only be measured by a photomultiplier.

Visible light interacting with a particle induces an oscillating dipole which emits scattered light.

With vertically polarized incident light, the dipoles oscillate in the z direction (Fig. 13.1).

With horizontally polarized incident light, the dipoles oscillate in the y direction (Fig. 13.1).

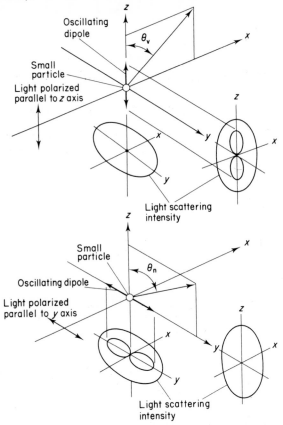

Fig. 13.1 Light-scattering profile of small particles with vertically (v) and horizontally (h) polarized incident light

The following relation exists between the angles θ_v and θ_h:

$$\sin^2\theta_v + \sin^2\theta_h = 1 + \cos^2\theta, \qquad (13.1)$$

where θ is the angle between the scattered and the incident light rays.

If the scattering particle is isotropic,

(i) vertically polarized incident light gives only vertically polarized scattered light;

(ii) horizontally polarized incident light gives only horizontally polarized scattered light.

If the scattering particle is anisotropic, vertically (or horizontally) polarized incident light will give both vertically and horizontally polarized scattered light. This effect is called DEPOLARIZATION OF THE SCATTERED LIGHT. The scattering is thus greater than that expected theoretically. This effect can be corrected by application of the so-called CABANNES FACTOR. With macromolecular solutions, the Cabannes factor is usually very close to one.

Bibliography: 236.
References: 2351, 3221, 4130.

13.1 WIDE-ANGLE LIGHT SCATTERING

13.1.1 Scattering in Solution by Small Particles

With scattered particles which are smaller than $\lambda/20$, the intensity of the scattered light (I_θ) is the same in all directions (spherically symmetrical) (Fig. 13.2).

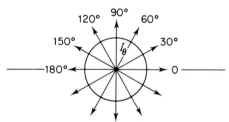

Fig. 13.2 Diagram of the light-scattering intensity I_θ for particles which are smaller than $\lambda/20$

The intensity of the scattered light (I_θ) is proportional to the polarizability (α) and molecular weight (M) of the scattering particles and is given by

$$I_\theta = 8\pi^4 \alpha^2 I_0 N(1+\cos^2\theta)/l^2 \lambda^4 V, \tag{13.2}$$

where
I_0 is the intensity of the incident light,
λ is the wavelength of the incident light,
l is the distance from the scattering source at which I_θ is measured,

$$N/V = cN_A/M, \tag{13.3}$$

N is the number of scattering particles,
V is the volume occupied by scattering particles,
c is the concentration of scattering particles,
N_A is the Avogadro number (cf. Section 40.3),
M is the molecular weight of the scattering particles,
θ is the angle between the scattered and the incident light,
α is the polarizability of the scattering particles,

$$\alpha = \frac{n_1 M}{2\pi N_A}\frac{dn}{dc}, \tag{13.4}$$

n_1 is the refractive index of the solvent,
dn/dc is the specific refractive index increment (cf. Chapter 11).
Combination of eqs (13.2) and (13.4) gives

$$R_\theta = I_\theta\, l^2/I_0 = K(1+\cos^2\theta)\, Mc, \qquad (13.5)$$

where
R_θ is known as the RAYLEIGH RATIO,
K is a constant defined by

$$K = \frac{2\pi^2\, n_1^2}{N_A\, \lambda^4}\left(\frac{dn}{dc}\right)^2. \qquad (13.6)$$

13.1.2 Determination of the Molecular Weight of Small Particles by Light Scattering

For small isotropic particles (with diameter $< \lambda/20$) the molecular weight (M) can be determined from the DEBYE EQUATION:

$$Kc/R_\theta = 1/M, \qquad (13.7)$$

where
K is the constant defined by eq. (13.6),
c is the concentration of scattering particles in solution,
R_θ is the Rayleigh ratio, given by eq. (13.5),
M is the molecular weight of the scattering particles.
The scattering of light by molecules is given by

$$I_0/I = e^{\tau l}, \qquad (13.8)$$

where
I_0 is the intensity of the incident light,
I is the intensity of the light transmitted through the sample,
l is the path length of the sample,
τ is the turbity of the sample.
The TURBIDITY (τ) is the Rayleigh ratio integrated over the solid angle:

$$\tau = 2\pi \int_0^\pi R_\theta \sin\theta\, d\theta, \qquad (13.9)$$

where
R_θ is the Rayleigh ratio, defined by eq. (13.5),
θ is the scattering angle.
For small, isotropic particles (with diameter $< \lambda/20$) the turbidity is equal to the Rayleigh ratio measured at a scattering angle $\theta = 90°$ and extrapolated to $\theta = 0°$:

$$\tau = \frac{16\pi}{3} R_{90°} = \frac{8\pi}{3} R_{0°}. \qquad (13.10)$$

The molecular weight (M) can be determined from the following DEBYE EQUATION:

$$Hc/\tau = (1/M) + A_2 c + ..., \qquad (13.11)$$

where
$H = 16\pi K/3$ (K is the constant defined by eq. (13.6)),
c is the concentration of scattering particles in solution,
A_2 is the second virial coefficient (cf. Section 2.10).

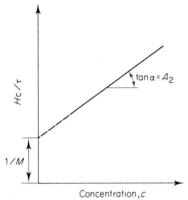

Fig. 13.3 Plot of Hc/τ versus concentration (c)

A plot of Hc/τ versus c is linear and yields $1/M$ as the intercept at the ordinate (Fig. 13.3). The slope, tan $\alpha = A_2$, yields the second virial coefficient.

Bibliography: 320, 1250.
References: 3106.

13.1.3 Scattering in Solution by Macromolecules

For macromolecules, where the coil diameter is larger than $\lambda/20$ (e.g. for vinyl polymers having a degree of polymerization larger than 500), the intensity of the scattered light (I_θ) is different with different observation angles (spherically unsymmetrical) (Fig. 13.4).

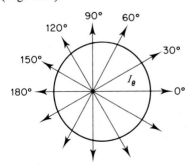

Fig. 13.4 Diagram of the light-scattering intensity I_θ for particles which are larger than $\lambda/20$

192

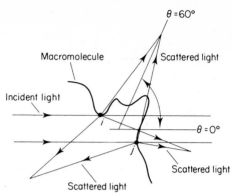

Fig. 13.5   Scattered light from different elements
i and j of a long macromolecule

The scattered light from two different elements (i and j) of a long macro-molecule (Fig. 13.5)

(i) does not interfere with each other at zero scattering angle,
(ii) interferes with each other in forward directions (narrow scattering angles) (cf. Section 13.3),
(iii) interferes more at wider scattering angles.

Considering the interferences of scatterings from different elements i and j for different shapes of the scattered macromolecule, it is necessary to introduce the scattering factor (P_θ) into the calculations.

The SCATTERING FACTOR (P_θ) is defined as the ratio of the measured scattered intensity at the angle θ (R_θ) to the scattered intensity at the angle θ extrapolated to $\theta = 0°$ ($R_{0°}$):

$$P_\theta = R_\theta / R_{0°}. \qquad (13.12)$$

For any shape, $P_\theta = 1$ at $\theta = 0$. As the θ increases, the P_θ decreases. The scattering factor (P_θ) is related to the radius of gyration ($\bar{s}^2$) and is dependent on the shape of the particles (e.g. spheres, thin discs, thin rods, circular cylinders, random coils, polydisperse random coils, and stiff coils). The mathematical presentation of the scattering factor (P_θ) for any desired shape of scattered particle is complicated and is not given here.

Bibliography: 166, 381, 382, 846, 1016.
References: 2020, 2021, 2293, 2311, 2312, 2563, 2653, 2654, 2757, 3103, 3104, 3106, 3222, 3346, 3361, 3489, 4101, 4151, 4371, 4377, 4533–4535, 4701, 5154, 5215, 5620, 5711, 5738, 5980, 6115, 6274, 6520, 6736, 6817, 6897, 6940, 6954, 7220, 7298.

13.1.4 Determination of the Weight-average Molecular Weight ($\bar{M}_w$) by Light Scattering

For large particles (macromolecules) (with diameter $> \lambda/20$) the weight-average molecular weight ($\bar{M}_w$) can be calculated from the equation

$$Kc/R_\theta = 1/\bar{M}_w P_\theta + 2A_2 c + ..., \qquad (13.13)$$

where

K is the constant defined by eq. (13.6),
R_θ is the Rayleigh ratio given by eq. (13.5),
P_θ is the scattering factor given by eq. (13.12),
$\overline{M}_w$ is the weight-average molecular weight,
c is the concentration of the scattering particles in solution,
A_2 is the second virial coefficient (cf. Section 2.10).

It can also be calculated from turbidities calculated from measurements of excess scattering at $\theta = 90°$, using

$$3Hc/8\pi R_{90°} = (1/\overline{M}_w P_{90°}) + 2A_2 c + ..., \qquad (13.14)$$

where

$H = 16\pi K/3$ (K is the constant defined by eq. (13.6)),
$R_{90°}$ is Rayleigh ratio measured at $\theta = 90°$,
$P_{90°}$ is the scattering factor for $\theta = 90°$.

There are two distinct methods for the application of the experimental data of light scattering to the determination of the weight-average molecular weight $(\overline{M}_w)$:

(i) the dissymmetry method (cf. Section 13.1.5),
(ii) the Zimm method (cf. Section 13.1.6).

13.1.5 The Dissymmetry Method

In order to obtain the scattering factor (P_θ) it is necessary to measure the dissymmetry coefficient (z).

The DISSYMMETRY COEFFICIENT (z) is defined as the ratio of scattering intensities at angles symmetrical about the 90° position:

$$z = I_\theta / I_{(180 - \theta)}. \qquad (13.15)$$

The value of z increases for any chosen value of θ with increasing molecule size.

The dissymmetry coefficient (z) is also related to the scattering factor (P_θ) by

$$z = P_\theta / P_{(180 - \theta)}. \qquad (13.16)$$

Measurements at two different angles (45° and 135°) yield values of the dissymmetry coefficient (z)

$$z = I_{45°} / I_{135°} = P_{45°} / P_{135°}, \qquad (13.17)$$

whereas the measured value of $I_{90°}$ is a direct measure of the Rayleigh ratio $R_{90°}$.

The Rayleigh ratio at 0° $(R_{0°})$ is given by

$$R_{0°} = R_{90°} / P_{90°} = I_{90°} / P_{90°}. \qquad (13.18)$$

The value $1/P_{90°}$ is called THE MOLECULAR WEIGHT CORRECTION FACTOR.

The dissymmetry coefficient is concentration dependent and it is necessary to extrapolate the experimental data to zero concentration. The plot of $1/z - 1$ versus c is linear (where c is the concentration of the solute). The value $z_{c=0}$ is used to find the molecular weight correction factor ($1/P_{90°}$) for converting $(Kc/R_{90°})_{c=0}$ to $Kc/R_{0°})_{c=0}$ (cf. eq. (13.13)), from which $\overline{M}_w$ can be calculated.

The value of $z_{c=0}$ gives a value for the characteristic dimension of the scattering particles, assuming one of the theoretical models for the polymer molecule, e.g. spheres, rods, coils, etc. Polymer molecules normally approximate to the random coil configuration.

Bibliography: 1016, 1250.
References: 2401, 3033, 3222.

13.1.6 The Zimm Method

The scattering factor (P_θ) for the random coil configuration is given by the ZIMM EQUATION:

$$1/P_\theta = 1 + (8\pi^2/9\lambda^2)\ \bar{r}^2\ \sin^2(\theta/2), \qquad (13.19)$$

where
λ is the wavelength of the incident light used,
$\bar{r}^2$ is the END-TO-END DISTANCE of a coiled polymer chain (cf. Section 2.11),
θ is the angle between the scattered and the incident light.
Substituting eq. (13.19) into eq. (13.13) we obtain

$$\frac{Kc}{R_\theta} = \frac{1}{\overline{M}_w}\left[\left(1 + \frac{8\pi^2}{9\lambda^2}\right)\bar{r}^2\sin^2\frac{\theta}{2}\right] + 2A_2 c + \dots \qquad (13.20)$$

or

$$Kc/R_\theta = (1/\overline{M}_w) + K'\sin^2(\theta/2) + 2A_2 c + \dots, \qquad (13.21)$$

where
K is the constant defined by eq. (13.6),
c is the concentration of the solute,
R_θ is the Rayleigh ratio given by eq. (13.5),
$\overline{M}_w$ is the weight-average molecular weight,
A_2 is the second virial coefficient (cf. Section 2.10),

$$K' = \frac{1}{\overline{M}_w}\left(\frac{8\pi^2}{9\lambda^2}\right)\bar{r}^2. \qquad (13.22)$$

(i) The plot Kc/R_θ versus c (at $\theta = $ const.) is linear (Fig. 13.6).
(ii) The plot Kc/R_θ versus $\sin^2(\theta/2)$ (at $c = $ const.) is also linear (Fig. 13.7).

In cases (i) and (ii) the measurements at different concentrations and over as large angles (θ) as possible (e.g. 45°, 60°, 75°, 90°, 105°, ...) yield a series of

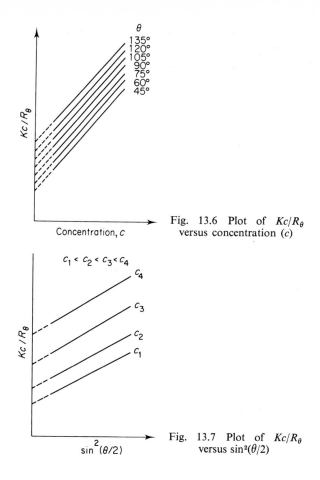

Fig. 13.6 Plot of Kc/R_θ versus concentration (c)

Fig. 13.7 Plot of Kc/R_θ versus $\sin^2(\theta/2)$

straight lines which can be extrapolated to the ordinate and therefore permit the determination of

$$(Kc/R_\theta)_{c=0} \qquad \text{(cf. Fig. 13.6),}$$

$$(Kc/R_\theta)_{\sin^2(\theta/2)=0} \quad \text{(cf. Fig. 13.7).}$$

(iii) The plot $(Kc/R_\theta)_{c=0}$ versus $\sin^2(\theta/2)$ is linear (Fig. 13.8).
(iv) The plot $(Kc/R\theta)_{\sin^2(\theta/2)=0}$ versus c is also linear (Fig. 13.9).

Considering eq. (13.21), it can be seen that

(v) for $c=0$, $(Kc/R_\theta)_{c=0} = (1/\overline{M}_w) + K' \sin^2(\theta/2)$;
(vi) for $\sin^2(\theta/2) = 0$, $(Kc/R_\theta)_{\sin^2(\theta/2)=0} = (1/\overline{M}_w) + 2A_2 c + \dots$.

Both straight lines (Figs 13.8 and 13.9) yield $1/\overline{M}_w$ as the intercept at the ordinate.

In Fig. 13.8 the slope is $\tan \alpha = K'$ and from this can be calculated the square average end-to-end distance $(\bar{r}^2)$ of a coiled polymer chain.

In Fig. 13.9 the slope is $\tan \alpha = 2A_2$ and from this can be calculated the second virial coefficient.

In general, the extrapolation shown in Figs 13.8 and 13.9 is not carried out separately, but rather in a single diagram (called the ZIMM PLOT) by plotting Kc/R_θ versus $100c + \sin^2(\theta/2)$ (Fig. 13.10).

(i) The lines of low slope (thin) show the dependence of Kc/R_θ values on $100c + \sin^2(\theta/2)$ for each constant concentration c. By extrapolating these lines to the intercept with each corresponding $100c$ value of the

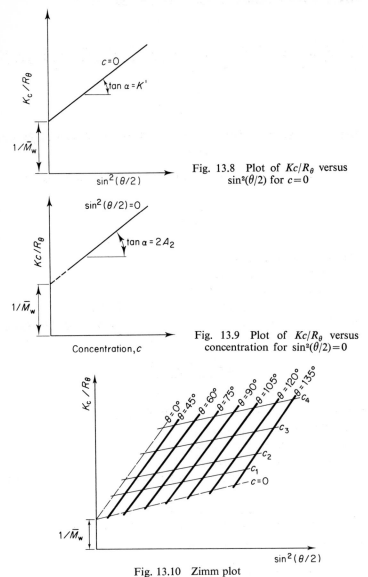

Fig. 13.8   Plot of Kc/R_θ versus $\sin^2(\theta/2)$ for $c=0$

Fig. 13.9   Plot of Kc/R_θ versus concentration for $\sin^2(\theta/2)=0$

Fig. 13.10   Zimm plot

abscissa ($\sin^2(\theta/2) = 0$), we obtain a series of points which together form the extrapolation of the line $Kc/R_\theta = f(c)$ at $\sin^2(\theta/2) = 0$.

(ii) The lines of high slope (thick) show the dependence of Kc/R_θ values on $100c + \sin^2(\theta/2)$ for constant scattering angles θ. By extrapolating these lines to the intercept with the corresponding abscissa value of $\sin^2(\theta/2)$ ($c = 0$), we obtain a series of points which together form the extrapolation of the line $Kc/R_\theta = f(\sin^2(\theta/2))$ at $c = 0$.

(iii) The two extrapolations must intercept the ordinate at the same point and this is a criterion for the validity of eq. (13.21).

Bibliography: 320, 776, 1016, 1350.
References: 2033, 2102, 2145, 2344, 2350, 2656, 2732, 2806, 2974, 3206, 3412, 3695, 3759, 3844, 4085, 4205, 4301, 4372, 4483, 4575, 4628, 4751, 4849, 5256, 5345, 5640, 5914, 5950, 6069, 6318, 6384, 6561, 6562, 6596, 6699, 6724, 6900, 7145, 7332, 7336.

13.1.7 Determination of the Second Virial Coefficient (A_2) from Light Scattering Measurements

The second virial coefficient (A_2) is expressed in light scattering terms as

$$A_2 = \frac{\rho_s}{\rho_p^2 M_s}\left(\tfrac{1}{2} - \chi\right) \qquad (13.23)$$

where

ρ_s is the density of the solvent,
ρ_p is the density of the polymer,
M_s is the molecular weight of the solvent,
χ is the Flory interaction parameter (cf. Section 2.9).

The second virial coefficient (A_2) can be calculated experimentally from values of c/R_θ measured at two concentrations c_1 and c_2:

$$A_2 = \frac{\tfrac{1}{2}K[(c/R_\theta)_{c_2} - (c/R_\theta)_{c_1}]}{c_2 - c_1} \qquad (13.24)$$

(in $\text{cm}^2 \text{ mol g}^{-2}$ (CGS) or $\text{m}^2 \text{ mol kg}^{-2}$ (SI)), where

K is the constant defined by eq. (13.6),
R_θ is the Rayleigh ratio given by eq. (13.5),
c is the concentration of the solution.

The second virial coefficient (A_2) can be measured from the slope $\tan \alpha$ in Fig. 13.9.

In general, the value of the second virial coefficient (A_2) obtained from light scattering measurements corresponds well with the value of A_2 obtained from osmotic measurements (cf. Section 5.2).

References: 4463, 4850, 5186, 5514, 6291, 6901.

198

13.1.8 Determination of the Square Average End-to-end Distance $(\bar{r}_2)$ from Light Scattering Measurements

The square average end-to-end distance $(\bar{r}^2)$ can be calculated from the Zimm method (cf. Section 13.1.6). Light scattering measurements give the dimensions of the Z-average value $(\bar{r}_Z^2)$.

In the interpretation of viscosity measurements (cf. Section 2.13) the number-average value of $(\bar{r}_n^2)$ should be used.

$$\bar{r}^2 = \bar{r}_Z^2 = \bar{r}_n^2 \qquad (13.25)$$

only for well-fractionated polymers. For linear polymers,

$$\bar{r}^2 = 6\bar{s}^2, \qquad (13.26)$$

where $\bar{s}^2$ is the square average radius of gyration (cf. Section 2.11).

Bibliography: 263.
References: 2145, 3721, 4174, 4301, 4302, 4463, 4696, 6105, 6338, 6897, 7332.

13.1.9 Light-scattering Instruments

Light-scattering instruments are built of the following parts (Figs 13.11 and 13.12):

(i) A light source. In older light-scattering photometers a high-pressure mercury lamp was used as a light source in conjunction with suitable filters in order to obtain a monochromatic beam (546 nm green mercury line or 436 nm blue mercury line). Recently, laser beams have been used (Fig. 13.12). Lasers are ideal light sources for light-scattering photometers. Lasers are monochromatic, perfectly collimated, intense,

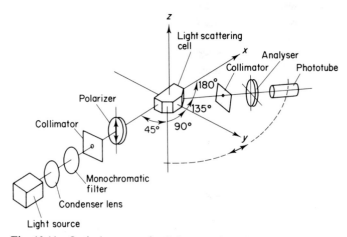

Fig. 13.11   Optical system of a light-scattering photometer designed for observation at 45°, 90°, and 135° angles

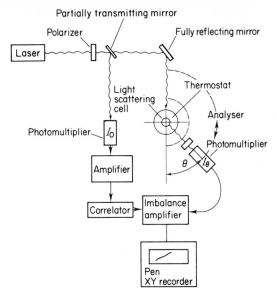

Fig. 13.12   Laser light-scattering photometer

can be completely polarized, and can be small in cross-section (cf. Section 10.6).

Bibliography: 846.
References: 3448, 4488, 4989, 4991, 7298.

(ii) A cell placed in a thermostat. Different types of cells are used in light-scattering measurements. A conical scattering cell helps to minimize reflections from the glass–liquid interface (Fig. 13.13). The volumes of cells vary from 8 to 30 ml in conventional photometers and can be

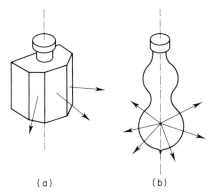

(a)   (b)

Fig. 13.13   Light-scattering cells. (a) For measurements at 45°, 90°, and 135° angles; (b) for measurements at any angle—this type of cell can be ultracentrifuged

smaller by a factor 10^3 or 10^4 in laser photometers. The cell is placed in a thermostat filled with a liquid having the same refractive index as the content of the scattering cell. A specially designed Dewar flask allows measurement of the light scattering over the temperature range from -40 to $200\,°C$.

Bibliography: 263.
References: 2375, 2573, 2749, 3049, 3238, 3362, 3447, 3724, 3908, 4312, 4488, 4626, 5633, 6832, 6898, 7332.

(iii) Scattered light detectors. The intensity of the scattered light is measured by photosensitive electrical devices, e.g. photomultipliers. The intensity of the scattered light must be compared with that of the primary beam. These two quantities differ by values of the order of 10^6, and it is inconvenient to measure both beams with the full sensitivity of the photomultiplier. A calibration procedure is necessary to determine the constants associated with any instrument, and is carried in two stages:

(a) A primary standard (12-tungstosilicic acid ($H_4SiW_{12}O_{40}$) $M = 2879$ in 1 M NaCl) is used to calibrate a permanent working standard. A permanent working standard (an opaque block of polymer or opal glass) used with a neutral filter (to reduce the light intensity) is mounted in the instrument or replaces the scattering cell.

(b) The permanent working standard is measured from time to time during the light-scattering measurement of the polymer solutions to provide the required calibration.

Modern laser ratio photometers compare the scattered and the incident light intensities by measuring the outputs of two photosensitive electrical devices (photomultipliers) (Fig. 13.12).

(iv) Data read-out and data handling. Automated read-out, computer data processing, and computer plotting permit easy acquisition and treatment of many data.

Bibliography: 412, 846.
References: 3032, 3909, 4989, 6802.

Many designs for light-scattering photometers are described in detail elsewhere.

Bibliography: 774.
References: 2179, 2297, 2375, 2507, 2573, 2733, 2838, 2877, 2989, 3105, 3362, 3447, 3724, 3759, 3908, 3909, 3989, 4030, 4312, 4502, 4714, 4720, 4778, 4874, 4991, 5196, 5618, 5633, 5642, 5844, 6283, 6388, 6825, 6898, 7136, 7298, 7332.

Computer analysis of light-scattering data provides more precise and accurate results. A number of computer programs for treating light scattering data have been published.

Bibliography: 733, 884.
References: 2576, 2612, 2953, 3333, 3807, 3909, 4533, 4823, 4989, 5096, 5267, 5289, 6069, 6109, 6237, 6387, 6802, 6996, 7024, 7205.

Extensive data on dissymmetries and particle-scattering factors are collected in The Polymer Handbook (B: 284).
Many tables of scattering functions have been published elsewhere (B: 366, 1200; R: 4989).

13.1.10 Sample Preparation for Light-scattering Measurements

Solutions used for light-scattering measurements should be absolutely dust-free. Dust can be removed through filtration through Millipore filters and/or high-speed centrifugation. Filtration is best carried out with cellulose membrane filters having pores of 0.2–0.5 μm in diameter. Membranes of this type are commercially available for use with most organic and inorganic solvents. In some cases porous glass or porcelain filters are used for such reactive solvents as strong acids and bases. A typical device for filtering polymer solutions through a membrane is shown in Fig. 13.14. The solvents used for light scattering should be spectrally pure in order to avoid fluorescence from impurities which may interfere with scattering measurements.

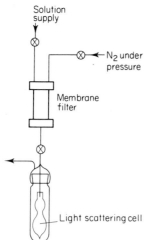

Fig. 13.14 Device for filtering polymer solutions through a membrane filter

The presence of dust is usually recognized from the marked deviations observed in the Zimm plot at angles below 45°.

Bibliography: 263, 1016, 1250.
References: 5091, 5732, 6226, 6283, 6782.

Scattering intensities of sufficient magnitude can be measured when $dn/dc >$ 0.05 cm^3 g^{-1}. For polymer solutions where $c = 0.01$ g cm^{-3} the difference in refractive index for solution and solvent is of the order of 0.002 unit. In order to determine the weight-average molecular weight ($\overline{M}_w$) to within $\pm 2\%$ it is necessary to know the refractive index increments to within $\pm 1\%$ and the difference in refractive indices with an accuracy of $\pm 2 \times 10^{-5}$.

Bibliography: 1016.
References: 2405, 2572, 2665, 3105, 3455, 3996, 4172, 6984.

13.1.11 Applications of Light Scattering in Polymer Research

Measurements of the intensity of light scattering by a dilute polymer solution give information about

(i) the weight-average molecular weight $(\overline{M}_w)$ (from $\sim 10^2$ to $\sim 10^7$);
(ii) the shape of the polymer molecule;
(iii) interactions between polymer and solvent;
(iv) the particles in suspension;
(v) impurities, flaws, and phase separations;
(vi) the total optical behaviour of single scatters (homogeneous or layered spheres and cylinders).

Bibliography, 99, 125, 126, 263, 264, 320, 425, 500, 650, 734, 735, 776, 797, 846, 899, 943, 994, 995, 1016, 1250, 1289, 1308, 1331.
References: 2573, 3063, 3179, 3191, 3206, 3694, 3869, 3909, 4085, 4205, 4301, 4302, 4370, 4424, 4848, 4989, 5860, 6336, 6452, 6540, 6963.

13.2 PULSE-INDUCED CRITICAL SCATTERING

PULSE-INDUCED CRITICAL SCATTERING (PICS) is a method for following the light scattering of polymer solutions isothermally. This scattering is induced by a preceding thermal pulse in special apparatus (Fig. 13.15).

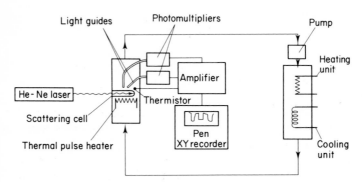

Fig. 13.15   Pulse-induced critical scattering spectrometer

A few microlitres of polymer solution is placed in a thin-walled capillary cell of 1 mm diameter. The He–Ne laser beam passes axially along the cell. The light scattered by the polymer solution after a fast temperature step is recorded at angles of 30° and 90°. The light-guides pass the light to photomultipliers. Temperatures are measured by thermistor.

The cell is placed in a water flow system with a small heater situated upstream from the cell to maintain it at a temperature above the cloud point. The temperature step is produced by switching off the heater. The cell contents then cool down rapidly to the temperature of the flowing water.

The pulse-induced critical scattering method can be applied in the study of

(i) phase separation phenomena in liquid–liquid demixing systems;
(ii) phase diagrams and the measurement of spinoidal points;
(iii) the compatibility of mixtures of polymers;
(iv) crystallization processes.

Bibliography: 367, 731.
References: 3720, 4631.

13.3 LOW-ANGLE LASER LIGHT SCATTERING

In low-angle laser light scattering the angles employed range from $2°$ to $10°$. Because data are obtained at low angles, measurements need be made at only one angle unless $\overline{M}_w$ exceeds 10^6. The complex angular extrapolations (Zimm plots) used in wide-angle light scattering are not required. Because the scattering volume is geometrically defined, the $\overline{M}_w$ values obtained in low-angle scattering are absolute. Sample clarification problems are minimized because of the small scattering volume $(0.1~\mu l)$ employed.

The weight-average molecular weight $(\overline{M}_w)$ can be calculated from the equation

$$Kc/R_\theta = (1/\overline{M}_w) + 2A_2 c \qquad (13.27)$$

or

$$M_w = \frac{1}{(Kc/R) - 2A_2 c}, \qquad (13.28)$$

K is the constant defined by eq. (13.6),
c is the concentration of scattering particles in the solution,
R_θ is the Rayleigh ratio defined by eq. (13.5),
A_2 is the second virial coefficient (cf. Section 2.10).
To determine $\overline{M}_w$, measurements of R_θ are obtained for a few solutions of different values of c. A plot of Kc/R_θ versus concentration (c) yields $1/\overline{M}_w$ as the intercept at the ordinate, and the slope is $\tan \alpha = 2A_2$.

If A_2 is known, then $\overline{M}_w$ can be calculated from a single observation of the Rayleigh ratio (R_θ) by use of eq. (13.27). This facilitates the continuous monitoring of solute molecular weight, such as in a GPC effluent.

The major features provided by the low-angle laser light scattering are:

(i) non-destructive method;
(ii) provides absolute calibration for molecular-weight calibration;
(iii) small scattering volume minimizes sample clarification requirements;
(iv) high sensitivity allows small and dilute samples;
(v) sample cell design allows elimination of background scatter;
(vi) high speed response for kinetic studies;
(vii) long-wavelength laser light source allows elimination of fluorescence;
(viii) rapid convenient operation yields molecular weight determinations in minutes.

Applications of low-angle laser light scattering may be found in many diverse scientific fields involving studies such as those listed below:

(i) molecular weight determinations;
(ii) molecular weight distribution (with GPC)—can effectively eliminate the need for band broadening corrections in certain GPC chromatographic methods (cf. Section 25.8);
(iii) macromolecular kinetics;
(iv) single particle fluorescence and nephelometry;
(v) molecular interactions (virial coefficients);
(vi) molecular conformations (radius of gyration);
(vii) molecular anisotropy (depolarization);
(viii) Rayleigh spectroscopy (diffusion coefficients);
(ix) Brillouin spectroscopy (acoustic modulation);
(x) refractometry.

Bibliography: 889.
References: 4486, 4488, 5636, 5637.

13.4 LIGHT-BEATING SPECTROSCOPY

LIGHT-BEATING SPECTROSCOPY (also called DOPPLER-SHIFT SPECTROSCOPY or RAYLEIGH LINEWIDTH SPECTROSCOPY) is used for measuring the shape (linewidth) of the Rayleigh peak.

The principle of Doppler-shift spectroscopy is as follows. The moving molecules in solution may be slightly shifted in frequency by the Doppler effect on the spectrum of the incident light. The Doppler shift is detected by optical mixing of the scattered radiation with a reference beam from the laser (Fig. 13.16). The resultant HETERODYNE or BEAT FREQUENCY is equal to the difference in frequencies of the reference and scattered beams. Knowledge of this frequency shift and the geometry permits direct computation of the velocity of the molecules in solution.

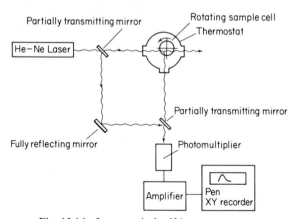

Fig. 13.16   Laser optical self-beat spectrometer

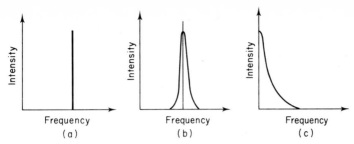

Fig. 13.17 Spectra of: (a) incident light from a laser; (b) scattered light from a rotating sample; (c) a photomultiplier current spectrum

The OPTICAL SELF-BEAT PROCESS allows one to shift the original spectrum from one centred at the incident light frequency down to frequencies of the order of the original linewidth, as is shown schematically in Fig. 13.17. The half-width at the half-height ($\Delta w_{1/2}$) is related to the translational diffusion coefficient (D) according to the equation

$$\Delta w_{1/2} = 2K^2 D, \tag{13.29}$$

where

$K = (4\pi/\lambda) n \sin (\theta/2)$ is the wave-vector change in the quasielastic scattering,
λ is the wavelength of the incident laser light,
n is the solution refractive index,
θ is the scattering angle.

Measurement of the translational diffusion coefficient (D) by this method is rapid and accurate, and is performed on solutions in equilibrium, thus giving valuable information on polymer conformations and solution hydrodynamics.

Light-beating spectroscopy can be applied to measurements of

(i) the weight-average molecular weight of polymers ($\overline{M}_w$);
(ii) the size of macromolecules;
(iii) concentration fluctuations;
(iv) rotational relaxation and diffusion;
(v) diffusion constants—translational, rotational, self-, and mutual;
(vi) the Doppler shift.

Bibliography: 43, 44, 108, 138, 344, 345, 745, 1019–1021, 1033.
References: 2655, 2697, 2840, 3032, 3237, 3244, 3245, 3443–3445, 3497, 3506, 3532, 3537, 3538, 4149, 4150, 4284, 4563, 4703, 4704, 5064, 5490, 5739, 5740, 5907, 6000, 6002, 6231, 6232, 6331–6335, 6684, 6773, 6774, 6981, 7263.

13.5 RAYLEIGH–BRILLOUIN SPECTROSCOPY

In inelastic scattering the scattered light has a different frequency to the incident light and produces two side-bands called BRILLOUIN PEAKS (Fig. 13.18).

The relative intensities of the various peaks in Rayleigh–Brillouin spectra are dependent on the structure and the relaxing processes in liquids. The shape of Rayleigh–Brillouin spectra can be measured using a scanning Fabry–Perot interferometer (Fig. 13.19), where the shift from the incident light frequency is determined by the mirror separation in the cavity. The distance between mirrors can be changed, for example, by piezoelectric methods (cf. Section 13.5.1).

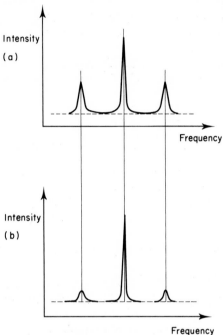

Fig. 13.18   Rayleigh–Brillouin spectra of:
(a) a pure solvent; (b) a polymer solution

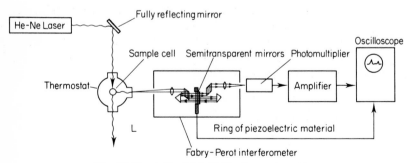

Fig. 13.19   A Rayleigh–Brillouin spectrometer

Rayleigh–Brillouin spectra are interpreted by using the Landau–Placzek ratio (J).

The LANDAU–PLACZEK RATIO (J) is defined as the intensity of the central Rayleigh peak divided by the intensity within the two Brillouin peaks.

The weight-average molecular weight $(\overline{M}_w)$ can be determined by measuring the Landau–Placzek ratio (J) for pure solvents and dilute solutions of macromolecules of various concentrations. $\overline{M}_w$ can be calculated from

$$BKc/(J_1 - J_2) = (1/\overline{M}_w) + 2A_2 c + \ldots, \qquad (13.30)$$

where

B and K are constants related only to the solvent,

J_1 and J_2 are the Landau–Placzek ratios of the pure solvent and of the solution, respectively,

c is the concentration of the scattering particles,

A_2 is the second virial coefficient.

Bibliography: 287, 745, 967.
References: 2906, 4981, 5272, 5273, 5737, 6684.

Rayleigh–Brillouin spectroscopy can also be applied to measurements of

(i) the size of macromolecules in solution;
(ii) branched polymers in solutions;
(iii) diffusion and diffusional processes in solutions;
(iv) rotational relaxation processes in solutions;
(v) polymer networks;
(vi) shear waves, hypersonic velocity, and sonic attenuation in solutions.

Bibliography: 43, 287, 288, 343, 745, 1033, 1045, 1074, 1199, 1230.
References: 2017, 2320, 2449, 2585, 2616, 2624, 2652, 2717, 2718, 3509–3511, 4017, 4018, 4268, 4287, 4550, 4668, 4669, 4875, 4935, 5163, 5255, 5275, 5304, 5373, 5374, 5384–5386, 5542, 5712–5720, 5907, 5945, 5804, 6001, 6090, 6439, 6529, 6722, 6773, 6774, 7181, 7238.

13.5.1 Fabry–Perot Interferometers

The FABRY–PEROT INTERFEROMETER shown in Fig. 13.20 is comprised of two pieces of glass or two fused quartz plates. The two inner surfaces are polished flat to within 0.05–0.005 of a wavelength over the working region, and they are coated with high-reflectivity metallic films (silver, gold, or aluminium) or alternating layers of high- and low-index dielectric materials. The glass plates are made with a small (0.1°) wedge angle between the major surfaces, in order to limit the reflections.

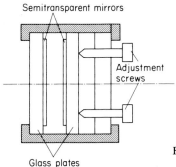

Semitransparent mirrors

Adjustment screws

Glass plates

Fig. 13.20 A Fabry–Perot interferometer

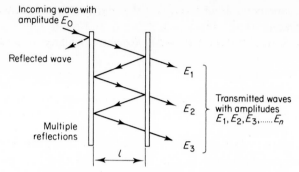

Fig. 13.21 Multiple reflections which occur between two highly reflecting films separated by a distance l

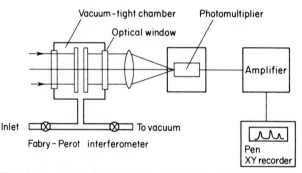

Fig. 13.22 A scanning Fabry–Perot interferometer (older type), where change in pressure produce changes in the index of refraction and the optical path between the plates

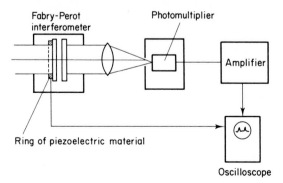

Fig. 13.23 A scanning Fabry–Perot interferometer (newer type), where a ring made from a piezoelectric material produces a change in the distance between plates when subjected to an electric field

Multiple reflections occur between the two highly reflecting films separated by a distance (l) in air (Fig. 13.21).

Normally the useful range of a Fabry–Perot interferometer is restricted to about the first ten fringes.

Two types of scanning Fabry–Perot interferometer are used:

(i) The old type, in which the optical path between the plates can be changed by varying the index of refraction (n) of the atmosphere placed between the plates (Fig. 13.22). The index of refraction for a vacuum is $n_0 = 1.000\ 000$, whereas the index of refraction for air is $n = 1.000\ 290$. The difference between n and n_0 is large enough so that the interferometer is swept through several free spectral ranges.

(ii) The new type, which works by varying the distance between the plates (l). One plate is mounted on a ring of piezoelectric material (Fig. 13.23). When such a material is subjected to an electric field it expands or contracts, changing the distance (l). The sweep frequency of the interferometer is synchronized with the scan frequency of the oscilloscope. When the photomultiplier output is connected to the input of the oscilloscope, the spectrum can be observed directly.

Bibliography: 1220, 1257.

Chapter 14

Ultraviolet and Visible Spectroscopy of Polymers

General bibliography: 71, 89, 192, 513, 672, 730, 1093, 1182, 1213, 1262, 1393.

The absorption of ultraviolet or visible radiation by polymers leads to transition among the electronic energy levels of the macromolecules, and as a result of this we obtain a typical ELECTRONIC ABSORPTION SPECTRUM (Fig. 10.4).

ULTRAVIOLET and VISIBLE SPECTROSCOPY yield information on multiple-bond and aromatic conjugation within macromolecules. The non-bonding electrons on oxygen, nitrogen, and sulphur may also be involved in extending the conjugation of multiple-bond systems in polymers.

Polymers which absorb light of wavelength between 400 and 800 nm (visible light) appear coloured to the human eye. Progressive absorption from 400 nm upwards leads to progressive darkening through yellow, orange, red, green, blue, violet, and black.

The absorption of light by macromolecules occurs only if the difference between two energy levels is exactly equal to the energy of a quantum:

$$\Delta E = h\nu = E_2 - E_1, \tag{14.1}$$

where

h is the Planck constant (6.626×10^{-27} erg s(CGS) or 6.626×10^{-34} J s (SI)),

ν is the frequency (in s^{-1}).

E_2 and E_1 are the energies of a single macromolecule in the final (higher level) and initial (lower level) states, respectively.

CHROMOPHORES are functional groups that absorb electromagnetic radiation (whether or not a colour is thereby produced, e.g. the carbonyl group C=O is a chromophore which absorbs in the region of 280 nm, but ketones are colourless). Important examples of organic chromophores which can be found in polymers are given in Table 14.1. Applications of ultraviolet-visible spectroscopy in polymer studies are mainly concerned with the presence of some of these chromophoric groups in polymer molecules.

AUXOCHROMES are groups (e.g. —OH, —OR, —NH$_2$, etc.) which extend the conjugation of a chromophore by sharing of the non-bonding electrons and become part of a new extended chromophore.

When a macromolecule absorbs ultraviolet or visible light of a particular energy, only one electron is shifted to a higher energy level, and all other electrons are unaffected (to a first approximation). The EXCITED state produced in this way has a short lifetime of the order of 10^{-6}–10^{-9} s. The

Table 14.1 Typical chromophores and their characteristics

Chromophore	Wavelength, λ_{max} (nm)	Molar absorptivity, ε_{max}
C=C	175	14 000
	185	8000
C≡C	175	10 000
	195	2000
	223	150
C=O	160	18 000
	185	5000
	280	15
C=C–C=C	217	20 000
⌬	184	60 000
	200	4400
	255	204

most probable ΔE transitions involve the shift of one electron from the highest occupied molecular orbital to the lowest available unfilled orbital. In many cases several transitions can be observed, giving several absorption bands in the polymer spectrum. Not all transitions from filled to unfilled orbitals are allowed. Where a transition is 'forbidden', the probability of that transition is low and the intensity of the associated absorption band is also low (cf. Chapter 16).

The relative energies of the molecular orbitals involved in electronic spectroscopy of polymers are shown in Fig. 14.1. Many factors influence the relative energies of molecular orbitals in polymers and a knowledge of these is necessary for an understanding of the electronic spectroscopy of macromolecules.

When a macromolecule absorbs ultraviolet or visible light an ABSORPTION BAND appears because vibrational and rotational effects are superimposed on the electronic transitions.

In some cases a solvent may influence for the intensity and the position of an absorption band and shift it to a longer or a shorter wavelength. According to the old system of nomenclature (which is progressively being discouraged):

(i) BATHOCHROMIC SHIFT (or RED SHIFT) is a shift to longer wavelength;

(ii) a HYPSOCHROMIC SHIFT (or BLUE SHIFT) is a shift to shorter wavelength;

(iii) a HYPERCHROMIC EFFECT is an effect which increases the intensity of absorption;

(iv) a HYPOCHROMIC EFFECT is an effect which decreases the intensity of absorption.

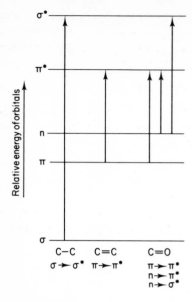

Fig. 14.1 Relative energies of molecular orbitals and different types of electronic transitions involved in electronic spectroscopy

As an aid in finding ultraviolet spectra of organic compounds of low molecular weight, the following publications are useful: B: 290, 478, 626, 792, 1037, 1472, 1476.

Application of ultraviolet spectroscopy in polymer research is limited to those polymers which have chromophoric groups (e.g. poly(enes), polymers having aromatic rings, or carbonyl groups).

References: 2295, 3143, 4407, 4803, 6167, 7134.

14.1 QUANTITATIVE ANALYSIS IN ULTRAVIOLET AND VISIBLE SPECTROSCOPY

In spectrophotometric practice, quantitative analysis is based on the application of the BEER–LAMBERT LAW, which is given by

$$A = \log_{10}(I_0/I) = \varepsilon cl \qquad (14.2)$$

(dimensionless), where

I_0 is the intensity of the incident light (in quanta per second),

I is the intensity of the light transmitted through the sample solution (or sample film) (in quanta per second),

Log_{10} (I_0/I) is called the ABSORBANCE (A) of the solution also called the OPTICAL DENSITY (D) or EXTINCTION (E)),

ε is the MOLAR ABSORPTIVITY (also called the MOLECULAR EXTINCTION COEFFICIENT),

$$\varepsilon = A/cl \qquad (14.3)$$

(in litres mol^{-1} cm^{-1} (CGS) or m^3 kg^{-1} m^{-1} (SI)).

The MOLAR ABSORPTIVITY (ε) is constant for a particular compound at a given wavelength. Where values for ε are very large, it is convenient to express it as its logarithm ($\log_{10}\varepsilon$). The absorbance (extinction) is a cumulative property for a mixture of two or more components:

$$A = l(\varepsilon_1 c_1 + \varepsilon_2 c_2 + \ldots + \varepsilon_n c_n). \qquad (14.4)$$

The concentration of two components may be determined when four values of the molar absorptivity (molar extinction coefficient) are known, and when the measurements are made for two wavelengths:

$$A' = l(\varepsilon_1' c_1 + \varepsilon_2' c_2) \quad \text{for } \lambda', \qquad (14.5)$$

$$A'' = l(\varepsilon_1'' c_1 + \varepsilon_2'' c_2) \quad \text{for } \lambda''. \qquad (14.6)$$

The wavelengths λ' and λ'' are chosen such that one component strongly absorbs these wavelengths whereas the other component absorbs them much less (Fig. 14.2).

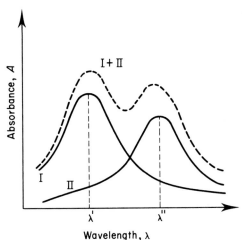

Fig. 14.2 Overlapping spectra of two components I and II and the spectrum of a mixture of the two components I+II

The following relationship is defined between absorbance (A) and transmittance ($T(\%)$) (Fig. 14.3):

$$A = \log_{10}(I_0/I) = -\log_{10} T = \log_{10}(100/T). \qquad (14.7)$$

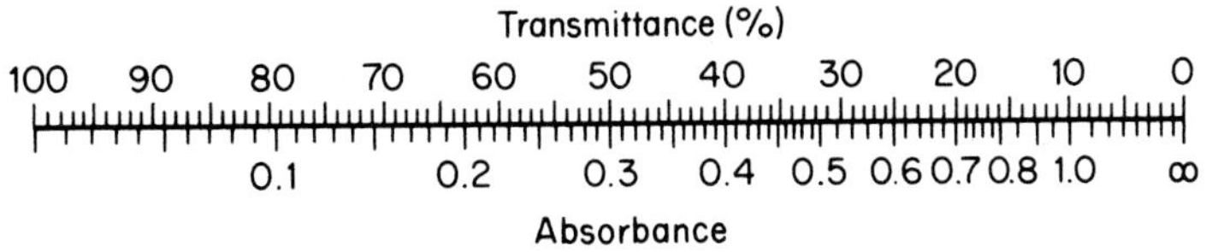

Fig. 14.3 Relation between a transmittance $T(\%)$ and an absorbance (A)

A conversion table between transmittance (T (%)) and absorbance (A) is given in Appendix C.

The absorbance (A) or transmittance (T) is recorded directly by all modern double-beam spectrometers. ABSORPTION SPECTRA are plotted with absorbance (A) (or ε, or $\log_{10}\varepsilon$) (Fig. 14.4 (a)) or transmittance (T (%)) (Fig. 14.4 (b)) on the ordinate and with wavelength (λ in nanometres or ångströms) on the abscissa.

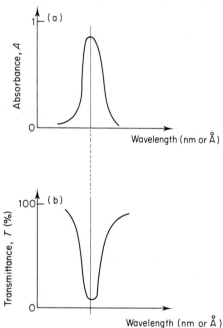

Fig. 14.4   Schematic absorption spectra: (a) expressed as absorbance A versus wavelength; (b) expressed as transmittance T versus wavelength

Other functions commonly used in spectrophotometry are given below:

(i) PERCENTAGE TRANSMISSION (per cent transmittance), $100\ I/I_0$;
(ii) PERCENTAGE ABSORPTION, $100(I_0 - I)/I_0$;
(iii) FRACTIONAL ABSORPTION, $(I_0 - I)/I_0$.

The concentration of any chromophoric compound can be quantitatively measured in solution by using the Beer–Lambert law (eq. (14.2)) if the molar absorptivity (molecular extinction coefficient) (ε), the path length (l) (the cell dimension), and the absorbance (A) are known.

The validity of the Beer–Lambert law should always be tested before using it for accurate quantitative analyses. To test eq. (14.2) it is necessary to measure absorbance (A) as a function of concentration (c) at a fixed wavelength (λ) and cell path (l) (Fig. 14.5). If the Beer–Lambert law is obeyed over the concentration range tested, a straight line should be obtained through the origin.

Deviations from the law are designated as positive or negative. Sometimes the Beer–Lambert law is obeyed at one wavelength (λ_1) but not at another (λ_2) (Fig. 14.6).

A number of common errors can occur in making quantitative absorption measurements which can be classified as chemical, instrumental, and personal.

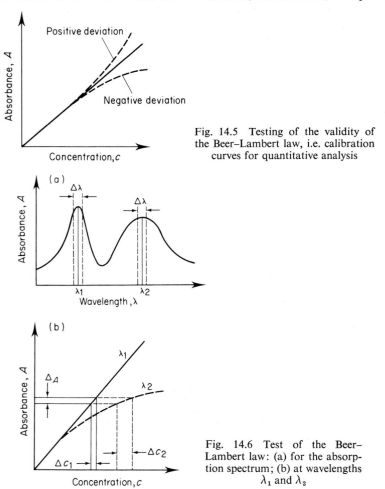

Fig. 14.5 Testing of the validity of the Beer–Lambert law, i.e. calibration curves for quantitative analysis

Fig. 14.6 Test of the Beer–Lambert law: (a) for the absorption spectrum; (b) at wavelengths λ_1 and λ_2

14.2 INSTRUMENTATION FOR ABSORPTION SPECTROSCOPY

14.2.1 Double-beam Spectrometer

The double-beam UV–VI spectrometer (Fig. 14.7) is built of the following parts:

(i) A light source, which consists of a tungsten filament lamp for wavelengths greater than 375 nm and a deuterium discharge lamp for wavelengths below 375 nm (cf. Section 10.5).

The original beam, after passing the grating or quartz prism monochromator, is divided into two equal beams by a rotating sector mirror. It alternately passes the beam (open sector) to one channel and reflects it (mirror sector) to the second.

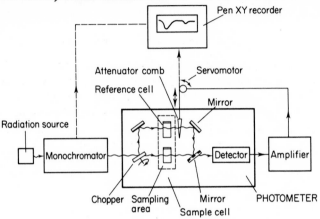

Fig. 14.7 Schematic diagram of an automatic, double-beam, optical null UV–VI spectrophotometer

(ii) A sampling area.

(iii) A photometer optical null readout system. In order to measure the absorption of a sample, intensities of sample and reference beams must be compared. The two chopped beams fall alternately on the detector (photomultiplier) and are amplified. If the intensities are identical, the amplifier has no output. Any difference in the intensities will result in an output signal of the frequency of chopping. This imbalance signal is further amplified and is used to drive an attenuator into or out of the reference beam. An attenuator is a thin, flat, metal comb with teeth whose separation increases linearly with distance. The fraction of open space a comb gives in a beam ($T\,(\%)$) can be varied linearly by careful positioning. Depending on the phase of the imbalance signal, the wedge is driven so as to increase or decrease the intensity of the reference beam until it equals that of the sample beam. Only when intensities in the beams are again identical is there no signal, and then the attenuator stops. The attenuator position is a measure of the relative absorption of the sample—by transmitting its position to a digital device the absorption is indicated.

For measurements of absorption with greater precision, especially at higher absorbances ($A > 1$), a ratio-recording readout has an advantage over the optical null method.

(iv) A data recording system. The ratio of reference beam to sample beam intensities (I_0/I) is fed to a pen recorder, which records absorbance (A) versus wavelength (λ) linearly over the range 0–2 absorbance units.

Absorption spectrometers are available for all ranges from the vacuum
ultraviolet to the far infrared (about 50 nm to 1000 μm).

Bibliography: 838, 991, 1213, 1275.
References: 5046, 5489, 5683, 7067.

14.2.2 Sample and Reference Cells

Different types of matched silica cell (Fig. 14.8) with path lengths of 0.1, 1.0,
and 10.0 cm are commonly used. The cells are expensive and fragile. They
must be thoroughly cleansed of polymer solution after each use. This can be
done by employing a simple device shown in Fig. 14.9. It is sometimes possible
to clean these cells with a soft tissue dipped in solvent. Cells must be properly
stored, and the optical surfaces must never be handled.

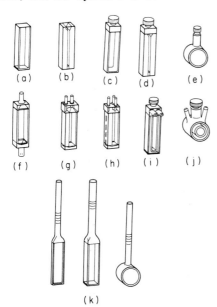

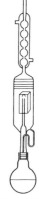

Fig. 14.8 Different types of cell used in
UV–VI spectroscopy. (a) Standard rectangular open-top cell; (b) standard open-
top microcell; (c) standard rectangular
stoppered cell; (d) standard stoppered
microcell; (e) standard cylindrical stoppered microcell; (f) flow-through cell;
(g) flow-through cell (automatic or continuous flow); (h) flow-through cell with
three tubes for inlet and vacuum connection; (i) standard rectangular constant
temperature cell; (j) standard cylindrical
constant temperature cell; (k) cells with
special vacuum-tight fused windows and
graded seal. (From B: 1473; reproduced
by permission of Hellma GmbH)

Fig. 14.9
Simple de-
vice for
cleaning a
spectro-
scopic cell

Table 14.2 Commercially produced solvents for UV-region spectroscopy (from B: 1471; reproduced by permission of BDH, England)

Solvent	Cut-off †	Minimum percentage transmittance in a 1 cm cell relative to distilled water at the following wavelengths (nm):														
		195	200	210	220	230	240	250	260	270	280	290	300	320	350	400
Acetonitrile ‡	<210	—	—	—	80	90	93	—	96	—	98	—	99	—	—	—
Benzene	278	—	—	—	—	—	—	—	—	—	25	80	90	95	99	—
Carbon tetrachloride	265	—	—	—	—	—	—	—	—	45	78	91	97	—	—	—
Chloroform	245	—	—	—	—	—	—	—	60	85	92	95	97	—	—	—
Cyclohexane ‡	200	—	—	15	50	75	90	97	—	—	—	—	—	—	—	—
Dimethylformamide	268	—	—	—	—	—	—	—	—	25	72	85	90	95	99	—
Dimethyl sulphoxide §	261	—	—	—	—	—	—	—	10 at 261	50	76	84	89	—	—	—
1,4-Dioxan ‡	215	—	—	—	20	35	47	59	70	(58)	(83)	(92)	(95)	—	—	—
Ethanol (95%) ‡	205	—	—	30	52	73	85	90	—	79	87	95	—	—	—	—
Ethanol (absolute)	205	—	—	30	52	73	85	90	—	—	—	—	—	—	—	—
Ethyl acetate	250	—	—	—	—	—	—	—	40	67	81	88	91	—	—	—
Hexane fraction from petroleum ‡	210	—	—	10	70	86	92	95	—	—	—	—	—	—	—	—
1,1,3,3,3,-Hexafluoropropan-2-ol ‡	<190	90	93	95	96	98	99	99	—	—	—	—	—	—	—	—
Methanol ‡	205	—	—	15	40	65	83	90	—	—	—	—	—	—	—	—
Propan-2-ol ‡	205	—	—	30	62	83	94	99	—	—	—	—	—	—	—	—
Tetrachloroethylene	290	—	—	—	—	—	—	—	—	—	—	10	75	85	—	—
Trifluoroacetic acid	260	—	—	—	—	—	—	—	10	80	90	93	95	95	—	95
2,2,2-Trifluoroethanol ‡	<190	70	90	92	94	95	96	97	98	—	—	—	—	—	—	—
2,2,4-Trimethylpentane ‡	200	—	10	30	60	85	95	97	—	—	—	—	—	—	—	—

§ The figures in parentheses apply after purging with nitrogen for 15 min.
‡ For use at short wavelengths the solvent should be purged with nitrogen immediately before use.
† The approximate wavelength at which the relative transmittance in a 1 cm cell falls at 10%.

14.2.3 High-pressure Cell for Optical Studies

Spectroscopic and other studies of solutions under 1000–10 000 atm pressure require specially constructed high-pressure cells. The main construction problem is that a large and variable birefrigence is formed by pressure in the optical windows. The special requirements of the materials to be used for high-pressure optical windows mean that Pyrex or quartz must be employed. In order to avoid solution leaks, a complicated pressure-locking system is used. A well-constructed high-pressure cell can be placed in conventional optical

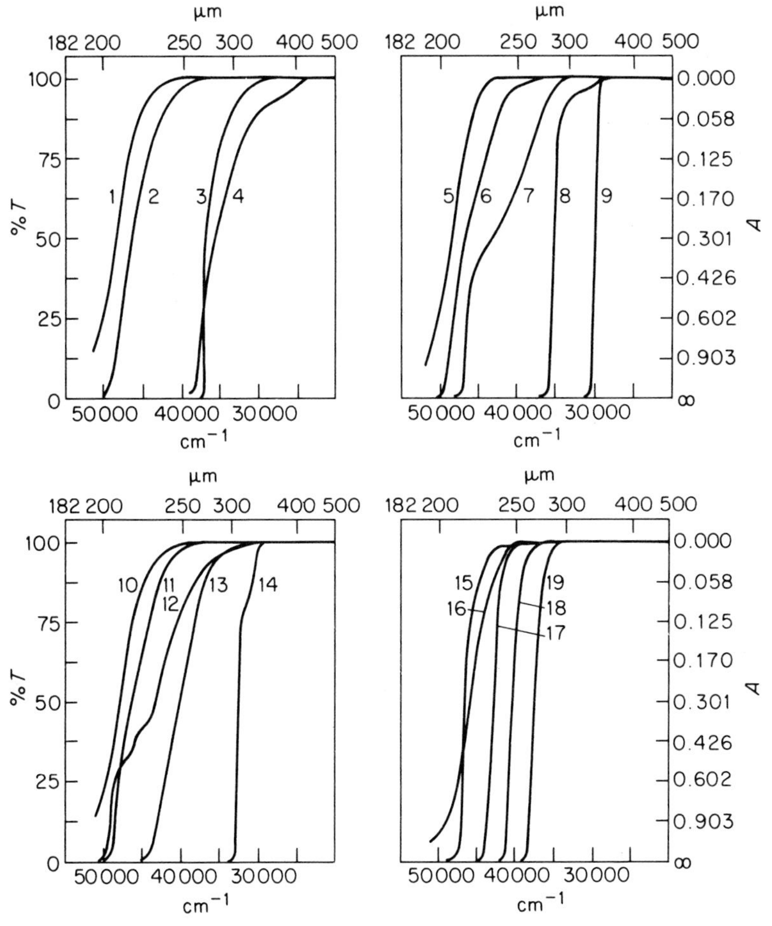

Fig. 14.10 Transmission spectra of different solvents (cell length 10 mm, versus water). Curves: 1, heptane; 2, methanol; 3, dimethylformamide; 4, dimethylsulphoxide; 5, hexane; 6, ethanol; 7, dioxane; 8, benzene; 9, acetone; 10, iso-octane; 11, isopropanol; 12, diethyl ether; 13, tetrahydrofuran; 14, pyridine; 15, acetonitrile; 16, cyclohexane; 17, dichloromethane; 18, chloroform; 19, carbon tetrachloride. (From B: 1472; reproduced by permission of Verlag Chemie)

instruments for measuring ultraviolet, visible, and infrared spectra, fluorescence, phosphorescence, light-scattering, etc.

References: 2868, 2874.

14.3 SOLVENTS FOR ULTRAVIOLET SPECTROSCOPY

Ultraviolet and visible spectra are usually measured on very dilute solutions (10^{-4}–10^{-6} mol l^{-1}). The solvent must be transparent within the wavelength range being measured. The lower-wavelength limits for common spectroscopic-grade solvents are given in Table 14.2. Below these limits, the solvents show excessive absorbance and sample absorbance will not be recorded linearly. The full ultraviolet spectra of some common solvents applied in spectroscopy are shown in Fig. 14.10.

Chapter 15

Infrared Spectroscopy of Polymers

General bibliography: 56, 71, 89, 127, 191, 274, 313, 316, 339, 359, 389, 431, 437, 554, 669, 730, 913, 965, 991, 1060, 1120, 1158, 1213, 1267, 1282, 1332, 1393.

When infrared light is passed through a polymer sample, some of the frequencies are absorbed while other frequencies are transmitted. The transitions involved in infrared absorption are associated with VIBRATIONAL changes within the molecule. Different bonds present in polymers (C—C, C=C, C—O, C=O, O—H, N—H, etc.) have different vibrational frequencies. The presence of these bonds in polymers can be detected by identifying characteristic frequencies as absorption bands in the INFRARED SPECTRUM.

Applications of infrared spectroscopy in polymers are concerned with the range 650–4000 cm^{-1} (15.4–2.5 µm). The 650–10 cm^{-1} frequency region is called the FAR INFRARED, and the 4000–12,500 cm^{-1} frequency region is called the NEAR INFRARED.

A conversion table between wavelengths in micrometres and wavenumbers in (cm^{-1}) is given in Appendix D.

As an aid in finding infrared spectra of low molecular weight organic-compounds the following publications are useful: B: 96, 97, 101, 378, 453, 580, 845, 911, 1061, 1283–1285, 1430, 1468, 1471, 1477–1479.

As an aid in finding infrared spectra of polymers, plastics, rubbers, and additives used in polymer processing the following publications are useful: B: 46, 124, 211, 239, 360, 405, 580, 582, 608, 652–657, 693, 701, 730, 982, 1213, 1268, 1324, 1441, 1445.

There are two general types of molecular vibration (Fig. 15.1):

(i) a STRETCHING VIBRATION is a rhythmical movement along the bond axis such that the interatomic distance increases or decreases;

(ii) a BENDING VIBRATION may consist of (a) a change in the bond angles between bonds with a common atom, or (b) movement of a group of atoms with respect to the remainder of the molecule without movement of the atoms in the group with respect to one another.

A molecule of n atoms has $3n$ vibrational degrees of freedom: three of the degrees of freedom describe rotation and three describe translation; the

222

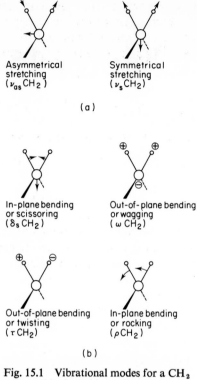

Asymmetrical stretching ($\nu_{as}CH_2$)

Symmetrical stretching ($\nu_s CH_2$)

(a)

In-plane bending or scissoring ($\delta_s CH_2$)

Out-of-plane bending or wagging ($\omega\, CH_2$)

Out-of-plane bending or twisting (τCH_2)

In-plane bending or rocking (ρCH_2)

(b)

Fig. 15.1 Vibrational modes for a CH_2 group. (a) Stretching vibrations; (b) bending vibrations. $+$ and $-$ indicate movement perpendicular to the plane of the page. (From B: 1213; reproduced by permission of John Wiley & Sons)

remaining $3n-6$ degrees of freedom are vibrational degrees of freedom of fundamental vibrations. The $3n-6$ rule does not apply to the CH_2 group, because the CH_2 represents only a portion of a molecule.

The theoretical number of fundamental vibrations is seldom observed because

(i) OVERTONES—multiples of a given frequency, and
(ii) COMBINATION TONES—sums of two other vibrations,

increase the number of bands, whereas other phenomena reduce the number of bands. These other phenomena include

(i) fundamental frequencies that fall outside the 650–4000 cm^{-1} (15.4–2.5 μm) region;
(ii) fundamental bands that are too weak to be observed;
(iii) fundamental vibrations that are so close that they coalesce;

(iv) the occurrence of a degenerate band from several absorptions of the same frequency in highly symmetrical molecules;

(v) the failure of certain fundamental vibrations to appear in the infrared because of the lack of the required change in dipole character of the molecule.

15.1 INSTRUMENTATION FOR INFRARED SPECTROSCOPY

15.1.1 Double-beam Spectrometer

The double-beam infrared spectrometer (Fig. 15.2) is built of the following parts:

(i) A radiation source, which is produced by electrically heating a Nernst filament (oxides of zirconium, thorium, and cerium) ($7100-1000 \text{ cm}^{-1}$) or a Globar (silicon carbide) ($5500-600 \text{ cm}^{-1}$) to $1000-1800°C$. The infrared radiation from the source is further divided into two beams and focused into the samples by a system of mirrors.

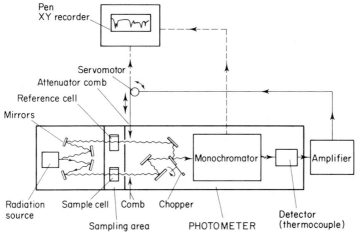

Fig. 15.2 Schematic diagram of an automatic, double-beam infrared spectrophotometer

(ii) A sampling area which allows accommodation for a wide variety of sampling accessories (cells). Reference and sample beams pass through the reference cell and the sample cell.

(iii) A photometer–optical null readout system. To measure the absorption of a sample, intensities of sample and reference beams must be compared. The reference beam and the sample beam pass through the attenuator and comb, respectively, and are reflected by a system of mirrors to the rotating sector mirror, which alternately reflects or transmits the beams to the monochromator slit. The two chopped beams then fall on the detector (thermocouple) and are amplified. If intensities are identical, the amplifier has no output. Any difference in intensities will result in an output signal of the frequency of the shopping. This imbalance signal is further amplified and used to drive an attenuator into or out of the reference beam in response to the signal created at the detector by the sample beam. The attenuator position is a measure of the relative absorption of the sample; by transmitting its position to a digital device the absorption is indicated.

(iv) A data recording system. The ratio of the intensities of the reference beam to the sample beam (I_0/I) is fed to a pen recorder, which records transmittance (T) versus wavenumber ($\bar{v}$) on a linear scale.

Infrared spectrometers are available for ranges from 4000 to 600 cm^{-1} (and even down to 200 cm^{-1}).

Automation of the recording of infrared spectra by digitization of the signal from a spectrometer is now a very common procedure. Details of circuits and programs providing band position, intensities, and half-width in the form of paper tape are given. An electronic programmer controls the cycles of sample feed, wavelength scanning, and digital readout. The last-mentioned, in the form of teletype and paper tape, is supplemented by an analogue output in the form of a conventional spectrum. The number of documented infrared spectra exceed 100 000. Tapes and programs are now commercially available for the ASTM and Sadtler collections of spectra (B: 1468, 1477–1479).

Bibliography: 257, 340, 926, 1060, 1120, 1438.
References: 2116, 3235, 3257, 3747, 4320, 4342, 4353, 4699, 5143, 5674.

15.1.2 Infrared Optical Materials

Infrared optical materials are available which transmit in the range of 10 000–10 cm^{-1} (1–1000 µm). The most common are as follows:

(i) SODIUM CHLORIDE (NaCl) used in the range 5000–625 cm^{-1} (2–16 µm). NaCl plates can be used up to 250 °C, but they are unsuitable for low-temperature work. They cannot be mechanically or thermally shocked. NaCl plates can easily be cut with a metal bandsaw or drilled. The salt plates can be ground on a metal or glass plate using a slurry of carborundum in water or a fine carborundum paper, and further polished on a pitch lap using a polishing compound mixed with a small amount of water. In the end a cloth polishing must be performed in a humidity-free atmosphere. Polished plates must be stored in a desiccator.

(ii) POTASSIUM BROMIDE (KBr) used in the range 5000–400 cm^{-1} (2–25 µm). KBr is softer and more plastic than NaCl. KBr has been used as windows in liquid cells and also as the matrix in the pressed disc technique.

(iii) SILVER CHLORIDE (AgCl) used in the range 5000–435 cm^{-1} (2–23 µm). AgCl is insoluble in water and organic solvents, is fairly soft, and does not cleave. It is extremely difficult to polish. AgCl can be rolled down to practically any thickness and can be moulded and fused. It darkens readily under exposure to light of wavelength under 500 nm, resulting in a reduction in its transmission. AgCl has been used as windows in low-temperature cells.

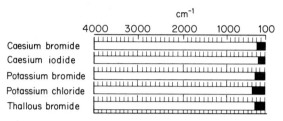

Fig. 15.3 Transparent regions of different commercially produced infrared optical materials. (From B: 1471; reproduced by permission of BDH, England)

Infrared spectra of some of commercially produced infrared optical materials are shown in Fig. 15.3 and in Appendix E.

Bibliography: 946.
References: 5178.

15.1.3 Care and Maintenance of Infrared Materials

Infrared optical materials are easily attacked by water and water vapour. When the water is absorbed on the salt plate surfaces there is a local dissolution and plates become 'fogged' and scatter the infrared radiation passing through them. NaCl and KBr plates can be exposed to a relative humidity of 30–40% for some hours without significant deterioration. Plates should be kept in a desiccator when not in use. They should never be handled with the fingers, only by using surgeon's gloves. Salt plates should be thoroughly washed after use with polymer solutions. After several flushings with solvent, the plates should be dried under a lamp (avoid for AgCl plates) in a dry atmosphere.

Bibliography: 946, 1066.

15.2 SAMPLING TECHNIQUES

15.2.1 Solid Polymers

The following techniques are applied for recording solid spectra:

(i) KBr PRESSED TABLET (DISC) TECHNIQUES. Polymer in the form of small particles is dispersed in a tablet of potassium bromide. KBr tablets are prepared by grinding the polymer sample (2 mg) with KBr (100–200 mg) and compressing the whole into a transparent tablet. The KBr must be completely dehydrated by drying at 105 °C. If possible the polymer sample and KBr should be heated to a temperature of 40 °C before grinding under an infrared lamp to avoid condensation of atmospheric moisture. The latest gives a broad absorption band between 3500 and 3400 cm^{-1}. The grinding is usually done with an agate mortar and pestle (10–15 min) or in a commercial vibration mill (1–2 min). Good dispersion is very important, because poorly ground mixtures in the tablet scatter more light than they transmit. The particle size must be less than 2 μm. Compression to a cohesive tablet requires high pressure (15 min at a pressure of 2 tonnes and a further 15 min at a pressure of 8–10 tonnes) and this is done by using an evacuable die (Fig. 15.4). The mould of the die is constructed in this way so that the

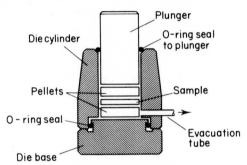

Fig. 15.4 An evacuable die for the preparation KBr tablets (discs). (From B: 1470; reproduced by permission of Beckman Instruments)

tablet is deaerated under pressure by a connected vacuum pump. Tablets produced by this technique commonly measure 13 mm in diameter and 0.3 mm in thickness. They can be handled by their edges. A well-made tablet is transparent. A certain skill is required to prepare the tablets correctly and the following suggestions may facilitate the procedure:

(a) A powdered fine grain KBr must be used; coarse grains may form white spots in the tablet.

(b) Moisture produces an uneven and rough surface.

(c) The lack of adequate transparency is frequently due to insufficient grinding of the polymer with powdered KBr, to too low a compression pressure, or to insufficient deaeration.

(d) Too long a grinding results in chalking of the tablet.

(e) Elastic polymers such as rubber should be ground in a vibration mill at a low temperature (−40 °C).

(ii) PREPARATION OF KBr PELLETS FROM MICROSAMPLES. Many microsamples originate from thin-layer chromatographic (TLC) separations (cf. Section 23.15).

The sample is separated by TLC, the adsorbent is removed from the area of the chromatogram containing the separated material, the sample is eluted with a suitable solvent, filtered to remove the suspended adsorbent, and finally mixed with KBr.

By using a porous triangle of pressed KBr (WICK-STICK elements) in a small glass vial, capped so that evaporation is restricted to the centre of the vial, the filtration of adsorbent and deposition of the sample on KBr can be accomplished in a single step. The adsorbent containing the sample is scraped from the thin-layer chromatographic plate and transferred to a glass vial containing a Wick-Stick using a thin-stemmed funnel to prevent the adsorbent from dusting on the top half of the Wick-Stick. A suitable eluting solvent is added and a vented cap is placed on the vial.

The Wick-Stick apparatus is shown in Fig. 15.5. The pressed KBr triangle is 2.5 cm high, 0.8 cm wide at the base, and 0.2 cm thick. The

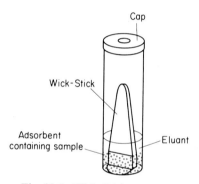

Fig. 15.5 Wick–Stick apparatus

solvent climbs the Wick-Stick by capillary action and evaporation takes place preferentially at the apex of the KBr triangle, deposing the sample. Evaporation of the solvent can be accelerated by maintaining the temperature 10–20 °C below the boiling point of the solvent and directing an air jet across the cap of the vial.

Reference: 3596.

(iii) The LYOPHILIZATION TECHNIQUE. In this method ground polymer (about 200 mg) is placed in a small flask containing about 2 g of KBr in 5 ml of water. The flask is connected with a special freeze-drying apparatus (Fig. 15.6). Several flasks with various samples can be placed in the apparatus simultaneously. Solutions in the flasks are then cooled to freezing point and a vacuum is applied. The sublimating ice is collected in a condenser filled with solid CO_2. After 6 h the samples are completely dehydrated and the polymer is perfectly coated with KBr. The sample is then ground in an agate mortar and the tablet is prepared.

Bibliography: 210.
References: 2089, 4102, 7290.

228

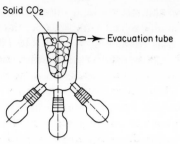

Solid CO$_2$

Evacuation tube

Fig. 15.6   Freeze-drying apparatus
for the lyophilization technique

(iv) The SUSPENSION TECHNIQUE. Polymer in the form of small
particles is dispersed in a drop of suspension liquid such as liquid
paraffin (Nujol) or hexachlorobutadiene or chlorofluorocarbon oil
(Kel-K oil). The suspension is then placed between two NaCl plates
to form a layer 0.1–0.01 µm thick.

These suspension liquids are transparent over a wide range (Fig. 15.7)
and Appendix G.

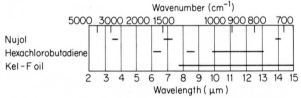

Fig. 15.7   Transparent regions of different suspension liquids.
(From B: 1471; reproduced by permission of BDH, England)

15.2.2 Isotropic Films

The following techniques are applied for the preparation of isotropic films:

(i) DRY CASTING. This involves evaporation of the solvent from a
solution of the polymer spread on a horizontal surface between two
wires (Fig. 15.8). A piece of glass plate must be thoroughly cleaned and
wiped with soft tissue dipped in a solution of dichlorodimethylsilane in

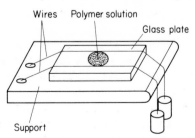

Wires   Polymer solution

Glass plate

Support

Fig. 15.8   Casting of polymer film
on a glass plate

carbon tetrachloride. When the solvent has evaporated, the plate is washed in water. After this treatment a plate has a water-repelling surface from which most polymers can easily be separated. Better uniformity of thickness can be obtained when drying is carried out slowly, e.g. by covering the plate with polymer solution with a small inverted beaker. The traces of solvent can be removed by heating the film in a vacuum.

Casting on cellophane is a very useful method for preparing films of uniform thickness. In this purpose the end of a glass tube of 2–3 cm diameter is covered with wet cellophane which becomes stressed during drying and obtains a perfectly smooth surface (Fig. 15.9). The tube is placed on a table and the measured amount of polymer solution (1–10 wt %) is poured on the cellophane. The evaporation rate of the

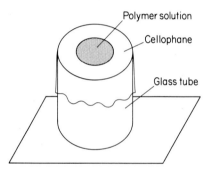

Fig. 15.9   Casting of polymer film on a
cellophane surface

solvent can be reduced by covering the cellophane with a watch-glass, obtaining in this way a smooth and uniform surface of the film. After evaporation of the solvent the polymer film is separated from the cellophane, which is eventually moistened to facilitate the separation. The polymer film must be dried in a vacuum before determination of the absorption spectrum.

Solid polymer films can be deposited directly on the NaCl plate by allowing a solution to evaporate drop by drop on the surface of the plate.

(ii) WET CASTING. This technique involves precipitation of the polymer film from a solution on to a glass plate; the plate is immersed in a solvent in which the polymer is insoluble but which is miscible with the solvent used to form the polymer solution.

(iii) MELT CASTING involves melting the polymer between two flat polished metal plates separated by spacing pins under high pressure at the required temperature (Fig. 15.10). Quick cooling may give a sample with little or no crystallinity.

The required thickness of films for infrared spectroscopy is between 5 and 200 µm. Films can be measured with a dial gauge micrometer down to between 0.05 and 0.100 mm thickness.

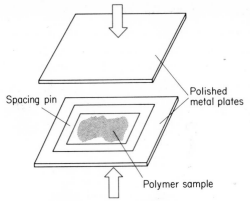

Fig. 15.10   Melt casting of polymer between two
flat polished metal plates

15.2.3 Oriented Films

Oriented films for infrared spectroscopy can be made by drawing or rolling the polymer film sample:

(i) Drawing. Films which are to be drawn should be of uniform thickness and width, and should have edges which have been cut cleanly with a sharp knife, razor-blade, or scalpel. Any tear in the edges will develop further under the drawing. Some polymers may be drawn without difficulties at room temperature; others, which break, must be drawn at a higher temperature or in the presence of a softening or plasticizing agent. Sometimes a small quantity of a solvent for the polymer may act as a plasticizer. Wet steam may also be effective.

(ii) Rolling. Mechanical rolling can be done between the cylindrical rollers of a small mill. It is difficult to carry out this operation on the very thin films required for infrared spectroscopy, but it is easier if the films are cast on a deformable material (e.g. lead, cadmium, or silver chloride sheets) which can be rolled thin along with the film.

The most commonly used sheet is silver chloride sheet, which has mechanical properties similar to those of lead and is transparent to infrared radiation. Because of this property it is necessary to remove the AgCl sheet from the polymer film. If required, the free polymer film can be obtained by dissolving away the silver chloride in aqueous sodium thiosulphate solution.

(iii) Many polymers solutions will give films with some orientation of the molecules if they are continually sheared during the drying process (e.g. by stroking the solution in a particular direction with a flat blade).

Polymers of high molecular weight are usually more easily oriented than are those of low molecular weight.

15.2.4 Cutting of Polymer Samples for Spectroscopic Examination

Some polymer samples are in pieces too thick for spectroscopic examination, and must be reduced in thickness without changing their physical state by solution or melting. There are two methods for reducing thickness:

(i) Cutting sections with a microtome. The sample must be embedded in a small block (Fig. 15.11) and then cut with a glass knife, i.e. the edge of a freshly broken piece of plate glass. It is not possible to define the angle of attack of the knife blade as this is too dependent on the nature of the

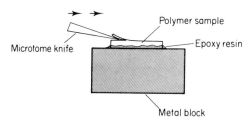

Fig. 15.11 Cutting a polymer sample with a microtome knife

polymer sample. Lubrication of the blade is by dilute aqueous detergent or organic solvents such as alcohols, acetone, or petroleum naphtha. Samples may be cemented to the block with an epoxy resin, but there is a possibility of such resins interacting with the specimen. After cutting, the sections are mounted in suitable frames or on salt plates.

(ii) Grinding with a fine aluminium oxide hone.

Bibliography: 70, 618.
References: 2106, 2969–2971, 3899, 3923, 4262, 6484.

15.2.5 Elimination of Fringes from Film Samples

Interference fringes may be observed on spectra obtained from unsupported films. These may be eliminated by

(i) casting the polymer film directly on the salt plate;
(ii) coating the film with immersion liquid and pressing it on the salt plate;
(iii) a slight delustring of the film by a very fine grinding paper.

15.2.6 Reducing Scatter from the Surface of Samples

Some polymers, if they are finely powdered, drawn, rolled, or cut, may display surface scattering of a great deal of infrared radiation. This effect can be reduced by putting the specimen in a suitable immersion liquid, e.g. liquid paraffin (Nujol) or hexachlorobutadiene or chlorofluorocarbon oil (Kel-F oil). These immersion liquids are transparent over a wide range (Fig. 15.7 and Appendix G), and they do not affect most polymers. The polymer specimen

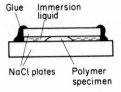

Fig. 15.12 A polymer specimen mounted in an immersion fluid to reduce scattering from the surface

is mounted between two salt plates, as is shown in Fig. 15.12. A liquid protein glue (Secotine) is recommended for sealing. Glues which contain phenol (recognizable by the smell) should be avoided.

Bibliography: 405, 618.

15.2.7 Preparation of Polymer Gel Samples for Spectroscopic Examination

Insoluble polymer gels can be examined by squeezing the gel between salt plates or pressing between silver chloride plates. The sandwich obtained is seldom uniform in thickness, but adequate spectra can be obtained. It is also recommended that the gel be swollen in a suitable solvent and only then should the swollen gel be squeezed between the plates. The solvent bands can be compensated for by using a variable path cell in the reference beam compartment of an infrared spectrometer.

Bibliography: 618.
References: 5306, 6183.

15.2.8 Preparation of Fibres for Spectroscopic Examination

Techniques for the preparation of natural and synthetic fibres are divided into two groups:

  (i) techniques which destroy the fibre form
    (a) potassium bromide discs
    (b) solvent cast films
    (c) mulls
    (d) hot-pressed films
    (e) cold-pressed discs
    (f) pyrolysis and hydrolysis
  (ii) techniques which retain the fibre form:
    (a) microspectroscopy of a single fibre
    (b) grid or pad of fibres
    (c) attenuated total reflection.

Bibliography: 76, 405, 608, 926, 1249, 1390, 1441, 1445, 1462.

233

15.2.9 Microspectrometry of a Single Fibre

A single fibre may be examined using a reflecting microscope attachment. The diameter of the fibre is usually too great for this technique, and the thickness must be reduced by microtoming, grinding (which does not affect orientation), or rolling (which affects the orientation).

For microtoming the fibre is embedded in a suitable wax, e.g. paraffin wax, collodion, ester wax, or polymethylmethacrylate, and is then further cut with a glass knife with a 90° cutting edge.

Bibliography: 70, 1249.
References: 2106, 2451, 3310, 3496, 3899, 4262, 6637.

15.2.10 Grid or Pad of Fibres

To obtain a parallel assembly of fibres one uses a simple U-shaped metal clamp (Fig. 15.13(a)) or alternatively the fibres can be packed between two salt plates (Fig. 15.13(b)). The dry fibres give satisfactory spectra, but the

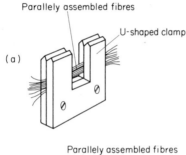

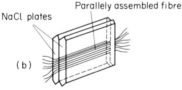

Fig. 15.13 Method for clamping fibres for infrared measurements: (a) in U-shaped metal clamps; (b) between two salt plates

surface scatters the infrared radiation. This effect can be reduced by wetting the fibres with an immersion liquid.

Bibliography: 1249.
References: 3031, 3456, 5282, 5509, 6128, 6469, 6470.

15.2.11 Liquid Polymers

Liquid polymers (e.g. silicone oils) are investigated in the form of liquid films. Drops of the liquid to be examined are placed between two salt plates which

234

are then inserted in a metallic frame. A salt plate of double thickness is placed in the reference beam. For low viscosity, liquid cells are used—as in the case of solutions.

15.2.12 Solutions

The selection of the correct solvent for infrared spectroscopy is difficult because all solvents exhibit a specific absorption. Spectra of the most common commercially produced solvents for infrared spectroscopy are shown in Fig. 15.14 and Fig. 15.15, and also in Appendix G.

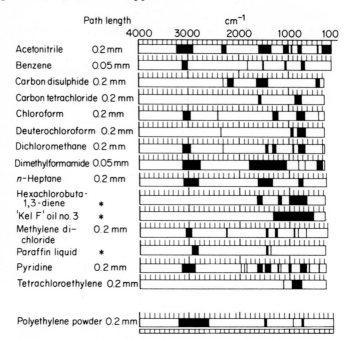

Fig. 15.14 Transparent regions of different commercially produced solvents. (The white windows in the chart show the regions in which solvents transmit more than 25% of the incident light.) (From B: 1471; reproduced by permission of BDH, England.) The chemicals indicated by an asterisk in the 'path length' column were analysed using a liquid film between caesium iodide plates. The data for n-heptane and methylene dichloride have been added by the author

Double-beam infrared spectrophotometers provide the possibility of automatic elimination of a part of the solvent absorption. To this end a cell filled with pure solvent is used in order to compensate for the solvent's specific absorption. Such compensation is not possible in the range of very strong absorption. It is most important to use anhydrous solvents because water absorbs strongly at 3710 cm^{-1} (2.7 µm) and 1630 cm^{-1} (6.15 µm).

The concentration of the polymer in the solution is 3–10% (sometimes even as high as 20%).

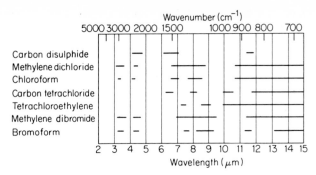

Fig. 15.15 Transparent regions of halogen-containing solvents. (The open regions between lines transmit more than 25% of the incident light at 1 mm thickness.) (From B: 1213; reproduced by permission of John Wiley & Sons)

There are three types of cell in general use:

(i) sealed permanent cells (seldom used for polymer solutions);
(ii) dismantlable cells (Fig. 15.16);
(iii) variable thickness cells (Fig. 15.17).

Liquid cells consist of two windows made from salt plates separated by a spacer of suitable thickness (0.1–0.01 mm) which also limit the volume of the

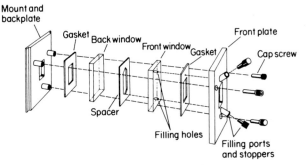

Fig. 15.16 Diagrams of a typical commercial dismantable liquid cell. (From R: 1470; reproduced by permission of Beckman Instruments)

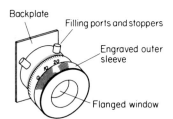

Fig. 15.17 Variable pathlength cell. (From B: 1470; reproduced by permission of Beckman Instruments)

sample. A variety of different cells are commercially available. The solution is placed in the cell with a capillary dropper or hypodermic syringe.

The variable thickness cell has no spacer. One window is fixed in the cell body and the other is moved in a piston seal by an accurately graduated micrometer screw. Variable thickness cells are specially designed to facilitate solvent compensation and to obtain exactly the required thickness of solution samples. Particular care must be taken in cleaning this type of cell after use and to avoid damage of the cell windows by overclosing.

If infrared cells are not carefully handled and maintained, they deteriorate rapidly (cf. Section 15.1.3).

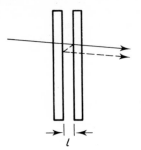

Fig. 15.18 Formation of interference fringes between closely spaced cell windows

The path length of a cell (l) can be determined as follows:

(i) From measurement of the thickness of the spacer with a dial gauge micrometer.
(ii) From interference fringes (Fig. 15.18) using

$$l = n/2\Delta\bar{v} \qquad (15.1)$$

(in centimetres), where
  n is the number of complete fringes in the wavenumber interval, $\Delta\bar{v}$,
  $\Delta\bar{v}$ is the wavenumber interval (in cm^{-1})—this value is constant for a given cell path length.

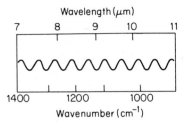

Fig. 15.19 Infrared spectrum of an empty cell

Figure 15.19 shows the interference pattern for an empty cell. To obtain distinct interference fringes, the cell windows must be flat; measurements must be made immediately after repolishing of the windows. Since cell thicknesses change with time because of gradual compression

237

of the spacer or erosion of the windows, they must be remeasured periodically.

(iii) By use of the interferometric method.

Bibliography: 1066.
References: 4281, 6633.

15.2.13 Multipass Cell

Multipass cells are used for infrared analysis of fractions collected from gas chromatography. Three mirrors in the cell (Fig. 15.20) give up to 16 passes (a distance of 1 m) through the vaporized sample. The mirrors must be gold

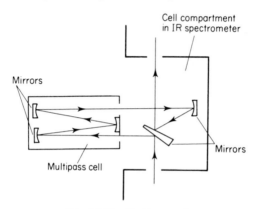

Fig. 15.20   Multipass cell

coated for maximum reflectivity and minimum deterioration over a period of time. The cell is electrically heated to 250 °C in order to vaporize the sample. Using this type of a cell, a 10 ml sample containing 1 % of polymer component may be examined. The cell can be connected directly to the gas chromatograph and, further, can be accommodated in an infrared spectrometer.

References: 3023, 7066.

15.3 QUALITATIVE APPLICATIONS OF INFRARED SPECTROSCOPY IN POLYMER RESEARCH

The applications of infrared spectroscopy to polymer characterization include the following:

(i) Structure determination. The most useful region of the infrared for structure determination is the characteristic GROUP-FREQUENCY REGION. There are several ways in which group-frequency data can be represented and tabulated. Figure 15.21 presents a comprehensive chart of wavelength of absorption organized according to the class of compound. After the infrared spectrum has been obtained, the strongest bands are selected and an attempt is made to identify the functional

238

Fig. 15.21(a)

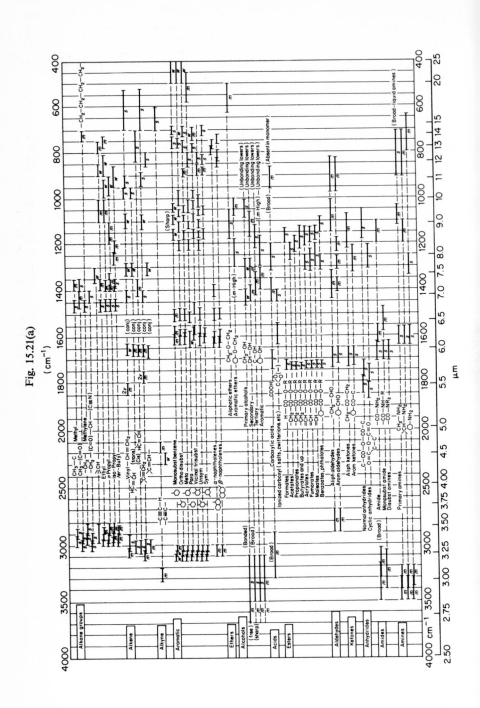

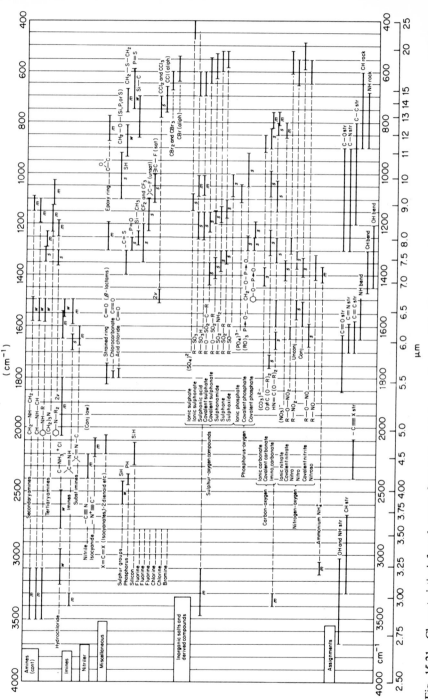

Fig. 15.21 Characteristic infrared group frequencies listed according to class of compound. Intensity of band: s = strong, m = medium, w = weak. Overtones bands are marked 2ν. (From **R**: 2940; reproduced by permission of the American Institute of Physics)

groups corresponding to those bands. This can result in several problems since characteristic frequencies often overlap. Therefore other information must be used, such as the relative intensities of various bands, other spectroscopic data, or any other chemical or physical data that are available. After the strongest bands have been assigned, an attempt should be made to assign the bands of moderate or even weak intensity. This can also cause a number of problems because of such complications as overtones and combination vibrations. It is useful to examine spectra of related model compounds. Finally, a series of possible models should be assumed by selection rules and compared with those observed. A complete vibrational analysis is difficult. In Table 15.1 are collected references to papers devoted to the vibrational analysis of different polymers; these references can sometimes help to provide information about polymer configuration and conformation.

(ii) Structure determination in copolymers. Infrared spectroscopy can sometimes provide information about the sequence distribution and tacticity in copolymers. Interpretation of these spectra is difficult because the absorptions lie very close together and often overlap. It is also difficult sometimes to decide whether the differences observed in a given

Table 15.1 Vibrational analysis of different polymers by infrared spectroscopy

Polymer	Published in
Poly(acrylonitryl)	R: 2150, 4908, 4915, 6297
Poly(butadiene)	B: 565; R: 2407, 2409, 3370, 3702, 4678, 6400
Poly(1-butene)	R: 2882, 4996, 5445, 6891, 7247
Poly(chloroprene)	R: 3364, 5156, 5477
Poly(ethylene)	B: 780; R: 2204, 2227, 2228, 2605, 4007, 4054, 4742, 4932, 5209, 5281, 5316, 5485, 5574, 6353, 6461, 6553, 6745, 6792, 6793, 7305
Poly(ethylene-2, 6-naphthalene)	R: 5639
Poly(ethylene terephthalate)	B: 780; R: 2191, 2462, 3776, 4909, 5281, 5309
Poly(isoprene)	R: 2408, 2410, 3036, 3077, 3701, 6044
Poly(methyl acrylate)	R: 3850
Poly(methyl methacrylate)	R: 2299, 5855
Poly(1-pentene) and poly(4-methyl-1-pentene)	R: 3566
Poly(propylene)	B: 780; R: 3438, 3743, 3746, 4212, 4910–4916, 4995, 5188, 5318, 5319, 5767, 6468, 6671, 6673, 6793
Poly(styrene)	B: 780; R: 4907, 5560, 5673, 6673, 6673, 6675, 6676, 7157
Poly(tetrafluoroethylene)	R: 2598, 4906, 5102, 5378, 7320
Poly(urethanes)	R: 2524, 6344, 6725
Poly(vinyl acetate)	R: 3851, 3852
Poly(vinyl alcohol)	B: 780; R: 4741, 4913, 5408, 6665, 6666, 6674, 6678, 6679, 6681
Poly(vinyl chloride)	B: 780; R: 2205, 3319, 3777, 4733, 4736–4740, 5307, 5604, 5344, 6297, 6364, 6373, 6746, 6748
Poly(vinylidene chloride)	B: 780; R: 4740, 5439, 5440

region in the spectra of a series of copolymers are due only to changes in copolymer composition or also to sequence distribution effects.

Bibliography: 578.
References: 2615, 3235, 3318, 3624.

15.4 QUALITATIVE ANALYSIS IN INFRARED SPECTROSCOPY

In infrared spectrophotometric practice, quantitative analysis is based on the application of the BEER–LAMBERT LAW, which is given by

$$A = \log_{10}(I_0/I) = acl \qquad (15.2)$$

(dimensionless), where

A is the absorbance,

I_0 is the intensity of the incident infrared radiation (or the radiation intensity passing through a reference cell),

I is the infrared radiation transmitted through the sample,

a is the ABSORPTION COEFFICIENT (in litres g^{-1} cm^{-1}),

c is the concentration of the solute (in grams per litre),

l is the path length of the sample (cm).

In order to correct for errors arising from partial scattering of the infrared radiation and for the overlap of neighbouring absorption peaks, a BASELINE

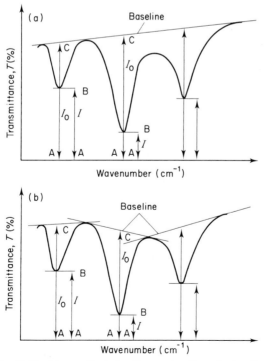

Fig. 15.22   Two methods for the construction of a baseline

must be constructed, as shown in Fig. 15.22, and I_0 determined as the transmitted intensity at the baseline.

The absorbance (A) at a given frequency (in cm^{-1}) is given by

$$A = \log_{10} (I_0/I) = \log_{10} (AC/AB). \tag{15.3}$$

For quantitative analysis one should choose a band in the region where the transmittance of the sample is constant and is not less than 25%.

There are several factors which can affect the infrared spectrum:

  (i) band overlapping;
 (ii) variation in the crystalline-amorphous state of the polymer sample;
(iii) variation in the configuration of the monomer units in the polymer chain, e.g. the different stereochemical forms (syndiotactic and isotactic);
(iv) formation of hydrogen bonds;
 (v) intermolecular interaction between constituents in polymer blends or graft-copolymers.

These factors should be considered before calibration procedures are carried out. The standard and model systems used for calibration should be similar to the samples examined.

For comparative quantitative measurements it is necessary to use films of the same thickness, and this condition is difficult to satisfy. A METHOD OF INTERNAL STANDARD is frequently applied to eliminate the differences in thickness. A known quantity of substance, the absorption band of which is easy to determine (e.g. KIO_4, $Pb(CNS)_2$), is added to the powdered polymer or its solution. Comparison of the absorption of the examined polymer at a determined wavelength with that of the internal standard makes the quantitative determination possible without the necessity of measuring the layer thickness.

Bibliography: 1324, 1447.

15.5 MEASUREMENT OF THE DEGREE OF CRYSTALLINITY IN POLYMERS BY INFRARED SPECTROSCOPY

From infrared spectrophotometric measurement the degree of crystallinity (χ_c) is defined as

$$\chi_c = A_c/A_a, \tag{15.4}$$

where A_a and A_c are absorbances at a given frequency (in cm^{-1}) of the amorphous and crystalline peaks (Fig. 15.23).

The frequencies of useful absorption peaks of several polymers are given in Table 15.2.

Bibliography: 16, 713, 1427.
References: 2536, 3047, 3351, 4020, 5187, 5281, 5574, 5615, 6663, 6796, 6797, 7305, 7318.

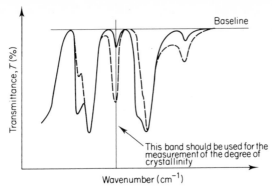

Fig. 15.23 Schematic amorphous (——) and crystalline
(- - -) infrared of a given semicrystalline polymer

Table 15.2 Absorption frequencies used for crystallinity determination by infrared methods
(From B: 1427; reproduced by permission of Academic Press)

Polymer	Crystalline (cm^{-1})	Amorphous (cm^{-1})
Poly(ethylene)	71, 730, 1050, 1178, 1889	1300, 1352, 1268
Poly(tetrafluoroethylene)	—	770
Poly(propylene), isotactic	809, 842, 894, 997	790, 1158
Poly(propylene), syndiotactic	866, 977	1131, 1199, 1230
Poly(vinyl alcohol)	1146	—
Poly(vinyl chloride)	955, 1226, 1254, 1373, 1428	2920
Poly(chlorotrifluoroethylene)	440, 490, 1290	754
Poly(2-chlorobutadiene)	952	—
cis-(2-Methylbutadiene)	1130	840
trans-(2-Methylbutadiene), α	800, 870, 890	840
trans-(2-Methylbutadiene), β	750, 800, 890	840
Poly(ethylene sebacate)	806, 970	720, 847
Poly(ethylene) terephthalate)	972, 848	790, 898, 1042
Nylon 6	835, 936, 964, 970, 1029, 1201	980, 990, 1076, 1118, 1170
Nylon 6.6	936	1139
Nylon 6.10	852, 940	1126
Cellulose	1372	2900
Poly[3, 3-bis(chloromethyl) oxacyclobutanel], α	525, 1318	—
Poly[3, 3-bis(chloromethyl) oxacyclobutane], β	525, 1310	—

15.6 INFRARED SPECTROSCOPY OF THE SURFACES OF POLYMER CRYSTALS

Infrared spectroscopy can sometimes be employed to study the nature of
chain folding at the surfaces of polymer crystals prepared from various
solutions. The ratio of the infrared absorbance at an amorphous band to that
at a crystalline band varies with the solvent used and with the thermal history
of the polymer.

The average number of monomer units per fold in polymers having double bonds can be calculated from the infrared expoxidation data.

References: 2226, 4016, 4633, 4639, 4735, 6255, 6564, 6565.

15.7 MICROSPECULAR REFLECTANCE MEASUREMENT

For quality control of films on reflective surfaces such as protective plastic coatings, varnishes, lacquers on cans, or epoxial coatings it is common to apply the microspecular reflectance measurement. The equipment (Fig. 15.24) fits directly into the normal cell slide, which is screwed to the base of the spectrophotometer sample compartment. The transmission obtained is generally

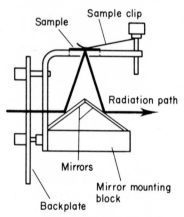

Fig. 15.24 Microspecular reflectance accessory. (From B: 1470; reproduced by permission of Beckman Instruments)

high enough for measurements on samples with a minimum diameter of 2 mm. Radiation is directed on to the sample at 20° incidence by a simple roof prism of surface-aluminized plane mirrors. The sample rests, with its reflective surface downwards, on an aperture plate covered with a protective coating of polytetrafluoroethylene. Samples may be gravity-held or retained in place by the spring clip attached to the sample plate. Typical transmission values for the various aperture plates for a surface-aluminized mirror are as follows:

> 2 mm diameter aperture, 28–30% transmission;
> 5 mm diameter aperture, 68–70% transmission;
> 10 mm diameter aperture, 80–81% transmission.

15.8 INTERNAL REFLECTION SPECTROSCOPY

INTERNAL REFLECTION SPECTROSCOPY (also called ATTENUATED TOTAL REFLECTANCE SPECTROSCOPY—ATR) is a technique which is based on internal reflection. The incident radiation is directed first into a material of high refractive index (n_1) at an angle (θ) greater than the critical angle (θ_c), and after slight penetration beyond the surface is reflected from the sample—which has lower refractive index (n_2) (Fig. 15.25).

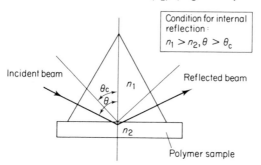

Fig. 15.25 Condition for a single internal reflection

When a polymer sample which selectively absorbs radiation is placed in contact with a reflecting surface, some frequencies of the incident beam will be absorbed while other frequencies are transmitted and reflected. This attenuated radiation is measured and plotted as a function of wavelength by a spectrometer, and will give an absorption spectrum of the sample, called an INTERNAL REFLECTION SPECTRUM.

It is possible to multiply the amount of absorption by multiplying the number of reflections (Fig. 15.26). An internal reflector plate is approximately 50 mm × 50 mm × 2 mm thick, with a 45° angle of incidence (θ), giving approximately 25 reflections.

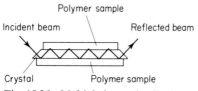

Fig. 15.26 Multiple internal reflections

Internal reflection spectroscopy can be applied to the ultraviolet, visible, and infrared regions. In Tables 15.3 and 15.4 are listed materials used for internal reflection plates applied to different spectral regions.

Most infrared spectrophotometers can be used with a MULTIPLE INTERNAL REFLECTION (MIR) ATTACHMENT (also called an ATR ATTACHMENT or ATR UNIT) which is placed in the sampling compartment. There are many different optical systems used in MIR (ATR) attachments:

(i) The two-mirror optical system (Fig. 15.27(a)): nine reflections with standard crystal; angle of incidence 45°. The crystal remains in the same position relative to the optical path irrespective of the sample thickness. Simplified design makes this unit most useful for routine laboratory work.

Table 15.3   Characteristics of materials for internal reflection
spectroscopy used in the visible and ultraviolet regions

Material	Transmission range (μm)	Refractive index (n_1) at 400 μm
Calcium tungstate	300–3500	1.91
Barium titanate	400–6500	2.40
Strontium titanate	400–4000	2.65
Crystal quartz	200–2500	1.56
Flint glass	200–1500	1.45
Sapphire	200–6500	1.78

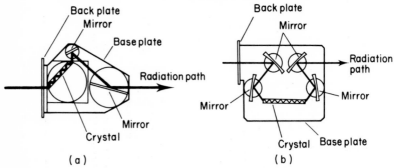

(a)   (b)

Fig. 15.27   Multiple internal reflection (MIR/ATR) attachments. (a) A two-mirror
optical system; (b) a four-mirror optical system. (From B: 1470; reproduced by
permission of Beckman Instruments)

(ii) The four-mirror optical system (Fig. 15.27(b)): 25 reflections with
standard crystal; angles of incidence 30°, 45°, and 60°. This system
offers three simply altered depths of penetration and considerably
intensifies the spectra of low-absorption samples.

Bibliography: 76, 571, 572, 768, 1369, 1374, 1389.

Internal reflection spectroscopy is used in

(i) qualitative identification of a wide variety of polymer samples, e.g.
films, adhesives, paper and paper coatings, powders, paints, fibres,
and foams;

(ii) studies of monomolecular layers;

(iii) molecular orientation studies (polarized internal reflection spectroscopy) in polymer films and in drawn fibres;

(iv) determination of optical constants;

(v) studies of oxidation and/or decomposition of polymer surfaces;

(vi) studies of contamination of surfaces by machine process conditions,
by human handling, or by a container;

(vii) studies of diffusion into a polymer sample and exudation of components to the polymer surface;

(viii) quantitative analysis of polymer samples.

Bibliography: 18, 76, 334, 572, 1389.
References: 2684, 2780, 3112, 4564, 6062, 6687, 6689, 7069.

Table 15.4 Characteristics of materials for internal reflection spectroscopy used in the infrared region

Material	Transmission range (mμ)	Refractive index, n_1, at 10 mμ	Critical angle θ (deg)	Desirable characteristics	Undesirable characteristics
Germanium (Ge)	2 –11.5	4.00	22	It has the highest refractive index of all materials and is completely insoluble	Brittle and fractures easily under pressure
Silicon (Si)	0.5– 6.2	3.42	26	It has the second highest refractive index and is a very hard and inert crystal	Oxidizes in air
Thalium-bromide–iodide (TlBr–TlI) (KRS–5)	0.5–35	2.37	39	It does not fracture under high pressure	High toxicity. It should not be used with water solutions
Silver chloride (AgCl)	0.4–23	1.98	49	Reflector plates are easily prepared from the rolled sheet using only a razor blade. It is soft and easily deformed	It should not be left in contact with a metal surface because of corrosion
Zinc selenide (ZnSe) (Irtran–4)	1.0–18	2.4		It is insoluble, and thus ideal for all liquid sampling applications.	Inordinately expensive
Cadmium telluride (CdTe) (Irtran–6)	2.0– 2.28	2.8			Inordinately expensive

15.9 REFLECTION–ABSORPTION INFRARED SPECTROSCOPY

REFLECTION–ABSORPTION SPECTROSCOPY is a technique which is based on infrared radiation polarized parallel to the plane of incidence. If the angle of incidence (θ) (Fig. 15.28) is only a few degrees less than 90°, the incident and reflected rays combine to establish a standing wave with appreciable amplitude at the surface of the metal mirror; this wave is capable of interacting with adsorbed film molecules.

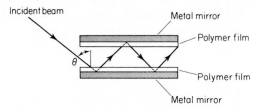

Fig. 15.28 Sample arrangement for obtaining multiple reflection at an angle θ from opposed coated mirrors

This type of infrared spectroscopy is very different from conventional infrared spectroscopy because of the high absorption intensity for parallel polarized radiation and the high angles of incidence which are not based on the geometric path length of the infrared radiation in the organic film.

The FRACTIONAL CHANGE IN REFLECTIVITY (ΔR) due to the presence of a polymer film on a reflecting metal mirror is given for $0° \leqslant \theta \leqslant 80°$

$$\Delta R = \frac{\varepsilon c d}{n^3}\left(\frac{9.212 \sin^2 \theta}{\cos \theta}\right), \tag{15.5}$$

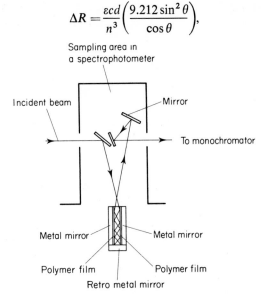

Fig. 15.29 Sample arrangement for reflection–absorption spectroscopy using multiple reflections from metal mirrors

where

 ε is the molar absorptivity (also called the molar extinction coefficient) for the polymer,

 c is the concentration of the absorbing polymer,

 d is the thickness of the film,

 n is the refractive index of the absorbing polymer,

 θ is the angle of incidence.

Sample arrangement for reflection–absorption infrared spectroscopy is shown in Fig. 15.29. The reflection attachment can be mounted in any common infrared spectrophotometer. Infrared radiation polarized parallel to the plane of incidence can be obtained by an AgBr wire grid polarizer which is placed in front of the entrance slit of the monochromator.

References: 2465, 3475, 3764–3766.

15.10 FOURIER TRANSFORM (FT) INFRARED SPECTROSCOPY

FOURIER TRANSFORM (FT) INFRARED SPECTROSCOPY is a technique which uses an interferometer (instead of a monochromator), e.g. a Michelson interferometer (Fig. 15.30). The interferometer consists of two plane mirrors at a right angle to each other and a beam splitter at an angle of 45° to the mirrors. One mirror is fixed in a stationary position and the other can be moved in a direction perpendicular to its front surface at a constant velocity.

The beam splitter divides the incoming light from the source: 50% is transmitted and 50% is reflected. It consists of a thin film of coating deposited on an optically flat support material (Table 15.5). A second equal thickness of this

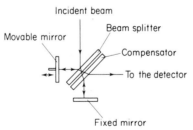

Fig. 15.30 Diagram of the Michelson interferometer

Table 15.5 Mid- and near-infrared beam splitters

Coating	Substrate	Wavelength range (μm)
Fe_2O_3	Quartz	0.65–2.5
Fe_2O_3	CaF	1.0– 5.0
Ge	KBr	2.7–25.0
Ge	CsI	10.0–50.0

support material (called the compensator) is placed in one arm of the interferometer to equalize the optical path lengths in both arms.

If the incoming radiation is monochromatic, the detector signal (or interferogram) goes through a series of maxima (the two light beams will be in phase when they return to the beam splitter) and minima (the two light beams will be out of phase when they return to the beam splitter). If the mirror is continuously moved the signal will oscillate from maximum to minimum for each quarter-wavelength movement of the mirror (Fig. 15.31(a)).

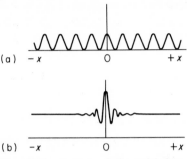

Fig. 15.31 Output of the Michelson interferometer as a function of mirror displacement (x) for: (a) a monochromatic source; (b) a broadband source. Zero refers to the mirror position for equal optical length in both arms of the interferometer

If the incoming radiation is polychromatic, the detector signal or interferogram is composed of the resultant signal for each frequency present in the incoming radiation. Each input frequency can be treated independently and hence the output (Fig. 15.31(b)) will be the sum of all cosine oscillations caused by all the optical frequencies in the incoming polychromatic radiation.

The interferogram also contains information on the intensity of each frequency in the spectrum.

The output information from the detector is digitalized in a computer and transformed to the FREQUENCY DOMAIN—each individual frequency is filtered out from the complex interferogram (FOURIER TRANSFORMS). Then the signals are converted into a conventional infrared spectrum. A block diagram of the basic components of a FT infrared spectrometer is shown in Fig. 15.32.

An entire spectrum can be recorded, computerized, and transformed in a few seconds. (In normal infrared measurements, spectra require several minutes to be recorded.)

Fourier transform (FT) infrared spectroscopy is used in

(i) detection of weak signals;
(ii) studies on samples at very low concentrations (0.5% concentration, 20 μg sample);

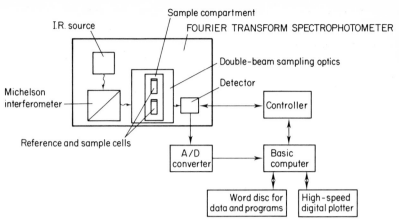

Fig. 15.32 Block diagram of the basic components of a FT infrared spectrometer

(iii) studies on adsorbed monolayers (e.g. a trace of ink on paper);
(iv) studies of infrared spectra of a single crystal (e.g. a crystal of benzene 300 μm in diameter);
(v) studies in aqueous solutions in the region between 950 and 1550 cm^{-1};
(vi) vibrational analysis (Table 15.6).
(vii) study of conformation-sensitive infrared bands.

Table 15.6 Vibrational analysis of different polymers by infrared Fourier transform spectroscopy

Polymer	Published in
Poly(acrylonitryle)	B: 302
Poly(butadiene)	R: 2927, 3916, 4417, 5741
Poly(chloroprene)	B: 302
Poly(ethylene)	B: 302; R: 5660, 5662
Poly(ethylene terephthalate)	R: 3141
Poly(propylene)	B: 302
Poly(styrene)	B: 302, 668; R: 5661
Poly(vinyl chloride)	B: 302

Bibliography: 73, 95, 302, 542–544, 639, 659, 668, 834, 840, 841, 883, 1330.
References: 2925–2928, 4129, 4632, 5661, 5663, 6663.

15.11 NEAR INFRARED SPECTROSCOPY

The NEAR INFRARED region is defined as the region from 12 500 to 4000 cm^{-1}. Absorption bands in the near infrared are principally overtones of low-frequency infrared bands (Fig. 15.33). Because of the complex overlapping bands, it is not easy to make assignments to the near infrared.

For measuring near infrared spectra it is possible to modify conventional infrared spectrophotometers by replacing the NaCl/KBr prism with a prism made from fused silica, quartz, lithium fluoride, or calcium fluoride and adding

Wavelength (μm)

		0.8	1.0	1.2	1.4	1.6	1.8	2.0	2.2	2.4	2.6	2.8	3.0
C—H absorptions	Terminal —CH₂ {Vinyloxy (—OCH=CH₂), Other}	⊢	⊢ ⊢		⊢0.02	⊢	⊢0.3 ⊢0.3		⊢0.2 ⊢0.2–0.5				
	Terminal —CH(O)CH₂				⊢	⊢	⊢0.2		⊢1.2				
	Terminal —CH(CH₂)CH₂				⊢	⊢			⊢				
	Terminal ≡CH		⊢			⊢1.0							⊢50
	cis —CH=CH		⊢					⊢0.15					
	>C(CH₂)(CH₂)O (oxetane)			⊢	⊢		⊢		⊢ ⊢		⊢⊢		
	—CH₃	⊢		⊢0.02	⊢		⊢0.1		⊢0.3				
	>CH₂			⊢0.02	⊢		⊢0.1		⊢0.25				
	>C—H			⊢		⊢	⊢						
	—CH aromatic	⊢			⊢0.1	⊢0.1		⊢					
	—CH aldehyde		⊢						⊢0.5				
	—CH (formate)							⊢1.0					
N—H absorptions	—NH₂ amine {Aromatic, Aliphatic}	⊢	⊢	⊢0.4	0.2⊢⊢1.4 ⊢0.5			⊢1.5 ⊢0.7				30⊢ ⊢30 1–5⊢ ⊢2	
	>NH amine {Aromatic, Aliphatic}	⊢ ⊢	⊢ ⊢			⊢0.5 ⊢0.5		0.5				⊢20 ⊢1	
	—NH₂ amide				0.7⊢ ⊢0.7			3⊢⊢0.5				100⊢ ⊢100	
	>NH amide					⊢0.5		⊢0.4				⊢50–100	
	>NH imide		⊢			⊢							
	—NH₂ hydrazine		⊢		0.5⊢⊢0.5			⊢					⊢
O—H absorptions	—OH alcohol				⊢2			⊢			⊢50		
	—OH hydroperoxide {Aromatic, Aliphatic}				1⊢⊢1 ⊢2			⊢1.3 ⊢0.8			30⊢⊢30 ⊢80		
	—PH phenol {Free, Intramolecularly bonded}				⊢3 ⊢			⊢			⊢200 Variable ⊢		
	—OH carboxylic acid				⊢							⊢10–100	
	—OH glycol {1.2, 1.3, 1.4}				⊢ ⊢ ⊢					50⊢⊢50 20–50⊢⊢20–100 50–80⊢⊢5–40			
	OH water				⊢0.7		⊢1.2				30⊢⊢7		
	≡NOH oxime				⊢						⊢200		
	HCHO (possibly hydrate)										⊢		
Other functional group absorptions	—SH							⊢0.05					
	>PH							⊢0.2					
	>C=O							⊢					⊢3
	—C≡N							⊢0.1					

Fig. 15.33 Summary of functional group correlations in the near infrared region. (Numbers alongside band locations are molar absorptivities.) (From B: 519, reproduced by permission of John Wiley & Sons)

a more sensitive detector. Many commercially produced ultraviolet and visible spectrophotometers are designed to cover the near infrared region.

Sample cells are made of a special grade of silica or quartz and they have path lengths between 0.1 and 10 cm.

The selection of the correct solvent for near infrared spectroscopy is difficult because it is necessary to avoid those containing O—H or N—H and even C—H groups. In Fig. 15.34 are shown the transmittance characteristics of some solvents suitable for near infrared spectroscopy. The most common are carbon tetrachloride (this is an ideal near infrared solvent), carbon disulphide, and methylene chloride.

Polymer films to be examined must be relatively thick.

The applications of near infrared spectroscopy to polymer characterization include the following:

(i) Quantitative organic functional-group analysis. The more important functional groups that have been studied are: C—H groups—terminal methylene and methyne groups give especially well-developed near infrared spectra; N—H groups—the fundamental stretching region at

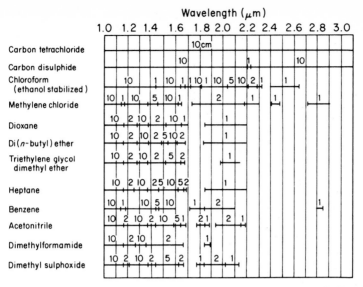

Fig. 15.34. Transmittance characteristics of some near infrared solvents. (Solid lines indicate usable regions; numbers indicate maximum desirable path lengths in centimetres.) (From B: 519; reproduced by permission of John Wiley & Sons)

2.8–3.0 µm, the first overtone region, near 1.5 µm, and the second overtone region, near 1.0 µm; O—H groups—the fundamental stretching region at 2.7—3.0 µm; C═O groups—the first overtone of carbonyl groups at 2.8—3.0 µm; S—H, P—H, nitriles, and other various groups.

(ii) Hydrogen-bonding studies.

(iii) Solute–solvent interaction studies.

Bibliography: 519, 716, 717, 1324.
References: 3247, 3457, 3472, 3473, 5276, 7115.

15.12 FAR INFRARED SPECTROSCOPY

The FAR INFRARED region is strictly defined as the region from 200 to 10 cm^{-1}. Many spectroscopists recognize the region below 650 cm^{-1} as the far infrared, but it can be defined as part of the mid-infrared (4000–200 cm^{-1}).

The 650–200 cm^{-1} region can be routinely recorded with commercially available spectrometers. Beyond 200 cm^{-1} specially designed spectrometers using gratings and transmission filters must be applied.

The windows in the sample cells are made of

(i) caesium iodide—used to about 50 µm;

(ii) crystal quartz—used beyond 50 µm (it is opaque from about 4.5 to 45 µm);

(iii) polyethylenes, which in most cases are free of bands beyond 25 µm, but they have weak absorption at about 50 and 150 µm;

(iv) diamond—which is completely transparent throughout the far infrared.

Water vapour should be completely removed from samples and the cell by flushing with dried air or nitrogen or by evacuation; widespread absorption due to the rotational bands of water occurs in the region of 50 μm.

The application of far infrared spectroscopy to polymer characterization includes the region 650–200 cm^{-1}, which is especially important for aromatic-, halogen-, phosphorous-, and organometallic-containing polymers. Bands in this region are sensitive to overall changes in molecular structure, i.e. conformation and tacticity, and molecular motion.

Bibliography: 101, 654, 693, 938.
References: 2108, 2226, 2385, 2791–2793, 3073, 3099, 3393, 3485–3488, 3697, 3698, 3793, 4178, 5320, 6107, 6592, 6744, 6747, 6673, 7114.

Chapter 16

Emission Spectroscopy

General bibliography: 75, 92, 133, 179, 180, 212, 242, 274, 328, 451, 463, 550, 610, 705, 791, 897, 952, 991, 1009, 1086, 1105, 1170, 1362, 1367, 1380, 1384.

The energy emitted as LUMINESCENCE (fluorescence and/or phosphorescence) is derived from the energy absorbed from the incident light.

Most molecules have a SINGLET GROUND STATE (S_0) (Fig. 16.1). The only excited states which may be reached directly by light absorption are EXCITED SINGLET STATES ($S_1, S_2, S_3, ..., S_n$). S_1 differs from S_0 in many

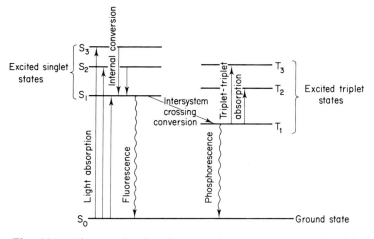

Fig. 16.1. Diagram showing the most important processes involving electronically excited states

ways and should be regarded as a chemically distinct species. The spins of electrons are paired in excited singlet states ($S_1, ..., S_n$). The lifetime of an excited singlet state (S_1) is about 10^{-8}–10^{-10} s.

By the INTERSYSTEM CROSSING CONVERSION (a non-radiative process occurring with spin inversion) a molecule from excited singlet state (S_1) can reach the FIRST TRIPLET STATE (T_1). The lifetime of a triplet state (T_1) is from minutes to 10^{-3} s.

The HIGHER TRIPLET STATES ($T_2, T_3, ..., T_n$) may be formed only when a molecule in its lowest triplet state (T_1) absorbs a new photon. Such a process is called TRIPLET–TRIPLET ABSORPTION.

An electronically excited molecule can lose its energy by emission of radiation which is known as luminescence. There are the following kinds of emission:

(i) FLUORESCENCE is a radiative emission process occurring from the lowest excited singlet state (S_1) to the singlet ground state (S_0) (Fig. 16.1);

(ii) PHOSPHORESCENCE is a radiative emission process occurring from the lowest excited triplet state (T_1) to the singlet ground state (S_0) (Fig. 16.1);

(iii) EXCIMER FLUORESCENCE is a radiative emission process occurring as a result of the decomposition of the EXCIMER COMPLEX formed between the lowest excited singlet state (S_1) and the singlet ground state (S_0) (Fig. 16.1) (cf. Section 16.1);

(iv) DELAYED FLUORESCENCE is a radiative emission process occurring from the decomposition of the EXCIMER COMPLEX formed between excited triplet states (T_1) (cf. Section 16.1).

Relations between absorption and emission spectra are shown in Fig. 16.2.

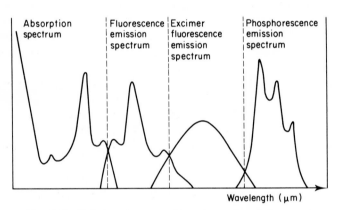

Fig. 16.2 Relations between absorption and emission spectra

An electronically excited molecule can also lose its energy by non-radiative processes such as the following:

(i) INTERNAL CONVERSION is a non-radiative process between two different electronic states of the same multiplicity, i.e. $S_n \rightarrow S_1$, or $T_n \rightarrow T_1$. In this process an electronically excited molecule can lose its energy by conversion to heat.

(ii) NON-RADIATIVE ENERGY TRANSFER is a process of transformation of energy from a singlet (or triplet) excited molecule (DONOR) to a molecule in the singlet ground state (ACCEPTOR).

(iii) DISSOCIATION is a process whereby excitation energy is used to stimulate the decomposition of a molecule into free radicals.

The UNIMOLECULAR LIFETIME (τ) of an excited state is defined in general as the time in seconds required for the concentration of molecules in that state to decay to $1/e$ of its initial value; e is the base of natural logarithms.

The INHERENT RADIATIVE LIFETIME (τ_0) of an excited state is defined as the time in seconds required for the concentration of molecules in that state to decay to $1/e$ of its initial value if no radiationless processes occur.

(i) for fluorescence emission,

$$\Phi_f \, \tau_{0(f)} = \tau_f; \tag{16.1}$$

(ii) for phosphorescence emission,

$$\tau_p \left(\frac{1 - \Phi_f}{\Phi_p} \right) = \tau_{0(p)}; \tag{16.2}$$

where

τ is the unimolecular lifetime (measured time) (in seconds,)
τ_0 is the inherent radiative lifetime (in seconds),
Φ is the quantum yield of the emission process,
subscripts f and p mean fluorescence and phosphorescence, respectively.

QUANTUM YIELD OF LUMINESCENCE is defined as the number of quanta emitted per exciting quantum absorbed.

The QUANTUM YIELD OF FLUORESCENCE (Φ_f) is defined as

$$\Phi_f = \frac{\text{Number of fluorescence quanta emitted}}{\text{Number of quanta absorbed to a singlet excited state}}. \tag{16.3}$$

The QUANTUM YIELD OF PHOSPHORESCENCE (Φ_p) is defined as

$$\Phi_p = \frac{\text{Number of quanta phosphorescence emitted}}{\text{Number of quanta absorbed to a singlet excited state}}. \tag{16.4}$$

The quantum yield is the most difficult characteristic to measure.

16.1 EXCIMERS AND EXCIPLEXES

An EXCIMER is a molecular dimer aggregate formed between an excited molecule in the lowest excited singlet state (S_1) and a molecule in the singlet ground state (S_0). Excimers are unstable in their ground state but are stable under electronic excitation:

$$S_1 + S_0 \rightarrow (S_1 \ldots S_0). \tag{16.5}$$

During the decomposition of an EXCIMER COMPLEX one may observe EXCIMER FLUORESCENCE, which differs from 'normal' fluorescence:

$$(S_1 \ldots S_0) \rightarrow S_0 + S_0 + \text{EXCIMER FLUORESCENCE}. \tag{16.6}$$

Excimer fluorescence shows a broad characteristic spectrum without vibrational structure (Fig. 16.2) and is dependent on temperature.

Excimers may also result from reactions of two excited triplet states (T_1) (BIMOLECULAR TRIPLET–TRIPLET INTERACTION):

$$T_1 + T_1 \rightarrow (T_1 \ldots T_1). \tag{16.7}$$

During the decomposition of such an excimer one may observe DELAYED FLUORESCENCE, which also differs from normal fluorescence:

$$(T_1 \ldots T_1) \rightarrow S_0 + S_0 + \text{DELAYED FLUORESCENCE.} \tag{16.8}$$

DELAYED FLUORESCENCE is an emission which has the spectral properties of fluorescence but with a much longer rise and decay time than ordinary fluorescence. Frequently the decay time is of the same order of magnitude as the phosphorescence decay time. Delayed fluorescence depends upon the square of the incident light intensity because a triplet state (T_1) arises only from an initial excitation via an excited singlet state (S_1) and two such events are required to obtain fluorescence of this type. Delayed fluorescence is sensitive to oxygen, which quenches the triplet state (T_1) with great efficiency.

Excimers are formed in concentrated polymer solutions or in the solid state. In order for an excited molecule to form an excimer with another molecule, its counterpart should approach to within a distance of 3–4 Å from the excited molecule during the period of excitation. The probability of formation of an excimer is also dependent upon the spatial alignment of a pair of molecules and the parallel alignment for rigid planar aromatic rings in polymers.

An EXCIPLEX is a well-defined complex which exists only in electronically excited states. Exciplexes are formed between excited donor molecules (D^*) and acceptor molecules (A) or excited acceptor molecules (A^*) and donor molecules (D):

$$D^* + A \rightarrow (D^- A^+)^* \leftrightarrow (D^+ A^-)^* \rightarrow D^* A \leftrightarrow DA^*; \tag{16.9}$$

$$A^* + D \rightarrow \underbrace{(A^- D^+)^* \leftrightarrow (A^+ D^-)^*}_{\substack{\text{Charge-transfer} \\ \text{complex}}} \rightarrow A^* D \leftrightarrow AD^*. \tag{16.10}$$

The fluorescence emission from an exciplex is dependent on solvent polarity. With increasing solvent polarity there is a decrease in the fluorescence quantum yield.

Triplet exciplexes can be identified in several cases by measuring phosphorescence and by triplet–triplet absorption.

Bibliography: 138, 464, 465, 528, 974, 1242.
References: 2041, 2414, 3019, 3035, 3066–3069, 3451, 3465, 3466, 3479, 3630, 3910, 4065, 4142, 4244–4246, 4611, 4612, 5095, 5195, 5507, 5835, 6015, 6016, 6482, 6753–6756, 6905, 7002, 7240, 7271.

16.2 FLUORESCENCE

FLUORESCENCE is a radiative emission process occurring from the lowest excited singlet state (S_1) to the singlet ground state (S_0) (Fig. 16.1).

The FLUORESCENCE SPECTRUM should be plotted with fluorescence intensity (I_f) (in arbitrary units) on the ordinate and with wavelength (λ) (in nanometres or ångströms) on the abscissa (Fig. 16.3).

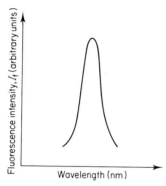

Fig. 16.3 Fluorescence emission spectrum for a given molecule

The FLUORESCENCE INTENSITY (I_f) from the solution can be calculated from the Beer–Lambert law:

$$I_f = I_0(1 - 10^{-\varepsilon cl})\,\Phi_f, \qquad (16.11)$$

where

I_f is the fluorescence intensity (in quanta per second),
I_0 is the intensity of the incident light,
ε is the molar absorptivity (molecular extinction coefficient),
c is the concentration of solute (in moles per litre),
l is the path length of the sample (in centimetres),
Φ_f is the quantum yield of fluorescence.

The absolute quantum yield of fluorescence is very difficult to determine because the fluorescence quanta are not monochromatic; they are unequally distributed in direction and are liable to reabsorption within the solution. Extra complications are due to polarization and refraction.

Measurement of the RELATIVE FLUORESCENCE YIELD is simple, especially when it is possible to compare the substance to be examined with a standard fluorescence compound which has a known quantum yield of fluorescence (Φ_{fs}) (Table 16.1). Quinine sulphate and quinine bisulphate dissolved in aqueous 1 N H_2SO_4 are widely used fluorescence standards.

The fluorescence yield (Φ_f) of an unknown compound can be calculated from the following equation:

$$\Phi_f = \frac{I_f\,\varepsilon_f\,c_s}{I_{fs}\,\varepsilon c}\,\Phi_{fs}, \qquad (16.12)$$

where

I_f and I_{fs} are the fluorescence intensities of the unknown compound and the standard compound, respectively,
ε and ε_s are the molar absorptivities of the unknown compound and the standard compound, respectively,

c and c_s are the concentrations of the unknown compound and the standard compound in solution.

Φ_{fs} is the fluorescence yield of the standard compound.

Table 16.1   The fluorescence standard compounds (from B: 1086; reproduced by permission of John Wiley & Sons)

Compound	Solvent	ϕ_{fs}
Fluorescein	Aqueous carbonate–bicarbonate buffer, pH 9.6 at low concentration	0.85
Rhodamine B	Ethanol at low concentration	0.69
Quinine bisulphate	0.1 N H_2SO_4, low concentration	0.55 (0.46)
2-Naphthol	Aqueous solution, 10^{-3} M in 0.05 M borate buffer, pH = 10	0.21
Proflavine	Aqueous solution, 10^{-4} M in 0.05 M acetate buffer, pH = 4	0.27

Measurements of relative fluorescence yield must be done in the same cell and with the same monochromatic beam of incident light. The geometry of the irradiation region, the positioning of the spectograph slit, and the optical limit of the fluorescence light entering into the detector system must be the same.

Fluorescence spectra obtained in the form of the response of a phototube which is dependent upon frequency are termed APPARENT FLUORESCENCE SPECTRA. A correction must be made to the observed phototube response to obtain values which are proportional to the number of quanta per second. (This is a major problem in the determination of the relative fluorescent quantum yield.)

TRUE FLUORESCENCE SPECTRA (proportional to the quanta per unit frequency interval) are derived from the apparent spectra by dividing the ordinates at each frequency of the apparent spectra by the sensitivity factor of the detector system for that frequency.

The areas under the true fluorescence spectral curves of the two solutions (unknown compound and standard compound) are proportional to the total quanta of fluorescence (I_f and I_{fs}).

The fluorescence quantum yield of the unknown compound can be obtained directly from the equation

$$\Phi_f = \Phi_{fs} \frac{I_f}{I_{fs}} = \frac{\text{Area under the true fluorescence emission curve for unknown compound}}{\text{Area under the true fluorescence emission curve for standard compound}} \cdot \Phi_{fs}. \quad (16.13)$$

Many types of modern commercial fluorescence spectrophotometer measure true fluorescence spectral curves.

The following molecular processes may compete with fluorescence:

(i) Internal conversion ($S_1 \to S_0$).
(ii) Intersystem crossing conversion ($S_1 \to T_1$).

(iii) Non-radiative energy transfer.
(iv) Quenching by impurities. It should be noted here that fluorescence measurements must be made with extremely pure solvent and solutes.
(v) Quenching by oxygen. For that reason the fluorescent intensity should be determined for a given solution both in an inert gas (e.g. nitrogen or argon) and when saturated with oxygen.
(vi) Concentration quenching and excimer formation.

The fluorescence intensity decays after withdrawal of the exciting source according to a first-order rate equation (exponentially with time):

$$I_f = I_{f(0)}\, e^{-t/\tau}, \tag{16.14}$$

where
$I_{f(0)}$ is the initial fluorescence intensity at time $t = 0$,
I_f is the fluorescence intensity at some later time (t),
t is the time,
τ is the mean lifetime (unimolecular lifetime) of the excited singlet state,
e is the base of natural logarithms.
After integration of eq. (16.14), we obtain

$$\ln I_f = \ln I_{f(0)} - (t/\tau). \tag{16.15}$$

The plot of $\ln I_f$ versus time (t) yields the mean lifetime (χ) (Fig. 16.4).

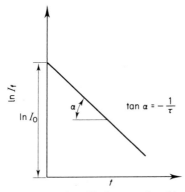

Fig. 16.4   Plot of $\ln I_f$ versus time (t)

Bibliography: 92, 179, 435, 549, 796, 923, 997, 1086.

16.3 INSTRUMENTATION FOR FLUORESCENCE EMISSION SPECTROSCOPY

16.3.1 Fluorescence Spectrophotometers

The fluorescence spectrophotometer for measuring true fluorescence spectral curves (Fig. 16.5) is built of the following parts:

(i) The light source, which consists of a deuterium discharge lamp for wavelengths below 375 nm and a xenon lamp for the region from 300 to 300 nm. Instead of a xenon lamp, some spectrophotometers use a medium- or high-pressure mercury lamp.

(ii) Monochromators. High brightness and resolution are required for monochromators applied in fluorescence spectrometers.

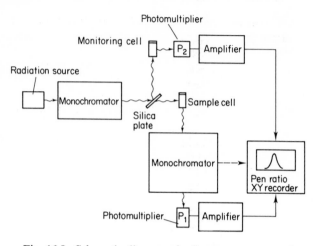

Fig. 16.5   Schematic diagram of a fluorescence spectrophotometer

(iii) An optical-monitoring system.

    (a) A small portion of the high-intensity, monochromatic excitation light beam is reflected by the clear matched silica plate on the monitoring cell, which contains fluorescence solution (e.g. rhodamine B).

    The fluorescence solutions chosen for the monitoring cell should, ideally, have a constant fluorescence yield and absorb substantially the whole beam to a depth of about 1 mm or less over the whole spectral range concerned.

    The fluorescence emitted is further monitored by the photomultiplier (P_2). After amplifying, the output of P_2 is fed to a ratio recorder.

    (b) The main portion of the initial light beam passes through the clear matched silica plate and through the cell at right angles to that part of the apparatus which measures the fluorescence emission.

    The fluorescence emission measuring device from the sample cell consists of the analysing monochromator (the same type as for the excitation monochromator) fitted with a photomultiplier (P_1). After amplification, the output of the photomultiplier (P_1) is fed to a ratio recorder.

(iv) A data recording system. The signal from a ratio recorder is fed to a pen which records the corrected true fluorescence intensity (I_f) against wavelength (λ) and gives true fluorescence spectra.

Bibliography: 242, 836, 949, 1008, 1086, 1380.
References: 3017, 3286, 4810, 4965, 6087, 7137, 7289.

16.3.2 Fluorescence Lifetime Measuring Methods

Fluorescence lifetime (fluorescence decay) measurements of the order of microseconds or milliseconds do not lead to serious difficulties. Most common fluorescent substances have lifetimes ranging from 10^{-8}–10^{-10} s and measurements of fluorescence decay in this time region is difficult.

The techniques for the measurement of fluorescence decay times of the order of nanoseconds are divided into two groups:

(i) pulse methods;
(ii) phase shift methods.

Bibliography: 949, 1361.

16.3.3 The Single-photon Method

The MONOPHOTON or SINGLE-PHOTON TIME CORRELATION METHOD measures the time of emission of individual fluorescence photons. The reference for zero time is the initial rise of the flash-lamp or an electrical pulse related in time to the flash-lamp discharge. The photon counting rate is in the range 100–200 counts s^{-1}. The time of arrival of each photon, with reference to a fixed time zero, is measured electronically by a time-to-amplitude

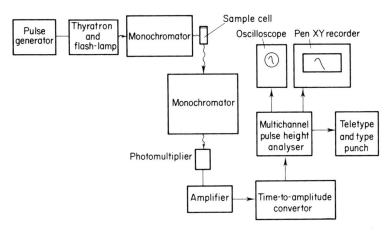

Fig. 16.6 Schematic diagram of a single-photon time correlation instrument used for fluorescence lifetime measurements

264

(TAC) circuit, and the resultant pulse height information is stored in a multi-channel pulse height analyser (MCPHA). The contents of the MCPHA memory can be read on to a punched tape or an XY recorder or typewriter. A block diagram of a typical single-photon instrument is shown in Fig. 16.6. The equipment required is of the type used in nuclear spectroscopy and fast coincidence measurements.

Bibliography: 484, 1361, 1435.
References: 2475, 4834, 5017.

16.3.4 The Pulse-sampling Oscilloscope Method

The PULSE-SAMPLING OSCILLOSCOPE METHOD measures the fluorescence lifetime by using a nanosecond flash-lamp, a fast photomultiplier, and a pulse-sampling oscilloscope (Fig. 16.7). The instrument functions by measuring

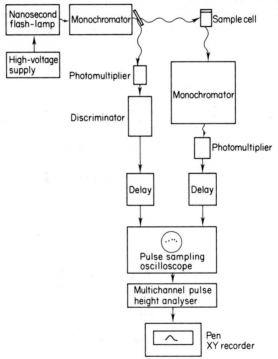

Fig. 16.7 Schematic diagram of a pulse-sampling oscilloscope used for fluorescence lifetime measurements

the amplitude at a series of times measured from a fixed time zero. Each time the transient to be measured triggers the system, it measures the amplitude at a time Δt later than the previous measurement. The result is displayed on a much slower time base as a series of dots on the oscilloscope screen.

Bibliography: 1361.
References: 4184, 6556.

16.3.5 The Phase-shift Method

The PHASE-SHIFT METHOD for determining fluorescence lifetime is based on the fact that a molecule excited using a sinusoidally modulated light signal will yield fluorescence of the same frequency, but phase-shifted with respect to the exciting light.

The fluorescence lifetime (τ) is determined from the measured phase difference between the exciting and emitted light at a known angular frequency (ω) of the exciting light, using

$$\tan \theta = \omega\tau, \tag{16.16}$$

where
θ is the phase angle,
ω is the frequency of the modulated exciting light,
τ is the fluorescence lifetime.

The phase-shift method requires a modulated light source and a phase-sensitive detector. A block diagram of a typical phase-shift instrument is shown in Fig. 16.8.

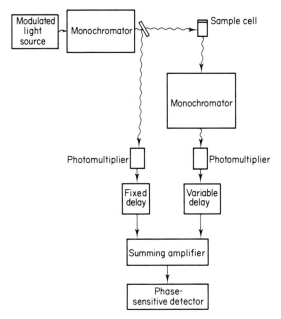

Fig. 16.8 Schematic diagram of a phase-shift instrument for fluorescence lifetime measurements

The modulated light source is provided either by modulation of the applied potential across a discharge lamp or by passing a steady light beam through a device with variable optical density. The following devices or methods are used for light modulation:

(i) Kerr cells;
(ii) electro-optic (solid state) light modulators;
(iii) ultrasonic modulators;
(iv) amplitude-modulated RF gaseous discharge;
(v) RF discharges operated at the desired frequency.

The following methods are used as for phase detection:

(i) methods based on phase shift or comparison at the modulating frequency;
(ii) methods based on a phase comparison after frequency conversion.

The phase-shift method permits measurement of fluorescence lifetimes in the range 10^{-9}–10^{-10} s.

Bibliography: 132, 179, 1160, 1361.
References: 2194, 2415, 2416, 2562, 2570, 2811, 4400, 4988, 5243, 6227–6229, 6494, 6591, 7011.

16.4 POLARIZED FLUORESCENCE

The polarization of fluorescence can be characterized by an optical coordinate system (x, y, z) (Fig. 16.9):

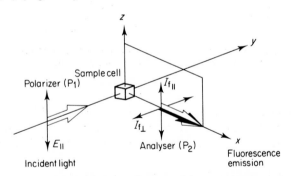

Fig. 16.9 Optical coordinate system for measurement of the degree of fluorescence polarization (P_f)

(i) the linearly polarized exciting light passes along the y axis and has its electric vector (P_1) in the direction of the z axis;
(ii) the polarized component of fluorescence is measured through an analyser in the direction of the y axis with the electric vector of the component (P_2) lying in the z–x plane.

The DEGREE OF FLUORESCENCE POLARIZATION (P_f) is defined as follows:

$$P_f = \frac{I_{f_\parallel} - I_{f_\perp}}{I_{f_\parallel} + I_{f_\perp}}, \qquad (16.17)$$

where

$I_{f_\parallel}$ is measured when the polarizer (P_1) and the analyser (P_2) are set parallel ($P_1 \| P_2$),

$I_{f_\perp}$ is measured when the polarizer (P_1) and the analyser (P_2) are set perpendicular ($P_1 \perp P_2$).

The polarized fluorescence may be measured by the following optical system (Fig. 16.10), which consists of the following components:

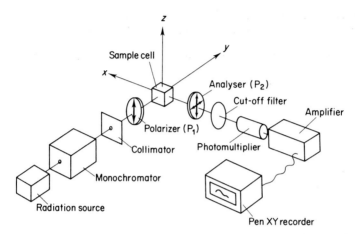

Fig. 16.10   Optical system for measuring the degree of polarization of fluorescence of a polymer solution or melt

  (i) A light source—a xenon mercury lamp. The original beam, after passing a quartz condensing lens, is monochromated by the grating or the quartz prism monochromator and polarized with the ultraviolet polarizer (P_1).
 (ii) The sample cell for the polymer solution or melt.
(iii) Fluorescence emitted from the sample passes through the analyser (P_2) and a cut-off filter, and is further measured by a photomultiplier. After amplification, the output of the photomultiplier is fed to a recorder.
 (iv) The polarization axes of the polarizer (P_1) and the analyser (P_2) are aligned either parallel or perpendicular for the measurement of $I_{f_\parallel}$ or $I_{f_\perp}$, respectively.

The fluorescence polarization method is applied to the study of

  (i) molecular orientation in polymer solids (cf. Section 35.11);
 (ii) relaxation times associated with the motion of polymer molecules (cf. Section 16.5).

Bibliography: 796, 974.
References: 2029, 5498, 5503, 5512.

16.5 STUDY OF MOLECULAR MOTIONS BY THE FLUORESCENCE METHOD

The rotational relaxation time for micro-Brownian motion in polymer solutions and melts is in the range 10^{-12}–10^{-5} s.

The lifetime of excitation for a fluorescent molecule is usually in the range 10^{-10}–10^{-7} s.

The lifetime of excitation can be utilized as a reference time scale for the measurement of relaxation times of molecular motions in polymer systems. The fluorescent groups can be attached to the polymer molecule or simply dispersed into a polymer system depending upon the purpose of the investigation.

If the relaxation time (ρ) of the molecular motions in the system becomes comparable to the lifetime of excitation (τ), the directions of the molecular axes of the excited molecules will be randomized by the rotational Brownian motion during the lifetime of excitation, and partial depolarization of fluorescence occurs.

For a system in which optically anisotropic fluorescent molecules are randomly dispersed, the extent of this rotational depolarization of fluorescence can be related to the rotational motions of the molecule by the following equation:

$$P_{f0}/P_f = 1 + A(\tau/\rho), \qquad (16.18)$$

where

P_f is the degree of polarization defined by eq. 16.17,

P_{f0} is the limiting degree of polarization attainable when the rotational motion of the molecule has completely ceased ($\rho \to \infty$),

A is a constant, dependent on the optical system used during the measurement,

τ is the lifetime of the fluorescence (in seconds),

ρ is the relaxation time of the molecular motion (in seconds).

The fluorescence yield (Φ_f) of molecules capable of internal rotation is markedly dependent upon the viscosity and temperature of the medium. This phenomenon can be interpreted as the INTERNAL QUENCHING OF FLUORESCENCE due to the rotation of a portion of the molecule with respect to the rest of the molecule (deviating from the planar conformation during the lifetime of excitation). For such molecules, the variation of the fluorescence yield (Φ_f) can be expressed in terms of the relaxation time of the internal rotation (ρ') by

$$\Phi_{f0}/\Phi_f = 1 + B(\tau/\rho'), \qquad (16.19)$$

where

Φ_f is the fluorescence yield,

Φ_{f0} is the limiting fluorescence yield attainable when the internal rotation has ceased ($\rho' \to \infty$),

B is the proportionality constant,

τ is the lifetime of the fluorescence (in seconds),

ρ' is the internal rotation time (in seconds).

The measurement of the degree of polarization is described in Section 16.4, whereas measurement of the fluorescence yield is described in Section 16.2.

Bibliography: 974.
References: 5499–5501, 5619.

16.6 PHOSPHORESCENCE

PHOSPHORESCENCE is a radiative emission process occurring from the lowest excited triplet state (T_1) to the singlet ground state (S_0) (Fig. 16.1).

Phosphorescence emission is generally observed only in viscous media or low-temperature glasses. In fluid media the phosphorescence emission is very low because of two reactions:

(i) TRIPLET QUENCHING is a radiationless reaction between an excited triplet state (T_1) and a quenching molecule (Q) (e.g. oxygen, impurities),

$$T_1 + Q \rightarrow S_0 \text{ (or products).} \tag{16.20}$$

(ii) TRIPLET–TRIPLET ANNIHILATION is a reaction between two excited triplet states (T_1) (bimolecular triplet–triplet interaction) which produces a triplet excimer:

$$T_1 + T_1 \rightarrow (T_1...T_1). \tag{16.21}$$

During the decomposition of such an excimer one observes a delayed fluorescence,

$$(T_1...T_1) \rightarrow S_1 + S_0 + \text{DELAYED FLUORESCENCE,} \tag{16.22}$$

which decreases the phosphorescence quantum yield.

If extreme care is taken with solvent purification and outgassing, it is possible to measure phosphorescence decay in fluid media.

The PHOSPHORESCENCE SPECTRUM should be plotted with the phosphorescence intensity (I_p) (in arbitrary units) on the ordinate and with the wavelength (λ) (in nanometres or ångströms) on the abscissa (Fig. 16.11).

The phosphorescence intensity decays after withdrawal of the exciting source according to a first- and second-order rate equation:

$$\frac{dI_p}{dt} = k_1 I_p + k_2 I_p^2, \tag{16.23}$$

where
k_1 and k_2 are different rate constants for different processes,
I_p is the phosphorescence intensity.

Bibliography: 92, 242, 1086.
References: 4024, 4339, 6505.

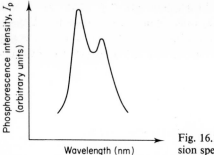

Fig. 16.11 Phosphorescence emission spectrum for a given molecule

16.7 INSTRUMENTATION FOR PHOSPHORESCENCE EMISSION SPECTROSCOPY

16.7.1 Phosphorescence Spectrophotometers

In order to separate phosphorescence and fluorescence from each other it is necessary to use a phosphoroscope (rotating slotted cylinder) (Fig. 16.12) or choppers with disc drive (Fig. 16.13) in which the beam of the incident light

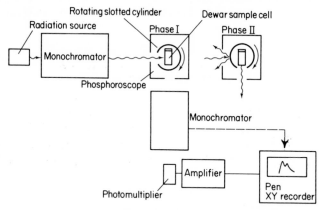

Fig. 16.12 Schematic diagram of a phosphorescence spectrophotometer in which a phosphoroscope is applied

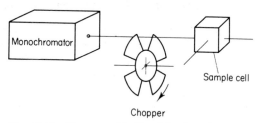

Fig. 16.13 Chopper with disc drive (rotating sector wheel)

and the beam of the emitted light are interrupted periodically. A phosphoroscope uses a rotating slotted cylinder which surrounds the Dewar cuvette containing the sample (Fig. 16.14). The sample is viewed by a detecting system only during darkness, when the short-lived fluorescence has completely decayed.

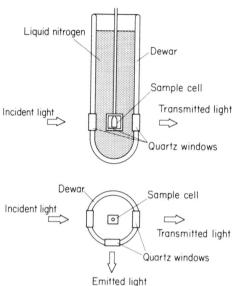

Fig. 16.14   Sample cell placed in a Dewar filled
with liquid nitrogen

The quartz Dewar flask which holds the sample has unsilvered strips which allow the passage of incident light and phosphorescence emission.

Other parts of the block system in a phosphorescence spectrophotometer (Fig. 16.12) are similar to those in a fluorescence spectrophotometer (cf. Section 16.3.1).

Bibliography: 1008, 1086, 1440.
References: 3950, 5749.

16.7.2 Measurement of the Phosphorescence Lifetime

Phosphorescence lifetime (phosphorescence decay) measurements in the time range of the order of $1-10^{-4}$ s do not cause any serious experimental difficulties.

Phosphorescence decay is commonly measured by the FLASH METHOD. A block diagram of a flash method instrument for phosphorescence lifetime measurements is shown in Fig. 16.15.

Commercial flash-lamps are available with decay times of less than 10^{-4} s and powers of a few joules. The intense flash of light passes into a sample cell and causes photoexcitation of molecules. The emitted phosphorescence light

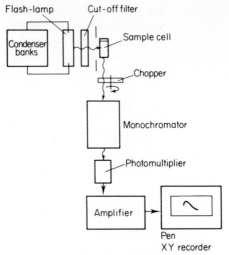

Fig. 16.15 Schematic diagram of a flash-lamp apparatus used for phosphorescence lifetime measurements

passes through a monochromator and is measured by a photomultiplier. The output of a photomultiplier is fed to an XY recorder after amplification.

Bibliography: 179, 242, 791, 1086.

16.8 FLASH KINETIC SPECTROSCOPY

FLASH KINETIC SPECTROSCOPY is a method for the study of triplet states and flash photolysis of molecules. When a sudden flash of high-intensity light is absorbed by reactant molecules, a relatively large concentration of free radicals or other transient species is formed.

Flash kinetic spectroscopy can also be applied to the study of the primary processes in photochemical systems (e.g. triplet–triplet absorption, decay of triplet excited state (T_1), etc.).

A typical flash photolysis apparatus with measuring spectrophotometer is shown in Fig. 16.16. The new species, formed by the flash of light in the reaction

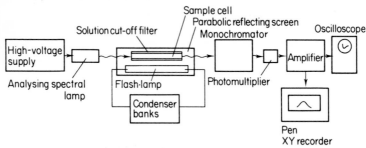

Fig. 16.16 Schematic diagram of a flash-photolysis apparatus

vessel, can be studied by recording the absorption spectra on photographic plates or by oscilloscope monitoring. The necessary light for analysing after flash exposure is produced by an additional spectrograph-analysing lamp. The photolysis lamp and the analysing lamp operate in a coupled system by which it is possible to control the time interval between the excitation from the photolytic flash and the appearance of the analysing light beam.

The DURATION OF THE FLASH (τ) is the time interval from $1/e$ of the maximum light intensity on the rising part of the curve to the corresponding point on the descending part of the transient absorption trace (Fig. 16.17).

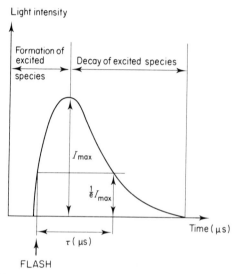

Fig. 16.17 Light intensity during a flash as a function of time

Considering a simple *RCL* circuit (Fig. 16.18), where

R is the resistance due to the lamp after discharge has started (it depends upon the length of the lamp, generally R is of the order 5 Ω or less),

C is the capacitance, which is variable but generally ranges from 1 to 20 μF,

L is the inductance due to the capacitor—for very short duration times the inductance must be very small to make it possible to maintain a large value for the capacitance (C),

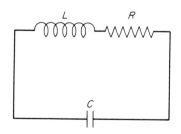

Fig. 16.18 *RLC* circuit

the duration of the flash (τ) is given by

$$\tau = RC. \tag{16.24}$$

For a 5 Ω resistance and a capacitor charged to 2 μF, the duration of the flash (τ) corresponds to 10 μs. For that reason the shortest lived species which can be studied under these conditions must have a lifetime somewhat in excess of 10 μs.

The energy obtained from the pulse is given by

$$E = \tfrac{1}{2}CV^2 \tag{16.25}$$

(in joules), where

C is the capacitance,

V is the voltage applied across the capacitor.

For a 2 μF capacitor charged to 10 kV, $E = 100$ J. For the same capacitor charged to 20 kV, $E = 400$ J. Typically the energies used in flash photolysis apparatuses range from 100 to 500 J.

Analysis of the flash kinetic data is based on the assumption that the photomultiplier response is linear with respect to the incident light intensity.

The absorption spectra of the new species formed are photographed at different time intervals after the photolysis flash on photographic plates mounted in the spectrograph or by photographic recording of oscilloscope traces. Absorption spectra may also be stored using magnetic memory.

The typical absorption spectrum traced on an oscilloscope during the formation and decay of a new species after a flash, measured only at one wavelength (λ), is shown in Fig. 16.19. The oscillograph deflection (x) is proportional to the light intensity (I):

$$I_0 = kx_0 \quad \text{and} \quad I = kx, \tag{16.26}$$

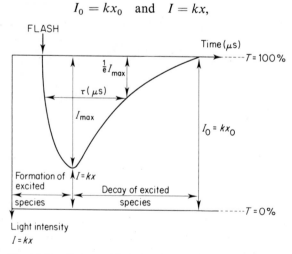

Fig. 16.19 Change of light intensity (oscilloscope deflection trace x) during the formation and decay of a new excited species measured at one wavelength (λ)

where k is a constant which depends on the sensitivity of the detector. However, k does not need to be evaluated because

$$A_\lambda = \log_{10}(I_0/I) = \log_{10}(x_0/x) = \varepsilon_\lambda cl, \qquad (16.27)$$

where

A_λ is absorption measured at wavelength λ,
ε_λ is the molar absorptivity at wavelength λ,
c is the concentration of the species formed,
l is the path length of the sample.

From eq. (16.27) the relative or absolute concentration of a new species which is formed and decays after a flash can be determined and used for kinetic calculations.

Bibliography: 1026, 1054, 1086.
References: 2342, 2564, 2586, 2804, 2866, 2870–2872, 2879, 3390, 4679, 5045, 5871, 7059.

16.9 NANOSECOND FLASH SPECTROSCOPY

The practical lower limit for the duration of the flash (τ) for the normal flash kinetic apparatus is about 1 µs. In order to study singlet state processes as well as other very reactive intermediates, it is necessary to apply nanosecond flash spectroscopy, which uses a Q-switched ruby laser (cf. Section 10.6.3). A diagram of such an apparatus is shown in Fig. 16.20. The light pulse from the Q-switched

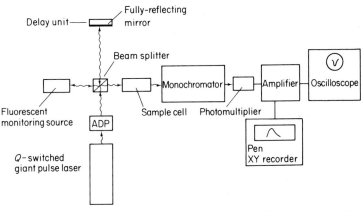

Fig. 16.20 Schematic diagram of the nanosecond flash spectrographic apparatus

ruby laser (vanadyl cyanine) passes through certain types of crystal (such as ammonium dihydrogen phosphate, ADP), resulting in frequency doubling. About 20% of the 694 nm radiation that enters the frequency doubler will emerge at a wavelength of 347 nm. The light then travels through a series of filters and a lens to a beam splitter. Part of the pulse is directed to the sample cell, the remainder travels through the beam splitter to a mirror and from there back. This light is then delayed relative to that which was split off to the sample cell, and the extent of the delay corresponds to twice the distance of the mirror from the beam splitter. After returning again to the beam splitter a portion of the light is reflected to a solution of a highly fluorescent compound (1,1′,4,4′-tetraphenylbuta-1,3-diene), which emits in continuously at a longer wavelength (400–600 nm) than the exciting pulse. This fluorescence pulse then passes through the beam splitter into the sample cell, and finally into the photomultiplier. The delay of this monitoring pulse relative to the exciting pulse can be up to 100 ns.

References: 3356, 3357, 3449, 3630, 4971, 5872, 5873.

16.10 CHEMILUMINESCENCE AND THERMOLUMINESCENCE

CHEMILUMINESCENCE is the emission of light resulting from certain chemical reactions which occur without light excitation. Detailed understanding of the excitation mechanism is restricted to a few systems, mainly those involving species in the gas phase and in solutions. In many cases chemiluminescence can be considered as an emission from vibrationally rather than electronically excited states.

Chemiluminescence can be observed during oxidative degradation of polymers as a result of a termination step involving recombination of free radicals.

THERMOLUMINESCENCE is the emission of light observed during the heating of a sample and resulting from some of the associated chemical reactions, e.g. thermal oxidation.

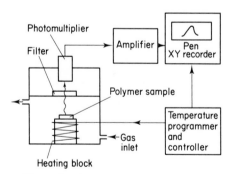

Fig. 16.21 Schematic diagram of an apparatus used for measuring chemiluminescence and/or thermoluminescence

Apparatus for studying both chemiluminescence and thermoluminescence is shown in Figure 16.21.

A polymer sample is placed on the surface of a thermostatically controlled heating block inside a light-tight box. The sample can be heated under various gaseous atmospheres, e.g. oxygen. The light emitted from the polymer sample passes through a glass filter to a photomultiplier. After amplification, the output of the photomultiplier is fed to a recorder.

Bibliography: 1086, 1103, 1104, 1201, 1362, 1372.
References: 2154, 2244, 3733, 3834, 4937, 5233, 5346, 5817, 5946, 5947, 6037, 6215, 7135.

16.11 APPLICATIONS OF EMISSION SPECTROSCOPY IN POLYMER RESEARCH

The applications of emission spectroscopy to polymer characterization include

(i) the study of molecular motions of macromolecules in solutions;
(ii) the study of the natural fluorescence of polymers and biopolymers;
(iii) the study of the interactions of polymers with dyes;
(iv) the study of the impurities present in commercial polymers;
(v) the study of the photodegradation and photostabilization of polymers;
(vi) the study of energy transfer processes in polymers;
(vii) the study of photosensitization processes (sensitized photopolymerization, photodegradation, photocrosslinking).

Bibliography: 22, 451, 469, 796, 977, 996, 1086, 1228.
References: 2067–2082, 2317, 2395, 2659, 2810, 3048, 3118, 3119, 3458, 3901, 4141, 4238, 4592, 4593, 5094, 5195, 5462, 5562, 5688, 6481, 6889, 6890.

Chapter 17

Raman Spectroscopy

General bibliography: 36, 71, 150, 191, 193, 212, 274. 313, 377, 431, 477, 514, 602, 604–606, 669, 693, 694, 761–763, 832, 835, 1045, 1154, 1286, 1296, 1297.

RAMAN SPECTROSCOPY deals with the phenomenon of a change in frequency when light is scattered by molecules. The RAMAN EFFECT arises from an exchange of energy between the scattering molecule and a photon of the incident radiation. In this process a transition of the molecule from one of its energy states to another state occurs, and a compensating change in the energy follows. The fundamental equation is

$$h\nu_0 + E_1 = h\nu_r + E_2, \qquad (17.1)$$

where

h is the Planck constant,

ν_0 is the frequency of the incident light (photon),

ν_r is the frequency of the scattered light (photon),

E_1 and E_2 are the initial and the final energies of the molecule.

The frequency shift $\nu_r - \nu_0 = \Delta\nu$ may be either positive or negative in sign. Its magnitude is referred to as the RAMAN FREQUENCY. The set of Raman frequencies of the scattering species constitutes its RAMAN SPECTRUM.

A Raman line of frequency $\nu_r < \nu_0$ is called a STOKES–RAMAN LINE.

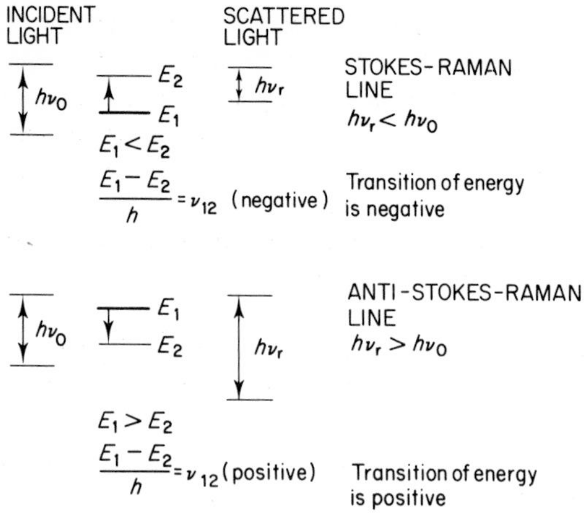

Fig. 17.1   Diagram for the most important processes involved in light scattering by molecules

A Raman line of frequency $v_r > v_0$ is called an ANTI-STOKES–RAMAN LINE (Fig. 17.1).

The STOKES LINES correspond to transitions in which the molecule is raised from a lower to a higher energy state and the photon loses its energy. The ANTI-STOKES LINES correspond to transitions in which the molecule drops from an excited state to a lower energy level and the photon increases its energy.

The RAMAN FREQUENCY SHIFT is expressed in terms of wavenumbers instead of frequencies:

$$ v = \frac{v_0 - v_r}{c} = \frac{E_2 - E_1}{hc}, \qquad (17.2) $$

where c is the velocity of light.

When the incident light consists of photons with energy $h v_0$, the photons colliding with a molecule may be

(i) ELASTICALLY SCATTERED, i.e. without change of energy, and produce the RAYLEIGH LINE;

(ii) INELASTICALLY SCATTERED.

(a) In an inelastic collision the molecule undergoes a quantum transition to a higher energy level, with the result that the photon loses energy and is scattered at a lower frequency (Δv negative).

(b) When the energy level of the molecule exceeds its lowest level, an encounter with a photon may produce a transition to a lower energy; in this case the photon is scattered at increased frequency (Δv positive).

RAMAN SHIFTS (RAMAN FREQUENCIES) are independent of the exciting frequency v_0 and are equivalent to the energy changes involved in the transition of the scattering species, and are therefore characteristic for this transition. The pattern on the low-frequency side of the exciting line (Δv negative) is 'mirrored' by an identical pattern on the high-frequency side (Δv positive). Intensities of Δv negative are greater than those of Δv positive, and the latter show a rapid decrease as $|\Delta v|$ increases. At temperature equilibrium the population of a higher level is consequently lower than that of a lower level and decreases exponentially with the energy.

RAMAN SHIFTS due to scattering phenomena correspond to vibrational or rotational transitions of the scattering molecules and can be observed in the visible region.

In Raman spectroscopy the photon is, on the whole, never observed, but it perturbs the molecules and induces them to undergo a vibrational or rotational transition.

The initial energy level (E_n) of a molecule (Fig. 17.1) must not be confused with the ground state level of a molecule (E_0).

When $E_n = E_m$ and hence $v_{nm} = 0$, the state of the scattering molecule remains unchanged and Rayleigh scattering is observed.

When $E_n \neq E_m$ we have the Raman effect. The necessary condition for Raman scattering is that the energy $h\nu_0$ of the incident photon must be greater than the energy difference $E_m - E_n$ between the final and the initial states of the actual transition.

17.1 INSTRUMENTATION FOR RAMAN SPECTROSCOPY

17.1.1 Raman Spectrometers

For liquid samples and light in the visible region, the total intensity of molecular scattering (including Rayleigh scattering) is of the order of 10^{-5} of the incident intensity, and of this only about 1 % may contribute to the Raman spectrum.

For gases, where the molecular density is lower, the intensity is correspondingly lower and, in general, larger sample volumes must be used. This imposes a requirement for intense sources of monochromatic light and spectrographic instruments of high luminosity.

The Raman spectrometer (Fig. 17.2) is built of the following parts:

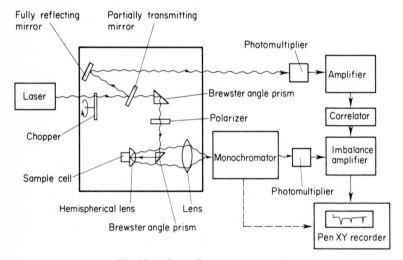

Fig. 17.2   Laser Raman spectrometer

(i) A light source. Polychromatic light sources cannot be applied, because each line excites its own Raman spectrum, and these could overlap each other. Application of filters is also limited due to the imperfect discrimination of most filters.

The Raman light source should emit strong, monochromatic radiation. For that reason specially designed mercury-arc lamps are in use, emitting the intense blue line at 4358 Å and the strong green line at 5461 Å. The chosen exciting line should not be absorbed by a sample and give fluorescence emission which would mask the Raman spectrum or cause photodecomposition of a sample under investigation. For that reason helium discharge lamps are sometimes used; these give intense lines at 5876 Å and 6678 Å which are used for the investigation of photosensitive samples.

In some new commercially produced Raman spectrometers, lasers are used as a light source (LASER RAMAN SPECTROSCOPY).

Most lasers generate one, or only a few, exciting lines which are highly monochromatic and can be emitted in a well-defined direction (Section 10.6).

The most powerful laser sources for Raman spectroscopy are: the He–Ne ion laser, with a strong emission line at 6328 Å; the argon ion laser, with strong emission lines at 4880 and 5145 A and weaker emission lines at 4579, 4765, 4965, and 5017 Å; the krypton ion laser, with emission lines at 6471, 5682, 5308, and 5208 Å. The application of lasers has greatly simplified the technique of Raman spectroscopy.

(ii) A sample optics system. In order to obtain a very high illumination of the sample and efficiently collect Raman radiation emitted from a small volume, a 90° (more common) or a 180° sample optics system is used (Figs. 17.4 and 17.6). The emitted radiation collected from the sample is passed to the monochromator entrance slit.

(iii) Monochromators. Almost all commercially produced Raman spectrometers use one or two grating monochromators (the Czerny–Turner type) in order to discriminate adequately between the weak Raman lines and the exciting radiation which enters the instrument.

(iv) A data recording system. The determination of line frequencies is best carried out on a photographic plate, but it is necessary to expose the sample for many hours. A photomultiplier is used for the determination of the intensities of the lines together with an amplifier and a recorder. For line intensity measurements when the intensity of incident light is fluctuating, it is necessary to record the ratio of the Raman signal and a reference signal. Most commercially made Raman spectrometers use either photographic or photoelectric recording systems. The spectra are recorded on a linear wavenumber scale.

Bibliography: 514, 603, 673, 832, 1297, 1420.
References: 4561.

17.1.2 Sampling Techniques

The following sampling techniques are applied for recording Raman spectra:

(i) Gases and vapours. Gas cells are designed as multiple-reflection cells (Fig. 17.3) in which the light beam can bounce back and forth, e.g. 100 times in the 25 mm long cell. The two plane reflectors are oriented at a slight angle ($\varphi \sim 5°$) with respect to each other. The incident beam makes an angle θ with the normal to the bottom reflector, and the first

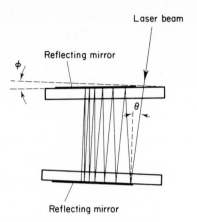

Fig. 17.3 Multiple reflection cell

reflected beam makes an angle of $\theta - \varphi$ with the normal to the top reflector. On the next reflection from the top plate, the angle is $\theta - 3\varphi$, etc. At each traverse, therefore, the beam gets closer to its path from the previous traverse and to the normal to one of the reflector plates. When the beam finally is perpendicular to either the top or bottom plate, it is reflected back on itself, retraces its path back through the cell, and exits at a point very close to its entrance point. In this way some 100 traverses, corresponding to about 1.5 m path length, occur within the cell.

(ii) Liquids. Liquid samples in large amounts can be examined directly in bottles or even in flasks using a 90° (Fig. 17.4(a)) or a 180° (Fig. 17.4(b))

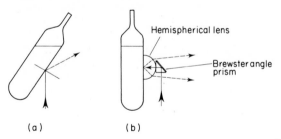

(a) (b)

Fig. 17.4 Examination of liquid in ampules under:
(a) a 90° and (b) a 180° illumination system

illumination system. Small liquid samples can be examined in glass or quartz capillaries in both 90° and 180° illumination systems (Fig. 17.5).

(iii) Solids. Solid samples can be examined using either a 90° or a 180° illumination system (Fig. 17.6). It is difficult to obtain good Raman spectra of solids. The solid scatters the exciting line along with the Raman scattering. Many advances have been made towards eliminating these problems. Factors involving the sample tube, sample crystal size,

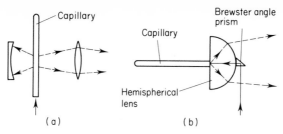

Fig. 17.5 Examination of liquids in capillaries under:
(a) a 90° and (b) a 190° illumination system

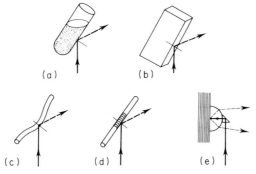

Fig. 17.6 Examination of solid samples under 90°
and 180° illumination systems. (a) Powder in a
bottle; (b) solid polymer block; (c) single fibre;
(d) fibre wound round a glass rod; (e) fibre in a
bundle

sample thickness, sample position, and sample preparation should be carefully considered. A single crystal scatters better than a fine powder. In the case of polymers, solid rods can be used. The KBr pellet technique is suitable for use with powders. Fibres, block specimens, and films can be studied without preparation.

It is possible to study the anisotropy of the Raman scattering from oriented specimens, e.g. drawn fibres or film.

Bibliography: 514, 602–604.
References: 4010.

17.2 POLARIZATION OF THE RAMAN EFFECT

Scattered light can be resolved into components $I_y(I_\perp)$, which is polarized perpendicular to the x–z plane, and $I_z(I_\parallel)$, which is polarized parallel to the x–z plane (Fig. 17.7).

The DEGREE OF POLARIZATION (ρ) of the scattered light is given by the ratio I_y/I_z or $I_\perp/I_\parallel$. ρ-Values always imply scattering at right angles to the direction of incident light.

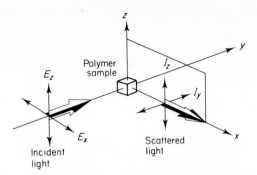

Fig. 17.7 Optical coordinate system for measurement of the degree of polarization of scattered light

When the scattering molecule is isotropic (e.g. carbon tetrachloride) the ρ-value is zero, i.e. the Rayleigh line is completely polarized.

The TOTAL INTENSITY OF THE SCATTERED LIGHT (I) is given by

$$I = I_y + I_z. \tag{17.3}$$

For study of the polarization of the Raman effect the incident light can be used either unpolarized ($E_z \neq 0$ and $E_x \neq 0$) or plane-polarized ($E_x = 0$). In this case the degree of depolarization is denoted ρ_n and ρ_p, respectively.

17.3 APPLICATIONS OF RAMAN SPECTROSCOPY IN POLYMER RESEARCH

Each scattering polymer species gives its own characteristic vibrational Raman spectrum which can be used for quantitative identification. Raman spectroscopy is not affected by the presence of water. In systems where chemical interactions occur, the presence of new molecular species can be detected by the appearance of new Raman lines. The intensity of a characteristic Raman line is proportional to the volume concentration of the species, which allows quantitative analysis. For the interpretation of an individual spectrum in terms of the molecular properties of the particular species, it is necessary to look more closely at the quantum-mechanical theory of the Raman effect.

Application of Raman spectroscopy to the study of polymers is partially limited by

(i) the low intensity of Raman scattering ($\sim 10^{-8}$ intensity of incident light) for many polymer samples;

(ii) the occurrence in many cases of a high background emission other than Raman, Tyndall, and Rayleigh scattering;

(iii) the presence of fluorescing impurities.

The applications of Raman spectroscopy to polymer characterization include

(i) study of the configuration and conformation of polymer and copolymer chains;

(ii) study of helix formation;

(iii) study of polymer crystals and interlamellar forces in crystals;

(iv) study of crystalline and amorphous orientation in polymers;

(v) study of texture (especially by using low-frequency Raman spectroscopy);

(vi) study of chain motions in solution;

(vii) study of polymer melts;

(viii) study of crosslinked polymers and gels;

(ix) study of the effect of stress applied to polymers;

(x) study of degradation processes.

In Table 17.1 are collected examples of the Raman vibrational analyses of different polymers.

Table 17.1 Vibrational analysis of different polymers by Raman spectroscopy

Polymer	Published in
Poly(butadiene)	R: 2927, 2975, 3332
Poly(ethylene)	R: 2463, 2597, 3344, 3490, 3491, 3582, 4006, 4007, 5486, 6463, 6600,
Poly(ethylene oxide)	R: 3267, 3917, 5161
Poly(ethylene terephthalate)	R: 2462, 5928
Poly(isoprene)	R: 2976
Poly(oxymethylene)	R: 5943
Poly(propylene)	R: 5878, 6216, 6793, 6930
Poly(vinyl chloride)	R: 2265, 4638, 4925, 6073–6075
Poly(vinylidene fluoride)	R: 4624
Poly(tetrafluoroethylene)	R: 2464, 3886, 4635, 5731, 5944
Ethylene–propylene copolymer	R: 3492

Bibliography: 42, 74, 149, 151, 211, 492, 762, 898.
References: 2198, 2463, 2466, 2523, 2739, 2766, 2772, 2972, 3173, 3267, 3335, 3493, 3494, 3626, 3727, 3918, 3919, 3949, 4009, 4627, 5113, 5160, 5158, 5159, 5162, 5225, 5592, 5746, 5878, 5927, 5929, 6217, 6463, 6464, 6465, 6489, 6490, 6599, 6612, 6777, 6778, 6930, 7286, 7319.

17.4 COMPARISON OF INFRARED AND RAMAN SPECTROSCOPY

Infrared and Raman spectroscopy are both techniques for investigating the vibrational spectra of polymers. However, these two spectroscopic methods are fundamentally different processes. There are some investigations for which each technique is uniquely applied; in other cases, infrared and Raman spectroscopy are nicely complementary. As a result, any laboratory involved in polymer characterization should be equipped with both infrared and Raman spectro-

Table 17.2 Common intensity differences between infrared and Raman spectra (from B: 211; reproduced by permission of Marcel Dekker Inc.)

Strongly infrared active †		Strongly Raman active ‡		Strong in both	
Vibration	Frequency range (cm^{-1})	Vibration	Frequency range (cm^{-1})	Vibration	Frequency range (cm^{-1})
C = O stretch	1600–1800	Aromatic C—H stretch	3000–3100	Aliphatic C—H stretch	2800–3000
C—O stretch	900–1300	C = C stretch	1600–1700	C ≡ N stretch	2200–2300
O—H stretch (H-bonded)	3000–3400	C ≡ C stretch	2100–2250	Si—H stretch	2100–2300
Aromatic CH out-of-plane deformation	650– 850	S = S stretch (Se–Se, etc.)	400– 500	C–Halogen stretch	500–1400
N—H stretch (H-bonded)	3100–3300	C—S Stretch (C–Se, etc.)	600– 700		
Si—O—Si antisymmetric stretch	1000–1100	Aromatic ring breathing	950–1050		
		Aromatic C = C in-plane vibration	1500–1700		
		N = N stretch	1575–1630		
		Si—O—Si symmetrical stretch	450– 550		

† In general this group includes vibrations of asymmetric groups, bending modes, and stretching of polar bonds.
‡ In general this group includes vibrations of symmetrical groups and stretching of bonds, particularly non-polar or slightly polar bonds.

meters so that the maximum amount of information about polymer structure may be obtained from the vibrational spectra.

A few differences between the two spectroscopic methods should be taken into consideration (Table 17.2):

(i) The most intense infrared bands are usually those arising from polar bonds such as O—H, N—H, and C=O, while the most intense Raman bands are usually those arising from bonds having nearly symmetrical charge distribution, such as C—C, C=C, and S—S, usually found in the substituents. Thus, in practice, infrared spectroscopy is usually preferred for studying the substituents in a polymer, while Raman spectroscopy is preferred for studying the conformation of a polymer.

(ii) The infrared absorption of water is quite intense, whereas Raman scattering of water is minimal, and for that reason Raman spectroscopy is especially useful for the study of aqueous solutions (e.g. the study of changes in the conformation of biological macromolecules in aqueous solution as a function of pH, ionic strength, and temperature).

(iii) Sample handling in Raman spectroscopy is much simpler than in infrared spectroscopy.

Infrared and Raman spectroscopy are complementary techniques.

Bibliography: 74, 211, 1224.
References: 2462, 2927, 3030, 3492, 4624, 6492, 7007.

17.5 INTRACHAIN VIBRATIONS OF POLYMERS

Any one helical polymer molecule has an infinite number of vibrations which may be classified into groups of finite numbers of vibrations, and each group may be specified by the PHASE DIFFERENCE (δ) between the vibrational displacement of corresponding atoms in adjacent units.

The selection rules for infrared absorption and Raman scattering are related to the angle (θ) of rotation around the chain axis per repeat unit.

(i) Infrared absorption bands arise from
  (a) non-degenerate vibrations with the phase difference $\delta = 0$;
  (b) degenerate vibrations with the phase difference $\delta = \theta$.

(ii) Raman absorption bands arise from
  (a) non-degenerate vibrations with the phase difference $\delta = 0$;
  (b) degenerate vibrations with the phase difference $\delta = \theta$;
  (c) vibrations with the phase difference $\delta = 2\theta$.

If the polymer chain has a twofold axis intersecting the helix at right angles, the chain belongs to the dihedral factor group and the non-degenerate vibrations with the phase difference $\delta = 0$ (called A vibrations) are separated into

(i) A_1 vibrations—infrared inactive but Raman active;
(ii) A_2 vibrations—infrared active but Raman inactive.

288

A_1 and A_2 vibrations are symmetric and antisymmetric, respectively, with respect to the twofold axis.

For neutron scattering (cf. Section 18.3), there is no selection rule related to helical symmetry.

For example, in polyethylene, which has a zigzag chain, there are three atoms per repeat unit of CH_2 and nine vibrational degree of freedom. Five (v_1-v_5) are symmetric and four (v_6-v_9) are antisymmetric:

v_1 refers to CH_2 symmetric stretching,
v_2 refers to CH_2 scissoring,
v_3 refers to CH_2 wagging,
v_4 refers to C—C stretching,
v_5 refers to C—C—C bending,
v_6 refers to CH_2 antisymmetric stretching,
v_7 refers to CH_2 twisting and rocking,
v_8 refers to CH_2 rocking and twisting,
v_9 refers to C—C internal rotation.

The relations between vibrational frequencies and phase differences (δ) are shown in Fig. 17.8. Vibrational modes of the polyethylene chain are shown in Fig. 15.1.

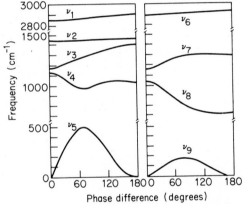

Fig. 17.8 Frequency–dispersion curves of the isolated chain of polyethylene. (From R: 6749)

For example, as the phase difference (δ) varies from 0 to π, the CH_2 rocking vibration (v_7) $(\delta = 0)$ becomes the twisting vibration (v_7) $(\delta = \pi)$, while the CH_2 twisting vibration (v_8) $(\delta = 0)$ becomes the rocking vibration (v_8) $(\delta = \pi)$.

The polyethylene chain has a twofold screw axis and the angle of rotation around the chain axis per repeat unit CH_2 is $\theta = \pi$. Infrared bands and Raman lines arise from the chain vibrations with the phase differences $\delta = 0$ and $\delta = \pi$.

Bibliography: 74, 752, 1287, 1290, 1443.
References: 2385, 3099, 3824, 4053, 4624, 4734, 4742, 4906, 4932, 5122, 5123, 5316, 5318, 5320, 5484–5486, 5841, 6112, 6462, 6463, 6466, 6613, 6661, 6667, 6670, 6671, 6682, 6744, 6747, 6749, 6750, 7320.

Chapter 18

Neutron Scattering Analysis

General bibliography: 62, 64, 168, 392, 752, 1135, 1442.

18.1 PROPERTIES OF NEUTRONS

The NEUTRON is an unstable particle with a mass given by

$$m_n = 1.674\,82 \times 10^{-24} \text{ g (CGS)}$$
$$= 1.008\,665 \text{ u} = 1.674\,82 \times 10^{-27} \text{ kg (SI)}$$

and a half-life of about 12 min, which decays into a proton, an electron, and a neutrino. Because the neutron carries no charge, it is not influenced by magnetic or by electrostatic fields. It can be deflected only by means of collisions with other particles. Lack of a charge also accounts to a large extent for its great penetrating power. It is stopped only by very thick barriers.

Neutrons are classified according to their velocity:

(i) THERMAL NEUTRONS are so named because their average energy is equal to the average kinetic energy of molecules at room temperature; their mean velocity $= 2200$ m s^{-1}, and their energy is 0.025 eV. These slow thermal neutrons are formed by the slowing down of fast neutrons by a process involving numerous collisions with nuclei.

(ii) INTERMEDIATE NEUTRONS have energy in the range 0.5 eV–10 keV.

(iii) FAST NEUTRONS have energy in the range 10–20 keV.

(iv) RELATIVISTIC NEUTRONS have energy >20 keV.

Because the neutron is electrically neutral, there is no barrier to prevent access of even slow neutrons to the atomic nucleus. Neutrons may interact with nuclei by one of the following methods:

(i) ELASTIC SCATTERING (elastic collision) is primarily responsible for the MODERATION (slowing down) of neutrons. In an elastic collision the total kinetic energy and the total momentum of the neutron and the nucleus with which it collides remain constant; that is, no energy is lost as electromagnetic radiation. Elements most often used as moderators are hydrogen and carbon.

COHERENT SCATTERING is a special type of elastic scattering (diffraction) of neutrons by crystals. Coherent scattering requires consideration of the neutrons as waves.

(ii) INELASTIC SCATTERING (inelastic collision) results in a loss in the total energy of the colliding systems.

(iii) A CAPTURE REACTION is the absorption of a neutron by a nucleus with an increase in energy and the formation of a high-energy state of the nucleus. The excess absorbed energy is re-emitted by emission of a particle or photon.

In general, equations for neutron scattering are complex and for that reason they are not cited here.

The NUCLEAR CROSS-SECTION (σ) is the probability of occurrence of a reaction. The TOTAL CROSS-SECTION (σ_t) is equal to the sum of the ABSORPTION CROSS-SECTION (σ_a) (also called the REACTION CROSS-SECTION (σ_r)) and the SCATTERING CROSS-SECTION (σ_s):

$$\sigma_t = \sigma_a + \sigma_s. \tag{18.1}$$

The unit of cross-section is the BARN: 1 barn $= 10^{-24}$ cm^2 (CGS) $= 10^{-28}$ m^2 (SI).

The scattering cross-section for hydrogen is approximately an order of magnitude larger than the cross-sections for most other elements (Table 18.1).

Table 18.1   Bound-nucleus cross-sections in barns (1 barn $= 10^{-24}$ cm^2 (CGS)) for neutron scattering

Nucleus	H	D	C	N	O	F	Cl
Incoherent	79.7	2.2	0.03	0.4	0.04	0.2	2.9
Coherent	1.8	5.4	5.5	11.0	4.2	3.8	12.2

18.2 INSTRUMENTATION FOR NEUTRON SCATTERING ANALYSIS

18.2.1 Neutron Sources

Neutrons are produced by two general processes:

(i) nuclear bombardment in accelerators;
(ii) fission in a reactor.

Experimental accuracy or resolution of neutron scattering spectra depends upon the flux intensity of the neutron beam. The higher the flux, the better the resolution.

18.2.2 Neutron Scattering Spectrometers

There are two classes of neutron scattering spectrometer:

(i) TIME-OF-FLIGHT SPECTROMETERS (Fig. 18.1). A beam of neutrons from a reactor passes through a beryllium–bismuth filter placed in liquid nitrogen. This filter transmits a neutron beam of 5.3 MeV

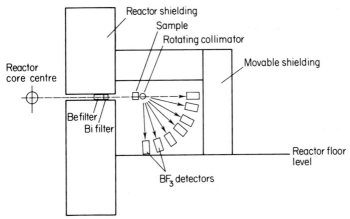

Fig. 18.1 Schematic diagram of a neutron time-of-flight spectrometer

energy, which impinges on a specimen and is scattered in all directions. The scattered neutrons are chopped by a rotating collimator and the distribution of neutron energies is measured by the BF_3 proportional counters. The times taken by the neutrons to go from the chopper to the detectors are recorded in an electronic channel time-of-flight analyser.

The difference between the energies of the scattered neutrons and the initial energy of the incident beam gives the vibrational energy of the scattering specimen.

(ii) TRIPLE-AXIS CRYSTAL SPECTROMETERS (Fig. 18.2). A beam of neutrons from a reactor passes through collimator and impinges on a monochromating crystal. This crystal selects monochromatic neutrons of wavelength λ_1 according to the Bragg equation,

$$n\lambda_1 = 2d_M \sin \theta_M, \tag{18.2}$$

where

n is any integer 0, 1, 2, 3, ..., called the order,

λ_1 is the wavelength of the monochromatic neutrons scattered by the monochromatic crystals,

d_M is the distance between adjacent planes in the monochromatic crystal,

θ_M is one half the angle of deviation of the scattered neutrons from the direction of the incident neutrons.

292

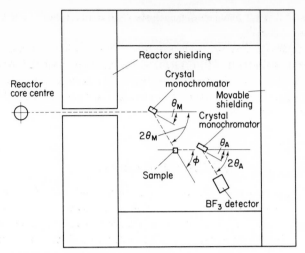

Fig. 18.2　Schematic diagram of a neutron triple-axis spectrometer

The scattered neutrons impinge on the specimen and are scattered through an angle φ and then analysed by scanning with the analysing crystal according to the Bragg equation,

$$n\lambda_2 = 2d_A \sin \theta_A, \qquad (18.3)$$

where

n is any integer 0, 1, 2, 3, ..., called the order,

λ_2 is the wavelength of the neutrons scattered by the analysing crystal,

d_A is the distance between adjacent planes in the analysing crystal,

θ_A is one half the angle of deviation of the scattered neutrons from the direction of the neutron beam scattered by the specimen.

Triple-axis spectrometers have a higher resolution and are generally applied for the study of the molecular vibrations of crystals and the direction of polarization.

Bibliography: 62, 64, 208, 214, 300, 392, 670, 752, 854, 1135.
References: 2363, 3665, 4193, 6230.

18.2.3 Neutron Detectors

Neutron detectors depend upon the production of ions by incident radiation. In a proportional counter for thermal neutrons, boron, and especially the isotope ^{10}B, is applied for this purpose:

$$^{10}B + {^1}n \rightarrow {^7}Li + {^4}He \text{ (α-particle)}. \qquad (18.4)$$

One α-particle is produced for each neutron captured. α-Particles are powerful ionizers (*c.* 80 000 ion pairs for one 2.5 MeV α-particle), sufficient to produce a very substantial pulse.

A typical neutron detector consists of a long, hollow, metallic cylinder filled with boron trifluoride (BF_3), a gas, which has been enriched in the isotope ^{10}B. The cylinder acts as the cathode while a wire located coaxially serves as the anode. A neutron passing through the BF_3 gas is captured by the reaction of ^{10}B and an α-particle is released.

18.2.4 Sample Preparations

There are no requirements as to the shape, form, and size of sample. It may be a liquid or a solid—the latter as a film, as a powder, as fibres, etc. The samples are mounted between pieces of aluminium foil in a sample holder appropriately shielded with cadmium. In order to reduce air scattering the sample holder is placed in a vacuum case, in transmission geometry. The high-temperature spectrum can be obtained by using electric heaters embedded in the sample holder. For low-temperature measurements the sample can be cooled by contact with a bath of liquid air.

18.3 APPLICATIONS OF NEUTRON SCATTERING ANALYSIS TO THE STUDY OF THE STRUCTURE OF POLYMERS

Neutron scattering analysis provides valuable information with regard to the following:

(i) Normal and interchain vibrations of polymers. Low-energy neutrons can be scattered by a polymer sample and lose a quantity of their energy which is equivalent to the characteristic molecular vibrational frequencies of the sample.

The incident neutrons should possess a narrow distribution of energies and an average energy comparable to that of the low-frequency motions of the molecules of the scattering specimen. At these energies, the wavelengths of the neutrons are comparable to atomic spacings.

(ii) Neutron scattering cross-sections of polymers.

(iii) Conformations of polymers in glass, rubbers, and in solutions (specially by small-angle neutron scattering (SANS)).

(iv) Characteristics of polymer networks.

There are two important differences between neutron scattering analysis and absorption spectroscopy (or Raman spectroscopy):

(i) neutron scattering spectra are not practically limited by quantum rules, and all molecular vibrational frequencies can be observed;

(ii) neutron scattering analysis is most sensitive to motions involving hydrogen atoms, because the scattering cross-section for hydrogen is approximately an order of magnitude larger than the cross-sections of most other elements (cf. Table 18.1).

Bibliography: 752, 854, 1135, 1382.
References: 2028, 2055, 2064, 2066, 2095, 2180, 2214–2216, 2345, 2346, 2363, 2790, 2984, 2985, 3054, 3055, 3252, 3384, 3665, 3879, 4021, 4052, 4573, 4574, 4586–4589, 4658, 4751, 4927, 5001, 5124, 5401, 5402, 5538, 5796, 5826, 6038, 6141, 6142, 6150, 6151, 6220–6223, 6477, 6492, 6620, 6621, 6833–6835, 6894, 7077, 7078, 7273–7275.

Chapter 19

Positron Annihilation Analysis

General bibliography: 520, 562, 583, 916, 1266, 1377.

19.1 PROPERTIES OF POSITRONS

There are two types of electron:

(i) the negative electron (NEGATRON) (e^-);
(ii) the positive electron (POSITRON) (e^+).

These two particles are identical except in respect of their charge, which is -1 for the negatron and $+1$ for the positron. One is the ANTIPARTICLE of the other.

The electron mass, $m = 9.1091 \times 10^{-28}$ g (CGS) $= 9.1091 \times 10^{-31}$ kg (SI). The charge on an electron, $e = 4.802 \times 10^{-10}$ esu (CGS) $= 1.602\,10 \times 10^{-19}$ A s (SI).

A positron interacting with a sample will be slowed down to thermal velocities within a very short time ($< 10^{-11}$ s).

THERMAL VELOCITIES are so named because the average velocity of such positrons is equal to the average velocity of molecules at room temperature.

After a period called the POSITRON LIFETIME, a positron will ANNIHILATE with an electron, and two γ-quanta named the ANNIHILATION PHOTONS are emitted:

$$e^+ + e^- \rightarrow 2\gamma. \tag{19.1}$$

Positron mean lifetimes in condensed samples are of the order of 0.1–5 ns (1 ns $= 10^{-9}$ s).

The annihilation photons yield information about the electron–positron state at the moment of annihilation, and hence measurement of annihilation photons can give useful information about the physical and chemical structure of the sample.

19.2 EXPERIMENTAL METHODS

There are three general methods used in positron annihilation analysis:

(i) The ANGULAR CORRELATION TECHNIQUE, which is based on the fact that the total momentum of the annihilating electron–positron pair is transferred to the two annihilation photons:

296

(a) if this momentum is zero, the two photons are emitted in exactly opposite directions;

(b) if this momentum differs from zero, there will be an angle (θ) ($\theta < 180°$) between the directions in which the photons are emitted, and the deviation will be proportional to the momentum of the annihilating pair.

In the angular correlation technique the angular distribution of the annihilation photons is measured (angle θ). The angles (θ) are of the order of a few milliradians.

(ii) The DOPPLER BROADENING TECHNIQUE, which is based on measurement of the energy distribution of the annihilation photons by applying the Doppler effect. Because of the Doppler effect, the energy distribution of the annihilation photons will broaden with increasing velocity of the annihilating pair. This method also measures the momentum distribution of the annihilating pair.

(iii) The LIFETIME TECHNIQUE is based on the measurement of a positron lifetime. In this method a positron source (^{22}Na) is applied in the form ^{22}NaCl. Simultaneously with the emission of a positron the source also emits a γ-quantum. One annihilation photon is recorded in one detector and the other annihilation photons in a second detector. The time difference between the signals from the two detectors is the positron lifetime.

Bibliography: 4, 520, 521, 562, 916, 1266.
References: 5012, 5013.

19.3 INSTRUMENTATION FOR POSITRON ANNIHILATION ANALYSIS

19.3.1 Positron Sources

Positrons are produced during the decay of two radioactive isotopes, ^{22}Na and ^{65}Zn.

A positron source can easily be obtained by evaporating the water from a droplet of ^{22}NaCl solution and sealing the residue between two thin foils made from metal or plastic. If the foils are sufficiently thin, not more than 20% of the positrons annihilate in the foils. The radioactivity of such a source is up to 50 μCi, and no special precautions are required for protection.

1 CURIE (1 Ci) is a unit of radioactivity defined as 3.70×10^{10} disintegrations per second.

19.3.2 Gamma Scintillation Counters

When gamma radiation interacts with certain substances called FLUORS (sometimes referred to as PHOSPHORS), a small flash of visible light (a

SCINTILLATION) is produced. Gamma scintillation uses crystals of sodium iodide containing a trace of thallium iodide as an activator (NaI(Tl) crystals). The PHOTOCELL used to detect these small scintillations must be extremely sensitive and must be connected intimately with the crystal so that the light is efficiently transmitted to the photosensitive cathode. The photocell used is of the type known as a PHOTOMULTIPLIER (cf. Section 10.7.1).

Bibliography: 781.

19.3.3 The Positron Lifetime Measuring System

In this procedure the signal from the detectors is transformed into an electrical signal whose amplitude is proportional to the time difference. This signal is fed into a multichannel analyser which registers the signal amplitude (i.e. the positron lifetime) and stores this information. When the annihilations of many positrons have been recorded, the results stored in the multichannel analyser will be a distribution of positron lifetimes, i.e. the number of positrons which have lived a certain time as a function of this time. This distribution is usually referred to as the LIFETIME SPECTRUM.

19.3.4 Sample Preparation

There are no requirements as regards the shape, form, and size of the sample. It may be a liquid or a solid—the latter as a film, as a powder, or as fibres. It is important that all positrons emitted from the source must stop and annihilate in the sample. For that reason the sample should be sufficiently thick to have an area density higher than 0.1 g cm^{-2}.

19.4 APPLICATIONS OF POSITRON ANNIHILATION ANALYSIS TO THE STUDY OF THE STRUCTURE OF POLYMERS

Positron annihilation occurs in all polymers for which data have been published. Before it annihilates, however, a positron forms a bound state with an electron which is called POSITRONIUM (and which has its own chemical symbol, Ps). There are two different positronium atoms:

(i) *ortho*-positronium (*ortho*-Ps), where the spins of the positron and the electron are parallel (a triplet state) (75% of all forms)—its lifetime is *c*. 140 ns and it decays into three photons;

(ii) *para*-positronium (*para*-Ps), where the spins of the positron and the electron are antiparallel (a singlet state) (25% of all forms)—its lifetime is *c*. 0.125 ns and it decays with emission of two photons.

The positron annihilation process depends significantly on polymer structure and can be used as a practical method for investigating polymer properties.

298

Positron annihilation has been applied to the study of radiation-induced solid-state polymerization. The lifetime spectrum is influenced by the polymerization.

Bibliography: 19, 113, 136, 194, 195, 1407.
References: 2850–2852, 2781, 3376, 3791, 4253, 4254, 4283, 4542–4544, 4966, 5549, 5550, 6573–6576, 6660, 6741, 6779, 6780.

Chapter 20

Nuclear Magnetic Resonance Spectroscopy

General bibliography: 1, 2, 39, 52, 54, 55, 71, 91, 120–122, 124, 174, 176, 177, 212, 213, 256, 274, 277, 321, 324, 373, 389, 391, 409, 413, 474, 506, 548, 561, 669, 671, 693, 730, 755, 789, 819, 824, 847, 878, 879, 903, 915, 960, 991, 1039, 1051, 1052, 1114, 1115, 1158, 1175, 1204, 1213, 1221, 1225, 1383, 1393.

NUCLEAR MAGNETIC RESONANCE (NMR) SPECTROSCOPY is a technique which records transitions between the energy levels of magnetic nuclei in an external magnetic field. NMR spectroscopy involves absorption of the energy of electromagnetic radiation in the radio-frequency region by a sample placed in an external magnetic field. Absorption is a function of the magnetic properties of some atomic nuclei in the molecule. A plot of the absorption of radio-frequency energy versus the external magnetic field gives an NMR SPECTRUM.

The atomic nuclei can be classified according to their NUCLEAR SPINS. The only nuclei which can absorb electromagnetic radiation are those for which the spin quantum number (M_I) is greater than zero.

The SPIN QUANTUM NUMBER (M_I) is associated with the MASS NUMBER and the ATOMIC NUMBER of the nuclei as follows:

Mass number	Atomic number	Spin quantum number (MI)
Odd	Odd or even	$\frac{1}{2}, \frac{3}{2}, \frac{5}{2}, \ldots$
Even	Even	0
Even	Odd	1, 2, 3, ...

There are a number of nuclei which have $M_I = 0$, e.g. ^{12}C, ^{16}O, and ^{32}S (these do not have angular momentum); others have $M_I \neq 0$, e.g.

$M_I = \frac{1}{2}$: ^1H (the proton), ^3H, ^{13}C, ^{15}N, ^{19}F, ^{31}P ⎫
$M_I = 1$: ^2H (deuterium), ^{14}N ⎬ Magnetic nuclei
$M_I > 1$: ^{10}B, ^{11}B, ^{17}O, ^{35}Cl ⎭

Properties of magnetic nuclei are shown in Table 20.1.

All nuclei that have spin quantum number $M_I \neq 0$ possess a MAGNETIC DIPOLE MOMENT or MAGNETIC MOMENT (μ_N) which arises from the motion (spinning) of a charged particle. Nuclei possessing electric quadrupoles can interact with both magnetic and electric field gradients.

Table 20.1   Properties of several magnetic nuclei (from B: 1213; reproduced by permission John Wiley & Sons)

Isotype	Natural abundance (%)	Spin number, I	Magnetic moment, μ	NMR frequency for a 10 kG field (MHz)	Relative sensitivity at constant field
^{1}H	99.9844	$\frac{1}{2}$	2.792 68	42.576	1.000
^{2}H	1.56×10^{-2}	1	0.857 39	6.5357	9.64×10^{-2}
^{3}H	—	$\frac{1}{2}$	2.9788	45.414	1.21
^{10}B	18.83	3	1.8005	4.575	1.99×10^{-2}
^{11}B	81.17	$\frac{3}{2}$	2.6880	13.660	0.165
^{12}C	98.9	0	—	—	—
^{13}C	1.108	$\frac{1}{2}$	0.702 20	10.705	1.59×10^{-2}
^{14}N	99.635	1	0.403 58	3.076	1.01×10^{-3}
^{15}N	0.365	$\frac{1}{2}$	0.283 04	4.315	1.04×10^{-3}
^{16}O	99.76	0	—	—	—
^{17}O	3.7×10^{-2}	$\frac{5}{2}$	1.8930	5.772	2.91×10^{-2}
^{19}F	100	$\frac{1}{2}$	2.6273	40.055	0.834
^{28}Si	92.28	0	—	—	—
^{29}Si	4.70	$\frac{1}{2}$	0.555 48	8.458	7.85×10^{-2}
^{30}Si	3.02	0	—	—	—
^{31}P	100	$\frac{1}{2}$	1.1305	17.236	6.64×10^{-2}
^{32}S	95.06	0	—	—	—
^{33}S	0.74	$\frac{3}{2}$	0.642 74	3.266	2.26×10^{-3}
^{34}S	4.2	0	—	—	—
^{35}Cl	75.4	$\frac{3}{2}$	0.820 91	4.172	4.71×10^{-3}
^{37}Cl	24.6	$\frac{3}{2}$	0.683 30	3.472	2.72×10^{-3}
^{79}Br	50.57	$\frac{3}{2}$	2.0991	10.667	7.86×10^{-2}
^{81}Br	49.43	$\frac{3}{2}$	2.2626	11.499	9.84×10^{-2}
^{127}I	100	$\frac{5}{2}$	2.7937	8.519	9.35×10^{-2}

The fundamental properties of a proton (^{1}H) are as follows:

(i) Mass: $m_p = 1.6730 \times 10^{-24}$ g (CGS) $= 1.007\,277$ u $= 1.6730 \times 10^{-27}$ kg (SI), where u is the universal unit of mass (u $= 1.661 \times 10^{-27}$ kg).

(ii) Charge.

(iii) Intrinsic angular momentum, known as the SPIN. This is a vector denoted by the symbol $\mathbf{I}$. The component of the spin vector $\mathbf{I}$ along any direction has only the value $\pm\frac{1}{2}\hbar$.

The symbol $\hbar = h/2\pi$ is read 'h bar' and

$$\hbar = 1.054 \times 10^{-27} \text{ erg s (CGS)} = 1.054 \times 10^{-34} \text{ J s (SI)},$$

where h is the Planck constant, and

$$h = 6.626 \times 10^{-27} \text{ erg s (CGS)} = 6.626 \times 10^{-34} \text{ J s (SI)}.$$

(iv) Magnetic moment. This is a vector denoted by the symbol $\boldsymbol{\mu}_N$. It is related to the intrinsic angular momentum ($\mathbf{I}$) by the following equation:

$$\boldsymbol{\mu}_N = +\frac{g_N \beta_N}{\hbar}\mathbf{I} \qquad (20.1)$$

where

g_N is called the NUCLEAR g-FACTOR and is dimensionless ($g_N = 5.5855$ for a proton),

β_N is the NUCLEAR MAGNETON, $\beta_N = 5.0509 \times 10^{-24}$ erg gauss^{-1} (CGS) $= 5.0509 \times 10^{-27}$ A m^2 (SI).

In an applied static magnetic field (H_0), magnetic nuclei like the proton (^1H) precess at a frequency (ω_0 or ν_0) given by the LARMOR EQUATION:

$$\omega_0 = -\gamma H_0, \tag{20.2}$$

$$\nu_0 = \omega_0/2\pi, \tag{20.3}$$

where

H_0 is the strength of the applied external static magnetic field (in gauss (CGS) or tesla (SI) (1 tesla (T) $= 10^4$ gauss (G)),

γ is the MAGNETOGYRIC RATIO (different for different nuclei),

$$\gamma = 2\pi\mu_N/hM_I, \tag{20.4}$$

μ_N is the magnetic moment of the proton,

h is the Planck constant,

M_I is the spin quantum number.

Typical approximate values for precessional frequencies (ν_0) for different nuclei are shown in Table 20.2.

Table 20.2 Precessional frequencies ν_0 (in megahertz) as a function of increasing field strength H_0 (in gauss)

Nucleus	H_0 (G)					
	14 000	21 000	23 000	51 000	58 000	71 000
	ν_0 (MHz)					
^{1}H	60	90	100	220	250	300
^{2}H	9.2	13.8	15.3	33.7	38.4	46.0
^{13}C	15.1	22.6	25.2	55.0	62.9	75.5
^{14}N	4.3	6.5	7.2	15.8	17.9	21.5
^{19}F	56.5	84.7	93.0	206.5	203.4	282.0
^{31}P	24.3	36.4	40.5	89.2	101.5	121.5
(Free electron)	3.9×10^4					

A proton (^1H) placed in an external static magnetic field, e.g. $H_0 = 14\,000$ G (CGS) $= 1.4$ T (SI) will precess at a frequency $\nu_0 = 60$ MHz (60 million times per second). For an external static magnetic field $H_0 = 71\,000$ G (CGS) $= 7.1$ T (SI), $\nu_0 = 300$ MHz (cf. Table 20.2).

Under the influence of an external static magnetic field (H_0), a magnetic nucleus can take up different orientations with respect to that field. The number of possible orientations is given by $2M_I + 1$, where M_I is a spin quantum number.

The proton (^{1}H) with spin quantum number $M_I = \frac{1}{2}$, placed in external static magnetic field (H_0), will have only two orientations (Fig. 20.1):

(i) aligned with the field (parallel orientation) (the lower energy state);
(ii) Opposed to the field (antiparallel orientation) (the higher energy state).

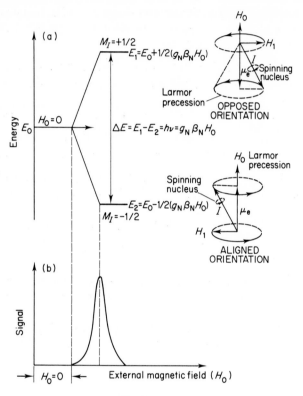

Fig. 20.1 (a) Energy level diagram showing Zeeman splitting for a proton system in crossed magnetic (H_0) and rf (H_1) fields. (b) NMR absorption line, (H_0) mode signal curve

The precessing proton (^{1}H) in the parallel orientation can absorb energy (ΔE) from the radio-frequency source (rf) and pass into the antiparallel orientation (this phenomenon is called ZEEMAN SPLITTING FOR A PROTON) under one particular condition: if the precessing frequency is the same as the frequency of the radio-frequency beam (such a condition is called NUCLEAR MAGNETIC RESONANCE). The energy absorbed is recorded in the form of an NMR SPECTRUM.

The energy absorbed ($\Delta E = h\nu$) is very small (of the order 10^{-4} kJ mol^{-1} (SI) in an applied external magnetic field $H_0 = 1.4$ T (SI)), but it can be re-emitted as 60 MHz energy, and this can be monitored by a radio-frequency detector as evidence of the resonance condition having been reached.

If the populations of the two states (the higher energy state and the lower energy state) approach equality, no further net absorption of energy will occur, and the observed resonance signal will fade out. This phenomenon is called SATURATION OF THE NMR SIGNAL.

In the normal measuring condition, the population in the two states is not equal because higher energy nuclei are constantly returning to the lower energy state. Two radiationless processes play important roles in the loss of (RELAXATION) energy by high-energy nuclei:

(i) in the SPIN–LATTICE RELAXATION PROCESS the energy difference (ΔE) is transferred to neighbouring atoms, either in the same molecule or in solvent molecules;

(ii) in the SPIN–SPIN RELAXATION PROCESS the energy difference (ΔE) is transferred to a neighbouring nucleus.

The rates of relaxation by these two processes are important and are given by

(i) The HALF-LIFE OF THE SPIN–LATTICE RELAXATION PROCESS (T_1) (commonly called the SPIN–LATTICE RELAXATION TIME) (T_1);

(ii) the HALF-LIFE OF THE SPIN–SPIN RELAXATION PROCESS (T_2) (in seconds) (commonly called the SPIN–SPIN RELAXATION TIME) (T_2).

The RELAXATION TIME is the time in which a non-equilibrium system completes 63%, or $1 - 1/e$, of its return to equilibrium; e is the base of natural logarithms.

Between the relaxation time (Δt) and the line broadening (Δv) is observed the following relationship (the UNCERTAINTY PRINCIPLE):

$$\Delta E . \Delta t \approx h/2\pi \approx \text{constant} \tag{20.5}$$

$$\Delta v . \Delta t \approx 1/2\pi \approx \text{constant.} \tag{20.6}$$

A general rule is that

(i) if Δt is small (in solid polymers), Δv is large (FAST RELAXATION) and broad absorption lines appear in the NMR spectrum;

(ii) if Δt is large (solutions of polymers in non-viscous solvents), Δv is small (SLOW RELAXATION) and narrow absorption lines appear in the NMR spectrum.

In other words:

(i) if T_1 and T_2 are small (in solid polymers), then the lifetime of an excited nucleus is short ($c.$ 10^{-5} s) and this will give very broad absorption lines in the NMR spectrum;

(ii) if T_1 and T_2 are large (solutions of polymers in non-viscous solvents), then the lifetime of an excited nucleus is long ($c.$ 1 s) and this will give sharp absorption lines in the NMR spectrum.

The spin–lattice relaxation process is very dependent on temperature, because the lattice vibration frequency increases with increasing temperature. In polymers the plot of T_1 versus temperature gives a curve which exhibits minima corresponding to the onset of characteristic modes of molecular motion (cf. Section 20.9).

20.1 INTERPRETATION OF NMR SPECTRA

There are four features of NMR spectra which are important for their interpretation:

   (i) line position (cf. Section 20.1.1);
  (ii) line intensity (cf. Section 20.1.2);
 (iii) line splitting (cf. Section 20.1.4);
 (iv) line width (cf. Section 20.10).

20.1.1 The NMR Position

The precessional frequency of all protons in a polymer sample (e.g. hydrogen atoms in CH_3, CH_2 and CH groups) placed in an external static magnetic-field (H_0) is not the same, and the precise value for any one proton depends on its chemical environment (i.e. the degree of shielding of a given proton on a carbon atom depends on the inductive effect of other groups of protons attached to the carbon). For that reason the shift in frequency is called the CHEMICAL SHIFT.

The CHEMICAL SHIFT is a difference in the absorption position of a given proton from the absorption position of a reference proton.

Two different groups of protons have different CHEMICAL SHIFT POSITIONS on the spectrum.

The chemical shift is dependent on the value of the magnetic field strength (H_0) and of the radio-frequency.

Measurement of the precessional frequency (absorption position) of a group of nuclei in absolute frequency units is difficult. Differences in frequency are usually measured from the chosen reference.

Reference substances for NMR spectroscopy may be

  (i) internal—dissolved in the solution under examination;
 (ii) external—sealed in a capillary immersed in the examined solution.

The most common reference is tetramethylsilane (TMS):

$$CH_3$$
$$|$$
$$CH_3\!-\!Si\!-\!CH_3$$
$$|$$
$$CH_3$$

TMS is chemically inert, magnetically isotropic, and volatile (b.p. 27 °C)—and because of that last property it can easily be removed from a sample. It is soluble in most organic solvents but insoluble in water and D_2O. It can be added to the sample solution (1 wt %) as an internal standard. It gives an intense single sharp absorption peak even at low concentration and absorbs at a higher field strength than almost all other protons. When water or D_2O is used as a solvent, TMS can be applied as an external reference.

For water and D_2O solutions it is normal to use the sodium salt of 3-(trimethylsilyl)propanesulphonic acid as a reference:

$$CH_3-\underset{\underset{CH_3}{|}}{\overset{\overset{CH_3}{|}}{Si}}-CH_2CH_2CH_2SO_3^-\,Na^+$$

Chemical shifts can be expressed in the following units (Fig. 20.2):

(i) In frequency units where the value of the frequency (v) is given in hertz (Hz). When chemical shifts are given in hertz the applied frequency must be specified. The results should be reported in frequency units in cases where complex spectra occur or where spin–spin coupling constants (J) are to be given.

(ii) In dimensionless units, where the value (δ) is independent of the applied frequency (v), by dividing v by the applied frequency and multiplying by 10^6 δ units are expressed in parts per million (p.p.m.).

(iii) In τ units, where

$$\tau = 10 - \delta. \tag{20.7}$$

It should be noted that δ is treated as a positive number.

Tetramethylsilane (TMS) protons resonate at exactly 60 MHz when the external static magnetic field (H_0) has field strength 14 092 G ($\delta = 0$ and $\tau = 10$

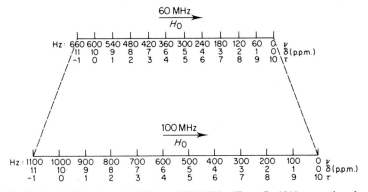

Fig. 20.2   NMR scale at 60 MHz and 100 MHz. (From B: 1213; reproduced by permission of John Wiley & Sons)

at that point, cf. Fig. 20.2). The position of signals from other protons is measured in relation to TMS in parts per million (p.p.m.).

Chemical shifts are influenced by several factors, such as

(i) shielding and deshielding effects (associated with varying electro-negativity in the molecule);

(ii) van der Waals deshielding effect (associated with groups which have sterically hindred positions, and in consequence the electron cloud of the hindering group will tend to repel, by electrostatic repulsion, the electron cloud surrounding the proton);

(iii) anisotropic effects (associated with the presence of π electrons, e.g. in alkene groups, carbonyl groups, aromatic rings, etc.).

Bibliography: 91, 409, 671, 697, 730, 934, 1085, 1213, 1270.

Predicting the NMR spectrum of an organic compound begins with predicting the chemical shift positions for the different hydrogens in the molecule using the CORRELATION DATA OF CHEMICAL SHIFTS and NMR SPECTRA CATALOGUES.

In Fig. 20.3 are shown (for example) approximate chemical shift positions for protons in some organic molecules.

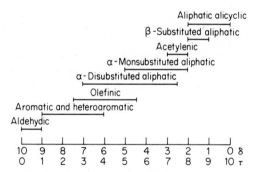

Fig. 20.3 Approximate chemical shift positions for protons in organic molecules. (From B: 1213; reproduced by permission of John Wiley & Sons)

As an aid in finding chemical shift data and NMR spectra, the following references are useful:

Bibliography: 2, 171, 176, 213, 615, 643, 1204, 1213, 1383, 1480, 1481.
References: 6428.

20.1.2 The NMR Line Intensity

The INTENSITY means the total strength of the signal. It is the total energy absorbed by the sample at resonance.

(Note: in optical spectroscopy the word 'intensity' means the value of the ordinate of the recorder trace of the signal. The intensity here is the height of line.)

The NMR LINE INTENSITY is the area under an NMR absorption curve.

The area under each NMR signal in the spectrum is proportional to the number of hydrogen atoms in that group. Measurement of the peak areas is carried out automatically on the NMR spectrometer by integration of each signal, and the integrated value is indicated on the spectrum in the form of a continuous line in which steps appear as each signal is measured (Fig. 20.4).

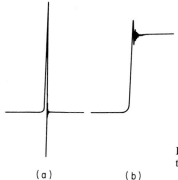

(a)　　　　　　(b)

Fig. 20.4 (a) NMR absorption spectrum. (b) Integral of absorption spectrum

The step height is proportional to the peak area. Broad peaks tend to give less accurate integrals than sharp peaks.

The integral trace of the proton NMR spectrum of a mixture can provide information on the relative amounts of the components. Quantitative analysis by this technique is especially useful when the components of the mixture are difficult or impossible to isolate.

20.1.3 Determination of Molecular Weight by NMR Spectroscopy

The integrated intensity of an absorption peak in the proton NMR spectrum depends only upon the molar concentration of the substance times the number of nuclei per molecule responsible for the peak.

The integrated intensity per nucleus per mole is the same for all substances in the sample.

When a known weight of a substance of known molecular weight (the standard—w_s/MW_s in moles) is added to a sample containing a known weight of a substance of unknown molecular weight (the unknown—w/MW), the following equality holds:

$$\frac{I_s/n_s}{w_s/MW_s} = \frac{I/n}{w/MW},$$　　　　(20.8)

where

I_s and I are the integrated intensities of the peak of the standard and the peak of the unknown, respectively,

n_s and n are the numbers of nuclei responsible for the peak of the standard
and the peak of the unknown, respectively,
w_s is the weight of standard substance added,
w is the weight of unknown substance used,
MW_s and MW are the molecular weights of the standard and the unknown
substance, respectively.

The molecular weight of the unknown substance (MW) may therefore be
calculated from eq. (20.8).

This method can be applied only when distinct reasonances are well separated
from the rest of the spectrum for both the unknown and the standard samples.
As a standard substance one can use iodoform (CHI_3) or 1, 3, 5-trinitrobenzene.

Bibliography: 55, 769, 819.

20.1.4 NMR Line Splitting

SPLITTING in the energy levels of magnetic nuclei is the phenomenon which
results in an increase in the number of energy levels as a result of exposing a
system containing magnetic nuclei to a magnetic field.

Table 20.3   The various possibilities of the spin–spin splitting pattern of NMR
spectra (from B: 55; reproduced by permission of Holden-Day)

Spin System	Appearance of resonance		
A	A resonance		
A_2	Singlet		
A_3	Singlet		
⋮	⋮		
A_m	Singlet		
	A Resonance		X Resonance
AX	1 : 1 doublet		1 : 1 doublet
AX_2	1 : 2 : 1 triplet		1 : 1 doublet
AX_3	1 : 3 : 3 : 1 quartet		1 : 1 doublet
⋮	⋮		⋮
AX_n	$(n+1)$-membered multiplet		1 : 1 doublet
A_2X_2	1 : 2 : 1 triplet		1 : 2 : 1 triplet
A_2X_3	1 : 3 : 3 : 1 quartet		1 : 2 : 1 triplet
⋮	⋮		⋮
A_2X_n	$(n+1)$-membered multiplet		1 : 2 : 1 triplet
⋮	⋮		⋮
A_mX_n	$(n+1)$-membered multiplet		$(m+1)$-membered multiplet
	A resonance	M resonance	X resonance
AMX	Pair of doublets	Pair of doublets	Pair of doublets
AMX_2	Pair of triplets	Pair of triplets	Pair of doublets
AMX_3	Pair of quartets	Pair of quartets	Pair of doublets
⋮	⋮	⋮	⋮
$A_m M_p X_n$	$(p+1)$ of $(n+1)$	$(m+1)$ of $(n+1)$	$(m+1)$ of $(p+1)$

Splitting of NMR spectral lines is caused by a SPIN COUPLING INTER-ACTION between neighbour protons, and is related to the number of possible spin orientations that these neighbours can have. This phenomenon is called SPIN–SPIN SPLITTING or SPIN COUPLING.

The number of lines (MULTIPLICITY) observed in an NMR signal for a group of protons is not related to the number of protons in that group, but to the number of protons in neighbouring groups.

The $(n+1)$ RULE helps to find the multiplicity of the signal from a group of protons, where n is the number of neighbours. (There are some exceptions to this rule.)

The various possibilities for NMR spin–spin splitting patterns are shown in Table 20.3. The simplest spin–spin splitting patterns are:

(i) Spin system AX (which gives the AX spectrum) (Fig. 20.5). This system represents two vicinal protons, which have different chemical and magnetic environments and which resonate at different positions in the

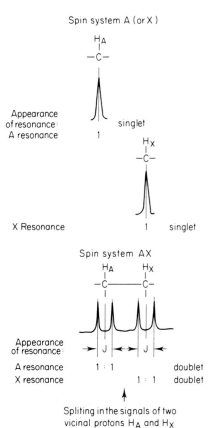

Fig. 20.5 Spin–spin splitting pattern for spin system A (or X) and spin system AX

NMR spectrum. They do not give single peaks (singlets), but doublets. The separation between the lines of each doublet is equal and this spacing is called the COUPLING CONSTANT (J) (measured in hertz).

Proton H_A appears as a doublet because of the two spin orientations (parallel (low energy) and antiparallel (high energy)) of proton H_X. Proton H_X also appears as a doublet because of the two spin orientations (parallel (low energy) and antiparallel (high energy)) of proton H_A.

(ii) Spin system AX_2 (which gives the AX_2 spectrum) (Fig. 20.6). In this system there are three different possible combinations of spin:

(a) the nuclear spin of the two X protons can have a parallel orientation to the proton A ($\uparrow\uparrow$);

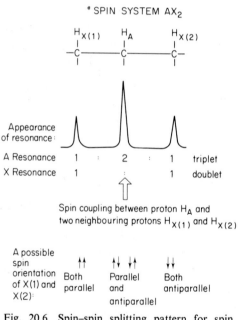

* SPIN SYSTEM AX_2

| | $H_{X(1)}$ | H_A | $H_{X(2)}$ |

Appearance of resonance:

| A Resonance | 1 | : | 2 | : | 1 | triplet |
| X Resonance | 1 | | : | | 1 | doublet |

Spin coupling between proton H_A and two neighbouring protons $H_{X(1)}$ and $H_{X(2)}$

A possible spin orientation of X(1) and X(2):

| $\uparrow\uparrow$ | $\uparrow\downarrow$ $\uparrow\downarrow$ | $\downarrow\downarrow$ |
| Both parallel | Parallel and antiparallel | Both antiparallel |

Fig. 20.6   Spin–spin splitting pattern for spin system AX_2

(b) the nuclear spin of the two X protons can have antiparallel orientations to the proton A ($\downarrow\downarrow$).

(c) One X proton can be parallel and the other X proton can be antiparallel ($\uparrow\downarrow$) or opposite ($\downarrow\uparrow$).

The sign $\uparrow$ means aligned (parallel) spin orientation. The sign $\downarrow$ means opposed (antiparallel) spin orientation. These three spin orientations ((a), (b), (c)) imply that proton A will appear as a triplet. The probability of the two spin orientations (a) and (b) is equal, but the third spin orientation (c) may occur in two different ways and the probability is two times higher: the intensity of the signal associated with this state is therefore twice that of the lines associated with the first two states and the relative line intensities in the triplet are $1:2:1$ (Fig. 20.6).

(iii) The number of the theoretical line intensities for triplets, quartets, quintets, etc., can be predicted from the PASCAL TRIANGLE (for the binominal expansion) (Fig. 20.7). The outer lines in multiplets can be of such low intensity that they may be not detectable unless that part of the spectrum is re-run on an expanded scale.

n	Relative intensity	
0	1	Singlet
1	1 1	Doublet
2	1 2 1	Triplet
3	1 3 3 1	Quartet
4	1 4 6 4 1	Quintet
5	1 5 10 10 5 1	Sextet
6	1 6 15 20 15 6 1	
7	1 7 21 35 35 21 7 1	
8	1 8 28 56 70 56 28 8 1	

Fig. 20.7   Pascal's triangle

There are several factors influencing the coupling constant (J):

(i) GEMINAL COUPLING, involving protons separated by only two bonds ($J \sim 10-18$ Hz);

(ii) VICINAL COUPLING, involving protons separated by three bonds ($J \sim 0-12$ Hz, usually around 8 Hz);

(iii) *CIS-* ($J \sim 5-14$ Hz) and *TRANS-* ($J \sim 11-19$ Hz) COUPLINGS;

(iv) AROMATIC COUPLINGS ($J_{ortho} \sim 7-10$ Hz $J_{meta} \sim 2-3$ Hz $J_{para} \sim 0-1$ Hz);

(v) ALLYLIC COUPLING ($J \sim 0-2$ Hz);

(vi) DEUTERIUM COUPLING, involving a proton (^{1}H) and a deuteron (^{2}H)—a proton coupled with a deuteron appears as a triplet and the relative line intensities are $1:1:1$;

(vii) HETERONUCLEAR COUPLING, involving a proton (^{1}H) and some other magnetic nucleus, e.g. ^{13}C, ^{15}N, ^{19}F, ^{31}P—the spectra are complicated.

FIRST-ORDER SPECTRA arise if the separation between multiplets (the chemical shift difference between signals) is much larger than the coupling constant (if $\Delta\delta > 6J$). In this case an unperturbed spectrum will be observed. (Fig. 20.8).

SECOND-ORDER SPECTRA arise if the separation between multiples (the chemical shift difference between signals) is small (if $\Delta\delta < J$) (Fig. 20.8).

312

In this case the spectra are almost distorted: the signals move together, the inner peaks become larger at the expense of the outer peaks, and the positions of the lines change. Such spectra are usually called AB SPECTRA (or STRONG COUPLED SPECTRA), indicating that the chemical shift values are closer than in AX cases.

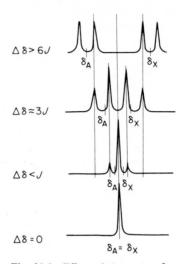

Fig. 20.8   Effect of the ratio $\Delta\delta/J$ on doublet appearance in the spin system AX

If two protons have the same chemical shift ($\Delta J = 0$) (e.g. the geminal protons is unhindered CH_2 groups) the interactions between spins is not observable in the NMR spectrum and a single peak is produced (Fig. 20.8).

Many second-order spectra have been analysed by calculation and can be found in the literature.

If a second-order spectrum is obtained a number of techniques can be applied to simplify its complexity (cf. Sections 20.5 and 20.6).

Bibliography: 1213, 1383.

20.2 INSTRUMENTATION FOR NMR SPECTROSCOPY

20.2.1 NMR Spectrometers

NMR spectrometers are built from the following parts (Fig. 20.9):

(i) A magnet. There are three types of magnet used in NMR spectrometers:
   (a) permanent magnets with field strengths of 14 000 G—instruments with a 14 000 G magnet are usually called 60 MHz instruments, and are used only for detecting proton (^{1}H) spectra;

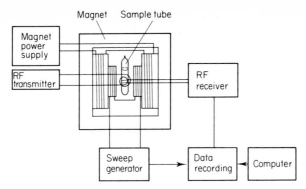

Fig. 20.9 A schematic diagram of a NMR spectrometer

(b) electromagnets with field strengths of 21 000 (23 000) G—instruments with a 23 000 G magnet are usually called 100 MHz instruments;

(c) superconducting magnets with field strengths of 51 000 G and 71 000 G —instruments with a 51 000 G magnet are usually called 220 MHz instruments and those with a 71 000 G magnet are called 300 MHz instruments—these superconducting magnets are produced by immersing electromagnets in liquid helium at 4 K, electrical resistance then vanishes and the metal becomes superconductor.

A homogeneous field across the sample is a fundamental requirement in NMR spectrometers. In order to compensate for any inhomogeneity in the main magnet's field it is necessary to use addition coils called Golay coils. These coils can be tuned to obtain specifically a contoured magnetic field.

The magnet must have pole pieces of large diameter relative to the width of the air gap. The air temperature for a permanent magnet and temperature of the cooling water for an electromagnet must be carefully controlled.

(ii) A source of radio-frequency power (RF transmitter). This is a crystal oscillator controlled at a single frequency, e.g.

$$60 \text{ MHz for a } 14\,000 \text{ G magnet,}$$
$$100 \text{ MHz for a } 23\,000 \text{ G magnet,}$$
$$220 \text{ MHz for a } 51\,000 \text{ G magnet,}$$
$$300 \text{ MHz for a } 71\,000 \text{ G magnet.}$$

Radio-frequency power is fed into a coil inside which is placed the sample.

(iii) A radio-frequency receiver (RF receiver), which detects, amplifies, and filters the NMR signal.

(iv) A sweep generator, which is applied for a changing strength of magnetic field (H_0). The output of the sweep generator is synchronized with the trace along the X-axis of an oscilloscope or XY-recorder. The variable electromagnetic coils are called sweep coils.

(v) A probe (double-coil type) fits into the magnet gap and holds the sample. The RF receiver coil is oriented with its axis perpendicular to the direction of the principal magnetic field and to the axis of the oscillator coil (Fig. 20.10). The sample tube is mounted on a light turbine, and a jet of air is adjusted to provide a steady spinning rate of around 30 Hz.

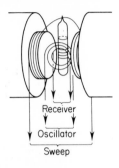

Receiver

Oscillator

Sweep

Fig. 20.10 Arrangement of sample and coils in the magnet gap

(vi) A data recording system. The oscilloscope and the XY-recorder are synchronized with the output of the sweep generator. A small computer attached to the NMR spectrometer can perform the operation called COMPUTER AVERAGING OF TRANSIENTS (CAT).

20.2.3 Computer Averaging of Transients (CAT)

With low polymer concentrations (less than 0.5 wt% of solute), the signal-to-noise ratio on the NMR spectrum becomes lower, since it is necessary to work at higher amplifications. It becomes difficult to distinguish true NMR peaks from the noisy baseline. Over a series of scans, therefore, the sum of all noise signals will be zero. The signals coming from the NMR spectrometer are stored in a computer memory disc. After averaging out all the transient noise signals to zero, all true NMR signals from the sample will appear at exactly the same place in the spectrum, and the algebraic summing of these signals gives signal enhancement. Instead of the CAT computer one can use a digital signal averager (DSA).

20.2.2 Sample Tubes for NMR Measurements

Sample tubes may have an influence on NMR measurements. Imperfections in the cylindrical symmetry of the sample and/or sample tube will give inhomogeneities in the magnetic field strength. These imperfections in symmetry can give significant variations in chemical shift and line-width values.

Several types of commercial microcell (Fig. 20.11) are in use. The microcell consists of a standard NMR sample tube, a Teflon holder, and a small glass bulb (30–100 µl capacity) in which the sample is placed. The sample bulb and Teflon plug are positioned in the NMR tube using a Kel-F rod, which is removed before the cell is placed into the instrument probe. The space around

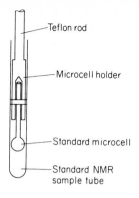

Fig. 20.11. Standard micro-cell and sample tube for NMR spectroscopy

the microsample bulb—below the Teflon holder—is filled with carbon tetra-chloride to minimize further any tendency for the sample bulb to wobble when the tube is rotated.

NMR tubes are expensive and they should be cleaned and re-used whenever possible. The best way to clean the tube is to rinse it with solvent immediately after use. Chromic acid cleaning solutions should never be used since it is easy to leave behind a trace of paramagnetic chromium ions that will result in line-broadening with subsequent samples.

References: 2539, 4993.

20.3 SOLVENTS USED IN NMR SPECTROSCOPY

The sharpest NMR spectra can only be obtained in non-viscous solutions. Choice of solvent for polymers is sometimes difficult. Solvents of low viscosity are highly volatile. In order to reduce dipolar broadening it is necessary to record NMR spectra at elevated temperatures (100–150° C). The typical solvents for polymers are normal organic solvents in which hydrogen has been replaced by deuterium (Table 20.4, Fig. 20.12). For deuterated solvents the isotopic purity should ideally be as high as possible. Good NMR spectra of polymers can be obtained at concentrations of 1–2%.

The NMR spectrum of a sample dissolved in one solvent may be slightly different from that measured in a more polar solvent, and it is important in all NMR work to quote the solvent used. The NMR signals for protons attached to carbon are, in general, shifted only slightly by changing the solvent. However, this is not the case for OH, NH, and SH protons, the signals for which move noticeably on changing to solvents of differing polarity (inter- or intramolecular hydrogen bonding). At high concentrations (strong hydrogen bonding), OH, NH, and SH protons appear at higher δ than in dilute solutions. The resonance positions for these protons are temperature dependent (higher temperatures—lower δ values) because increasing temperature reduces hydrogen bonding.

Table 20.4 Solvents for NMR work

Solvent	Formula	Approximate δ for ^1H equivalent as contaminant	Boiling point (°C)	Freezing point (°C)
Carbon disulphide	CS_2	—	46	−108.5
Carbon tetracholoride	CCl_4	—	77	− 23
Deuteroacetic acid	CD_3COOD	13 and 2	118	16.6
Deuteroacetone	CD_3COCD_3	2	56	− 95
Deuteroacetonitrile	CD_3CN	2	82	− 44
Deuterobenzene	C_6D_6	7.3	80	5.5
Deuterium oxide	D_2O	5	101.5	3.8
Deuterochloroform	$CDCl_3$	7.3	61	− 63
Deuteromenthanol	CH_3OD	3.4	65	− 98
Deuteropyridine	C_6D_5N	7.5	115	− 42
Deuterotoluene	$C_6D_5CD_3$	7.3 and 2.4	110	− 95
Deuterotrifluoroacetic acid	CF_3COOD	13	72	− 15
Deuterodimethyl sulphoxide	$(CD_3)_2SO$	2	189	18
Hexachloroacetone	CCl_3COCCl_3	—	203	− 2

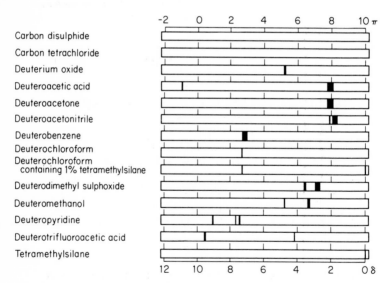

Fig. 20.12 Solvents used in NMR spectroscopy. The chart depicts regions in which weak bands due to minor impurities (e.g. incompletely deuterated analogues) are likely to be found. The exact positions of these bands are dependent on experimental conditions, especially temperature and concentration. (From B: 1471; reproduced by permission of BDH, England)

Polymer solutions used for NMR spectroscopy should be absolutely free from insoluble material (gel). Gel can be removed by filtration through Millipore filters and/or high-speed centrifugation (cf. Section 13.1.10).

20.4 DEUTERATION IN NMR SPECTROSCOPY

In polymers, deuterium oxide (D_2O) exchanges with labile protons such as those from OH, NH, and SH groups:

$$ROH + D\text{—}O\text{—}D \longrightarrow ROD + H\text{—}O\text{—}D, \qquad (20.9)$$

$$RCOOH + D\text{—}O\text{—}D \longrightarrow RCOOD + H\text{—}O\text{—}D. \qquad (20.10)$$

Peaks previously observed for the labile protons (^1H) disappear or are diminished and a new peak corresponding to H—O—D appears at around 5δ. The NMR spectrum can be recorded in a solvent other than D_2O and then a few drops of D_2O are shaken with the sample solution and the spectrum is re-recorded.

20.5 CONTACT-SHIFT REAGENTS

Contact-shift reagents spread our NMR absorption patterns without increasing the strength of the applied magnetic field (Fig. 20.13). The contact-shift

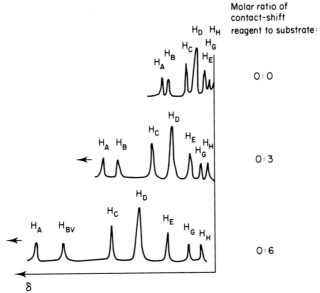

Fig. 20.13 Effect of the contact-shift reagent on the NMR spectrum

reagents are ions in the rare earth (lanthanide) series coordinated to organic ligands e.g.:

(i) tris-(dipivaloylmethanato)europium ($Eu(DPM)_3$);

(ii) tris-1,1,1,2,2,3,3,-heptafluoro-7,7-dimethyl-3,5-octanedionato europium ($Eu(FOD)_3$);

(iii) tris-2,2,6,6-tetramethyl-3,5-heptane-4,6-dionato europium ($Eu(FHD)_3$);

318

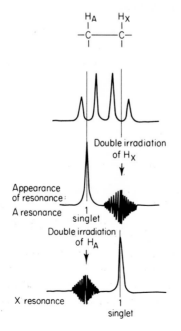

Eu(DPM)₃: $R_1 = R_2 = Me_3$
Eu(FOD)₃: $R_1 = CMe_3, R_2 = C_3F_4$
Eu(FHD)₃: $R_1 = R_2 = C_2F_5$

The most effective solvents for contact-shift reagents are carbon tetrachloride, deuterobenzene, and deuterochloroform.

It is necessary to report the mole ratio of shift reagent to substrate, concentration, solvent, etc., for a given shift reagent experiment.

Bibliography: 1211, 1213.
References: 2686, 4070, 4229, 5577.

20.6 SPIN–SPIN DECOUPLING TECHNIQUES

20.6.1 Double Resonance

DOUBLE RESONANCE (DOUBLE IRRADIATION) is a technique which uses two radio-frequency sources in the basic NMR spectrometer. This technique is applied for simplifying NMR spectra. In the case of two neighbouring protons, H_A and H_X, multiplicity of signals arises (Fig. 20.14). Proton H_A appears as a doublet because of the two spin orientations (parallel (low energy) and antiparallel (high energy)) of proton H_X. Proton H_X also appears as a

Fig. 20.14 Double irradiation (spin decoupling) of an AX system

doublet because of the two spin orientations (parallel (low energy) and anti-parallel (high energy)) of proton H_A. If proton H_X is irradiated with the correct strong radio-frequency at its resonance frequency, the transitions between the two spin states (parallel and antiparallel orientations) of proton H_X are very rapid, so that the lifetime of proton H_X in any one spin state is too short to resolve the coupling with proton H_A. Thus proton H_A will resonate only once, and the previously observed doublet collapses to a singlet (Fig. 20.14). This will also happen when proton H_A is irradiated instead of proton H_X. This process is called SPIN–SPIN DECOUPLING.

To perform this operation one requires a basic NMR spectrometer with two radio-frequency sources. One proton is irradiated with a strong radio-frequency signal at its resonance frequency, while the other nuclei are scanned to detect which ones are affected by decoupling from the irradiated proton.

In addition to complete spin–spin decoupling (total collapse of multiples) and selective spin–spin decoupling (partial collapse of multiples) there are three types of partial spin–spin decoupling:

(i) spin tickling (splitting of lines into multiples);
(ii) the nuclear Overhauser effect (increasing of the absorption signal);
(iii) internuclear double resonance (INDOR) (a change in the more sensitive nucleus signal while irradiating the less sensitive nucleus).

Bibliography: 67, 91, 177, 631, 730, 1115, 1149, 1213.

20.6.2 Spin Tickling

SPIN TICKLING involves irradiating a nucleus with a much less intense radiation than is necessary for complete or selective decoupling. The effect is to increase the number of lines in the coupled absorption. This technique is seldom used in NMR spectroscopy of polymers.

Bibliography: 91.

20.6.3 The Nuclear Overhauser Effect (NOE)

The NUCLEAR OVERHAUSER EFFECT (NOE) occurs when two protons (H_A) and (H_B) (or two carbon atoms $(^{13}C_A)$ and $(^{13}C_B)$) are close enough to each other to allow interaction of their fluctuating magnetic vectors (each contributes to the other's spin–lattice relaxation process).

The double irradiation at the H_A signal will give rise to absorption and emission processes for H_A. This stimulation will be transferred via the fluctuating magnetic vectors to the relaxation mechanism of proton H_B. The spin–lattice relaxation of proton H_B will be speeded up, providing an increase in the NMR absorption signal of H_B (10–50%).

NOE experiments should be carried out in an oxygen-free atmosphere because molecular oxygen is paramagnetic and can contribute in the spin–relaxation processes (cf. Section 20.9).

Bibliography: 91, 144, 975, 1001, 1002, 1241.
References: 3971, 3974, 4334, 4335, 4760, 4830, 5168, 6202, 6205, 6209, 6214, 6225, 6479.

(Note: several papers discussing application of NOE technique in NMR spectroscopy of polymers are included in references in Sections 20.7 and 20.9.)

20.6.4 Internuclear Double Resonance

INTERNUCLEAR DOUBLE RESONANCE (INDOR) is a technique applied for examination of coupling between two nuclei of very different relative sensitivities (e.g. ^1H and ^{13}C, ^1H and ^{15}N).

Normally, if such nuclei are coupled, the spectrum of the less sensitive nucleus cannot be obtained. By applying the INDOR technique this spectrum can be registered indirectly by observing the height of a peak in the spectrum of the high-sensitivity nucleus while irradiating the low-sensitivity nucleus with a narrow band frequency. Irradiation of the low-intensity peak results in a proportionate decrease in the intensity of the high-sensitivity peak. A plot of these changes versus sweep frequency (in hertz) gives an image of the low-intensity spectrum. The INDOR spectra are recorded directly on the same trace as the normal NMR spectrum in such a way that the INDOR peaks line up with the original transitions that are being affected. The INDOR technique is difficult and time-consuming.

Bibliography: 91, 730, 1213.

20.7 ^{13}C NMR SPECTROMETRY

^{13}C has a nuclear spin $I = \frac{1}{2}$ and can be observed by NMR at a precessional frequency v_0 (cf. Table 20.2):

$v_0 \approx 15.1$ MHz at a field $H_0 = 14\,000$ G (60 MHz instrument);

$v_0 \approx 25.2$ MHz at a field $H_0 = 23\,000$ G (100 MHz instrument).

The relative abundance of ^{13}C is only 1.1% in comparison to ^{12}C. The ^{13}C resonance has only 1.6% of the sensitivity of the proton (^1H) resonance, and the relaxation time for ^{13}C is longer than for the proton (^1H) (i.e. of the order of minutes). The proton NMR signal from a ^{13}C—^1H bond is split into a doublet (symmetrically astride the main proton ^1H peak). These peaks are known as ^{13}C SATELLITE PEAKS.

The coupling constants (J) for ^{13}C—^1H and ^{13}C—^{12}C—^1H are ~ 100 MHz and ~25 MHz, respectively (i.e. they are large) and thus interpretation of

^{13}C spectra can be difficult because of overlapping of ^{13}C—^{1}H multiplets. In order to simplify the ^{13}C spectrum, ^{13}C nuclei can be completely decoupled from all of the protons (^{1}H) by use of the double resonance technique (cf. Section 20.6.4). The spectrum is then simply a series of singlets corresponding to each variety of carbon atom present.

Since decoupling can interfere with relaxation times, the application of the nuclear Overhauser effect (NOE) (cf. Section 20.6.3) may lead to signal enhancement of ^{13}C peaks.

One solution to the low natural abundance and small magnetic moment of ^{13}C is to use the technique of computer averaging of transients (CAT) (cf. Section 20.2.2) to scan a large number of successive runs. A tenfold signal-to-noise enhancement requires 100 stored spectra, and this takes a lot of time, because the time of one scan is about 3 min.

Bibliography: 61, 200, 201, 299, 690, 814, 815, 947, 1091, 1271.

20.7.1 Applications of ^{13}C NMR Spectroscopy in Polymer Research

The application of ^{13}C NMR spectroscopy in polymer research includes uses in the study of

(i) stereochemistry of macromolecules (Table 20.5), e.g. structural isomerism, steric isomerism, conformation of macromolecules, and helical conformation;
(ii) short and long chain branching;
(iii) the structure of crosslinked gels;
(iv) mechanisms of polymerization;
(v) mechanisms of polymer oxidation and degradation processes.

Table 20.5 ^{13}C NMR spectroscopy of polymers

Polymer	Published in
Hydroxyethyl cellulose	R: 3127
Poly(N-acryliminoalkanes)	R: 4258
Poly(acrylonitrile)	R: 2213
Poly(alkene sulphides)	R: 3358
Poly(buta-1,2-dienes)	R: 3298, 3290, 3561, 5223
Poly(buta-1,4-dienes)	R: 2448, 2880, 2881, 2949, 2950, 3128, 3239, 3289– 3292, 3298, 3561, 3562, 3563, 5323, 5960, 5971, 5974, 6617, 6727, 6731, 6732, 6771
Poly(but-1-ene)	R: 5151
Poly(chloroprene)	R: 3262
Poly(cyclic butadienes)	R: 5933
Poly(ethylene)	R: 6498
Poly(ethylene oxide)	R: 5125
Poly(isoprene)	R: 3186, 3239, 3789, 5348, 6728, 6729, 6733, 6734

Table 20.5 (*cont*)

Polymer	Published in
Poly(methacrylonitrile)	R: 4230, 4231
Poly(3-methylbut-1-ene)	R: 6730
Poly(4-methylhex-1-ene)	R: 2948
Poly(methyl methacrylate)	R: 4326
Poly(α-methylstyrene)	R: 3299
Poly(N-formylpropylene-imines)	R: 3868
Poly(pentadiene)	R: 3293, 3599
Poly(pentamer)	R: 2726
Poly(propylene)	R: 2152, 3008, 3182, 4232, 4234, 4326, 5968, 5972, 5973, 7156, 7309, 7310, 7313
Poly(propylene imine)	R: 3867
Poly(propylene oxide)	R: 6204
Poly(styrene)	R: 3263, 4233, 4326, 4831, 5132, 5969, 5973
Poly(styrene peroxide)	R: 2670
Poly(vinyl alcohol)	R: 5534, 7178, 7180
Poly(vinyl acetate)	R: 7179
Poly(vinyl carbazole)	R: 4480, 6843
Poly(vinyl chloride)	R: 2001, 2124, 2721, 2724, 2725, 4228, 5813, 6205
Poly(vinyl methylether)	R: 4326
Poly(vinyl pyridine)	R: 2575, 3629, 4987, 5128
Butadiene–isoprene copolymer	R: 4956
Butadiene–methacrylonitrile copolymer	R: 3297
Butadiene–styrene copolymer	R: 6328
Butadiene–propylene copolymer	R: 2369
Ethylene oxide–propylene oxide copolymer	R: 7061
Ethylene–propylene copolymer	R: 7308
α-Methylstyrene–methacrylo-nitrile copolymer	R: 3296
Propylene–butadiene copolymer	R: 2722, 5975
Propylene–ethylene copolymer	R: 2517, 2723, 2727, 3008, 3294, 4326, 5982, 6181, 6726, 6814, 7090, 7311
Styrene–acrylonitrile copolymer	R: 6205, 6557

References: 3029, 3216, 3629, 3810, 3971, 3974, 3988, 4258, 4545, 4608, 4722, 4760, 4956, 5125, 5126, 5131, 5151, 5362, 5556, 5968–5971, 6061, 6181, 6205, 6208, 6211, 6214, 6502, 6558, 6729, 7266, 7321.

20.8 FOURIER TRANSFORM (FT) NMR SPECTROSCOPY

FOURIER TRANSFORM (FT) NMR SPECTROSCOPY (PULSED NMR) is a technique which uses a series of short radio-frequency (rf) pulses ($\approx 30\mu s$) instead of a continuous signal (as in normal NMR measurements).

For proton (^{1}H) FT NMR, the sample is irradiated in a fixed magnetic field with a strong pulse of radio-frequency energy containing all the frequencies over the proton (^{1}H) range. The protons in each environment absorb their frequencies from the pulse and couple with one another to form all the coupling energy sublevels.

When the pulse is switched off all the protons undergo relaxation processes and re-emit the absorbed energies and coupling energies simultaneously.

These re-emitted energies give a TIME DOMAIN—a complex interacting pattern—which decays rapidly (Fig. 20.15(a)). This output is digitalized in a computer and transformed to the FREQUENCY DOMAIN—each individual frequency is filtered out from the complex interacting pattern (Fig. 20.15(b)) (FOURIER TRANSFORMS). Then the signals are converted into a conventional NMR spectrum (Fig. 20.15(c)).

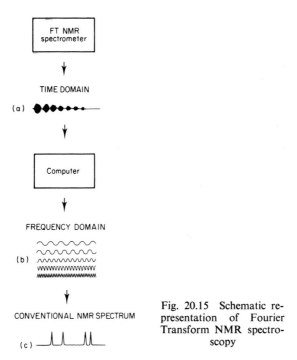

Fig. 20.15 Schematic representation of Fourier Transform NMR spectroscopy

An entire spectrum can be recorded, computerized, and transformed in a few seconds (in normal NMR measurements, spectra require several minutes to record).

Fourier transform (FT) NMR spectroscopy is used in

(i) detection of weak signals;
(ii) studies on samples at very low concentrations (10–50 µg);
(iii) studies on samples of low natural abundance and with small magnetic moments (e.g. ^{13}C);

(iv) determination of spin–lattice relaxation times (cf. Section 20.9).

Bibliography: 187, 414, 415, 429, 430, 475, 624, 947.
References: 2353, 3944, 4235, 5710, 6205, 6206.

20.9 NUCLEAR MAGNETIC DIPOLE INTERACTIONS

Any magnetic nucleus in a molecule supplies an instantaneous magnetic dipole, which is proportional to the magnetic moment of the nucleus.

This fluctuating magnetic dipole field depends on

  (i) the magnitudes of the nuclear moments;
 (ii) the distance between the nuclei in a solid state;
(iii) the concentration of molecules possessing magnetic nuclei in solution;
(iv) the frequency distribution of the molecular motion.

The CORRELATION TIME (τ_c) is a measure of the average time that two nuclei remain in a given relative orientation:

  (i) for the rotational process (related to an INTRAMOLECULAR RELAXATION) the correlation time (τ_c) is the average time that the molecule requires to rotate through an angle of one radian;
 (ii) for the translational process (related to an INTERMOLECULAR RELAXATION) the correlation time (τ_c) is the average time that the molecule takes to move one molecular diameter.

Both intramolecular and intermolecular relaxations depend on the viscosity and on the size and shape of the macromolecules.

The dependence of T_1 upon τ_c is illustrated in Fig. 20.16:

  (i) for rapid molecular motion,

$$1/\tau_c \geqslant 2\pi v_0 \quad \text{and} \quad 1/T_1 \sim \tau_c; \tag{20.11}$$

 (ii) for very slow motion,

$$1/\tau_c \leqslant 2\pi v_0 \quad \text{and} \quad 1/T_1 \sim 1/\tau_c; \tag{20.12}$$

(iii) a correlation time of

$$\tau_c = 1/2\pi v_0, \tag{20.13}$$

where v_0 is the Larmor frequency (cf. equation 20.3), leads to the most effective spin–lattice relaxation, with T_1 increasing for shorter or longer τ_c (Fig. 20.16).

The dependence of T_2 upon τ_c is also illustrated in Fig. 20.16:

  (i) for short τ_c (rapid molecular motion) T_2 coincides with T_1;
 (ii) for long τ_c (slow molecular motion), T_2 differs from T_1—long correlation time permits dipole–dipole interactions to become effective, leading to broad lines.

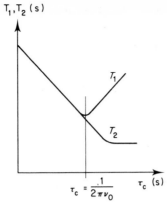

Fig. 20.16 The variation of T_1
and T_2 with correlation time (τ_c)

The dipole–dipole interaction for a pair of identical nuclei of spin $I = \frac{1}{2}$ (e.g. protons) will yield a static local magnetic field (H_{loc}) given by

$$H_{loc} = 3\mu/r^3 \, (\cos^2 \theta - 1), \tag{20.14}$$

where
  μ is the magnetic moment of the proton,
  r is the distance between the two protons (cf. Fig. 20.17),
  θ is the angle between the line joining the two protons and the direction of the applied magnetic field H_0.

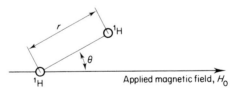

Fig. 20.17. A pair of protons ^1H in an
applied magnetic field (H_0)

In general we observe the following:

(i) In solid polymers where the angle θ is fixed, the static local field (H_{loc}) can be quite large ($H_{loc} \sim 14\,G$ for a proton at a distance of 1 Å).

(ii) In solid polymers where macromolecules are in motion, the angle θ has an average value of $\bar{\theta}$. Incomplete averaging of θ leads to broader lines in the NMR spectrum.

(iii) If the motion of macromolecules is random and rapid (short τ_c) $\cos^2 \theta$ can be averaged over all space to give $\overline{\cos^2\theta} = \frac{1}{3}$ and in this case the local field is $H_{loc} = 0$. For that reason, measurements of NMR spectra of polymers in solution are usually carried out, when possible, at elevated temperatures in order to decrease τ_c and thus sharpen the resonance lines.

T_1 and T_2 are related to the width, shape, and intensity of the NMR absorption line. These quantities can be derived as follows:

(i) T_1 can be found directly by using rapid scans and observing the growth of the signal immediately after inserting the sample in the magnetic field.
(ii) T_2 may be obtained from the line-width so long as it is less than about 1 s:

$$\Delta v_{1/2} T_2 \approx 1/\pi \qquad (20.15)$$

where $\Delta v_{1/2}$ is the width of the line at half maximum intensity. For liquids $T_2 \simeq T_1$ and this method cannot be employed.
(iii) When $T_1 < 2-3$ s or $T_2 > \sim 1$ s the pulsed NMR technique (cf. Section 20.8) is commonly used.
(iv) The true value can be determined by the SPIN ECHO method (not discussed in this book).

Bibliography: 91, 146, 174, 177, 318, 831, 885, 1150, 1223.
References: 2087, 2119–2121, 2168, 2186, 2195, 2506, 2518, 2522, 2600, 2661, 2670–2672, 2769, 2920, 2981, 3041, 3079, 3219, 3220, 3224, 3225, 3256, 3435, 3436, 3524–3530, 3616, 3628, 3629, 3745, 3786–3788, 3790, 3942, 3943, 3970, 3971, 3976, 4025, 4074, 4128, 4229, 4235, 4251, 4333–4335, 4440, 4556, 4650–4652, 4821, 4822, 4830, 4831, 4950, 4951, 4953, 4998–5000, 5070, 5169–5171, 5173, 5367, 5426, 5494, 5497, 5541, 5972, 6107, 6108, 6110, 6209, 6225, 6402, 6426, 6440, 6441, 6500, 6501, 6514, 6559, 6584, 7010, 7101, 7346 (cf. also references in Section 20.11).

20.10 THE NMR SECOND MOMENT

The NMR SECOND MOMENT $(\overline{\Delta H^2})$ is a function of the shape and the width of the NMR resonance signal and is given by

$$\overline{\Delta H^2} = \int_{-\infty}^{+\infty} (H - H_0)^2 f(H) \, dH \Big/ \int_{-\infty}^{+\infty} f(H) \, dH, \qquad (20.16)$$

where

$f(H)$ is the line shape, i.e. the amplitude of the resonance signal as a function of the magnetic field H,

$(H-H_0)$ are measures (in gauss) of the departure of each point in the spectrum from the central resonance position H_0.

The NMR resonance signal can be presented as an absorption spectrum or first-derivative spectrum (Fig. 20.18). The half-height width $(\Delta v_{1/2})$ (Fig. 20.18(a)) and the peak-to-peak (width Δv_{pp}) (the distance between extrema) (Fig. 20.18(b)) are equal to $1/\pi T_2$, where T_2 is the half-life of the spin–spin relaxation process (the spin–spin relaxation time).

The width of the NMR signal decreases with increasing temperature (Fig. 20.19).

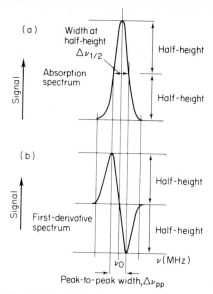

Fig. 20.18   NMR signals. (a) Absorption
spectrum, ν mode signal curve; (b) first
derivative curve of absorption spectrum

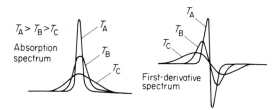

Fig. 20.19   Decrease of signal widths with increasing
temperature

The NMR SECOND MOMENT ($\overline{\Delta H^2}$) is related to the molecular chain
orientation in rigid solid polymers by

$$\overline{\Delta H^2} = \frac{G}{N}\sum_{ij} r_{ij}^{-6}\overline{(3\cos^2\theta - 1)^2} \tag{20.17}$$

(in square gauss (CGS) or square tesla (SI)), where

$$G = \tfrac{3}{2}I(I+1)g_N^2\beta_N^2 \tag{20.18}$$

I is the spin quantum number of the determined nuclei being examined,
g_N is the nuclear g factor (cf. page 301),
β_N is the nuclear magneton (cf. page 301),
N is the number of nuclei over which the sum is taken,
r_{ij} is the distance between the nuclei i and j being examined,

θ is the angle between the line joining the i and j nuclei and the direction of the applied magnetic field H_0.

The effect of molecular motion can be observed as a reduction in the second moment $(\overline{\Delta H^2})$ when the frequency associated with the motion exceeds the width of the resonance spectrum (in frequency units).

Experimentally a reduction in the second moment is observed as the temperature of the sample increases and the frequencies associated with the molecular motion increases (Fig. 20.20).

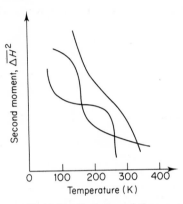

Fig. 20.20 Variation of the second moment $(\overline{\Delta H^2})$ with temperature for different polymers

The decrease in the second moment $(\overline{\Delta H^2})$ with increasing temperature is caused by the thermal motion of the atoms leading to a reduction in the average value of the $(3\,\overline{\cos^2\theta} - 1)$ and r^{-6} terms in eq. (20.17).

Bibliography: 145, 146, 885, 1347.
References: 2013, 3038, 3260, 3830, 3836, 4015, 4442, 4939 (cf. also references in Section 20.9).

20.11 APPLICATIONS OF BROAD-LINE NMR SPECTROSCOPY

The main applications of measurements of proton relaxation by direct observations of broad-line NMR spectra or by pulse NMR methods are

(i) the study of polymer morphology;
(ii) the study of molecular motions.

The NMR signal of rigid molecules in a crystal or in the amorphous phase in a polymer sample below the glass transition temperature consists of a comparatively broad line. The half-height width of such a line is about 10–20 G. The broadening of the line is caused by the local magnetic fields of the protons in the sample. With increasing mobility of molecules, the line-width decreases because the average time that two nuclei remain in a given relative orientation decreases. Therefore, for example, the signal of the molecules in an amorphous polymer sample has a half-height width of only 0.1 G or less.

As a consequence of the influence of mobility on line-width, the NMR signal of a partially crystalline polymer sample above the glass transition temperature consists of two components (Fig. 20.21):

(i) a narrow component caused by the chains in the amorphous regions which are able to perform segmental motions;
(ii) a broad component caused by the rigid chains in the crystalline regions and by a rigid chain fraction which can sometimes be found in amorphous regions.

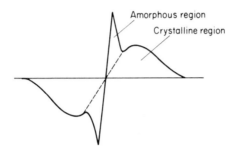

Fig. 20.21 Broad-line NMR signal of a
partially crystalline polymer

The study of broad-line NMR spectra can also give information on the structure in the melt, e.g. the formation of bundles of parallel chains causes deviations from the Lorentzian shape of the NMR line.

Bibliography: 20, 174, 461, 581, 885, 1063, 1145, 1222.
References: 2196, 2221, 2368–2370, 3065, 3171, 3277, 3279, 3382, 3703, 4127, 4441, 4557, 4583, 4675, 4745, 4840, 4958, 5175, 5361, 5543, 5591, 5814, 5885, 6211, 6235, 6261, 6326, 6422, 6424, 6444, 6445, 6968, 7126, 7302–7304 (cf. also references in Section 20.9).

20.12 APPLICATIONS OF HIGH RESOLUTION NMR SPECTROSCOPY IN POLYMER CHEMISTRY

The applications of high resolution NMR spectroscopy to polymer characterization include

(i) study of the configuration of polymer chains (the polymer shape formed by primary valency bonding);
(ii) study of the conformation of polymer chains (the polymer shape due to rotation around primary valence bonds);
(iii) the study of sequence distribution and tacticity in polymers and copolymers;
(iv) differentiation between polymer mixtures, block copolymers, alternating copolymers, and random copolymers;

(v) the study of helix-coil transitions;
(vi) studies of molecular interactions in polymer solutions;
(vii) the study of diffusion in polymer films (cf. Section 37.6);
(viii) the study of polymer compatibility and polymer blends;
(ix) the study of crosslinking;
(x) investigation of the mechanism of propagation in vinyl polymerization.

Some selected examples of the NMR study of polymer microstructure are given in Table 20.6.

Table 20.6 ^{1}H NMR spectroscopy of polymers

Polymer	Published in
Poly(acetaldehyde)	R: 3709
Poly(acryl amide)	R: 7265
Poly(acrylate esters)	R: 2905, 5133, 7284
Poly(acrylic acid)	R: 5127
Poly(acrylonitrile)	R: 5394
Poly(butadiene)	R: 3946, 3947, 4348, 6184, 6185, 6604, 6731, 6735, 7350
Poly(chloroprene)	R: 3363, 3365, 3369, 3597, 5573
Poly(cyclic butadienes)	R: 5933
Poly(esters)	R: 5768
Poly(ethers)	R: 2602, 3189, 4952, 5446, 5617, 5959
Poly(ethylene)	B: 753; R: 3321, 3368, 3728
Poly(ethylene oxide)	R: 6738
Poly(isoprene)	R: 3186, 6207, 6264, 6735
Poly(methacrylic acid)	R: 4607
Poly(methyl acrylate)	R: 5134
Poly(methyl acrylonitrile)	R: 4243, 5130, 6327, 6645, 7207
Poly(methyl methacrylate)	B: 173; R: 2520, 3366, 3367, 3369, 3521, 3945, 5957, 6497, 6499, 6500
Poly(α-methylstyrene)	R: 4883, 5404, 5492
Poly(perfluorobutadiene)	R: 6824
Poly(propylene)	R: 3366, 3368, 3409, 3418, 3520, 3975, 3977, 4464, 4961, 5567, 5568, 6198, 6326, 6530, 6531, 6787, 7167, 7168, 7312, 7314, 7315
Poly(propylene oxide)	R: 2945, 5557, 5558
Poly(propylene sulphide)	R: 4259
Poly(styrene)	R: 2516, 2601, 3549, 3972, 6329, 6619, 7276
Poly(urethanes)	R: 6618
Poly(N-vinyl acetamide)	R: 5393
Poly(vinyl acetate)	R: 5958
Poly(vinyl alcohol)	R: 3126, 5360, 7178
Poly(vinyl amine)	R: 5393
Poly(vinyl carbazole)	R: 3994, 6843, 7103, 7283

Table 20.6 (*cont.*)

Polymer	Published in
Poly (vinyl chloride)	R: 2240, 2461, 2514, 2515, 2521, 2761, 2854, 3319, 3940, 3973, 4318, 4319, 5803, 5956, 6197, 6365, 6785, 6786, 7089, 7285
Poly(vinyl chloride sulphone)	R: 2673
Poly(vinyl fluoride)	R: 7128
Poly(vinyl ketone)	R: 5129, 5236
Poly(vinyl pyridine)	R: 4987, 5128, 7031
Poly(vinylidene fluoride)	R: 7125
Methyl methacrylate–methyl acrylate acid copolymer	R: 4317
Methyl methacrylate–chloroprene copolymer	R: 3261

Bibliography: 20, 170, 172–175, 177, 178, 280, 341, 683, 778, 831, 887, 1084, 1113, 1165, 1190, 1402, 1416.
References: 3295, 3417, 5357, 5377, 5577, 7055, 7226.

20.13 MEASUREMENT OF THE DEGREE OF CRYSTALLINITY IN POLYMERS BY NMR SPECTROSCOPY

From nuclear magnetic resonance measurement the degree of crystallinity (χ_c is defined (Fig. 20.21) as

$$\frac{\chi_c}{1-\chi_c} = \frac{\text{Area of broad component}}{\text{Area of narrow component}}.$$

(20.19)

Bibliography: 555, 713, 885, 1222, 1225, 1427.
References: 2369, 2938, 3278, 3350, 5842, 5889, 5890, 6425, 6427, 6966, 7126, 7127.

Chapter 21

Electron Paramagnetic Resonance Spectroscopy

General bibliography: 33, 50, 51, 60, 71, 111, 147, 212, 274, 473, 505, 506, 573, 669, 693, 794, 799, 842, 904, 959, 988, 1005, 1050, 1071, 1088, 1095, 1157, 1166, 1175, 1193, 1248, 1279, 1326, 1375.

ELECTRON PARAMAGNETIC RESONANCE SPECTROSCOPY (EPR), also known as ELECTRON SPIN RESONANCE SPECTROSCOPY (ESR), is a technique which records transitions between spin levels of molecular unpaired electrons in an external magnetic field. EPR (ESR) spectroscopy involves absorption of energy from electromagnetic radiation in the microwave frequency region by a sample placed in an external magnetic field. Absorption is a function of unpaired electrons present in the molecule. A plot of the absorption of microwave energy versus the external magnetic field gives an EPR (or ESR) SPECTRUM.

The electrons in atoms and molecules form pairs. For each electron in a certain orbital with the SPIN QUANTUM NUMBER $M_S = -\frac{1}{2}$, there is another electron in the same orbital with the spin quantum number $M_S = +\frac{1}{2}$. The paired electrons do not give an ESR spectrum. An unpaired electron has no other electron as partner in the same orbital, and gives an ESR spectrum.

The fundamental properties of an electron are

(i) Mass, $m_e = 9.112 \times 10^{-28}$ g (CGS) $= 5.486 \times 10^4$ u $= 9.112 \times 10^{-31}$ kg (SI) where u is the universal unit of mass (u $= 1.661 \times 10^{-27}$ kg).

(ii) Charge, $e = 4.803 \times 10^{-10}$ esu (CGS) $= 1.60210 \times 10^{-19}$ A s (SI).

(iii) Intrinsic angular momentum, known as the SPIN. This is a vector denoted by the symbol **S**. The component of the spin vector **S** along any direction may only take the values $\pm\frac{1}{2}\hbar$.

The symbol $\hbar = h/2\pi$, is read 'h bar' and

$$\hbar = 1.054 \times 10^{-27} \text{ erg s (CGS)} = 1.054 \times 10^{-34} \text{ J s (SI)},$$

where h is the Planck constant and

$$h = 6.626 \times 10^{-27} \text{ erg s (CGS)} = 6.626 \times 10^{-34} \text{ J s (SI)}.$$

(iv) Magnetic moment. This is a vector denoted by the symbol $\boldsymbol{\mu}_e$. It is related to the intrinsic angular momentum (**S**) by the following equation:

$$\boldsymbol{\mu}_e = -\frac{g\beta}{\hbar}\mathbf{S}, \qquad (21.1)$$

where

g is the g-factor (also called the LANDE g-factor) (dimensionless)— $g = 2.002\,319$ for unbound electrons,

β is the BOHR MAGNETON—$\beta = 9.2740 \times 10^{-21}$ erg gauss^{-1} (CGS) $= 9.2740 \times 10^{-24}$ A m^2 (SI) (it is also commonly denoted by the symbol μ_0).

The minus sign in eq. (21.1) shows that the magnetic moment vector ($\boldsymbol{\mu}_e$) of the electron is opposite in direction to the angular momentum $\mathbf{S}$.

In addition to its spin, an electron can have additional angular momentum (called ORBITAL ANGULAR MOMENTUM) as it moves not only around its own axis but also in an orbit.

The electron with spin quantum number $M_S = \frac{1}{2}$, placed in an external magnetic field (H_0) will have only two orientations (Fig. 21.1(a)):

(i) aligned with the field (parallel orientation) (the lower energy state);
(ii) opposed to the field (antiparallel orientation) (the higher energy state).

The precessing electron in the parallel orientation can absorb energy (ΔE) from the microwave-frequency source and pass into the antiparallel orientation (this phenomenon is called ZEEMAN SPLITTING FOR AN ELECTRON) under one particular condition: if the precessing frequency is the same as the frequency of the microwave beam (such a condition is called a ELECTRON SPIN RESONANCE). The absorbed energy is recorded in the form of an ESR (or EPR) SPECTRUM (Fig. 21.1(b) or (c)).

If the populations of the two states (the higher energy state and the lower energy state) approach equality, no further net absorption of energy will occur, and the observed resonance signal will fade out. This phenomenon is called SATURATION OF THE ESR (or EPR) SIGNAL.

Under normal measurement conditions, the populations in the two states are not equal because higher energy electrons are constantly returning to the lower energy state. Two radiationless processes play an important role in the loss (RELAXATION) of energy by high-energy electrons:

(i) the SPIN–LATTICE RELAXATION PROCESS, in which the energy difference (ΔE) is transferred to neighbouring atoms, either in the same molecule or in other molecules;
(ii) the SPIN–SPIN RELAXATION PROCESS, in which the energy difference (ΔE) is transferred to neighbouring electrons.

The rates of relaxation by these two processes are important and are given by

(i) the HALF-LIFE OF THE SPIN–LATTICE RELAXATION PROCESS (T_1) in seconds (commonly called the SPIN–LATTICE RELAXATION TIME) (T_1);

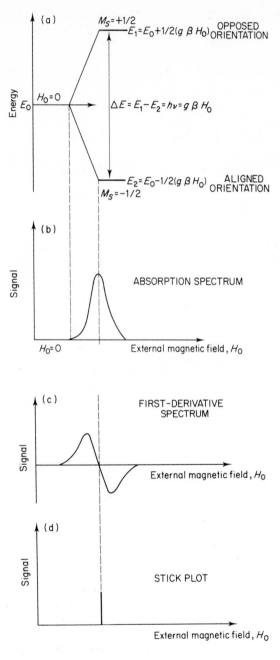

Fig. 21.1 (a) Energy level diagram showing Zeeman splitting for an electron system in crossed (H_0) and microwave magnetic fields. (b) ESR (EPR) absorption line; H_0 mode signal curve. (c) ESR (EPR) first-derivative line. (d) Stick plot of a first derivative line

(ii) the HALF-LIFE OF THE SPIN–SPIN RELAXATION PROCESS (T_2) in seconds (commonly called the SPIN–SPIN RELAXATION TIME) (T_2).

The RELAXATION TIME is the time in which a non-equilibrium system completes 63%, or $1 - 1/e$, of its return to equilibrium; e is the base of natural logarithms.

The spin–lattice relaxation time (T_1) can be determined by observing the return of the signal to its equilibrium intensity after quickly reducing the incident microwave power.

The spin–spin relaxation time (T_2) can be determined from the line-width according to

$$\frac{1}{T_2} = \frac{g\beta\Delta H_{1/2}}{\hbar} \tag{21.2}$$

(in s^{-1}), where

g is the g-factor (dimensionless),

β is the Bohr magneton,

$\Delta H_{1/2}$ is the line-width (in gauss) measured between the two points of the absorption peak (not the derivative plot) which have half the maximum height (Figs 21.2 or 21.3),

$\hbar$ is a constant, $\hbar = h/2\pi$.

Between the relaxation times (Δt) and the line broadening (Δv) is observed the following relationship (the UNCERTAINTY PRINCIPLE):

$$\Delta E . \Delta t \simeq h/2\pi \simeq \text{constant} \tag{21.3}$$

or

$$\Delta v . \Delta t \simeq h/2\pi \simeq \text{constant} \tag{21.4}$$

or

$$g\beta\Delta H_{1/2} . \Delta t \simeq h/2\pi \simeq \text{constant.} \tag{21.5}$$

A general rule is that

(i) if Δt is small, Δv is large (FAST RELAXATION) and broad absorption lines appear in the ESR (or EPR) spectrum;

(ii) if Δt is large, Δv is small (SLOW RELAXATION) and narrow absorption lines appear in the ESR (or EPR) spectrum.

Calculation of the relaxation times T_1 and T_2 is necessary for the study of transition probabilities between the two spin levels and provides information about molecular movement in the crystalline regions of semicrystalline polymers.

21.1 INTERPRETATION OF EPR (OR ESR) SPECTRA

There are four features of EPR (or ESR) spectra which are important for their interpretation:

(i) line shape (cf. Section 21.1.1);

(ii) line intensity (cf. Section 21.1.2);
(iii) line position (cf. Section 21.1.3);
(iv) line splitting (cf. Section 21.1.4).

21.1.1 The EPR Line Shape

The EPR (or ESR) SPECTRUM can be registered (Figs 21.2 and 21.3) as

(i) an absorption spectrum;
(ii) a first-derivative spectrum;
(iii) a second-derivative spectrum.

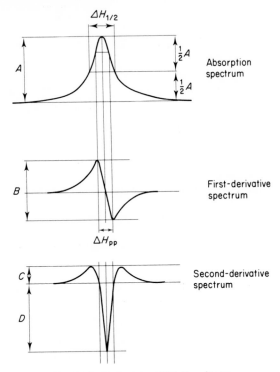

Fig. 21.2   Lorentzian ESR line shapes

Note: in contrast to NMR spectrometers, which usually record the NMR absorption spectrum directly, ESR spectrometers immediately plot the first derivative of the energy absorption curve.

The shapes of ESR lines are usually described by comparison with the Lorentzian and the Gaussian line shapes. Analytical expressions for these lines are as follows:

(i) Lorentzian line type (Fig. 21.2),

$$y = a/(1+bx^2);$$ (21.6)

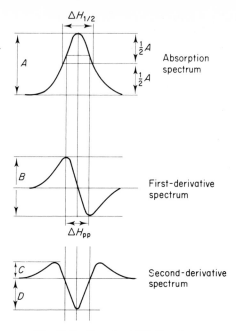

Fig. 21.3   Gaussian ESR line shapes

(ii) Gaussian line type (Fig. 21.3),

$$y = a^{-bx^2}. \tag{21.7}$$

Measurable experimental parameters characterizing the two types of spectral line are given in Table 21.1.

Table 21.1   The most important measurable parameters characterizing Lorentzian and Gaussian lines

Parameter	Lorentzian line	Gaussian line
Width at half-height	$\Delta H_{1/2}$	$\Delta H_{1/2}$
Peak-to-peak width	$\Delta H_{pp} = \dfrac{H_{1/2}}{\sqrt{3}}$	$\Delta H_{pp} = \left(\dfrac{2}{\ln 2}\right)^{1/2} \dfrac{\Delta H_{1/2}}{2}$
Peak amplitude	$A = \dfrac{2}{\pi \Delta H_{1/2}}$	$A = \left(\dfrac{\ln 2}{\pi}\right)^{1/2} \dfrac{2}{\Delta H_{1/2}}$
Peak-to-peak amplitude	$B = \dfrac{3\sqrt{3}}{\pi} \dfrac{1}{(\Delta H_{1/2})^2}$	$B = \left(\dfrac{2}{\pi e}\right)^{1/2} \dfrac{2 \ln 2}{(\Delta H_{1/2})^2}$
Peak amplitude of positive lobe	$C = A\left(\dfrac{2}{\Delta H_{1/2}}\right)$	$C = A\left(\dfrac{16 e^{3/2} \ln 2}{(\Delta H_{1/2})^2}\right)$
Peak amplitude of negative lobe	$D = -A\left(\dfrac{8}{\Delta H_{1/2}}\right)$	$D = -A\left(\dfrac{8 \ln 2}{\Delta H_{1/2}}\right)$

338

In several cases the shape of an absorption line is a mixture of the Lorenztian and Gaussian types. The mathematical calculation of the exact shape of an ESR (or EPR) line is very complicated. There are two methods which simplify this treatment and distinguish the two types:

(i) The SLOPE OF THE DERIVATIVE CURVE METHOD, presented in Fig. 21.4. For a Gaussian line the ratio of slopes $m : n = 2.2$ and for a Lorentzian line the ratio of slopes $m : n = 4$.

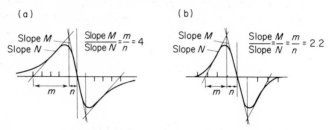

Fig. 21.4 The normalization technique for identification of (a) Lorentzian and (b) Gaussian first-derivative curves

(ii) The NORMALIZATION PLOT, shown in Fig. 21.5. In this method, the baseline is divided into units of a, where a is the point on the abscissa where the curve reaches a maximum, and the ordinates are normalized to unity.

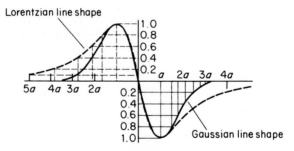

Fig. 21.5 The slope method for identification of Lorentzian and Gaussian first-derivative curves

The experimental measurement of a line-width is described in Section 21.2.4.

Bibliography: 646, 1088, 1375.
Reference: 3501.

21.1.2 The EPR Line Intensity

The INTENSITY means the total strength of the signal. It is the total energy absorbed by the sample at resonance.

(Note: in optical spectroscopy the word 'intensity' means the value of the ordinate of the recorder trace of the signal. The intensity here is the height of line.)

The EPR (or ESR) LINE INTENSITY is the area under an EPR absorption curve.

The area under each EPR signal is proportional to the number of unpaired spins in the sample (per gram, per millilitre, or per millimetre length of sample). In order to relate the number of unpaired spins in different samples through their ESR signals it is necessary to integrate the absorption signals because the line shapes can be very different. Unfortunately this measurement cannot be made to a high degree of accuracy (not better than 30–50%).

The measurement of the absolute number of unpaired spins is described in Section 21.3.

21.1.3 The EPR Line Position

The position of an EPR (or ESR) line is usually determined as the point where the derivative spectrum crosses the zero level. For asymmetric lines in powder spectra the points of maximum and minimum derivative are also often determined. The precision of the measurements depends on the width of the line and the noise level.

The position in the spectrum is measured in gauss (or tesla). Because the field depends on the frequency of the spectrometer, the position is usually characterized by a g-value.

The g-VALUE is defined as the constant of proportionality between the frequency and the field at which resonance occurs, and is given by

$$\Delta E = hv = g\beta H_0, \tag{21.8}$$

where

ΔE is the energy absorbed from the radio-frequency source which will induce transitions between the two states of the electron,

h is the Planck constant,

v is the frequency of the radio waves (rf) (MHz),

g is the g-factor ($g = 2.002\,319$ for unbound electrons),

β is the Bohr magneton,

H_0 is the strength of the applied external magnetic field.

The g-value is proportional to the magnetic moment of the molecule being studied.

In general, the g-value of bound unpaired electrons in a molecule is not the same as the g-value of free (unbound) electrons.

The magnetic interactions involving the orbital angular momentum of the unpaired electron are responsible for the deviation of the g-value from $2.002\,319$. Since the orbital angular momentum of the unpaired electron depends upon its chemical environment in an atom, molecule, or crystal, thus the g-value calculated from eq. (21.8) must likewise depend upon the chemical environment. As a consequence, deviations of the g-value from 2.002 319 are much like the chemical shifts of nuclear magnetic resonance (NMR) (cf. Section 20.1.1).

In single crystals the g-value is dependent on the direction of the external magnetic field in relation to the crystallographic axes x, y, z. One must use g_x, g_y, and g_z to indicate the three possible g-values obtained by making measurements on a crystal oriented successively in three mutually perpendicular directions with respect to the external field (this phenomenon is called ANISO-TROPY OF THE g-VALUE).

For crystals having axial symmetry the following g-value expressions are used:

(i) the quantity $g_\perp$ is the g-value measured when a particular crystal axis is perpendicular to the direction of the external magnetic field H_0;
(ii) the quantity $g_\parallel$ is the g-value measured when a particular crystal axis is parallel to the direction of the external magnetic field H_0.

The anisotropy of g-values causes asymmetry in the spectral line (Figs 21.6 and 21.7). In solutions, with rapid tumbling, we can observe only an average g-value.

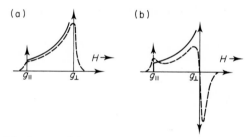

Fig. 21.6   Axial asymmetry, $g_\parallel \neq g_\perp$. (a) Absorption spectrum; (b) first-derivative spectrum

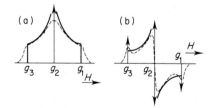

Fig. 21.7   An asymmetric g-tensor, $g_1 < g_2 < g_3$. (a) Absorption spectrum; (b) first-derivative spectrum

The g-values of organic radicals are all close to the value for a free electron, i.e. 2.002 319. The deviation of the g-value of radicals from this value may yield very important information on the structure of a free radical.

A compilation of g-values is given in B: 447.

The method for determination of the g-values is described in Section 21.4.

Bibliography: 111, 1088, 1375.
References: 4123, 4124, 4615.

21.1.4 EPR Line Splitting

SPLITTING in the energy levels of an electron is the phenomenon resulting in an increase in the number of energy levels as a result of exposing a system with unpaired electrons to a magnetic field.

By placing the unpaired electrons in a magnetic field, the number of energy levels increases from one ($E = E_0$) to two (Fig. 21.1(a)):

(i) $\hspace{2cm} E_1 = E_0 + \tfrac{1}{2}g\beta H_0;$ $\hspace{2cm}$ (21.9)

(ii) $\hspace{2cm} E_2 = E_0 - \tfrac{1}{2}g\beta H_0.$ $\hspace{2cm}$ (21.10)

If there is a magnetic nucleus (e.g. a proton, ^{1}H) in the region near the unpaired electron, the magnetic moment of the electron is affected by the orientation of the magnetic moment of the nucleus. As a result of this interaction, each magnetic energy level for the electron is split into a number of sublevels (Figs 21.8 and 21.9). This interaction of the electron and the magnetic

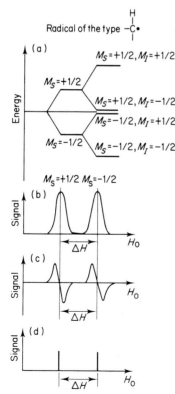

Fig. 21.8 (a) Energy level diagram showing hyperfine splitting for a free electron by a nucleus of spin $M_I = \tfrac{1}{2}$ (^{1}H proton). (b) Absorption line. (c) First-derivative line. (d) Stick plot of a first-derivative line—the relative intensities of the lines in a derivative spectrum are shown by the heights of the lines in the diagram. ΔH is the hyperfine splitting constant

342

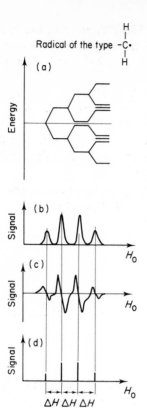

Radical of the type $-\overset{\text{H}}{\underset{\text{H}}{\text{C}}}\cdot$

(a)

Energy

(b)

Signal

H_0

(c)

Signal

H_0

(d)

Signal

H_0

ΔH ΔH ΔH

Fig. 21.9 (a) Energy level diagram showing a hyperfine splitting for a free electron by two nuclei of spin $M_I = \frac{1}{2}$ (^{1}H protons). (b) Absorption line. (c) First-derivative line. (d) Stick plot of a first-derivative line—the line intensities are proportional to the coefficient of the expansion of $(1+x^n)$. ΔH is the hyperfine splitting constant.

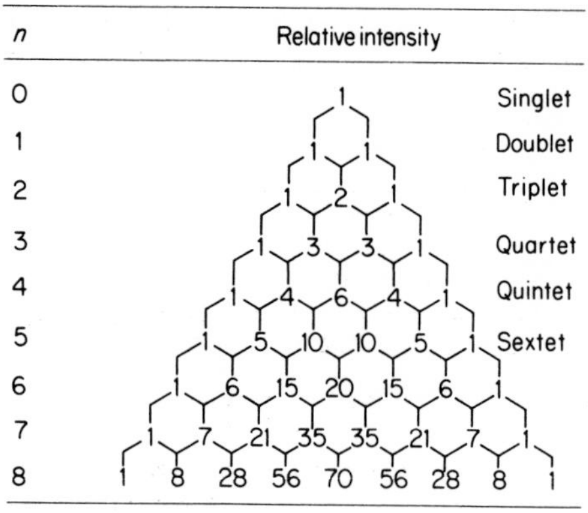

n	Relative intensity								
0					1				Singlet
1				1		1			Doublet
2			1		2		1		Triplet
3		1		3		3		1	Quartet
4	1		4		6		4		1	Quintet
5	1	5		10		10		5	1	Sextet
6	1	6	15	20	15	6	1		
7	1	7	21	35	35	21	7	1	
8	1	8	28	56	70	56	28	8	1

Fig. 21.10 Pascal's triangle: the binominal triangle representing the coefficients in the expansion of $(1+x)^n$. Note that the sum across any row is 2^n

nucleus is called the HYPERFINE INTERACTION, and the splitting in the energy levels is called HYPERFINE SPLITTING. There are as many sub-levels due to hyperfine splitting as there are possible orientations of the nuclear magnetic moment.

The HYPERFINE SPLITTING CONSTANT (ΔH) (measured in gauss) is the separation between the two hyperfine lines of the spectrum (Figs 21.8 and 21.9).

The interaction of the unpaired electron with n equivalent protons results in $n+1$ lines whose relative intensities are proportional to the coefficient of the binominal expansion of $(1+x)^n$ (Fig. 21.10).

A compilation of the hyperfine splitting constants (ΔH) is given in B: 447.

The experimental measurement of a hyperfine splitting constants (ΔH) is described in Section 21.2.4.

Bibliography: 111, 1088, 1375.

21.2 INSTRUMENTATION FOR EPR (ESR) SPECTROSCOPY

21.2.1 EPR (ESR) Spectrometers

ELECTRON SPIN (PARAMAGNETIC) RESONANCE SPECTROMETERS (ESR or EPR SPECTROMETERS) are built of the following parts (Fig. 21.11):

(i) A magnet. There are three types of magnet used in EPR spectrometers:
   (a) air core solenoids of low-frequency spectrometers with a field strength of 100–200 G (not used frequently);
   (b) iron core magnets with a field strength of 3000–18 000 G;
   (c) superconducting magnets with a field strength of 60 000 G—an electromagnet is immersed in liquid helium at 4 K with the result

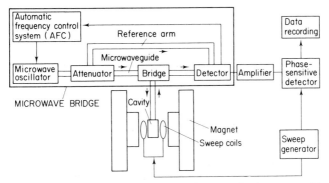

Fig. 21.11   Schematic diagram of an electron spin (paramagnetic) resonance spectrometer

that electrical resistance vanishes and the metal becomes a super-conductor.

Magnetic field homogeneity across the sample is a fundamental requirement in EPR spectrometers. Stability of the magnet is often obtained by regulation with a Hall crystal in the magnet gap. This crystal gives a voltage proportional to the field and any changes can be corrected with a feedback circuit.

The magnet must have pole pieces of large diameter relative to the width of the air gap. The air temperature for a permanent magnet and the temperature of the cooling water for an electromagnet must be carefully controlled.

(ii) A microwave bridge. This can be subdivided into different units:

(a) A microwave oscillator, which is a source of microwave power. The most suitable oscillator is the klystron, which can only be tuned over a small frequency region. The propagation of micro-waves occurs in waveguides. EPR spectrometers have been constructed for the following bands:

Band S for 3 GHz (wavelength 90 mm) and for an 1100 G magnet;
Band X for 9 GHz (wavelength 30 mm) and for a 3300 G magnet;
Band K for 24 GHz (wavelength 12 mm) and for an 8500 G magnet;
Band Q for 35 GHz (wavelength 8 mm) and for a 12 500 G magnet;
Band E for 70 GHz (wavelength 4 mm) and for a 25 000 G magnet.

The large majority of EPR spectrometers are designed for a frequency around 9.5 GHz (Band X).

The klystron frequency is tuned to the resonance frequency of the microwave cavity with the sample in place. The power of the klystrons used in EPR spectrometers is usually a few hundred milliwatts. The heat generated by the klystron is commonly dissipated by circulating water.

(b) The attenuator (parametric amplifier) is applied in order to attenuate the microwave power reaching the cavity.

(c) The bridge (microwaveguide), where the propagation of micro-waves occurs. In highly sensitive spectrometers the bridge uses a reflection cavity and is designed as the 'magic T' (Fig. 21.12). The power from the klystron is divided into the two arms of the T seen from the klystron. If both arms are perfectly matched, half of the power goes to the cavity and the other half is absorbed in the load. When resonance absorption occurs there is mismatching, and a signal is reflected from the cavity. The signal power is divided into the arms of the T seen from the cavity. Half of the signal power is absorbed in the isolator in the klystron arm, the other half goes to the detector.

(d) The microwave cavity is generally designed as the reflection type, where the power enters and leaves by the same hole, called the iris.

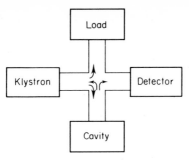

Fig. 21.12 Schematic diagram of the
'magic T'

Cavities can be designed in different shapes and sizes for various purposes. The most common cavities are:

TE_{102} rectangular cavity—useful for large samples and especially for liquid samples (Fig. 21.13).

TE_{011} cylinder cavity—useful for gaseous systems and liquid samples in capillaries (Fig. 21.14).

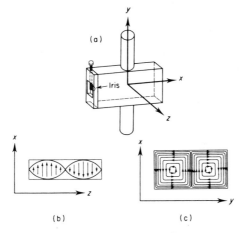

Fig. 21.13 (a) Rectangular cavity operating in the TE_{102} mode. (b) Electric field contours. (c) Magnetic field contours

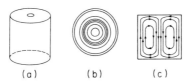

Fig. 21.14 (a) Cylindrical cavity operating in the TE_{011} mode. (b) Electric field contours. (c) Magnetic field contours

Cavities are designed by the pattern of the internal electromagnetic field as either transverse electric $TE_{l,m,n}$ modes. The subscripts l, m, n are integers which define the field patterns by the number of repetitions in the E (electric) and H (magnetic) fields along the coordinates x, y, and z.

(e) The detector, which is a microwave receiver. It is a crystal diode or a tunnel-effect diode which works as a microwave rectifier. The signal is further amplified and fed to the phase-sensitive detector; after filtration it is recorded as an EPR signal.

(f) The reference arm. This is a waveguide that channels some microwave power from the klystron directly to the detector.

(g) The automatic frequency control system (AFC) is applied for the correction of the exact cavity resonance frequency.

(iii) A sweep generator, which is applied to change the strength of the magnetic field (H_0). The output of the sweep generator is synchronized with the trace along the X axis of an oscilloscope or XY-recorder. The variable electromagnetic coils are called sweep coils and are on each side of the cavity.

(iv) Data recording. The oscilloscope and the XY-recorder are synchronized with the output of the sweep generator. A small computer attached to the EPR spectrometer can perform the storage of information.

The sensitivity of ESR spectrometers is much higher than that for NMR spectrometers because the magnetic moment of an electron is much higher than that for a proton. From the BOLTZMANN DISTRIBUTION EQUATION,

$$n_i/n_j = e^{-\Delta E/kT}, \qquad (21.11)$$

where

n_i and n_j are upper- and lower-level spin-state populations, respectively,

ΔE is the energy required for level splitting,

k is the Boltzmann constant (cf. Section 40.4),

T is the thermodynamic temperature (in kelvins),

it can be seen that there is a difference of only about 7 p.p.m. in upper- and lower-level spin-state populations using the NMR method, whereas there is a population imbalance of about 1600 p.p.m. by the ESR method, giving ESR spectrometers a higher sensitivity by about 230-fold. The energy required for level splitting (ΔE) is directly proportional to the magnetic field strength and thus greater sensitivity can be obtained at higher field strengths and resonance frequencies. Lowering the temperature will also increase the sensitivity, but the effect of temperature on molecular motion limits this approach.

Bibliography: 17, 1050, 1088.

21.2.2 Sampling Techniques

Gases, liquids, and solids may be examined by EPR (ESR) spectroscopy by the use of different types of arrangement.

(i) Gases. There are three groups of gases which can be studied by ESR:
   (a) paramagnetic molecules: O_2, NO, NO_2, ClO_2, etc.;
   (b) reactive atoms: H, N, O;
   (c) free radicals in a gas phase.
   The main experimental problem is to produce a sufficient concentration
   of free radicals in the gas to obtain a good signal-to-noise ratio. Measurements at a pressure of 10–100 torr (10^{17}–10^{18} molecules cm^{-3}) sometimes give line broadening. Pressures of 0.1–1 torr (10^{15}–10^{16} molecules
   cm^{-3}) give spectra with better resolution. For that reason it is necessary
   to have a gas handling system which permits pressure adjustment so
   that one can optimize signals and resolutions. Two methods are commonly used:
   (a) differential pumping to remove excess molecules;
   (b) dilution with inert diamagnetic gases such as helium or argon.
(ii) Liquids. The studies of free radicals in liquids are devoted to the
   following problems:
   (a) Radicals generated by radiolysis. These radicals can be obtained
       by two main techniques: first there are those radicals generated in a
       radiation source and then pumped through the resonance cavity;
       second there are those radicals formed in the cavity by direct intro-
       duction of a high-energy electron beam into the resonance cavity.
       From an experimental point of view, both methods are complicated.
   (b) Radicals generated by photolysis. These radicals are obtained by
       direct irradiation of the sample with ultraviolet light in the resonance
       cavity, equipped with a quartz window in one wall (Fig. 21.15).

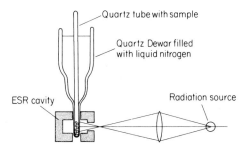

Fig. 21.15   Schematic diagram of arrangement
for measuring ESR spectra of radicals by ultraviolet irradiation of a sample in the cavity

   (c) Negative and positive radical ions produced in solutions.
   (d) Intermediate species produced in chemical reactions. These free
       radical species are observed mainly in flow systems (Fig. 21.16), e.g.
       study of redox reactions as indicators of free radical polymerization.
       For that reason special mixing cells have been designed, e.g. the
       Dixon–Norman mixing cell (Fig. 21.17). This cell consists of two tubes
       with a flat cell on top. The inner tube contains a spray head from

which one solution (e.g. a reductant) is injected into the second solution (e.g. an oxidant). Free radicals are immediately formed after mixing the two solutions in the resonance cavity. The flow rate of the solvents should be selected experimentally; rates between 1 and 10 cm^3 s^{-1} are commonly used.

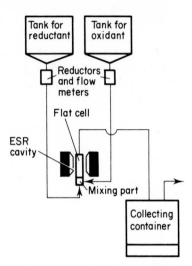

Fig. 21.16 Schematic diagram of a flow system for the production of a high concentration of free radicals

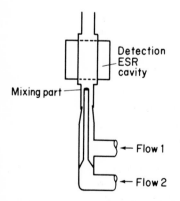

Fig. 21.17 The Dixon–Norman type mixing cell for ESR study of free radicals formed in a flow system

The samples dissolved in water or sulphuric acid, because of the large dielectric constants of these solvents, have very low detection signals. For that reason a fine sample tube must be used.

(iii) Solids. Free radicals in solids can be produced directly in the resonance cavity by radiolysis, photolysis, pyrolysis, mechanical stretching of the sample, etc.

21.2.3 Stabilization of Free Radicals

The lifetimes of many organic and polymer free radicals and ion radicals are too short for detection at room temperature by EPR spectroscopy. For that reason it is very important to stabilize free radicals and radical ions by one of the following methods:

(i) Cooling to low temperatures, e.g. that of liquid nitrogen ($-196\,^\circ$C $=$ 77 K) or liquid helium ($-268.8\,^\circ$C $= 4.2$ K).

Measurement at the temperature of liquid nitrogen requires a simple Dewar container in which, after it has been filled with liquid nitrogen, one places the sample-containing tube. Inserting the sample directly into liquid nitrogen frequently causes problems because the bubbles of nitrogen gas produce excessive noise during the recording of the spectrum. This effect may be suppressed by blowing helium gas into the liquid nitrogen.

Measurements of EPR spectra at different temperatures (77–300 K) are made by applying a temperature variation through use of a temperature control device (Fig. 21.18). This device is a heat exchanger. Dry

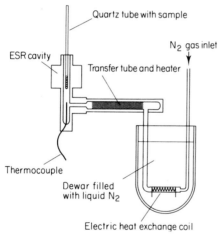

Fig. 21.18   Gas flow apparatus for varying
and controlling the temperature of the sample

nitrogen passes through a copper coil immersed in liquid nitrogen and is transferred through a tube heater into a resonance cavity. Temperature control is achieved by varying the rate of flow of the cooling gas and controlling the heater current.

Measurements at the temperature of liquid helium (1.2–4.2 K) necessitate the use of a special low-temperature continuous-flow cryostat. This kind of experiment is difficult to perform because it is difficult to handle the liquid helium.

(ii) Stabilization of free radicals in organic crystalline matrices, e.g. alcohols.

(iii) Trapping free radicals by the SPIN TRAPPING TECHNIQUE. This technique is used for the detection and identification of short-lived free radicals by detecting the EPR spectra of nitroxyl addition products (SPIN ADDUCTS) of free radicals (R$^{\cdot}$) to nitroso (e.g. 2-nitroso-2-methyl propane, t-butyl nitroxide, or nitrosobenzene) or nitrone (phenyl-N-tert-butyl nitrone) compounds called SPIN TRAPS:

$$R^{\cdot} + CH_3-\underset{\underset{CH_3}{|}}{\overset{\overset{CH_3}{|}}{C}}-N=O \longrightarrow CH_3-\underset{\underset{CH_3}{|}}{\overset{\overset{CH_3}{|}}{C}}-\underset{\underset{R}{|}}{\overset{\overset{O^-}{|}}{N}}{}^+ \qquad (21.12)$$

$$R^{\cdot} + \langle\!\!\!\bigcirc\!\!\!\rangle-N=O \longrightarrow \langle\!\!\!\bigcirc\!\!\!\rangle-\underset{\underset{R}{|}}{\overset{\overset{O^-}{|}}{N}}{}^+ \qquad (21.13)$$

$$R^{\cdot} + \langle\!\!\!\bigcirc\!\!\!\rangle-\underset{\underset{\underset{H_3C\diagup\diagdown CH_3}{CH}}{|}}{\overset{\overset{H}{|}}{C}}=\overset{\overset{O^-}{|}}{N}{}^+ \longrightarrow \langle\!\!\!\bigcirc\!\!\!\rangle-\underset{\underset{\underset{H_3C\diagup\diagdown CH_3}{CH}}{\underset{|}{R}}}{\overset{\overset{H}{|}}{C}}-\overset{\overset{O^-}{|}}{N}{}^+ \qquad (21.14)$$

Bibliography: 1088.
References: 6192–6195.

21.2.4 Field Sweep Calibration

For measurements of line-widths and hyperfine splittings, an accurate calibration of the field sweep is necessary. This can easily be carried out using a standard sample in which the line positions have been determined using accurate methods. The standard sample can be inserted in a double cavity together with the unknown sample, and their spectra recorded by a dual-channel recorder.

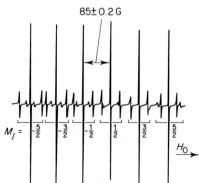

85±0.2 G

$M_I = $ $\begin{array}{cccccc} -\frac{5}{2} & -\frac{3}{2} & -\frac{1}{2} & \frac{1}{2} & \frac{3}{2} & \frac{5}{2} \end{array}$

H_0

Fig. 21.19 ESR spectrum of Mn^{2+} in MgO at room temperature, with $\nu \approx 9.3$ GHz. (From B: 1375; reproduced by permission of McGraw-Hill)

Various standard samples are used:

(i) Mn^{2+} ions in SrO powder give a six-line spectrum (Fig. 21.19) which is 1.6 G wide, over 420 G, with a hyperfine splitting constant (ΔH) of 85 ± 0.2 G;

(ii) The tetracyanoethylene anion radical (TCNE radical) in solution (Varian standard sample) gives nine lines, 100 mG wide, over 13 G, with a hyperfine splitting constant (ΔH) of 1.574 G.

References: 5978, 6099, 6322.

21.3 MEASUREMENTS OF THE ABSOLUTE NUMBER OF UNPAIRED SPINS

Spin concentration measurements are made by comparison of the integrated intensity of the signal from the unknown sample with that from a standard sample with a known number of unpaired electron spins.

A standard sample should have the following properties:

(i) a line-width and a line shape similar to that of the unknown sample;

(ii) the number of spins should be similar to that in the unknown sample;

(iii) physical shape and dielectric loss similar to those of the unknown sample;

(iv) stability both with time and temperature—the number of spins, the line-width, and the g-value should remain constant;

(v) a short-spin–lattice relaxation time T_1 to avoid easy saturation of the signal.

Since line-widths and spin concentration vary over several orders of magnitude, it is necessary to prepare a number of standard samples.

Various secondary standard samples are used:

(i) synthetic ruby crystal, which can be placed permanently in the resonance cavity, permitting simultaneous observation of standard and unknown sample;

(ii) charred dextrose, which does not change its magnetic properties in the temperature range 4–500 K;

(iii) powdered coal (pitch) diluted with KCl, supplied commercially;

(iv) Mn^{2+} ions in MgO;

(v) freshly recrystallized 1,1-diphenyl-2-picrylhydrazine (DDPH).

The dual cavity allows the use of samples with similar filling factors and positions in the radio-frequency field. The change of sample is also simple. Differences in the two signal measuring channels are eliminated by exchange of samples.

Spin concentration measurements made by relative comparison with a secondary standard can be carried out with an accuracy of 5–10%.

Bibliography: 17, 558, 1279.
References: 2759, 4133, 6412, 7204, 7242.

21.4 ACCURATE DETERMINATION OF g-VALUES

The effective g-value can normally be determined to two significant places of decimals by applying a Hall crystal field controller. This accuracy is not sufficient, however, to identify many organic and polymeric free radicals.

In order to improve the accuracy to four or five places of decimals for narrow lines, it is necessary to measure the frequency separately with a wavemeter or a frequency counter. The field is simultaneously measured with an NMR gaussmeter or a Hall crystal field controller calibrated with an NMR gaussmeter. The g-value is further determined from eq. (21.8).

The g-value can also be measured by comparison of the unknown sample with the reference sample. Both samples are inserted into the double cavity and their spectra are recorded by a dual-channel recorder. The separation of the centres of the two spectra represents the magnetic field difference (ΔH) at the

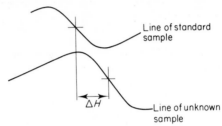

Fig. 21.20 Determination of the separation of the two centres of the standard and the unknown samples

two positions of the sample (Fig. 21.20). The g-value of the unknown sample (g) can be calculated from

$$g = g_s - \frac{\Delta H}{H_0} g_s, \qquad (21.15)$$

where
g_s is the g-value of the reference sample,
ΔH is the separation of the centres of the two spectra (in gauss) (cf. Fig. 21.20),
H_0 is the strength of the applied external magnetic field (in gauss).
Common standard samples are

(i) tetracene positive ion, $g = 2.002\,604 \pm 0.000\,007$;
(ii) perylene positive ion, $g = 2.002\,583 \pm 0.000\,006$.

References: 2086, 4594, 6219, 6321.

21.5 FREE RADICAL REACTIONS

A FREE RADICAL is defined as an atom, a group of atoms, or a molecule in a certain state containing one unpaired electron which occupies an outer orbital, e.g. atomic hydrogen ($H^{\cdot}$), the hydroxyl radical ($HO^{\cdot}$), or the methyl radical ($H_3C^{\cdot}$).

A BIRADICAL is a species containing two unpaired electrons in outer orbitals, e.g. molecular oxygen ($O_2^{\cdot \cdot}$) in its ground or excited state.

A RADICAL ION is a free radical with a positive or negative charge, e.g. a protonated amine radical ($H_3N^{+\cdot}$) or a naphtalene anion radical ($C_{10}H_8^{-\cdot}$).

Free radicals are usually very active on account of the strong tendency of their unpaired electrons to interact with other electrons and to form electron pairs (chemical bonds).

Typical free radical reactions are chain reactions which occur in three steps:

(i) THE INITIATION STEP, which is a radical formation process, e.g. the radicals are formed in pairs by homolytic cleavage of a two-electron bond.

(ii) THE PROPAGATION STEP, which is a transfer reaction of free radicals in which the site of free radical is changed. There are four types of propagation reactions:

   (a) ATOM TRANSFER REACTIONS, e.g. abstraction of hydrogen by a free radical with formation of a new free radical,

$$A^{\cdot} + RH \longrightarrow AH + R^{\cdot} ; \qquad (21.16)$$

   (b) ADDITION REACTIONS involving the addition of free radical to a double bond,

$$A^{\cdot} + C{=}C \longrightarrow A{-}C{-}C^{\cdot} ; \qquad (21.17)$$

   (c) FRAGMENTATION REACTIONS—a typical example being β-SCISSION, in which an unpaired electron in a molecule splits a bond in the β-position and produces a new free radical and a molecule containing a double bond,

$$\overset{..}{R}C{-}C^{\cdot} \longrightarrow R^{\cdot} + C{=}C ; \qquad (21.18)$$

   (d) REARRANGEMENT REACTIONS, in which a free radical changes position in a molecule, giving a new type of free radical,

$$R{-}\overset{\cdot}{\underset{\underset{R}{|}}{C}}{-}CH_2{-}R \longrightarrow R{-}\overset{\overset{R}{|}}{\underset{\underset{R}{|}}{C}}{-}\overset{\cdot}{C}H_2 ; \qquad (21.19)$$

(iii) TERMINATION REACTIONS, which occur in all systems where free radicals are present. There are two types of termination reaction:
   (a) COMBINATION of two radicals,

$$R^{\cdot} + R^{\cdot} \longrightarrow R{-}R ; \qquad (21.20)$$

(b) DISPROPORTIONATION—a reaction involving the transfer of hydrogen,

$$R^{\cdot} + RCH_2\text{—}\overset{\cdot}{C}HR \longrightarrow R\text{—}H + RCH\text{=}CHR . \qquad (21.21)$$

All of these reactions may be observed in free radical polymerization and free radical-initiated degradation.

21.6 APPLICATIONS OF ELECTRON PARAMAGNETIC RESONANCE SPECTROSCOPY IN POLYMER RESEARCH

The main applications of EPR (ESR) spectroscopy in polymer research are in the study of the free radicals formed during

(i) polymerization processes (radiation- and photopolymerization, free radical initiation, polymerization initiated by redox systems, ionic polymerization, copolymerization, etc.);
(ii) degradation processes in polymers (radiation- and photo-degradation; Table 21.2);

Table 21.2   ESR spectroscopy of polymers

Polymer	Published in
Cellulose acetate	R: 4105–4109
Poly(butadiene)	R: 2737, 4918 5212, 7344
Poly(ethylene)	B: 987, 1088, 1306; R: 2009, 2177, 2599, 2682, 2801, 2803, 3343, 3539, 3540, 3858, 4321, 4437–4439, 4782, 4843, 4918, 5437, 5563, 5564, 5568–5570, 5967, 6169, 6330, 6358– 6360, 6704, 6848–6852, 6854–6858, 7021
Poly(isobutylene)	R: 4980
Poly(isoprene)	R: 2599, 2736, 4290, 4466, 5564, 7344
Poly(propylene)	R: 2183, 2599, 3387, 3450, 3860, 4918, 5436, 5603, 5963, 5967, 6853, 6858, 7281
Poly(styrene)	R: 2009, 3070, 3388, 3399, 4221, 6236, 6657, 6942, 7118
Poly(tetrafluoroethylene)	R: 2808, 3980, 4888, 6152, 6393, 6819
Poly(vinyl alcohol)	R: 7163
Poly(vinyl chloride)	R: 2856, 3952, 4318, 4842, 4979, 4926, 5265, 5564– 5566, 5938, 5965, 6656, 6818
Poly(vinylidene chloride)	R: 5566

(iii) oxidation of polymers;

(iv) molecular fracture in polymers (mechanical degradation);

and in the study of

(v) stable polymer radicals;

(vi) mobility of macromolecules (spin labelling).

Bibliography: 107, 203, 204, 233, 234, 245, 440–446, 448, 711, 743, 987, 1088, 1130, 1306.
References: 4123, 4124, 4265, 5409–5412, 5488, 5834, 5934, 6157, 6203, 6357, 6361, 7287.

21.7 SPIN LABELLING AND SPIN PROBE TECHNIQUES

Both these techniques make use of stable nitroxy radicals, e.g. 4R-2,2,6,6-tetramethylpiperidine:

R = OH, OCOPh

SPIN LABELLING is a technique which involves covalent bonding of different nitroxide radicals to diamagnetic polymers (which do not contain unpaired electrons and do not give ESR spectra). Line-width measurement of the ESR spectra carrying stable nitroxide radicals may give information on molecular transitions.

Bibliography: 665, 666, 821, 1088, 1130.
References: 2184, 2628–2636, 3781, 4104, 4799, 4809, 5955, 6007, 6008, 6820, 6820, 6821, 6935, 6936, 7009.

The SPIN PROBE technique is a method whereby different nitroxide radicals are mixed with a polymer (i.e. unbound) and their tumbling behaviour during the relaxation and transition processes in the polymers is studied. In the spin probe technique the nitroxide radicals are present in the polymer matrix at concentrations of 10–100 p.p.m.

Bibliography: 186, 219, 1088, 1130.
References: 2545, 2628, 2635, 3792, 4698, 4765, 4769–4781, 4783, 5413, 5942, 6007, 6203, 6820, 6937.

21.8 ELECTRON SPIN DOUBLE RESONANCE SPECTROSCOPY

21.8.1 Electron Nuclear Double Resonance Spectroscopy

ELECTRON NUCLEAR DOUBLE RESONANCE (ENDOR) SPECTROSCOPY is a technique which detects changes in a monitored ESR signal due to the interaction of nuclear magnetic resonance (NMR) of nuclei coupled to the paramagnetic centre.

The fundamental processes which lead to an ENDOR absorption are as follows:

(i) A partial saturation of some particular electron transitions by a microwave frequency produces an ESR signal of a given amplitude.

(ii) When the second field having an NMR radio-frequency is slowly altered, a change takes place in the spin population of the electron levels.

(iii) This change affects the amplitude of the original ESR signal, and it is the magnitude of this 'difference in amplitudes' which is observed as a function of the NMR radio-frequency. This effect is commonly known as ENDOR.

A plot of change in the ESR signal amplitude versus NMR radio-frequency is called an ENDOR SPECTRUM.

ENDOR spectroscopy is especially useful when the hyperfine lines are not well resolved in the ESR spectrum.

Bibliography: 739, 1375.
References: 4189.

21.8.2 Electron–electron Double Resonance Spectroscopy

ELECTRON–ELECTRON DOUBLE RESONANCE (ELDOR) SPECTROSCOPY is a technique which detects the reduction in the intensity of one hyperfine transition when a second hyperfine transition is simultaneously being saturated.

Simultaneous electron spin resonance (ESR) in one magnetic field for two different transitions requires irradiation of the sample with two microwave frequencies.

ELDOR spectroscopy is mainly applied to the study of spin–spin relaxation processes, and is useful for the study of molecular motions with correlation times from 10^{-10} to 10^{-3} s.

Bibliography: 660, 739, 1375.
References: 2608, 3214, 3215, 3502, 4190, 4191, 5372, 7232.

Chapter 22

Mass Spectrometry

General bibliography: 12, 103, 117–119, 123, 158, 212, 220, 221, 227, 274, 424, 480, 564, 658, 692, 750, 894, 900, 902, 908, 930, 984, 1012, 1055, 1096, 1097, 1119, 1188, 1202, 1213, 1247, 1316, 1354, 1381, 1393, 1395.

When an organic substance is exposed to an impact from an electron the following reactions may occur:

(i) Production of POSITIVE MOLECULAR IONS (RADICAL CATIONS) (also called PARENT IONS):

$$M + e^- \longrightarrow M^{\cdot +} + 2e^-. \tag{22.1}$$

If some of the molecular (parent) ions remain intact long enough to reach the detector (about 10^{-6} s), the PARENT PEAK is observed. It is important to recognize the parent peak because this gives the molecular weight of the compound. This molecular weight is the molecular weight to the nearest whole number.

(ii) Production of MULTIPLY CHARGED IONS:

$$M + ne^- \longrightarrow M^{\cdot n+} + (n+1)e^-. \tag{22.2}$$

(iii) Formation of POSITIVE FRAGMENT IONS and radicals, produced by molecular (parent) ion disintegration 10^{-10}–10^{-3} s after the impact:

$$\left. \begin{aligned} M^{\cdot +} &\longrightarrow M_1^+ + R_1^{\cdot}, \\ M^{\cdot +} &\longrightarrow M_2^+ + R_2^{\cdot}, \\ &\quad\vdots \\ M^{\cdot +} &\longrightarrow M_n^+ + R_n^{\cdot}. \end{aligned} \right\} \tag{22.3}$$

(iv) Formation of METASTABLE IONS. These ions are formed by a metastable transition of positive ion fragments as they travel from the electrostatic field to the magnetic field:

$$M_n^+ \longrightarrow M_i^+ + M^0, \tag{22.4}$$

where M_i^+ is the metastable ion and M^0 is the neutral fragment. If an ion (M_n^+) which fragments after acceleration but before entering the magnetic field has a mass m_1, but is dispersed in the magnetic field as a mass m_2, it will be carried out by an ion current and recorded as a low-intensity broad peak at an apparent mass m^*, given by

$$m^* = (m_2)^2/m_1. \tag{22.5}$$

The peak caused by the ion current corresponding to mass m^* is called a METASTABLE PEAK.

Measurement of the mass of the metastable peak gives the information that the mass m_2 is derived directly from mass m_1 by loss of a neutral fragment (M^0).

(v) Production of ISOTOPIC IONS in the case of organic substances which possess naturally occurring isotopes (Table 22.1). The presence of

Table 22.1 Different elements and their isotope abundances (from B: 1213; reproduced by permission of John Wiley & Sons)

Elements			Abundance			
Carbon	^{12}C	100	^{13}C	1.08		
Hydrogen	^{1}H	100	^{2}H	0.016		
Nitrogen	^{4}N	100	^{15}N	0.38		
Oxygen	^{16}O	100	^{17}O	0.04	^{18}O	0.20
Fluorine	^{19}F	100				
Silicon	^{28}Si	100	^{29}Si	5.10	^{30}Si	3.35
Phosphorus	^{31}P	100				
Sulphur	^{32}S	100	^{33}S	0.78	^{34}S	4.40
Chlorine	^{35}Cl	100	^{37}Cl	32.5		
Bromine	^{79}Br	100	^{81}Br	98.0		
Iodine	^{127}I	100				

isotopes leads to the production of different m/e ions, which appear as peaks $M+1$ and $M+2$ (QUASIMOLECULAR IONS).

The isotopic abundance ratios for various combinations of carbon, hydrogen, nitrogen, and oxygen are collected in special tables.

(vi) Formation of REARRANGED IONS. These ions arise by the transfer to another bonding location of one or more atoms in an ion moiety before subsequent cleavage occurs.

(vii) Production of NEGATIVE IONS by one of the following reactions:

(a) DISSOCIATIVE RESONANCE CAPTURE,

$$AB + e^- \longrightarrow A + B^-; \tag{22.6}$$

(b) RESONANCE CAPTURE,

$$AB + e^- \longrightarrow AB^-; \tag{22.7}$$

(c) ION-PAIR PRODUCTION,

$$AB + e^- \longrightarrow A^+ + B^- + e^-. \tag{22.8}$$

The efficiency of production of positive ions is from three to many times greater than that of negative ions in the 50–100 eV region employed in commercial mass spectrometers. A spectrum of positive ion fragments gives a mass spectrum.

The MASS SPECTRUM of a sample is the plot of the experimentally determined ion intensities for a set of ion masses (m) divided by charges (e), i.e. m/e (Fig. 22.1).

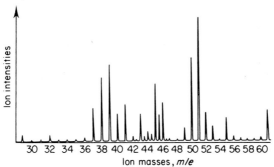

Fig. 22.1   Fragment of the mass spectrum

The ION INTENSITIES are detected as peak intensities of vertical distance (in millimetres) from the horizontal axis. Peak heights are proportional to the number of ions of each mass.

The MASS-TO-CHARGE RATIO (m/e) is detected at a distance from a zero or scan starting point along the horizontal axis.

Mass spectral data are usually presented in one of the following forms:

(i) PERCENTAGE RELATIVE ION INTENSITY ($\%$ RI). To obtain the percentage relative ion intensity ($\%$ RI) from the raw peak height of the ith ion, the ion of largest intensity in the spectrum is selected. The intensity of this ion peak is called the BASE PEAK ION INTENSITY (p_b). All of the raw peak heights (p_i) are multiplied by 100 and divided by p_b to obtain the PERCENTAGE RELATIVE ION INTENSITY ($\%$ RI) (relative to the largest ion):

$$\% \, RI = (p_i/p_b) \times 100 \qquad (22.9)$$

(in per cent). The calculated data are presented in the form of a bar graph (Fig. 22.2). The bar graph has no ability to show ions with low intensities (e.g. metastable ions).

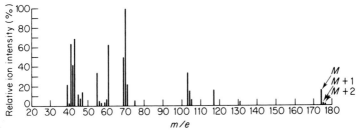

Fig. 22.2   Bar graph of the relative ion intensity ($\%$)

In the standard method a tabulation of relative ion intensities is given by the increasing order of the mass-to-charge (m/e).

Mass spectral data catalogues do not usually include bar graphs.

(ii) PERCENTAGE TOTAL ION INTENSITY ($\% \sum_i$). To obtain the percentage total ion intensity ($\% \sum_i$) the row peak heights (p_i) are summed to obtain the TOTAL ION INTENSITY ($\sum p_i$). Then the various peaks (p_i) are multiplied by 100 and divided by the sum of the p_i ($\sum p_i$) to obtain the PERCENTAGE TOTAL ION INTENSITY ($\% \sum_i$):

$$\sum_i = (p_i / \sum p_i) \times 100 \qquad (22.10)$$

(in per cent). The calculated data are presented in the form of a bar graph (Fig. 22.2).

Mass spectra should include information on sensitivity. The SENSITIVITY is usually given for the ion of largest intensity or for the base peaks, and is expressed as microamperes of ion current for a selected peak per microgram of sample. The sensitivities of various compounds may vary considerably.

It is important to check the BACKGROUND PEAKS before the sample mass spectrum is obtained. These peaks may arise from trace quantities of samples which were not completely pumped out of the instrument in measurements made previously. The peaks observed in such a 'background' spectrum should be subtracted from the sample peaks.

As an aid in interpretation of mass spectra of organic compounds the following publications are useful: B: 117, 326, 564, 802, 803, 901, 902, 1213, 1261, 1316, 1461, 1463, and R: 4530.

22.1 INSTRUMENTATION FOR MASS SPECTROMETRY

Mass spectrometers can be classified according to the method of separating the charged particles:

(i) Magnetic field deflection mass spectrometers, which can be subdivided into
   (a) single-focusing (magnetic field only) (low–middle resolution) mass spectrometers (cf. Section 22.1.1), and
   (b) double-focusing (electrostatic field before magnetic field) (high resolution) mass spectrometers (cf. Section 22.1.3);
(ii) time-of-flight mass spectrometers (cf. Section 22.1.4);
(iii) quadrupole mass spectrometers (cf. Section 22.1.5);
(iv) cyclotron resonance mass spectrometers;
(v) double-beam mass spectrometers.

Mass spectrometers can also be classified according to their resolution:

(i) low-resolution instruments are instruments giving resolutions of 200/1 or less—these include older commercial instruments;

(ii) medium-resolution instruments are instruments giving resolutions of from 500/1 to about 5000/1—these include modern single-focusing commercial instruments;

(iii) high-resolution instruments are instruments giving resolutions from 10 000/1 to about 100 000/1 or above—these include most of the double-focusing intruments of Mattauch–Herzog or Nier–Johnson design and some other special instruments.

The resolution (R) of an instrument can be calculated from

$$R = M/\Delta M, \qquad (22.11)$$

where

M is the higher mass number of the two selected peaks,

ΔM is the difference between the two mass numbers.

Bibliography: 117, 231, 564, 658, 750, 809.

22.1.1 Single-focusing Mass Spectrometers

The single-focusing mass spectrometer (Fig. 22.3) is built of the following parts:

(i) A sample handling system. This consists of
  (a) a device for introducing the sample (cf. Section 22.1.2);

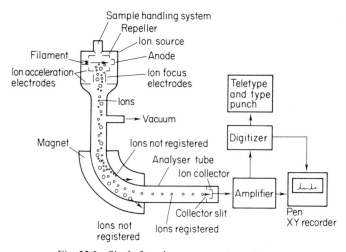

Fig. 22.3   Single-focusing mass spectrometer

(b) a micromanometer for determining the amount of sample introduced;

(c) a molecular leak for releasing the sample into the ionization chamber;

(d) a pumping system which produce a vacuum of 10^{-1}–10^{-3} torr (1 torr = $\frac{1}{760}$ atm = 133.322 N m^{-2} (SI)).

(ii) Ionization and accelerating chambers. The gas stream from the molecular leak enters the ionization chamber (operated at a pressure of about 10^{-6}–10^{-5} torr), in which it is bombarded at right angles by an electron beam emitted from a hot filament.

The positive ions produced by interaction with the electron beam are forced through the first accelerating slit by a small electrostatic field between the repellers and the first accelerating slit. A strong electrostatic field between the first and second accelerating slits accelerates the ions to their final velocities. Additional focusing of the ion beams is provided between the accelerating slits. To obtain a spectrum, either the magnetic field applied to the analyser tube, or the accelerating voltage between the first and second ion slits, is varied. Thus, the ions are successively focused at the collector slit as a function of mass/charge. In modern instruments, a scan from mass 12 to mass 500 may be performed in 1–4 min or even less.

(iii) An analyser tube and magnet. The analyser tube has a 180° curve; it is made from metal and is evacuated to 10^{-7}–10^{-8} torr. The ion beam passes from ion source to collector through the analyser tube. A homogeneous magnet field across the analyser tube is a fundamental requirement in mass spectrometers.

(iv) An ion collector and an amplifier. A typical ion collector consists of one or more collimating slits and a Faraday cylinder. The ion beam impinges axially into the collector, and the signal is amplified by a vacuum-tube electrometer or an electron multiplier.

(v) A recorder. A widely used recorder employs five separate galvanometers that record simultaneously on photographic paper. The sensitivity levels of a five-element galvanometer decreases from top to bottom in the ratio $1 : 3 : 10 : 30 : 100$. Peak heights from the baseline are read on the most sensitive trace remaining on the scale, and are multiplied by the appropriate sensitivity factor.

Most modern mass spectrometers have digitizer units which convert the ion intensity signals and the m/e to produce a digital output, which can be printed on paper type, punched on computer tape or cards, and/or recorded on magnetic tape.

Bibliography: 117, 564, 658, 750, 902, 1213.

22.1.2 Sampling Techniques in Mass Spectrometry

Gases, liquids, and solids may be examined by mass spectroscopy by the use of ordinary sampling devices. Sample handling depends upon

(i) the type of inlet system in a given mass spectrometer;
(ii) the mechanism for introducing samples;
(iii) the requirement of physical and chemical separation techniques.

Usually every different laboratory uses its own characteristic accessories, inlet systems, and sample-handling techniques for the introduction of samples to the mass spectrometer:

(i) Introduction of gas samples. A number of gas-tight syringes are commercially produced.

(ii) Introduction of liquid samples. The wide variety of liquid microlitre-size syringes ($c.$ 2 μl) provides many different methods of introducings amples. The sample is drawn into the needle, injected through a septum into the inlet system, and the syringe is withdrawn. The syringes permit introduction of constant volumes, or they may be calibrated to introduce fixed amounts.

(iii) Introduction of solid samples. Samples ($c.$ 10 μg) may be introduced in capillaries, glass cups, metal tubules, etc. The amount of sample to be used varies greatly, usually depending upon the volatility of the substance.

Bibliography: 117, 123, 564, 750.
References: 3075, 4048, 5259, 6403, 7241.

22.1.3 High-resolution Mass Spectrometers

Modern double-focusing mass spectrometers apply an electrostatic field ahead of the magnetic field. A positive ion in an electrostatic field experiences a force in the direction of the field. The path of an ion moving through the field is thus curved. In a radial electrostatic field (always perpendicular to the direction of flight of the ions) the radius of curvature of the ion path is dependent on the energy of the ion and the strength of the electrostatic field. The electrostatic field is an energy analyser, instead of a mass analyser, and serves to limit the energy spread of the ion beam before it enters the magnetic field. The energy spread is one of the major factors in decreasing the resolution of the single-focusing mass spectrometer.

There are two types of double-focusing mass spectrometer:

(i) The MATTAUCH–HERZOG double-focusing mass spectrometer (Fig. 22.4). In this instrument are incorporated a 31°51′ electrostatic field analyser and a 90° magnetic field analyser. The multimass ion beam is

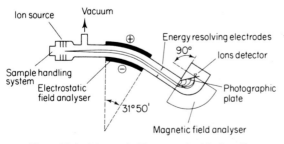

Fig. 22.4 Mattauch–Herzog double-focusing mass spectrometer

uniform in energy as it leaves the electrostatic analyser, and the magnetic field produces the desired mass dispersion while focusing each unimass ion beam to a different point.

(ii) The NIER–JOHNSON double-focusing mass spectrometer (Fig. 22.5). In this instrument the electrostatic and magnetic field analysers are both 90°, and all ions are focused to the same point in the detector.

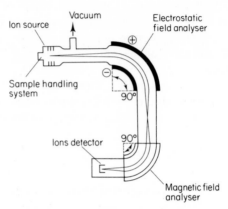

Fig. 22.5 Nier–Johnson double-focusing mass spectrometer

Bibliography: 407, 564, 685, 750, 880, 881.
References: 2387–2389, 5135.

22.1.4 Time-of-flight Mass Spectrometers

The high-resolution mass spectrometer known as the TIME-OF-FLIGHT MASS SPECTROMETER makes use of a drift tube and employs no magnetic field (Fig. 22.6). Ions leave the acceleration field with the same kinetic energies,

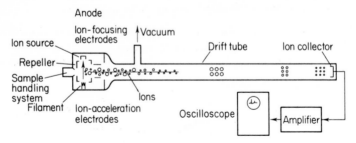

Fig. 22.6 Time-of-flight mass spectrometer

but with different velocities depending on their masses. They travel in a straight line through a field-free drift tube. Different ions will travel a given distance in a drift tube in different times. The measurement of this time of flight (in microseconds) is the basis for the determination of the m/e value of different ions. A control grid placed in the ion beam produces ions in the form of pulses lasting only about 0.25 μs at a frequency of 10 000 times per second. If the ions were allowed to enter the drift tube continuously, there would be no possibility of measuring the time required for a given ion. The accelerating grid and collector device must also be pulsed in sequence. An oscilloscope is commonly used to display the spectrum.

Reference: 5996.

22.1.5 Quadrupole Mass Spectrometers

The high-resolution mass spectrometer known as the QUADRUPOLE MASS SPECTROMETER employs four electric poles (called QUADRUPOLES) (Fig. 22.7) and no magnetic field (Fig. 22.8). Ions entering from the ion source

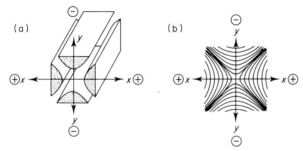

Fig. 22.7 (a) The quadrupole electrode structure, which produces the equipotential lines shown in (b)

travel with constant velocity in the direction parallel to the poles (z direction), but acquire oscillation in the x and y directions. This is accomplished by application of both a DC voltage and a radio-frequency voltage to the poles. There is a stable oscillation that allows an ion to pass from one end of the

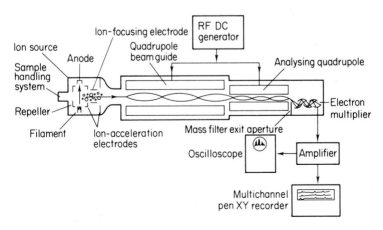

Fig. 22.8 Quadrupole mass spectrometer

quadrupole to the other without striking the poles. This oscillation is dependent on the mass-to-charge (m/e) ratio of an ion. Therefore only ions of a single m/e value will traverse the entire length of the analyser. All ions will have unstable oscillations and will strike the poles. Mass scanning is carried out by varying the DC voltage and the radio-frequency while keeping their ratios constant. Commercial instruments recently developed have a resolution up to 8000.

Bibliography: 361, 564, 750, 1213.
References: 2603, 3392, 5728, 6437, 6967.

22.2 GAS CHROMATOGRAPHY MASS SPECTROMETRY

Several commercial instruments are built as a gas chromatograph coupled to a mass spectrometer through an interface that enriches the concentration of the

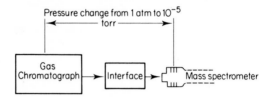

Fig. 22.9   Gas chromatograph–mass spectrometer analysing system

sample in the carrier gas (Fig. 22.9). Scan times are rapid enough so that several mass spectra can be obtained during the elution of a single peak from the gas chromatograph.

Bibliography: 547, 564, 895, 1202.
References: 3780, 4602, 4982, 5239, 5333, 6705.

22.3 ELECTROSPRAY MASS SPECTROSCOPY

ELECTROSPRAY MASS SPECTROSCOPY is a technique in which a dilute solution of polymers is electrosprayed from a hypodermic needle at voltages of about 10 kV into nitrogen at atmospheric pressure. A supersonic molecular beam of the resultant mixture of ions and neutral molecules is formed on passage of the gas mixture through a nozzle–skimmer system and energy is analysed by means of a repeller grid–Faraday cage ion collector system. Estimating the velocity of the beam (c. 500–1000 m s^{-1}), it is possible to calculate the M/z ratio of the macroions (M is the mass of the ion and z is the number of elementary charges).

Bibliography: 424.
References: 3185, 5011, 6762.

22.4 APPLICATIONS OF MASS SPECTROMETRY TO THE CHARACTERIZATION AND ANALYSIS OF POLYMERS

The main applications of mass spectrometry are
  (i) study of the stereoregularity of polymers by thermal decomposition of macromolecules and analysis of the gas evolved;
 (ii) study of the thermal degradation of polymers (pyrolysis) by analysis of the gaseous products evolved (cf. Section 34.14);
(iii) determination of gas permeability constants.

Bibliography: 658, 1356.
References: 2604, 2607, 2668, 3251, 3633, 3873, 3898, 4076, 4179–4181, 4267, 4275, 4276, 4723, 4853, 4882, 4982, 4983, 5099, 5333, 5518, 5924, 5996, 6707, 6136, 6263, 7085, 7086.

Chapter 23

Chromatography

General bibliography: 15, 202, 265, 349, 379, 417–419, 509, 556, 595, 645, 703, 738, 749, 798, 830, 905, 924, 1072, 1073, 1176, 1216, 1264, 1436, 1451, 1456.

23.1 GENERAL CHROMATOGRAPHIC THEORY

All types of CHROMATOGRAPHY may be defined as differential migration processes where sample components are selectively retained by a stationary phase.

The STATIONARY PHASE may be an active solid or it may be an immobile liquid.

CHROMATOGRAPHIC TECHNIQUES are separation processes, and they can be divided into two main branches:

(i) gas chromatography;
(ii) liquid chromatography.

A CHROMATOGRAM is a record obtained from a chromatographic analysis in the form of plot of the detector response versus the retention time or the retention volume (cf. Fig. 23.1).

The RETENTION TIME (ELUTION TIME) (t_R) is the time from injection to the peak maximum (in seconds or minutes).

The RETENTION VOLUME (ELUTION VOLUME) (V_R) is given by

$$V_R = t_R v = V_m + KV_s = t_m v + KV_s, \qquad (23.1)$$

where
t_R is the retention time,
v is the flow rate,
V_m is the mobile phase interstitial volume (void volume),
K is the distribution coefficient,
V_s is the stationary phase volume,
t_m is the retention time of unretained components (solvent front).

Retention time or retention volume is a characteristic property of a given compound on a specific column under a given set of conditions such as flow rate, temperature, pressure, stationary phase, mobile phase.

The area under a particular peak is proportional to the concentration of a particular compound injected.

The resolution of chromatographic peaks on chromatograms is related to two factors:

(i) COLUMN EFFICIENCY, which is measured by the number of theoretical plates, or the height equivalent to a theoretical plate (HETP);
(ii) COLUMN SELECTIVITY, which is measured by the relative separation of peaks.

The NUMBER OF THEORETICAL PLATES (N) is given by (Figs 23.1 and 23.2)

$$N = 16(a/b)^2, \tag{23.2}$$

where

a is the distance from injection to peak maximum and is equivalent to the retention time (t_R) or the retention volume (V_R),

b is the length of the peak baseline limited by the two tangents.

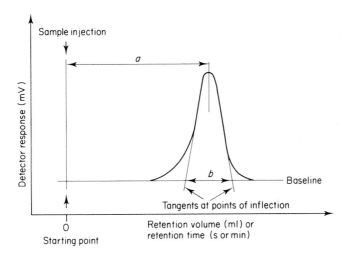

Fig. 23.1   One-peak chromatogram and method of calculation of number of theoretical plates (column efficiency) (N)

The HEIGHT EQUIVALENT TO A THEORETICAL PLATE (HETP) is related to the number of theoretical plates (N) by

$$HETP = L/N \tag{23.3}$$

where L is the length of the chromatographic column (in centimetres).

The HETP measure of column efficiency allows for comparisons between columns of different lengths.

The VAN DEEMTER EQUATION is an equation relating the HETP to flow velocity:

$$\text{HETP} = A + B/v + C/v, \qquad (23.4)$$

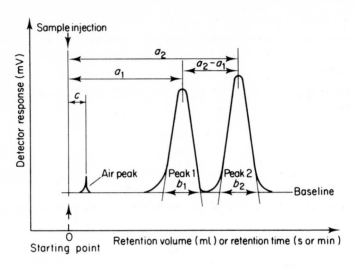

Fig. 23.2 Two-peak chromatogram and method of calculation of column efficiency (N), column selectivity (x), and column resolution (R)

where
  A is the EDDY DIFFUSION, i.e. the contribution to band broadening because of the paths of varying length that the sample molecules may take through a system (Fig. 23.3)—molecules that take a short path emerge earlier than the main peak, whereas molecules that travel by a long path emerge later than the main peak,
  B is the molecular diffusion,

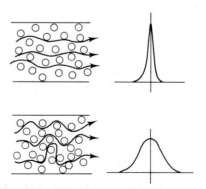

Fig. 23.3 Schematic representation of the influence of Eddy diffusion on band broadening

C is the MASS TRANSFER, i.e. the movement of a sample through a phase to an interface,

v is the flow rate.

The Van Deemter equation can be presented in the form of a VAN DEEMTER PLOT (Fig. 23.4).

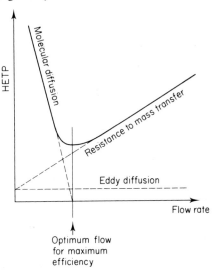

Fig. 23.4   Van Deemter plot

The COLUMN SELECTIVITY (COLUMN EFFICIENCY) (α) is the ratio of two corrected retention times or retention volumes or distribution coefficients:

$$\alpha = \frac{t_{R_2} - t_m}{t_{R_1} - t_m} = \frac{V_{R_2} - V_m}{V_{R_1} - V_m} = \frac{K_2}{K_1}, \tag{23.5}$$

where

t_{R_1} and t_{R_2} are corrected retention times of compounds 1 and 2, respectively,

V_{R_1} and V_{R_2} are corrected retention volumes of compounds 1 and 2, respectively,

K_2 and K_1 are distribution coefficients of compounds 1 and 2, respectively,

t_m is the retention time of unretained components (solvent front),

V_m is the mobile phase interstitial volume (void volume).

The effect of band spreading on peak resolution is shown in Fig. 23.5.

The COLUMN SEPARATION FACTOR (α_s) is the ratio of two uncorrected retention times (t_R) or retention volumes (V_R):

$$\alpha_s = t_{R_2}/t_{R_1} = V_{R_2}/V_{R_1}. \tag{23.6}$$

The DISTRIBUTION COEFFICIENT (K) is defined as

$$K = \frac{\text{Concentration of solute in the stationary phase}}{\text{Concentration of solute in the mobile phase}}. \tag{23.7}$$

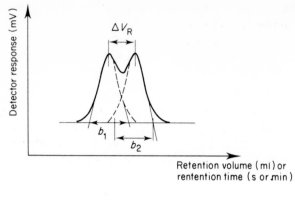

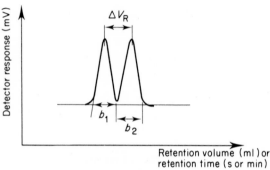

Fig. 23.5 Effect of band spreading on peak resolution

The distribution coefficient is also called:

(i) the PARTITION COEFFICIENT, in liquid partition chromatography;
(ii) the PERMEATION COEFFICIENT, in exclusion chromatography;
(iii) the ADSORPTION COEFFICIENT, in liquid adsorption chromatography;
(iv) the DISTRIBUTION COEFFICIENT, in ion-exchange chromatography and gel permeation chromatography.

The general rule is that

(i) the distribution coefficient (K) is high when most of the substrate is retained in the stationary phase;
(ii) separation between two compounds is possible if their distribution coefficients are different;
(iii) the greater the difference in K values, the fewer the plates or the shorter the column required to achieve a separation;
(iv) the distribution coefficient (K) decreases with increasing temperature.

The COLUMN CAPACITY FACTOR (k) is given by

$$k = KV_s/V_m = (V_R - V_m)/V_m, \qquad (23.8)$$

where

K is the distribution coefficient,

V_m is the stationary phase volume,

V_s is the mobile phase interstitial volume,

V_R is the retention volume.

The general rule is that

    (i) small values of k show that the compounds are poorly retained by the column and elute close to the unretained peak (poor separations);

    (ii) large values of k show that separation is improved, but the analysis time is increased, with the formation of wide peaks which are difficult to detect;

    (iii) optimum values of k are between 1.5 and 4.0.

The COLUMN RESOLUTION (R_c) is the true separation of two consecutive peaks and is given by

$$R_c = \frac{2(t_{R_2} - t_{R_1})}{b_1 + b_2} = \frac{2(V_{R_2} - V_{R_1})}{b_1 + b_1}. \tag{23.9}$$

The column resolution (R_c) is a combined measure of column efficiency (term A), column selectivity (term B), and column capacity (term C):

$$R_c = \tfrac{1}{4}\sqrt{(N)} \underbrace{\left(\frac{\alpha - 1}{\alpha}\right)}_{} \underbrace{\left(\frac{k}{k+1}\right)}_{}. \tag{23.10}$$

                         (A)   (B)   (C)

Separation can be improved by adjusting any one of these three factors.

References: 4427, 5539, 6912.

23.2 QUANTITATIVE CHROMATOGRAPHIC ANALYSIS

In chromatographic methods, the area produced for a particular peak is proportional to the concentration of the particular component. There are many ways of relating the peak shape to the sample concentration (cf. Section 23.3). The weight per cent composition can be obtained by the following means:

    (i) The AREA NORMALIZATION METHOD. In this method the weight per cent composition can be obtained by measuring the area of each peak and dividing the individual areas by the total area (Fig. 23.6):

$$A_x(\%) = (A_x / \textstyle\sum A_1) \times 100, \tag{23.11}$$

$$A_1 + A_2 + \ldots + A_i + \ldots + A_n = \textstyle\sum A_i, \tag{23.12}$$

where

$A_x(\%)$ is the normalized peak area of component x expressed as a percentage of the sum of the individual peak areas.

$A_1, \ldots, A_n$ are the areas of the individual peaks

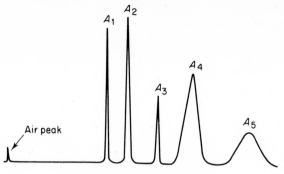

Fig. 23.6   Area normalization plot

This method is valid only when all peaks are eluted and when every compound in the mixture has the same detector response. Normally, areas of compounds are not directly proportional to the percentage composition because different compounds have different detector responses and it is necessary to determine correction factors (f_i).

The peak area can be expressed as

$$A_i = f_i\, C_i \qquad (23.13)$$

(in per cent) where
  A_i is the raw peak area,
  C_i is the adjusted peak area,
  f_i is the detector response factor (in mole per cent, volume per cent, or weight per cent).

The DETECTOR RESPONSE FACTOR (f_i) is the proportionality factor which depends on the chemical nature of the substance and on the type of detector used. Usually, it is not given as an absolute amount (e.g. peak area by unit weight), but rather relative to a given standard, i.e. as a RELATIVE DETECTOR RESPONSE FACTOR. If these factors are known, the concentration of the individual component in the sample can be calculated from the raw peak areas.

The determination of relative response factors can be carried out using the following treatment. From eq. (23.13) the peak area for the first two components can be expressed as

$$A_1 = f_1\, C_1, \qquad (23.14)$$

$$A_2 = f_2\, C_2. \qquad (23.15)$$

Thus

$$C_1 = A_1/f_1, \qquad (23.16)$$

$$C_2 = A_2/f_2. \qquad (23.17)$$

If the concentration of both components is the same, then $C_1 = C_2$ and therefore

$$A_1/f_1 = A_2/f_2 \qquad (23.18)$$

or

$$f_2 = (A_2/A_1)f_1. \qquad (23.19)$$

Usually, f_1 is assigned a value of 100 and thus

$$f_2 = (A_2/A_1) \times 100. \qquad (23.20)$$

The relative response factors of the other components can be expressed similarly.

The relative response factors are used to help in the calculation of true concentration values from the peak areas. In this calculation, the so-called REDUCED PEAK AREA VALUES (A/f) are first established. Then, the concentrations of the individual components (c_x) are calculated as the ratios of their respective reduced peak areas and their sum:

$$(A_1/f_1)+(A_2/f_2)+...+(A_i/f_i)+...+(A_n/f_n) = \Sigma(A_i/f_i), \qquad (23.21)$$

$$C_x(\%) = (A_x/f_x)/\Sigma(A_i/f_i) \times 100. \qquad (23.22)$$

Bibliography: 417, 419, 905.
References: 3153, 3154, 3329, 3782, 4288, 4289, 4356, 5242, 6101.

(ii) The EXTERNAL STANDARD METHOD. This method allows one to obtain a correct value of the weight per cent composition. In this method exact amounts of the pure calibration samples are injected (A_1, A_2). The values of the peak areas are then plotted against the known weight of sample injected in order to give the calibration curve, which should be linear and should pass through the origin (Fig. 23.7).

An exact amount of the unknown composition is further injected (A_x). The peak area is then measured and, from the calibration curve, the amount of sample present in the unknown composition can be calculated (Fig. 23.7).

Bibliography: 419, 556, 905.

(iii) The INTERNAL STANDARD METHOD. This method allows one to obtain a correct value of the weight per cent composition from one analysis. The detector response factors do not have to be determined. Since the determination is relative, analytical conditions do not have to be reproduced exactly.

In this method a known weight ratio of the sample and a standard are prepared and chromatographed. The peak areas are measured and area ratios are plotted against weight ratios to obtain a graph (Fig. 23.8). An accurately known amount of the internal standard is further added to the unknown sample and this mixture is chromatographed. Area ratios are measured and, from the calibration graph, the weight ratio of the unknown to the standard can be obtained. Since the amount of

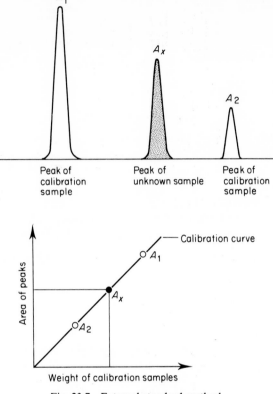

Fig. 23.7   External standard method

standard added is known, it is a simple calculation to determine the amount of the unknown compound present.

Bibliography: 419, 556, 905.

23.3 CONVERSION OF STRIP CHART RECORDER DATA INTO NUMERICAL DATA

There are several methods for relating the peak shape to the sample concentration:

(i) The PEAK HEIGHT MEASUREMENT METHOD. Peak height (in millimetres) is measured as the distance from the baseline to the peak maximum, as shown in Fig. 23.9. Baseline drift is compensated for by interpolation of the baseline between start and finish, as shown for peaks A, B, and C. Peak heights and widths are dependent on the sample size and the sample feed volume.

(ii) PEAK AREA MEASUREMENT METHODS are less dependent on operating conditions than peak height methods.

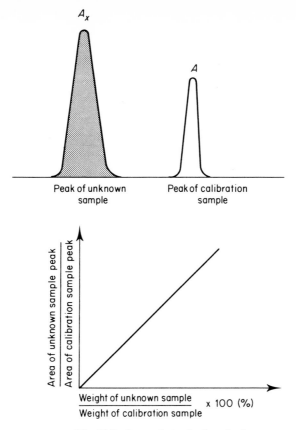

Fig. 23.8   Internal standard method

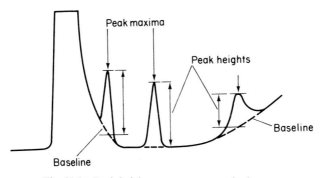

Fig. 23.9   Peak height measurement method

(a) HEIGHT TIMES WIDTH AT HALF-HEIGHT METHOD. The area can be calculated by the triangle formula (Fig. 23.10):

$$A = \tfrac{1}{2}HW, \tag{23.23}$$

where
H is the peak height (in millimetres);
W is the width (in millimetres) at half-height.

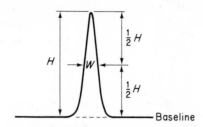

Fig. 23.10   Height times width at half-height method

This method should be applied only to symmetrical peaks or peaks which have similar shapes. The area measured is several per cent less than the true area, but it is proportional to the sample size.

(b) TRIANGULATION METHOD. In this technique, height is measured from the baseline to the intersection of the two tangents (Fig. 23.11). The area can be calculated by the triangle formula:

$$A = \tfrac{1}{2}bH. \tag{23.24}$$

This method should be applied only to symmetrical peaks (i.e. of Gaussian shape).

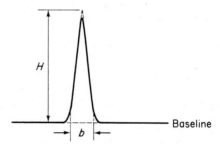

Fig. 23.11   Triangulation method

(c) PLANIMETRY. In this method the peak is traced manually with a planimeter. A planimeter is a mechanical device which measures area by tracing the perimeter of the peak. The area is presented digitally on a dial. Planimetry is less precise than other methods. Precision can be improved by tracing each peak several times and taking an average value.

(d) CUTTING AND WEIGHING PAPER TECHNIQUE. In this method peak areas are determined by cutting out the peak and weighing the paper on an analytical balance. This method is fairly precise, particularly for asymmetrical peaks. The homogeneity, moisture content, and weight of the paper, as well as the skill exercised in cutting the peaks, are important factors.

(e) DIGITAL INTEGRATORS are devices which automatically measure the peak area, but do not make the calculation or data interpretation. The combination of a digital integrator with a teletype output and with time-sharing on a computer is very useful.

Bibliography: 88, 493, 556, 905.

23.4 GAS CHROMATOGRAPHY

GAS CHROMATOGRAPHY is a technique for separating volatile substances by percolating a gas stream over a stationary phase.

Gas chromatography can be divided into two main branches:

(i) GAS–LIQUID CHROMATOGRAPHY (GLC). In this technique the stationary phase is a thin film of a liquid over an inert solid.

(ii) GAS–SOLID CHROMATOGRAPHY (GSC). In this technique the stationary phase is a solid (silica gel, molecular sieve, charcoal).

Bibliography: 338, 410, 568, 1168, 1176, 1450.

(iii) MASS CHROMATOGRAPHY is a modified form of gas chromatography that measures not only the retention data but also gas density, thus providing additional information to calculate the molecular weight and absolute weight of each component. Peak identification is given by combined retention data and molecular weight.

Bibliography: 1269.
References: 3565–3568.

23.5 GAS–LIQUID CHROMATOGRAPHY

In GAS-LIQUID CHROMATOGRAPHY (GLC) the components to be separated are carried through the column by an inert gas called a CARRIER GAS. The sample mixture is partitioned between the carrier gas and a non-volatile solvent (STATIONARY PHASE) supported on an inert size-graded solid (SOLID SUPPORT). The solvent selectively retards the sample components according to their distribution coefficient until they form separate bands in the carrier gas. These component bands leave the column in the gas stream and are recorded as a function of time by a detector.

The solvent selectively retards the sample components because interaction forces exist between solute and solvents. There are four types of interaction force:

(i) ORIENTATION FORCES (KEESOM FORCES) resulting from the interactions between two permanent dipoles (e.g. a hydrogen bond);

(ii) INDUCED DIPOLE FORCES (DEBYE FORCES) resulting from the interactions between a permanent dipole in one molecule and an induced dipole in a neighbouring molecule—these forces are usually very small;

(iii) DISPERSION FORCES (NON-POLAR FORCES) (LONDON FORCES) resulting from synchronized variations in the instantaneous dipoles of the two interaction species—these forces are very weak;

(iv) SPECIFIC INTERACTION FORCES resulting from chemical bonding and complex formation between solute and solvents.

Gas-liquid chromatography is a very useful technique for analysing low molecular weight compounds (e.g. monomers, additives, solvents). The problem in applying gas–liquid chromatography to polymer analysis is that most macromolecules have too low a vapour pressure even at elevated temperatures to pass through a column of this type. This problem can be partially solved by using pyrolysis gas chromatography (cf. Section 34.15).

23.5.1 Gas Chromatographs

These are built of the following parts (Fig. 23.12):

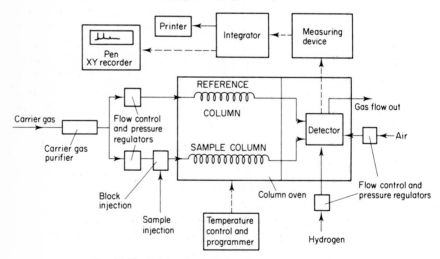

Fig. 23.12 Schematic diagram of a gas chromatograph

(i) A cylinder of carrier gas (hydrogen, helium, or nitrogen) with flow control and a pressure regulator. The optimum flow rate can easily be determined experimentally from a van Deemter plot of HETP versus linear gas velocity (Fig. 23.4). The most efficient flow rate is at the HETP minimum or the maximum number of plates.

The presence of oxygen and water in the carrier gas can damage the chromatographic liquid phase at elevated temperatures, causing tailing and loss of peak resolution. Adsorbents such as molecular sieves, silica gel, and alumina deteriorate as they pick up water from the carrier gas. Carbon supports are attacked by O_2 at temperatures above 200 °C. If water is present in the carrier gas, it will accumulate at the head of the column during the cooling period in temperature programming work and, when the column is heated, the eluted water peak causes an erratic baseline and an interfering peak.

Many of the ultrasensitive detectors used in gas chromatography are affected by O_2 and H_2O impurities in the carrier gas. Competition of O_2 in the carrier gas for free electrons lowers the response of the detector for the analysed species. Flame ionization and thermal conductivity detectors will give negative peaks with traces of O_2 and H_2O in the carrier gas.

In order to remove traces of O_2 and H_2O one must make use of a high capacity purifier which will reduce the O_2 and H_2O content to less that 1 p.p.m. of each. Commercially produced purifiers contain a tube filled with a molecular sieve pretrap heated by an electric furnace.

(ii) A sample injection system. Gases are usually introduced by gas-tight syringes or by a gas sampling valve. Liquids are also introduced with syringes. Solid samples are injected in solution in a solvent which does not interfere with the samples being analysed. Samples are injected through a self-sealing septum using a hypodermic syringe needle. The injected volume is determined from the volume of the syringe (0.2–µl for a regular analytical column and 0.004–0.5 µl for a capillary column).

(iii) A column. Columns are made from copper, stainless steel, aluminium, and glass in straight, bent, or coiled forms; they are 1–3 m in length and 0.25–4 mm in inner diameter. A longer length gives more theoretical plates and better resolution but requires very high inlet pressure. Typical gas chromatographic column characteristics are given in Table 23.1.

Use of wall-coated open tubular columns (WCOT), support-coated open tubular columns (SCOT), or micropacked capillary columns requires careful attention to dead volume and sample size, as well as the mechanical hookups.

DEAD VOLUME is the total volume of the system that is not in the path of, and swept by, the mobile phase. For the highest efficiency the dead volume should be minimized.

In the use of capillary columns, special consideration must be given to the good connection of these columns with the inlet system of a chromatograph.

A column is packed with the solid support (Chromosorb) to provide a uniform, inert surface area for distributing the liquid phase. Chromosorbs are composed of the skeletons of diatoms (microscopic unicellular algae). The diatom support surfaces are covered with silanol (Si—OH) and siloxane groups (Si—O—Si) which can hydrogen bond with solvents and solutes. Particle sizes of 100/120, 80/100, 60/80, and 40/60 mesh are recommended.

Table 23.1 Typical gas chromatograph column characteristics (in CGS units) (from B: 1469; reproduced by permission of Alltech)

	Column type			
	Wall-coated open tubular (WCOT) column (Golay)	Porous-layer open tubular (PLOT) or support-coated open tubular (SCOT) column	Micropacked capillary	Conventionally packed column
Graphic description				
Coil diameter	130 mm	130 mm	130 mm	Variable
Outer diameter	0.75 mm	1.0 mm	1.5 mm	3.175–6.35 mm
Inner diameter	0.25 mm and 0.50 mm	0.50 mm	0.6–0.8 mm	2 mm and 4 mm
Typical length	20–100 m	15–30 m	3–6 m	1–3 m
Total efficiency per column	100 000	50 000	28 000	4 000
Type of sample injection system	Splitting	Splitting or splitless	Splitting or splitless	Conventional
Sample size	10^{-4}–10^{-3} µl	10^{-3}–10^{-2} µl	10^{-2}–10^{-1} µl	1 µl and larger
Flow rate	0.05–5 cm³ min⁻¹	2–8 cm³ min⁻¹	2–8 cm³ min⁻¹	20–100 cm³ min⁻¹
Speed of analysis	Fast	Fast	Good	Slow
Temperature stability	Good	Good	Good	Good
Chemical inertness	Best ──→ Poorest			
Uniform flow velocity	Best ──→ Poorest			
Linear flow velocity	20–30 cm s⁻¹	20–30 cm s⁻¹	2–8 cm³ min⁻¹	10–70 cm³ min⁻¹

The MESH NUMBER refers to the number of openings per linear inch. Particles that will pass through 100 mesh but not 120 mesh are referred to as 100/120 mesh (Table 23.2).

Table 23.2   Particle sizes and mesh ranges

Particle diameters (μm)	US standard mesh range
590–420	30/40
420–297	40/50
297–210	50/70
210–149	70/100
149–105	100/140
105–74	140/200
74–37	200/400
37–20	400/625
20–5	625/2500

Correct selection of the partitioning solvent is based mainly on experience and/or trial and error. The liquid phase chosen (Table 23.3) depends on the composition of the sample. For efficient separation, the liquid phase should be similar in chemical structure to the components of the mixture, e.g. hydrocarbon compounds are best separated with a hydrocarbon solvent, polar compounds with a polar solvent, etc. If the components of the mixture are of different chemical structure, liquid phases of different polarity must be used. According to EWELL'S CLASSIFICATION all solutes can be divided into five groups (Table 23.4). Chemical structure and properties of the most common liquid phases are given in Appendixes I and J.

The amount of liquid phase used should be enough to coat the particles with a thin uniform layer, and between 2 and 10 wt% is usual. Column efficiency decreases drastically for liquid loadings exceeding 30 wt%. Several methods of coating the solid support with liquid phase can be employed. The most useful is a rotating evaporator method. In this method the correct amount of liquid phase (cf. Appendix J) is dissolved in a suitable solvent and placed in a round-bottom flask in a rotating evaporator. The weight amount of solid support is added. After connection of the rotating evaporator with the vacuum, the flask is rotated until all of the solvent is evaporated.

The column is filled with a coated support by using a funnel and by being vibrated until no further packing can be added. It may then be coiled or bent to fit the configuration of the chromatograph oven. Columns should be conditioned for at least 2 h at 25 °C above the maximum temperature at which the column will be used, but below the maximum temperature limit for the liquid phase. A small carrier gas

Table 23.3 Liquid phase classification (from B: 905; reproduced by permission of Varian Associates) (cf. Appendixes I and J)

Classification	Liquid phase
Class A (I)	FFAP
	20M–TPA
	Carbowaxes
	Ucons
	Versamid 900
	Hallcomid
	Quadrol
	Theed
	Mannitol
	Diglycerol
	Castorwax
Class B (II)	Tetracyanoethyl panterythritol
	Zonyl E–7
	Ethofat
	β,β-Oxydipropionitrile
	XE–60
	XF–1150
	Amine 220
	Epon 1001
	Cyanoethyl sucrose
Class C (III)	All polyesters
	Dibutyl tetrachlorophthalate
	SAIB
	Tricresyl phosphate
	STAP
	Benzyl cyanide
	Lexan
	Propylene carbonate
	QF–1
	Polyphenylether
	Dimethylsulfolane
	OV–17
CLASS D (IV and V)	SE–30
	SF–96
	DC–200
	Dow 11
	Squalane
	Hexadecane
	Apiezons
	OV–1

flow (5–10 ml min^{-1}) should be used during conditioning. The exit end of the column should be left disconnected from the detector to avoid contamination of the detector.

(iv) Detectors indicate the presence and measure the amount of components in the column eluent. The two most popular detectors are

(a) A thermal conductivity cell (TC detector) (Fig. 23.13), which employs a tungsten filament which is heated by passing a constant

Table 23.4 Ewell's classification of solutes (from B: 905; reproduced by permission of Varian Associates)

Classification	Solute	Character
Group I	Water Glycol, glycerol, etc. Amino alcohols Hydroxy acids Polyphenols Dibasic acids	High polar
Group II	Alcohols Fatty acids Phenols Primary and secondary amines Oximes Nitro compounds with α–H atoms Nitriles with α–H atoms NH_3, HF, N_2H_4, HCN	Polar
Group III	Ethers Ketones Aldehydes Esters Tertiary amines Nitro compounds with no α–H atoms Nitriles with no α–H atoms	Medium polar
Group IV	$CHCl_3$ CH_2Cl_2 CH_3CHCl_2 CH_2ClCH_2Cl $CH_2ClCHCl_2$ etc. Aromatic hydrocarbons Olefinic hydrocarbons	Low polar
Group V	Saturated hydrocarbons CS_2 Mercaptans Sulphides Halocarbons CCl_4	Non-polar

current through it. When the molecules mixed with the carrier gas pass over the hot filament, the rate of heat loss is reduced and the resistance of the filament increases. The resistance change is measured by using a Wheatstone bridge and the signal is fed to a recorder, where it appears as a peak.

(b) A flame ionization detector (FID) (Fig. 23.14), where hydrogen and air are used to produce a flame. A collector electrode with a DC potential applied is placed above the flame and measures the conductivity of the flame. With pure hydrogen, the conductivity is quite low, whereas organic compounds are burnt with a resulting

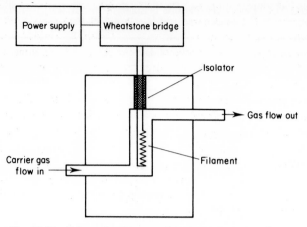

Fig. 23.13 Schematic diagram of a thermal conductivity cell (TC detector)

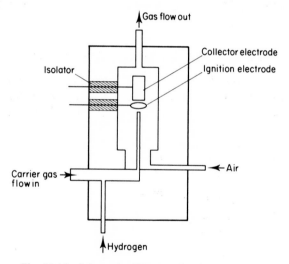

Fig. 23.14 Schematic diagram of a flame ionization detector (FID detector)

conductivity increase and the current which flows is amplified and fed to a recorder.

A summary of detector parameters is given in Table 23.5.

Bibliography: 356, 891, 905.

(v) A recorder. The potentiometric strip chart XY recorder is a commonly used instrument. More advanced gas chromatographs use automatic devices to convert the analogue signal from the chromatograph to a digital value. Electronic digital integrators integrate the area under the peak and print out peak areas and retention times.

Gas chromatographs can be connected to computers, which not only can integrate the peak area, but can also perform the necessary calculations to convert area to concentration of sample component.

(vi) Temperature control is required for the injection chamber, the column, and the detector. These three component parts require three different temperature control systems:

(a) The injection chamber should be at a sufficiently high temperature to vaporize the sample rapidly with no loss in efficiency. Too high a temperature may cause thermal decomposition of the sample.

(b) The column temperature should be high enough to obtain the desired degree of separation. The best method is to employ temperature programming. Temperature programming is an increase in the column temperature during analysis in order to provide a faster and more versatile analysis.

(c) The detector temperature must be high enough to eliminate condensation of the sample and/or liquid phase.

Bibliography: 379, 703, 1216.
References: 2669, 4046, 5801, 5335.

23.6 GAS–SOLID CHROMATOGRAPHY

In GAS–SOLID CHROMATOGRAPHY (GSC), the gas components are separated by adsorption on an active solid, e.g. silica gel, a molecular sieve, charcoal. A typical gas–chromatogram of a mixture of permanent gases is shown in Fig. 23.15.

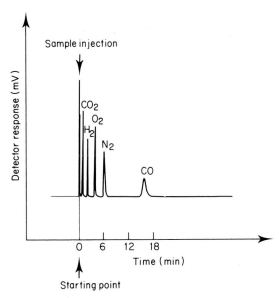

Fig. 23.15 Typical chromatogram of a gas mixture

Table 23.5 Summary of detector parameters (from B:

Detector	Principle of operation	Selectivity	Sensitivity $(g\ s^{-1})$	Line ran
Thermal conductivity	Measures thermal conductivity of gases	Universal—responds to all compounds	6×10^{-10}	10^4
Gas density balance	Difference of molecular weight of gases	Universal—responds to all compounds whose MW differs from carrier gas	Variable in range of TCD	10^3
Flame ionization	H_2–O_2 flame 2000 °C plasma	Responds to organic compounds, but not to fixed gases or water	9×10^{-13} for alkanes	14^7
Electron capture ^3H ^{63}Ni	$N_2 + \beta \rightarrow e^-$ $e^- + sample \rightarrow$ loss of I	Response to electron-absorbing compounds, especially halogens, nitrates, and conjugated carbonyls	2×10^{-14} for CCl_4 5×10^{-14} for CCl_4	$5 \times$ 50
Alkali flame P compound N compound	Alkali modified H_2–O_2 flame 1600 °C plasma	Enhanced response to phosphorus compounds Enhanced response to nitrogen compounds	4×10^{-14} 7×10^{-12}	10^3 10^3
Helium	$He + \beta \rightarrow He^*$ $Sample \xrightarrow{He^*} 1_0$	Universal—responds to all compounds	2×10^{-14} for methane	$5 \times$
Cross-section	$\beta + sample \rightarrow 1_0$	Universal—responds to all compounds	10^{-9}	10^4

roduced by permission of Varian Associates)

Minimum detectable quantity (g)	Stability	Temperature limit (°C)	Carrier gas	Remarks
$^{-5}$ g of CH_4 per millilitre of detector effluent	Good	450° C	He H_2 N_2	Non-destructive—requires good temperature and flow control. Simple, inexpensive, and rugged
	Good	Better sensitivity 150 °C	N_2 CO_2 Ar	Good for analysis of corrosive compounds. Non-destructive
$\times 10^{-11}$ g for alkanes	Excellent	400 °C	He N_2	$H_2O + CS_2$ good solvents because no response. Destructive
$^{-13}$ g for lindane $\times 10^{-12}$ for lindane	Fair Fair	225 °C 350 °C	N_2 or Ar + 10% CH_4	Detector is easily contaminated and easy to clean. Sensitive to water; carrier gas must be dry. Can be operated in pulsed or DC mode. Non-destructive
$\times 10^{-12}$ g parathion $\times 10^{-10}$ azobenzene	Fair Fair	300 °C 300 °C	N_2 He	Destructive—requires flow controller for hydrogen and air Destructive—requires flow controller for hydrogen and air. High sensitivity operating in starved O_2 mode
$^{-12}$ g for fix gases	Poor	100 °C	He	High-sensitivity to bleed precludes its use with columns other than active solids. Non-destructive
$^{-5}$ g	Good	225 °C	H_2 or He + 3% CH_4	

The most useful solid phase for gas separation are molecular sieves (synthetic zeolites having the structure $Na_{12}(AlO_2)_{12}(SiO_2)_{12} \cdot 27H_2O$). These compounds are capable of separating materials based on their molecular size and configuration, and they adsorb molecules that have polar or polarizable properties.

23.7 SUPERCRITICAL FLUID CHROMATOGRAPHY

SUPERCRITICAL FLUID CHROMATOGRAPHY is a technique in which the mobile phase is a gas—above its critical temperature under pressure—which is able to dissolve the substance to be chromatographed. This dense fluid state has the advantage of better mass transfer when compared to the liquid state of the same substance.

Supercritical fluid chromatography has been applied to the separation of oligomers on an analytical and on a preparative scale.

References: 3731, 3732, 3922, 4309, 4310, 4605, 4606.

23.8 ISOLATION OF COMPONENTS OF THE EFFLUENT FROM GAS CHROMATOGRAPHY

If further analysis of the gas chromatograph effluent is required it is necessary to separate and isolate one component from another. This can be carried out by using a series of small tube-traps equipped with a four-way stopcock and refrigerated with liquid nitrogen (Fig. 23.16). The effluent gas flow is

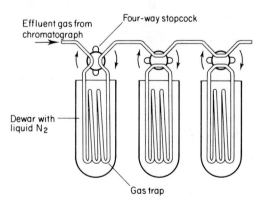

Fig. 23.16 Traps for gas chromatograph effluents

diverted, by means of the four-way stopcock, through the trap when the frontal zone of a component reaches the trap. The time delay between the recorder signal and the arrival of the component at the trap can be estimated from the flow rate and the distance between the katharometer and the trap.

23.9 APPLICATIONS OF GAS CHROMATOGRAPHY TO CHARACTERIZATION AND ANALYSIS OF POLYMERS

The main applications of gas chromatography are

(i) Determination of the purity of monomers, solvents, and additives.
(ii) Combined with other techniques as follows:
- (a) heating of the sample in order to analyse for the volatile materials in polymers, e.g. monomers, plasticizers, antioxidants, antistatic agents, and solvents in adhesives and coatings;
- (b) extraction of the sample in order to analyse for the same products as in (a) above;
- (c) pyrolysis in characterization of polymer structures (cf. Section 34.15);
- (d) hydrolysis in characterization of polymer structures;
- (e) hydrogenation in characterization of polymer structures;
- (f) micro-ozonolysis in characterization of polymer structures (cf. Section 39.6).

References: 3838, 4182, 4183, 4628, 5572, 6212–6213.

23.10 LIQUID CHROMATOGRAPHY

LIQUID CHROMATOGRAPHY can be divided into four main branches:

(i) LIQUID–LIQUID CHROMATOGRAPHY (LLC), which is a form of partition chromatography or solution chromatography. The sample is partitioned between the mobile liquid (usually water) and the stationary liquid (usually an organic solvent). The mobile liquid cannot be a solvent for the stationary liquid.
(ii) LIQUID–SOLID CHROMATOGRAPHY (LSC), which is a form of adsorption chromatography. Adsorbents such as silica gel, alumina, a molecular sieve, or porous glass are packed in a column and the sample components are displaced by the mobile phase (cf. Section 23.11). To this group belongs THIN LAYER CHROMATOGRAPHY (TLC) where instead of using a column the adsorbent is spread out on a flat glass plate (cf. Section 23.15).
(iii) ION-EXCHANGE CHROMATOGRAPHY is a technique which uses zeolites and synthetic organic and inorganic resins to perform chromatographic separations by an exchange of ions between the sample and the resins. Compounds which have ions with different affinities for the resin can be separated (cf. Section 23.20).
(iv) EXCLUSION CHROMATOGRAPHY is a technique which uses uniform, highly porous, non-ionic gels to separate materials according to their molecular size. The largest molecules of the solute cannot penetrate the pores within the cross-linked gel beads, and thus elute first. Smaller macromolecules of the solute are retained in interstices

within the gel beads and therefore require more time to elute. Exclusion chromatography (Chapter 25) includes

(a) gel filtration chromatography (cf. Section 25.2);
(b) gel permeation chromatography (cf. Section 25.3);
(c) hydrodynamic chromatography (cf. Section 25.15).

23.11 LIQUID–SOLID CHROMATOGRAPHY

In LIQUID–SOLID CHROMATOGRAPHY (LSC) the components to be separated are carried through the column in a mobile liquid phase. The interaction between the solute molecule and the stationary phase occurs on the surface of the stationary phase. It is governed by the difference in the strength of the adsorption forces between the stationary phase (ADSORBENT) and the solute molecules (ADSORBATES). In general, polar molecules are more strongly adsorbed by the polar stationary phase than are non-polar molecules. Adsorption of polar molecules is enhanced in a non-polar medium, but is reduced in a polar medium due to increased competition of the mobile phase for the surface.

An ADSORPTION ISOTHERM describes the equilibrium concentration relationship between the amounts of adsorbed and unadsorbed solute at a given temperature. It is a plot of the concentration of solute in the adsorbed phase versus its concentration in the unadsorbed phase. There are four basic types of adsorption isotherm between a liquid and a solid surface (Fig. 23.17). The

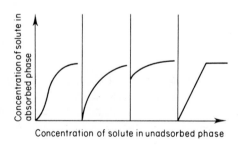

Concentration of solute in unadsorbed phase

Fig. 23.17   Basic types of adsorption isotherm
between liquid and solid surfaces

shape of the adsorption isotherm depends on the forces involved in the adsorption mechanism. There are four types of interaction force:

(i) ORIENTATION FORCES (KEESOM FORCES) resulting from the interaction between two permanent dipoles (e.g. hydrogen bonds);
(ii) INDUCED DIPOLE FORCES (DEBYE FORCES) resulting from the interaction between a permanent dipole in one molecule and the induced dipole in a neighbouring molecule—these forces are usually very small;

(iii) DISPERSION FORCES (NON-POLAR FORCES) (LONDON FORCES) resulting from synchronized variations in the instantaneous dipoles of the two interaction species—these forces are very weak.

(iv) SPECIFIC INTERACTION FORCES resulting from chemical bonding (chemisorption) and complex formation (e.g. charge transfer complexes) between solute and solvent.

There is an unfortunate tendency to use the terms 'high performance liquid chromatography' (HPLC) and 'high pressure liquid chromatography' (HPLC) interchangeably.

HIGH PERFORMANCE LIQUID CHROMATOGRAPHY (HPLC) is liquid–solid chromatography carried out at moderate to low pressure (e.g. at about 500 p.s.i. $\simeq 3.5 \times 10^7$ dyne cm^{-2} (CGS) $\simeq 3.5 \times 10^6$ Pa (SI)).

HIGH PRESSURE LIQUID CHROMATOGRAPHY (HPLC) is liquid–solid chromatography carried out at up to 5000 p.s.i. ($\simeq 3.5 \times 10^8$ dyne cm^{-2} (CGS) $= 3.5 \times 10^7$ Pa (SI)).

HPLC is a valuable separation and analytical technique because separation results can be obtained easily and rapidly. HPLC can generally be applied to separate compounds which are dissolved in some solvent.

Bibliography: 269, 363, 380, 546, 556, 590, 595, 684, 746, 828, 1029, 1183, 1184, 1238, 1239, 1265. References: 3655, 4517, 6458.

23.12 SELECTION OF THE COLUMN PACKING IN LSC

Liquid–solid separations can be carried out on the basis of one or more of the following mechanisms:

(i) adsorption,
(ii) partition,
(iii) ion exchange,
(iv) exclusion.

In Fig. 23.18 is presented a guide to LSC mode selection for a particular separation on the basis of minimal information about the sample. After the mode has been selected a particular column must be chosen. When ordering columns from manufacturers it is necessary to supply the following information:

(i) column packing, i.e. description, part number, part sizes;
(ii) ion form desired (for ion exchange);
(iii) length (in centimetres);
(iv) internal and external diameters of column (in millimetres);
(v) jacketed or unjacketed.

The column length is commonly 25, 30, 50, or 100 cm. Columns are packed with different materials dependent on the mode selected:

(i) Liquid–solid mode. These columns are packed with small particles of porous support material, e.g. silica (5–10 µm), alumina (5–10 µm), or charcoal.

394

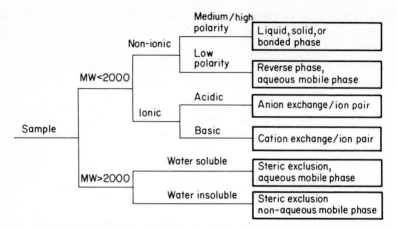

Fig. 23.18   Guide to mode selection for LSC column packing

(ii) Bonded phase mode. These columns are packed with both polar and non-polar phases. A stable bonded phase is prepared by chemical reactions forming the siloxane (—Si—O—Si—) bond.

(iii) Reverse phase mode.

(iv) Ion exchange mode. These columns are packed with two distinct types of ion exchange resin: porous and superficially porous (called often PELLICULAR). Ion exchange columns are offered as strong anion $(R\,N(CH_3)_3^+)$ and strong cation (RSO_3^-) (cf. Section 23.20). When ordering ion exchange columns the ion form must be specified: possible cations include H^+, NH_4^+, Na^+; possible anions include Cl^-, CH_3COO^-, $HCOO^-$, BO_3^-, PO_4^- and citrate ions.

(v) Steric exclusion mode. These columns are packed with semi-rigid cross-linked gels filled with tetrahydrofuran. Each gel has a distinct pore size which excludes molecules of larger size. To choose the correct gel for a particular application one must know the molecular weight range of the sample (cf. Section 25.9).

Selection of column packing and operating conditions requires considerable experimentation.

GUARD COLUMNS are used for the protection of the analytical column and in order to extend its useful life. The guard column removes sample and mobile-phase impurities. It should be packed with a similar material to the analytical column.

Bibliography: 143.
Reference: 5038.

23.13 SELECTION OF THE MOBILE PHASE IN LSC

Liquid–solid chromatography (LSC) is carried out with a polar adsorbent with mobile phases ranging from the non-polar (pentane) to the very polar (alcohols).

The relative polarities of the mobile phase and the solute govern the adsorption equilibria. A polarity scale can be established by empirically rating solvents in the order of their strength of adsorption on an adsorbent such as silica. A solvent high in polarity will displace one lower in the polarity scale. Such a scale is called an ELUOTROPIC SERIES (Table 23.6).

Table 23.6 Elutropic series and characteristics of pure solvents (from B: 1456; reproduced by permission of Regis Chemical Co.)

Solvent strength $\varepsilon°$	Solvent	Viscosity (cP at 20°C)	Boiling point (°C)	UV cut-off (nm)
−0.25	Fluoroalkanes	—	—	—
0.00	n-Pentane	0.23	36	210
0.01	1-Octane	0.50	99	210
0.01	Hexane	0.3	69	210
0.04	n-Decane	0.92	174	210
0.04	Cyclohexane	1.0	81	210
0.05	Cyclopentane	0.47	50	210
0.15	Carbon disulphide	0.37	46	380
0.18	Carbon tetrachloride	0.97	77	265
0.26	Xylene	0.7	—	290
0.28	1-Propyl ether	0.37	68	220
0.29	Toluene	0.59	111	285
0.32	Benzene	0.65	80	280
0.38	Ethyl ether	0.23	35	220
0.40	Chloroform	0.57	62	245
0.42	Methylene chloride	0.44	40	245
0.43	Methyl isobutyl ketone	0.54	118	330
0.45	Tetrahydrofuran	0.55	66	220
0.49	Ethylene dichloride	0.79	57	230
0.51	Methyl ethyl ketone	0.4	80	330
0.56	Acetone	0.32	56	330
0.56	Dioxane	1.54	107	220
0.58	Ethyl acetate	0.45	77	260
0.60	Methyl acetate	0.37	57	260
0.62	Dimethyl sulphoxide	2.24	189(d)	—
0.62	Aniline	4.4	184	—
0.63	Diethylamine	0.38	56	275
0.64	Nitromethane	0.67	101	380
0.65	Acetonitrile	0.37	82	210
0.71	Pyridine	0.94	116	305
0.82	1-Propanol	2.3	82	210
0.82	n-Propanol	2.3	97	210
0.88	Ethanol	1.20	78	210
0.95	Methanol	0.60	64	210
1.11	Ethylene glycol	19.9	198	210
Large	Acetic acid	1.26	118	—

The SOLVENT STRENGTH PARAMETER ($\varepsilon°$) is a measure of the polarity of a solvent (solvent strength) for liquid–solid adsorption chromatography. It is based on the free energy of adsorption on a standard surface.

TUNED SOLVENTS are binary solvent mixtures carefully blended to 0.05 increments of solvent strength ($\varepsilon°$) (Table 23.7).

Table 23.7 Tuned solvents (from B: 1456; reproduced by permission of Regis Chemical Co

Solvent strength, $\varepsilon°$	Components	Viscosity (cP at 20°C)	Boiling points, low/ high (°C)	UV cut-off (nm)
0.00	n-Pentane	0.23	35	210
0.05	Cyclohexane/ carbon tetrachloride	1.00	76/82	265
0.10	Cyclohexane/ carbon tetrachloride	1.00	76/82	265
0.15	Cyclohexane/ carbon tetrachloride	1.00	76/82	265
0.20	Carbon tetrachloride/ chloroform	0.95	60/77	265
0.25	Carbon tetrachloride/ chloroform	0.91	60/77	265
0.30	Carbon tetrachloride/ chloroform	0.83	60/77	265
0.35	Carbon tetrachloride/ chloroform	0.73	60/77	265
0.40	Carbon tetrachloride/ methylene chloride	0.53	40/77	265
0.45	Chloroform/ ethylene dichloride	0.68	60/84	245
0.50	Ethylene dichloride/ ethyl acetate	0.76	60/78	260
0.55	Ethylene dichloride/ ethyl acetate	0.58	60/78	260
0.60	Ethyl acetate/ acetonitrile	0.44	77/82	260
0.65	Ethyl acetate/ isopropanol	0.58	82/83	260
0.70	Acetonitrile/ isopropanol	0.70	77/83	210
0.75	Acetonitrile/ isopropanol	1.20	77/83	210
0.80	Acetonitrile/ isopropanol	1.89	77/83	210
0.85	Acetonitrile/methanol	0.43	64/77	210
0.90	Acetonitrile/methanol	0.48	64/77	210
0.95	Methanol	0.60	64	210

Selection of the correct mobile phase strength is generally a matter of trial and error. It is best to start with a mobile phase of intermediate polarity and then increase the solvent polarity until components start to elute.

Bibliography: 143, 546, 1238.
Reference: 4520.

23.14 HPLC CHROMATOGRAPHS

High pressure liquid chromatographs (HPLC) are built of the following parts (Fig. 23.19):

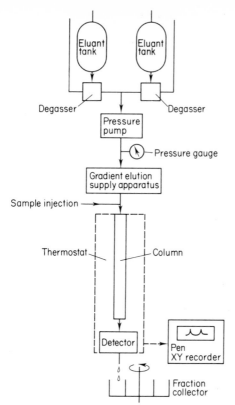

Fig. 23.19 A schematic diagram of a high
performance liquid chromatograph

(i) A pumping system, including solvent containers, degasser, pressure
pump, and pressure gauge. There are two main classifications of pumping
systems:
(a) constant pressure;
(b) constant flow rate (more advantage).
In the latter the solvent is pumped at a constant flow rate (0.01–10 ml
min^{-1}) at pressures up to 5000 p.s.i. ($\simeq 3.5 \times 10^8$ dyne cm^{-2} (CGS) $=$
3.5×10^7 Pa (SI)). Flow rates must be minimized because they affect
retention and resolution and can cause undesirable changes in the
baseline.

References: 4271, 5204, 5355, 5311.

(ii) Gradient elution supply apparatus is used for the preparation of solvent
mixtures of progressively stronger elution power. (Gradient elution in
liquid chromatography is an analogue of temperature programming
in gas chromatography.) Solvent programming devices can be divided
into two main categories:

398

(a) devices which continuously mix two streams in changing proportions;

(b) devices which employ a prefilled mixing chamber.

(iii) A sample injection system. There are two main systems for injection of the sample:

    (a) The hypodermic injection system. In this system the sample solution is added from the syringe, which is inserted through the neoprene membrane into the flowing solvent at the head of the column.

    (b) Multiport waves. In this system the sample loop has interchangeable loops of different volumes which can be filled with the sample solution of desired concentration while the solvent is flowing through the column. The polymer sample is injected by guiding the solvent through the valve into the sample loop. This system may be completely automated and allows the introduction of reproducible volumes of sample solution every time.

(iv) HPLC chromatographic columns. These are made from steel or glass in straight or coiled forms of length 100–600 cm and inner diameter 1–6 mm. Longer lengths give more theoretical plates and better resolution but require a very high inlet pressure.

Bibliography: 49, 858, 1238.
References: 3853, 4131, 4156, 4308, 5037.

(v) Detectors indicate the presence and measure the amount of every component of the column eluent. There is no ideal detector for liquid chromatography. Different types of continuous detectors are available:

    (a) The ultra violet detector (the most common) is based on the double-beam photometer fitted with a flow cell (Fig. 23.20). A standard

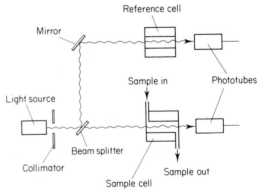

Fig. 23.20 Ultraviolet absorbance detector with a
flow cell

analytical cell is 1 cm long and 1 mm in internal diameter. Photometers function at 254 nm, a wavelength emitted by a low-pressure mercury lamp.

    (b) Refractive index detectors are based on the differential refractometer (cf. Section 11.2). These detectors must be thermostatted to within 10^{-3} °C for maximum sensitivity. They are applied in the analysis of compounds which do not absorb light at any convenient wavelength.

(c) Fluorescence detectors fitted with a flow cell (Fig. 23.21), in which the primary filter is chosen to pass only the exciting wavelength and

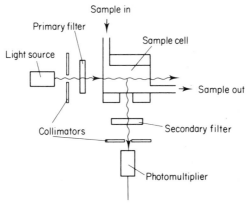

Fig. 23.21 Fluorescence detector with a flow cell

the secondary filter is chosen to cut off the exciting wavelength but to pass the emitted wavelength.

(d) A flame ionization detector, which uses a travelling wire or chain to pick up column effluent (Fig. 23.22). The wire then travels

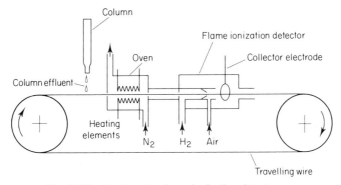

Fig. 23.22 Moving-wire flame ionization detector

through an oven where the solvent is evaporated. The residue of sample left on the wire is then transported to the detector for combustion and measurement.

Bibliography: 235, 317, 411, 948, 1029, 1184, 1237, 1342.
Reference: 4347.

(vi) Recorder. The potentiometric-type strip chart recorder is commonly used. More advanced liquid chromatographs use automatic devices to convert the analogue signal from the chromatograph to a digital value.

Electronic digital integrators integrate the area under the peak and print out peak area and retention time.

Liquid chromatographs can be connected to computers which not only integrate the peak area, but perform the necessary calculations to convert the area to the concentration of sample component.

(vii) Temperature control is required for the column and for the detector. Temperature can also be changed to increase solubility, decrease the reaction with the column packing, lower viscosity, and to facilitate recovery of a volatile sample.

Bibliography: 372, 411, 556, 736, 1011, 1029.
Reference: 6659.

23.15 THIN-LAYER CHROMATOGRAPHY

In THIN-LAYER CHROMATOGRAPHY (TLC), 2–10 µl portions of a sample are applied in the form of a 0.1–1% solution near the bottom of a glass plate coated with a thin layer—usually 100–300 µm thick—of stationary phase. The plate is placed in a chamber containing a depth of a few millimetres of the appropriate mobile phase, which travels through the layer by capillary action. Sample components travel at different speeds through the layer depending on their adsorption coefficients. Developing is stopped by removing the plate from the chamber and evaporating the mobile phase. The position of the separated solutes is located by using normal or ultraviolet light, or by spraying with a chromogenic agent.

The position of the solute spot is described by measuring its R_f VALUE, which is defined (Fig. 23.23) by

$$R_f = \frac{\text{Distance moved from point of application by solute}}{\text{Distance moved from point of application by mobile phase}} = \frac{c}{d}. \quad (23.25)$$

The R_f values of all adsorbed solutes are therefore less than 1.00. For convenience, hR_f values are sometimes quoted, hR_f is defined as $100R_f$. Tables of R_f values are published regularly in the *Journal of Chromatography*.

The NUMBER OF THEORETICAL PLATES (N) is given (Fig. 23.23) by

$$N = 16(a/b)^2, \quad (23.26)$$

where

a is the distance from the spot to the starting line (dipline) and is equivalent to the retention time (t_R) or retention volume (V_R) (cf. Section 23.1),
b is the spot diameter.

The HEIGHT EQUIVALENT TO A THEORETICAL PLATE (HETP) is related to the number of theoretical plates (N) by

$$\text{HETP} = L/N \quad (23.27)$$

where $L = a$, and therefore

$$\text{HETP} = b^2/16a. \quad (23.28)$$

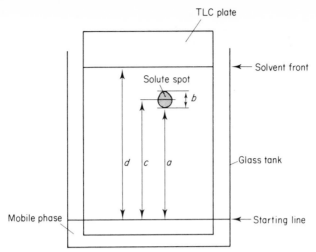

Fig. 23.23 Thin layer chromatogram (TLC) and methods
of calculation of the R_f value and the number of theoretical
plates (N)

Bibliography: 498, 663, 870, 972, 1092, 1196, 1236, 1251, 1303, 1304, 1386, 1449.

23.16 TECHNIQUE OF THIN-LAYER CHROMATOGRAPHY

23.16.1 Preparation of TLC Plates

TLC plates consist of thin layers of adsorbent (0.1–10 mm thick) spread on a supporting plate made of glass or aluminium foil. Several standard TLC plates (20 × 20 cm, 10 × 20 cm, and 5 × 20 cm) are readily available commercially.

Commercial adsorbents for TLC (in particular silica gel or alumina) have particle diameters in the range 1–40 µm. They contain additives, e.g. gypsum (5–20 %) or starch (2–5 %) as binders, which aid in the production of a uniformly coated plate. Additional additives can be fluorescent indicators, such as zinc silicate, which aid the detection of fluorescent materials when viewed under an ultraviolet lamp.

The plates should be cleaned by washing with water, immersing for 24 h in a tank of distilled water containing 1 % of an aqueous surfactant detergent, and finally washing in distilled water and leaving to dry.

The adsorbent is applied to the plate as an aqueous slurry. The best and most reproducible method of coating a plate is by using a commercially produced mechanical spreading device. The freshly coated plate should be left undisturbed for 15–30 min in order to evaporate water from the surface. The plates are then inserted into a well-ventilated drying oven at 105–100 °C for 60 min. At the end of plate preparation, they should be stored in a box containing drying agents.

23.16.2 Application of the Sample

Samples are applied in the form of 0.1–1% solutions in a non-polar volatile solvent. The solvent should be non-polar to minimize the spreading sample at the point of application. The technique for applying the sample is shown in Fig. 23.24.

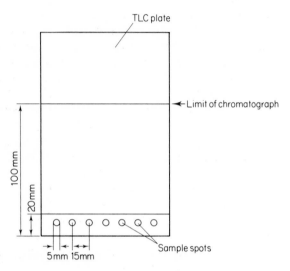

Fig. 23.24   Method of spotting a TLC plate

In order to spot a plate, 2–10 µl of the sample solutions are applied from a microsyringe. The diameters of the spots should be between 2–4 mm and the spot centres 10–15 mm apart. The distance between the point of application and the solvent front is usually 10 cm. The adsorbent should be removed with a sharp point above the limit line of the chromatogram, so that solvent cannot travel past this line. For quantitative work spot sizes must be identical.

The separation of a sample on a preparative scale (greater than 50 µg) requires that the sample solution should be applied as a narrow band and that thicker layers of adsorbent should be employed.

23.16.3 Development of the Chromatogram

Development of the chromatogram is normally carried out in a glass tank. Mobile phase is poured into the tank to a depth of 5 mm. The bottom of the plate is then dipped into the mobile phase. It is important to see that the spot at the starting point should be above the level of the solvent in the tank, otherwise the sample will diffuse away.

23.16.4 Detection of Solutes

Colourless solutes on a developed chromatogram are detected by the following methods:

(i) Viewing the plate under ultraviolet light.
(ii) Revealing the spots as coloured zones by spraying the plate with one of the following chromogenic reagents:
 (a) Iodine vapour. The plate is placed in a closed jar containing a few crystals of iodine. The iodine vapour dissolves in most organic compounds, which show up as brown spots on a pale yellow background. Spots revealed with iodine fade rapidly when left in the open atmosphere. By spraying the plate with 7,8-benzoflavone (0.3 g dissolved in 95 ml ethanol+5 ml 30% sulphuric acid), one can produce an intense blue colouration with iodine.
 (b) Dichromate–sulphuric acid. The reagent is prepared by dissolving 5 g of potassium dichromate in 100 ml of 40% sulphuric acid. Spraying the solvent-free plate and then briefly heating it at 110 °C produces light blue spots on an orange background with most organic substances. Heating at higher temperatures (200 °C) chars the spots, i.e. forms black zones on a colourless background.
 (c) Phosphomolybdic acid. Spraying the chromatogram with a 5% solution of the acid in isopropanol and then heating it briefly at 110 °C gives blue-black spots on a yellow background with a large variety of organic substances.
 (d) Rhodamine B. About 50 mg of dye are dissolved in 100 ml of ethanol. The reagent produces mauve spots on a pink background with a large variety of organic substances. The spots are intensified by blowing sufficient bromine vapour on the sprayed chromatogram to decolorize the background. Inspection of the plate under a long-wave ultraviolet lamp shows orange fluorescent spots on a dark background.
 (e) Antimony trichloride or antimony pentachloride. These reagents are used as 10–20% solutions in chloroform or carbon tetrachloride. Heating the sprayed plates to 100 ° C for a short period produces spots of various colours on a white background with most organic substances.
(iii) Radioactive methods.

Bibliography: 69, 663, 1029, 1092, 1196, 1251, 1386.
References: 2565, 3150, 3339, 5480–5481, 5745, 6294, 6459, 6460, 6467, 6515, 7073.

(iv) The sample concentration as a function of a distance along its elution path can be measured directly by densitometric scanning of developed chromatograms by a spectrodensitometer. Instruments for this purpose are commercially available.

Bibliography: 426, 525, 614, 696, 1092, 1251.
References: 4614, 5625.

23.17 THE TLC PROGRAMMED MULTIPLE DEVELOPMENT TECHNIQUE

TLC MULTIPLE DEVELOPMENT is a method for improving the resolution of the normal TLC technique. In this method the chromatogram is developed, removed from the chamber, dried, and developed again in the same solvent. Using such a multiple development, two 10 cm developments take less time than one 20 cm development. Separation is most improved if the R_f values of the components to be separated are below 0.5. Spot separation decreases when the average R_f of the components to be separated is above 0.7.

TLC PROGRAMMED MULTIPLE DEVELOPMENT is a special kind of multiple development. In this method thin-layer plates are automatically and repeatedly developed a preset number of times. With each succeeding development the solvent advances on the plate. Following each development, controlled evaporation causes the solvent front to recede, usually to or below the point of initial spot deposition. After the last preset cycle, continued controlled evaporation prevents further development on the plate.

The development times are relatively short—the development time for the first cycle is between 10 and 100 s. A large number of development cycles are automatically carried out, e.g. 100 cycles. Each successive development cycle is longer than its predecessor, i.e. the solvent advances farther with each cycle and the spots do not catch up with the solvent front. The solvent is removed from the plate while the plate remains in contact with the solvent reservoir (Fig. 23.25). Solvent is evaporated by heat and/or flow of inert gas, again for a

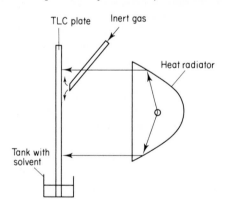

Fig. 23.25 Programmed multiple development device. (From B: 1456; reproduced by permission of Regis Chemical Co.)

preset period. The solvent front recedes, usually to or beyond the point of initial spot deposition. After this period of solvent removal, the solvent is again allowed to advance, this time for a longer period. The solvent is then evaporated again. Each time the development cycle is repeated, the solvent advances farther.

Commercially produced programmed multiple development devices (Regis Chemical Co.) have an electronic programmer controlling several developers.

Bibliography: 1456.
References: 4350–4352, 5773–5779.

23.18 APPLICATIONS OF TLC IN POLYMER RESEARCH

Thin-layer chromatography (TLC) may be applied to

  (i) determination of a molecular weight distribution (MWD);

 (ii) separation of low molecular weight additives from polymers;

(iii) fractionation of polymers;

 (iv) study of the compositional heterogenity of copolymers;

  (v) separation of copolymers from homopolymers;

 (vi) separation of polymer blends;

(vii) study of the percentage of grafting and the separation of grafted polymers from homopolymers.

Bibliography: 663, 1111.
References: 2667, 3208, 3209, 4125, 4126, 4214, 4216–4218, 4381, 4402, 4403, 4686, 5311, 5312, 6144, 6683, 6685, 7062.

23.18.1 Determination of Molecular Weight Distribution by TLC

Thin-layer chromatography (TLC) is a useful method for the rapid and accurate determination of the molecular weight distribution for polymers.

The PHASE RATIO, which is defined as the weight of solvent per unit weight of adsorbent, is a decreasing function of distance from the dip line. This results in a continuous increase in the concentration of polymer in the mobile phase as a band travels away from its starting point. At some point depending on the molecular weight of the polymer, the solubility limit is exceeded and the polymer precipitates.

If mixed solvents are employed as eluents, the most polar component of the eluent is preferentially adsorbed and the composition of the mobile phase, as well as that of the phase ratio, becomes a function of distance from the dip line.

The polymer fractionation mechanism also depends on adsorption and desorption rates and equilibrium coverage on polymer molecular weight.

The application of porous adsorbents (in TLC and GPC) can introduce an additional fractionation mechanism as a result of molecular size (GPC fractionation mechanism).

The fraction results can be presented as the R_f value as a function of molecular weight (Fig. 23.26).

References: 2334, 2335, 3857, 4215, 4404, 4686, 5625, 5628, 5630, 5631, 6356.

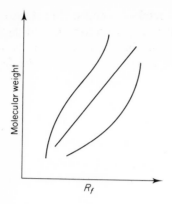

Fig. 23.26 Typical shapes of chromatographic mobilities of polymer fractions on TLC adsorbents

23.19 HIGH PERFORMANCE RADIAL CHROMATOGRAPHY

HIGH PERFORMANCE RADIAL CHROMATOGRAPHY (HPRC) is a separation technique which includes many of the advantages of the stationary, mobile, and gas phases. The results from HPRC correlate to high performance liquid chromatography (HPLC) and to a thin-layer chromatography (TLC).

In HPRC a small amount of solvent (100–200 µl) is pumped on to the centre of a plate through a small-bore Pt/Ir capillary (Fig. 23.27). As the solvent is delivered to the plate, it spreads in a radial fashion as shown in Fig. 23.28.

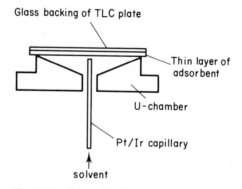

Fig. 23.27 Schematic diagram of the developing chamber for HPRC chromatography

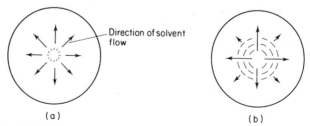

Fig. 23.28 TLC plates. (a) A spotted plate; (b) a developed plate

The solvent delivery is constant and is adjustable to any value between 1 and $10 \mu l s^{-1}$ by use of a pumping system connected to a syringe reservoir (Fig. 23.29). A controlled delivery of mobile phase eliminates variations associated with capillary action. Samples are applied to the layer in a circular fashion,

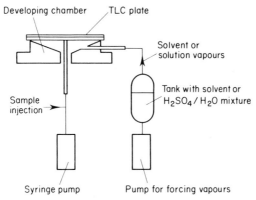

Fig. 23.29 High performance radial chromatography (HPRC) development device

the centre of which is positioned exactly above the solvent-feeding capillary. As solvent is pumped on the layer, it flows over and at right angles to the spots, compressing ensuing zones into tight, compact bands, and thus increasing the degree of resolution.

Preconditioning with solvent vapours and/or sulphuric acid/water mixtures (used to control relative humidity and activity) sometimes improves the resolution of a specific separation system. Conditioning may be carried out before and during chromatographic development.

Bibliography: 1346.

23.20 ION EXCHANGE CHROMATOGRAPHY

In ION EXCHANGE CHROMATOGRAPHY the components to be separated are carried through the column in the mobile phase. Chromatographic separation occurs by an exchange of ions between the sample in the mobile phase and resins in the stationary phase. Compounds which have ions with different affinities for the resin can be separated.

Resins used as the stationary phase in ion exchange columns must be ionic in nature and highly permeable. Synthetic ion exchangers are crosslinked polyectrolytes prepared from styrene–divinylbenzene copolymers. There are two general types of ion exchange resins (Fig. 23.30):

(i) Anion ion exchangers;
(ii) Cation ion exchangers.

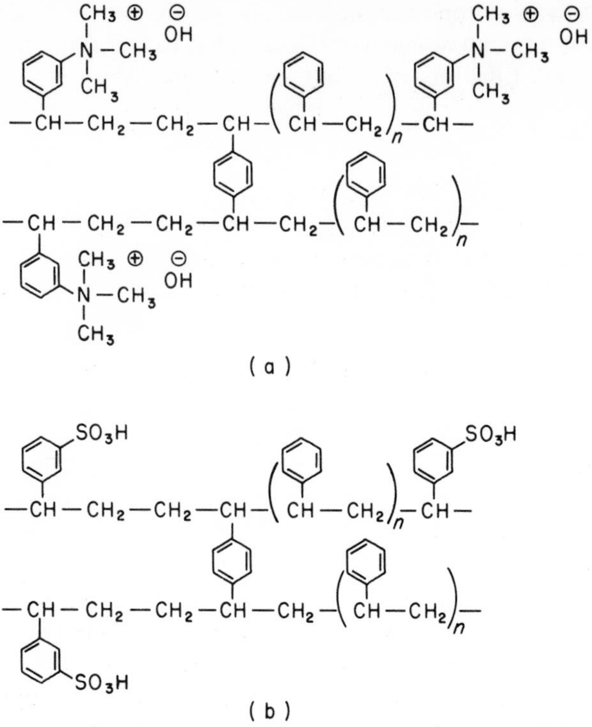

Fig. 23.30 Ion exchanger structures. (a) Anion ion exchanger;
(b) cation ion exchanger

The CAPACITY OF AN ION EXCHANGER is defined as the number of functional groups available for ion exchange and is expressed in milliequivalents per gram of dry resin in the H^+ (cation) or Cl^- (anion) form.

The capacity of weakly acidic and basic resins shows a marked dependence and generally has a small range of maximum capacity depending on the pK_a of the functional group. The strongly acid and basic resins have a much wider range of maximum capacity.

The ion exchange process may be carried out in aqueous or non-aqueous media. The mobile phase usually contains a counter-ion, opposite in charge, to the surface ionic group, which is in equilibrium with the resin in the form of an ion pair.

The DISTRIBUTION COEFFICIENT (K) is defined by the equilibrium of the ion exchange process:

(i) for an anion exchange process,

$$K = k\left(\frac{V_m}{V_s}\right) = \frac{[-N(CH_3)_3^+ \, X^-][Cl^-]}{[-N(CH_3)_3^+ \, Cl^-][X^-]};$$ (23.29)

(ii) for a cation exchange process,

$$K = k\left(\frac{V_m}{V_s}\right) = \frac{[-COO^- X^+][Na^+]}{[-COO^- Na^+][X^+]};\qquad(23.30)$$

where

k is the column capacity factor (cf. eq. (23.7)),

V_m is the mobile phase interstitial volume (void volume),

V_s is the stationary phase volume (an ion exchange capacity).

The distribution coefficient (K) is a function of many experimental parameters, e.g.

  (i) pH value (the distribution coefficient (K) can be affected by altering the pH);
  (ii) ionic charge;
  (iii) ionic radius;
  (iv) resin porosity;
  (v) ionic strength and solvent;
  (vi) temperature.

The counter-ions in eqs (23.29) and (23.30) are usually dependent on a buffer. A decrease in the buffer concentration will shift the equilibrium in favour of the solute in the resin phase, and increase the chromatographic retention of the solute. An increase in the buffer concentration will have the opposite effect.

For product separation one can make use of simple open columns with diameters of 1–10 cm and lengths of 20–200 cm or automatic commercially produced liquid chromatographs with closed columns (cf. Section 23.14).

Bibliography: 34, 105, 411, 600, 601, 664, 1105, 1359.
Reference: 5037.

23.21 CHEMICAL SEPARATIONS WITH OPEN-PORE POLY(URETHANE)

Open-pore poly(urethane) is composed of agglomerated spherical particles (1–10 μm in diameter) bonded to each other in a rigid, highly permeable structure (Fig. 23.31).

Open-pore poly(urethane) columns are prepared in glass or metal columns by in situ precipitation-polymerization of isocyanate and a polyol in 60–40 vol.% toluene–carbon tetrachloride solutions. By variation of the monomer concentrations and reaction conditions during the preparation of open-pore poly(urethane), it is possible to alter the density, porosity, and surface area of the polymer. The diameters of the polymer spheres can be changed by altering the temperature fo the reaction mixture or by using catalysts. The crosslinked composition of open-pore (polyurethane) makes it resistant to chemical attack. Open-pore poly(urethane) is compatible with a number of solvents and dilute

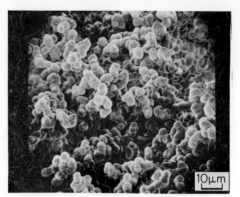

Fig. 23.31 Electron micrograph of
open-pore polyurethane. (Reproduced
by courtesy of R. E. Sievers)

acids. It has no cation-exchange properties, but exhibits low-capacity, weak-
base anion exchange characteristics.

Open-pore poly(urethane) can be utilized in liquid chromatography (LC) and
also in thin-layer chromatography (TLC).

Bibliography: 969, 1129.
References: 2557, 3895, 4056, 5002, 6103.

Chapter 24

Inverse Gas Chromatography

General bibliography: 104, 198, 551, 892, 970.

INVERSE GAS CHROMATOGRAPHY is a chromatographic technique which employs a polymer as the stationary phase and its interaction with a known, volatile solute is then measured. From the magnitude and temperature dependence of this interaction, properties of the polymer–solute and pure polymer can be evaluated.

Solute retention data are determined by the SPECIFIC RETENTION VOLUME (V_g) corrected at $0\,°C$ and given by

$$V_g = \frac{273.2}{T}\left(\frac{V_R}{w}\right) = \frac{273.2}{T}K \tag{24.1}$$

(in cubic centimetres per gram (CGS) or cubic metres per kilogram (SI)), where

V_R is the retention volume at temperature T,

w is the mass of stationary phase in the column,

K is the distribution (partition) coefficient.

The RETENTION DIAGRAM is obtained by plotting the logarithm of the specific retention volume (V_g) versus the reciprocal of the absolute temperature (Fig. 24.1). The retention diagram is linear and its slope is related to the enthalpy of the solution in the stationary liquid phase (GLC) or adsorption on the solid

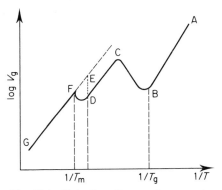

Fig. 24.1   Retention diagram for a semi-crystalline polymer sample

surface (GSC) according to

$$\frac{\partial \ln V_{\mathrm{g}}}{\partial (1/T)} = -\frac{\Delta H}{R},$$
(24.2

where

ΔH is the corresponding enthalpy,

R is the universal gas constant (cf. Section 40.2).

Studies of transition of the polymer stationary phase are called MOLE CULAR PROBES. The plot of the specific retention volume (V_{g}) versus the reciprocal of the absolute temperature ($1/T$) (Fig. 24.1) gives several importan pieces of information regarding transition of the polymer stationary phase

(i) Segment AB. In this temperature region the polymer is below it glass transition temperature (T_{g}) (cf. Chapter 32) and penetration o the solute molecules into the bulk of the polymer phase is precluded Retention occurs mainly by surface absorption and the retentio diagram is linear in this region and gives information on the surfac properties of the polymer.

(ii) Point B. This point corresponds to the glass transition temperature (T_{g}) Penetration of the solute into the bulk of the polymer begins, causin, an increase of the retention volume with temperature. Due to a initially slow rate of diffusion of the solute into and out of the stationar polymer phase, non-equilibrium conditions prevail.

(iii) Segment BC. In this temperature region, the diffusion co-efficient of the solute increases with an increase in the temperature.

(iv) Point C. This point corresponds to the equilibrium conditions of the diffusion process.

(v) Segment CD. In this temperature region—which is below the meltin; point (T_{m}) of the polymer (cf. Section 32.7)—retention proceeds by bul sorption, but polymer–solute interaction is restricted to the amorphou domains of the stationary polymer phase.

(vi) Segment DF. In this temperature region (upon melting), the fractio of the amorphous phase increases, leading to an increase in retentio volume.

(vii) Point F. This point corresponds to the melting point (T_{m}) (cf. Sectior 32.7).

(viii) Segment FG. In this region—which is above the melting point (T_{m})– the bulk sorption into the completely amorphous polymer phase occur and the retention diagram is linear in this region. By extrapolation o this line to lower temperatures (dashed line FE), the crystalline conten of the stationary phase can be determined by comparison of th experimental retention volume with the extrapolated value.

In the region FG the polymer is completely amorphous and th properties of the polymer–solute can be investigated.

Bibliography: 104, 198, 551, 892, 970.
References: 2551, 2554, 3580, 3870, 4250, 4921, 4922.

24.1 APPARATUS FOR INVERSE GAS CHROMATOGRAPHY

Any type of commercial chromatograph can be utilized for inverse gas chromatography (Fig. 24.2). A column is packed with an inactive solid support containing the polymer dispersed as a thin film on the surface. In some cases

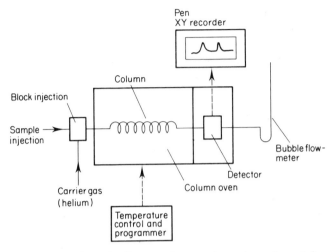

Fig. 24.2 Schematic diagram of a gas chromatograph used for inverse gas chromatography

the column may also be packed directly with the polymer in either film, fibre, or powder form. A uniform flow of inert gas is maintained through the column, and a pulse of probe molecules is injected at one end and detected at the other by a suitable detector. A small pulse of non-interacting gas can also be injected with the probe molecules to aid in detection of the carrier gas front. The usual data recorded are the temperature and pressure drop across the column and the extent of the retention times.

Bibliography: 198.
References: 2549, 2552, 3753, 4921, 6534.

24.2 PREPARATION OF CAPILLARY COLUMNS

Stainless steel columns (0.075 cm diameter, 20 m long) should be carefully cleaned with several solvent washings before coating. In order to coat the inner walls of the capillary column with a polymer film of thickness in the range 10^4–10^5 Å a dynamic method is commonly used (Fig. 24.3). In this method about 10 ml of a dilute polymer solution (6–10 wt %) are placed in the reservoir and pushed through the column with nitrogen at about 0.5 atm pressure. Continued nitrogen flow dries the polymer, which adheres to the tube's inner wall. A 10 ml portion of the coating solution is put through the column as many times as it takes to obtain a film thickness of the order of

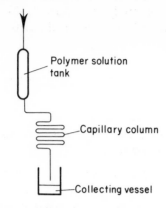

N₂ under pressure

Polymer solution tank

Capillary column

Collecting vessel

Fig. 24.3 Schematic diagram of a dynamic method for coating capillary columns with the liquid phase

10^4–10^5 Å. The amount of polymer in the column is determined by careful weighing of the column before and after coating using a high-precision balance.

24.3 APPLICATIONS OF INVERSE GAS CHROMATOGRAPHY TO THE CHARACTERIZATION AND ANALYSIS OF POLYMERS

The main applications of inverse gas chromatography are

   (i) study of the retention mechanism in solid polymers;
  (ii) determination of the glass transition temperature (T_g) and the melting point (T_m) (Section 24.3.1);
 (iii) study of polymer solution thermodynamics;
  (iv) determination of the crystallinity of a polymer stationary phase (Section 24.3.2);
   (v) determination of diffusion coefficients (Section 24.3.3);
  (vi) study of polymer surface properties (Section 24.3.4);
 (vii) determination of polymer solubility parameters;
(viii) determination of the Flory–Huggins (χ) interaction parameter (cf. Table 2.4).

Bibliography: 198, 551, 892, 970.
References: 3167, 3168.

24.3.1 Determination of the Glass Transition Temperature by Inverse Gas Chromatography

The determination of the glass transition temperature (T_g) for a polymer stationary phase is based on the change from surface adsorption below T_g to bulk sorption above T_g (Fig. 24.1). The change in retention mechanism at T_g is a result of increased molecular mobility of the polymer segments at and above T_g, allowing for the penetration of the solute molecules into the bulk of the polymer.

In the detection of the glass transition, the solute must be non-solvent of the polymer in order to observe reversal from the linear behaviour. To maximize reversal, the bulk partition coefficient should be as large and the surface partition coefficient as small as possible.

Bibliography: 198.
References: 2547, 2548, 2550, 2551, 2555, 2676, 2768, 2992, 3138, 3572, 3579, 3756, 3757, 4836, 4924, 4941, 4948, 5428, 5772, 6435, 6995, 7223.

24.3.2 Determination of the Crystallinity of Polymers by Inverse Gas Chromatography

The determination of the crystallinity of polymers in the stationary phase is based on the differential solubility of the solute in crystalline and amorphous domains (phases). The polymer–solute interaction is restricted to the amorphous domains, leading to an increase in the specific retention volume (V_g) with decreasing crystallinity.

Retention diagrams through the melting transitions of semicrystalline polymers (Fig. 24.4) can be analysed quantitatively to yield the crystallinity

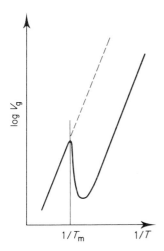

Fig. 24.4 Retention diagram through the melting transition of a semicrystalline polymer sample

and melting curve of the polymer. Above the melting point (T_m) the polymer is completely amorphous and a linear retention diagram is obtained. By extrapolating this straight line to lower temperature the retention volume for the theoretically amorphous polymer can be obtained.

The DEGREE OF AMORPHOUS FRACTION (χ_a) of the polymer stationary phase can be calculated from

$$\chi_a = V_g/V_g',\qquad (24.3)$$

where V_g and V_g' are the experimental and extrapolated retention volumes at the same temperature.

416

The DEGREE OF CRYSTALLINITY (χ_c) of the polymer stationary phase can be calculated from

$$\chi_c = 100[1 - (V_g/V_g')] \tag{24.4}$$

(in per cent).

The DEGREE OF MAXIMUM POSSIBLE CRYSTALLINITY ($\chi_{c(max)}$) at a given temperature is obtained from

$$\chi_{c(max)} = 100[1 - (V_{g(eq)}/V_g')] \tag{24.5}$$

(in per cent), where $V_{g(eq)}$ is the retention volume of the polymer once equilibrium between crystalline and amorphous phases has been reached.

The DEGREE OF CRYSTALLINITY (χ_c) at any time (t) can be obtained from

$$\chi_c = 100\left(\frac{V_g' - V_{g(t)}}{V_{g(t)} - V_{g(eq)}}\right) \tag{24.6}$$

(in per cent), where $V_{g(t)}$ is the retention volume measured at time t.

The melting curve is obtained by determining the crystallinity at several temperatures.

Bibliography: 198.
References: 2049, 2549, 2553, 2991, 3753, 3816, 4160, 4941, 6534.

24.3.3 Determination of Diffusion Coefficients by Inverse Gas Chromatography

The diffusion coefficient of the solute in the polymer stationary phase can be determined from the width of a symmetrical eluted peak. The peak broadening as a function of column properties is expressed by

$$HETP = A + B/v + Cv \tag{24.7}$$

where
HETP is the height equivalent to a theoretical plate (cf. Section 23.1),
v is the flow rate of carrier gas in the column,
A is the empirical constant taking into account instrumental broadening,
C is the constant pertaining to the finite time required to achieve equilibrium in the stationary phase, and is given by

$$C = \frac{8d^2}{\pi^2 D_1}\left(\frac{K}{(1+K)^2}\right), \tag{24.8}$$

d is the thickness of the stationary phase,
D_1 is the diffusion coefficient of the solute in the polymer,
K is the distribution coefficient (partition coefficient),

$$K = (t_R - t_M)/t_M, \tag{24.9}$$

t_R and t_M are the retention times for the solute and the non-interacting marker.

The HETP value can be determined from the width of the eluting solute peaks for several carrier gas flow rates. The constant C can be obtained from the slope of plot of HETP versus v, which becomes linear at high velocities ($B/v \rightarrow 0$). From the known values of d and K the diffusion coefficient D_1 can be calculated.

Bibliography: 198.
References: 2556, 3273, 3755, 5270.

24.3.4 Determination of Surface Properties of Polymers by Inverse Gas Chromatography

The following data can be evaluated from the magnitude of the retention volume and the shape of the eluted solute peak obtained in the polymer stationary phase at temperatures below the glass transition temperature, where retention proceeds exclusively by surface adsorption:
(i) An ADSORPTION ISOTHERM for the polymer, which describes the variation of the amount of solute adsorbed with its concentration (i.e. pressure) in the bulk phase, at constant temperature (Fig. 24.5).

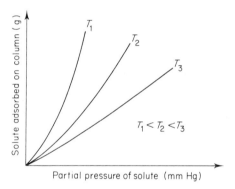

Fig. 24.5 Adsorption isotherms for a given polymer at different temperatures

The amount of solute adsorbed (a) on a mass of polymer (m) is given by

$$a = \frac{1}{m} \int_0^c V_R \, dc, \tag{24.10}$$

where
  a is the amount of solute adsorbed,
  m is a mass of polymer,
  V_R is the retention volume,
  c is the gas phase concentration.
(ii) A HEAT OF ADSORPTION which can be obtained from the experimental adsorption isotherms. For non-linear adsorption isotherms the

heat of adsorption varies with surface coverage and results are expressed as an ISOSTERIC HEAT OF ADSORPTION at a specified coverage (a):

$$(-\Delta H_s)_a = -R\left(\frac{\partial \ln p_1}{\partial(1/T)}\right)_a, \qquad (24.11)$$

where

$-\Delta H_s$ is the heat of adsorption (equal to but of opposite sign to the heat of desorption),

p_1 is the partial pressure of solute in the gas phase,

R is the universal gas constant (cf. Section 40.2).

(iii) The SURFACE AREA OF THE POLYMER STATIONARY PHASE can be determined by applying the B-E-T theory (which uses a model of multimolecular adsorption) from

$$\frac{p_1/p_{1(0)}}{v(1-p_1/p_{1(0)})} = \frac{1}{v_m C} + \left(C - \frac{1}{v_m C}\right)\left(\frac{p_1}{p_{1(0)}}\right), \qquad (24.12)$$

where

p_1 is the partial pressure of solute in the gas phase,

$p_{1(0)}$ is the saturated vapour pressure of the solute,

v is the volume of the solute on the surface,

v_m the volume of the solute on the surface corresponding to monolayer formation,

C is a constant.

Bibliography: 198, 1127.
References: 2540, 2548, 2580, 2817, 3265, 3754, 4135, 4158, 4293, 5326, 6168, 6623, 6586, 6830, 6831.

Chapter 25

Exclusion Chromatography

General bibliography: 31, 32, 269, 371, 449.

25.1 LIQUID EXCLUSION CHROMATOGRAPHY

LIQUID EXCLUSION CHROMATOGRAPHY (LEC) is a chromato-graphic technique which separates products by molecular size using silica gels having very large pores (100–50 000 Å) (Table 25.1). The separation mechanism

Table 25.1 Characteristics of different types of silica gel applied in liquid exclusion chromatography (from R: 6413; reproduced by permission of John Wiley & Sons)

Packing type	Mean pore diameter (Å)	Packing size (mesh)
Fractosil 25 000	30 000	120–230
(E. Merck, Darmstadt)		
Fractosil 10 000	14 000	120–230
Fractosil 5000	4900	120–230
CPG-10-1250	1100	120–200
(Corning Glass Works, Corning, N.Y.)		
BioGlass 2500 ⎫ in equal	2500	100–200
BioGlass 1500 ⎬ proportions	1500	100–200
BioGlass 1000 ⎭	1000	100–200
(Corning Biological Products, Medfield, Mass.)		
Glass ⎫ in equal	400–800	75–125 μm
Glass ⎬ proportions	200–400	75–125 μm
Glass ⎭	< 100	75–125 μm
(Water Associates, Inc.)		

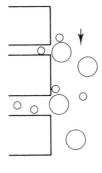

Fig. 25.1 Schematic representation of the mechanism of parti-cle separation by liquid exclusion chromatography

occurs as a result of steric exclusion from the pores of the packing material (Fig. 25.1). Some authors have pointed out that it is not possible to make a phenomenological distinction between the separation mechanism of liquid exclusion chromatography (LEC) and hydrodynamic chromatography (HDC) (cf. Chapter 25.16) because both mechanisms are operable with or without porous packing.

For product separation by liquid exclusion chromatography we can utilize a simple open chromatographic column. The main application of LEC chromatography is for the separation of polymer latex particles.

References: 2746, 3831, 4718, 6413.

25.2 GEL FILTRATION CHROMATOGRAPHY

GEL FILTRATION CHROMATOGRAPHY (GFC) is a chromatographic technique which separates products according to their molecular size using hydrophilic macrogels swelled in aqueous systems.

Two types of macrogel may be employed:

(i) The crosslinked poly-(dextran) gels (Sephadex), e.g.
Sephadex G-10—fractionation range 0–700,
Sephadex G-200—fractionation range 5 000–800 000;
(ii) the cross-linked poly(acrylamide) gels (Bio-Gel P), e.g.
Bio-Gel P2—fractionation range 200–1800,
Bio-Gel P300—fractionation range 60 000–400 000.

The separation mechanism occurs as a result of

(i) the molecular sieve properties of macrogels;
(ii) adsorption;
(iii) ion exchange;
(iv) ion exclusion.

For product separation one can utilize simple open columns with diameters of 1–5 cm and lengths of 20–200 cm or automatic commercially produced chromatographs with closed columns (cf. Section 23.14).

The main applications of gel filtration chromatography (GFC) are

(i) separation of macromolecules from low molecular weight salts;
(ii) buffer exchange in macromolecular solutions;
(iii) fractionation of macromolecules, especially biopolymers, in aqueous solutions;
(iv) the study of molecular association, e.g. to measure free and bound small molecules in macromolecular solutions;
(v) the study of equilibrium constants in macromolecular solutions and polyelectrolytes.

Bibliography: 5, 244, 346–348, 454, 1411, 1458, 1460.
References: 2127, 3396–3398, 3617, 3744, 4832, 5314, 5867, 7044.

25.3 GEL PERMEATION CHROMATOGRAPHY

General bibliography: 28, 31, 32, 125, 130, 148, 216, 250, 260, 267–269, 371, 411, 422, 449, 481, 570, 599, 687, 688, 790, 940, 943, 998–1000, 1310, 1315, 1401, 1433.

GEL PERMEATION CHROMATOGRAPHY (GPC) is a chromatographic technique which uses highly porous, non-ionic gel beads for the separation of polydisperse polymers in solution.

Present theories and models of GPC fractionation indicate that the hydrodynamic volume of the molecule governs the separation, not the molecular weight.

The general concept of the fractionation mechanism is that the largest macromolecules of the solute cannot penetrate the pores within the crosslinked gel beads, and thus elute first (their retention volume is smaller). Smaller macromolecules of the solute are retained in interstices within the gel beads and therefore require more time to elute (their retention volume is larger).

A general chromatographic theory given in Section 23.1 is also valid for gel permeation chromatography (GPC)

The DISTRIBUTION COEFFICIENT (PARTITION COEFFICIENT) in GPC is defined as

$$K = V_{s_i}/V_s \qquad (25.1)$$

where

V_{s_i} is the pore volume of the beads accessible to permeation by the ith component with a specific molecular size,

V_s is the volume of the stationary phase (the volume of all the pores in the beads accessible to permeation).

The value of K is dependent on molecular size and ranges from zero (the molecular size is larger than the largest pore size) to unity (the molecular size is smaller than the smallest pore size): $0 \leqslant K \leqslant 1$.

The COLUMN RESOLUTION (R_c) (cf. eq. (23.9)) in GPC for complete separation must be equal to or greater than 1.

For high molecular weight, polydisperse materials the column resolution (R_c) decreases with increasing molecular weight because fewer pores are available to the large molecules than are available to the small molecules.

References: 2450, 2459, 2460, 2688, 2728–2731, 2742, 2743, 2745, 2748, 2753, 3084, 3149, 3643, 3831, 3906, 4026, 4045, 4526, 4832, 5047, 5074, 5252, 5336, 5337, 5629, 5866, 6456, 6862, 6868, 7022, 7255, 7256.

25.4 THE GPC CHROMATOGRAM

The GPC CHROMATOGRAM is presented as a plot of detector response versus retention volume (V_R) (Fig. 25.2). Instead of detector responses and retention volumes (V_R), however, it is usual to measure heights (H_i) above the baseline and counts, respectively.

The **GPC** chromatogram can be normalized, so that

$$\int_0^\infty H_i(V_{\mathrm{R}i})\,dV_{\mathrm{R}} = 1. \tag{25.2}$$

The GPC chromatogram of a sample should be normalized before its shape is compared with the standard chromatogram because it is almost impossible to inject exactly the same amount of sample into the chromatograph each time.

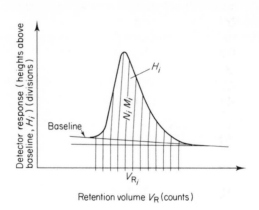

Fig. 25.2 A typical GPC chromatogram

The following schedule should be followed in comparing sample and reference standard chromatograms:

(i) First the baseline must be constructed. Normally this is done by joining portions of the chromatograms occurring prior to the sample appearance and after the final peaks have been eluted (Fig. 25.2).

(ii) Each 2.5 or 5 ml count is divided into between two and five equal parts and the chromatogram peak height (H_i) at each division is measured and recorded.

(iii) The chromatogram height (H_i) for each equal division of retention volume $(V_{\mathrm{R}i})$ is divided by the sum of all such heights.

$$H_i' = H_i \bigg/ \sum_{i=1}^{n} H_i = \frac{H_i}{H_1 + H_2 + H_3 + \ldots + H_n}. \tag{25.3}$$

This procedure is called NORMALIZATION OF THE GPC CHROMATOGRAM. The resulting value represents the weight fraction of macromolecules eluting at the retention volume.

(iv) The plot of the normalized chromatogram of the polymer sample is usually presented together with the normalized chromatogram of the reference standard sample.

25.5 GPC CALIBRATION METHODS

Three calibration methods are generally used:

(i) the narrow molecular weight distribution standards method;
(ii) the universal calibration method;
(iii) the broad molecular weight distribution (polydisperse) standards method.

25.5.1 The Narrow Molecular Weight Distribution Standards Method

The main disadvantage of this method is the difficulty of preparing standards of narrow molecular weight distribution for most polymers. Only polystyrene is commercially available over the whole range of molecular weights. Reliance on a molecular weight (or size) calibration performed with a given polymer to interpret the chromatograms of other polymers can lead to serious errors.

Bibliography: 1164.
References: 2360, 2708, 2715, 2837, 3905, 4014, 4837, 5043, 5339, 5824, 6453.

25.5.2 The Universal Calibration Method

The limiting viscosity number (intrinsic viscosity) $[\eta]$ is related to the hydrodynamic volume of the macromolecules through the following equation (sometimes called the EINSTEIN VISCOSITY LAW) (cf. Section 2.13):

$$[\eta] \, \overline{M}_v = \Phi(h) \, V_h \tag{25.4}$$

(in decilitres per gram (CGS) or cubic metres per kilogram (SI), where
  $[\eta]$ is the limiting viscosity number (intrinsic viscosity) of the polymer sample,
  $\overline{M}_v$ is the viscosity-average molecular weight of the polymer,
  $\Phi(h)$ is a function related to the hydrodynamic behaviour of the macromolecules (constant for a particular solvent and a given temperature),
  V_h is the hydrodynamic volume of the macromolecule.
The value $[\eta] \, \overline{M}_v$ from eq. (25.4) is a direct measure of the hydrodynamic volume of the macromolecule. It is a UNIVERSAL CALIBRATION PARAMETER for GPC because an approximately linear relation is generally obtained between $\log[\eta] \, \overline{M}_v$ and the retention volume (V_R) (Fig. 25.3).

The interpretation of data in gel permeation chromatography requires evaluation of the molecular weight distribution of a particular polymer on the basis of known molecular weight distributions of a series of standard samples. Standard samples are not available for most polymers. This requires transformation of the primary calibration curve so that it can be used for polymers structurally different from the standard.

Calibration standards should be

(i) monodisperse with a polydispersity index ($\overline{M}_w/\overline{M}_n$) of unity;
(ii) available over a very wide range of molecular weights (10^2–10^7).

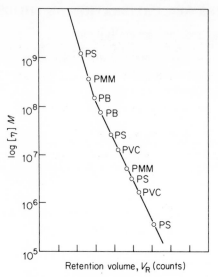

Fig. 25.3 Universal calibration curve:
PB = poly(butadiene), PMM = poly
(methyl methacrylate), PS = poly
(styrene), and PVC = poly(vinyl chloride). (From R: 3801; reproduced by
permission of John Wiley & Sons)

Typical commercially available calibration standards are, for example,
poly(styrene) ($\overline{M}_w/\overline{M}_n = 1.02$–$1.20$, molecular weight range (MWR) = 600–2×10^6);
poly(vinyl chloride) ($\overline{M}_w/\overline{M}_n = 2.4$–$2.7$, MWR = 68 000–132 000);
poly(butadiene) ($\overline{M}_w/\overline{M}_n = 1.05$–$1.48$, MWR = 17 000–423 000);
poly(ethylene) ($\overline{M}_w/\overline{M}_n = 1.7$–$3.7$, MWR = 7000–41 700).
When two monodisperse polymers possess the same hydrodynamic volume
in solution, their molecular weights are related through their limiting viscosity
numbers (intrinsic viscosities):

$$\overline{M}_{v_1}/\overline{M}_{v_2} = [\eta]_2/[\eta]_1. \qquad (25.5)$$

This relationship cannot be used directly to evaluate the molecular weight
distributions of polydisperse samples.

For polydisperse samples the limiting viscosity numbers are determined
by the Mark–Houwink–Sakurada relation (cf. Section 9.1.2):

$$[\eta]_1 = K_1 \, \overline{M}_{v_1}^{a_1}, \qquad (25.6)$$

$$[\eta]_2 = K_2 \, \overline{M}_{v_2}^{a_2}, \qquad (25.7)$$

where K and a are Mark–Houwink–Sakurada constants. Combination of eqs (25.6) and (25.7) yields

$$\log \overline{M}_{v_2} = \frac{1}{1+a_2} \log \frac{K_1}{K_2} + \frac{1+a_1}{1+a_2} \log \overline{M}_{v_1}. \qquad (25.8)$$

A primary poly(styrene) calibration curve can be used for other polymers through eq. (25.8).

When K and a are known, $\log[\eta] \overline{M}_v$ can be written as

$$\log \overline{M}_v^{1+a} + \log K, \qquad (25.9)$$

and the retention volume (V_R) can be directly related to the molecular weight of the polymer (M) (Fig. 25.4).

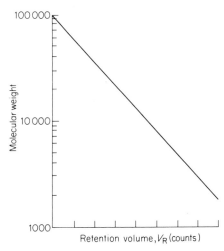

Fig. 25.4   A semilogarithmic plot of molecular weight versus the retention volume

Bibliography: 260, 261, 304, 1069, 1070, 1126.
References: 2097, 2104, 2173, 2348, 2349, 2485, 2611, 2707, 2934–2936, 3080–3083, 3085, 3087, 3088, 3160, 3161, 3555, 3799, 3800, 3801, 3897, 4458, 4465, 4768, 5253, 5678, 5911, 6120, 6941, 7034, 7080.

25.5.3 The Broad Molecular Weight Distribution Standards Method

This method seems to be the most useful in practical applications because broad molecular weight distributed samples are always readily available.

Two methods are used:

(i) a method which utilizes polydisperse calibrating samples for which molecular weight averages are known, but which neglects column peak spreading;

(ii) a method which determines the calibration and column spreading for the GPC column, but which requires calibrating samples with fairly narrow molecular weight distributions for which both the weight- and number-average molecular weights must be known.

References: 2211, 2212, 2546, 2708, 3484, 4148, 5033, 5181, 5926, 6651, 6655, 6980, 7035, 7068, 7258.

25.6 OVERLOADING EFFECT

The OVERLOADING EFFECT (non-linear response) in GPC is characterized by a large loss in resolution and column efficiency due to an excessively high sample molecular weight. This effect may be caused by

(i) concentration dependence of the solute hydrodynamic volume;
(ii) non-equilibrium partition of the solute between the stationary gel and the solvent mobile phases.

The sample load is restricted by two limitations:

(i) the sample viscosity should not differ from the solvent viscosity by a factor greater than 2;
(ii) the sample volume must be small since the width of a zone, i.e. the peak width, increases linearly with sample volume.

GPC columns should not be overloaded; the sample size must be limited to about 15 mg per 100 ml column volume.

Bibliography: 29, 941, 999.
References: 2706, 4807, 5338, 5339 5634, 6116.

25.7 CALCULATION OF MOLECULAR WEIGHT AVERAGES

The basic data from GPC are recorded in the form of a chromatogram (Fig. 25.2). The peak heights (H_i) as a function of the retention volumes (V_{R_i}) are proportional to some value $N_i M_i$, where N_i is the number of molecules of the ith kind with molecular weight M_i.

Using corresponding values of $M_i = f(V_{R_i})$ from the calibration curve (Fig. 25.3), it is easy to calculate the values N_i and $N_i M_i^2$ (cf. Table 25.2) and to obtain the sums necessary to calculate $\overline{M}_n$ and $\overline{M}_w$ from

$$\overline{M}_n = \frac{\sum N_i M_i}{\sum N_i}, \tag{25.10}$$

$$\overline{M}_w = \frac{\sum N_i M_i^2}{\sum N_i M_i} \tag{25.11}$$

and the polydispersity index from $\overline{M}_w / \overline{M}_n$.

Two types of graph of molecular weight distribution (MWD) are used:

(i) The INTEGRAL MOLECULAR WEIGHT DISTRIBUTION CURVE is obtained by accumulating recorder divisions ($N_i M_i$) and then normalizing them (cf. Table 25.3). The resulting data expressed as the sum of the

Table 25.2 Calculation of molecular weight averages

Retention volume, V_{R_i} (ml or counts)	Recorder division, $H_i = N_i M_i$ (mm)	Molecular weight, M_i, from calibration curve	Number of molecules of the ith kind, $N_i = H_i/M_i$	Value of $N_i M_i$
V_{R1}	H_1,	M_1	N_1	$N_1 M_1$
.	.	.	.	.
.	.	.	.	.
.	.	.	.	.
	$\sum N_i M_i$		$\sum N_i$	$\sum N_i M_i$

Table 25.3 Calculation of molecular weight distribution from GPC data

Molecular weight, M_i, from calibration curve	Recorder division, $H_i = N_i M_i$ (mm)	Recorder division normalized, W_i (%)	Sum up to given M_i, W_i (%)
M_1	$N_1 M_1$	$W_1 = \dfrac{N_1 M_1}{\sum N_i M_i}$	W_1
M_2	$N_2 M_2$	$W_2 = \dfrac{N_2 M_2}{\sum N_i M_i}$	$W_1 + W_2$
M_3	$N_3 M_3$	$W_3 = \dfrac{M_3 N_3}{\sum N_i M_i}$	$W_1 + W_2 + W_3$
.	.	.	.
.	.	.	.
.	.	.	.
M_n	$N_n M_n$	$W_n = \dfrac{N_n M_n}{\sum N_i M_i}$	$W_1 + W_2 + W_3 + ... + W_n$
	$\Sigma N_i M_i$	W_i (%) = 100	W_i (%) = 100

mass concentrations ($\sum W_i$) are plotted against molecular weight (M) (Fig. 25.5) or against log M (Fig. 25.6).

(ii) The DIFFERENTIAL MOLECULAR WEIGHT DISTRIBUTION CURVE is obtained by differentiating the curve drawn through the points of Fig. 25.5, as is shown in Fig. 25.7, or by plotting the recorder readings directly against molecular weight (M).

Accurate calculations of molecular weight and molecular weight distribution (MWD) curves require corrections for

(i) band broadening (cf. Section 25.8);
(ii) Eddy diffusion (cf. Section 23.1);

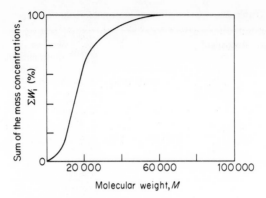

Fig. 25.5 Integrated molecular weight distribution curve. Plot of $W_i\,(\%)$ versus M

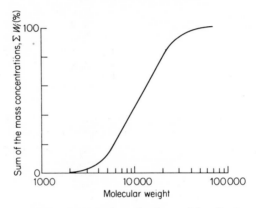

Fig. 25.6 Integrated molecular weight distribution curve. A semilogarithmic plot of $W_i\,(\%)$ versus molecular weight

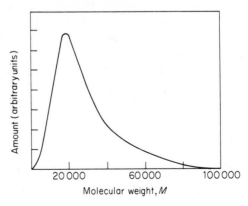

Fig. 25.7 Differential molecular weight distribution curve

(iii) viscous drag;
(iv) mixing in various parts of the GPC chromatograph;
(v) sample overloading.

25.8 BAND BROADENING CORRECTION

The raw GPC chromatogram does not represent the true molecular weight distribution of the sample. It includes band spreading which results in a chromatogram broader than the true sample molecular weight distribution (Fig. 25.8).

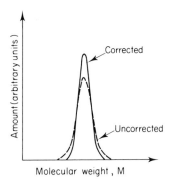

Fig. 25.8 Comparison of an instrumental spreading-corrected distribution with an uncorrected distribution

The error due to instrumental band spreading becomes more significant as the molecular weight distribution of the sample narrows. In general, instrumental spreading becomes less important for samples having a polydispersity index $\overline{M}_w/\overline{M}_n > 2$.

In order to assess properly the importance of zone broadening correction one must take the following into account:

(i) the width of the molecular weight distribution of the sample;
(ii) the resolution of the fractionating column used in the separation.

Careful selection of columns and column combinations must be made before correction for band broadening. Poor column selection can give errors much higher than $\pm 8\%$ for the molecular weight average and the MWD, even for samples with a polydispersity index $\overline{M}_w/\overline{M}_n > 4$.

There are several methods for correcting band broadening effects.

Bibliography: 305, 384, 509, 729, 999, 1312, 1313.
References: 2211, 2402, 2404, 3093, 3242, 3640, 3645, 3866, 4013, 4045, 4156, 4159, 4525, 4527, 4684, 5825, 5831, 6390, 6454, 6862, 6863, 6865, 6869.

25.9 GEL PERMEATION CHROMATOGRAPHS

Gel permeation chromatographs are built of the following parts (Fig. 25.9):

(i) A pumping system, including a solvent container, a degasser, and a high pressure pump. The solvent is pumped at a constant flow rate (0.1–5 ml min^{-1}) at pressures up to 250 p.s.i. ($\simeq 1.72 \times 10^7$ dyne cm^{-2}

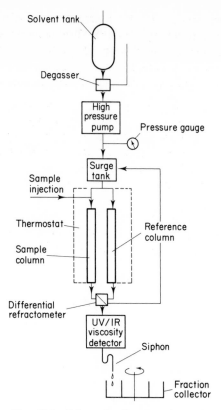

Fig. 25.9 Schematic diagram of a gel
permeation chromatograph

(CGS) $= 1.72 \times 10^6$ Pa (SI)). Pressure fluctuations must be minimized
because they are registered in the detector as noise and distort the
chromatogram.

Bibliography: 79.
References: 4306, 4324, 6102.

(ii) A sample injection system. There are two main systems for injection
of the sample:

(a) The hypodermic injection system. In this system the polymer
solution is added from the syringe, which is inserted through the
Neoprene membrane into the flowing solvent at the head of the
column. This system is not automated and difficulties are encountered
as regards the introduction of reproducible volumes of polymer
solution each time. Sample injection is usually additionally complicated by the risk of introducing gas bubbles, which are registered
in the detector as noise and distort the chromatogram.

(b) The Multiport valve system. In this system the sample loop has
interchangeable loops of different volumes which can be filled with

polymer solution of the desired concentration while the solvent is flowing through the column. The polymer sample is injected by guiding the solvent through the valve into the sample loop. This system may be completely automated and allows the introduction of reproducible volumes of polymer solution each time.

The balance between sample volume and concentration for optimum resolution is a complex function depending on the molecular weight of the polymer sample and the flow properties of two liquids of different densities (flow rate) and other factors. The introduction of a known amount of sample into the column with a minimum loss of resolution is one of the most important operations.

Bibliography: 79, 829.
References: 2404, 2478, 4307, 4944, 4945.

(iii) GPC chromatographic columns. Columns are made from steel and they have standard dimensions: length 30–100 cm and inner diameter 8–10 mm

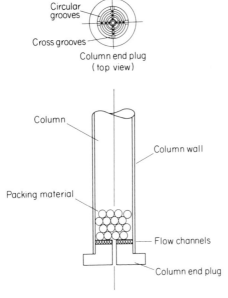

Fig. 25.10   Cross-section of a gel permeation
column and detail of end plug

(Fig. 25.10). The following types of GPC column packing are usually applied:

(a) Beads of crosslinked copolymer and divinylbenzene (STYRAGEL). These semirigid gels are very fragile and must be carefully handled and maintained. Before packing into the column the polystyrene gel should be swollen in the same solvent as that used as the mobile phase. Swollen beads are stirred up in the solvent and poured on the top of the column. The bottom fitting should be connected to a vacuum line through a trap holding the excess of swelling solvent. After filling the column with polystyrene beads, the excess solvent is drawn off under vacuum. It is necessary to avoid the removal of too much solvent, and to prevent the access of air to the beads. The formation of an air pocket in the column results in lost resolution. GPC columns must be protected from gas bubbles because when air is introduced into the packing it may destroy the packing of the column completely. Changing solvents during the life of the column is *not* recommended. Styragel columns should not be filled with water, alcohols, acetone, methyl ethyl ketone, and dimethyl sulphoxide (DMSO) as solvents. Poly(styrene) gel columns may operate at up to 100 °C. Commercially made columns filled with Styragel are available. Table 25.4 is given as an example of the characteristics of a μ-STYRAGEL column produced by Waters Associates.

Table 25.4 μ-Styragel specifications (from B: 1482; reproduced by permission of Waters Associates)

Column size:	
Length	30 cm
Diameter	7.8 mm i.d. (columns with new end fittings); 9.5 mm o.d. (columns with old end fittings)
Flow rate/pressure	Up to 3.0 ml min^{-1} depending on solvent used for 500, 10^3, 10^4, 10^5, 10^6 Å columns; 2000 p.s.i. maximum for 100 Å columns.
Packing material	Fully porous highly crosslinked styrene–divinyl-benzene copolymer
Particle size	$\simeq$ 10 μm in diameter
Pore size	100, 500, 10^3, 10^4, 10^5, or 10^6 Å
Separation method	Separates molecules by size
Separation range (approx.)	From 20–2×10^7 molecular weight (MW)
Approximate column range for column selection	100 Å: 20–700 MW
	500 Å: 50–10^4 MW
	10^3 Å: 500–2×10^4 MW
	10^4 Å: 4×10^3–2×10^5 MW
	10^5 Å: 10^5–8×10^6 MW
	10^6 Å: 10^6–2×10^7 MW
Plates per column (minimum)	4000 plates per 30 cm column for 100 Å
	3000 plates per 30 cm column for 500, 10^3, 10^4, 10^5, and 10^6 Å

(b) Porous glass (BIOGLASS), porous silica (POROSIL), and porous alumina. Before being placed in the column these rigid packings should be mixed with 5% hexamethyldisilazane in *n*-hexane. The careful removal of air from the pores is carried out by boiling the slurry under a light vacuum for a few hours and then pouring it into the column.

(c) Semirigid and soft packing materials include poly(vinyl acetate) gels, dextran gels (SEPHADEX), and poly(acrylamide) gels.

(d) Lignin gels.

Bibliography: 322, 569, 599, 829.
References: 2256, 2482, 2699, 2700, 2954–2960, 3056, 3086, 3092, 3094, 3855, 3856, 3985, 4456, 4461, 4462, 4885, 4929, 4936, 4945–4947, 5037, 5043, 5336, 5371, 5689, 5735, 6956, 7254.

(iv) Sample concentration detectors. Normal samples of polymer contain 2–10 mg of polymer which is eluted from the column in 25–75 ml of solvent. For that reason the sample concentration detectors must be very sensitive and different types of continuous detectors are applied:

  (a) Differential refractometers measure directly the deflection of a light beam due to the difference in refractive indices of the sample and reference liquids (cf. Section 11.2). This method allows one to measure small changes of refractive index. The refractive index Δn is proportional to the change of density $\Delta\rho$ (at a polymer concentration of Δc) (cf. Chapter 11).

  (b) Spectrophotometric detectors (ultraviolet and infrared spectrometers).

  (c) Viscometric detectors (cf. Section 9.1.6).

  (d) Low-angle laser light scattering (cf. Section 13.3).

  (e) Other types such as conductivity detectors, or calorimetric detectors.

Bibliography: 79, 920, 1315.
References: 2404, 2413, 2362, 2698, 2946, 3089, 3360, 3685, 3803, 3855, 4157, 4285, 4322, 4323, 4431, 5253, 5254, 5297, 5336, 5635, 5684, 6077, 6102, 6201, 6339, 6765.

(v) Eluant volume detectors. The time dependence of the volume of eluant flowing out of a column is usually recorded by the siphon system. When the siphon is full and overflows the signal from the photocell registering system is sent to the recorder, producing a blip.

Bibliography: 79.
References: 3758, 4944, 7259.

(vi) A data recording system. The signal from the polymer concentration detector and the eluant volume detector are recorded on a standard potentiometer or on a digital curve translator.

Bibliography: 79.
References: 2398, 2476, 2478, 3808.

The commercial GPC instruments now available provide extensive aids to automation, such as facilities for automatic sequential injection of several polymer samples, for rapid changeover of solvents and columns, and for collection of fractions of eluted polymer solution.

As a result of new advances in high-pressure solvent transport systems, high-sensitivity detectors of low cell volume, and polystyrene gel bead processing technology, high-speed gel permeation chromatography (GPC) is a method which allows one to obtain data in less than 20 min, in contrast to the 2–4 h normally required using a conventional GPC chromatograph.

Bibliography: 1315.
References: 3808, 4462, 4929, 4945, 4947, 5354.

A number of computer programs for treatment of GPC data have been published. The input data include the GPC data recording, the Mark–Houwink–Sakurada exponent (a), and the calibration curve. The output data are the five molecular weight averages, differential and cumulative MWD and MWD curve plots with standard deviations, and the absolute MWD with correction for instrumental axial spreading.

Bibliography: 540, 1035, 1313.
References: 2212, 3242, 3665, 3666, 4045, 4455, 5155, 5824, 5825, 6454, 6648, 6863, 6957.

25.10 SAMPLE PREPARATION FOR GPC CHROMATOGRAPHY

The polymer sample must be completely dissolved in the chosen solvent. Partial solution will not give a real distribution curve because of the prefractionation of the sample. Before injection into the column the solution must be filtered under pressure through a micropore filter.

The choice of solvent is determined by:

(i) polymer type;
(ii) GPC column type;
(iii) the type of sample concentration detector.

When a diffractional refractometer is used the refractive index of the polymer must differ considerably from that of the solvent (e.g. THF). Application of spectrophotometric detectors requires chlorinated solvents for infrared spectrometers and spectral-grade solvents for ultraviolet spectrometers. (Ultraviolet spectrometers are very sensitive to the presence of traces of additives in commercial polymers.)

The polymer concentration should be 0.1% or less to avoid errors due to viscosity effects in columns. Recommended sample concentrations versus molecular weights are shown in Table 25.5. When μ-STYRAGEL columns

Table 25.5 Molecular weight versus sample concentration

Molecular weight range	Sample concentration (%)
Up to 20 000	0.25
34 000–200 00	0.10
400 000–2 000 000	0.05
2 000 000	0.01

are used for the determination of molecular weight distributions, the recommended injection volume in 0.5 ml or less. For certain application the volume can be increased to 1.0 ml, but care should be taken to ensure that the distribution does not change when 1 ml is used instead of 0.5 ml. Injection of a volume larger than that recommended may damage the column.

Polymer solutions are very easily oxidized by air when antioxidants are not present. It is recommended that polymer fractions are stored with the anti-oxidant (e.g. Ionol) present until the fraction is to be examined by GPC chromatography.

25.11 PREPARATIVE GPC

Preparative GPC is a magnification of analytical GPC. The same packing, solvents, and detectors are used. The methods of sample injection, solvent handling and fraction collection differ. The sample concentration seldom

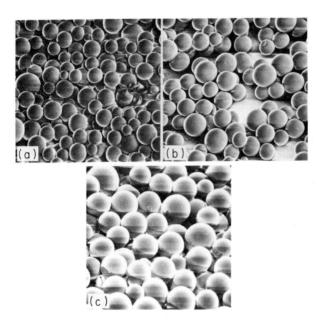

Fig. 25.11 Crosslinked poly(styrene) gels: (a) 16–20 μm; (b) 20–24 μm; (c) 24–28 μm. (From R: 3092; reproduced by courtesy of J. F. Dawkins, T. Stone, and G. Yeadon with the permission of IPC Science and Technology Press

exceeds 1 %. For a total fractionation of 100 g of polymer, 10 litres of solution are required. The injection system is constructed as an automatic system on a timed basis. The usual sample injection system consists of repeated loop injection. The collection of fractions is usually done by the selection procedure. Electronic switches are set to divert the eluant to a selected collection bottle via an automatic multiport valve. The separation of the polymer selected from a large amount of solvent is carried out in a vacuum rotary evaporator designed to add the fractionated solution continuously under a nitrogen atmosphere. The evaporator must be equipped with a low-temperature condenser to recover the solvent. The evaporation container should be no larger than 50 ml, otherwise the small amount of isolated fraction will be spread over a large glass surface.

Bibliography: 13.
References: 2250, 4459, 4838, 5331, 5735, 5810, 5811, 6653, 6934.

25.12 HIGH-RESOLUTION GPC

High resolution in gel permeation chromatography (GPC) can be obtained by increasing the column efficiency, i.e. the number of theoretical plates (N) in a column. Instead of using long columns and the high pressure required to pump the solvent through the column, high-resolution gel permeation chromatography may be attained in commercial GPC chromatographs by a recycling operation through a reciprocating pump (recycle GPC).

The number of cycles necessary depends on the sample distribution (Fig. 25.12).

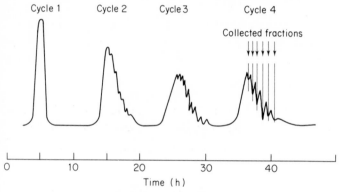

Fig. 25.12   A typical recycling GPC chromatogram

The required cycle is collected and analysed.

Recycle GPC is an effective method for preparative separation of small molecules and low molecular weight polymers.

Bibliography: 157, 1315.
References: 2397, 2398, 2479, 2480, 2481, 2762, 3253, 3802, 3987, 4012, 4014, 4460, 4461, 5336, 5340, 7023.

25.13 DIFFERENTIAL GPC

DIFFERENTIAL GPC is a chromatographic technique in which the positions of solvent and polymer solution are reversed, i.e. pure solvent is injected as a sample. The chromatogram obtained is a mirror image of the normal GPC chromatogram of the polymer sample (Fig. 25.13). If the injected sample is a polymer which differs in molecular weight distribution from the polymer in

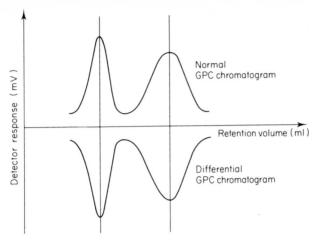

Fig. 25.13 Comparison of two chromatograms: a normal GPC chromatogram and a differential GPC chromatogram

solution, the chromatogram obtained is a difference chromatogram of these two polymers.

References: 2849, 5047.

25.14 APPLICATIONS OF GEL PERMEATION CHROMATOGRAPHY TO THE CHARACTERIZATION AND ANALYSIS OF POLYMERS

Gel permeation chromatography (GPC) can be successfully applied for the following investigations:

(i) Determination of the molecular weight (MW) of homopolymers and copolymers.
(ii) Determination of the molecular weight distribution (MWD) curves of homopolymers and copolymers.
(iii) Analysis of copolymer compositions. The analysis of copolymers by GPC as a function of molecular weight and chemical composition is much more complex than that of homopolymers.

The molecular weight distribution of copolymers or a mixture of homopolymers can be measured when
(a) the composition of these does not vary with molecular weight;
(b) the calibration curves are available for specific compositions.
Frequently, the composition does not vary with molecular weight. When the composition varies the quantitative data are difficult to obtain because a differential refractometer is too sensitive to changes in composition.
(iv) Determination of polymer branching (short chain branching distribution (SCBD) and long chain branching distribution (LCBD)) with

438

parallel measurements such as viscosity of solutions or infrared spectroscopy (cf. Section 25.15).

(v) Study of preferential solvation of polymers.

(vi) Qualitative and quantitative analysis of additives in polymers.

Bibliography: 7, 26, 876, 964, 1132.
References: 2035, 2088, 2098, 2100, 2261, 2266, 2354, 2355, 2361, 2403, 2443, 2687, 2689, 2694, 2784, 3207, 3236, 3272, 3371, 3372, 3809, 3883, 3986, 4089, 4115, 4116, 4838, 4902, 5003, 5044, 5067, 5068, 5234, 5294–5296, 5327, 5332, 5342, 5382, 5418, 5419, 5552, 5621, 5622, 5638, 5742, 5950, 6126, 6312, 6323, 6420, 6920, 6922, 7027.

25.15 DETERMINATION OF CHAIN BRANCHING IN POLYMERS

The SHORT CHAIN BRANCHING DISTRIBUTION (SCBD) and the LONG CHAIN BRANCHING DISTRIBUTION (LCBD) can be measured using GPC with other methods, e.g. viscosity measurements.

The intrinsic viscosity and the average viscosity molecular weights of a branched ($\overline{M}_{vB}$) and a linear ($\overline{M}_{vL}$) structure are related by

$$[\eta]_B \overline{M}_{vB} = [\eta]_L \overline{M}_{vL}. \qquad (25.12)$$

The viscosity-average molecular weight of branched molecules ($\overline{M}_{vB}$) may be calculated when a calibration plot of elution volume against $[\eta]_L \overline{M}_{vL}$ is available and $[\eta]_B$ is known. A molecular weight distribution of branched macromolecules based on a calibration plot established with a linear unbranched polymer involves a significant error. For branched macromolecules the limiting viscosity number (intrinsic viscosity) is lower than that of a linear polymer of the same molecular weight.

A BRANCHING FACTOR (G) under these conditions is defined by

$$G = [\eta]_{0B}/[\eta]_{0L} \qquad (25.13)$$

(dimensionless). It is related to the NUMBER OF BRANCH POINTS (g):

(i) for lightly randomly branched polymers and star branched polymers by

$$G = g^{1/2}; \qquad (25.14)$$

(ii) for highly randomly branched polymers by

$$G = g^{3/2} \qquad (25.15)$$

When the relationship between the number of branch points per macromolecule and the molecular weight is known it is possible to calculate the degree of branching and true molecular weight distribution. The necessary information is

(i) the limiting viscosity number (intrinsic viscosity);
(ii) the GPC chromatogram;
(iii) the theoretical long chain branching distribution as a function of molecular weight.

Short chain branching reduces the limiting viscosity number (intrinsic viscosity) in comparison to that of a linear polymer of the same molecular weight. To avoid this error short chain branching is usually measured by collecting fractions from GPC chromatography and values from infrared measurements to determine the degree of branching based on calibration with known samples.

Bibliography: 109, 110, 532, 538, 1226, 1274, 1315.
References: 2099, 2101–2104, 2249, 2263, 2871, 2946, 2983, 3234, 4457, 4715, 5232, 5431, 5606, 5609, 5632, 5678, 5685, 5952, 5953, 6339, 6378, 6864, 7057, 7081–7083, 7113, 7334, 7338.

25.16 HYDRODYNAMIC CHROMATOGRAPHY

HYDRODYNAMIC CHROMATOGRAPHY (HDC) is a chromatographic technique which uses packed beads for the separation of colloidally suspended particles in solution. A bed packed with porous or non-porous packing presents the particles suspended in the carrier solvent with a tortuous path through a large number of capillary-like tunnels (Fig. 25.14). The separation is due to the

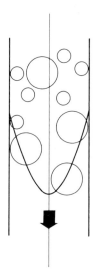

Fig. 25.14 Schematic presentation of the mechanism of particle separation by hydrodynamic chromatography (HDC)

velocity profiles, in the interstitial volume of the column. Larger particles are excluded faster from regions near the capillary wall, where axial velocities are small, and hence experience on the average a higher velocity and small retention time. Porous packing presents an additional force for size separation by steric exclusion from the pores (cf. liquid exclusion chromatography, Section 25.1).

The rate of elution of the colloid through the packed bed is affected by

   (i) the diameter of the colloidal particles;
  (ii) the diameter of the packing beds;
  (iii) the flow rate of the fluid eluant.

440

For characterization one employs the elution volume of R_f VALUE, defined as

$$R_f = \frac{\text{Rate of movement of the colloid peak}}{\text{Rate of movement of the fluid front}}. \qquad (25.16)$$

The RATE OF MOVEMENT OF THE COLLOID PEAK is obtained by measuring the transit time of the peak through the columns.

The RATE OF ELUANT FLOW is obtained by measuring the transit time of a marker species (a small ionic or low molecular weight species to which the detector is sensitive).

R_f values increase with a reduction in the bead diameter of the packing. The smaller the packing diameter the better is the separation.

25.17 THE UNIVERSAL CALIBRATION PLOT FOR HDC

The universal calibration plot for HDC involves determination of the elution volumes (or R_f values) of a series of poly(styrene) latexes of known particle size (Fig. 25.15). The particle sizes may be measured by electron microscopy.

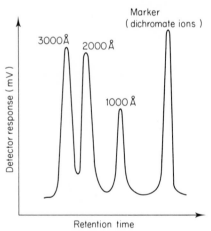

Fig. 25.15 Calibration elution of a latex of known size and a small marker species (dichromate ions) enables one to construct a calibration plot of R_f versus particle diameter (Å)

The calibration plot of latex diameter versus R_f is then used to give size information on the other colloids examined.

25.18 HYDRODYNAMIC CHROMATOGRAPHS

HDC instruments are composed of the following parts (Fig. 25.16):

(i) A pumping system, including a solvent container, a degasser, and a high pressure pump (cf. Section 25.9).

(ii) A sample injection system.

(iii) HDC chromatographic columns. These have the following dimensions: length 100 cm and inner diameter 9 mm. The following types of HDC packing are usually applied:
  (a) strong acid cation exchange resins in sodium form (20–40 µm);
  (b) styrene divinylbenzene copolymer beds (which do not suffer from shrink–swell sensitivity to the level of the electrolyte in the solution, as do cation exchange resins);
  (c) solid glass beads.

(iv) A sample concentration detector. A single-wavelength flow-through spectrometer is commonly used.

(v) An eluant volume detector and a data recording system, which are similar to those applied in GPC chromatographs (cf. Section 25.9).

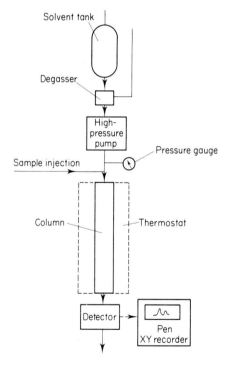

Fig. 25.16 Schematic diagram of a hydro-dynamic chromatograph

Water is commonly used as a solvent. The elution time for an injected sample is 100 min, but this can be reduced by up to 10 min without much loss in resolution by increasing the flow rate.

442

25.19 APPLICATIONS OF HYDRODYNAMIC CHROMATOGRAPHY TO THE CHARACTERIZATION AND ANALYSIS OF POLYMERS

The main applications of hydrodynamic chromatography (HDC) are

(i) the determination of particle size distribution—Fig. 25.17 shows the chromatogram of a latex having two distinct groups of particles;

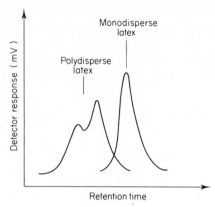

Fig. 25.17 Hydrodynamic chromatogram
of two kinds of latex: polydisperse latex
and monodisperse latex

(ii) the determination of particle agglomeration in polymer latexes;
(iii) the study of emulsion polymerization kinetics in order to determine the particle size of the final emulsion, its polydispersity, and the rate of particle growth;
(iv) the study of the effect of environmental factors on the size of latex particles (e.g. the pH, ionic strength, or surfactant type and concentration in the continuous phase);
(v) the study of the kinetics of colloid shrinking, swelling, and aggregating.

References: 6413, 6430, 6431, 6589.

Chapter 26

Overview of Polymer Morphology

General bibliography: 19, 85, 127, 237, 308, 352, 432, 433, 438, 471, 503, 585, 641, 667, 725, 22, 862, 865, 869, 927, 1047, 1081, 1098, 1117, 1169, 1194, 1195, 1350, 1396, 1424, 1427, 1434.

MORPHOLOGY OF POLYMERS is a field of science devoted to study of he arrangement, form, and structure of polymer molecules in crystalline and amorphous regions.

Polymers can be divided into two classes:

   (i) AMORPHOUS—described by the ideal melt.

  (ii) SEMICRYSTALLINE, which consists of two categories of region:

     (a) AMORPHOUS REGIONS; and

     (b) CRYSTALLINE REGIONS—described by the CRYSTAL-DEFECT STRUCTURE CONCEPT—which allow for the occurrence of such phenomena as polymer single crystals, chain folding, lamellar crystalline growths, and intermediate crystalline objects such as axialites and dendrites, lattice dislocations, etc.

The FRINGED-MICELLE MODEL (Fig. 26.1) assumes that an individual macromolecule runs through several crystalline regions dispersed in an amorphous region (domain).

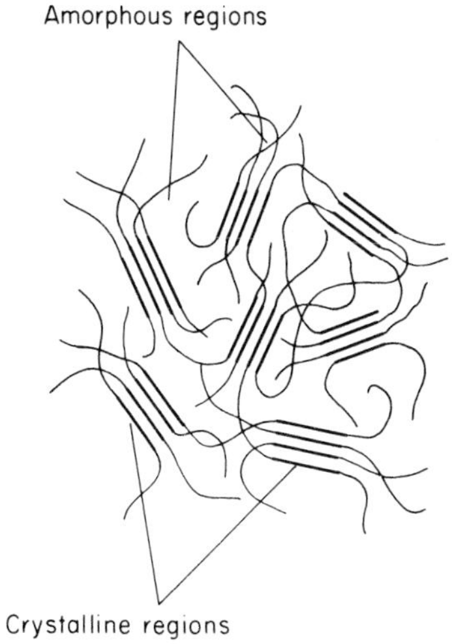

Amorphous regions

Crystalline regions

Fig. 26.1 Fringed micelle model

Polymers which crystallize normally have regular configurations. Polymer with bulky side groups and irregular configuration do not crystallize, or crystallize in very limited manner.

High crystallinity is evident in polymers that are dense, stiff, hard, tough and resistant to solvents and chemical reactions.

A high proportion of amorphous domains increases softness, flexibility, and processing.

The rate and extent of crystallization in a given polymer determines

(i) the degree of crystallinity;

(ii) the size and shape of crystalline regions.

A SINGLE CRYSTAL is a crystal in which all the parts can be related to the same microscopic translation lattice.

A TWINNED CRYSTAL is a crystal in which the lattice undergoes an abrupt change at one or more well-defined planes (twin planes)

A POLYCRYSTALLINE SAMPLE is a sample made up of larger number of single crystals or twinned crystals.

PARACRYSTALLINITY is the intermediate state between liquid and crystalline states.

POLYMORPHISM is the occurrence of different crystal modifications of the same polymer possessing the same monomeric unit. The various modifications are characterized by different crystallographic parameters such as different lattice constants and/or lattice angles, and consequently different unit cells. Polymorphism can result from conformational differences in the chain molecule or different packings of molecules with the same conformation.

ISOMORPHISM is the phenomenon by which various monomer units can replace each other in the lattice. Isomorphism is possible in copolymers if the corresponding monomer units show similar lattice constants and/or lattice angles, and the same helical type.

ANNEALING is a technique in which the samples are heated to the required temperature for a specified time under vacuum in order to change their morphological structure. The film samples should be placed between two aluminium frames to prevent films from curling during annealing at temperatures in excess of the T_g. Vacuum annealing is recommended to prevent oxidation of the sample at the elevated annealing temperature. Samples removed from the oven should be allowed to cool to room temperature immediately after the specified annealing time.

References: 2270, 2679, 3406, 3581, 3689, 4164, 4508, 5063, 7030.

26.1 POLYMER CRYSTALLIZATION

In unorientated polymer solutions and melts crystallization occurs by nucleation processes.

The PRIMARY NUCLEATION PROCESS can be defined as origination of a crystal phase from a parent phase. It can be divided into three types:

(i) SPONTANEOUS HOMOGENEOUS NUCLEATION. This is nucleation which occurs under no other influence besides supercooling or supersaturation of the parent phase (Fig. 26.2). It is rarely observed in bulk polymers.

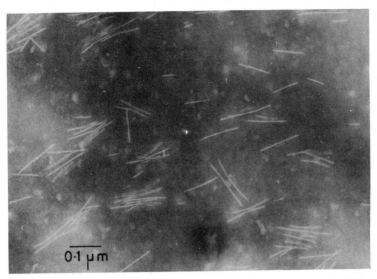

Fig. 26.2 Homogeneous nucleation in *cis*-poly(isoprene) (From R: 3270; reproduced by courtesy of B. C. Edwards and P. J. Phillips with the permission of John Wiley & Sons)

Fig. 26.3 Heterogeneous nucleation in *cis*-poly(isoprene). (From R: 3270; reproduced by courtesy of B. C. Edwards and P. J. Phillips with the permission of John Wiley & Sons)

446

(ii) ORIENTATION-INDUCED NUCLEATION. This is nucleation caused by some degree of alignment of liquid molecules which reduces the difference between the liquid and crystalline arrangements of the molecules. It is very important in the processing of thermoplastics.

(iii) HETEROGENEOUS NUCLEATION. This is nucleation at the interface with a foreign phase (solid impurities, purposely added nucleating agents, or an extended substrate on which crystallization may occur) (Fig. 26.3).

SECONDARY NUCLEATION is a surface nucleation on a growth plane. TRANSCRYSTALLIZATION is a crystallization process in which the formation of a large number of adjacent nuclei occurs on the surface of the melt.

Bibliography: 131, 862, 1025, 1140, 1141, 1194, 1195, 1424, 1427, 1444.
References: 2043, 2044, 2200, 2207, 2229, 2230, 2234, 2271, 2318, 2319, 2324, 2677, 2962, 3101 3269–3271, 3353, 3483, 3581, 3610, 3682, 3688, 3783, 3991, 4081, 4082, 4260, 4269, 4389, 4444 4623, 4833, 4881, 4974, 4975, 5031, 5032, 5051, 5054, 5055, 5192, 5615, 5789, 5790, 5797, 5818 5819, 6079, 6178, 6179, 6254, 6255, 6266, 6270, 6348, 6767, 6794, 6795, 7184, 7198, 7300, 7301.

CRYSTALLIZATION KINETICS are commonly studied by dilatometry (cf. Chapter 31.4). The fraction of amorphous material remaining (λ) at time t is given by

$$\lambda = \frac{h_\infty - h_t}{h_\infty - h_0},$$ (26.1)

where

h_0 is the height of the mercury in the dilatometry capillary at the beginning of crystallization,

h_∞ is the height of the mercury column at a later stage in the crystallization at which further change is imperceptible,

h_t is the height of the mercury column at time t,

This simple equation is valid only when crystallization proceeds until practically the full mass of material is transformed to the crystalline state.

The values of the amorphous content (λ) at given times can be plotted as:

(i) λ versus $\log t$ (Fig. 26.4);

(ii) $\log(-\log \lambda)$ versus $\log t$ (Fig. 26.6), called the AVRAMI PLOT.

The crystallization rate constant (k) can be calculated from

$$\lambda = \exp(-kt^n),$$ (26.2)

where

λ is the value of the amorphous content,

k is the crystallization rate constant (in s^{-n}),

t is the crystallization time (in seconds),

n is a function of the type of nucleation and growth (dimensionless).

Bibliography: 588, 862.
References: 2181, 2182, 2232, 2255, 2322, 2390, 2489, 2490, 2660, 2763, 3585, 3681, 3690, 3692, 3826, 3953, 3956, 3957, 3969, 4058, 4314, 4798, 5022, 5061, 5062, 5548, 5941, 6337, 6407, 6442, 6511, 6767.

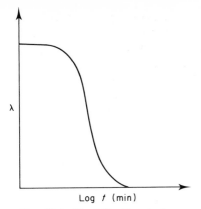

Fig. 26.4   Crystallization isotherm

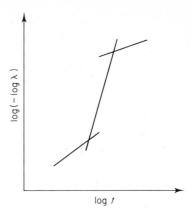

Fig. 26.5   Avrami plot for isothermal crystallization

Extensive data regarding the rate of crystallization of polymers are collected in The Polymer Handbook (B: 856).

26.2 LAMELLAR SINGLE CRYSTALS

Macromolecules in polymer single crystals crystallize with chain folding (Fig. 26.6), giving molecular LAMELLAE.

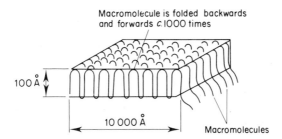

Fig. 26.6   Schematic representation of chain folding in lamellae

The thicknesses of the lamellae are uniform from one polymer to another, being of the order of 100 Å, whereas the dimensions in the plane of the lamella are very much larger—typically of the order of several micrometres.

The polymer single crystals are about 75–85% crystalline, whereas 100% crystallinity can be obtained if the surface layers are decomposed with fumic nitric acid. (cf. Section 26.14).

Single crystals of many different polymers crystallize from dilute solutions in the form of different crystallographic forms (Figs 26.7 and 26.8).

There are three proposed models for the lamellar crystals crystallized from the melt (Fig. 26.9):

448

Fig. 26.7   Single crystals of poly(ethylene) in the form of molecular lamellae in polarized light. (Reproduced by courtesy of B. Wunderlich)

(a)

5 μm

(b)

Fig. 26.8 Single crystals of polymers other than poly(ethylene). (a) Poly(oxymethylene). (From R:2270; reproduced by courtesy of D. C. Bassett, F. R. Damont, and R. Salovey with the permission of IPC Science and Technology Press) (b) Multilayer of poly(oxymethylene) lamellae. (From R:2251; reproduced by courtesy of W. J. Barnes and F. P. Price; with the permission of IPC Science and Technology Press). (c) Poly(4-methylpentene). From R:2270; reproduced by courtesy of D. C. Bassett, F. R. Damont and R. Salovey with the permission of IPC Science and Technology Press.) (d) Cellulose triacetate (magnification ×1300). (From R:2794; reproduced by courtesy of D. H. Chanzy and E. J. Roche with the permission of John Wiley & Sons)

(i) the REGULAR MODEL, in which adjacent re-entry folds have the same thickness;

(ii) the IRREGULAR MODEL, in which adjacent re-entry folds have different thicknesses;

(iii) the SWITCHBOARD MODEL, in which non-adjacent re-entry folds occur and in which a non-ordered amorphous layer is present on each side of the lamellae.

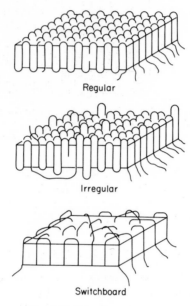

Regular

Irregular

Switchboard

Fig. 26.9 Lamellae models

The fold length (or lamellar height) increases with increasing crystallization temperature and pressure. The fold length for many polymers increases linearly with the reciprocal of the difference between the melting point and the crystallization temperature (the SUPERCOOLING EFFECT).

When crystallization is carried out from a concentrated solution, or from the melt, the same macromolecule will crystallize into more than one lamellae (Fig. 26.10).

The number of INTERLAMELLAR LINKS (also called TIE MOLECULES) increases with increasing molecular weight. A sample crystallized from the melt always has a relatively high content of amorphous region.

The characterization of lamellar single crystals can be carried out using

(i) electron microscopy (cf. Chapter 27);

(ii) X-ray diffraction methods (cf. Chapter 28);

(iii) electron-diffraction methods (cf. Chapter 29).

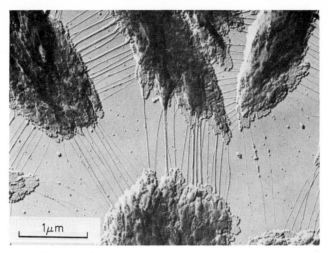

Fig. 26.10 Interlamellar links between poly(ethylene) lamellae. (From R: 4497; reproduced by courtesy of F. H. Keith, F. J. Padden Jr, and R. G. Vadimsky with the permission of John Wiley & Sons)

Bibliography: 137, 434, 501, 627, 640, 724–726, 823, 862–864, 1030, 1047, 1144, 1195, 1417, 1427. References: 2027, 2042, 2251, 2270, 2294, 2430, 2431, 2455–2457, 2625, 2794, 2904, 3074, 3174, 3272, 3379, 3383, 3406, 3481, 3586–3589, 3610, 3614, 3773, 3926, 4338, 4469, 4474, 4478, 4484, 4496–4499, 4501, 4507, 4511, 4513–4516, 4518, 4610, 4643, 4694, 4753, 4933, 4969, 5050, 5077, 5219, 5288, 5313, 5366, 5483, 5657, 5675, 5698, 5757, 5786, 5792, 5820, 5900, 5903, 6022, 6081, 6199, 6375, 6784, 6812, 7063, 7079, 7112, 7183, 7202, 7261.

26.3 POLYMER SPHERULITES

SPHERULITES are supermolecular polygonal structures of anisotropic aggregates of crystals. The spherulite is made up of fibrils which are arranged in a radial pattern (cf. surface replica of a ring spherulite, Fig. 26.11). The FIBRILS

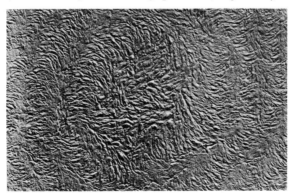

Fig. 26.11 Electron micrograph of surface replica of a ring spherulite of a high density poly(ethylene) (magnification × 8000). (From R: 4753; reproduced by courtesy of M. Kryszewski, A. Galeski, T. Pakula, and J. Grebowicz with the permission of *The Journal of Colloid and Interface Sciences*)

are made up of crystallites with the chain folded at right angles to the fibril length.

Spherulites crystallize from the melt or from relatively concentrated solutions (above 1 wt %). Spherulites grow from a heterogeneous nucleus at their centre by a two-stage process (Figs 26.12 and 26.13):

(i) PRIMARY CRYSTALLIZATION. The original lamellae that grew from the primary nucleus branches at points of imperfection (DISLOCA-TIONS) in the crystal lattice to yield secondary lamellae, which bend and twist as they grow, and in turn branch to produce tertiary lamellae, and so on. The final result is the production of a multitude of fibrous lamellae

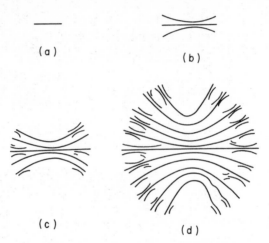

(a)

(b)

(c)

(d)

Fig. 26.12 Schematic representation of the different stages of spherulitic growth. (a) Single lamellar crystal; (b) Spawning of new lamellar crystals; (c) and (d) formation of radial spherulites by the spawning mechanism

that develop radially outward in all directions from the nucleus until stopped by the interaction of neighbouring spherulites on one another.

(ii) SECONDARY CRYSTALLIZATION. This process proceeds within each spherulite, transforming a portion of the remaining amorphous material.

The size of spherulites ranges from 1 μm to a few millimetres, depending upon the crystallization conditions.

The characteristics of spherulites can be obtained by using one of the following methods:

(i) Electron microscopy (cf. Chapter 27).
(ii) X-ray diffraction methods (cf. Chapter 28).
(iii) Birefringence measurements (cf. Section 35.3). In polarized light, spherulites show the typical MALTESE CROSS (Fig. 26.14) which is caused by a

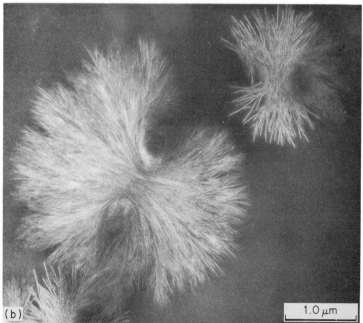

Fig. 26.13   Growth of lamellar spherulites during the crystallization of *cis*-poly (isoprene): (a) and (b) show two successive stages. (From R: 3268, 5820; reproduced by courtesy of B. C. Edwards and P. J. Phillips with the permission of John Wiley & Sons)

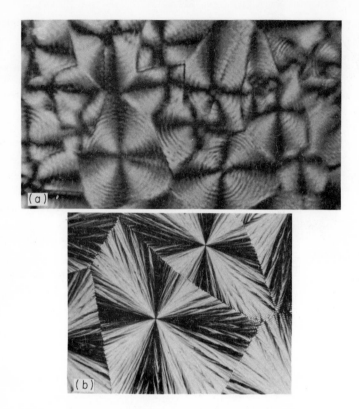

Fig. 26.14 The Maltese cross observed in spherulites in polarized light: (a) in poly(ethylene) (reproduced by courtesy of B. Wunderlich); (b) in L-poly(propylene oxide) (reproduced by courtesy of J. H. Magill)

birefringence effect (the speed of light varies within the different spherulitic regions). Different types of spherulites give different birefringences: (a) POSITIVE SPHERULITES show the highest refractive index in the radial direction; (b) NEGATIVE SPHERULITES show the highest refractive index in the tangential direction. Study of spherulite birefringence gives information on their structure.

(iv) Light scattering methods. The orientation of the molecular axis in the spherulites can be measured by a light scattering method (cf. Section 35.10). Different types of spherulites give different scattering patterns: (a) POSITIVE SPHERULITES have their optical axis along the radial direction; (b) NEGATIVE SPHERULITES have their optical axis perpendicular to the radial direction.

In the PERFECT SPHERULITE the crystal orientation with respect to the radius is the same everywhere within the spherulite. During the growth of spherulites, chain ends and non-crystalline material may be included in the spherulite, and the spherulite formed may be regarded as a DISORDERED SPHERULITE.

Spherulites can be classified into following categories:

(i) type I (α-form crystallinity (monoclinic)), which exhibit a well-defined Maltese cross and can be differentiated only on the basis of birefringence, which is positive;

(ii) type II (α-form crystallinity (monoclinic)), which exhibit a well-defined Maltese cross and can be differentiated only on the basis of birefringence, which is slightly negative;

(iii) type III (β-form crystallinity (hexagonal)), which has more highly negative birefringence,

(iv) type IV (β-form crystallinity (hexagonal)), which has much more negative birefringence and shows a distinctly banded structure;

(v) type V (γ-form crystallinity (possessing a triclinic unit cell)), but this has been obtained in bulk only at high pressure.

Spherulites make films and foils opaque when their diameters are greater than half the wavelength of the light and when inhomogeneities in the density and refractive index exist within the spherulite.

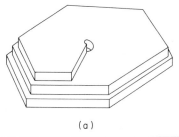

(a)

Fig. 26.15 Hedrites. (a) Orientation of lamellae;
(b) Electron micrograph of surface replica from
poly(methylene oxide) (magnification × 3000).
(From R: 3574, 3575; reproduced by courtesy
of M. Kryszewski and A. Galeski with the
permission of John Wiley & Sons)

456

Bibliography: 1032, 1127, 1139, 1194, 1195, 1422, 1427.
References: 2030, 2241, 2260, 2272, 2273, 2411, 2412, 2644, 2695, 2711, 2785, 3027, 3028, 3268, 3322, 3379, 3480, 3649, 3687, 4008, 4199, 4278, 4425, 4491–4495, 4500, 4506, 4510, 4519, 4603, 4880, 4890, 4904, 4973, 4974, 5021, 5030, 5059, 5299, 5450, 5656, 5675, 5690, 5806, 5819, 5901, 5916, 6071, 6177, 6870, 6964, 7203.

26.4 AXIALITES

AXIALITES are aggregates of crystal lamellae and they exhibit the different characteristics of polymer single crystals or spherulites depending on the angle of view. They can crystallize in a wide variety of supermolecular structures, such as

- (i) HEDRITES, which have hexagonally oriented fibrous lamellae which may incorporate a screw dislocation (Fig. 26.15);
- (ii) OVOIDS, which have radially oriented fibrous lamellae (Fig. 26.16);
- (iii) SPIRAL OVOIDS, which have spirally oriented fibrous lamellae (Fig. 26.17).

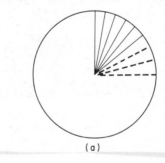

(a)

(b)

Fig. 26.16   Ovoid. (a) Orientation of lamellae; (b) electron micrograph of surface replica from poly(methylene oxide) (magnification ×3000). (From R: 3574, 3575; reproduced by courtesy of M. Kryszewski and A. Galeski with the permission of John Wiley & Sons)

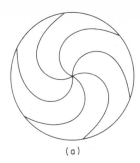

(a)

3 μm

(b)

Fig. 26.17 Spiral ovoid. (a) Orientation of lamellae; (b) Electron micrograph of surface replica from poly(methylene oxide) (magnification × 5600). (From R: 3574, 3575; reproduced by courtesy of M. Kryszewski and A. Galeski with the permission of John Wiley & Sons)

They crystallize during the supercooling of the polymer melt.

Bibliography: 1427.
References: 2275, 3574, 3575, 3611, 4511, 4752, 5747.

26.5 DENDRITES

DENDRITES are branched crystals (Fig. 26.18) formed if the overall crystallization rate is directionally dependent. They grow from dilute solutions at increasingly lower temperatures. A regular dendrite of poly(ethylene) photographed in polarized light is shown in Fig. 26.19.

Irregular dendrites are called HEDGEHOG DENDRITES (Fig. 26.20).

Bibliography: 1143, 1427.
References: 2795, 2797, 3386, 3615, 4491, 4492, 4546, 5349, 6199, 6450, 7196, 7197, 7203.

458

Fig. 26.18   Photomicrograph of a dendric crystal of pentaerythrityl tetrabromide (Magnification × 500). (From R: 6450; reproduced by courtesy of A. J. Pennings and P. Smith with the permission of John Wiley & Sons)

26.6 EXTENDED CHAIN CRYSTALS

EXTENDED CHAIN CRYSTALS are crystals of long-chain molecules which have no sign of folding (Fig. 26.21). This type of ordering can take place with chains of molecular weight greater than 10 000 and whose extended lengths are greater than 1000 Å. Sometimes extended chain crystals may form well-developed needle-like crystals (Figs 26.22 and 26.23). They can be formed

  (i) by very slow crystallization at temperatures close to the melting point;

  (ii) during the crystallization of melts under high pressure (5000 atm);

  (iii) by orientation in polymers.

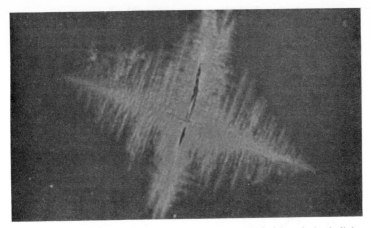

Fig. 26.19 Regular poly(ethylene) dendrite crystal in polarized light. (Reproduced by courtesy of B. Wunderlich)

Fig. 26.20 Hedgehog poly(ethylene) dendrite crystals in polarized light (scale bar = 10 μm). (Reproduced by courtesy of B. Wunderlich)

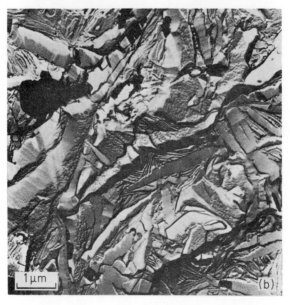

Fig. 26.21 Extended chain crystals of poly(ethylene).
(a) (From B: 82; reproduced by courtesy of D. C. Bassett
with the permission of IPC Science and Technology Press.)
(b) (From B: 83; reproduced by courtesy of D. C. Bassett
with the permission of *High Temperature–High Pressure
Journal*, Pion Ltd)

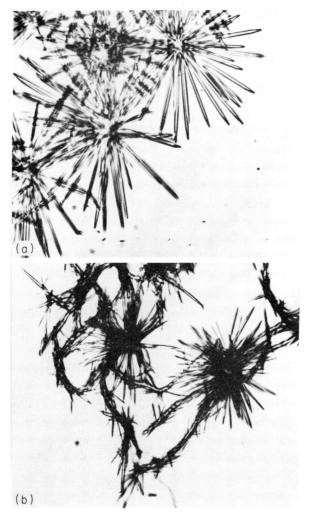

Fig. 26.22 Needle-like extended chain crystals of poly
(oxymethylene): (a) neat and (b) branched. (From R: 4202;
reproduced by courtesy of M. Iguchi and I. Murase with the
permission of *Makromolekulare Chemie*)

Bibliography: 82, 83, 258, 727, 805, 1333, 1424, 1427.
  References: 2117, 2174, 2268, 2269, 2273, 2276–2281, 2296, 2678, 2710, 3044, 3232, 3233, 3610, 3804, 4198–4200, 4202, 4203, 4426, 4540, 4855, 5013, 5027, 5194, 5365, 5593, 5754, 5756, 5904, 5905, 6003, 6394, 7186, 7193, 7200.

26.7 FIBRILLAR POLYMER CRYSTALS

FIBRILLAR POLYMER CRYSTALS are formed by alignment of segments
and partial crystallization of several polymer molecules. The formation of these
fibrillar crystals may be initiated by BUNDLE-LIKE NUCLEI (Fig. 26.24),

462

Fig. 26.23 Needle-like extended chain crystals of poly(oxymethylene): (a) in a normal light and (b) in a polarized light. (From R: 4203; reproduced by courtesy of M. Iguchi, I. Murase, and K. Watanabe with the permission of the Society of Chemical Industry, London)

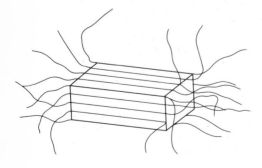

Fig. 26.24 Schematic illustration of a bundle-like nucleus (ciliated or fringed micellae nucleus)

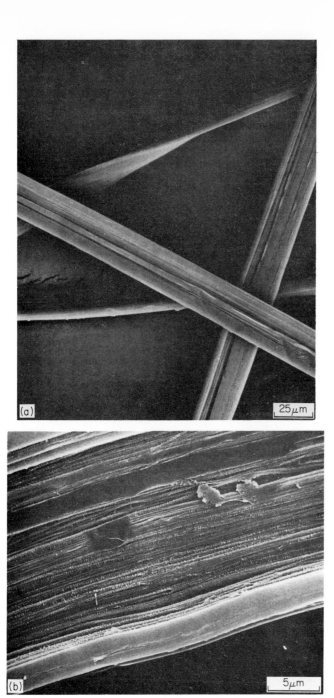

Fig. 26.25   Scanning electron micrographs of poly(ethylene) macro-fibrils: (a) magnification × 750, and (b) magnification × 600 of portion of (a). (From R : 7349; reproduced by courtesy of A. J. Pennings and A. Zwijnenburg; with the permission of John Wiley & Sons)

also referred to as CILIATED or FRINGED MICELLE NUCLEI. They are formed by

  (i) oriented crystallization in flowing solutions;
 (ii) oriented crystallization in oriented melts undergoing raw nucleation;
(iii) oriented crystallization in stretched networks;
 (iv) complexation between polymers through van der Waals bonds, Coulombic forces, hydrogen bonds, and hydrophobic interactions;
  (v) eutectic crystallization (cf. Section 26.8).

Two examples of macrofibrils are shown in Figs 26.25 and 26.26.

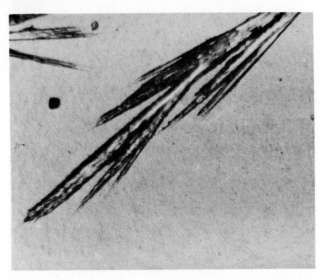

Fig. 26.26   Fibrous isotactic poly(styrene) crystal (magnification × 1200). (From R: 4879; reproduced by courtesy of P. J. Lemstra and G. Challa with the permission of John Wiley & Sons)

Bibliography: 727, 728, 1025, 1047, 1141, 1427.
References: 2242, 2437, 2438, 2795, 3507, 3622, 3664, 4099, 4199, 4247, 4505, 4879, 5078, 5193, 5545, 5754, 5756, 5758, 6449, 6533, 6664, 6823, 6840, 6842, 6951, 7260, 7348, 7349.

26.8 EUTECTIC CRYSTALLIZATION IN POLYMER SOLUTIONS

EUTECTIC CRYSTALLIZATION is the simultaneous generation of at least two crystalline phases yielding a fine-grained and intimately mixed solid. Such a phase transition occurs in general at a composition of the liquid mixture and below a temperature which is determined by the intersection of the liquidus curves of the phase diagram (Fig. 26.27).

A general requirement for the occurrence of a EUTECTIC POINT is that

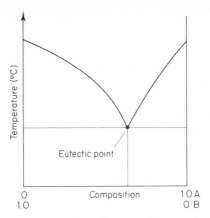

Fig. 26.27 Phase diagram of a eutectic
system composed of components A and B

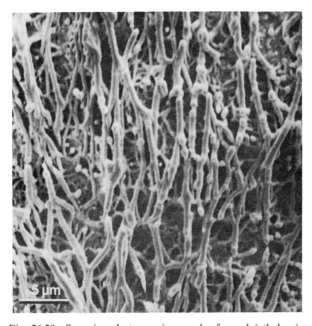

Fig. 26.28 Scanning electron micrograph of a poly(ethylene)
fracture surface after sublimation of the 1, 2, 3, 4, 5-tetra-
chlorobenzene from the eutectic mixture (scale bar = 3 μm).
(From R: 5755; reproduced by courtesy of A. J. Pennings and
P. Smith with the permission of the Society of Chemical
Industry, London)

the components are completely miscible in the liquid state and incompletely
miscible or immiscible in the solid state.

As an example of a eutectic system we can take a composition of polyethylene
and 1,2,4,5-tetrachlorobenzene. When the eutectic mixture is quenched to room

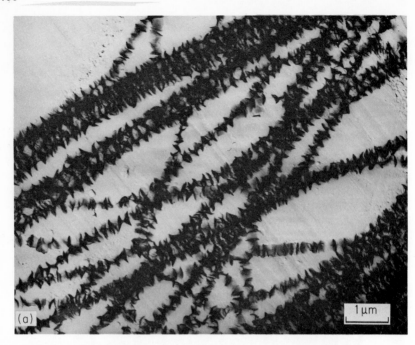

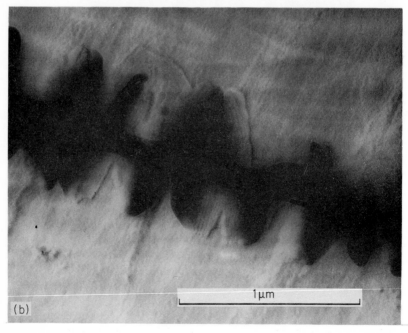

Fig. 26.29 (a) Typical shish-kebab structure found in specially crystallized poly (olefins). (b) Magnified detail of (a). (From R: 5414; reproduced by courtesy of T. Nagasawa and Y. Shimomura with the permission of John Wiley & Sons)

temperature and the solid tetrachlorobenzene is subsequently removed by sublimation, the remaining polyethylene has a fibrillar structure (Fig. 26.28).

References: 5755, 6447.

26.9 EPITAXY

EPITAXY is the oriented growth of one crystalline substance on another. Epitaxially grown crystals are frequently oriented with the chain axis parallel to the substrate surface. Epitaxial growth of one polymer may occur on

(i) oriented polymer surfaces of the same or another polymer;
(ii) non-polymeric crystals, e.g. alkali halide surfaces;
(iii) other surfaces, e.g. glass, quartz, metals.

A typical example of epitaxial growth is the formation of SHISH-KEBAB structures (Fig. 26.29) during crystallization from stirred solutions.

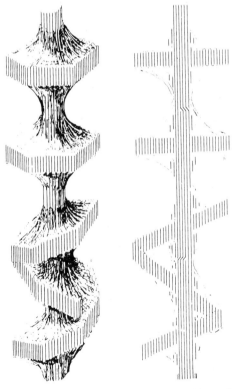

Fig. 26.30   Model of polymeric shish-kebab illustrating the interconnection between the chains in the lamellar overgrowth and in the backbone. (From R: 1025; reproduced by permission of John Wiley & Sons)

468

The macromolecules are oriented along the direction of flow and settle out parallel to each other to form folded-chain lamellae (folded-chain crystals) (KEBABS). These lamellae overgrow the extended chain fibril (the SHISH) epitaxially (Fig. 26.30).

Polymers may crystallize as SHISH-KEBAB structures from the melt as well as from stirred solutions.

Bibliography: 783, 823, 882, 1024, 1025, 1392, 1427.
References: 2094, 2157, 2411, 2626, 2627, 2734, 3170, 3380, 3381, 3613, 3621, 3622, 4201, 4204, 4513, 4523, 4524, 4580, 4688–4690, 4750, 4776, 4970, 5006, 5073, 5149, 5150, 5224, 5269, 5414, 5415, 5751–5757, 6047, 6048, 6254, 6448, 6693, 6694, 7047, 7079, 7097, 7099, 7100, 7139, 7193, 7199, 7201.

26.10 TEXTURE

TEXTURE is the arrangement of the crystalline regions, or crystallites, with respect to each other and to possible intervening amorphous regions or voids (Fig. 26.31).

The presence of voids may be observed by using scanning microscopy (Fig. 26.32).

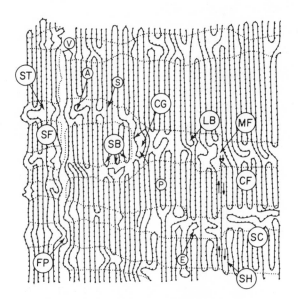

Fig. 26.31 Statton's model of fine texture of stretched linear poly(ethylene). Key: A, amorphous region; CF, clustered fibrils (hot stretched); CG, part corresponding to crystal growth in the bulk material; E, end of molecular chain; FP, structure giving four-point diffraction pattern; LB, long backfolding of molecular chain; MF, migrating fold; P, Paracrystalline layer lattice; S, straight chains; SB, short backfolding of molecular chains; SC, lamellar crystals (single crystals); SF, single fibrils (cold stretched); SH, shearing region; ST, V, void.

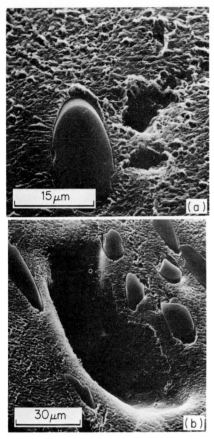

Fig. 26.32. Scanned electron micrographs of small (a) and large (b) voids in glass fibre-reinforced poly(propylene). (From R: 3057; reproduced by courtesy of M. W. Darlington and G. R. Smith with the permission of IPC Science and Technology Press)

Texture is most significant in polymers in which chains and crystallites have a preferred orientation as the result of mechanical treatment such as drawing, rolling, or stretching.

Bibliography: 642, 1427.
References: 3057, 4064, 5856, 7017, 7208.

26.11 CRYSTAL LATTICE DEFECTS

Crystal lattice defects in semicrystalline polymers can be divided into two groups:

   (i) POINT DEFECTS, which are conformational defects (Fig. 26.33):

      (a) a KINK DEFECT occurs when the displacement is smaller than the interchain distance;

      (b) a JOG DEFECT occurs when the displacement is larger than the interchain distance;

      (c) a FOLD DEFECT occurs when the displacement is associated with a change in the relative position of the chain fragments;

(d) a RENEKER DEFECT is a defect which passes along the polymer chain without causing any change in the relative position of the chain in the crystal aggregate.

(ii) NETWORK DEFECTS occur when the lattice atom positions are randomly displaced from their ideal lattice positions.

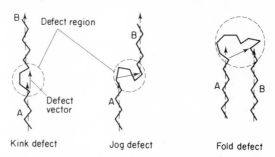

Kink defect   Jog defect   Fold defect

Fig. 26.33   Schematic representation of point defects

Bibliography: 1144, 1427.
References: 5188, 5805, 6021, 6695.

26.12 DETERMINATION OF THE MELTING POINT OF A CRYSTALLINE POLYMER

The MELTING POINT (T_m) (cf. Section 32.7) of a crystalline polymer can be measured by one of the two following methods:

(i) the dilatometric method (cf. Section 31.4). The volume change is plotted versus the increasing (Fig. 26.34 (a)) or decreasing (Fig. 26.34 (b) temperature. The melting point is then given by the temperature at which the heating and cooling curves diverge. The rate at which temperature is increased or decreased affects the determination of the melting point. The apparent melting point is decreased if the heating rate is very high. A rate of about 1 °C per hour is recommended.

For accurate determination of T_m it is necessary to plot the apparent melting point (observed) $(T_m)_{obs}$ versus the crystallization temperature (T_c) (Fig. 26.35). The true melting point (T_m) can be found as the point at which the line $T_m = T_c$ crosses with the line $(T_m)_{obs}$ versus T_c.

(ii) The birefringence method. The melting point (T_m) is determined as the observed temperature at which spherulite birefringence (cf. Section 35.3) disappears on heating.

The T_m values obtained by the birefringence method are always lower than the dilatometrically determined values (even up to 10 °C).

Bibliography: 501, 1324, 1396, 1427.

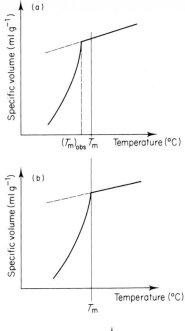

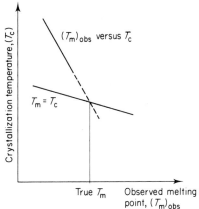

Fig. 26.34 Specific volume change versus temperature. (a) Crystallization during cooling; (b) melting during heating

Fig. 26.35 Method for determination of true T_m

26.13 DEGREE OF CRYSTALLINITY

The ABSOLUTE DEGREE OF CRYSTALLINITY cannot be unambiguously defined, because of existing crystal defect structures in crystalline regions of polymers (cf. Section 26.11).

The definition of the apparent degree of crystallinity assumes that a given polymer sample can be considered as a two-phase system:

(i) an ideal perfectly ordered crystalline phase (region);

(ii) an ideal disordered liquid-like amorphous phase (region).

The APPARENT DEGREE OF CRYSTALLINITY (commonly called the DEGREE OF CRYSTALLINITY or the INDEX OF CRYSTALLINITY (χ_c) (expressed in per cent) can be measured by the following methods:

   (i) IR spectroscopy (cf. Section 15.5);
  (ii) NMR spectroscopy (cf. Section 20.13);
 (iii) inverse chromatography (cf. Section 24.3.2);
 (iv) X-ray diffraction (cf. Section 28.11);
  (v) density measurements (cf. Section 31.8);
 (vi) DTA–DSC methods (cf. Section 34.10).

The APPARENT DEGREE OF CRYSTALLINITY of a given polymer sample, measured by different physical methods, such as density, X-ray diffraction, infrared absorption, nuclear magnetic resonance (NMR), specific heat, specific enthalpy, etc., can differ very considerably. For that reason experimental details should be quoted, e.g. 'the degree of crystallinity from NMR crystallinity'.

Bibliography: 714, 1109, 1396, 1427.

26.14 ETCHING TECHNIQUES

The ETCHING TECHNIQUE, used to reveal defects in crystals, is a method based on the greater chemical reactivity or solubility of less perfectly crystallized material, and even of amorphous parts of samples.

There are several different etching techniques:

   (i) gas discharge in oxygen or other gases at pressures of 10^{-3}–10^{-4} torr (1 torr = $\frac{1}{760}$ atm = 133.322 N m^{-2} (SI)) (cf. Section 39.4);
  (ii) ozonolysis (cf. Section 39.6);
 (iii) oxidation by nitric acid (90%), fumic·nitric acid (containing $\sim 15\%$ free NO_2), or permanganic reagent;
 (iv) Solvent–non-solvent mixtures;
  (v) hydrolysis;
 (vi) enzymic attack.

By properly adjusting the reaction conditions, it is frequently possible to reveal details of the defect structure of a crystalline sample by progressive removal of amorphous or poorly crystallized material.

The solvent etching technique using a boiling mixture of solvents is the only method which removes molecules as a whole without breaking the covalent bonds.

Bibliography: 86, 1427.
References: 2197, 2264, 2273, 2292, 2619, 3152, 3258, 3334, 3349, 3622, 4209, 4354, 4512, 4515, 4518, 4521, 4539, 4633, 4759, 4905, 5018, 5308, 5675, 5748, 5754, 5792, 5794, 5902, 5981, 6503, 6770, 7008, 7027, 7028, 7112, 7189.

Chapter 27

Electron Microscopy

General bibliography: 309, 557, 559, 589, 597, 679, 875, 981, 1108, 1151, 1280, 1412.

27.1 PRINCIPLES OF MICROSCOPY

The unaided eye is incapable of perceiving detail finer than about 0.1 mm in size.

The MICROSCOPE is an instrument which produces enlarged images of fine detail. The wave nature of light sets a fundamental limit to the fineness of detail which the light microscope can resolve.

The RESOLVING POWER is the finest detail which a microscope can resolve.

The IMAGE RESOLUTION is the finest detail which is reproduced in a given image.

For various reasons, the image resolution can be larger than the resolving power of the instrument (microscope). Thus the resolving power, as the term suggests, expresses the capability, theoretical or practical, of the instrument to resolve, while the resolution describes the result actually obtained.

The basic optical system for the microscope contains a series of condenser lenses (Fig. 27.1), which produces a series of images of progressively increased size.

The BRIGHTNESS of the final image depends upon the intensity of the radiation from the object entering the objective lens, and the total magnification (M). Since the brightness falls in proportion to $1/M^2$ it is important to collect as much of the illuminating radiation as possible to give a visibly bright image. To achieve this end the ANGULAR APERTURE OF THE OBJECTIVE, that is, the angle of the cone of radiation accepted by the lens, must be large. This can be defined in terms of the semi-angle (Θ) of the cone of light leaving each point in the object and entering the objective lens (the ANGLE OF ACCEPTANCE OF THE LENS).

The resolving power of a microscope is limited by the wave nature of the radiation which it employs. The minimum spacing (d) which can be resolved by a good microscope is of the order of the wavelength of the radiation, and is given by

$$d = \frac{\lambda}{n \sin \Theta},$$ (27.1)

474

where

λ is the wavelength of the radiation in the space between object and objective lens,

Θ is the angle of acceptance of the lens,

n is the refractive index of the medium between specimen and lens,

$n \sin \Theta$ is called the NUMERICAL APERTURE OF THE OBJECTIVE.

For the best resolution, large values of the angle of acceptance of the lens (Θ) and short wavelength radiation are required.

Figure 27.1 illustrates image formation in three basic microscopes: optical, transmission electron, and scanning electron.

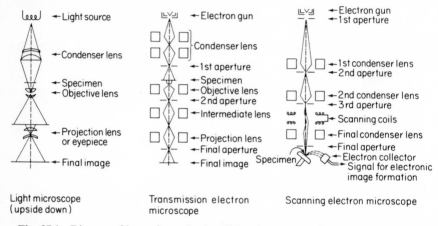

Fig. 27.1   Diagram of image formation in a light microscope, and a transmission electron microscope, and a scanning electron microscope. (From B: 66; reproduced by permission of John Wiley & Sons)

27.2 TRANSMISSION ELECTRON MICROSCOPY

TRANSMISSION ELECTRON MICROSCOPY (TEM) is a technique analogous to that of visible light microscopy but which applies an electron beam instead of a light beam, and where electrostatic and/or electromagnetic lenses replace the usual glass ones.

The transmission electron microscope is built of the following parts (Fig. 27.1):

(i) The electron gun, usually made of a tungsten filament, which produces thermal electrons. These electrons are accelerated by applying a high voltage (50–100 kV, and even 1 mV) and the electron beam is focused on the sample with aid of a series of electrostatic and/or electromagnetic lenses.

(ii) The electron beam penetrates the specimen, which is placed in a special holder. The area of the sample is limited to 2 mm or less in diameter and its thickness to a few thousand ångstroms (for instruments with an accelerating voltage of 100 kV).

(iii) The image is formed by two or more additional electrostatic (Fig. 27.2) and/or electromagnetic lenses, and is observed on a fluorescent screen or recorded photographically. The strength of the lenses can be controlled by varying the current through them; in this way it is possible easily and rapidly to change the magnification.

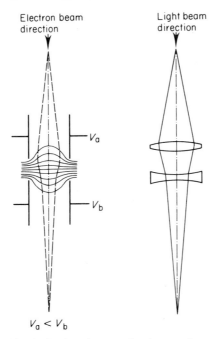

Electron beam direction

Light beam direction

V_a

V_b

$V_a < V_b$

Fig. 27.2 An electrostatic electron lens and its optical analogue. Cylinders are at different potentials V_a and V_b

(iv) The microscope column (from the electron gun to the fluorescent screen) is evacuated to 10^{-5} torr (1 torr $= \frac{1}{760}$ atm $= 133.322$ N m^{-2} (SI)) or less by rotational and oil diffusion pumps because of strong scattering of electrons by air.

In modern commercial electron microscopes there is the possibility of observing both the image of the sample and its diffraction pattern by changing the power of the intermediate lens.

The resolving power (d) of an electron microscope can easily be calculated by applying the equation

$$d = \frac{\lambda}{n \sin \Theta},$$ (27.2)

where
λ is the wavelength of the electron beam (in ångstroms),

$$\lambda = (1.5/V)^{1/2},$$ (27.3)

V is the applied high voltage for electron acceleration (in volts),

n is the refractive index of the medium between specimen and lens (for a vacuum $n = 1.000$),

Θ is the angle of acceptance of the lens.

The best resolution in electron microscopy is of the order of 2–5 Å.

The magnification is determined by photographing a diffraction grating with known spacing of the ruled lines, or a monodisperse polystyrene latex of known particle diameter.

Bibliography: 309, 502, 557, 559, 563, 597, 715, 788, 1218, 1280, 1412.

27.3 SAMPLE PREPARATION FOR TRANSMISSION ELECTRON MICROSCOPY

Samples for examination in the electron microscope have to be mounted on very thin films, because of the low penetrating power of electrons in the 50–100 kV range. These films must be made from materials with a high transparency to electrons and should not be more than about 200 Å thick. The films are mounted on discs usually made of copper or nickel and containing a number of apertures. These discs are known as SPECIMEN SUPPORT GRIDS, and are available in different designs and sizes (Fig. 27.3). The standard grids are

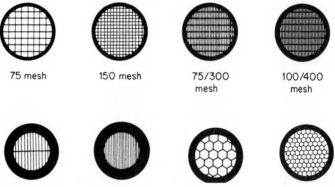

| 75 mesh | 150 mesh | 75/300 mesh | 100/400 mesh |

Fig. 27.3 Metal grids. (From B: 1474; reproduced by permission of Ladd)

3 mm in diameter, 25–30 μm thick, and contain a square mesh (100–200 per inch or 4–8 per millimetre) with windows 90 μm square and 35 μm bars. Most grids have one side shiny and the other matt in appearance. Support films are mounted on the matt surface for maximum adhesion. Figure 27.4 shows a specimen resting on a support film and carried on a metal grid.

Specimen support films can be prepared from many different materials, such as polymers (Formvar—polyvinyl formal, and collodion—nitrocellulose), or evaporated carbon, or silicon oxide (SiO_2). The specimen support film should be extremely thin (less than 100 Å), amorphous, and structureless.

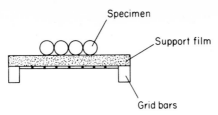

Fig. 27.4   Specimen resting on a support
film carried on a metal grid

The most simple method of preparation of plastic film coating on a grid is shown in Fig. 27.5. This technique uses a Petri dish fitted with a drain-tube and tap. A dilute solution of polymer in a suitable solvent is spread out on a water surface. After solvent evaporation, a thin polymer film is placed on the grid by

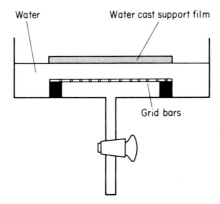

Fig. 27.5   A device for the preparation
of a support film on a grid

removing the water. Plastic films are now being superseded by evaporated carbon films.

The apparatus used for the carbon evaporation process is shown in Fig. 27.6. Two pointed carbon rods are arranged in a vacuum bell-jar, one fixed and the other sliding in a silica tube under a spring pressure. An alternating current (20–50 A) from a small transformator is passed through the rods and results in intense heating in the region of contact and evaporation on the target platform. The carbon is evaporated on to a substrate which can easily be stripped from a glass side. Suitable substrates are Bedacryl 122X (supplied as a 40% solution in xylene), boron oxide, or glycerol.

Almost all polymer specimens are too thick for direct observation in the electron microscope. The technique of cutting thin sections is applicable to all types of polymer materials and fibres. Ultramicrotomes of a variety of designs are commercially available which can cut sections as thin as 200–300 Å. When the microtome is operated the specimen is brought slowly past the glass or diamond knife and is then returned to its original position for the next cut.

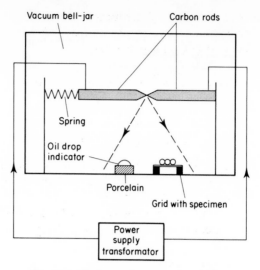

Fig. 27.6   Carbon rod evaporation device

The speed of cutting can be varied by adjusting the motor drive. The knife should be adjusted at the right clearance angle (Fig. 27.7).

Rubbery samples can be cut by cooling the knife and the sample to liquid air temperatures, where the rubbery phase has become glassy. A miniature water

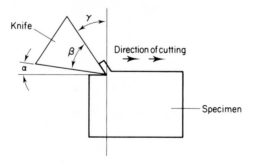

Fig. 27.7   Cutting thin sections of specimen by the knife and definition of the angles: α is the clearnace angle, β is the knife angle, γ is the rake angle

bath is fixed to the back of the cutting edge, so that the sections are drawn off on the surface of the water (or ice) as they are cut. They are then placed on the specimen-supporting film and grid under a binocular microscope. It is also possible to 'shadow' the specimen with a heavy metal such as chromium, tungsten, gold, platinum, or uranium in order to increase contrast.

Unsaturated polymer particles (e.g. polydienes) can be stained by treating cut specimens with osmium tetroxide (OsO_4) vapour or 2% osmic acid (for 1–2 days) in order to improve the electron contrast. This method allows visualization of details of the internal structure of unsaturated polymers. Osmium tetroxide reacts with the unsaturated double bonds and oxidizes them; reduced metallic osmium is deposited and gives image contrast.

Bibliography: 189, 516, 557, 592, 715, 1018, 1106, 1169.
References: 2126, 2425, 2466, 2836, 3006, 3270, 3345, 3863, 4064, 4236, 4534, 4408, 4446, 4450–4453, 4539, 4755, 5115, 5117, 5544, 6056, 6131, 6258, 6828, 7117.

27.4 SHADOW CASTING

In order to shadow a specimen on a supporting film and grid, one must place it in a bell-jar at a distance of 10–20 cm from the evaporating source, which may be a V-shaped filament or a tantalum boat (Fig. 27.6).

A shadow (free of metal) is formed on the lee-side of a specimen, whilst the front side collects a greater amount of metal (Fig. 27.8). The angle of shadowing

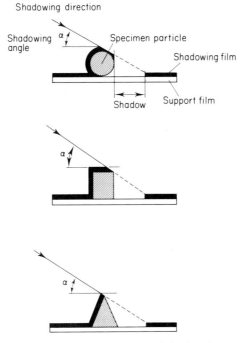

Fig. 27.8 Schematic representation of shadowed particles of different shape

varies according to the shape of the specimen. For large objects the angle of shadowing is of 45°, whereas for small objects it should be 10–12°.

A shadow-cast specimen, when photographed, is usually projected or printed as negative; the shadows appear dark and the region covered by metal are light (Fig. 27.9).

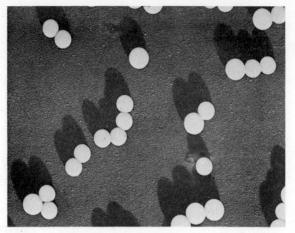

Fig. 27.9 Poly(styrene) latex in solution after being freeze-dried and shadowed. (Reproduced by courtesy of S. Claesson)

Macromolecular solutions can be studied by rapid freezing and use of a freeze-drying method at low temperatures, employing the very simple arrangement shown in Fig. 27.10. The specimen can be stabilized at any temperature from that of liquid nitrogen to room temperature, measured by a thermocouple.

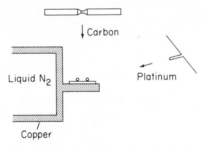

Fig. 27.10 Apparatus for shadowing frozen specimens. (From R: 2865; reproduced by permission of *Makromolekulare Chemie*)

The solvent sublimes after evacuation of the bell-jar, and the frozen macromolecules maintain the same position which they had in solution. The frozen specimen is then shadowed, thus allowing one to obtain three-dimensional pictures of the frozen macromolecule (Fig. 27.11).

Bibliography: 189, 557, 589, 715.
Reference: 2865.

27.5 REPLICA TECHNIQUES

Bulk specimens cannot be observed directly in the transmission electron microscope and so replica techniques must be applied.

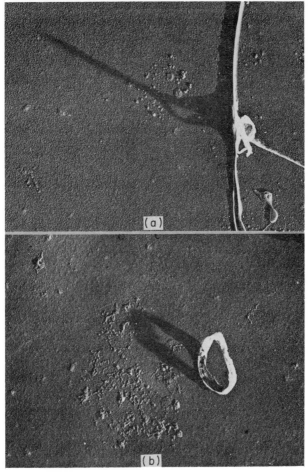

Fig. 27.11 (a) Odd-shaped particle from chill haze prepared by the freeze-drying and shadowing technique. (b) A chill haze particle standing on edge. (From R: 2865; reproduced by courtesy of S. Claesson with the permission of *Makromolekulare Chemie*)

The principle of the replica technique is to transfer the surface topography of the solid body to a thin film which can be observed in a transmission electron microscope. There are many variants of the replica technique, but generally they produce one of two types of replica (Fig. 27.12):

(i) The single-stage method. In this method the replicas are made by depositing a replicating material directly on the specimen, separating the two, and examining the replica in the electron microscope.

(ii) The two-stage method. This method consists of making a preliminary replica of the specimen surface in plastic material, coating the surface of this replica with the final replicating material, separating the two, and examining the final replica in the electron microscope.

482

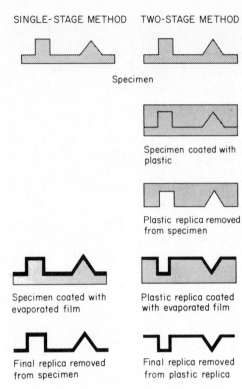

SINGLE-STAGE METHOD   TWO-STAGE METHOD

Specimen

Specimen coated with
plastic

Plastic replica removed
from specimen

Specimen coated with   Plastic replica coated
evaporated film   with evaporated film

Final replica removed   Final replica removed
from specimen   from plastic replica

Fig. 27.12   Two types of replica method

The contrast produced by many replicas in the electron microscope is often
very low and it may be necessary to increase it by means of the shadow-casting
technique.

Bibliography: 189, 557, 589, 715.
References: 6256, 6257.

27.6 DETERMINATION OF MOLECULAR WEIGHT DISTRIBUTION (MWD) CURVES BY ELECTRON MICROSCOPY

The molecular weight distribution of various polymers can be estimated from
electron micrographs of their molecules. The number of molecules needed to
construct the distribution is between 200 and 300. Micrographs of a single
molecule can be improved by shadowing with a platinum–carbon mixture
(Fig. 27.13). From the shadow dimensions it is possible to calculate the volume
of the molecule. Knowing the macroscopic density of the polymer, the correct
mass of a single molecule can be obtained. The lower limit of an exact determination of the dimensions is a molecular weight of $c.\ 10^5$.

References: 2252–2254, 3034, 4680–4683, 5935, 5999, 6040, 6127, 6392.

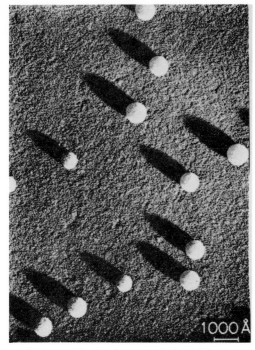

Fig. 27.13 Typical shadows given by poly(styrene) latex single molecules observed under the electron microscope, (From R: 4680; reproduced by courtesy of G. Koszterszitz, W. K. R. Barnikol, and G. V. Schulz with the permission of *Makromolekulare Chemie*)

27.7 APPLICATIONS OF TRANSMISSION ELECTRON MICROSCOPY IN POLYMER RESEARCH

Transmission electron microscopy (TEM) can be applied to the study of

(i) internal micromorphology;
(ii) polymer lattices;
(iii) polymer networks;
(iv) pore size distribution;
(v) molecular weight distribution (cf. Section 27.6).

Bibliography: 309, 592, 1427.
References: 2273, 2305, 3006, 3173, 3434, 3863, 4161, 4354, 4408, 4481, 4570, 4610, 4749, 4852, 5117, 5245, 5544, 5798, 5837, 6768, 6828, 6917, 7065.

27.8 SCANNING ELECTRON MICROSCOPY

SCANNING ELECTRON MICROSCOPY (SEM) is a technique which forms an image of a microscopic region of the specimen surface. An electron beam from 5 to 10 nm in diameter is scanned across the specimen. The interaction of the electron beam with the specimen produces a series of phenomena (Fig. 27.14) such as

   (i) backscattering of electrons of high energy;
  (ii) secondary electrons of low energy;
 (iii) absorption of electrons, measurable as specimen current;
 (iv) X-rays;
  (v) visible light (cathodoluminescence).

Any one of these signals can be continuously monitored by detectors. The detector signal is amplified and used to modulate the brightness of a cathode ray tube, the beam of which is scanned in synchrony with the electron beam impinging upon the specimen. A correspondence between each scanned point at the specimen surface and a corresponding point on the cathode ray tube screen is thus established. The area scanned upon the specimen is very small in comparison with the corresponding area on the cathode ray tube screen. The magnification of the image on the screen (or photograph) is the ratio of a distance on the screen and the corresponding distance on the specimen.

The scanning electron microscope is built of the following parts (Fig. 27.1):

   (i) The electron gun, usually made of a tungsten filament, which produces a flow of thermal electrons. These electrons are accelerated by applying a high voltage (1–30 kV) and the electron beam is focused with the aid of a series of electromagnetic lenses.

  (ii) The system of two-stage electromagnetic lenses is used to demagnify the electron beam diameter to 5–10 nm at the specimen. A set of electromagnetic scanning coils, located within the bore of the final condenser lens, scans the beam in a rectangular raster on the specimen surface.

 (iii) Detectors may detect electrons, X-rays, or cathodoluminescent light (photons) (Fig. 27.14). One such detector (depending on the manufacturer) is placed in the specimen chamber (the secondary electron emission is the principal mode of imaging). The signal from a detector is amplified and fed to a cathode ray tube. As the electron beam scans the specimen surface, the information collected modulates the raster of a cathode ray tube, which is scanned in synchrony with the electron beam. Each point on the cathode ray tube raster corresponds to a point on the specimen surface, and the cathode ray tube intensity varies with the intensity of the signal generated by the electrons impinging on the specimen surface.

 (iv) The microscope column (from the electron gun to a specimen) is evacuated to 10^{-5} torr (torr $= \frac{1}{760} = 133.322$ N m^{-2} (SI)) by rotational and oil-diffusion pumps because of strong scattering of electrons by air.

The MAGNIFICATION is defined as the ratio of the length of one side of the cathode ray tube screen relative to that of the area scanned on the specimen surface. Most scanning electron microscopes have a magnification range of $\times 20$ to $\times 100\,000$, with the optimum useful magnification being $\times 20\,000$ to $\times 50\,000$ depending on the type of specimen and the construction of the instrument. Low magnifications allow easy specimen orientation.

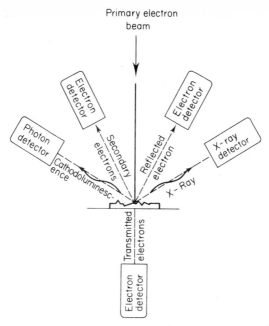

Fig. 27.14 Information that can be generated in the SEM by an electron beam striking the sample. (From B: 66; reproduced by permission of John Wiley & Sons)

The RESOLUTION of the scanning electron microscope is dependent on two parameters:

(i) the thickness of the electron beam;
(ii) the specimen area from which secondary electrons are emitted.

Commercially available instruments have a resolution of 6–10 nm (60–100 Å).
 The DEPTH OF FOCUS (DEPTH OF FIELD) (ΔF) of the scanning electron microscope is given by

$$\Delta F = \pm \frac{d}{2\alpha}, \tag{27.4}$$

where
 ΔF is the depth of focus,
 d is the resolution of the electron optical system,
 2α is the aperture angle.
The depth of focus is usually 300–600 times better than that of an optical microscope. The large depth of focus, together with a great mobility of the specimen stage, allows one to obtain a three-dimensional image.

Bibliography: 66, 163, 309, 593, 598, 983, 1294.
Reference: 4558.

27.9 SAMPLE PREPARATION FOR SCANNING ELECTRON MICROSCOPY

If the specimen is not a good conductor, it should be coated with a thin layer (100–500 Å) of conducting material (gold, silver, carbon, and gold–palladium). This coating is carried out by placing the specimen in a high-vacuum evaporator and vaporizing a suitable metal held in a heated tungsten basket. Charging phenomena can be reduced by coating the specimen with organic antistatic agents from solution.

27.10 ELECTRON CHANNELLING CONTRAST

Electrons which penetrate a crystal travel along certain CHANNELLING DIRECTIONS (Fig. 27.15). In these particular directions, a small fraction

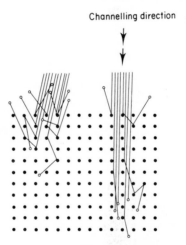

Fig. 27.15 Schematic representation of the electron channeling process in a crystal

(5% or less) advance up to 100 nm before being scattered. The remainder of the electrons scatter at normal distances. If the electron beam is made to rock through all directions within a conical solid angle, the re-emerging backscattered and secondary electrons produce a pattern, with contrast up to 5%, known as an ELECTRON CHANNELLING PATTERN.

The electron channelling contrast technique can be used to obtain information on the orientation and degree of perfection of the crystalline lattice.

Bibliography: 163 598.

27.11 APPLICATIONS OF SCANNING ELECTRON MICROSCOPY IN POLYMER RESEARCH

Scanning electron microscopy (SEM) can be applied to the study of

(i) the morphology of polymers, copolymers, block copolymers, and polyblends;
(ii) the microstructure of two-phase polymers;
(iii) polymer networks;
(iv) rough surfaces;
(v) fractured surfaces;
(vi) adhesives, and especially failures thereof;
(vii) filled and fibre-reinforced plastics;
(viii) organic coatings (pigment dispersion, flow, and adhesion of binder to pigment and substrate, weathering because of mildew, chalking, blistering, and cracking, or swelling of paint films in water);
(ix) foamed polymers;
(x) the extrusion or moulding performance of plastics.

Bibliography: 266, 388, 531, 541, 592, 1068, 1363, 1427.
References: 2289, 2305, 2421, 3057, 3060, 3287, 3377, 3391, 3683, 3706, 3771, 3967, 4072, 4316, 4570, 4786, 4854, 5353, 5687, 5700, 5906, 6089, 6374, 6450, 6616, 7175.

Chapter 28

X-ray Diffraction Analysis

General bibliography: 16, 57, 298, 309, 342, 669, 704, 758, 820, 846, 979, 1023, 1131, 1254, 1255, 1328.

28.1 PROPERTIES OF X-RAYS

X-RAYS are generated when high-energy electrons impinge on a metal target (iron, copper, or molybdenum). The X-ray beam possesses a spectrum (Fig. 28.1). The X-RAY SPECTRUM consists of two parts (cf. Table 28.1):

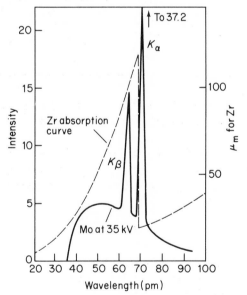

Fig. 28.1 Zirconium absorption curve (broken line) superimposed by the 35 kV molybdenum emission line (continuous line). (From B: 758; reprinted by permission of John Wiley & Sons)

(i) A broad band responsible for the continuous radiation.

(ii) Characteristic lines, called K_α and K_β. The K_α line is composed from two components K_{α_1} and K_{α_2}, separated by a very small wavelength interval.

Table 28.1 Wavelengths of K emission series and K absorption edges (from B: 758; reproduced by permission of John Wiley & Sons)

Element	Z	K_{α_2}	K_{α_1}	K_{β_1}	K_{β_2}	K absorption edge
Fe	26	1.939 91	1.935 97	1.756 53		1.743
Cu	29	1.544 33	1.540 51	1.392 17	1.381 02	1.380
Mo	42	0.713 54	0.709 26	0.632 25	0.620 99	0.619

K_α and K_β lines can be separated by absorbing filters made from metals such as zircon, nickel, or manganese (Table 28.2). Nickel-filtered CuK_α radiation (1.541 78 Å) is the most useful X-ray radiation for polymer studies.

Monochromatic radiation can be obtained by inserting a crystal monochromator in the direct X-ray beam.

Table 28.2 β-Filters for target elements (from B: 758; reproduced by permission of John Wiley & Sons)

Target element	Filter	Length (mm)	Density (g cm^{-2} (CGS))	Per cent loss KP
Mo	Zr	0.081	0.053	57
Cu	Ni	0.015	0.013	45
Fe	Mn	0.011	0.008	38

The amount of X-ray radiation absorbed by a sample is determined by the absorption coefficient of the substance concerned for X-rays of the given wavelength.

The LINEAR ABSORPTION COEFFICIENT (μ) is defined as the natural logarithm of the ratio of the intensity of the incident X-ray beam on a specimen to the intensity of the X-ray beam after it has penetrated 1 cm of the sample material, and is given by

$$I/I_0 = e^{-\mu l}, \tag{28.1}$$

where

I_0 is the intensity of the X-ray incident beam,
I is the intensity of the transmitted X-ray beam,
μ is the linear absorption coefficient,

$$\mu = \frac{1}{l} \log \frac{I_0}{I}, \tag{28.2}$$

l is the thickness of the specimen.

The MASS ABSORPTION COEFFICIENT (μ_m) is obtained by dividing the linear absorption coefficient (μ) by the density of the material (ρ):

$$\mu_m = \mu/\rho \tag{28.3}$$

(in square centimetres per gram (CGS) or square metres per kilogram (SI)). The mass absorption coefficient (μ_m) is independent of the physical and chemical state of the substance, which is not true of μ:

$$\mu_m = CNZ^4 \lambda^n / A \qquad (28.4)$$

where

N is the Avogadro number (cf. Section 40.3),

Z is the atomic number,

A is the atomic weight of the absorbing element,

λ is the wavelength,

n is the exponent—between 2.5 and 3.0,

C is a constant, which is approximately the same for all elements.

The mass absorption coefficient (μ_m) determined by eq. (28.4) is called the CROSS-SECTION of an atom.

28.2 DIFFRACTION AND SCATTERING

When an X-ray monochromatic beam impinges on a sample, two processes can be observed:

 (i) If the sample has a periodic structure (crystalline regions) the X-ray is scattered COHERENTLY. This process occurs without change of wavelength and without loss of phase relationship between the incident and scattered rays. This process is called a DIFFRACTION X-RAY EFFECT and is measured by wide-angle X-ray diffraction.

(ii) If the sample has an unperiodic structure which possesses different electronic densities (crystalline and amorphous regions), the X-ray is scattered INCOHERENTLY (COMPTON SCATTERING). This process occurs with change of wavelength and change of phase relationship between the incident and scattered rays. This process is called a DIFFUSE X-RAY DIFFRACTION EFFECT (simply SCATTERING) and is measured by small-angle X-ray scattering.

28.3 GEOMETRY OF DIFFRACTION

A crystal lattice can be considered as a family of planes extending through the atoms of the lattice and defined by a triplet (hkl) of whole numbers which are called MILLER INDICES. The incident waves are assumed to be reflected by these planes. The spacing $d_{(hkl)}$ between adjacent planes with Miller indices (hkl) can be calculated from the angle 2θ using the BRAGG equation:

$$n\lambda = 2d_{(hkl)} \sin \theta, \qquad (28.5)$$

where

n is an integer $0, 1, 2, 3, \ldots$, called the order,

λ is the X-ray wavelength,

$d_{(hkl)}$ is the distance between adjacent planes in the crystal,

θ is one-half the angle of deviation of the diffracted rays from the incident X-rays.

The geometry of the diffraction of X-rays in a crystal is given in Fig. 28.2.

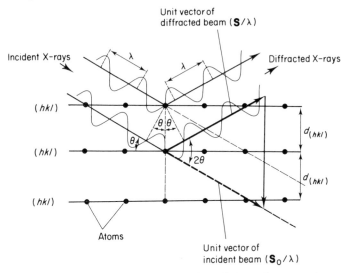

Fig. 28.2   Geometry of the diffraction of X-rays in a crystal

The ANGULAR DISTANCE between several reflections in the diffraction pattern can be evaluated with the help of the reciprocal lattice transformation.

The RECIPROCAL LATTICE can be constructed from the data of the real crystal lattice (Fig. 28.3) and has the following properties:

(i) each point in the reciprocal lattice represents a family of net planes (hkl) in the crystal lattice;

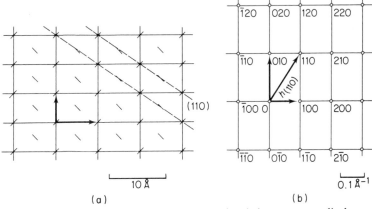

Fig. 28.3   (a) Crystal lattice of (001) plane. The chains are perpendicular to the drawn plane. The dotted lines represent the lines of intersection of the (110) plane with the (001) plane. (b) Reciprocal lattice (001) plane. (From B: 439; reproduced by permission of John Wiley & Sons)

(ii) the vector $\mathbf{h}(hkl)$ from the origin to the point (hkl) of the reciprocal lattice is normal to the (hkl) plane of the crystal lattice;

(iii) the length of the vector $\mathbf{h}(hkl)$ is equal to the reciprocal of the spacing $d_{(hkl)}$ between the planes of the crystal lattice.

When the reciprocal lattice has been constructed, the directions of the diffracted ray can be found by the following geometrical construction (Fig. 28.4):

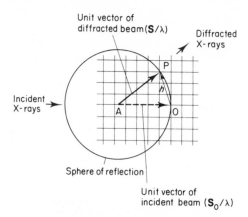

Fig. 28.4 Geometrical condition for diffraction in reciprocal space. (From B: 439; reproduced by permission of John Wiley & Sons)

(i) From the origin (O) of the reciprocal lattice the vector $\mathbf{s}_0/\lambda$ is drawn along the same line as the unit vector $\mathbf{s}_0$ of the incident wave, but in the direction opposite to it.

(ii) The origin of the vector $\mathbf{s}_0/\lambda$ is assumed to be situated at a certain point A, not necessarily a lattice point, and a sphere of radius $1/\lambda$ is described around this point.

(iii) If this SPHERE OF REFLECTION which passes through O also passes through a lattice point P, the vector $\overrightarrow{OP}$ is a radius vector $\mathbf{h}$ of the reciprocal lattice (cf. Fig. 28.4).

(iv) The direction of the diffracted beam is given by the vector $\overrightarrow{AP}$.

The sphere of reflection allows one to determine the system of planes giving an interference maximum if the wavelength of the incident radiation is known.

Around every lattice point there exists a so-called INTENSITY REGION, which depends on the following:

(i) The size and shape of the crystal are determined by two factors:

    (a) the STRUCTURE FACTOR, which describes the arrangement of the scattering atoms within a unit cell of the crystal lattice;

    (b) the LATTICE FACTOR, which describes the arrangement of the single cells within the whole crystal.

(ii) The degree and kind of lattice distortions in the crystal.

There are two general methods for measuring X-ray diffraction and scattering:

(i) photographic techniques, which utilize photographic films;
(ii) a diffractometry (counter) technique, which utilizes different types of X-ray detectors (counter diffractometers).

Bibliography: 16, 165, 223, 439, 541, 704.
References: 2282, 2340, 2417, 2488, 2815, 2921, 3140, 3337, 3341, 3663, 4047, 4176, 4225, 4261, 4470, 4590, 4625, 4649, 4776, 4777, 5166, 5228, 5229, 5324, 5369, 6161, 6669, 6697, 6698, 6721, 7091, 7129, 7267, 7269.

Extensive crystallographic data for various polymers are collected in The Polymer Handbook (B: 928).

28.4 WIDE-ANGLE X-RAY DIFFRACTION (WAXS)

Bulk and moulded polymers may be considered as randomly oriented semi-crystalline specimens. Wide-angle X-ray diffraction (WAXS) in such samples will yield a typical powder-diffraction pattern (Fig. 28.5).

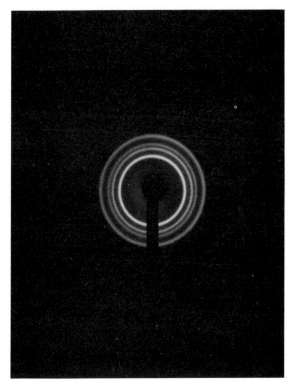

Fig. 28.5 Powder diffraction pattern of partially crystalline isotactic poly(propylene). (Reproduced by courtesy of M. Takayanagi)

Many small crystallites with many different crystal plane orientations yield diffraction in the form of concentric cones (Fig. 28.6). The diffraction effects recorded on a flat film consist of a number of relatively sharp concentric circles superimposed on a background of incoherent scatter and at least one noticeable

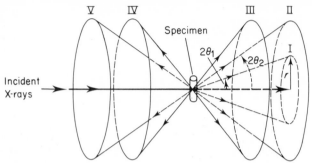

Fig. 28.6 Diffraction of X-rays by a randomly oriented poly-crystalline sample. (From B: 758; reproduced by permission of John Wiley & Sons)

amorphous halo (Figs 28.5 and 28.7). The diffraction pattern of an amorphous sample differs significantly from that of a crystalline sample (Fig. 28.7). A well-crystallized sample shows two regions of reflection (Fig. 28.6):

(i) forward-reflection ($0° < 2\theta < 90°$)—concentric cones I, II, and III;
(ii) back-reflection ($90° < 2\theta < 180°$)—concentric cones IV and V.

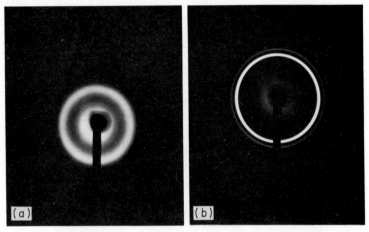

Fig. 28.7 Diffraction pattern of poly(ethylene): (a) amorphous, (b) partially crystalline. (Reproduced by courtesy of M. Takayanagi)

There are two beam transmission methods for measuring the wide-angle diffraction of powder patterns:

(i) The forward-reflection measuring method (the LAUE TECHNIQUE) (Fig. 28.8). In this method a plane film is placed a few centimetres (~ 5 cm)

WIDE-ANGLE X-RAY
DIFFRACTION

SMALL-ANGLE X-RAY
SCATTERING

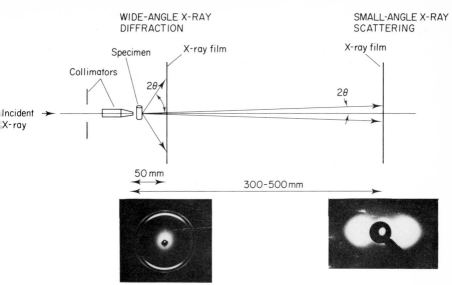

Fig. 28.8 Differences between wide-angle X-ray diffraction (WAXS) and small-angle X-ray diffraction (SAXS) techniques and their patterns. (From B: 309; reproduced by permission of John Wiley & Sons.) (WAXS and SAXS patterns reproduced by courtesy of T. Hashimoto)

from the specimen, perpendicularly to the incident monochromatic X-ray beam. Exposures of several hours are required. Instead of photographic recording, an X-ray detector is used which counts the separate X-ray photons in a diffracted beam. In this method the intensity of the scattered X-ray beam versus the scattering angle 2θ is measured.

(ii) The forward- and back-reflection measuring method (the DEBYE–SCHERRER TECHNIQUE) (Fig. 28.9). In this method a narrow beam of monochromatic X-rays irradiates a small cylindrical sample, and short

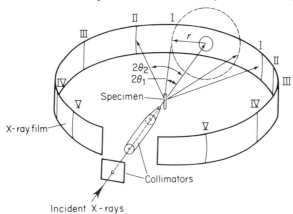

Fig. 28.9 Schematic representation of the Debye–Scherrer technique. (From B: 758; reproduced by permission of John Wiley & Sons)

sections of the diffraction cones are intercepted by a strip of film (Fig. 28.10) which is placed cylindrically around the sample (Fig. 28.9). This method is especially useful for qualitative analysis, measurement, and comparison of interplanar spacing ($d_{(hkl)}$). From measurements of the line

Fig. 28.10 Strip-film of pattern of poly(ethylene) obtained by the Debye–Scherrer technique. (This sample is the same as in Fig. 28.7.) (Reproduced by courtesy of M. Takayanagi)

position on properly calibrated films, the interplanar spacing ($d_{(hkl)}$) can be calculated using the Bragg equation (eq. (28.5)):

$$d_{(hkl)} = \frac{\lambda}{2}\frac{1}{\sin\theta}. \qquad (28.6)$$

Instead of photographical recording one can use an X-ray detector to determine the line intensity.

Single crystals and drawn fibres and films can be investigated by the CRYSTAL ROTATION METHOD (FIBRE-DIFFRACTION TECHNIQUE) (Fig. 28.11). In this method the principal crystallographic axis is perpendicular to an

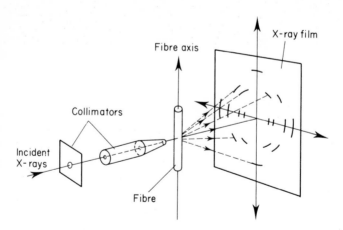

Fig. 28.11 Geometry of diffraction by a specimen with axial (fibre) orientation. (From B: 16; reproduced by permission of John Wiley & Sons)

incident monochromatic X-ray beam. The diffraction pattern (called the FIBRE PATTERN) (Fig. 28.12) is registered on cylindrical film coaxial with the crystal rotation axis. In the case of fibres no sample rotation is required, because many crystallites are already oriented.

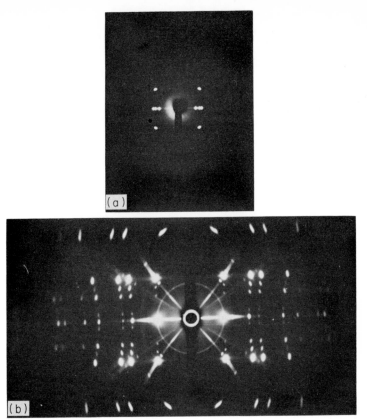

Fig. 28.12 Rotating X-ray diffraction pattern of: (a) axially oriented isotactic poly(propylene), (b) poly(tetraoxane). (Reproduced by courtesy of M. Takayanagi)

28.5 DUAL WAVELENGTH X-RAY DIFFRACTION

The dual wavelength X-ray diffraction technique employs a composite X-ray tube which simultaneously emits CuK_α (1.544 33 Å) and AlK_α (8.339 16 Å) radiation. A sample is irradiated with dual wavelength X-ray radiation, and a diffraction pattern for both wavelengths is registered on the photographic plate.

The diffraction of CuK_α radiation yields information on $d_{(hkl)} < 100$ Å, whereas the diffraction of AlK_α radiation gives data on $d_{(hkl)} > 100$ Å.

The dual wavelength X-ray diffraction technique provides valuable information for the study of fibres.

Bibliography: 611.

28.6 SMALL-ANGLE X-RAY SCATTERING (SAXS)

Small-angle (less that 2°) X-ray scattering (SAXS) has no dependence on the inhomogeneities of atomic dimensions that give rise to wide-angle X-ray diffraction (WAXS).

498

Small-angle X-ray scattering (SAXS) depends only on

(i) the alteration of crystalline and amorphous regions, which possess different electronic densities;
(ii) the presence of microvoids dispersed throughout the solid polymer matrix.

The intensity of small-angle scattering increases with the degree of contrast between the electron densities of the two or more kinds of region that produce the heterogeneity, e.g. in swollen polymers, where X-ray scatter intensity depends on the difference in the electron densities of the particles and solvent. A small-angle X-ray scattering (SAXS) pattern is shown in Fig. 28.8.

Bibliography: 16, 215, 309, 552, 704, 1053.
References: 4710, 5859, 6508.

28.7 INSTRUMENTATION FOR X-RAY DIFFRACTION AND SCATTERING MEASUREMENTS

28.7.1 X-ray Cameras

There are three basic types of X-ray camera:

(i) Flat-film cameras (transmission cameras) where film is placed a few centimetres (~ 5 cm) from the specimen (Fig. 28.8). The film is held in a light-tight cassette that is loaded in the dark room and then replaced in precisely its original position on the supporting bracket in the camera. Commercially produced flat-film cameras provide the possibility for mounting of the collimator, specimen, and film cassette separately.
(ii) Cylindrical-film cameras are required for measuring forward- and back-reflections using the Debye–Scherrer technique (cf. Section 28.4), and for detecting the diffraction pattern by the crystal rotation method (cf. Section 28.4).

In Debye–Scherrer cameras only a narrow film strip is used, in one of the following ways (Fig. 28.13):

(a) regular position,
(b) precision position,
(c) Straumanis asymmetric position,
(d) Wilson asymmetric position.

The Straumanis asymmetric position (c) permits the calculation of accurate values of the scattering angle (2θ) and the distance between adjacent planes in the crystal (d) for every line measured.

In the crystal rotation method a wide film is placed inside a cylindrical camera for registration of the complete diffraction pattern. Commercially produced X-ray spectrometers have a cylindrical camera fitted directly with a goniometer head in which an oriented sample is placed.

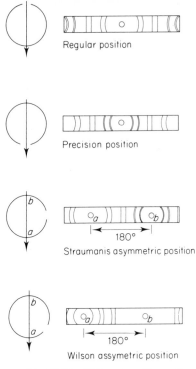

Fig. 28.13 Different positions of film in the Debye–Scherrer technique. (From B: 16; reproduced by permission of John Wiley & Sons)

(iii) Small-angle cameras, where a flat film is placed at fixed distances of 300–500 mm from a specimen (Fig. 28.14). Measurement of an undistorted small-angle scattering pattern requires collimation of the incident X-ray beam with a pair of fine pinholes. Because of very weak incident beam, exposures of several hours are required. In order to avoid air-scattered X-rays the cameras have to be evacuated.

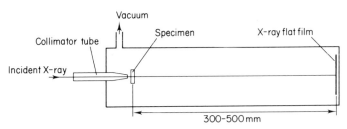

Fig. 28.14 Schematic representation of a small-angle scattering camera

The quantitative measurement of intensities from photographic patterns is made by microdensitometer (microphotometer), which

(i) scans the photographic pattern radially in order to measure the intensity distribution as a function of the scattering angle (2θ);
(ii) scans the photographic pattern circularly about the $2\theta = 0°$ point in order to determine the intensity distribution azimuthally.

Most of the commercially produced microdensitometers allow for enlargement of the pattern, easy positioning of the film in XY translational coordinates, and rapid choice of several aperture shapes and sizes, and are completely automated.

For the measurement of high resolution of the details of a diffraction pattern, fine-grained films (e.g. Eastman type KK, Eastman No Screen, and Ilford type G) are required.

Bibliography: 16, 57, 222, 644, 704, 758, 1023, 1254, 1255.
References: 2591, 2623, 3246, 3309, 3312, 3569, 4096, 4434, 4547, 4562, 4565, 4705–4707, 4711, 4712, 5069, 5359, 6267, 7124.

28.7.2 X-ray Diffractometer

An X-ray diffractometer (X-ray spectrometer) (Fig. 28.15) is built of the following parts:

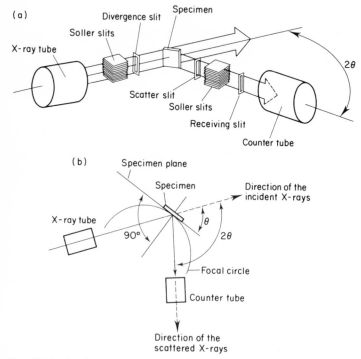

Fig. 28.15   (a) Schematic representation of a X-ray diffractometer. (b) Geometry of an X-ray diffractometer

(i) An X-ray source, which can be used in the form of a sealed X-ray tube or X-ray tubes with rotating anticathodes. The latest type of source produces higher intensity X-rays.

(ii) Lenses and mirrors cannot be used with X-rays. A collimated beam can be obtained from an X-ray tube with an extended target by passage through a bundle of metal tubes (collimator) or, if collimation in one plane is required, through the space between a stack of parallel metal sheets (Soller slits) (Fig. 28.15).

(iii) Crystal monochromators. A monochromatic beam can be obtained by reflecting (diffracting) the direct X-ray beam from the surface of a flat crystal (rock salt, fluorite, urea nitrate, or pentaerythritol) plate that has been cleaved or cut with its surface parallel to the diffracting planes.

(iv) A goniometer for mounting a specimen (Fig. 28.16). Goniometer heads can be divided into two types:
- (a) cross-type goniometer heads (normal type) permit parallel movement for adjustment of the orientation of the specimen;

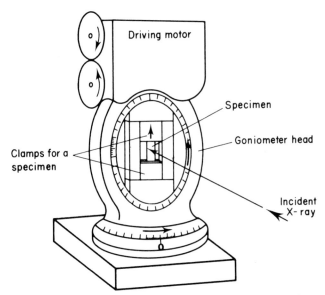

Fig. 28.16 Schematic representation of goniometer head

- (b) eucentric goniometer heads permit adjustment in such a way that the centres of the two axis arcs exactly coincide on the axial line of specimen rotation.

(v) X-ray counters. There are three types of counters for X-ray intensity measurements:
- (a) Geiger–Müller counters for measuring weak X-ray scattering intensity in small-angle X-ray scattering.

(b) Proportional counters, which produce an electrical pulse output proportional to the energies of the exciting X-ray photons.

(c) Scintillation counters, which consist of a scintillation counter (fluorescent crystal made of a sodium iodide activated with 1% thalium) and a photomultiplier. When an X-ray photon strikes the crystal, it generates a quantity of fluorescent light that is proportional to the energy of the X-ray photon. The emitted fluorescent light is measured by a photomultiplier.

(vi) An amplifier, pulse-height analyser, and recording system.

Bibliography: 16, 704.
References: 2488, 4942, 6268.

28.7.3 Sample Preparation

The preparation and mounting of the specimen is one of the most important experimental elements in the X-ray analysis of polymers:

(i) Unoriented specimens (randomly oriented crystalline and amorphous material) can be mounted with simple techniques.

For the Laue technique (forward-reflection measuring technique) the sample can be prepared in the form of a sheet (more convenient) or a cylinder. Polymer powder can be pressed (under light pressure to avoid deformation of the microstructure) into coherent pellets of 5 mm in diameter. Sometimes amorphous binders such as gum tragacanth, collodion, or Duco cement are used, but they produce a diffuse X-ray diffraction effect.

For the Debye–Scherrer technique (forward- and back-measuring technique) specimens should be prepared in the form of cylinders of about 1 mm diameter. They can be prepared by the following methods:

(a) filling a thin-walled plastic or glass tube with bulk polymer powder;
(b) moulding bulk polymer into a cylindrical shape;
(c) mechanically cutting or turning a moulded specimen into a cylindrical shape;
(d) building up a sheet of many thin lamellae and then trimming it to a roughly cylindrical cross-section.

(ii) Oriented specimens. For studies of polymer sheets, films, and fibres samples have to be mounted in a special specimen holder (goniometer), which is shown schematically in Fig. 28.16. The 360° azimuthal circle permits the sample axis to be tilted at any desired azimuth with the vertical. The horizontal setting circle varies the inclination of the specimen with respect to the incident X-ray beam.

(iii) Long fibres may be wound tightly on a small rectangular frame (Fig. 28.17(a)), parallel to one another and several layers deep. Short fibres may be mounted individually by glueing the ends (Fig. 28.17(b)).

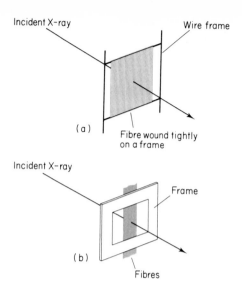

(a)

Incident X-ray — Wire frame

Fibre wound tightly on a frame

Incident X-ray

Frame

(b)

Fibres

Fig. 28.17   Fibre sample preparation for X-ray study

Bibliography: 16, 704, 1255, 1272.
References: 2606, 4031, 6324.

28.8 INFORMATION OBTAINED FROM DIFFRACTION AND SCATTERING MEASUREMENTS

Typical examples of wide-angle diffraction (scattering) (WAXS) patterns for different polymers are shown in Figs 28.18 and 28.19.

Qualitative analysis (Fig. 28.20) of WAXS patterns (Fig. 28.21) may supply the following information:

(i) morphology of specimen (crystalline or amorphous);
(ii) approximate amount of crystalline fraction;
(iii) preferred orientation of crystallites;
(iv) degree of alignment;
(v) perfection of crystalline regions;
(vi) degree of orientation;
(vii) translation period along the fibre axis.

The quantitative analysis of measurements of the X-ray scattered beam versus the scattering angle (2θ) may supply information necessary for

(i) determination of the crystal unit cell;
(ii) calculation of interatomic distances and bond angles;
(iii) determination of the degree of crystallinity in polymers (χ_c) (cf. Section 28.11).

504

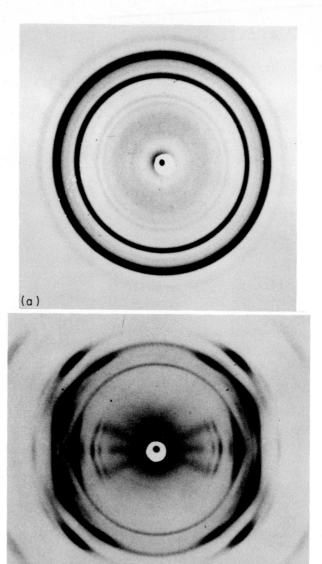

Fig. 28.18   WAXS diffraction pattern of poly(ethylene oxide):
(a) bulk crystallized sample; (b) drawn fibre. (From R: 4975;
reproduced by courtesy of A. J. Lovinger, Lau Chi-Ming, and
C. C. Gryte with the permission of IPC Science and Technology
Press)

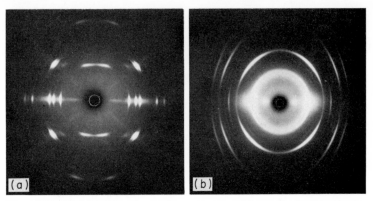

Fig. 28.19 WAXS diffraction pattern of: (a) poly(propylene); (b) celanese acetal copolymer (Celcon®). (From R: 5517; reproduced by courtesy of H. D. Noether and W. Whitney with the permission of Steinkopft Verlag)

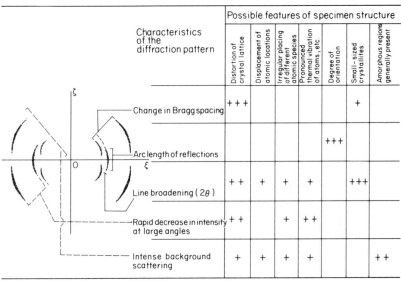

	Possible features of specimen structure						
Characteristics of the diffraction pattern	Distortion of crystal lattice	Displacement of atomic locations	Irregular placing of different atomic species	Pronounced thermal vibration of atoms, etc.	Degree of orientation	Small-sized crystallites	Amorphous regions generally present
Change in Bragg spacing	+ + +					+	
Arc length of reflections					+ + +		
Line broadening (2θ)	+ +	+	+	+		+ + +	
Rapid decrease in intensity at large angles	+ +		+	+ +			
Intense background scattering	+	+	+	+			+ +

Fig. 28.20 Correlation between diffraction pattern and structural elements of the diffracting substance. (From B: 704; reproduced by permission of Elsevier–Kodansha)

28.9 APPLICATIONS OF WIDE-ANGLE X-RAY DIFFRACTION (WAXS) TO THE STUDY OF THE STRUCTURE OF POLYMERS

Wide-angle X-ray diffraction (scattering) (WAXS) provides valuable information for

(i) the identification of polymer crystals (size and perfection);
(ii) the analysis of crystallite orientation;
(iii) the determination of type and the degree of crystallite orientation;

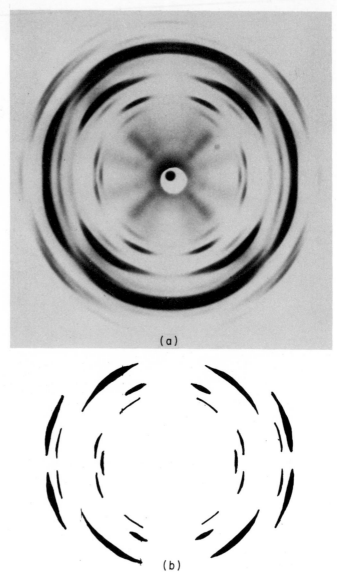

Fig. 28.21 WAXS diffraction pattern of poly(ethylene oxide). (a) Photographic pattern; (b) the major inner reflections (redrawn). (From R:4975; reproduced by courtesy of A. J. Lovinger, Lau Chi-Ming, and C. C. Gryte with the permission of IPC Science and Technology Press)

(iv) study of the degree of crystallinity (χ_c) (cf. Section 28.11);

(v) the study of polymer conformation (especially helical conformation);

(vi) the study of the deformation and annealing of polymers;

(vii) the study of molecular motions in polymer crystals;

(viii) the study of molten polymers.

Bibliography: 16, 327, 704, 1047, 1139, 1260, 1360.
References: 2011, 2032, 2134, 2153, 2171, 2205, 2247, 2283, 2394, 2675, 2756, 2910, 2939, 2947, 3047, 3164, 3173, 3218, 3240, 3340, 3598, 3654, 3686, 3729, 3849, 3854, 3928, 4047, 4061, 4064, 4143, 4260, 4327, 4369, 4440, 4470, 4559, 4581, 4583, 4590, 4591, 4643, 4763, 4777, 4930, 4962, 4972, 4975, 5097, 5166, 5229, 5230, 5325, 5451, 5469, 5511, 5517, 5528, 5529, 5615, 5641, 5648, 5654, 5814, 5863, 5930, 6165, 6267, 6533, 6695, 6697, 6718, 6721, 6965, 7051, 7076, 7261, 7272, 7292.

28.10 APPLICATIONS OF SMALL-ANGLE X-RAY SCATTERING (SAXS) TO THE STUDY OF THE STRUCTURE OF POLYMERS

Small-angle X-ray scattering (SAXS) provides valuable information for the study of

(i) external dimensions of such morphological domains as lamellae, spherulites, separated phases, and voids;

(ii) macromolecules in solutions (analysis of particle size and shape);

(iii) dilute or dense systems of colloidal particles;

(iv) swollen polymers;

(v) deformation and annealing of polymers;

(vi) branched polymers.

Bibliography: 16, 748, 773, 846, 1047, 1254, 1260, 1360.
References: 2112, 2153, 2159–2161, 2394, 2458, 2538, 2595, 2675, 2839, 2844, 2910, 3021, 3022, 3090, 3107, 3117, 3170, 3218, 3240, 3243, 3434, 3551, 3592, 3598, 3623, 3686, 3900, 3928, 3931, 3936, 3961–3963, 3981, 3983, 3984, 4031, 4143, 4176, 4192, 4327, 4368, 4409, 4433–4436, 4469, 4541, 4553, 4554, 4559, 4576, 4577, 4583, 4643, 4653, 4725, 4732, 4903, 4957, 5040, 5097, 5144, 5157, 5197, 5245, 5305, 5325, 5488, 5536, 5590, 5649, 5781–5783, 5814, 5845, 5856, 5863, 5864, 6023, 6024, 6033, 6039, 6080, 6164, 6165, 6171, 6269, 6515, 6516, 6518, 6528, 6601, 6609, 6267, 6718, 6955, 6965, 7051, 7052, 7257, 7261, 7274, 7345.

28.11 MEASUREMENT OF THE DEGREE OF CRYSTALLINITY IN POLYMERS BY X-RAY DIFFRACTION ANALYSIS

The degree of crystallinity can be determined if the crystalline and amorphous scattering in the diffraction pattern can be separated from each other.

The DEGREE OF CRYSTALLINITY (χ_c) is equal to the ratio of the integrated crystalline scattering to the total scattering, both crystalline and amorphous, and is given by

$$\chi_c = \int_0^\infty s^2 I_c(s)\, ds \bigg/ \int_0^\infty s^2 I(s)\, ds, \tag{28.7}$$

where

s is the magnitude of the reciprocal-lattice vector and is given by

$$s = (2 \sin \theta)/\lambda, \tag{28.8}$$

θ is one-half the angle of deviation of the diffracted rays from the incident X-rays (cf. Fig. 28.2),

λ is the X-ray wavelength,

$I(s)$ is the intensity of coherent X-ray scatter from a specimen (both crystalline and amorphous),

$I_c(s)$ is the intensity of coherent X-ray scatter from the crystalline region.

The degree of crystallinity calculated from eq. (28.7) tends to be smaller than the true crystalline fraction, because part of the X-ray intensity that is scattered by the crystalline regions is lost from the peaks and appears as diffuse scatter in in the background as a result of atomic thermal vibrations and lattice imperfections.

Differentiation between crystalline and amorphous scattering in coherent scattering is very difficult, and sometimes provides errors which seriously influence the measurement of the degree of crystallinity. There are two basic methods for differentiation between crystalline and amorphous scattering:

(i) For samples where completely amorphous or completely crystalline specimens cannot be obtained. In this case (Fig. 28.22) a line is drawn

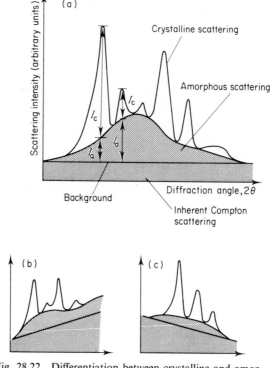

Fig. 28.22 Differentiation between crystalline and amorphous scattering for different semicrystalline polymer samples: (a), (b) and (c) show different positions of background lines)

which should connect the minima between the crystalline peaks. The scatter intensity above this line (I_c) is from a crystalline region, whereas the scatter intensity below this line (I_a) is from an amorphous region. The degree of crystallinity (χ_c) can then be calculated from eq. (28.7).

(ii) For samples where completely amorphous or completely crystalline specimens can be obtained. In this case (Fig. 28.23),

(a) an amorphous reference specimen,

(b) a crystalline reference specimen, and

(d) a specimen of unknown crystallinity

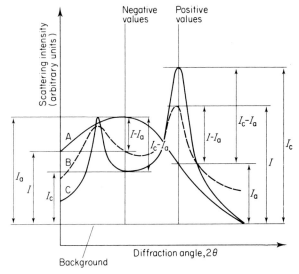

Fig. 28.23 Diffraction intensity curves for: curve A, amorphous; curve B, unknown, and curve C, crystalline specimens

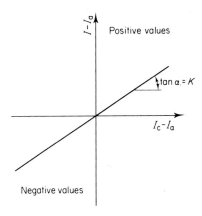

Fig. 28.24 Plot of scattering intensity differences obtained from Fig. 28.23

are examined separately. The degree of crystallinity (χ_c) can be calculated from

$$\chi_c = \frac{(I - I_a)_i - K}{(I_c - I_a)_i},$$ (28.9)

where
  I is the scatter intensity from an unknown sample,
  I_a is the scatter intensity from an amorphous reference specimen,
  I_c is the scatter intensity from a crystalline reference specimen,
  K is a constant which can be calculated from the slope of a curve obtained by plotting $I - I_a$ versus $I_c - I_a$ (Fig. 28.24).

Bibliography: 16, 541, 614, 704, 713, 927, 1155, 1427.
References: 2024, 2037, 2537, 2771, 2855, 2978, 3340, 3729, 3825, 4020, 4033–4038, 4059, 4060, 4062, 4373, 4374, 4743, 4744, 4934, 5098, 5136, 5211, 5477, 5857, 5895, 5896, 6122–6125, 6325, 6517, 6988, 7141, 7272.

Chapter 29

Electron Diffraction Analysis

General bibliography: 41, 438, 439, 1043, 1327.

29.1 PROPERTIES OF ELECTRON BEAMS

Electron radiation has a typical wave character. The wavelength (λ) of an electron beam accelerated by an electrical potential (U) is given by the DE BROGLIE EQUATION:

$$\lambda = h/\sqrt{(2meU)} = 12.25/\sqrt{U}, \qquad (29.1)$$

where
  λ is the wavelength of the electron beam (in ångstroms),
  h is the Planck constant,
  m is the mass of an electron,
  e is the charge on an electron,
  U is an electrical potential (in volts).
With a voltage $U = 40$ kV, the wavelength $\lambda = 0.06$ Å. This is much shorter than the 1.54 Å CuK_α radiation usually used in X-ray diffraction studies (cf. Section 28.1).

The geometry of diffraction of electron radiation is analogous to that for X-ray diffraction described in Section 28.3, and for the purposes of calculation can be applied to the BRAGG equation (eq. (28.5)), and the reciprocal lattice construction.

29.2 ELECTRON DIFFRACTION EQUIPMENT

Most diffraction studies of polymers are carried out using the transmission electron microscope (cf. Section 27.2). In all modern electron microscopes it is possible to change from imaging to diffraction by a simple commutation of the lenses. The interference pattern produced in the focal plane of the objective lenses is magnified and projected on the screen (Fig. 29.1).

Studies of texture orientation can only be carried out with a specimen holder which can be tilted around one or two axes.

An exact evaluation of the diffraction pattern requires the internal calibration of the apparatus. The experiment yields only the distance $r_{(hkl)}$ of the diffraction spots from the primary beam on the photographic plate. The relationship

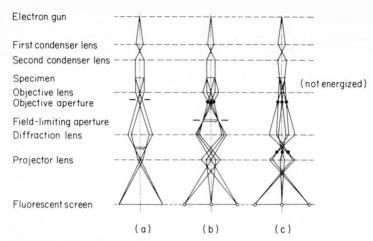

Fig. 29.1 Formation of images in the electron microscope: (a) bright-field image; (b) wide-angle electron diffraction; (c) small-angle electron diffraction. (From B: 439; reproduced by permission of John Wiley & Sons)

between the distance $(r_{(hkl)})$ and the diffraction angle (2θ) is given by

$$L \tan 2\theta = r_{(hkl)}, \tag{29.2}$$

where:

L is the effective distance between the specimen and the photographic plate,
θ is the diffraction angle,
$r_{(hkl)}$ is the distance of the diffraction spots from the primary beam.

The relation between the measured $r_{(hkl)}$ and the plane distance $(d_{(hkl)})$ can be obtained by comparison with a diffraction pattern of a calibration material (LiF or $TiCl_3$) in which the distance between lattice planes is well known.

Bibliography: 14, 439.

29.3 SAMPLE PREPARATION FOR ELECTRON DIFFRACTION ANALYSIS

The optimal thickness of a specimen for electron diffraction is in the range of several hundred ångstrom units. For that reason special methods for sample preparation must be applied (cf. Section 27.3):

(i) production of thin film by casting from solution;
(ii) precipitation of small particles from dilute solutions;
(iii) sectioning;
(iv) splitting and disintegration of bulk material;
(v) surface detachment replicas;
(vi) interfacial polycondensation and pyrolysis;

It should be noted that electron radiation induces the formation of radicals in polymers and these can cause chain scission and/or crosslinking. The radiation is less effective at a higher voltage of acceleration. As a consequence of additional electron radiation effects, a change of lattice constants and an increase of the number of lattice imperfections are observed.

Bibliography: 439.
Reference: 5610.

29.4 SMALL-ANGLE ELECTRON SCATTERING

Small-angle electron scattering is caused by the different electronic densities in a specimen. From the positions of reflections, an average distance between regions of equal density can be calculated.

This method has been used to investigate the crystalline–amorphous macro-lattice of drawn material and fibres. The main advantage of small-angle electron scattering compared with X-ray techniques is the possibility of combining diffraction experiments and morphological studies in the electron microscope.

Bibliography: 439.
Reference: 5034.

29.5 APPLICATIONS OF ELECTRON DIFFRACTION ANALYSIS TO STUDY THE STRUCTURE OF POLYMERS

Electron diffraction analysis provides valuable information for the study of

(i) single crystals (identification of size and perfection);
(ii) the degree of crystallinity;
(iii) textured and unoriented polycrystalline structures;
(iv) drawn and oriented polymers.

Bibliography: 438, 439.
References: 2209, 2267, 2274, 3378, 3379, 3381, 3482, 3612, 3978, 3979, 4112, 4509, 4519, 4548, 5675, 5682, 5788, 6048, 6304, 6305, 6784, 7064, 7097, 7098, 7100.

29.6 DIFFERENCES BETWEEN ELECTRON AND X-RAY ANALYSIS

There are several important differences between electron and X-ray diffraction analysis of polymers:

(i) As a result of the short wavelength of electrons (e.g. 0.06 Å at $U = 40$ kV), the electron interference maxima appear at very small diffraction angles (θ), so that a single-crystal pattern obtained by electron diffraction contains many more reflections than does an X-ray pattern (Fig. 29.2).

(ii) A line broadening obtained by electron diffraction becomes observable with particles smaller than 50–100 Å in comparison to about 1000 Å for X-ray diffraction.

(iii) The strong scattering of electrons passing through the specimen results in diffraction intensities greater by a factor of 10^6–10^8 than those produced by X-rays under the same conditions.

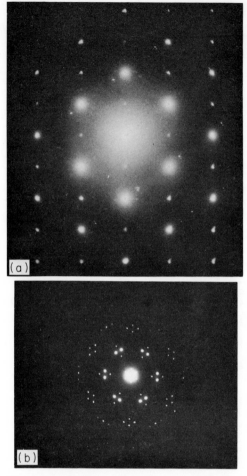

Fig. 29.2 Electron diffraction pattern of: (a) a poly(ethylene) single crystal (reproduced by courtesy of M. Takayanagi); (b) a poly(oxymethylene) single crystal. (From R22.67; reproduced by courtesy of D. C. Bassett with the permission of *Philosophical Magazine*, Taylor and Francis Ltd)

(iv) The high electron diffraction intensities, even from very thin layers, allow one to observe the diffraction pattern on a fluorescent screen.

(v) The exposure time required to take the photograph using the electron diffraction method is of the order of a few seconds by comparison with several hours in the X-ray diffraction method.

(vi) The quantity of sample required to produce a photograph by the electron diffraction method is of the order of 10^{-18} g. The optimal thickness of a polymer sample for electron diffraction is of the order of several hundred ångstroms, whereas using the X-ray diffraction method it is of the order of several millimetres.

(vii) Electron diffraction analysis can be combined with electron microscopic observation of the same specimen, and this allows the determination of the arrangement of the unit cells and the orientation of the molecular chains within the morphological structure units.

Chapter 30

Electron Spectroscopy for Chemical Application

General bibliography: 206, 292–295, 297, 669, 1207–1209.

ELECTRON SPECTROSCOPY FOR CHEMICAL APPLICATION (ESCA) involves the measurement of binding energies of electrons ejected by interactions of a molecule with a monoenergetic beam of soft X-rays.

When a soft X-ray beam interacts with core electrons the following processes occur (Fig. 30.1):

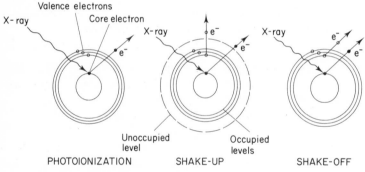

Fig. 30.1   Interaction of soft X-rays with electrons in a molecule

(i) a PHOTOIONIZATION PROCESS—this process occurs with the removal of a core electron (emission of photoelectron);

(ii) a SHAKE-UP PROCESS, which is the photoionization process accompanied by simultaneous excitation of a valence electron from an occupied to an unoccupied level (shake-up);

(iii) a SHAKE-OFF PROCESS, which is the photoionization process accompanied by ionization of a valence electron (shake off).

Monitoring of the energies of these emitted electrons gives an ELECTRON CORE LEVEL SPECTRA (Fig. 30.2). When an electron is emitted from the core, a core hole state is formed. The lifetimes of core hole states are typically in the range 10^{-13}–10^{-15} s. De-excitation of the hole state can occur via both fluorescence and Auger processes. The removal of a core electron (which is

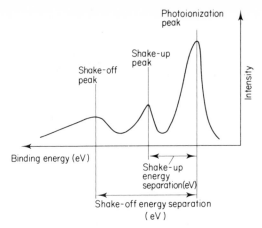

Fig. 30.2 Electron core level spectra for a given sample

shielded by the valence electrons, c.f. Fig 30.1) is accompanied by substantial reorganization of the valence electrons in response to the effective increase in nuclear charge. This perturbation is responsible for the photoionization process and can be accompanied by shake-up and shake-off processes.

Valuable information can be obtained from core level spectra (Fig. 32.2), e.g.

(i) the binding energy for core electrons;
(ii) the separation energy for the shake-up and the shake-off processes;
(iii) the energy levels for valency electrons;
(iv) the distribution of unpaired electrons, spin states;
(v) identification of structural features;
(vi) angular-dependent studies of a single core level gives information concerning inhomogeneities in the surface and subsurface of a sample.

30.1 ESCA SPECTROMETERS

The ESCA spectrometer (Fig. 30.3) is built of the following parts:

(i) An X-ray generator consisting of a high-power electron gun with a well-focused beam hitting the rotating water-cooled aluminium anode. AlK$_\alpha$ radiation (1486.6 eV) is emitted at a small angle ($\sim 5°$) and diffracted by one or several spherically bent quartz crystals before being focused on the sample.
(ii) Because of strong scattering of electrons by air, the spectrometer is evacuated to 10^{-7} torr (1 torr $= \frac{1}{760}$ atm $= 133.322$ N m^{-2} (SI)) or less by two-stage rotary pumps and oil diffusion pumps.
(iii) Samples may be introduced from the atmosphere into the spectrometer by using pre-pumped insertion locks and valves. Samples can be cooled or heated *in situ* and additional equipment for sample treatment (e.g.

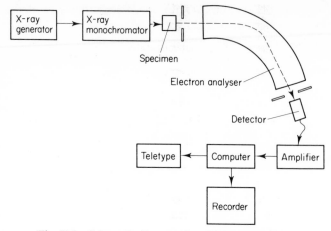

Fig. 30.3   Schematic diagram of an ESCA spectrometer

argon ion bombardment, plasma synthesis, electron bombardment, ultraviolet irradiation, chemical treatment) can be mounted on the spectrometer.

(iv) The electron analyser is a spherical condenser. In its focal plane there is an extended multidetector system, consisting of two closely spaced channel plate detectors in series, which results in a high amplification (10^8) of each electron beam focused on the detector in the focal plane. All pulses are transferred to optical signals by means of a phosphor and the focal plane is continuously scanned by a photomultiplier (or TV camera). After amplification the output of the photomultiplier is fed to a computer for further processing to form the final electron core level spectra.

The time required to record a spectrum is typically *c.* 10 min.

Bibliography: 1207–1209.

30.2 SAMPLE PREPARATION FOR ESCA SPECTROSCOPY

Polymer samples may be studied directly in the form of films or powder. Powders can be mounted on double-sided Scotch tape. Surface coatings can be investigated without the need to separate the coating from the surface. The minimum thickness of the sample is in the range of 100 Å and area 0.2 cm². The ESCA technique is essentially non-destructive.

30.3 APPLICATIONS OF ESCA SPECTROSCOPY IN POLYMER RESEARCH

ESCA spectroscopy can be applied to

(i) the identification of polymers, copolymers, and polymer blends;

(ii) the study of structural isomerism of polymers and copolymers (e.g. alternating or random nature);

(iii) the study of valence bands of polymers;

(iv) the study of polymer films produced on surfaces;

(v) the study of surface treatments, e.g. plasma modification;

(vi) the study of chemical degradation of polymers, e.g. oxidation, nitration, etc.;

(vii) the study of thermal and photodegradation of polymers;

(viii) the study of photoconductivity in polymers;

(iv) the study of the statics and dynamics of sample charging;

(x) the study of triboelectric phenomena in polymers.

Bibliography: 291–297, 388, 1240, 1394.
References: 2018, 2774, 2853, 2884–2901, 2993, 4067, 6473, 7249.

Chapter 31

Density Measurements

General bibliography: 309, 1324, 1427.

31.1 THE ABSOLUTE DENSITY

The ABSOLUTE DENSITY (ρ) is defined as

$$\rho = m/V \qquad (31.1)$$

(in grams per cubic centimetre (CGS) or kilograms per cubic metre (SI)) where
m is the mass of the sample (in grams (CGS) or kilograms (SI)),
V is the unit volume of the sample (in cubic centimetres (CGS) or cubic
metres (SI)).

The volume depends upon temperature (and pressure) and for that reason the
temperature at which the density is measured is indicated in tables of physical
data—usually in degrees Celsius—or by a subscript, e.g. ρ_{20}.

The most common methods for measuring the absolute density (ρ) of macromolecular substances are

(i) the pycnometric method;
(ii) the density gradient column;
(iii) the dilatometric method.

31.2 THE PYCNOMETRIC METHOD

Density (ρ) is measured by determining the weight of a volume-calibrated
pycnometer (Fig. 31.1) filled with a liquid of known density in which a certain
quantity of the polymer sample is immersed. The sample volume equals the
pycnometer volume minus the undisplaced volume of liquid of known density.
The accuracy of this method is ± 0.004 g cm^{-3}.

Bibliography: 1324.
References: 2159, 2161, 2430, 3386, 4475, 5090, 5096, 6347.

31.3 THE DENSITY GRADIENT COLUMN METHOD

Density (ρ) is measured by the flotation level after dropping a polymer sample

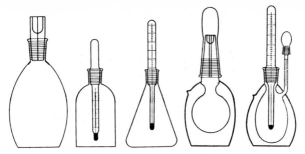

Fig. 31.1 Different types of volume-calibrated pycnometers

into the density gradient column. A device for the preparation of a linear density gradient column by mixing heavy and light liquids in various proportions is shown in Fig. 31.2.

A typical calibration curve for a linear density gradient column is shown in Fig. 31.3. After a gradient is established, a density gradient column may be used for months. Heights of flotation can be measured up to ± 0.5 mm. The accuracy of this method is ± 0.0002 g cm^{-3}.

Bibliography: 1324, 1427.
References: 2430, 3352, 3862, 6347.

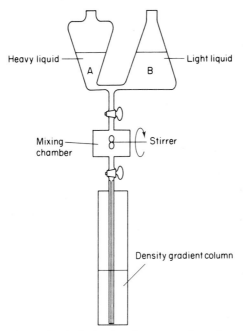

Fig. 31.2 A device for the preparation of a
linear density gradient column

522

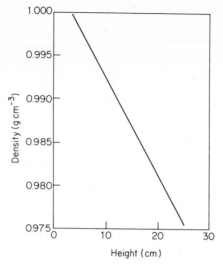

Fig. 31.3   Calibration curve for a linear
density gradient column (CGS units)

31.4 THE DILATOMETRIC METHOD

This method is commonly used for measuring the density change as a function
of temperature. The change in volume (ΔV) of a polymer sample is calculated
from the equation

$$\Delta V = \pi r^2 \Delta H + \Delta V_1 - \Delta V_2, \qquad (31.2)$$

where
   r is the dilatometer capillary radius,
   ΔH is the change in height of the mercury column,
   ΔV_1 is the change in volume of the glass dilatometer,
   ΔV_2 is the change in volume of the mercury.
All changes ΔH, ΔV_1, and ΔV_2 are related to a reference temperature and
temperature t (in degrees Celsius).

Figure 31.4 shows a typical dilatometer for polymer density determinations.
The volume expansion or contraction as a function of temperature is measured
by recording the level change of mercury in the dilatometer capillary. The
accuracy of this method is ± 0.001 g cm^{-3}.

Bibliography: 309, 1427.
References: 2159, 2161, 2333, 2952, 3280, 3421, 5022, 5090, 6046, 6767, 7130.

31.5 THE RELATIVE DENSITY

The RELATIVE DENSITY (d) is defined as the ratio of the absolute density of
a polymer sample to the absolute density of a specified reference material
(e.g. mercury). The relative density is dimensionless.

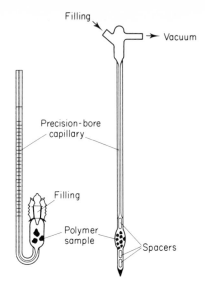

Fig. 31.4 Different types of dilatometer

31.5.1 Mercury as a Reference Material

Before making any measurements, the mercury must be purified. The most widely used method of purifying mercury is treatment by dilute nitric acid (1 part by volume of concentrated acid in 20 parts of water), followed by washing with water and by distillation. The nitric acid treatment is more effective if the mercury is shaken for a few minutes with the acid rather than dropped through a column of the acid. The purified mercury is best stored in poly(ethylene) bottles (in order to avoid electrostatic charges) in a vacuum or an air-free atmosphere. Small amounts of impurity on the surface of the mercury may be removed by filtration, or a pure sample may be drawn off through a tube dipping below the surface.

31.6 AUTOMATIC DIGITAL DENSITOMETER

The principle of this type of densitometer is based on the measurement of the vibration of a glass U-tube filled with a liquid (or a gas) of unknown density. The vibration period (T) is given by

$$T = 2\pi \left(\frac{I_0 + v\rho}{C}\right)^{1/2},$$ (31.3)

where
I_0 is the moment of inertia of the empty tube,
v is the inner volume of the tube,

ρ is the density of the investigated liquid,
C is the elastic constant of the tube.

The square of the vibration period (T) is a linear function of the density of the liquid:

$$T^2 = A + K\rho, \tag{31.4}$$

where A and K are constants depending upon the geometry of the tube. In order to know the absolute value of the density, it is necessary to determine the constants A and K starting from calibration of the apparatus with liquids of known densities.

The vibrating tube is sealed in a thermostat (Fig. 31.5). The total length of tube is 2–3 cm with inner diameter 1 mm. The electronic measurement system

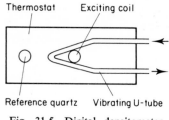

Fig. 31.5 Digital densitometer
cell

compares the frequency of tube vibrations with that of reference quartz of 1 MHz. This type of densitometer can be applied to detect low variations in concentration of a solute and small volume changes, and is very useful in macromolecular studies.

References: 3478, 4708, 6187.

31.7 DETERMINATION OF THE PARTIAL SPECIFIC VOLUME OF THE SOLUTE FROM THE DILATOMETRIC METHOD

The PARTIAL SPECIFIC VOLUME OF THE SOLUTE (POLYMER) $(\bar{v})$ is defined as the ratio of the volume (v_2) and the mass (m_2) of the ploymer in solution:

$$\bar{v} = v_2/m_2 \tag{31.5}$$

(in cubic centimetres per gram (CGS) or cubic metres per kilogram (SI)).

It can be determined by accurate measurements of the density of the solvent and of the polymer solution in a pycnometer (cf. Section 31.2).

The volume of the solute (polymer) (v_2) is given by

$$v_2 = v - v_1 \tag{31.6}$$

(in cubic centimetres (CGS) or cubic metres (SI)), where
v_2 is the solute volume,

v_1 is the solvent volume,
v is the solution volume.
The volume of the solvent (v_1) is given by

$$v_1 = m_1/\rho_1, \tag{31.7}$$

where
   m_1 is the solvent mass (in grams (CGS) or kilograms (SI)),
   ρ_1 is the solvent density (in grams per cubic centimetre (CGS) or kilograms
      per cubic metre (SI)).
The mass of the solvent (m_1) is given by

$$m_1 = m - m_2, \tag{31.8}$$

where
   m is the mass of solution,
   m_2 is the mass of solute (polymer).
Combination of eqs (31.5)–(31.8) yields the equation for the PARTIAL
SPECIFIC VOLUME OF THE SOLUTE (POLYMER) in the form

$$\bar{v} = \frac{1}{m_2}\left(v - \frac{1}{\rho_1}(m - m_2)\right) = \frac{1}{c_v} - \frac{1}{\rho_1}\left(\frac{1}{c_w} - 1\right), \tag{31.9}$$

where
   c_v is the volume concentration (in grams per cubic centimetre (CGS) or
      kilograms per cubic metre (SI)),
   c_w is the weight concentration (in grams per gram (CGS) or kilograms per
      kilogram (SI)).

References: 4429, 6218.

31.8 MEASUREMENT OF THE DEGREE OF CRYSTALLINITY IN POLYMERS BY DENSITY MEASUREMENTS

From the density measurement the degree of crystallinity (χ_c) is defined in two
forms:

   (i) weight fraction crystallinity (w_c);
   (ii) volume fraction crystallinity (v_c).

The TOTAL VOLUME OF SEMICRYSTALLINE POLYMER SAMPLE (V)
is given by

$$V = \frac{W}{\rho} = \frac{W_a}{\rho_a} + \frac{W_c}{\rho_c}, \tag{31.10}$$

where
   $V = V_a + V_c$ is the total volume of a sample,
   V_a and V_c are volumes of amorphous and crystalline components, respectively,
   $W = W_a + W_c$ is the total weight of a sample,

526

W_a and W_c are weights of amorphous and crystalline components, respectively, ρ_a and ρ_c are densities of amorphous and crystalline components, respectively.

The WEIGHT FRACTION CRYSTALLINITY (w_c) is defined as

$$w_c = W_c/W = \rho_c V_c/\rho V, \qquad (31.11)$$

but

$$V_c/V = 1 - (V_a/V) \qquad (31.12)$$

and

$$V = \rho_a V_a + \rho_c V_c \qquad (31.13)$$

Combining eqs (31.11)–(31.13) with eq. (31.10) gives a final equation for calculation of the weight fraction crystallinity (w_c):

$$w_c = \frac{\rho_c}{\rho}\left(\frac{\rho - \rho_a}{\rho_c - \rho_a}\right). \qquad (31.14)$$

The VOLUME FRACTION CRYSTALLINITY (v_c) is given by

$$v_c = \frac{V_c}{V} = \frac{\rho - \rho_a}{\rho_c - \rho_a}. \qquad (31.15)$$

Between the weight fraction crystallinity (w_c) and the volume fraction crystallinity (v_c) is the following relation:

$$w_c = \frac{\rho_c}{\rho} v_c \qquad (31.16)$$

In order to calculate v_c or w_c from eqs (31.15) and (31.16), it is necessary to determine

(i) the density of the crystalline phase (ρ_c), calculated from the crystal structure;
(ii) the density of the sample at completely amorphous conditions (ρ_a) and the density of the unknown sample (ρ) from pycnometric measurements (cf. Section 31.2) or dilatometric measurements (cf. Section 31.4).

The density increases whereas the specific volume decreases if the amount of crystalline fraction increases.

Bibliography: 1427.
References: 2141, 2142, 2800, 2823, 3324, 3351, 3352, 3806, 4020, 4185, 4208, 4227, 4647, 5487, 6043, 6491, 6649, 7119.

Chapter 32

The Glass Transition Temperature and the Melting Point

General bibliography: 48, 127, 309, 436, 508, 524, 586, 909, 1298, 1336, 1345, 1350, 1396, 1419.

The SPECIFIC VOLUME (in millilitres per gram (CGS) or cubic metres per kilogram (SI)) of an amorphous polymer changes linearly with increasing temperature up to the TRANSITION REGION, where a change of the slope occurs (steeper gradient). At the GLASS TRANSITION TEMPERATURE (T_g), the rate increases and a discontinuity is formed in the specific volume curve (Fig. 32.1). The GLASS TRANSITION TEMPERATURE (T_g) is usually defined as the point at which the tangents of the two curves intersect. Such curves are obtained by dilatometric measurements (cf. Section 31.4).

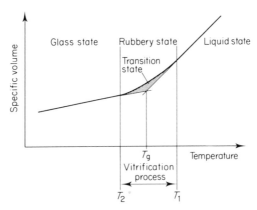

Fig. 32.1   The specific volume as a function of temperature in the transition range

When a liquid polymer (melt) is cooled, the transition from highly viscous supercooled melt to the rigid glass begins at temperature T_1 (Fig. 32.1). At temperature T_2 the solidification (vitrification) process is terminated. This phenomenon is called the GLASS TRANSITION because amorphous polymers exhibit a change from soft, elastic (rubbery) behaviour above T_g to glass-like behaviour below T_g.

If the polymer is crystallizable the specific volume curve shows a discontinuity at the melting point. A typical behaviour of a semicrystalline polymer forming both a glassy and a crystalline state is shown in Fig. 32.2. T_m represents the melting point and $T_{g_1}, T_{g_2}, \ldots$ the glass transition temperatures obtained at various cooling rates. In the region between T_m and T_g supercooling occurs and sudden crystallization takes place. Below T_g the rapid crystallization is not

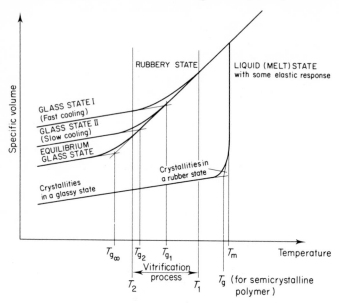

Fig. 32.2 The specific volume as a function of temperature of a glass-forming polymer cooled at various rates

possible by reason of an increase of viscosity, and the polymer remains in the disordered glassy state. When the cooling rate is slower the supercooling persists to lower temperatures and the transition T_{g_2} occurs at a lower temperature than T_{g_1}. If an infinite time were allowed, a limiting value $T_{g\infty}$ would be reached. The glassy states obtained by different cooling rates differ in T_g and also in density. Completely crystalline polymers should not exhibit a glass transition temperature (Fig. 32.3).

Below T_g, chain segments are frozen in fixed positions in a disordered quasi-lattice. Some molecular movements of chain segments take place in the form of vibrations about a fixed position. A diffusional rearrangement of the segmental position is less probable. With increasing temperature, the amplitude of segmental vibrations increases. In the transition state, chain segments have sufficient energy to overcome the secondary intramolecular bonding forces. Chain segments or chain loops may perform rotational and translational motions (Fig. 32.4) which are called SEGMENTAL MOTION or SHORT-RANGE DIFFUSIONAL MOTION. In the rubbery state the segmental motions are very rapid, whereas the MOLECULAR MOTION (the motion of entire molecules) is restricted by chain entanglements. With increasing temperature the degree of entanglement decreases and the molecular slip increases, but

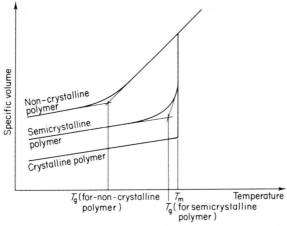

Fig. 32.3 The specific volume for non-crystalline, semi-crystalline, and crystalline polymers

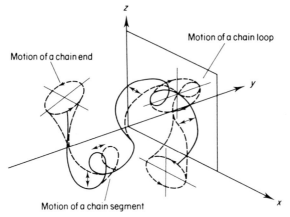

Fig. 32.4 Motion of chain ends, chain loops, and chain segments in the glass transition zone

the sample still shows some elasticity. In the liquid state the slip of entire molecules is predominant, and the elasticity of the sample disappears.

In semicrystalline polymers the same mechanism is involved in the amorphous region. In the crystalline region segments incur short-range vibrations about their equilibrium lattice positions up to the melting point (melting temperature) (T_m).

All polymers exhibit a glass transition at a particular temperature or range of temperatures. The glass transition temperature is well marked in the amorphous polymers (e.g. poly(styrene), poly(methyl methacrylate), and poly(vinyl chloride)), whereas in semicrystalline polymers (e.g. poly(ethylene)) it is less conspicuous because it only occurs in the non-crystalline amorphous parts of the polymer.

Typical glass transition temperatures in various polymers are given in Table 32.1. The value of T_g depends on the experimental time scale and the method of measurement. Values of T_g reported in the literature for any particular polymer may vary from 10 to 30 °C.

Table 32.1   Glass transition temperatures (T_g) for different polymers (from B: 807; reproduced by permission of John Wiley & Sons)

Polymer	T_g (°C)
Poly(dimethyl siloxane)	−120
Poly(butadiene)	−85
Poly(ethylene)	−80
Poly(isoprene)	−73
Poly(isobutylene)	−70
Poly(propylene)	−19
Poly(methyl acrylate)	9
Poly(caproamide) (6 Nylon)	50
Poly(vinyl chloride)	83
Poly(styrene)	100
Poly(tetrafluoroethylene)	126

Bibliography: 129, 183, 186, 218, 308, 396, 397, 634, 686, 712, 771, 806, 826, 1099, 1197.
References: 2047, 2285, 2333, 2531, 2534, 2986, 2987, 3158, 3548, 3603, 3604, 3634, 3635, 3723, 3760, 4079, 4188, 4443, 5056, 5080, 5200, 5278, 5807, 5989, 6117, 6234, 6405, 6423, 6532, 6719, 6875, 6972, 7158, 7188.

T_g values for a large number of polymers are available in The Polymer Handbook (B: 807) and other publications (B: 818, 1121, 1459).

32.1 RELAXATION PROCESSES IN THE REGION OF T_g

The RELAXATION PROCESSES in polymers can be considered as the movement of molecular segments of polymer molecules.

For polymers in the amorphous state, in a temperature region on either side of T_g, three well-defined molecular relaxation processes are commonly observed:

(i) α-RELAXATION is the process above T_g, considered as the primary process and referred to as the glass–rubber transition process. This relaxation process results from large-scale conformational rearrangements of the polymer chain backbone, which occur by a mechanism of hindered rotation around main-chain bonds.

(ii) β-RELAXATION is the process below T_g, considered as the secondary process and referred to as the glass transition process. This relaxation process results from hindered rotations of side groups independent of the polymer chain backbone.

(iii) γ-RELAXATION is the process below T_g associated with the disordered regions of the polymer.

Bibliography: 184, 185, 503, 628, 634, 893, 973, 1217.
References: 2528, 2567, 3002, 3960, 4858, 5814, 5918, 6367, 6710.

32.2 FREE VOLUME THEORY OF GLASS TRANSITION

The FREE VOLUME (V_f) of liquids is defined by

$$V_f = V_T - V_0, \tag{32.1}$$

where

V_T is the total volume of the liquid at temperature T (in kelvins),

V_0 is the theoretical molar volume of the most dense packing of the liquid molecules at 0 K.

The TOTAL VOLUME of the liquid (V_T) is the sum of the free volume (V_f) and of the occupied volume (V_0):

$$V_T = V_f + V_0. \tag{32.2}$$

The occupied volume (V_0) includes not only the van der Waals radii but also the FLUCTUATION VOLUME, which is associated with the thermal vibrational motion.

The TEMPERATURE COEFFICIENT OF EXPANSION OF FREE VOLUME (α_f) is given by

$$\alpha_f = \frac{1}{\bar{v}_g} \left(\frac{dV}{dT} \right), \tag{32.3}$$

where

$\bar{v}_g$ is the specific volume at T_g,

dV/dT is the observed volume change as a function of the temperature change. The temperature coefficient of expansion (α_f) is greater than that of the occupied volume (V_0) above T_g (Fig. 32.5). The GLASS TRANSITION TEMPERA-TURE (T_g) based on the free volume theory is that temperature at which the free volume (V_f) reaches a constant value, and does not decrease any more when

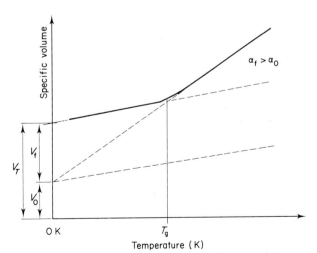

Fig. 32.5 Definition of free volume

532

the sample is cooled below T_g (Fig. 32.5). Below T_g the hole mobility is restricted and the only movement occurs in the occupied volume (V_0).

The FREE VOLUME FRACTION (f) is given by

$$f = \frac{V_f}{(V_f + V_0)}.$$ (32.4)

The CRITICAL FREE VOLUME ($V_f)_{crit}$ is measured by extrapolating the liquid state line to absolute zero:

$$(V_f)_{crit} = V_0 \quad \text{at } 0 \text{ K.}$$ (32.5)

Bibliography: 48, 186, 285, 436, 585, 587, 634, 698, 804, 826.
References: 2238, 2483, 2569, 2919, 3211–3213, 3468, 3470, 4691, 4692, 6046, 6405, 6406, 6972, 7111, 7130, 7343.

32.3 THE HOLE THEORY

The HOLE THEORY is based on a lattice model. Generally a system is considered in which N segments are situated in M lattice sites. The NUMBER OF EMPTY SITES or HOLES is given by $M - N$, and the CELL VOLUME by

$$v = M/N.$$ (32.6)

At the glass transition temperature the number of holes and the conformation of the chain backbone are frozen and do not change on further cooling. Above the glass transition temperature (T_g) the expansion of the hole size occurs simultaneously with the increase in the number of holes. The lattice itself expands and contracts above and below T_g, whereas the number of holes varies only above T_g.

References: 5530, 7342.

32.4 POLYMERIC GLASSES

POLYMERIC GLASSES in the transition range exhibit a behaviour which is similar to a first-order thermodynamic transition. The characteristic effect appears when slowly cooled or annealed glassy polymers are heated through the transition range, showing an abrupt, almost discontinuous, increase in volume (V) and enthalpy (ΔH) before approaching the steady-state conditions above T_g. The thermal expansion coefficient (α) and the specific heat (c_p) pass through a maximum before reaching, at high temperature, values characteristic of the supercooled liquid state (Fig. 32.6).

The temperature at which the peak appears, and the corresponding increase of volume (V) and enthalpy (ΔH), depend on the rate of heating and on the initial state of the glass, i.e. on its thermal history.

Bibliography: 629, 951, 1336, 1419.
References: 2048, 2129, 3155–3159, 3696, 4188, 5199, 6155, 6404, 7041, 7045.

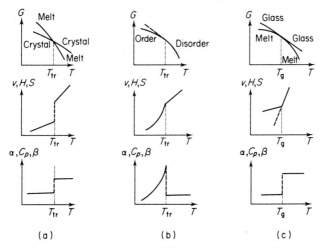

(a)   (b)   (c)

Fig. 32.6 A comparison of different types of transition processes: (a) first-order; (b) second-order; and (c) vitrification process. Notation: G is the free energy, H is the enthalpy, S is the entropy, α is the coefficient of expansion, β is the compressibility, C_p is the specific heat, T_g is the glass transition temperature, T_{tr} is the transition process temperature

32.5 FACTORS INFLUENCING T_g

32.5.1 Chain Microstructure

Chain flexibility and intermolecular packing distances (chain stiffness), bulkiness, flexibility of side chains, and polarity of the chain are major parameters influencing T_g.

CHAIN FLEXIBILITY is determined by the ease with which rotation occurs about primary valence bonds. A rotation involves an energy barrier that is of the same order as the molecular cohesive forces (1–5 kcal mol^{-1} (4.2–20.9 J mol^{-1}))

The decreasing chain flexibility increases the T_g temperatures by increasing steric hindrance. STERIC HINDRANCE is dependent on the size, shape, and constitution of the backbone group. The most common backbone hindrance group is the p-phenylene group.

Rigid and bulky side groups decrease the flexibility of the chain (steric hindrance is increased) and the value of T_g is reduced; for example, for poly-(styrene) $T_g = 100\,^{\circ}$C, for poly(α-methylstyrene) $T_g = 175\,^{\circ}$C, and for poly-(acenaphthalene) $T_g = 264\,^{\circ}$C.

Introducing flexible side groups results in an increase of intermolecular distances; consequently free volume (cf. Section 32.2) predominates and the T_g value is lowered.

Increased chain symmetry lowers T_g values; for example, for poly(vinyl chloride) $T_g = 83\,^{\circ}$C, and for poly(vinylidene chloride) $T_g = -17\,^{\circ}$C.

Cis–trans isomerization influences T_g temperatures because of differences in the freedom of rotation of chain atoms on either side of the double bond; for example, for *cis*-poly(butadiene) $T_g = -96\,°C$, and for *trans*-poly(butadiene) $T_g = -83\,°C$.

Polar substituents increase chain cohesion and T_g are raised; for example, for *cis*-poly(butadiene) $T_g = -96\,°C$, and for poly(vinyl chloride) $T_g = 83\,°C$.

Bibliography: 183, 218, 397, 691, 1197.

32.5.2 Crystallinity

In crystalline polymers T_g increases proportionally to the amount of crystallinity, because crystallites tend to reinforce or stiffen the structure. It has also been observed that T_g may decrease with increasing crystallinity. For that reason the effect of variation in the degree of crystallinity on T_g should be measured and reported.

Bibliography: 218.
References: 4248, 5964, 6100, 6224.

32.5.3 Tacticity

The glass transition temperatures (T_g) of mono(vinyl)-substituted polymers are independent of tacticity, whereas those of di(vinylidene)-substituted polymers $(-CH_2CXY-)_n$ are dependent on tacticity, e.g. for poly(methyl methacrylate) ($T_{g\,(atactic)} = 104–108\,°C$, $T_{g\,(isotactic)} = 42–45\,°C$, and $T_{g\,(syndiotactic)} = 105–120\,°C$. Variation in T_g temperatures arises as a result of the extent of polar interaction between the pendant ester groups in the polymeric configuration. In general, for any pair of stereoisomers,

$$T_{g(syndiotactic)} - T_{g(isotactic)} = 0.59(\Delta E/k), \tag{32.7}$$

where

ΔE is the difference in the flex energy between the syndiotactic and isotactic isomers,

k is the Boltzmann constant (cf. Section 40.4).

Bibliography: 183, 218, 691.
References: 2128, 2521, 3707, 3772, 4419, 4423, 5473, 5964, 5990, 6354, 6786, 7056.

32.5.4 Molecular Weight

For linear polymers the T_g value is an increasing function of the molecular weight. T_g varies linearly with the reciprocal of the number-average molecular weight ($\bar{M}_n$) of polymers; for example, the T_g of poly(styrene) with $\bar{M}_n = 3000$ is 43 °C, whereas with $\bar{M}_n = 300\,000$ it is 99 °C. This dependence is a result of the contribution of chain-end segments in molecular motions. This motion increases the free volume. As the number of chains increases ($\bar{M}_n$ decreases) the free volume as a whole increases faster with temperature, and the glass transition occurs at lower temperature. This becomes

$$T_g = T_{g\infty} - (K/\bar{M}_n),\qquad(32.8)$$

where

$T_{g\infty}$ is the glass transition temperature of an infinite molecular weight polymer,
$\bar{M}_n$ is the number-average molecular weight,
K is a constant, given by

$$K = 2V_c \rho N_A/\alpha,\qquad(32.9)$$

V_c is the free volume contributed by a chain end,
ρ is the polymer density,
N_A is the Avogadro number (cf. Section 40.3),
α is the thermal expansion coefficient.

Equation (32.8) gives a straight line when T_g is plotted against $\bar{M}_n^{-1}$ (Fig. 32.7). The free volume contributed by a chain end (V_c) can be calculated from the

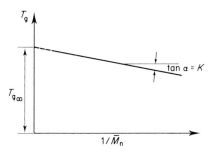

Fig. 32.7 Change of glass transition temperatures versus $\bar{M}_n^{-1}$

slope of the line, since α and ρ are known. For polystyrene V_c was found to be 80 Å³, which is about half the size of a styrene monomer unit, and K is about 2×10^5.

Bibliography: 218, 771, 1197, 1318.
References: 2330, 2532, 3004, 3354, 3355, 3468, 3470, 4291, 5213, 5809, 6046, 6434, 6880, 6883.

32.5.5 Branching

The change of T_g with increase in branching is the result of two effects:
  (i) the increased number of end groups will increase the chain mobility and free volume;
  (ii) the introduction of branch points will reduce chain mobility and free volume.

In general, the presence of end groups has more influence than the presence of branch points, and lowering of T_g is always observed.

Bibliography: 218.

32.5.6 Intermolecular Bonds

Polymers with strong intermolecular bonds, e.g. chemical crosslinks or hydrogen bonds, exhibit higher values of T_g. Crosslinking reduces free volume and increases T_g. When the degree of crosslinking is sufficiently high the T_g temperature completely disappears.

The glass transition temperature for crosslinked polymers can be expressed by a semi-empirical equation:

$$T_g = T_{g\infty} + (K\Gamma/N_A), \tag{32.10}$$

where

$T_{g\infty}$ is the glass transition temperature of a polymer of infinite molecular weight,

K is a constant,

Γ is the crosslink density (cf. Section 1.6),

N_A is the Avogadro number (cf. Section 40.3).

Bibliography: 218, 1318.
References: 2107, 2952, 3471, 6051, 6052, 6106.

32.5.7 Effects of Solvents and Plasticizers

The presence of low molecular weight compounds (such as solvents, water, and plasticizers) in the amorphous phase of a semicrystalline polymer lowers T_g. The nature of the solvent (plasticizer) can affect the degree to which T_g is decreased.

The glass transition temperature of the 'plasticized' polymers is given by

$$T_g = \frac{Kw_1 T_{g_1} + w_2 T_{g_2}}{Kw_1 + w_2}, \tag{32.11}$$

where

T_{g_1} and T_{g_2} are the glass transition temperatures of solvent (plasticizer) and polymer, respectively,

w_1 and w_2 are the weight fractions of solvent (plasticizer) and polymer, respectively ($w_1 + w_2 = 1$),

K is a constant given by

$$K = (\alpha_r - \alpha_g)_2/(\alpha_r - \alpha_g)_1, \tag{32.12}$$

α_1 and α_2 are the thermal expansion coefficients of the solvent (plasticizer) and the polymer, respectively—the subscript r referring to the rubbery state and the subscript g to the glassy state.

Equation (32.11) is valid only if the solvent (plasticizer) is molecularly distributed in the amorphous phase.

Bibliography: 218, 314, 1197.
References: 2013, 2065, 3723, 4277, 4298, 4645, 5177, 6019, 6709.

32.5.8 Copolymerization and Polyblend Effects

Amorphous random copolymers exhibit a single T_g which can be predicted by

$$T_g = \frac{Kw_1 T_{g_1} + w_2 T_{g_2}}{Kw_1 + w_2}, \qquad (32.13)$$

where
T_{g_1} and T_{g_2} are glass transition temperatures of two homopolymers, w_1 and w_2 are weight fractions of two homopolymers ($w_1 + w_2 = 1$),
K is a constant given by

$$K = (\alpha_r - \alpha_g)_2 / (\alpha_r - \alpha_g)_1, \qquad (32.14)$$

α_1 and α_2 are the thermal expansion coefficients of the homopolymers—the subscript r referring to the rubbery state and the subscript g to the glassy state.

For a series of homogeneous copolymers or compatible polyblends, the T_g value is in the range between the values of the parent homopolymers (Fig. 32.8).

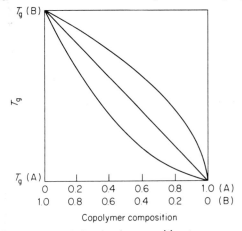

Fig. 32.8   Variation in glass transition temperature with a given copolymer composition

In semicompatible systems, one of the components tends to be distributed as droplets in a continuous matrix of the other component. The droplet size and the degree of separation depend on composition, molecular weight, the method of sample preparation, and the degree of compatibility of the homopolymer sequences.

Where one component is completely incompatible with the other, two glass transition points are observed.

If all segments in either block or graft copolymers are fully compatible, the T_g change is analogous to that of homogeneous copolymers and compatible polyblends.

Bibliography: 218, 661, 662.
References: 2327, 2331, 2593, 3467, 3727, 4328, 4329, 4410, 4412, 4859, 5079, 6810, 7060, 7165.

32.5.9 Drawing of Polymer Samples

Glass transition temperatures change during the drawing of polymer samples. The variation of T_g is a result of changing the configurational entropy and the rate of molecular motion in local-mode relaxation.

References: 3605, 4249, 5105, 5531, 6590, 6993.

32.6 THE DETERMINATION OF T_g

Experimental methods of measuring T_g are based on the change in properties of polymers converted from the glassy to the rubbery state. Two methods are applied:

(i) the STEADY-STATE METHOD, in which polymer properties are measured under static isothermal equilibrium conditions over a temperature range including T_g;

(ii) the DYNAMIC METHOD, in which polymer properties are measured during the heating of the material above T_g—in this method T_g is measured by an extrapolation procedure in order to obtain isothermal conditions.

Dynamic methods are more popular, on account of their convenience. The two methods may sometimes give different T_g values. T_g depends not only upon the experimental conditions, such as the heating rate, but also on the thermal history of polymer sample, e.g. rate of cooling. At any finite rate of cooling the

Table 32.2   Methods of T_g measurements

Method	Published in
Dilatometry	B: 309; R: 2333, 3993, 4094, 4196, 5932, 6046, 6532, 7130, 7174
Buoyancy	R: 3285, 5429
Expansivity	R: 5146, 6532
Optical, dynamic IR spectroscopy	R: 2092, 2132, 2534, 3887, 4007, 4153, 4693, 5559, 5560, 7155
Refractometry	R: 2330, 5029, 7084
Light scattering	B: 499; R: 3814, 6090
Radiothermoluminescence	R: 2802, 5375
NMR	R: 2185, 3528, 5174, 6441, 6444
ESR	B: 186, 1088; R: 2544, 4765, 5941, 6203, 6822
X-ray diffraction	B: 741; R: 2135, 3243, 3385
DSC–DTA measurements	(cf. Section 34.5)
Electric measurements	R: 5478, 6138, 7012
Penetrometry	R: 5769, 6763
Brittle temperature measurements	B: 1110; R: 5007
Softening point measurements	B: 1464; R: 2235
Dynamic mechanical properties	B: 309; R: 3284, 6166, 6532, 7174
Dielectric constant	B: 309; R: 6137, 6139
Ultrasonic	R: 2144, 3395, 3921, 4897, 6985, 7020
Chromatography	R: 4836, 5428 (cf. also Section 24.3.1)

polymer never reaches thermodynamic equilibrium. After an infinitely long time under equilibrium conditions the configurational entropy becomes zero at T_g.

Polymer samples for T_g measurements should be amorphous, free from internal stress, and unoriented. T_g measurements should be made on samples which have been melted, rapidly quenched, and annealed for at least 30 min at some 20 °C above T_g, but below the crystallization temperature range. It is important that the T_g should be shown to be reproducible after repeated annealing and that the period between measurements at all temperatures, including room temperatures, should be recorded.

In Table 32.2 are collected references to examples of different methods of T_g measurement.

32.7 THE MELTING POINT (T_m)

The MELTING POINT (T_m) of a semicrystalline polymer is that temperature at which the liquid crystalline polymer repeat units are in equilibrium with respect to the composition of the liquid phase. T_m is the temperature at which the last traces of crystallinity disappear when melting is carried out according to an extremely slow heating schedule. The condition of equilibrium may be expressed as follows:

$$\mu_u^c - \mu_u^\circ = \mu_u - \mu_u^\circ, \qquad (32.15)$$

where

μ_u is the chemical potential of the liquid polymer repeat unit,
μ_u^c is the chemical potential of the polymer in crystalline form,
μ_u° is the chemical potential in the standard state.
The left side of eq. (32.15) may be expressed as

$$\mu_u^c - \mu_u^\circ = -\Delta H_u[1 - (T/T_m)], \qquad (32.16)$$

where

T is the thermodynamic temperature (in kelvins),
T_m is the melting point of pure polymer,
ΔH_u is the heat of fusion of the polymer repeat unit.

At the melting point (T_m) all changes follow a first-order thermodynamic transition and include the free enthalpy and discontinuous changes in volume, density, enthalpy, entropy, and coefficients of expansion, compressibility, specific heat, refractive index, birefrigence, and transparency. Any one of these changes can be used to determine T_m.

The T_m depends on the thermal history of the polymer sample, because the crystal perfection and crystallite size are dependent on the rate of crystallization.

The T_m of branched polymers are slightly dependent on T_c (the crystallization temperature).

Bibliography: 375, 686, 861, 1024, 1099.
References: 2158–2160, 2324, 2469, 3111, 3426, 3954, 4208, 5020, 5022, 5081, 5596, 6070, 6627, 7192.

540

32.8 EFFECT OF SOLVENT ON THE MELTING POINT (T_m)

For a semicrystalline polymer the equilibrium melting point of the pure polymer (T_m°) will be lowered to a new value (T_m)

(i) if the amorphous phase contains a solvent but the diluent does not enter the crystalline phase,

$$\frac{1}{T_m} - \frac{1}{T_m^\circ} = \frac{RV_2}{\Delta H_2 V_1}(v_1 - \chi v_1^2), \tag{32.17}$$

where

T_m is the melting point of the semicrystalline polymer whose amorphous phase contains solvent,

T_m° is the melting point of the pure semicrystalline polymer,

R is the universal gas constant (cf. Section 40.2),

V_1 and V_2 are molar volumes of solvent and polymer, respectively,

v_1 is the volume fraction of solvent,

χ is the Flory interaction parameter (cf. Section 2.9),

ΔH_2 is the heat of fusion of the repeat unit of the polymer;

(ii) if the amorphous phase contains a solvent but the diluent enters the crystalline phase,

$$\frac{1}{T_m} - \frac{1}{T_m^\circ} = \frac{RV_2}{\Delta H_2 V_1}\left(v_1 - \frac{BV_1}{RT_m}v_1^2\right), \tag{32.18}$$

where

B is a parameter that characterizes the interaction energy density of the polymer solvent pair.

From comparison of eqs (32.17) and (32.18) it can be seen that

$$\chi = BV_1/RT_m. \tag{32.19}$$

The heat of fusion of the repeat unit of the polymer (ΔH_2) can be calculated from the plot of $(1/T_m - 1/T_m^\circ)/v_1$ versus v_1/T_m (Fig. 32.9) if the ratio V_2/V_1 is

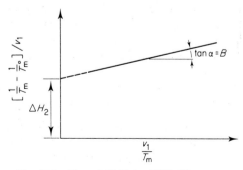

Fig. 32.9   Plot of $[(1/T_m)-(1/T^\circ{}_m)]/v_1$ versus v_1/T_m for a given polymer

known. The plot yields ΔH_2 as the intercept at the ordinate and the slope is given by $\tan \alpha = \chi$ (or B).

Bibliography: 456, 861, 862, 1024.
References: 2207, 4080, 4270, 4430, 4629, 5056, 5417, 6068.

32.9 RELATION BETWEEN T_g AND T_m

In linear semicrystalline homopolymers T_g and T_m values cannot be independently controlled. As T_g is raised or lowered in a homologous homopolymer series, T_m is similarly raised or lowered. The ratio T_g/T_m changes from 0.5 for symmetrical polymers, e.g. poly(ethylene), to 0.75 for unsymmetrical polymers, e.g. poly(styrene) (Table 32.3).

Table 32.3 T_g, T_m, and T_g/T_m values for some selected polymers

Polymer	T_g (°C)	T_m (°C)	T_g/T_m†
Poly(amide) (Nylon)	47	225	0.64
Poly(ethylene)	− 68	135	0.50
Poly(isoprene)	− 70	28	0.67
Poly(propylene)	− 18	176	0.57
Poly(styrene)	100	230	0.75
Poly(vinyl chloride)	82	180	0.78
Poly(vinylidene chloride)	− 39	210	0.48
Silicone rubber	− 123	− 58	0.70

† This ratio is calculated using absolute temperature values of T_g and T_m.

Bibliography: 183, 806, 1396.
References: 2307, 2530, 2643, 5596.

32.10 PRACTICAL IMPLICATIONS OF T_g AND T_m

The thermal behaviour of a high polymer is one of its most important properties. For crystalline polymers it is determined by the melting point (T_m) and for amorphous polymers by the glass transition temperature (T_g). Knowledge of these transition temperatures is important for the full characterization of physical and mechanical properties and for the processing of polymers and the manufacture of plastics.

According to the importance of the transition temperature in engineering design, polymeric materials may be divided into five classes:

(i) Elastomers. These materials are used above T_g in order to maintain the high, local-segmental mobility required in such materials.

(ii) Amorphous polymers. These materials are used at temperatures below T_g.

(iii) Tough, leather-like polymers. These polymers are limited to use in the immediate vicinity of their glass transition temperatures.

(iv) Semicrystalline polymers. Polymers which have 50% crystallinity can be used at temperatures between T_g and T_m, where the material exhibits moderate rigidity and a high degree of toughness.

(v) Highly crystalline and oriented polymers. These materials are used at temperatures below T_m (of the order of 100 °C), since changes in crystal structure occurs above T_g as T_m is approached.

Bibliography: 127, 909, 1337, 1396.

Chapter 33

Dielectric Properties of Polymers

General bibliography: 167, 309, 350, 483, 591, 594, 625, 686, 710, 877, 893, 1234, 1341, 1351.

Polymers which contain polar groups exhibit dielectric properties. The dielectric properties of a polymer are characterized by the DIELECTRIC CONSTANT (ε), defined as

$$\varepsilon = 4\pi lC/A \qquad (33.1)$$

(dimensionless), where

l is the thickness of the polymer sample (in centimetres),
A is the area of the sample (in square centimetres),
C is the capacitance of a condenser,

$$C = q/V \qquad (33.2)$$

(in centimetres in the electrostatic system of units),
q is the charge on the condenser plates,
V is the potential between the condenser plates.
(Note: the dielectric constant for a vacuum is $\varepsilon = 1.0000$ and for air $\varepsilon = 1.0005$ at 20 °C.)

The DIELECTRIC CONSTANT is much more conveniently expressed as the ratio of the capacitance of the condenser with the polymeric material in place to its capacitance with a vacuum.

$$\varepsilon = C_{\text{polymer}}/C_{\text{vacuum}}. \qquad (33.3)$$

For polymers there is no adequate theory predicting dielectric behaviour and all data from experiments must be correlated with physical, mechanical, and molecular properties by empirical means.

In dielectric experiments an alternating high-frequency electric field produces an alternating electric polarization which, in the case of polar solids, will lag behind the applied field by some phase angle δ.

The complex dielectric constant in this case consists of two components:

(i) ε', which is the real dielectric constant of the sample;
(ii) ε'', which is the DIELECTRIC LOSS FACTOR.

The ratio of these two components is called the LOSS TANGENT or DISSIPATION FACTOR (tan δ)

$$\tan \delta = \varepsilon''/\varepsilon'. \qquad (33.4)$$

The components ε' and ε'' and tan δ plotted against the applied high-frequency electric field (f) for one relaxation time are shown in Fig. 33.1.

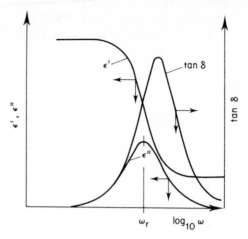

Fig. 33.1 ε' and ε'' and tan δ plotted versus log ω

Both factors ε' and ε'' describe the dielectric relaxation processes by the following equations:

$$\varepsilon' = \varepsilon_0 + (\varepsilon_0 - \varepsilon_\infty)/(1 + \omega^2 \tau^2), \qquad (33.5)$$

$$\varepsilon'' = \omega\tau + (\varepsilon_0 - \varepsilon_\infty)/(1 + \omega^2 \tau^2), \qquad (33.6)$$

where

ε_0 is the static dielectric constant related to the actual dipole moment of the macromolecule,

ε_∞ is the dielectric constant measured at high frequencies,

ω is the angular frequency,

τ is the DIELECTRIC RELAXATION TIME (in seconds), i.e. the time required for orientation of the permanent dipoles in the applied electric field.

In a relaxation region ε' decreases with frequency from a value of ε_0 to ε_∞. This decrease is called the DIELECTRIC DISPERSION ($\Delta\varepsilon$) and it corresponds to some relaxation processes in polymers (cf. Section 32.1). The dielectric dispersion can be calculated from the area under a plot of ε'' versus $1/T$ on the basis of the equation

$$\Delta\varepsilon = \frac{2E}{\pi R} \int_{-\infty}^{+\infty} \varepsilon'' \, d(1/T), \qquad (33.7)$$

where

E is the activation energy,

R is the universal gas constant (cf. Chapter 40.2).

The difference $(\varepsilon_0 - \varepsilon_\infty)$ is known as the magnitude of the relaxation and is a measure of the orientation polarization. In the relaxation region ε'' passes through a maximum at a frequency ω_r (Fig. 33.1).

The α-, β-, and γ-relaxation processes associated with various molecular motions of macromolecules (cf. Section 32.1) can be identified by measuring the dielectric dispersion ($\Delta\varepsilon$) from the area under peaks of ε'' versus $1/T$ at different frequencies (Fig. 33.2).

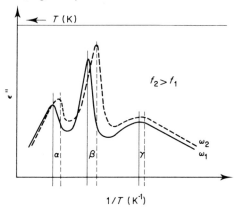

Fig. 33.2 Plot of ε versus $1/T$ for a polymer sample which may exhibit α-, β-, and γ-relaxation processes. The shape of curves changes from one polymer to another; f is the applied frequency of measurement

A diagram in which ε'' is plotted against ε', each point corresponding to one frequency, is called a COLE–COLE DIAGRAM (Fig. 33.3). For given values of ε' and ε'' the curve should be completely defined and uninfluenced by the frequency range in which the relaxation appears (i.e. independent of ω).

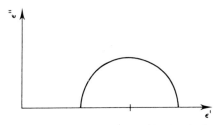

Fig. 33.3 Cole–Cole plot for a single relaxation time (τ)

When relaxation processes are not described by one relaxation time, then a semicircular arc is not obtained (Fig. 33.4).

The study of dielectric relaxation processes is useful in the analysis of dynamic mechanical data.

The glass transition temperature T_g is generally considered as the temperature at which the dielectric α-relaxation process occurs. The disadvantage of dielectric

546

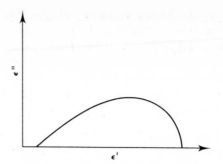

Fig. 33.4 Cole–Cole plot for processes which are described by different relaxation times (τ)

measurement for the determination of T_g is that the dielectric relaxation for many polymers shifts with an increase in the applied frequency of measurement (f) (Fig. 33.5).

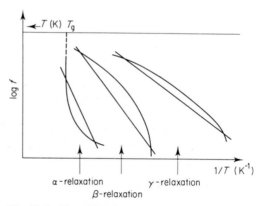

Fig. 33.5 Plot of $\log f$ versus $1/T$ showing shifting of α-, β-, and γ- relaxation processes with increasing applied frequency of measurement (f)

The EFFECTIVE DIPOLE MOMENT can be calculated from the ONSAGER EQUATION:

$$\frac{\bar{\mu}_s^2}{n^s} = \frac{3kT}{4\pi N}\left(\frac{(\varepsilon_0 - \varepsilon_\infty)(2\varepsilon_0 + \varepsilon_\infty)}{3\varepsilon_0}\right)\left(\frac{3}{\varepsilon_\infty + 2}\right)^2, \tag{33.8}$$

where

  $\bar{\mu}_s$ is the average dipole moment per effective chain segment,
  n_s is the number of monomer units per segment,
  k is the Boltzman constant (cf. Section 40.3),
  T is the absolute temperature,
  N is the number of dipoles per unit volume,

ε_0 is the static dielectric constant related to the actual dipole moment of the macromolecule,

ε_∞ is the dielectric constant measured at high frequency.

33.1 INSTRUMENTATION FOR THE MEASUREMENT OF THE DIELECTRIC CONSTANT AND THE LOSS FACTOR

The dielectric constant (ε) and the loss factor (ε'') are measured using equipment based on the Schering bridge design (Fig. 33.6). The Schering bridge operates

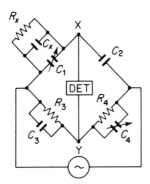

Fig. 33.6 Schematic diagram of a Schering bridge

on alternating current in the frequency (f) range 10^1–10^6 Hz. Its elements are impedances instead of resistances, and the balance is achieved by adjusting separately the capitance and resistance of the variable inpedance:

(i) measurement of the capacitance (C_x) yields the dielectric constant (ε);
(ii) measurement of the resistance (R_x) yields the loss tangent (tan δ).

The sample takes the form of a thin disc and is placed between two electrodes of the condenser; these electrodes must adhere directly to the sample surfaces and can be made

(i) from aluminium foil;
(ii) by painting with special conducting paints containing colloidal silver or graphite;
(iii) by vacuum deposition of metal or graphite.

The samples are used in the form of film, typically 5 cm in diameter and 0.5 mm in thickness. They can be prepared by moulding, calendering, or casting from a solvent.

The data from the Schering bridge can be computerized. The input data include details of measured capacitance and resistance, sample thickness, cell area at constant temperature, and applied frequency. The computer then calculates the series and parallel capacitance and the true dissipation factor.

These quantities are combined with the sample area and thickness data to calculate the dielectric constant and the loss factor.

Bibliography: 309, 310, 877, 893, 1351.
References: 4239, 4241, 5994.

33.2 APPLICATIONS OF DIELECTRIC MEASUREMENTS TO THE STUDY OF THE STRUCTURE OF POLYMERS

Dielectric measurements provide valuable information for the study of
  (i) the dipole moments in polymers,
  (ii) the α-, β-, and γ-relaxation processes in polymers,
  (iii) the conformation of polymers in the glassy state,
  (iv) the stereoregularity of polymers,
  (v) the determination of a glass transition temperature (T_g).

Bibliography: 584, 591, 594, 710, 825, 857, 921, 1010.
References: 2014, 2015, 2090, 2122, 2133, 2156, 2202, 2203, 2384, 2445–2447, 2568, 2638, 2783, 2834, 2951, 3564, 3619, 3815, 3823, 3884, 3951, 4068, 4071, 4078, 4144, 4145, 4206, 4207, 4240, 4242, 4252, 4263, 4264, 4357, 4800, 4806, 4828, 4829, 4960, 4984, 5041, 5103, 5104, 5120, 5266, 5292, 5298, 5416, 5432–5435, 5476, 5519, 5526, 5527, 5614, 5680, 5787, 5821, 5822, 5846–5851, 5861, 5869, 5870, 5994, 6034, 6035, 6064, 6153, 6188–6190, 6196, 6313, 6368–6370, 6372, 6514, 6579, 6581, 6713–6717, 6791, 6887, 6910, 7033, 7104, 7105, 7173, 7234–7239.

33.3 THERMALLY STIMULATED DISCHARGE TECHNIQUE

The THERMALLY STIMULATED DISCHARGE TECHNIQUE is a method used to investigate the dielectric properties of polymers by measuring the depolarization current of the polarized sample.

The THERMAL SAMPLING TECHNIQUE is a modification of the thermally stimulated discharge technique and is applied to investigate molecular motions (relaxation processes) in polymers.

In both techniques a polymer film is placed between vacuum-evaporated gold electrodes and polarized by an electric field. The depolarization current is further measured as a function of temperature.

Bibliography: 1338.
References: 2384, 2805, 3389, 3815, 4063, 4754, 4800, 6701, 6918, 6919, 7328–7330.

Chapter 34

Thermal Analysis of Polymers

General bibliography: 124, 127, 142, 243, 309, 315, 351, 460, 620, 719, 851, 855, 1320, 1372, 1373, 1419.

THERMAL ANALYSIS can be defined as the measurement of a property of a polymer sample as a function of temperature.

The following types of thermal analytical techniques are used:

(i) Static methods associated with weight change:
   (a) ISOBARIC WEIGHT-CHANGE DETERMINATION is a technique which records the equilibrium weight of a substance as a function of temperature (T) at a constant partial pressure of the volatile products;
   (b) ISOTHERMAL WEIGHT-CHANGE DETERMINATION is a technique which records the dependence of the weight of a substance on time (t) at constant temperature.
(ii) Dynamic methods associated with weight change:
   (a) THERMOGRAVIMETRY (TG) is a technique which records the weight of a substance in an environment heated or cooled at a controlled rate as a function of time or temperature;
   (b) DERIVATIVE THERMOGRAVIMETRY (DTG) is a technique which yields the first derivative of the thermogravimetric curve with respect to either time or temperature.
(iii) Methods associated with energy change:
   (a) HEATING CURVES ANALYSIS is a technique which records the temperature of a substance against time in an environment heated at a controlled rate;
   (b) HEATING-RATE CURVES ANALYSIS is a technique which records the first derivative of the heating curve with respect to time, i.e. dT/dt, plotted against time or temperature;
   (c) INVERSE HEATING-RATE CURVES ANALYSIS is a technique which records the first derivative of the heating curve with respect to temperature, i.e. dt/dT, plotted against either time or temperature;
   (d) DIFFERENTIAL THERMAL ANALYSIS (DTA) is a technique which records the difference in temperature between a substance and a reference material against either time or temperature as the two specimens are subjected to identical temperature conditions in an environment heated or cooled at a controlled rate;

(e) DERIVATIVE DIFFERENTIAL THERMAL ANALYSIS is a technique which yields the first derivative of the differential thermal curve with respect to either time or temperature;

(f) DIFFERENTIAL SCANNING CALORIMETRY (DSC) is a technique which records the energy necessary to establish zero temperature difference between a substance and a reference material against either time or temperature as the two specimens are subjected to identical temperature conditions in an environment heated or cooled at a controlled rate.

(iv) Methods associated with evolved volatiles:

(a) EVOLVED GAS DETECTION (EGD) is a technique which detects whether or not a volatile product is formed during thermal analysis;

(b) EVOLVED GAS ANALYSIS (EGA) is a technique which determines the nature and/or amount of volatile products formed during thermal analysis.

The above definitions are recommended by the Council of the International Confederation for Thermal Analysis (ICTA).

Bibliography: 124, 351, 851, 1372.
References: 5014–5016, 5164, 5165, 6577.

34.1 THERMOGRAVIMETRY

THERMOGRAVIMETRY (TG) is a dynamic method in which the weight loss (w) of a sample is measured continuously as

(i) a function of temperature (T) at a constant rate;

(ii) a function of time (t) at a constant temperature (ISOTHERMAL or STATIC THERMOGRAVIMETRY):

$$w = f(T \text{ or } t). \tag{34.1}$$

An experimental plot of weight loss versus temperature is called a THERMO-GRAM (Fig. 34.1) and exhibits a thermogravimetric curve which has

(i) a first step which represents a small initial weight loss ($w_0 - w_1$) and which results from desorption of solvent—if it occurs near 100 °C, this may be attributed to loss of water;

(ii) a second step ($w_1 - w_2$)—and sometimes a third step ($w_2 - w_3$)—which are results of thermal decomposition of the sample.

DERIVATIVE THERMOGRAVIMETRY (DET) is a dynamic method in which the derivative of the change of mass with respect to time (dw/dt) is measured as

(i) a function of temperature (T) at a constant rate;

(ii) a function of time (t) at a constant temperature (ISOTHERMAL or STATIC DERIVATIVE THERMOGRAVIMETRY):

$$\frac{dw}{dt} = f(T \text{ or } t). \tag{34.2}$$

The curve obtained is the FIRST DERIVATIVE OF THE MASS-CHANGE CURVE. A curve contains a series of peaks instead of the stepwise curve. The areas under the peaks are proportional to the total change in the mass of the sample.

A comparison between a TG and a derivative (DTG) mass-loss curve is given in Fig. 34.1. The derivative (DTG) curve may be obtained from the TG curve

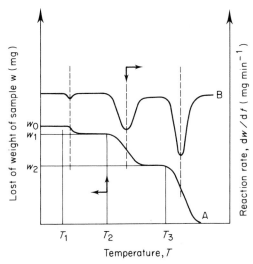

Fig. 34.1   Typical thermograms: curve A, a thermo-gravimetric (TG) curve; curve B, a derivative thermodynamic (DTG) curve

either by manual differentiation methods or by electronic differentiation of the TG signal. The DTG curve, whether derived mathematically or recorded directly, contains no more information than does an integral TG curve obtained under the same experimental conditions. Accessory equipment is available for most thermobalances so that the DTG curve can be easily recorded along with the TG curve.

The main application of thermogravimetric analysis is in the study of thermal kinetics. For reactions which involve a decomposition of the type

$$A_{(solid)} \xrightarrow{k} B_{(solid)} + C_{(gas)}, \tag{34.3}$$

the rate of loss of weight (k) is given by the Arrhenius equation:

$$k = \frac{dW}{dt} = AW^n e^{-E/RT}, \tag{34.4}$$

552

where

A is the pre-exponential factor,
W is the weight of the remaining sample,
n is the order of reaction,
dW/dt is the reaction rate (in milligrams per second (CGS) or kilograms per second (SI)),
E is the activation energy,
R is the universal gas constant (cf. Section 40.2),
T is the temperature.

Equation (34.4) can be presented in its logarithmic form as

$$\ln k = \ln\left(\frac{dW}{dt}\right) = \ln A + n \ln W - E/RT. \tag{34.5}$$

There are several methods for determination of the pre-exponential factor, the order of reaction (n), and the activation energy (E):

(i) DIFFERENTIAL METHOD. In this method eq. (34.5) is applied at two different temperatures and the resulting expressions are substracted from one another, and then eq. (34.5) can be expressed as a difference equation:

$$\Delta \ln\left(\frac{dW}{dt}\right) = n \Delta \ln W - (E/R)\Delta(1/T). \tag{34.6}$$

The n and E/R can be calculated from the plot on Fig. 34.2. The plot yields E/R as the intercept at the ordinate and the slope is $\tan \alpha = n$.

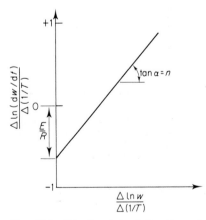

Fig. 34.2 Kinetic plot of the thermal degradation of a given polymer at constant $\Delta(1/T)$

Bibliography: 459, 479, 812, 1102, 3512.
References: 2118, 2175, 3431, 6238.

(ii) MULTIPLE HEATING RATES METHOD. In this method the value of the constant heating rate is changed from run to run but other conditions remain the same. A set of different thermogravimetric curves is obtained (Fig. 34.3). Equation (34.6) can be applied directly to the determination of the pre-exponential factor (A) and the activation energy (E) from the plot on Fig. 34.4. The plot yields A as the intercept at the ordinate and the slope is $\tan \alpha = E$.

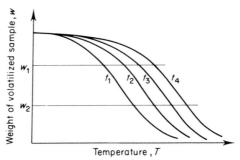

Fig. 34.3 Thermograms for a given polymer sample heated at various heating rates

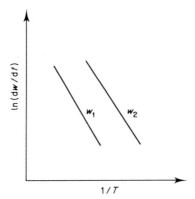

Fig. 34.4 Kinetic plot of ln (dw/dt) versus $1/T$ for a given polymer at the indicated values of w_1 and w_2 obtained from Fig. 34.3

In order to evaluate the order of reaction (n) one can use eq. (34.6) applied at $\ln k = 0$, i.e.

$$E/RT_0 = \ln A + n \ln W. \qquad (34.7)$$

In this case, a plot of E/RT_0 versus $\ln W$ (Fig. 34.5) should give a linear reaction whose slope is $\tan \alpha = n$.

Bibliography: 459, 479, 812, 1102.
References: 2118, 2175, 3431, 3512, 6238.

554

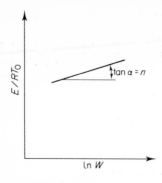

Fig. 34.5 Determination of the reaction order (n) for a given polymer

(iii) MAXIMIZATION OF RATE METHOD. See B: 1102; R: 3554, 6013.
(iv) VARIABLE HEATING RATE FOR A SINGLE THERMOGRAM METHOD. See B: 1102; R: 6013.
(v) APPROXIMATE INTEGRATION OF RATE EQUATION METHOD. See B: 459, 1102; R: 2332, 2680, 2916, 3229, 3230, 3430, 5005, 5651, 5653, 6010, 6011, 6340, 6408, 6419.

In polymer chemistry, thermogravimetry is mainly applied to the study of

(i) thermal degradation of polymer samples (kinetics and mechanisms);
(ii) thermal stability of polymer samples;
(iii) oxidative degradation;
(iv) solid-state reactions;
(v) determination of moisture, volatiles, and ash;
(vi) absorption, adsorption, and desorption processes;
(vii) plasticizer volatility;
(viii) composition of plastics and composites;
(ix) identification of polymers by analysis thermograms.

Bibliography: 309, 383, 459, 460, 812, 1101, 1102, 1104, 1180, 1372, 1421.
References: 2115, 2560, 3228, 3503, 4926, 4928, 5025, 6014, 6438.

34.1.1 Thermobalances

The most important problems in thermal analysis are a constant rate of temperature change and temperature measurement. A temperature sensor should be in contact with the sample even if this causes some inconvenience. Commercial thermobalances have a programmed heating temperature and a gaseous atmosphere. Most of the newer systems display derivative TG curves and several of them have a computer for temperature programming and for retrieval and analysis of data. The most common thermobalances are as follows:

(i) The horizontal balance. An instrument of this type is shown in Fig. 34.6. The sample (0.5–500 mg) is placed in a furnace on the panel of one arm of a precision balance. The change in sample weight is automatically recorded as the sample is maintained at a constant high temperature or

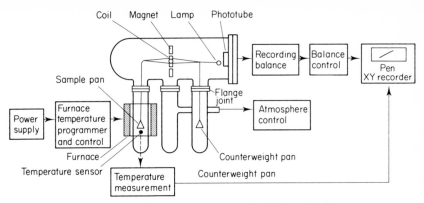

Fig. 34.6 Schematic representation of a horizontal balance

dynamically heated by programmed heating (maximum temperature of 1200 °C). The floating thermocouple which is not in direct contact with the sample should be placed as near to the sample as possible. The gaseous laminar flow should be preheated before entering the furnace tube.

(ii) The hangdown balance (vertical balance). An instrument of this type is shown in Fig. 34.7. Hangdown balances have very high sensitivity up to

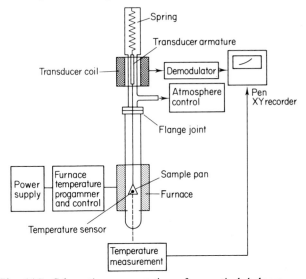

Fig. 34.7 Schematic representation of a vertical balance

10^{-7} g. The main disadvantage of this type of balance is the inconvenience in the placement of a temperature thermocouple in contact with the sample.

(iii) The substitution balance. An instrument of this type operates on the substitution principle with built-in counterweights and electromagnetic

556

compensation. The temperature thermocouple is in direct contact with the sample during the weight-loss measurement. A special reactive gas inlet allows for a rapid change of gaseous atmosphere in the sample chamber. The balance mechanism operates in the inert gas atmosphere, and is protected from corrosive gases.

Most commercially produced balances can be evacuated to 10^{-5}–10^{-8} torr (1 torr = $\frac{1}{760}$ atm = 133.322 N m^{-2} (SI)).

Bibliography: 38, 142, 351, 459, 496, 722, 850, 860, 1102, 1370–1373.

34.2 THERMAL CHROMATOGRAPHY

THERMAL CHROMATOGRAPHY is a technique which continuously measures and records the weight loss of a polymer sample as its temperature is raised at a steady rate. The sample temperature is· programmed. Instead of recording the weight loss, thermal conductivity of the evolved gases is measured, which is then followed by gas chromatography. Nanogram quantities can be detected by this method.

Bibliography: 1269.
References: 2341, 3565, 3567, 3568, 5722, 6654.

34.3 DIFFERENTIAL THERMAL ANALYSIS METHODS

DIFFERENTIAL THERMAL ANALYSIS (DTA) is a technique which records the difference in temperature between a sample and a reference material against either time or temperature as the two specimens are subjected to identical temperature conditions in an environment heated or cooled at a controlled rate.

DIFFERENTIAL SCANNING CALORIMETRY (DSC) is a technique which records the energy necessary to establish zero temperature difference between a sample and a reference material against either time or temperature as the two specimens are subjected to identical temperature conditions in an environment heated or cooled at a controlled rate.

Schematic DTA or DSC curves are plotted as a function of time or temperature at a constant rate of heating. The ordinate represents ΔT (the difference between the sample and reference temperatures) for DTA measurements or $d\Delta Q/dt$ (the power difference between the sample and reference cells) for DSC measurements. Schematic DTA or DSC curves illustrating many of typical phenomena are shown in Fig. 34.8. The peak area between the curve and a baseline is proportional to the enthalpy change in the sample. A shift in the baseline results from a change in the heat capacity (or mass) of the sample.

As a general rule:

(i) First-order transitions should give narrow peaks. Physical transitions in polymers are often from structures of varied size and configuration, and

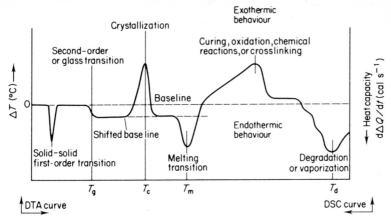

Fig. 34.8 Schematic DTA or DSC curve

yield spectra much broader than similar transitions in low molecular weight compounds.

(ii) Second-order transitions or glass transitions cause abrupt changes in curve shape. The sample absorbs more heat because of its higher heat capacity.

(iii) Chemical reactions such as polymerization, curing (which may be either endothermic or exothermic), oxidation, or crosslinking (which are always endothermic) give broad peaks.

(iv) The curves for cooling scans are inverse to those for heating to the extent that the transitions and processes are reversible. Cooling transitions often either occur at lower temperatures due to supercooling or tend to be spread out over a wider temperature interval.

The interpretation of curves is difficult and often depends on subsidiary information from other analytical techniques.

Bibliography: 78, 90, 309, 351, 358, 459, 496, 532, 718, 751, 850, 852, 867, 1058, 1174, 1181, 1233, 1368, 1370–1373, 1425, 1428.

A computer program for analysis of polymer cooling curves by DTA is published in R: 3425, 6042.

34.3.1 Differential Thermal Analysis Instruments

Differential thermal analysis instruments (differential thermographs) measure a temperature difference (ΔT) between a sample and a reference material. They are built of the following basic components (Fig. 34.9):

(i) A measuring cell, which consists of the heating block which is programmed to increase in temperature linearly with time. There are two types of cell:

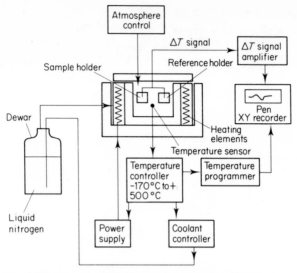

Fig. 34.9   Low- and high-temperature operating DTA instrument

(a) the classical DTA cell (Fig. 34.10). The sample and the reference are placed in separated holders. Thermocouples placed in the centres of the sample and reference material (thermally inert materials, e.g. fused quartz, porcelain, glass beads, or MgO) measure their temperature and also the temperature difference between them. The main

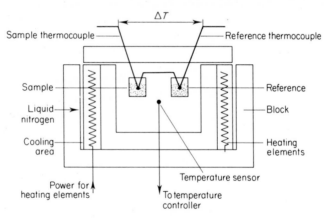

Fig. 34.10   Essential elements of the DTA cell

disadvantage of this type of cell is that the temperature difference depends on the density of the sample, the thermal conductivity of the sample, the specific heat of the sample, the thermal diffuseness of the sample, and the reference material, and the geometry of the system applied. Therefore, poor agreement is often obtained upon comparison of DTA curves from different instruments.

(b) The calorimetric DTA cell (Fig. 34.11). The sample and reference holders are isolated from each other and the thermocouples are

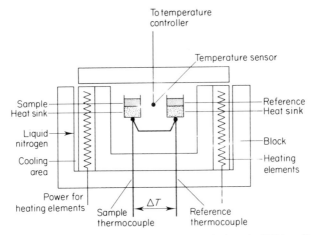

Fig. 34.11 Essential elements of the calorimetric DTA cell

placed in the heat flow path. The heat sink is of sufficient thermal inertia that it limits the heat flow through the thermocouples, unlike the DTA measuring cell in which thermal properties and the geometry of the sample affect the magnitude of heat flow.

The sample holder assembly and heating block are designed to achieve uniform temperature throughout sample and reference material and to operate under steady-state conditions.

(ii) A temperature programmer. It is impossible to evaluate heats of transition or decomposition without a linear temperature increase or decrease. The heating rate is an important parameter of the measurement. With a higher heating rate, the glass transition temperature (T_g) tends to increase slightly, whereas the melting point (T_m) is little influenced; the temperature of crystallization (T_c) and the temperature of decomposition of the sample (T_d) are also influenced significantly.

(iii) A heating–cooling system. Modern differential thermographs are designed to work at temperatures ranging from $-170\,^{\circ}C$ to $+500\,^{\circ}C$.

(iv) A ΔT signal amplifier.

(v) A recorder. In strip-chart recording the ΔT signal is recorded as a function of time (t), and the temperatures of the sample (T_s) or of the system (T_a) are recorded simultaneously with a separate recorder or pen on a comparable time base.

(vi) An atmosphere control system. When low-temperature ($-170\,^{\circ}C$ to $20\,^{\circ}C$) measurements are made, dry nitrogen or helium must be passed through the sample holder assembly to prevent condensation of water vapour. Up to $600\,^{\circ}C$, nitrogen is used as a carrier when water vapour,

CO_2, or other products given off during transitions are to be swept through the thermocouple.

Bibliography: 78, 309, 496, 852, 1372.
References: 2140, 2259, 2467, 6314, 7050.

34.3.2 Sampling Techniques for DTA

DTA instruments can analyse solid and liquid samples. Solid samples can be in foil, powder, crystal, or granular form.

The general recommendations are as follows:

(i) The sample should be small (20–50 mg) and it is sometimes advisable to dilute the sample with a small amount of diluent (material identical with the reference sample).

(ii) Maximum sensitivity and resolution are obtained by minimizing the sample and increasing the efficiency of heat transfer from the sample to its thermocouple.

Table 34.1   Aluminium oxide heat capacity standards used for DTA and DSC calibration (CGS system) (from R: 3660; reproduced by permission of the American Chemical Society)

K	ΔH (cal g^{-1})	K	ΔH (cal g^{-1})	K	ΔH (cal g^{-1})	K	ΔH (cal g^{-1})
0	0.0	160	0.084 36	420	0.231 02	690	0.273 42
5	0.000 003	170	0.093 62	430	0.233 57	700	0.274 33
10	0.000 022	180	0.102 63	440	0.236 03	720	0.276 07
15	0.000 074	190	0.111 47	450	0.238 35	740	0.277 68
20	0.000 178	200	0.119 88	460	0.240 61	760	0.279 23
25	0.000 332	210	0.127 99	470	0.242 71	780	0.280 69
30	0.000 616	220	0.135 77	480	0.244 77	800	0.282 07
35	0.001 026	230	0.143 20	490	0.246 72	820	0.283 36
40	0.001 619	240	0.150 30	500	0.248 60	840	0.284 58
45	0.002 436	250	0.157 08	510	0.250 38	860	0.285 75
50	0.003 497	260	0.163 50	520	0.252 11	880	0.286 85
55	0.004 850	270	0.169 62	530	0.253 78	900	0.287 91
60	0.006 517	280	0.175 43	540	0.255 35	920	0.288 89
65	0.008 486	290	0.180 94	550	0.256 90	940	0.289 85
70	0.010 74	298.16	0.185 21	560	0.258 35	960	0.290 74
75	0.013 29	300	0.186 15	570	0.259 78	980	0.291 61
80	0.016 16	310	0.191 09	580	0.261 14	1000	0.292 41
85	0.019 33	320	0.195 74	590	0.262 47	1020	0.293 21
90	0.022 72	330	0.200 17	600	0.263 74	1040	0.293 93
95	0.026 31	340	0.204 36	610	0.264 95	1060	0.294 64
100	0.030 10	350	0.208 34	620	0.266 15	1080	0.295 31
110	0.038 23	360	0.212 19	630	0.267 30	1100	0.295 97
120	0.047 00	370	0.215 80	640	0.268 41	1120	0.296 58
130	0.056 17	380	0.219 20	650	0.269 46	1140	0.297 18
140	0.065 54	390	0.222 41	660	0.270 51	1160	0.297 73
150	0.074 99	400	0.225 45	670	0.271 50	1180	0.298 27
		410	0.228 32	680	0.272 48	1200	0.298 81

Note: 1 cal (CGS) = 4.1868 J (SI).

(iii) The small particle size increases the surface area and shifts the transition peaks to lower temperatures. Thermal conductivity and diffusity are influenced by sample density. Tighter packing increases heat conduction. Packing of the sample is especially important in cases where gaseous products are evolved or where samples are studied under a gaseous atmosphere.

(iv) The reference material (heat capacity standards) should be chosen to give as small a temperature difference as possible between sample and reference by balancing their heat capacities.

Table 34.2 Different heat capacity standards used for DTA and DSC calibration (from B: 1372; reproduced by permission of John Wiley & Sons)

Temperature (°C)	Standard	ΔH_f (cal g^{-1})	ΔH_t (cal g^{-1})
34.6	Azoxybenzene	21.6	
44.9	C_2H_6		2.59
47	CBr_4		4.81
48.2	Benzophenone	23.5	
62.5	Palmitic acid	51.2	
69	Stearic acid	47.5	
69.8	Biphenyl	28.7	
99.3	Phenanthrene	25.0	
114	o-Dinitrobenzene	32.3	
121.8	Benzoic acid	33.9	
125	NH_4NO_3		12.6
128	KNO_3		12.86
130	$BaCl_2.2H_2O$	116.6 ($-2H_2O$)	
137.2	NH_4Br		(882 cal mol^{-1})
150	$CuSO_4.5H_2O$	228.5 ($-4H_2O$)	
156.4	In	6.79	
177.0	KCNS	25.72	
179	Ag_2S		4.08
180	$CaSO_4.2H_2O$	157.2 ($-2H_2O$)	
183.1	NH_4Cl		(1873 cal mol^{-1})
187.8	Pentaerythritol	77.1	
212	$AgNO_3$	17.7	
231.9	Sn	14.4	
252	$LiNO_3$	88.5	
299.8	$KClO_4$		23.7
306.2	$NaNO_3$	44.2	
327.4	Pb	5.50	
337	KNO_3	28.1	
350	$CdCO_3$	134.5 ($-CO_2$)	
398	$K_2Cr_2O_7$		28.9
419.5	Zn	24.4	
498	$PbCl_2$	20.9	
553	LiBr	36	
575	$LiSO_4$	62	
588.8	Na_2WO_4		28.57
850	$CaCO_3$	427.1 ($-CO_2$)	

Note: 1 cal (CGS) = 4.1868 J (SI).

The standards used for calibration must be chemically stable and have low vapour pressures (the heat of vaporization should not contribute to the heat effect measured). Most standards used are metal oxides (Table 34.1), pure metals, or organic compounds (Table 34.2) of high purity.

Bibliography: 78, 309, 1233, 1425, 1428.
Reference: 2259.

34.4 DIFFERENTIAL SCANNING CALORIMETERS

Differential scanning calorimeters (DSC) use a method of heat flow measurement based on power compensation. They are built of the following parts (Fig. 34.12):

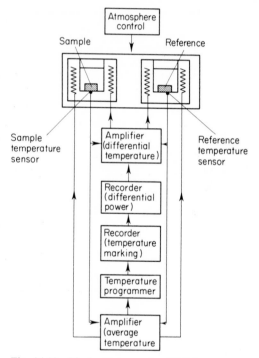

Fig. 34.12   Block diagram of a DSC instrument

The sample and reference are heated separately by individually controlled heater elements (Fig. 34.13). The power to these heaters is adjusted continuously in response to any thermal effects in the sample. In this way sample and reference are at identical temperatures. The differential power required to achieve this condition is recorded as the ordinate on the recorder, with the programmed temperature of the system as the abscissa.

The atmospheric control system is similar to that for DTA instruments (cf. Section 34.3.1).

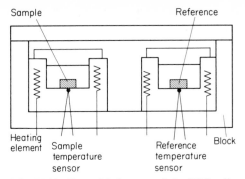

Sample   Reference

Heating
element Sample   Reference   Block
        temperature   temperature
        sensor   sensor

Fig. 34.13   Essential elements of the DSC cell

Bibliography: 78, 309, 1372, 1457.
References: 2364, 3071, 3217, 3336, 3348, 3498, 4900, 5301, 5358, 5597, 6042, 6772, 7025, 7049.

34.4.1 Sampling Techniques for DSC

DSC instruments can analyse solid and liquid samples. Solid samples can be in foil, powder, crystal, or granular form. Materials such as polymer films can be conveniently sampled by cutting out sections of the film with a standard paper punch or cork borer. Solid polymers can be sliced into thin sections with a razor blade or knife. A sample is placed in an aluminium pan which is sealed using a sample pan crimper. (The seal should be tight but not hermetic.)

The proper weight of sample depends upon the problem and may vary between 0.5 mg and 10 mg. Small samples

  (i) permit higher scan speeds;
 (ii) yield maximum resolution and thus better qualitative results;
(iii) yield most regular peak shapes;
 (iv) permit better sample contact with controlled atmospheres and better removal of decomposition products;
  (v) are recommended where the transition energy to be measured is very high.

Large samples:

  (i) permit observation of small transitions;
 (ii) yield more precise quantitative measurements;
(iii) produce larger quantities of volatile products for detection by the effluent analysis system.

Other recommendations are similar to those mentioned in discussing sampling techniques for DTA (cf. Section 34.3.2).

34.5 APPLICATION OF DTA–DSC TECHNIQUES IN POLYMER RESEARCH

DTA–DSC techniques are commonly used to measure
  (i) glass transition temperatures (T_g);

(ii) melting points (T_m);
(iii) decomposition temperatures (T_d).

DTA–DSC methods can also be applied in the direct measurement of the energy absorbed or evolved in studies of
   (i) heats of fusion:
  (ii) heats of vaporization;
 (iii) heats of crystallization;
 (iv) heats of reaction (including polymerization, oxidation, and combustion);
  (v) heats of decomposition (dehydration);
 (vi) heats of solution;
(vii) heats of adsorption (desorption);
(viii) specific heats;
 (ix) activation energies;
  (x) entropies of transition;
 (xi) solid-state transition energies.

The DTA–DSC curve (Fig. 34.14) illustrates how the various thermal processes appear during heating of a polymer sample.

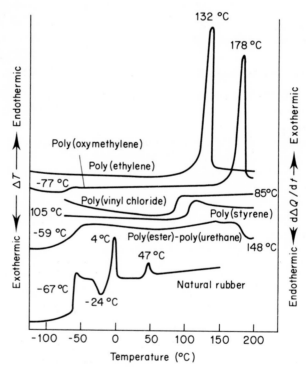

Fig. 34.14   DTA–DSC curves of some common polymers. (From B: 309; reproduced by permission of John Wiley & Sons)

Bibliography: 78, 719, 954–958, 1058, 1100, 1101, 1180, 1181, 1426.
References: 2013, 2016, 2258, 2262, 2372, 2447, 2501, 2765, 2835, 2937, 3020, 3164, 3307, 3308, 3323, 3340, 3373, 3503, 3531, 3556, 3602, 3622, 3652, 3653, 3730, 3826, 3903, 3914, 3955, 3958, 4070, 4293, 4314, 4346, 4380, 4428 4578, 4628, 4629, 4784, 4785, 4604, 4700, 4997, 5028, 5032, 5044, 5065, 5097, 5121, 5218, 5278, 5363, 5417, 5427, 5494, 5555, 5582, 5594, 5602, 5646, 5681, 5808, 5862, 5892, 6079, 6176, 6200, 6265, 6302, 6343, 6348, 6519, 6871, 6906, 7028, 7131, 7187, 7251.

34.5.1 T_g Determination by DTA–DSC Methods

DTA–DSC methods belong to the dynamic group of methods (cf. Section 34.3). The heat capacity changes are observed in a polymer sample with changes in temperature (Fig. 34.15).

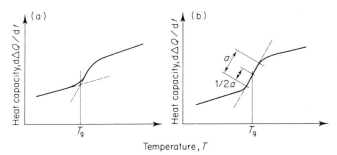

Fig. 34.15 (a) T_g taken as the intersection of the extrapolation of the baseline with the extrapolation of inflection. (b) T_g taken as the inflection point

Two methods are used for defining T_g from DTA–DSC data (Fig. 34.15):

(i) T_g taken as the intersection of the extrapolation of the baseline with the extrapolation of the inflexion;
(ii) T_g taken as the inflexion point.

Several factors may affect the reproducibility of T_g from DTA–DSC measurements:

(i) the effect of sample form, i.e. granules or a thin sheet moulded from the granules;
(ii) the effect of an imposed controlled thermal treatment;
(iii) the effect of thermally cycling the samples.

References: 2047, 2140, 2594, 2596, 2764, 2909, 2997, 3145, 3528, 3593, 3715, 3761, 3774, 4092, 4807, 4808, 5028, 5145, 5277, 5417, 5915, 6045, 6082, 6532, 6597, 6598, 6808, 7158, 7174.

34.5.2 T_g and T_m Determination by Differential Thermal Analysis Under High Pressure

The glass transition temperature T_g and the melting point T_m increase with increasing pressure. For the study of pressure effects on phase transition the high pressure DTA cell must be used (Fig. 34.16). The cell contains a chamber 100 mm long and 10 mm in diameter with openings at both ends. One of the

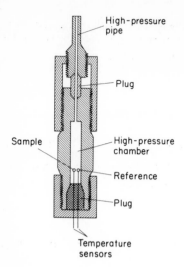

High-pressure pipe

Plug

Sample

High-pressure chamber

Reference

Plug

Temperature sensors

Fig. 34.16 High-pressure DTA cell

openings is connected with a pipe connected to the pressure intensifier. The other end is stopped with a plug containing two metal-sheathed chromel–alumel thermocouples. A sample of a few milligrams is fixed to the junction of the element wires at one end of one thermocouple and the reference sample is fixed to that of the other thermocouple. Silicone fluid is used as the pressure-transmitting fluid.

References: 3072, 4549, 6702.

34.6 QUANTITY OF HEAT

The word HEAT should be used only when referring to a method of energy transfer and, when the transfer is completed, to refer to the total amount of energy so transferred.

The QUANTITY OF HEAT has meaning only in the context of an interaction in which energy is transferred from one system to another as a result of a temperature difference.

The unit of quantity of heat is the CALORIE—defined as that amount of heat required to raise the temperature of 1 g of water by 1 K or 1 °C, from 14.5 °C to 15.5 °C.

A corresponding unit defined in terms of Fahrenheit degrees and British units is the BRITISH THERMAL UNIT, or Btu. By definition, 1 Btu is the quantity of heat required to raise the temperature of 1 lb of water from 63 °F to 64 °F.

A third unit in common use, especially in measuring the energies of chemical bonds, is the KILOCALORIE.

The relations among these units are

$$1 \text{ Btu} = 252 \text{ cal} = 0.252 \text{ kcal} = 1055 \text{ joules}.$$

The International Committee on Weights and Measures no longer recognizes the calorie as a fundamental unit, but recommends instead that the JOULE be used for quantity of heat as well as all other forms of energy.

The following relation holds:

$$1 \text{ cal} = 4.186 \text{ joules} = 4.186 \text{ J}.$$

34.7 ENTHALPY MEASUREMENTS BY DTA–DSC METHODS

The heat of transition or reaction (enthalpy) of a polymer sample (ΔH) can be determined from the area of the curve peak (A) using

$$\Delta Hm = KA, \tag{34.8}$$

where

ΔH is the heat of transition or reaction (enthalpy) (in calories per gram (CGS) or joules per kilogram (SI)),

m is the mass of the polymer sample (in grams (CGS) or kilograms (SI)),

K is the calibration coefficient—the expression used for K depends on the type of instrument used and the method of recording the DTA or DSC curves,

A is the curve peak area (in square centimetres (CGS) or square metres (SI)).

The calibration constant (K) is related to the geometry and thermal conductivity of the sample holder, and is usually determined by calibration of the system with compounds having known heats of transition (or reaction). The expression used for K depends on the type of instrument used and on the method of recording the DTA or DSC curves.

The sample curve peak can be integrated using one of the methods described in the Section 23.3. If overlapping peaks are present in the curve, the curve may be integrated in parts. For each of the various peak areas, a different K value must be used to calculate the ΔH (if DTA is used). The total ΔH is thus the sum of all the areas. In the DSC method, K is independent of the temperature and only one K value is needed. This is the advantage of DSC over DTA for quantitative measurements.

If the baseline undergoes a large displacement during the reaction or transition, the integration of the curve is difficult and leads to large errors in the ΔH calculations. The precision of DTA–DSC methods is 5–10% in most cases.

Bibliography: 78, 309, 537, 1058, 1181, 1372.
References: 2016, 2471, 5274, 5652, 5655, 5837, 5898, 6042, 6593, 6605, 6960, 7071, 7072.

34.8 DETERMINATION OF SPECIFIC HEAT CAPACITY BY DTA–DSC METHODS

The SPECIFIC HEAT CAPACITY under constant pressure (c_p) is defined as that heat quantity which is required to heat the unit mass (m) of a substance through 1 K or 1 °C (ΔT) under constant pressure (p), and is given by

$$c_p = \Delta Q/m\Delta T \qquad (34.9)$$

(in cal g^{-1} °C^{-1} (CGS) or J kg^{-1} K^{-1} (SI)), where

ΔQ is the required heat quantity (in calories (CGS) or joules (SI)),

m is the mass of the sample (in grams (CGS) or kilograms (SI)),

$\Delta T = T_2 - T_1$ is the temperature increase from T_1 to T_2 (in degrees Celsius (CGS) or kelvins (SI)).

The MOLAR HEAT CAPACITY under constant pressure (C_p) is defined as that heat quantity which is required to heat 1 mol (M) of a substance by 1 K or 1 °C (ΔT) under constant pressure (p), and is given by

$$C_p = Mc_p = \Delta Q/n\,\Delta T \qquad (34.10)$$

(in cal mol^{-1} °C^{-1} (CGS) or J mol^{-1} K^{-1} (SI)) where $n = m/M$ is the number of moles.

Specific heats are generally determined with calorimeters, but it is much more convenient to use the rapid DTA–DSC methods. In these latter methods, a curve for the empty sample container is first run (upper curve on Fig. 34.17).

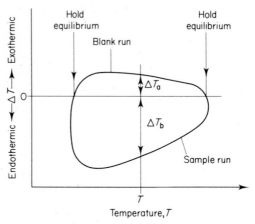

Fig. 34.17 Specific heat curves for a given polymer

The sample is then placed in the sample container and its curve recorded, using the same instrument adjustment. The specific heat capacity (c_p) at temperature (T) can be calculated from

$$(c_p)_T = (\Delta T_a + \Delta T_b)\,K/mq, \qquad (34.11)$$

where

ΔT_a is the absolute differential temperature for the empty container,

ΔT_b is the absolute differential temperature for the sample,

K is the calibration coefficient at temperature T (in cal s^{-1} °C^{-1} (CGS) or J s^{-1} K^{-1} (SI)),

m is the mass of the polymer sample,

q is the heating rate (in degrees Celsius per second (CGS) or kelvins per second (SI)).

Using the DTA–DSC method allows one to obtain a value for the specific heat (c_p) with an accuracy of about $\pm 2\%$ for the range from $-100\,^\circ\text{C}$ to $600\,^\circ\text{C}$.

Bibliography: 90, 309, 375, 537, 686, 993, 1372, 1419.
References: 2010, 2045, 2239, 2315, 2316, 2427, 2510, 2511, 2558, 2786, 2787, 2833, 2835, 2908, 3045, 3178, 3184, 3226, 3557–3560, 3606, 3607, 3659, 3679, 3680, 3769, 3778, 3925, 3955, 4003, 4091, 4092, 4195, 4418–4421, 4648, 4845, 4846, 4860, 4861, 4730, 5100, 5101, 5222, 5223, 5274, 5425, 5605, 5691, 5692, 5729, 6004–6006, 6041, 6082, 6086, 6156, 6190, 6436, 6471, 6472, 6566, 6751, 6859, 6961, 7014–7016, 7087, 7095, 7120–7123, 7161, 7162, 7164, 7172, 7185, 7187, 7194, 7195, 7261, 7282.

Extensive data on heat capacities for different polymers are collected in The Polymer Handbook (B: 1404) and in B: 818.

34.9 CALCULATION OF THERMODYNAMIC FUNCTIONS FROM SPECIFIC HEAT DATA

Thermodynamic functions for a given polymer can be computed with the aid of the specific heat (c_p) by graphical integration of the following equations:

(i) enthalpy,

$$\Delta H = H_T - H_0 = \int_0^T c_p \, dT \tag{34.12}$$

(in calories per gram (CGS) or joules per kilogram (SI));

(ii) entropy,

$$\Delta S = S_T - S_0 = \int_0^T \frac{c_p \, dT}{T} \tag{34.13}$$

(in cal $g^{-1}\,^\circ C^{-1}$ (CGS) or J $kg^{-1}\,K^{-1}$ (SI));

(iii) free energy,

$$\Delta G = G_T - G_0 = (H_T - H_0) - T(S_T - S_0) \tag{34.14}$$

(in calories per gram (CGS) or joules per kilogram (SI)).

References: 2036, 4422, 6045, 7013.

34.10 MEASUREMENT OF THE DEGREE OF CRYSTALLINITY IN POLYMERS BY DTA–DSC METHODS

From thermal methods (DTA–DSC) the degree of crystallinity (χ_c) is defined as follows:

(i) Using specific heat measurements,

$$\chi_c = \frac{(c_p)_a - c_p}{(c_p)_a - (c_p)_c}, \tag{34.15}$$

where

c_p is the specific heat of the unknown specimen,

$(c_p)_a$ and $(c_p)_c$ are the specific heats of the amorphous and crystalline standards, respectively.

(ii) Using enthalpies of fusion,

$$\chi_c = \frac{\Delta H_a - \Delta H}{\Delta H_a - \Delta H_c},$$ (34.16)

where

ΔH is the enthalpy of fusion of the unknown specimen,

ΔH_a and ΔH_c are the enthalpies of fusion of the amorphous and crystalline standards, respectively.

To obtain the crystallinity data, the enthalpy of fusion must be derived and compared with the enthalpy of fusion of 100% crystalline material. In order to measure an unknown density and crystallinity of a polymer sample, the area under the fusion curve per weight of sample must be measured through the fusion temperature range and compared with the calibration curve (Fig. 34.18). The calibration plot can be obtained by measuring samples of known density and crystallinity (Fig. 3.19).

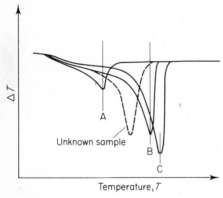

Fig. 34.18 Superimposed fusion curves of three polymer samples of different density and crystallinity. Sample A, density = 0.92, crystallinity = 30%; sample B, density = 0.95, crystallinity = 60%; sample C, density = 0.96, crystallinity = 73%

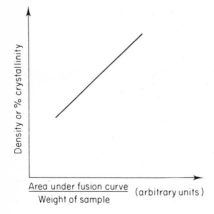

Fig. 34.19 Calibration curve of area under fusion curve per weight of sample versus density (or % crystallinity) of polymer samples

571

Bibliography: 78, 627, 713, 1426.
References: 2136, 2201, 2714, 2735, 2883, 2941, 2966, 3183, 3184, 3601, 3826, 3862, 4001, 4019, 4093, 4186, 4489, 4920, 5052, 5053, 5200, 5220, 5647, 5648, 5791, 5793, 5794, 5899, 5951, 6042, 6989, 7133, 7190, 7191, 7194.

34.11 HIGH-SPEED SCANNING CALORIMETRY

High-speed calorimetry is applied for the study of
(i) time-dependent thermal phenomena in polymers during decomposition or during transitions involving non-equilibrium states;
(ii) the behaviour of polymers heated at rates encountered during manufacturing and combustion processes.

Rapid heating of polymer samples (which usually have low diffusivity) leads to large thermal gradients and causes difficulties in finding correlation between the sample temperatures and the measured properties. For that reason heating speeds in common calorimetric measurements are of the order of 1 K s⁻¹.

Bibliography: 1423.
Reference: 7158.

34.12 THIN-FOIL CALORIMETRY

In high-speed thin-foil calorimetry the heating rates are up to 600 K s⁻¹. In this method thin polymer film samples (0.0005 cm) are used which minimize the path length for heat flow, and thus minimize the thermal gradient problem.

For measurement purposes it is necessary to prepare a laminated array of metal foils and sample sheets (Fig. 34.20). Two sample sheets are inserted

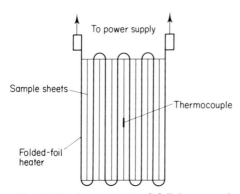

Fig. 34.20 Arrangement of foil heater and sample sheets for high-speed thin-foil calorimetry. (From R: 3848; reproduced by permission of John Wiley & Sons)

between each pair of neighbouring foils and a thin-foil thermocouple (0.001 cm copper–constantan) is inserted between the two sample sheets near the median plane. The sample is heated by passing an electric current through the foil. The thermocouple signal is monitored as a function of time on a chart recorder.

The rate of heating $(\mathrm{d}T/\mathrm{d}t)$ (in kelvins per second) can be determined from the curve.

References: 3846–3848.

34.13 EVOLVED GAS DETECTION METHODS

EVOLVED GAS DETECTION (EGD) methods involve the detection (by gas thermal conductivity or pressure) of the gases evolved during a chemical reaction as a function of temperature.

The combination of EGD methods with other thermal techniques such as DTA and TG would yield especially significant results in the study of polymers.

Bibliography: 732, 833, 1372.
References: 3375, 3595, 6931, 7048.

34.13.1 Thermal Volatilization Analysis

THERMAL VOLATILIZATION ANALYSIS (TVA) is one of the EGD methods which detect a change of pressure of the gases evolved during thermal decompositions of polymers.

A sample of a polymer is heated to degradation temperatures in a continuously evacuated vessel in which a liquid nitrogen trap is interposed between the hot reaction zone and the pumps (Fig. 34.21). A readily measurable transient pressure (10^{-5}–10^{-1} torr) (1 torr = $\frac{1}{760}$ atm = 133.322 N m^{-2} (SI)) is formed by those degradation products sufficiently volatile to distil molecularly to the liquid

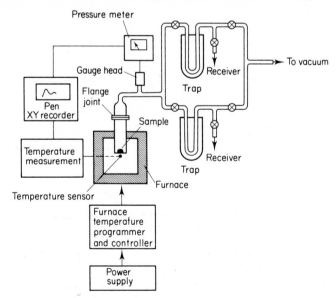

Fig. 34.21 Schematic representation of apparatus for thermal volatilization analysis

nitrogen trap. Up to pressures of the order of 0.1 torr, the transient pressure is linearly proportional to the rate of volatilization. The resulting thermogram shows one or more peaks corresponding to maxima in the rate of breakdown of the sample. Under isothermal degradation conditions, therefore, thermal volatilization analysis (TVA) gives the same data as a derivative thermogravimetry (DTG) curve, but with the added advantage of experimental simplicity.

Thermal volatilization analysis (TVA) thermograms provide information on the activation energy (E), the order of reaction (n), and the Arrhenius pre-exponential factor (A). Computer simulation allows the prediction of the TVA thermogram characteristics of reactions.

Bibliography: 906, 907.
References: 2680, 5205–5208, 6067.

34.13.2 Cold Traps

Various simple glass traps (Fig. 34.22) with external refrigerant baths (Table 34.3) are commonly used for collecting volatile samples. The efficiency of a trap can be increased by packing with glass beads or glass wool.

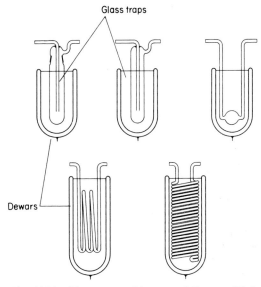

Fig. 34.22   Glass traps with external Dewars filled with liquid freezing bath

Adsorption traps packed with activated charcoal, silica gel, alumina, or molecular sieves can be used at room temperature.

Glass traps are commonly used for the collection of gaseous products formed during the degradation of polymers (e.g., thermal pyrolysis, photodegradation, etc.).

Table 34.3   Commonly used constant-temperature baths (from B: 553; reproduced by permission of Marcel Dekker)

Temperature (°C)	Slush†
−12	Methyl benzoate or *t*-amyl alcohol
−15.3	Benzyl alcohol
−23	Carbon tetrachloride‡
−30.6	Bromobenzene
−35	1,2-Dichloroethane‡
−42	Diethyl ketone
−45	Chlorobenzene
−48	*n*-Hexanol
−52	Benzyl acetate
−57	Octane (E)
−64	Chloroform‡
−71	Methyl hexanoate
−78	CO_2/acetone‡ or CO_2/trichloroethylene
−84	Ethyl acetate
−87	Trichloroethylene
−91	Heptane‡
−95	Toluene
−98	Methanol‡
−104	Cyclohexene‡
−107	2,2,4-Trimethylpentane
−112	Carbon disulphide (O)
	1-Bromobutane (E)
−117	2-Chlorobutane
	isoamyl alcohol‡
−119	Ethyl bromide
−123	1-Chloropropane‡
−126	Methylcyclohexane (S)
−127	*n*-Propyl alcohol‡
−131.5	*n*-Pentane‡
−136.4	3-Chloropropene (allyl) (O)
−139	Chloroethane (CV)
−147	Isoprene (EOS)
−158	Vinyl chloride (COV)
−161	2-Methyl-butane (isopentane)‡ (SV)
−169	3-Methyl-but-1-ene (CEOSV)
−185.6	Liquid argon‡ (or oxygen)
−196	Liquid nitrogen‡
−210 to −218	Solid nitrogen‡

† C = cylinder; E = expensive; O = odour; S = supercools;
V = volatile (at room temperature).
‡ Recommended.

34.14 EVOLVED GAS ANALYSIS

EVOLVED GAS ANALYSIS (EGA) is a technique for analysing the gases evolved during the heating and thermal decomposition (PYROLYSIS) of polymers.

Depending on their structure, polymers can be pyrolysed by one of the following processes (Fig. 34.23):

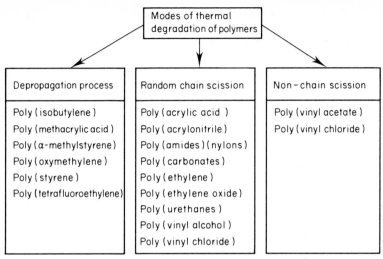

Fig. 34.23   Modes of thermal degradation of polymers

(i) DEPROPAGATION (commonly called chain unzipping), which is a free radical chain reaction which results in monomers as the main product;

(ii) RANDOM CHAIN SCISSION (random degradation) to smaller fragments—the monomer yield is very small.

(iii) NON-CHAIN SCISSION, which occurs when a facile elimination reaction can take place, e.g. evolution of HCl from poly(vinyl chloride).

The gases may be analysed in an intermittent or continuous manner by the various techniques shown in Fig. 34.24.

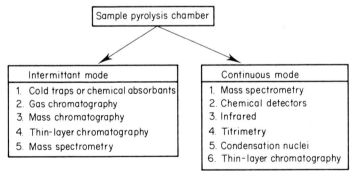

Fig. 34.24   Analytical methods for the analysis of evolved gases

EGA methods have been widely applied to the study of

(i) the kinetics of the degradation of polymers;

(ii) the mechanism of polymer stabilization;

(iii) the identities of polymers and copolymers;

576

(iv) differentiation of polymeric blends from true copolymers;
(v) stereoregularity in polymers;
(vi) the determination of trace amounts of solvent, residual monomers, or absorbed water;
(vii) the identities of solvents in coatings and adhesives;
(viii) the thermal stability of polymers;
(ix) the mechanism of curing.

Bibliography: 732, 793, 833, 855, 953, 1264, 1372, 1405.
References: 2780, 3748, 3275, 3513, 3794, 4567–4568, 5996, 6085.

34.15 PYROLYSIS–GAS CHROMATOGRAPHY OF POLYMERS

PYROLYSIS–GAS CHROMATOGRAPHY OF POLYMERS is a method which involves rapid heating of the polymeric material in the carrier gas stream at the entrance to the column. The high temperature causes that polymer sample is pyrolysed into fragments which have a sufficiently high vapour pressure to pass through the column and be analysed by gas chromatography.

A chromatogram obtained from pyrolysis of polymers is called a PYRO-GRAM. The shape of the pyrogram is dependent on the structural characteristics, such as

(i) the monomer sequence;
(ii) stereoregularity;
(iii) the degree of branching or crosslinking;
(iv) the length distribution in block and graft copolymers;
(v) crystallinity.

Comparing pyrograms in the form of FINGERPRINT FRAGMENTATION PATTERNS of unknown samples with those of known polymers and copolymers pyrolysed under identical conditions is one of the simplest methods of polymer identification. Figure 34.25 shows a pyrogram fingerprint in which retention

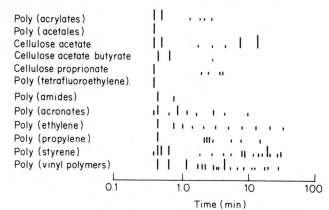

Fig. 34.25 Pyrograms in the form of a fingerprint fragmentation pattern. (From R: 5463)

times are plotted on a logarithmic scale with peak size denoted only as large, medium, or small.

The following pyrolysis techniques are used:

(i) The FURNACE CHAMBER TECHNIQUE. This method employs some type of chamber that can be heated by an external source to the desired reaction temperature (Fig. 34.26). The sample is introduced by

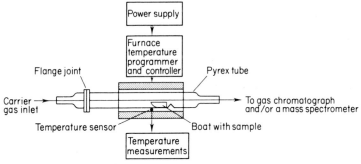

Fig. 34.26   Reactor chamber pyrolyser

injection (for liquids) or by movable boat or cup (for liquid or solid). The volatile pyrolysis products are carried by carrier gas directly to the chromatographic column or to the mass spectrometer.

References: 3330, 4110, 4871, 6650.

(ii) The FLASH PYROLYSIS TECHNIQUE. This method employs a high-resistance coil or filament made of platinum or Nichrome. The apparatus is shown schematically in Fig. 34.27. The sample is dissolved

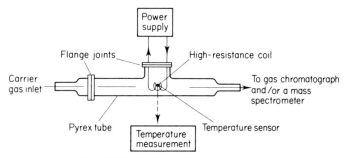

Fig. 34.27   Flash pyrolysis apparatus

and coated on the surface of a coil. After the solvent has been removed by drying, the coil is placed in the entrance to a gas chromatograph column and pyrolysed by a 'continuous mode' or 'pulse mode' by passing an electric current through the wire. Wire temperatures between 600 °C and 1200 °C are used for most organic polymers. The major disadvantages of this method are that

578

(a) the exact pyrolysis temperature is not measurable, although the temperature is reproducible;

(b) where coated filaments are used, it is impossible to get accurate weighings of sample and residue;

(c) the metal surface of the filament could have a catalytic effect on the degradation process;

(d) carbon deposits or other residues on the filament cause fluctuation in heat output so that frequent recalibration is neccessary.

Bibliography: 817, 1027.
References: 2176, 2236, 2245, 2658, 3144, 3162, 3327, 3347, 3817, 4336, 4870, 5910, 6513, 6603, 6931, 6939.

(iii) The LASER PYROLYSIS TECHNIQUE. This method employs a focused laser beam for pyrolysis of a sample placed in the entrance tube to a gas chromatograph column (Fig. 34.28). The main disadvantage of

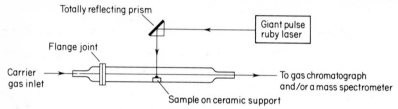

Fig. 34.28   Laser pyrolysis apparatus

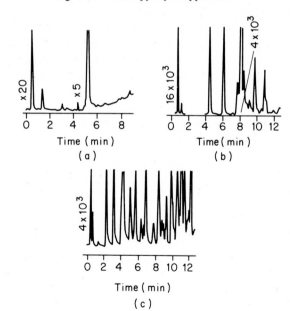

Fig. 34.29   Pyrolysis of poly(styrene) with different types of pyrolyser. (a) A laser pyrolysic chromatogram; (b) a filament pyrolysis chromatogram; (c) a tube furnace pyrolysis chromatogram. (From B: 462; reproduced with permission of Plenum Press)

this method is that only polymers which absorb the light of laser beams can be pyrolysed. The solution of the problem is the addition of carbon black to the sample. Mixing of the sample with carbon black affects the pattern of the fragment peaks produced, and samples should only be compared which have the same concentration of carbon black.

Bibliography: 462.
References: 3437, 3656, 3829, 4644.

The pyrogram patterns from laser pyrolysis are simpler than those from conventional pyrolysis methods such as the furnace chamber technique or the flash pyrolysis technique (Fig. 34.29).

One of the major difficulties in pyrolysis studies is in obtaining reproducible results, for that reason all operating conditions must be rigorously controlled.

Bibliography: 78, 196, 450, 620, 785, 817, 1027, 1029, 1264, 1348, 1372, 1385.
Reference: 5024.

34.15.1 Applications of Pyrolysis–Gas Chromatography in Polymer Research

The main applications of pyrolysis–gas chromatography in polymer research are
- (i) qualitative identification of polymers by comparison of pyrograms and mass spectra with those of known polymers;
- (ii) study of the stereoregularity of polymers;
- (iii) quantitative analysis of copolymers and their structures, e.g. detection of differences between random and block copolymers;
- (iv) differentiation of polymeric blends from true copolymers;
- (v) study of thermal stability and degradation of polymers and copolymers;
- (vi) study of the kinetics of degradation of polymers and copolymers;
- (viii) study of the thermal oxidative degradation of polymers and copolymers;
- (viii) determination of traces of monomers, solvent, additives, and absorbed water present in polymeric materials;
- (ix) identification of solvents in coatings and adhesives;
- (x) study of the curing of polymers.

Bibliography: 46, 197, 620, 732, 816, 953, 1264.
References: 2236, 2246, 2247, 2442, 2477, 2858, 2859, 2922, 3005, 3078, 3131, 3274, 3331, 3597, 3633, 3749, 3780, 3794, 3795, 3837, 4311, 4337, 4602, 4646, 4870, 4872, 4889, 5180, 5201, 5238, 5264, 5463, 5465, 5518, 5583, 5584, 6026, 6252, 6315–6317, 6355, 6594, 6844–6847, 6927, 6959, 7116, 7224, 7250.

34.16 THERMAL CONDUCTIVITY OF POLYMERS

The THERMAL CONDUCTIVITY of polymers (K) is defined as the ratio of the heat flow (dQ/dt) across unit area of a surface to the negative temperature gradient in the direction of flow (dT/dx):

$$K = \left(\frac{dQ}{dt}\right) \bigg/ \left(\frac{dT}{dx}\right) \tag{34.17}$$

(in cal cm^{-1} s^{-1} °C^{-1} (CGS) or J m^{-1} s^{-1} K^{-1} (SI)).

Thermal conductivity (K) of a crystalline polymer sample is greater than that of an amorphous sample, but its dependence on temperature is the same. It increases with molecular weight and crosslinking and increases in the direction of orientation, but decreases in the perpendicular direction as the result of molecular alignment.

There are two general methods for measuring the thermal conductivity of polymers (K):

(i) steady-state methods;

(ii) transient methods.

Bibliography: 37, 309, 394, 756, 759, 760, 1062, 1320, 1467.
References: 2123, 2314, 3282, 3283, 3693, 3892–3894, 3904, 4002, 4609, 6262, 6349, 6527, 6881.

34.16.1 Steady-state Methods of Measuring Thermal Conductivity

In this method, the thermal conductivity of the polymer sample (K) can be determined by comparing the time required to vaporize a given amount of a known liquid when heated by conduction through the sample and through a reference material in specially constructed glass apparatus (Fig. 34.30).

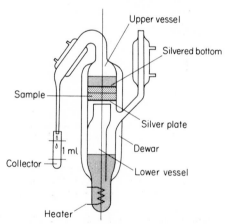

Fig. 34.30 Apparatus for measuring the thermal conductivity of polymers

The sample is placed between two vessels: lower and upper. A pure liquid is boiled in the lower vessel, heating the silver plate and returning after condensation to the lower vessel. An upper vessel has a silvered bottom and contains a pure liquid boiling 10–20 °C lower than the liquid in the lower vessel. Vapour from the liquid boiling in the upper vessel is condensed into a collector.

The temperature gradient between the lower vessel and the upper vessel (ΔT) is dependent on the sample thickness (l) and area (A). When the steady state is reached, the time (t) necessary to distil a fixed amount (1 ml) of the liquid into the collector is measured.

The thermal conductivity (K) is given by

$$K = \frac{\Delta H_v l}{A t \Delta T} = \frac{l}{AR},\qquad (34.18)$$

where
ΔH_v is the heat of vaporization of liquid from the upper vessel,
l is the thickness of the sample,
A is the area of the sample,
t is the time taken to distil a fixed amount of a liquid from the upper vessel,
ΔT is the temperature gradient,
$R = t\,\Delta T/\Delta H_v$ is the thermal resistance of the sample (a calibration parameter)
(in s °C cal^{-1} (CGS) or s K J^{-1} (SI)).

The thermal resistance parameter (R) is first determined for a standard material of known thermal conductivity (K) and different sample thickness (l) (e.g. glass) and plotted versus time (t) (Fig. 34.31). Using this calibration curve,

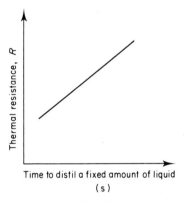

Fig. 34.31 Calibration curve for the measurement of the thermal conductivity of polymers

the thermal resistance parameter (R) for a given polymer can be read off for the observed distillation times for the polymer sample at several different values of the sample thickness (l). After reading the value (R) from the plot in Fig. 34.31, the thermal conductivity of the polymer sample can be calculated from eq. (34.18).

Bibliography: 309, 1467.
References: 3533, 4609, 6262.

Chapter 35

Measurements of Anisotropy of Polymers

General bibliography: 309, 368, 369, 633, 680, 909, 1139, 1345, 1350, 1360, 1396.

35.1 THE DEFINITION OF ORIENTATION IN A POLYMER

The single unit crystal cell can be oriented in a rectangular Cartesian system having coordinate axes x, y, z (Fig. 35.1).

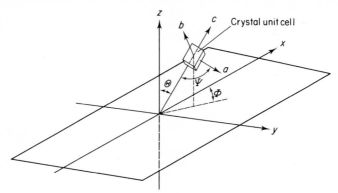

Fig. 35.1   The orientation of the c axis is defined by the polar angles
Θ and Φ. The angle Ψ refers to the projection of the a axis

The orientation of a single crystal unit cell can be defined in terms of the Eulerian angles Θ, Φ, and Ψ:

(i) the angle Θ defines the angle between the c axis and the z axis;
(ii) the angle Φ defines the angle between the projection of c axis in the x, y plane and the x axis.
(iii) the angle Ψ explains the projection of the a axis in the plane normal to the c axis.

The orientation distributions for various planes in the unit crystal cell can be obtained from X-ray data (cf. Section 28.3).

35.2 THE MOLECULAR CHAIN ORIENTATION FUNCTION

The HERMAN CHAIN ORIENTATION FUNCTION (f) is defined as

$$f = \tfrac{1}{2}(3\overline{\cos^2\Theta} - 1), \tag{35.1}$$

where Θ is the angle between the chain segment and the direction of preferred orientation of the sample (an orientation angle).

For parallel orientation, $\Theta = 0°$ and $f = +1$ (perfectly oriented fibres); for perpendicular orientation, $\Theta = 90°$ and $f = -\frac{1}{2}$ (all the molecular chains lie in the plane perpendicular to the fibre axis); for random orientation, $\overline{\cos^2 \Theta} = \frac{1}{3}$ and hence $f = 0$ (unoriented and isotropic sample).

The AVERAGE DEGREE OF MOLECULAR ORIENTATION ($\bar{f}$) for a semicrystalline polymer is given by

$$\bar{f} = v_c f_c + (1 - v_c) f_a, \tag{35.2}$$

where

v_c is the volume fraction crystallinity (cf. Section 31.8);

f_c is the crystalline chain orientation function, determined by X-ray diffraction or by the dichroism method;

f_a is the amorphous chain orientation function determined from birefringence data (cf. Section 35.3).

Bibliography: 309, 368, 369, 633.
References: 4029, 5026, 6078, 6173, 6535.

35.3 BIREFRINGENCE

BIREFRINGENCE (DOUBLE REFRACTION) is an optical phenomenon in which a polymer sample shows different refractive indices for light with plane polarization in two perpendicular directions.

ORIENTATION BIREFRINGENCE is a result of a physical ordering of optically anisotropic elements (e.g. chemical bonds) along some preferential direction.

For a uniaxially oriented sample, there will be two refraction indices, $n_{\parallel}$ and $n_{\perp}$, corresponding to polarization of light parallel to and perpendicular to the sample symmetry axis.

The BIREFRINGENCE (Δn) is defined as the difference between the two refraction indices:

$$\Delta n = n_{\parallel} - n_{\perp}, \tag{35.3}$$

where

$n_{\parallel}$ is the apparent parallel refractive index of the sample;

$n_{\perp}$ is the apparent perpendicular refractive index of the sample.

The birefringence in the Cartesian coordinate system is given by

$$\Delta n_x = n_z - n_y, \tag{35.4}$$

$$\Delta n_y = n_z - n_x, \tag{35.5}$$

$$\Delta n_z = n_x - n_y. \tag{35.6}$$

For uniaxial symmetry only a single birefringence value is necessary to describe the orientation:

$$\Delta n_x = \Delta n_y; \quad \Delta n_z = 0. \tag{35.7}$$

584

Completely amorphous polymers gave no birefringence. Semicrystalline and oriented polymers exhibit a birefringence.

The birefringence of a semicrystalline polymer is given by

$$\Delta_n = v_c \Delta n_c + (1 - v_c) \Delta n_a + \Delta n_f \tag{35.8}$$

or

$$\Delta_n = \chi_c \Delta n_c + (1 - \chi_c) \Delta n_a + \Delta n_f, \tag{35.9}$$

where

v_c is the volume fraction crystallinity determined from density measurements (cf. Section 31.8),

χ_c is the degree of crystallinity determined from X-ray diffraction data (cf. Section 28.11),

Δn_c is the crystalline phase birefringence,

Δn_a is the amorphous phase birefringence.

Δn_f is a part of birefringence which arises from the distortion of the electric field of the light wave at the phase boundaries.

The value of Δn_f depends upon the anisotropy of the shape of the phase boundary and upon the difference in the average refractive indices of the two phases. Because of the small refractive index difference between the amorphous and crystalline phases of a polymer, Δn_f is usually small and is often neglected. It may be estimated by swelling studies in which the oriented polymer is swollen with solvents of differing refractive index. The solvent enters only the amorphous phase and changes its refractive index. Thus the refractive index difference between the crystalline and swollen amorphous phase may be altered.

When a semicrystalline polymer is heated, the birefringence disappears gradually as the crystallites melt.

DEFORMATION BIREFRINGENCE arises as a result of a change of bond angles and/or bond length, or a change in packing (a change in the lattice spacing) by an external deformation (Fig. 35.2). This type of birefringence may occur in optically isotropic particles as well as in anisotropic systems.

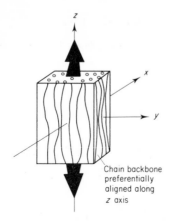

Chain backbone
preferentially
aligned along
z axis

Fig. 35.2 Formation of mole-cular orientation under external stretching of a sample along the z axis to produce orien-tation birefrigence

The deformation birefringence of a semicrystalline polymer is given by

$$\Delta n = v_c \Delta n_c^\circ f_c + (1 - v_c) \Delta n_a^\circ f_a + \Delta n_f, \qquad (35.10)$$

where

v_c is the volume fraction crystallinity,

Δn_c° is the intrinsic birefringence of a crystalline phase where the axis of the crystal is most nearly parallel to the chain axis—this axis generally is aligned along the stretch direction during orientation of the sample—in a biaxial orientation, planar orientation effects may influence the value of Δn_c°,

Δn_a° is the intrinsic birefringence of the amorphous phase defined as the maximum in orientation birefringence,

f_c is the crystalline chain orientation function determined by X-ray diffraction or by the dichroism method,

f_a is the amorphous chain orientation function determined from birefringence data.

The deformation birefringence for block or graft copolymers is given by

$$\Delta n = v_A \Delta n_A^\circ f_A + v_B \Delta n_B^\circ f_B + \Delta n_f, \qquad (35.11)$$

where

v_A and v_B are volume fractions of the individual components A and B, respectively,

f_A and f_B are chain orientation functions of the individual components A and B, respectively.

The birefringence arising from strain or orientation in the amorphous region of a semicrystalline polymer disappears at the glass transition temperature.

The correlation of molecular orientation behaviour with birefringence data requires an understanding of the origin of inherent molecular anisotropy.

Bibliography: 309, 369, 466, 912, 929, 971, 1138, 1139, 1258, 1260, 1387.
References: 2007, 2032, 2365, 2366, 2782, 2847, 2914, 3043, 3453, 3610, 3618, 3828, 3928, 3938, 3939, 4032, 4049, 4098, 4227, 4366, 4373, 4868, 4919, 5114, 5330, 5511, 5522, 5598, 5600, 5806, 5827, 5830, 5949, 5985, 6174, 6539, 6542, 6608, 6712, 6717, 6757, 6950.

35.4 BIREFRINGENCE MEASUREMENTS

There are four general methods applied for the determination of birefringence:

(i) the transmission method;
(ii) the compensator method;
(iii) the interference microscopy method;
(iv) the refractometry method.

35.4.1 Birefringence Measurements by the Transmission Method

This method depends on measuring the OPTICAL RETARDATION (R), which is defined as

$$R = (d/\lambda_0) \Delta n \qquad (35.12)$$

(dimensionless), where

d is the thickness of the polymer sample,

λ_0 is the incident light wavelength,

Δn is the birefringence.

The optical retardation (R) is related to the transmission (T) by

$$T = I/I_0 = \sin^2(\pi R), \tag{35.13}$$

where

I_0 is the intensity of the incident light,

I is the intensity of the light transmitted through the polymer sample.

The birefringence is given from eqs (35.12) and (35.13) as

$$\Delta n = \lambda_0/\pi d \arcsin(\sqrt{I/I_0}). \tag{35.14}$$

The single-beam spectrophotometer (Fig. 35.3) is built from the following parts:

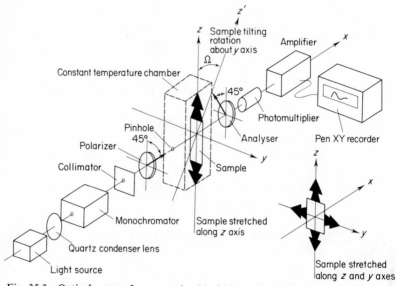

Fig. 35.3 Optical system for measuring birefrigence by the transmission method

(i) A light source, which consists of a xenon mercury lamp. The original beam, after passing a quartz condensing lens, is monochromated by a grating or a quartz prism monochromator and polarized with an ultraviolet polarizer. The light then passes through a pinhole to exclude stray light.

(ii) A polymer sample is set in a constant-speed stretcher in a constant-temperature chamber.

(iii) The light transmitted through the polymer sample passes through an analyser (second polarizer) and the final intensity of the light transmitted is detected with a photomultiplier. The photomultiplier output is fed to an amplifier and then to a recorder.

The polarizer and analyser (second polarizer) each have their electric vectors at 45° to the z axis but at 90° to each other (cf. Section 10.3.1).

This technique allows one to measure

(i) changes of birefringence simultaneous with stretching;
(ii) the relaxation of birefringence (e.g., short-time relaxation behaviour).

Bibliography: 1260, 1387.
References: 3326, 4121, 5830, 6174, 6547, 7093.

35.4.2 Birefringence Measurements by the Compensator Method

This method also depends on measuring the optical retardation (R) defined by eq. (35.12). The measuring device is built from the following parts (Fig. 35.4):

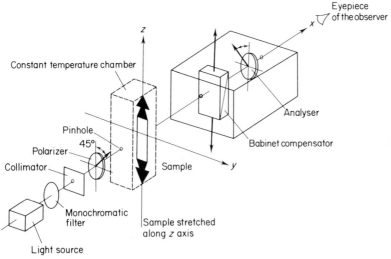

Fig. 35.4 Optical system for measuring birefrigence by the compensator method

(i) A light source consisting of a xenon mercury lamp. The original beam, after passing through a quartz condensing lens, is monochromated by a grating or a quartz prism monochromator and polarized with an ultra-violet polarizer. The light then passes through a pinhole to exclude stray light.

(ii) A polymer sample is set in a constant-speed stretcher in a constant-temperature chamber.

(iii) A Babinet compensator, through which the light is passed after trans-mission through the polymer sample. The compensator contains a pair of quartz wedges and an analyser (second polarizer). The optical axis of one wedge is perpendicular to that of the other, and both optical axes are orthogonal to the beam (x-axis). No retardation is induced when the two wedges are set in the original beam (phase angle $\delta = 0$). However, when the light transmitted through the polymer sample (which induces a

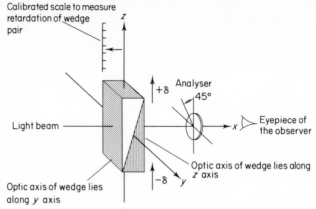

Fig. 35.5 Schematic diagram of the Babinet compensator

phase shift of $+\delta$) passes through the wedges it is possible—by movement of the wedges—to induce an offsetting phase shift $(-\delta)$ which an observer will see as a fringe or a black line when he looks along the x-axis. A calibrated scale allows one to measure retardation by changing the thickness of the wedge pair. This device replaces the photomultiplier, amplifier, and recorder in the single-beam spectrophotometer shown in Fig. 35.3.

Bibliography: 575, 1260, 1353, 1387.
Reference: 3266.

35.5 DICHROISM

DICHROISM is an optical phenomenon in which a polymer sample shows different absorbances for light with plane polarization in two perpendicular directions.

The DICHROIC RATIO is the ratio of sample absorbances measured in two mutually perpendicular directions.

For a uniaxially oriented sample, there will be two absorbances, $A_{\parallel}$ and $A_{\perp}$, corresponding to polarization of light parallel to and perpendicular to the sample symmetry axis.

In a uniaxially oriented sample, only one dichroic ratio is defined:

$$D = A_{\parallel}/A_{\perp}. \tag{35.15}$$

For a biaxially oriented sample, there will be three absorbances A_1, A_2, and A_3 measured along the x, y, and z axes, respectively (Fig. 35.6).

In biaxial orientation the three dichroic ratios can be defined as

$$D_{32} = A_3/A_2, \quad D_{31} = A_3/A_1, \quad D_{21} = A_2/A_1, \tag{35.16}$$

only two of which are independent since $D_{31} = D_{32} D_{21}$. The determination of absorbance A_1 is difficult—if not impossible. However, it can be evaluated from measurements on samples tilted with respect to the direction of the incident beam.

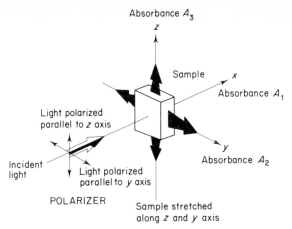

Fig. 35.6 Schematic representation of the three absorbance A_1, A_2, A_3 measured along the x, y and z axes, respectively

The dichroic method can be related quantitatively to certain orientation functions characteristic of the average orientation of the structural unit within the polymer sample, e.g. the Herman orientation function (f). For example,

$$f = \tfrac{1}{2}(3\overline{\cos^2\Theta} - 1) = (D-1)/(D+2), \qquad (35.17)$$

but this equation is valid only for the case where the dipole transition moment lies along the chain segment axis. Alternatively,

$$f = \tfrac{1}{2}(3\overline{\cos^2\Theta} - 1)\tfrac{1}{2}(3\overline{\cos^2\alpha} - 1) = (D-1)/(D+2), \qquad (35.18)$$

and this equation is valid for the case where the dipole transition moment makes an angle α with the chain segment axis.

The use of such relationships requires knowledge of the orientation of the dipole transition moments relative to defined axes, such as the local ploymer chain segment axis or the crystallographic axes, within the structural units.

Dichroic studies of oriented polymers assist in

(i) estimating the degree of orientation;
(ii) determining the chain conformation and structure;
(iii) detection of bands associated with amorphous and crystalline polymers.

Bibliography: 58, 182, 246, 405, 780, 912, 963, 1249, 1260, 1387, 1441.
References: 2325, 2798, 3411, 3495, 4560, 5520, 5710, 6536–6538, 6371, 7210.

35.6 DICHROISM MEASUREMENTS IN THE ULTRAVIOLET AND VISIBLE REGIONS

The single-beam spectrophotometer (Fig. 35.7) is built from the following parts:

(i) A light source consisting of a xenon mercury lamp. The original beam, after passing through a quartz condensing lens, is monochromatized by

590

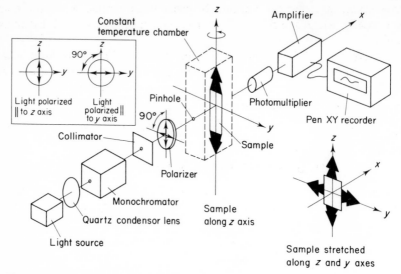

Fig. 35.7   Optical system for measuring ultra violet and visible-light dichroism

a grating or a quartz prism monochromator and polarized with an ultraviolet polarizer. The light is then passed through a pinhole to exclude stray light.

(ii) A polymer sample is set in a constant-speed stretcher in a constant-temperature chamber.

(iii) The intensity of the light transmitted through the polymer sample is detected with a photomultiplier. The photomultiplier output is fed to an amplifier and next to a recorder.

For uniaxially oriented polymers, the dichroic ratio is obtained from the measured intensities with the polarization direction successively orientated parallel and perpendicular to the sample reference axis (e.g. the draw direction).

Dichroic measurements for monofilaments utilize a polarizing microscope.

Bibliography: 1094, 1260, 1387.
References: 5710, 6371, 7094.

35.7 DICHROISM MEASUREMENTS IN THE INFRARED REGION

Dichroism measurements in the infrared region are made using the double-beam infrared spectrometer (cf. Section 15.1.1). In the sampling area are the polymer sample and the infrared polarizer. The infrared spectra are recorded with the polarizer successively orientated parallel and perpendicular to the sample reference axis.

The effects of scattering and overlap of neighbouring peaks may be corrected for by a suitable choice of baseline. The errors due to polarization effects can be eliminated by orienting the polarizer and sample at $\pm 45°$ to the monochromator entrance slit. Other errors arise from insufficient spectrometer resolution, beam convergence, and spectral dilution. In order to determine the absorbance A_1 for biaxially oriented specimens, absorbance measurements on tilted films are required.

Measurements of dichroic ratios require very thin films (2 μm to 0.1 mm) in order to avoid complete absorption of the radiation. Thicker samples may be used if measurements are limited to overtone or combination bands.

Bibliography: 405, 1094, 1139, 1441.
References: 2130, 3037, 3039, 3328, 3429, 3495, 3671, 3727, 4154, 4294, 4341, 4636, 4637, 4640, 5092, 5111, 5210, 5280, 5560, 5561, 5823, 5986–5988, 6233, 6342, 6395, 6703, 6837, 6888, 7154, 7209.

35.8 DYNAMIC INFRARED SPECTROSCOPY

DYNAMIC INFRARED SPECTROSCOPY is a technique which measures frequency shifts in the infrared spectra of stressed polymers (Fig. 35.8). The

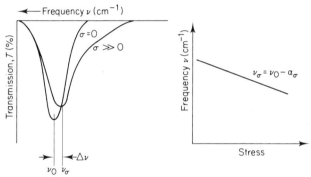

Fig. 35.8 Schematic effect of stress on a band associated with a backbone vibration. (From R: 7171; reproduced by permission of John Wiley & Sons)

frequency shift $\Delta \nu$ for a particular absorption band can be described by

$$\Delta \nu = \alpha \delta, \qquad (35.19)$$

where

α is a stress sensitivity factor,

δ is the applied stress in the polymer sample.

During the stressing of a polymer molecule there can occur three different modes of deformation:

(i) bond stretching;
(ii) angle bending;
(iii) torsional angle change.

The molecular orbitals and bonding energies are affected by these deformations and in turn can alter the vibrational spectrum of the molecule.

Pressure also affects infrared vibrational frequencies.

Dynamic infrared spectroscopy can also be used to measure the dichroic ratio of shifted frequencies.

References: 6027, 6161, 6943, 7169–7171, 7176, 7323–7325.

35.9 ELECTRO-OPTIC EFFECTS IN POLYMERS

Polymer molecules in solution placed in a homogeneous pulsed electric field (Fig. 35.9) show induced optical anisotropies such as

- (i) ELECTRIC FIELD BIREFRINGENCE (also called the KERR EFFECT), i.e. different refractive indices for light with plane polarization in two perpendicular directions;
- (ii) ELECTRIC FIELD DICHROISM, i.e. different absorbances for light with plane polarization in two perpendicular directions;
- (iii) ELECTRIC OPTICAL ROTATION (in the presence of asymmetric centres), i.e. optical rotatory dispersion and molecular circular dichroism.

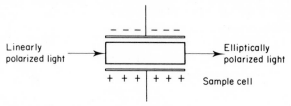

Fig. 35.9 Schematic representation of the formation-induced optical anisotropy

A schematic diagram of optical components of an apparatus for measuring electro-optical effects is shown in Fig. 35.10.

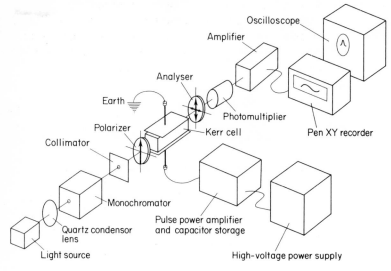

Fig. 35.10 Optical system for measuring electro-optic effects

Polymer molecules in solution yield large electro-optic effects because the interaction which orientates a macromolecule in an electric field depends upon its size as well as its electric asymmetry. By measuring induced electro-optic effects and their time dependence, it is possible to study the motion of macro-molecules in solution and their macromolecular structure. Polymer solids should also yield electro-optic effects in strong electric fields, but this effect in solids is relatively weak and difficult to measure.

For the production of a homogeneous pulsed electric field it is common to apply an electronic system which uses a hydrogen thyratron tube in a delay-line circuit employing coaxial cable as the delay-line element. Pulse rise and decay times of the order of a few hundredths of a microsecond allow for measuring birefringence decay times of the order of 10^{-7} s.

Bibliography: 303, 472, 625, 808, 989, 990.
References: 2146, 2326, 2343, 2352, 2371, 2590, 2775–2779, 2929, 2930, 3102, 3108, 3116, 3227, 3454, 3462, 3583, 4111, 4303–4305, 4528, 4716, 4862–4865, 4895, 4896, 5291, 5370, 5406, 5510, 5580, 5581, 5611, 5829, 5886–5888, 5931, 6049, 6539, 6580, 6602, 6813, 6998, 7225, 7341.

35.10 SMALL-ANGLE LIGHT-SCATTERING (SALS) METHOD FOR STUDYING ORIENTATION

Crystalline polymers scatter light as a result of

(i) fluctuations in density (e.g. the presence of microgels) (cf. Section 1.6);
(ii) orientation of correlated domains (spherulites) (cf. Section 26.3).

Analysis of the intensity and angular dependence of small-angle scattered light gives information about the morphology of polymer films.

Molten polymer samples give no scattering. As crystallization begins, characteristic scattering from spherulites can be observed.

To study orientation fluctuations, scattering intensity is measured as a function of the polarization directions of the incident and the scattered light. In Fig. 35.11 are shown different angles needed to specify scattering conditions. Polarizers are mounted in the incident and scattered beams.

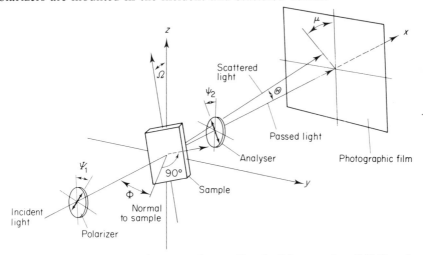

Fig. 35.11 Optical system for measuring small-angle light-scattering (SALS) and description of different angles needed to specify scattering conditions

Scattering intensity is usually measured as

(i) H_v SALS, in which the analyser has its plane of polarization perpendicular to that of the incident beam ($\Psi_1 = \Psi_2 + 90°$) (Fig. 35.12(a));

(ii) V_v SALS in which the analyser has its plane of polarization parallel to that of the incident beam ($\Psi_1 = \Psi_2$) (Fig. 35.12(b)).

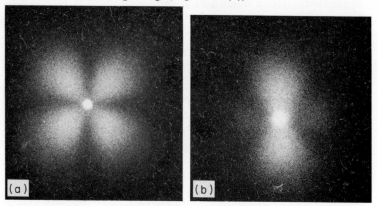

Fig. 35.12 Typical experimental SALS pattern from undeformed low-density poly(ethylene) film: (a) H_v pattern; (b) V_v pattern. (From R: 5665; reproduced by courtesy of T. Pakula and Z. Soukup with the permission of John Wiley & Sons)

The SALS patterns can be calculated theoretically.

During the deformation of a polymer sample the SALS patterns change in a characteristic way (Fig. 35.13).

For oriented polymer film samples, the scattering may be affected by birefringence. In such cases, the measurement conditions should be such that the orientation direction, making an angle Ω (Fig. 35.11) with the vertical, lies along the polarized or the analyser polarization direction ($\Psi_1 = \Omega$ or $\Psi_1 = \Omega + 90°$). For biaxially oriented films, the scattering pattern depends upon the angle Φ between the normal to the film and the incident beam. The sample stage must be able to rotate about its normal through the angle Ω and to be tilted through the angle Φ (Fig. 35.11).

Bibliography: 1139, 1259, 1260, 1360.

35.10.1 Light-scattering Apparatus

Evaluation of the entire small-angle light-scattering (SALS) pattern provides qualitative information about the nature and size of structures within the polymer, but quantitative information can only be obtained by determining the angular dependence of the scattering intensity with a photomultiplier.

In the light-scattering instrument the incident light beam should be exactly parallel in order to make accurate determinations of scattering angles. The light source should be monochromatic and stable.

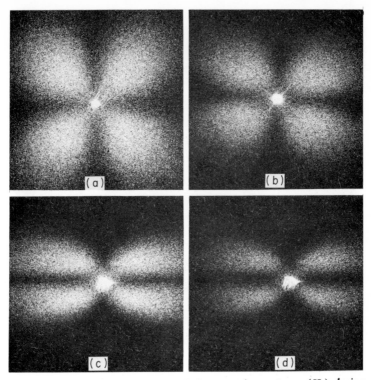

Fig. 35.13 Changes in the shape of the scattering patterns (H_v) during the deformation ($\updownarrow$ draw direction) of low-density poly(ethylene) film: (a) 0%; (b) 40%; (c) 60%; and (d) 110% elongation. (From R: 5665; reproduced by courtesy of T. Pakula and Z. Soukup with the permission of John Wiley & Sons)

As light sources it is possible to use laser sources with high-resolution optics. The most commonly employed is the He–Ne laser and a $\frac{1}{50}$ s exposure time. Weaker scattering patterns can be recorded using more intense lasers (argon ion or cadmium lasers), but increasing laser power may provide heat decomposition of a polymer sample.

The light-scattering apparatus is shown schematically in Fig. 35.14.

The film sample (0.02–0.1 mm thick) is mounted between Corning glass cover slips using an immersion fluid and set in a goniometer which permits tilting of the sample at an angle Φ relative to the incident beam and rotating it through an angle Ω about the sample normal (for oriented samples). Special cells for controlling sample temperature, for immersing sample in liquids, and for stretching polymer films are available.

Using a special high-speed cine camera it is possible to photograph patterns with very short exposure times, and this permits one to follow rapid changes with respect to time.

In obtaining photographic patterns it is necessary to eliminate surface scattering by using films having a very smooth surface, or by immersing the

596

sample between glass plates using a silicone fluid of matching refractive index which does not swell the polymer.

For measuring the intensity of scattering as a function of angle one can apply

(i) photometric systems containing a photomultiplier (Fig. 35.14);

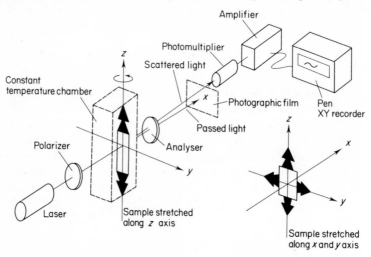

Fig. 35.14   Optical system for measuring small-angle light-scattering (SALS) using a photographic film or photomultiplier tube

(ii) electron scanning systems containing an optical multichannel analyser or a rapid scan spectrometer—these two devices conventionally employ a monochromator before the vidicon for recording the variation of intensity as a function of wavelength.

The detection system should have very high resolution at low scattering angles, and also high sensitivity and accuracy at large angles. Neutral filters allow for attenuation of the intensity of scattered light, which varies with angle. For absolute intensity measurements, apparatus calibration is necessary. The scattering from films must be corrected for reflection, refraction, and secondary scattering. To follow rapid changes in the intensity of scattered light, all requisite adjustments need to be made rapidly, and the time when a measurement is made should be accurately determined.

Bibliography: 1260.
References: 2845, 3575, 3934, 4503, 5664, 5665, 5977, 6171, 6543, 6546, 6548, 6549, 6908, 7010, 7092.

35.10.2 Applications of Small-angle Light Scattering (SALS) in Polymer Research

The small-angle light-scattering (SALS) method may be applied to the study of

(i) the orientation of spherulites (cf. Section 35.10);
(ii) polymer crystallization;

597

(iii) particle size analysis in multicomponent polymer systems;
(iv) polymer networks and gels.

Bibliography: 1139, 1259, 1260, 1360.
References: 2453, 2454, 2472, 2473, 2824, 2846, 2847, 2913, 3325, 3326, 3574–3576, 3678, 3684, 3928–3930, 3932, 3933, 3935, 3937, 4097, 4197, 4236, 4541, 4868, 4869, 4898, 5058, 5073, 5116, 5176, 5300, 5330, 5444, 5521, 5599, 5664, 5828, 5830, 5838, 5849, 5917, 5919–5923, 6171, 6174, 6175, 6345, 6541, 6543, 6544, 6548–6552, 6554, 6555, 6712, 6895, 6907, 6909, 6952, 6953, 7182, 7279, 7280.

35.11 STUDY OF MOLECULAR ORIENTATION BY POLARIZED FLUORESCENCE

Polarized fluorescence measurements can yield valuable information on the orientation distribution of polymer molecules.

When a film in which optically anisotropic fluorescent molecules (M) are dispersed is irradiated with linearly polarized light, the fluorescent molecules are selectively excited, depending on the angular alignments of the molecular axes with respect to the direction of the electric vector of the incident exciting light. The fluorescence emitted from the film possesses polarization characteristics which are related to the spatial distribution of the molecular axes (Fig. 35.15) of the molecules thus excited at the instant of emission of fluorescence.

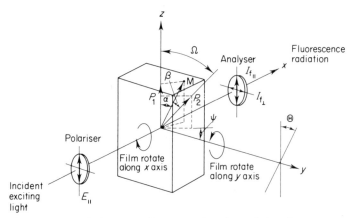

Fig. 35.15 Optical system for measuring the relative fluorescence polarization of dispersed fluorescence molecules, (M). The electric vector of the exciting light (E) is plane polarized in the z, x ($E_\parallel$) plane. Emitted polarized fluorescence intensity $I_{f\parallel}$ and $I_{f\perp}$ are measured along the x axis. Angles Ω, Ψ, and Θ define the position of a film in an x, y, z coordinate system.

The intensity of the polarized component of fluorescence (I_f) is proportional to $\cos^2\alpha\cos^2\beta$ for the molecule M, which orients at angles α and β with vectors P_1 and P_2 and is given by

$$I_f = I_0\cos^2\alpha\cos^2\beta, \tag{35.20}$$

where I_0 is the intensity of the incident light.

The choice of optically anisotropic fluorescent molecule depends on

  (i) chemical structure;
 (ii) place of location in polymer matrix, e.g. some fluorescent compounds may be located only in the amorphous regions, others only in crystalline regions;
(iii) photostability;
 (iv) high fluorescence efficiency;
  (v) interaction with macromolecules (the fluorescent molecule must not perturb the polymer structure).

Compounds which satisfy such criteria are uranine (**I**) and 2,2'-(vinylene-di-*p*-phenylene)*bis*benzoxazole (**II**) used in low concentration (100–200 p.p.m.).

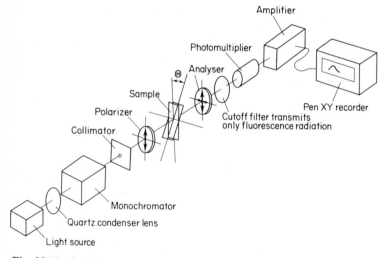

(I)

(II)

The polarized fluorescence of a film containing an optically anisotropic fluorescent molecule (M) is measured along the *x* axis (Fig. 35.16). The sample is mounted in a goniometer, which permits rotation at any chosen angle along

Fig. 35.16 Optical system for measurement of the angular distribution of polarized components in a polymer film

the x axis and the y axis. The cut-off filter transmits only fluorescent light. Further description of the measurement process is as given for Fig. 16.5.

The theoretical distribution of the polarized components of fluorescence obtainable by the polarized fluorescence method (cf. Section 16.4) are given for each pattern of orientation in Fig. 35.17.

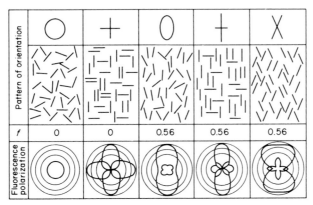

Fig. 35.17 Orientation patterns with corresponding orientation function (f) and the angular distributions of polarized components of fluorescence. $\bigcirc$, $I_\parallel$; $\bullet$, $I_\perp$. (From B: 974; reproduced by permission of John Wiley & Sons)

For fluorescent polymers containing fluorescent groups along the molecular chain (e.g. poly(amides) having monometic units of stilbene, partially dehydrochlorinated poly(vinyl chloride), and partially dehydrated poly(vinyl alcohol)), the molecular orientation behaviour can be investigated directly from the polarization characteristics of fluorescence emitted from the excited specimen.

The angular distribution of $I_{f\parallel}$ (Fig. 35.18) can be analysed as being composed of three basic types of orientation:

(i) axial orientation along the stretching direction;
(ii) orientation perpendicular to the stretching direction;
(iii) unoriented random distribution.

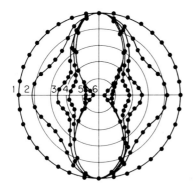

Fig. 35.18 Angular distribution of the polarized component of fluorescence emitted from the fluorescent compound dispersed in unixially stretched poly(vinyl alcohol) film at draw ratios of: (1) 1; (2) 1.08; (3) 1.3; (4) 1.6; (5) 2.0 and (6) 5.0. (From B: 974; reproduced by permission of John Wiley & Sons)

For non-fluorescent polymers, optically anisotropic fluorescent compounds such as (**I**) or (**II**) can be dispersed in the specimen as the probe to indicate molecular orientation behaviour.

Bibliography: 182, 974.
References: 4050, 5502, 5504, 5506, 5508, 5511.

Note that the polarized phosphorescence may also yield valuable information (R: 6134).

35.12 ACOUSTIC (SONIC) METHODS

Study of the velocity of sound (or sound pulse propagation) is a useful method for determination of the orientation in polymer fibres and films. The rate of sound propagation is much greater when directed along the chain than when it is directed perpendicular to the chain (Fig. 35.19).

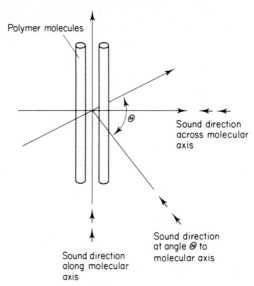

Fig. 35.19 Possible modes of sound transmission in macromolecules

The velocity of sound (C) in an oriented polymer sample is given by

$$\frac{1}{C^2} = \frac{1-\overline{\cos^2\Theta}}{C_\perp^2} + \frac{\overline{\cos^2\Theta}}{C_\parallel^2} = \frac{\overline{\sin^2\Theta}}{C_\perp^2} + \frac{\overline{\cos^2\Theta}}{C_\parallel^2}, \qquad (35.21)$$

where

Θ is the angle between the molecular chain axis and the direction of the sound propagation,

$C_\perp$ represents the sonic velocity across the molecular axis (90°) (perpendicular orientation),

$C_{\parallel}$ represents the sonic velocity along the molecular axis ($0°$) (parallel orientation).

For the polymer sample treated as a single-phase system, the SONIC MODULUS (YOUNG'S MODULUS) (E) can be used:

$$\frac{1}{E} = \frac{1}{\rho C^2} = \frac{(1 - \overline{\cos^2 \Theta})}{E°}, \tag{35.22}$$

where

C is the velocity of sound,

ρ is the density of the polymer,

Θ is the angle between the molecular chain axis and the direction of the sound propagation,

$E°$ is the INTRINSIC LATERAL MODULUS defined as the transverse Young's modulus for a perfectly oriented fibre—it is a function of the intermolecular forces in the fully oriented fibre.

For the polymer sample treated as a two-phase system (a semicrystalline polymer), the sonic modulus (E) may be identified with the average degree of molecular orientation ($\bar{f}$) (cf. Section 35.2) by

$$\frac{3}{2}\left(\frac{1}{\Delta E}\right) = \frac{v_c f_c}{E_c°} + \frac{(1 - v_c) f_a}{E_a°}, \tag{35.23}$$

where

$$\frac{1}{\Delta E} = \frac{1}{E_1} - \frac{1}{E_2}, \tag{35.24}$$

E_1 and E_2 are the sonic moduli for the unoriented and the oriented sample, respectively,

v_c is the volume fraction crystallinity (cf. Section 31.8),

f_c is the crystalline chain orientation function determined by X-ray diffraction or by the dichroism method,

f_a is the amorphous chain orientation function determined from birefrigence data (cf. Section 35.3),

$E_c°$ is the intrinsic lateral (transverse) modulus of the crystal region,

$E_a°$ is the intrinsic lateral modulus of the amorphous region.

Bibliography: 369, 1036, 1138, 1139, 1260, 1387.
References: 2799, 3139, 3734, 5368, 5525, 6050, 6170, 6172.

Note: the sonic method is also useful for the study of segmental motion in polymers (R: 3059).

35.12.1 Sonic Apparatus for Measuring Orientation

A piezoelectric crystal (Rochelle salt twister bimorph) is held in contact with the sample (Fig. 35.20). The transmitter crystal, when pulsed, transmits a sound

602

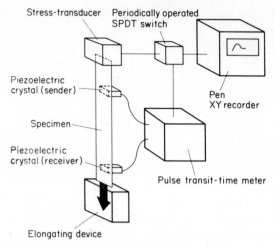

Stress-transducer Periodically operated
                    SPDT switch

Piezoelectric
crystal (sender)

Pen
XY recorder

Specimen

Piezoelectric
crystal (receiver)

Pulse transit-time meter

Elongating device

Fig. 35.20   Sonic apparatus for measuring orientation

wave through the sample and this propagating wave is detected by a second
piezoelectric crystal (receiver). Knowing the propagation distance and the
computed incremental time between pulse and signal detection allows calculation
of the sound velocity (C). The sample length–width ratio must be of the order
of 10 or more.

References: 3864, 5368.

Chapter 36

Surface Tension of Solid Polymers

General bibliography: 8, 205, 530, 539, 617, 699, 700, 827, 1014, 1049, 1075, 1336, 1364, 1448.

SURFACE TENSION is a result of unbalanced intermolecular forces (called COHESION FORCES) of molecules at the surface of a solid. These molecules have additional surface energy called COHESION ENERGY in comparison with the molecules inside the solid. The cohesion forces have a tendency to reduce the surface area to a minimum.

The INTERFACE TENSION is a result of intermolecular forces (called ADHESION FORCES) which are formed at the interface between two solids. The additional free energy formed is called ADHESION ENERGY.

The SURFACE TENSION of solids is most conveniently determined by CONTACT ANGLE MEASUREMENTS, which consist of measuring the contact angle (θ) of a homologous series of liquids of known surface tension (γ_L) on plane solid surfaces. The different values of the contact angle (θ) (Fig. 36.1) lead to different effects:

(i) $\theta = 0$ ($\cos \theta = 1$)—the liquid wets the solid and tends to spread right over the surface;

(ii) $0 < \theta < \pi/2$—the liquid spreads on the surface in a limited range;

(iii) $\theta > \pi/2$—the liquid does not wet the surface and has a tendency to shrink away from the solid.

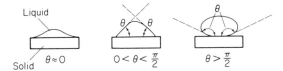

Fig. 36.1   Different contact angles of liquid
on a solid polymer

The plot of $\cos \theta$ versus γ_L yields the critical surface tension of the solid (γ_{crit}). The critical surface tension of a solid is not a well-defined value and it depends on the particular series of liquids used for its determination.

The ADHESION TENSION (equilibrium contact angles of liquids on solids) is determined by the YOUNG EQUATION:

$$\gamma_L \cos \theta = \gamma_S - \gamma_{SL} - \pi_L, \tag{36.1}$$

(in dynes per centimetre (CGS) or newtons per metre (SI)), where
$\gamma_L \cos \theta$ is the adhesion tension,
γ_L is the surface tension of the liquid,
γ_S is the surface tension of the solid,
γ_{SL} is the surface tension between liquid and solid,
θ is the contact angle,
π_L is the equilibrium pressure of the absorbed vapour of the liquid on the
solid.

If the vapour pressures are small, eq. (36.1) can be written as

$$\gamma_L \cos \theta = \gamma_S - \gamma_{SL}. \tag{36.2}$$

The rate of spreading of a liquid on a solid surface can be determined from

$$\frac{dA}{dt} = k\gamma_L(\cos \theta_s - \cos \theta_d), \tag{36.3}$$

where
dA/dt is the rate of increase in the liquid–solid contact area,
k is the rate constant,
γ_L is the surface tension of the liquid (cf. Fig. 36.2),

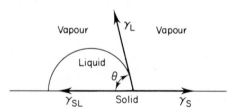

Fig. 36.2   Profile of a small liquid drop spreading on a plane solid
surface

θ_d is the dynamic contact angle,
θ_s is the static contact angle achieved at a prolonged time of contact (sometimes called the EQUILIBRIUM CONTACT ANGLE).

36.1 MEASUREMENT OF THE CONTACT ANGLE

There are several methods for measurement of the contact angle:

(i) the tilting plate method, giving a precision of $1°$;
(ii) liquid drop or a gas bubble method;
(iii) the slide technique, giving contact angles to a precision of $0.1°$.

In the slide technique method the polymer plate is immersed in the liquid
(Fig. 36.3). The meniscus on a partially immersed plate rises to a definite height
(h) if θ is finite, and can be measured by cathetometer (travelling microscope)

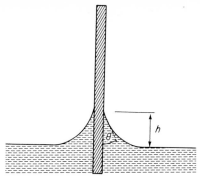

Fig. 36.3   Slide technique for contact angle measurement

under proper illumination. The contact angle (θ) can be calculated from

$$\sin \theta = 1 - (h/a)^2 \qquad (36.4)$$

where a is the capillary constant.

Bibliography: 8.
Reference: 3552.

Extensive data on critical surface tensions of polymers are collected in The Polymer Handbook (B: 1191).

36.2 APPLICATIONS OF SURFACE TENSION MEASUREMENTS IN POLYMER CHEMISTRY

The cohesion and adhesion forces play an important role in
   (i) spinning;
  (ii) polymer adhesion;
 (iii) stability of dispersions;
  (iv) wetting of polymers by liquids;
   (v) spreading of polymer melts on solid surfaces.

References: 2050, 2149, 2192, 2338, 2373, 2820, 2825, 3463, 3464, 3573, 3661, 3662, 3865, 4128, 4147, 4237, 4824, 4826, 4827, 4856, 4857, 4990, 5472, 5532, 5650, 5992, 6140, 6259, 6300, 6415, 6652, 7177, 7264.

Chapter 37

Diffusion in Polymers

General bibliography: 331–333, 490, 574, 621, 784, 910, 1031, 1124, 1252, 1321.

The DIFFUSION OF SMALL MOLECULES (organic liquids and vapours) occurs by movement of these small molecules from one 'hole' to another. The term 'HOLES' indicates, from the thermodynamic point of view, the probability of creating either sorption sites for the penetrant molecules or diffusion channels in the polymer through which the penetrant can move from one site to another.

Diffusion requires a concentration gradient of the penetrant in the polymer, and flux occurs from regions of high concentration to regions of low concentration.

The DIFFUSION FLUX (J) of a species is proportional to the concentration gradient measured normal to the unit area of cross-section through which a species diffuses (FICK'S FIRST LAW) and is given by

$$J = - D \frac{\partial c}{\partial x} \qquad (37.1)$$

where

J is the flux (on g cm^{-2} s^{-1} (CGS) or kg m^{-2} s^{-1} (SI)),

c is the concentration of penetrant molecules (diffusant) (in grams per cubic centimetre (CGS) or kilograms per cubic metre (SI)),

x is the space coordinate measured normal to the section (in centimetres (CGS) or metres (SI)),

D is the diffusion coefficient (in square centimetres per second (CGS) or square metres per second (SI)),

$\partial c/\partial x$ is the concentration gradient across a thickness ∂x.

The DIFFUSION COEFFICIENT (D) is a function of the concentration of diffusant (penetration molecules) (FICK'S SECOND LAW) and for Cartesian coordinates is given by

$$\frac{\partial c}{\partial t} = D \left[\frac{\partial^2 c}{\partial x^2} + \frac{\partial^2 c}{\partial y^2} + \frac{\partial^2 c}{\partial z^2} \right]. \qquad (37.2)$$

Polymer systems whose diffusion characteristics can be described by Fick's first and second laws are said to follow 'FICKIAN DIFFUSION BEHAVIOUR'.

The diffusion coefficient (D) depends on temperature (increasing with increasing temperature) according to the equation

$$D = D_0 \exp(-E_D/RT), \qquad (37.3)$$

where

D_0 is a pre-exponential constant,

E_D is the activation energy.

R is the universal gas constant (cf. Section 40.2),

T is the temperature.

A diffusion plot for a given polymer can be represented by one of the following:

(i) A plot of absorption (or desorption) of a penetrant versus the square root of the time of diffusion ($\sqrt{t}$) (Fig. 37.1).

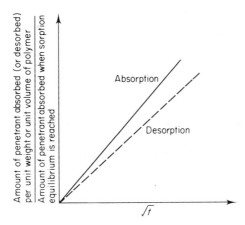

Fig. 37.1 Absorption and desorption curves
of a given penetrant by a given polymer
(Fickian diffusion behaviour)

(ii) A plot of absorption (or desorption) of a penetrant versus $\sqrt{(t)}/l$, where l is the thickness of the dry polymer film. Such a plot is called a RE-DUCED SORPTION (or DESORPTION) CURVE (Fig. 37.2). The rates of absorption and desorption are equal and no net gain or loss of penetrant occurs.

(iii) A series of absorption (or desorption) curves (under identical experimental conditions) for films of varying thickness can be superimposed on one single curve if each curve is plotted in the form of a reduced sorption (or desorption) curve.

The reduced desorption curve always lies between the reduced absorption curve and the $\sqrt{(t)}/l$ axis if the diffusion coefficient (D) is an increasing function of concentration.

Fickian diffusion behaviour is evidenced by a linear relationship between the absorption (or desorption) of penetrant and the square root of the time of diffusion ($\sqrt{t}$).

608

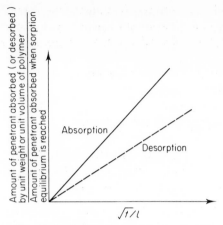

Fig. 37.2 Reduced sorption (or desorption) curves

The following factors affect the diffusion process:

(i) Factors which decrease the segmental mobility (crystallization, chain stiffness, and cross-linking) decrease the diffusion rate. The modified diffusion coefficient for the semicrystalline, two-phase (spherulic) system is given by

$$D = D_a(\Psi/B), \tag{37.4}$$

where

D_a is the diffusion coefficient for the completely amorphous system,

B is the BLOCKING FACTOR, which describes the blocking of the diffusion paths through the amorphous layers by contiguous impenetrable crystalline lamellae (Fig. 37.3),

Ψ is the IMMOBILIZATION FACTOR, which describes the reduction in chain mobility caused by anchoring of the amorphous chains between crystallites.

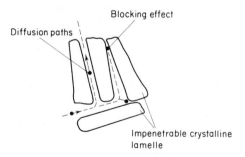

Fig. 37.3 Schematic representation of diffusion path and blocking effect in the amorphous phase of a semicrystalline solid

(ii) Factors which increase the fractional free volume, e.g. deformation. Elastic strain in a direction normal to crystalline lamellae will increase the thickness and the specific volume of the amorphous layers between the lamellae.

(iii) Factors affecting processes of immobilization of the penetrant in the polymer matrix.

References: 2356, 2592, 2832, 2843, 3009, 3010, 3241, 3321, 3514, 3515, 3518, 3545, 3722, 4004, 4113, 4595, 4597, 4598, 4619, 4655, 4795, 4801, 5008, 5009, 5039, 5260, 5262, 5263, 5723, 5725, 5795, 5993, 6084, 6319, 6418, 6493, 6737, 6790, 6798, 6971–6973, 6976, 7001, 7037–7040, 7106, 7246.

NON-FICKIAN DIFFUSION BEHAVIOUR (also called CASE II DIFFUSION) is an anomalous diffusion behaviour which cannot be described in terms of Fickian laws. It can exist in polymers

(i) below the glass transition temperature;

(ii) above the glass transition temperature in crosslinked and crystalline polymers.

Anomalous diffusion behaviour is evidenced by significant deviations from the linear relationship between absorption of penetrant and the square root of the time of diffusion ($\sqrt{t}$) (Fig. 37.4). The rates of absorption and desorption are not equal in non-Fickian diffusion.

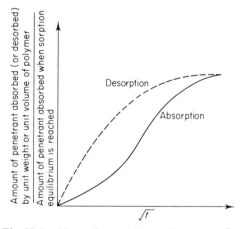

Fig. 37.4   Absorption and desorption curves of a given penetrant by a given polymer (non-Fickian diffusion behaviour)

Both types of diffusion, i.e. Fickian and non-Fickian (case II), can be superimposed. Some penetrants will show Fickian behaviour in a given polymer while others will show non-Fickian (case II) behaviour. By increasing the crosslink density of a given polymer, the diffusion tends to move from Fickian type to non-Fickian (case II) type.

Bibliography: 1007.
References: 2046, 2189, 2426, 2789, 2907, 3011, 3320, 3517, 3522, 4117–4119, 4279, 4796, 4797, 5261, 5438, 5784, 5785, 5853, 5993, 6139, 6418, 6970, 6974, 6975, 6977, 7245.

37.1 SORPTION MEASUREMENTS

The equilibrium determination of the uptake of vapour (or solute) by a solid polymer gives information on the surface concentration of a diffusant. When the vapour pressure (or solute concentration) is constant, the rate of approach to equilibrium gives a direct measure of the diffusion coefficient.

The sorption kinetic technique has many advantages over the permeation technique for obtaining the diffusion coefficient, because this method is not affected by some defects, e.g. holes in the polymer membrane, and can be applied to very thin membranes. The sorption technique is particularly valuable in the measurement of low diffusion coefficients.

The kinetics and equilibria of vapour sorption (desorption) are commonly monitored gravimetrically using quartz helical springs. Spring extension is calibrated as a function of weight gain to a precision of ± 2 μg. A fixed amount of solvent sufficient to maintain a specified partial pressure is placed into a temperature-controlled (± 0.1 °C) evacuated chamber (Fig. 37.5). The consequent spring extension is measured as a function of time to determine the

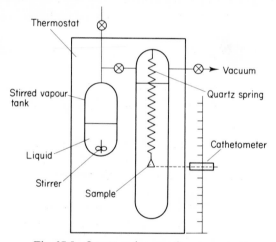

Fig. 37.5 Quartz spring sorption apparatus

sorption kinetics. Sorption equilibria are determined from the final weight of penetrant absorbed by the polymer sample. The time necessary to establish sorption equilibria varies from approximately 24 h to more than 1000 h depending upon film composition, temperature, and penetrant activity.

The kinetics and equilibrium data of liquid sorption are monitored by placing film samples in a liquid at a constant temperature (± 0.1 °C). After a specified period of time a sample is removed from the liquid, placed between several folds of tissue paper to remove liquid adhering to the film surface, quickly placed in a tarred and capped weighing bottle, and weighed to the nearest 0.0001 g (cf. Section 2.15).

Bibliography: 333.
References: 2486, 2487, 4226, 4280, 5891, 7262.

37.2 MEASUREMENT OF THE DIFFUSION COEFFICIENTS

There are several experimental methods for measuring diffusion coefficients (Table 37.1).

Table 37.1 Methods of diffusion coefficient measurement

Method	Published in
Interference microscopy	R: 2105, 2773, 3012, 3827, 3965, 3966, 5505, 6072
Holographic interference microscopy	R: 2990, 3311, 6032, 6924, 6925
IR spectroscopy	R: 4596, 4597
NMR spectroscopy	R: 5172, 6401, 6560, 6739

One of the most practical methods is by the application of radioactively tagged penetrants. A small amount of penetrant (c. 1 mg) is placed in contact with the lower surface of a disc-shaped sample. During diffusion the concentration of penetrants at the upper surface of a sample increases and is measured by radioactivity counts.

References: 2178, 2621, 2622, 2813, 2814, 4272, 5343, 6028–6030, 7058, 7159, 7160.

37.3 GAS PERMEABILITY OF POLYMER FILMS

General bibliography: 309, 333, 488, 490, 504, 609, 636, 800, 918, 919, 1031, 1122, 1123, 1125, 1252, 1253, 1329.

The permeation of a gas through a polymer film can be described in terms of the gas or vapour dissolving at one surface of the film, diffusing through the film under a concentration gradient, and evaporating from the other surface at the low concentration, i.e. low pressure, side.

If a constant pressure difference is maintained across the film, the gas will diffuse through the film at a constant rate. Under steady-state conditions the permeability process is described by Fick's first law:

$$J = -D\frac{\partial c}{\partial x} \qquad (37.5)$$

where

J is the flux (i.e. rate of gas transportation) or the amount of gas diffusing through unit area of the film in unit time, expressed at STP (standard temperature and pressure) composite unit system as

$$J = \frac{\text{(Amount of gas)}}{\text{(Membrane area)(Time)}} \qquad (37.6)$$

J is expressed in units: $(cm^3 (STP) cm^{-2} s^{-1})$ CGS, $(m^3 (STP) m^{-2} s^{-1})$ SI.

D is the diffusion coefficient: ($cm^2 s^{-1}$) CGS, ($m^2 s^{-1}$) SI.

$\dfrac{\partial c}{\partial x}$ is the concentration gradient across a thickness x.

After integration of eq. (37.5) across the total thickness of the film (l), the amount of gas diffusing through the polymer film (J) is given by

$$J = \frac{D(c_1 - c_2)}{l}, \tag{37.7}$$

where

c_1 and c_2 are the concentrations of gas in the two surfaces: (g vapour/g polymer) CGS, (kg vapour/kg polymer) SI or (cm^3 vapour/cm^3 polymer) CGS, (m^3 vapour/m^3 polymer) SI.

l is the thickness of polymer film: (cm) CGS, (m) SI.

Gas concentrations are normally measured in terms of the pressure (p) of the gas which is at equilibrium with the film. According to Henry's law,

$$c = Sp, \tag{37.8}$$

where

c is the gas concentration: (cm^3 vapour/cm^3 polymer) CGS,
(m^3 vapour/m^3 polymer) SI.

S is the SOLUBILITY COEFFICIENT of the gas in the polymer film, expressed by STP composite unit system as

$$S = \frac{\text{(Amount of gas)}}{\text{(Amount of polymer) (Differential pressure of gas)}}, \tag{37.9}$$

S is expressed in units: (cm^3 (STP) cm^{-3} atm^{-1}) CGS,
(m^3 (STP) m^{-3} Pa^{-1}) SI.

p is the pressure, expressed in units: (atm) CGS, (Pa) SI.

Note: $Pa = m^{-1} kg s^{-2} = N m^{-2}$ (SI system).

Hence

$$J = \frac{DS(p_1 - p_2)}{l}. \tag{37.10}$$

The PERMEABILITY CONSTANT (P) is defined as the product $D \times S$ and is given by

$$P = DS = \frac{Jl}{p_1 - p_2}. \tag{37.11}$$

Gas permeability constants of polymers (P) are generally expressed by STP composite units system as

$$P = \frac{\text{(Amount of gas)(Thickness of membrane)}}{\text{(Membrane area) (Time) (Differential pressure of gas)}} \tag{37.12}$$

where
P is expressed in units: (cm³ (STP) cm cm⁻² s⁻¹ atm⁻¹) CGS,
(m³ (STP) m m⁻² s⁻¹ Pa⁻¹) SI.

The diffusion (D), permeability (P), and solubility (S) quantities are dependent on:

(i) temperature;
(ii) concentration;
(iii) penetrant size and shape;
(iv) penetrant–polymer interactions;
(v) morphology and molecular motion of the polymer.

References: 2172, 3040, 3516, 4139, 4155, 4282, 4286, 4529, 4670–4672, 4757, 4758, 4994, 5694, 5695, 5724, 5726, 6385, 7570, 6636, 6800, 6801, 6841, 7243, 7244, 7248, 7327.

37.4 DETERMINATION OF THE PERMEABILITY CONSTANT

The most common method for measuring the permeability constant (P) is by measuring the increase of pressure with time on the low-pressure side of the film (Fig. 37.6) under conditions where the pressure difference remains constant.

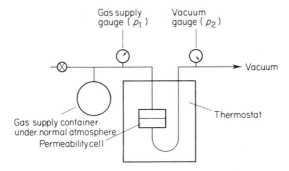

Fig. 37.6   Gas permeability apparatus

A typical pressure–time plot is shown in Fig. 37.7. From a plot of the increase in pressure at the low-pressure side versus time after the steady state one can determine $\Delta p/\Delta t$. The volume of the low-pressure side (V) may be determined experimentally. The amount of gas transmitted in unit time at STP is given by

$$\frac{\Delta p}{\Delta t} = \frac{V \times 273}{760(273 + T)}. \tag{37.13}$$

For an area of film (A) and a thickness (l), then

$$P = \frac{\Delta p}{\Delta t} \frac{V \times 273 \times l}{760(273 + T) p_1 A}. \tag{37.14}$$

Here $p_2 \approx 0$ and p_1 is the manometer pressure (in torr) on the high-pressure side, and T is the temperature (in degrees Celsius).

614

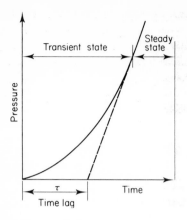

Fig. 37.7 Typical pressure–time
plot

Bibliography: 309, 333, 1125, 1465.
References: 3873, 4266, 4401, 5676, 5800, 6571.

37.5 DETERMINATION OF THE DIFFUSION CONSTANT

The diffusion constant (D) can be calculated directly from the pressure–time plot (Fig. 37.7) using

$$D = l^2/6\tau, \qquad (37.15)$$

where
D is the diffusion constant,
l is thickness of the polymer film,
τ is the intercept on the time axis of the extrapolated pressure–time plot (cf. Fig. 37.7).

When the diffusion constant is large, no time lag is observable, the steady-rate curve passes through the origin, and only P is obtainable.

Diffusion constants are reproducible within 10% when the time lags are larger than about 20 min.

The solubility coefficient (S) can be calculated from

$$S = P/D, \qquad (37.16)$$

where
S is the solubility coefficient,
P is the permeability constant,
D is the diffusion constant.

Bibliography: 309, 333, 1125, 1465.

37.6 STUDY OF THE CONCENTRATION AND PRESSURE DEPENDENCE OF THE DIFFUSION OF GASES IN POLYMERS BY NMR SPECTROSCOPY

Under pressure the gas dissolved in the polymer may plasticize the polymer matrix and increase the gas diffusion rate while the surrounding gas can compress the polymer, reduce the free volume, and decrease the gas diffusion rate.

In NMR measurement the polymer sample is exposed to the desired pressure of gas penetrant (contained H atoms). The sample is allowed to come to equilibrium, and then the concentration of gas dissolved in the polymer is measured by comparing its NMR signal intensity to that of pure water. A concentration gradient, necessary for both permeation and sorption experiments in the gas permeability method (cf. Fig. 37.7), is not necessary for the NMR measurement.

References: 2167, 2169.

Chapter 38

Degradation of Polymers

General bibliography: 169, 276, 278, 315, 355, 376, 507, 522, 535, 536, 674–678, 706, 855, 859, 932, 1059, 1077, 1086, 1104, 1161, 1162, 1273, 1406.

DEGRADATION OF POLYMERS is a process which occurs with rupture of the main chain backbone or the side group bonds.

Two of the most important types of degradation are as follows:

(i) The RANDOM DEGRADATION process of polymers is characterized by rupture of main-chain links according to the laws of probability. Polymers which have tertiary hydrogen atoms such as

$$
\begin{array}{c c c c}
H & H & H & H \\
| & | & | & | \\
-C- & C- & C- & C- \\
| & | & | & | \\
H & R & H & R \\
\end{array}
$$

have a tendency to degrade randomly and/or crosslink.

(ii) The DEPOLYMERIZATION PROCESS (UNZIPPING PROCESS) is a reversal of the polymerization process. Polymers which have structures such as

$$
\begin{array}{c c c c}
H & R_1 & H & R_1 \\
| & | & | & | \\
-C- & C- & C- & C- \\
| & | & | & | \\
H & R_2 & H & R_2 \\
\end{array}
$$

have a tendency to degrade via depolymerization reactions. The pure depolymerization process consists of the following steps: (a) chain end initiation, (b) depropagation, (c) second-order termination, such as disproportionation and/or combination.

Random degradation and depolymerization processes may occur if the polymer sample is affected by

(i) heat (THERMAL DEGRADATION);
(ii) light (PHOTODEGRADATION);
(iii) high-energy radiation (RADIATIVE DEGRADATION);

(iv) shearing forces, ultrasonic vibrations, repeated and rapid freezing of the solvent of a polymer solution, high-speed stirring (MECHANO-DEGRADATION);

(v) chemical agents (CHEMICAL DEGRADATION);

(vi) enzymes, bacteria, fungi (BIODEGRADATION).

38.1 RANDOM DEGRADATION OF POLYMERS

The chain scission reaction is characterized by the DEGREE OF DEGRADATION (α), which is defined as the fraction of the total amount of mers in the polymer sample which is scissioned:

$$\alpha = \bar{s}/(DP_0 - 1), \tag{38.1}$$

where

$\bar{s}$ is the average number of main-chain links which have been scissioned in one original chain at time t,

DP_0 is the original chain length or degree of polymerization of the monodisperse polymer sample.

If DP_0 is large, eq. (38.1) is reduced to

$$\alpha = \bar{s}/DP_0. \tag{38.2}$$

38.2 RANDOM CROSSLINKING PROCESSES

In many cases a random crosslinking process occurs simultaneously with random chain scission.

The crosslinking reaction is characterized by the DEGREE OF CROSSLINKING (CROSSLINKING DENSITY) (Γ), which is defined as the fraction of the total quantity of mers in the polymer sample which is crosslinked (cf. Section 1.6).

The CROSSLINKING INDEX (δ) is defined in terms of the weight-average molecular weight ($\bar{M}_w$) as

$$\delta = p \, DP_w, \tag{38.3}$$

where

p is the probability of one mer being crosslinked in each original chain,

DP_w is the weight-average chain length at time t.

The CROSSLINKING INDEX (γ) is defined in terms of the number-average molecular weight ($\bar{M}_n$) as

$$\gamma = p \, DP_0, \tag{38.4}$$

where DP_0 is the original chain length or degree of polymerization of the monodisperse polymer sample.

For a monodisperse sample; $DP_0 = DP_w$ and $\delta = \gamma$; for a sample of random initial distribution, $DP_w = 2DP_0$ and $\delta = 2\gamma$.

The NUMBER-AVERAGE MOLECULAR WEIGHT BETWEEN CROSS-LINKS $((\bar{M}_n)_c)$ is given by

$$(\bar{M}_n)_c = \frac{(\bar{M}_n)_0}{\Gamma} \tag{38.5}$$

where

$(\bar{M}_n)_0$ is the number average molecular weight of the primary chain in the polymer sample,

Γ is the crosslinking density.

38.3 PHOTODEGRADATION OF POLYMERS

There are two types of photodegradation of polymers:

(i) DIRECT PHOTODEGRADATION, when macromolecules absorb light directly, and free radicals are formed after photoexcitation;

(ii) SENSITIZED PHOTODEGRADATION, when the degradation of macromolecules is initiated by the free radicals which are formed from the photodecomposition of low molecular weight photo-initiators (sensitizers).

Most polymers contain only C—C, C—H, C—O, C—N, and C—Cl bonds, and they do not absorb light of wavelength longer than 190 nm. When polymers contain different types of chromophoric group (cf. Table 14.1), they can absorb light with wavelengths of 250–400 nm and higher.

A necessary condition for the dissociation of a particular bond into free radicals is the excitation of a molecule by the absorption of light of sufficient energy. The bond dissociation energies are listed in Table 38.1, whereas the

Table 38.1 Approximate bond dissociation energies of various chemical bonds (CGS units)

Chemical bonds	Bond dissociation energy (kcal mol^{-1})
C—H (primary)	99
C—H (tertiary)	85
C—H (allylic)	77
C—C	83
C=C	145
C≡C	191
C—Cl	78
C—N	82
C=N	153
C≡N	191
C—O	93
C=O	186
C—Si	78
Si—H	76
N—O	37
O—O	66

Note: 1 kcal = 4.1868×10^3 J (SI)

energies of light quanta (photons) of different wavelength are shown in Table 38.2.

Table 38.2   Energy of light
(CGS units)

Wavelength (Å)	Energy (kcal mol^{-1})
10 000	28.6
7000	40.8
6200	46.1
6000	47.6
5800	49.3
5300	53.9
5000	57.1
4700	60.8
4200	68.1
4000	71.4
3660	78.0
3530	81.0
3250	88.0
3000	95.3
2890	99.0
2537	112.7
2000	142.9

Note: 1 kcal = 4.1868×10^3 J (SI);
1 eV = $1.602\ 10 \times 10^{-19}$ J (SI).

The quantitative relationship between the number of polymer molecules which are scissioned and the number of photons absorbed per unit time is given by the QUANTUM YIELD OF CHAIN SCISSION (Φ_{cs}):

$$\Phi_{cs} = \frac{\text{Number of macromolecules undergoing scission reaction}}{\text{Number of quanta absorbed by the polymer}}. \quad (38.6)$$

The quantitative relationship between the number of polymer molecules which are crosslinked and the number of photons absorbed per unit time is given by the QUANTUM YIELD OF CROSSLINKING (Φ_{cr}):

$$\Phi_{cr} = \frac{\text{Number of macromolecules undergoing crosslinking reaction}}{\text{Number of quanta absorbed by the polymer}}.$$

$$(38.7)$$

Sometimes the quantum yield is determined for a gas of low molecular weight which is formed during the degradation of macromolecules. The QUANTUM YIELD OF GASES EVOLVED (Φ_{gases}) is defined as

$$\Phi_{gases} = \frac{\text{Number of molecules of low molecular weight gaseous product}}{\text{Number of quanta absorbed by the polymer}}.$$

$$(38.8)$$

The number of gaseous molecules formed per unit time can be measured by any convenient analytical kinetic method, e.g. chromatography, mass spectrometry, etc.

Bibliography: 451, 507, 1086, 1162, 1187.

38.4 DETERMINATION OF THE QUANTUM YIELD

The QUANTUM YIELD OF CHAIN SCISSION (Φ_{cs}) is calculated experimentally from the equation

$$\Phi_{cs} = \frac{1}{I_a}\left[\frac{1}{(\bar{M}_n)_t} - \frac{1}{(\bar{M}_n)_0}\right], \tag{38.9}$$

where

I_a is the number of quanta of light absorbed by the polymer,
$(\bar{M}_n)_0$ is the initial number-average molecular weight of the polymer,
$(\bar{M}_n)_t$ is the number-average molecular weight after absorption of I_a quanta of light.

By measuring the viscosity of the photodegraded polymer samples, the quantum yield of chain scission (Φ_{cs}) is calculated from

$$\Phi_{cs} = \frac{1}{I_a}\left[\frac{1}{(\bar{M}_v)_t} - \frac{1}{(\bar{M}_v)_0}\right] = \frac{1}{I_a}\left[\frac{(\bar{M}_v)_0}{(\bar{M}_v)_t} - 1\right], \tag{38.10}$$

where

I_a is the number of quanta absorbed by the polymer,
$(\bar{M}_v)_0$ is the initial viscosity-average molecular weight of the polymer,
$(\bar{M}_v)_t$ is the viscosity-average molecular weight after absorption of I_a quanta of light.

The plot of $[(\bar{M}_v)_0/(\bar{M}_v)_t] - 1$ versus the number of quanta of light absorbed (I_a) is almost linear (Fig. 38.1) and the slope is $\tan\alpha = \Phi_{cs}$.

If crosslinking and main-chain scission occur simultaneously, the following equation should be taken into consideration:

$$\Phi_{cs} - \Phi_{cr} = \frac{1}{I_a}\left[\frac{1}{(\bar{M}_n)_t} - \frac{1}{(\bar{M}_n)_0}\right]. \tag{38.11}$$

For the determination of the number of quanta absorbed one uses chemical actinometers. A liquid-phase actinometer containing a light-sensitive compound (A) absorbs light of a certain wavelength and gives a product (B) with the quantum yield (Φ_B). The fraction of incident light absorbed by compound (A) is given by

$$1 - (I_a/I_0) = 1 - 10^{-\varepsilon_A c_A l} \tag{38.12}$$

and is measured by a photometric method or calculated from known values:

I_a is the number of quanta of light absorbed,
I_0 is the number of quanta in the incident radiation,
ε_A is the molar absorptivity of compound A,

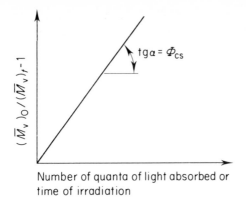

Fig. 38.1 Plot of $[(\bar{M}_v)_0/(\bar{M}_v)_t] - 1$ versus the number of quanta of light absorbed or the time of irradiation

c_A is the concentration of compound A,
l is the path length of the light.
The number of molecules of product B (n_B) formed during the irradiation time (t) can be determined by analysis. From these data, the intensity of the light beam (I_0) can be calculated from

$$I_0 = n_B/\Phi_B\, t(1 - 10^{-\epsilon_A c_A l}). \tag{38.13}$$

When the quantum yield Φ_B is known, the quantity I_0 can easily be determined.

A typical apparatus for the determination of quantum yield and also for quantitative photochemical studies is shown in Fig. 38.2.

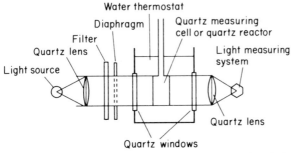

Fig. 38.2 Arrangement for quantitative photochemical studies

A number of liquid-phase actinometer systems are used for the determination of quantum yield:

(i) Potassium ferri-oxalate actinometers are very sensitive within a wide range of wavelengths from 2540 Å to 5780 Å. When a solution of $K_3Fe(C_2O_4)_3$ in aqueous sulphuric acid is irradiated by light, the Fe^{3+} ions are reduced to Fe^{2+}. The number of Fe^{2+} ions (n_B) formed during the photolysis can be calculated from

$$n_B = \frac{6.023 \times 10^{20} V_1 V_2 \log_{10}(I_0/I)}{V_2 l \varepsilon_B}, \tag{38.14}$$

where

V_1 is the volume of actinometer solution irradiated (in millilitres),
V_2 is the volume of aliquot taken for analysis (in millilitres),
V_3 is the final volume to which the aliquot V_2 is diluted (in millilitres),
$\log_{10}(I_0/I_a)$ is the measured optical density of the solution at 5100 Å,
l is the path length of the spectrophotometer cell (in centimetres),
ε_B is the experimental value of the molar absorptivity of the Fe^{2+} complex.

The quantum yield $\Phi_B = \Phi_{Fe^{2+}}$ for the different wavelengths is given in Table 38.3.

Table 38.3 Quantum yields of Fe^{2+} in a $K_3Fe(C_2O_4)_3$ chemical actinometer at 22 °C (from B: 242; reproduced by permission of John Wiley & Sons)

Wavelength (Å)	$[K_3Fe(C_2O_4)_3]$ (M)	Fraction of light absorbed (l = 15 mm)	Φ_B (i.e. $\Phi_{Fe^{2+}}$)
5770 5790	0.15	0.118	0.013
5460	0.15	0.061	0.15
5090	0.15	0.132	0.86
4800	0.15	0.578	0.94
4680	0.15	0.850	0.93
4360	0.15	0.997	1.01
	0.006	0.615	1.11
4050	0.006	0.962	1.14
3660	0.006	1.00	1.21
			1.26
	0.15	1.00	1.15
			1.20
3340	0.006	1.00	1.23
3130	0.006	1.00	1.24
2970 3020	0.006	1.00	1.24
2537	0.006	1.00	1.25

Bibliography: 242, 952, 1086.
Reference: 3948.

(ii) Uranyl oxalate actinometers are very sensitive within a wide range of wavelengths from 2080 Å to 4360 Å.

Bibliography: 242, 1086, 1291.
References: 3169, 4873, 5843, 5878, 5879, 6962.

(iii) Liquid-phase actinometers are used in the visible range.

Bibliography: 952, 1086.
References: 6301, 7003.

Chapter 39

Oxidation Reactions of Polymers

General bibliography: 708, 1077, 1086, 1104, 1409.

MOLECULAR OXYGEN exists in its ground state as a triplet $^3O_2$ ($^3\Sigma_g^-$), has a biradical nature, and reacts easily with other organic or polymer-free radicals (P˙) giving POLYMER PEROXY RADICALS (POO˙):

$$P˙ + O_2 \longrightarrow POO˙. \tag{39.1}$$

The peroxy radical can abstract hydrogen from another polymer molecule (PH) to form POLYMER HYDROXYPERIDES (POOH):

$$POO˙ + PH \longrightarrow POOH + P˙. \tag{39.2}$$

Two peroxy radicals may react with each other to produce POLYMER OXY-RADICALS (PO˙):

$$2 POO˙ \longrightarrow 2 PO˙ + O_2. \tag{39.3}$$

The polymer oxyradical can also be formed from the decomposition of polymer hydroperoxides (POOH):

$$POOH \longrightarrow PO˙ + ˙OH. \tag{39.4}$$

When the oxygen pressure is high (atmospheric pressure), termination occurs by the following reactions:

$$\left.\begin{array}{l} POO˙ + POO˙ \\ PO˙ + POO˙ \\ PO˙ + PO˙ \end{array}\right\} \longrightarrow \text{Inactive products} \tag{39.5}$$

Inactive products contain ether bridges or peroxide bridges depending on the reactions which participate in the termination reaction. When the oxygen pressure is low, the termination reaction may also occur by the following reactions

$$\left.\begin{array}{l} P˙ + POO˙ \\ P˙ + PO˙ \\ P˙ + P˙ \end{array}\right\} \longrightarrow \text{Inactive products.} \tag{39.6}$$

39.1 MEASURING OXYGEN UPTAKE

Measurement of oxygen uptake can be carried out using many types of microgas burette. One of the most practical is shown in Fig. 39.1. It allows the measurement of oxygen uptake as a change in volume at constant pressure and temperature. These steady conditions are achieved by using two manometers filled with

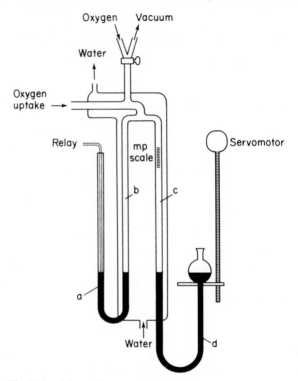

Fig. 39.1   Schematic representation of the oxygen uptake
instrument

mercury. Any decrease in volume causes a rise of mercury level in tubes b and c and a lowering in tube a, which breaks an electrical contact in a. A relay starts a servo motor which raises the mercury level in d, thereby also raising the level in tube a. Volume expansions are corrected in a similar way. When the mercury level in a starts to rise, a second contact in a is closed and causes the servo motor to lower the level in tube d. The accuracy of the instrument is determined by the difference of the two contacts in tube a.

Bibliography: 1086.
References: 2474, 2541, 3750, 6133.

A much more highly sensitive system for measuring the oxygen uptake is based on a reluctance transducer (Fig. 39.2). The pressure transducer is made entirely of welded stainless steel, with a ± 10 torr (1 torr $= \frac{1}{760}$ atm $= 133.322 \, \mathrm{N \, m^{-2}}$

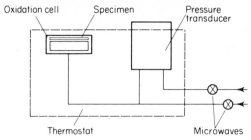

Fig. 39.2 Oxygen uptake instrument based on a
reluctance transducer

(SI)) linear operation range and a diaphragm displacement of only 0.005 cm³
at full deflection.

Reference: 3752.

39.2 DETERMINATION OF HYDROPEROXIDE GROUPS

The hydroperoxide group content in polymers can be measured by two sensitive
spectrophotometric methods:

(i) The IODOMETRIC METHOD. The polymer sample is diluted with a
mixture of acetic acid and chloroform, carbon tetrachloride, or chlorobenzene and treated with potassium iodide after deaeration. The iodine
liberated is measured spectrophotometrically at 360 nm in 1 cm cells.
Hydroperoxide group contents in the range 1–100 p.p.m. can be determined.

References: 2223, 5035.

(ii) The TRIPHENYLPHOSPHINE METHOD. The polymer sample is
treated in solution with triphenylphosphine, which is oxidized to triphenylphosphine oxide. The decreasing absorbance at 260 nm when
triphenylphosphine is oxidized is a quantitative measurement of the
hydroperoxide. Hydroperoxide group contents in the range 0.1–10 p.p.m.
can be determined.

References: 6535, 6568.

Hydroperoxide group contents in polymer films can be measured by using
infrared spectroscopy. The hydroperoxide absorption peak is located in the same
spectral range as that for hydroxide groups, i.e. at 3300–3500 cm⁻¹, and it is
difficult to determine one from the other.

The oxidized polymer is exposed to sulphur dioxide vapour at room temperature. Sulphate groups are formed rapidly and quantitatively by selective reaction
only with the hydroperoxide groups. Analysis is based on the strong infrared
absorption of sulphate groups at 1195 cm⁻¹. Hydroperoxide groups in the range
0.1–1 p.p.m. can be determined.

Reference: 5302.

626

39.3 DETERMINATION OF CARBONYL GROUPS IN POLYMERS

During the oxidation of polymers different types of carbonyl group are formed, e.g. monoketo systems, diketo systems, aldehyde systems, acid and ester systems, and cyclic systems, e.g. lactones.

Carbonyl absorptions are sensitive to both chemical and physical effects, and due allowance must be made for both in assessing the likely position of ν_{CO} in a given polymer structure. Many factors are involved in determining the precise frequency of a given carbonyl group. These include not only those, such as inductive, resonance, and field effects, which alter the force constant of the CO bond, but also physical factors, such as mass and angle effects, vibrational coupling, and changes in the force constant of the adjacent bonds. Phase changes, solvent effects, and hydrogen bonding also play a significant part in determining the final frequency.

In most oxidized polymers carbonyl frequencies lie in the range 1780–1640 cm^{-1}.

Bibliography: 96, 112, 1013, 1077, 1086.

39.4 OXIDATION OF POLYMERS BY SINGLET OXYGEN

SINGLET OXYGEN is the excited form of molecular oxygen and can exist in two forms:

(i) SINGLET OXYGEN $^1O_2$ $(^1\Delta_g)$;
(ii) SINGLET OXYGEN $^1O_2$ $(^1\Sigma_g^+)$.

These two forms of singlet oxygen differ one from another in electron configuration and in excitation energy: singlet oxygen $^1O_2$ $(^1\Delta_g)$ exceeds the ground state of molecular oxygen $(^3\Sigma_g^-)$ by 22.5 kcal mol^{-1} (0.98 eV); singlet oxygen $^1O_2$ $(^1\Sigma_g^+)$ exceeds the ground state of molecular oxygen $(^3\Sigma_g^-)$ by 37.5 kcal mol^{-1} (1.63 eV).

In the gas phase under low pressure singlet oxygen $^1O_2$ $(^1\Delta_g)$ is extremely long lived, c. 45 min. Singlet oxygen $^1O_2$ $(^1\Delta_g)$ has a shorter lifetime (c. 7 s) and is rapidly quenched in the presence of water vapour. The lifetime of singlet oxygen in solution is strongly affected by the nature of the solvent (Table 39.1). Singlet oxygen can be generated by the following methods:

(i) Microwave excitation in a specially constructed apparatus (Fig. 39.3). Oxygen is passed through a quartz tube which is a part of the discharge section which is connected to a 2450 MHz microwave generator. Such a discharge produces atomic oxygen, ozone, and singlet oxygen. The oxygen entering the discharge zone is saturated with mercury vapour, and after discharge reacts immediately with atomic oxygen and/or ozone according to the reactions

$$O + H_g \longrightarrow HgO, \qquad (39.7)$$

$$O_3 + H_g \longrightarrow HgO + O_2. \qquad (39.8)$$

Table 39.1 Lifetime of singlet oxygen $^1O_2$ ($^1\Delta_g$) in various solvents (from R: 5235; reproduced by permission of the American Chemical Society)

Solvent	$\tau^1 \Delta$ (μs)
H_2O	2
CH_3OH	7
50% D_2O–50% CH_3OH	11
C_2H_5OH	12
C_6H_{12}	17
C_6H_6	24
CH_3COCH_3	26
CH_3CN	30
$CHCl_3$	60
CS_2	200
CCl_4	700
Freon-113	1000

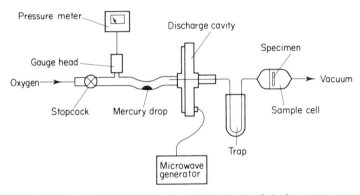

Fig. 39.3 Microwave generator for production of singlet oxygen

Bibliography: 47, 1086.
References: 2561, 4413, 5937.

(ii) Laser irradiation of a liquid solution of oxygen in Freon-113.

References: 4432, 5108, 5109.

(iii) The photosensitizing method. Singlet oxygen is formed in the energy transfer process from the excited sensitizer (S) molecule to the molecular oxygen:

$$S + h\nu \longrightarrow S^*, \tag{39.9}$$

$$S^* + {}^3O_2 \longrightarrow S + {}^1O_2. \tag{39.10}$$

Common dyes such as methylene blue, Rose Bengal, or rhodamine are used as sensitizers.

Bibliography: 720, 721, 740, 1076, 1086.
References: 5937, 7132.

(iv) Chemical methods, of which two are widely applied in research.

(a) Reaction of active chlorine from the decomposition of sodium or calcium hypochlorites with hydrogen peroxide,

$$Cl_2 + H_2O_2 \longrightarrow {}^1O_2 + 2HCl. \tag{39.11}$$

References: 3440–3442.

(b) Thermal decomposition of triphenylphosphite–ozone adduct,

$$(C_6H_5O)_3P\overset{O}{\underset{O}{\diamond}}O \xrightarrow{-35\,°C} {}^1O_2 + (C_6H_5O)_3P{=}O. \tag{39.12}$$

The adduct is formed between triphenylphosphite and ozone at $-70\,°C$.

References: 4414, 5396, 5397.

The reactivity of singlet oxygen with chemical compounds is determined by the β VALUE, which is defined as follows:

$$\beta = k_1/k_2, \tag{39.13}$$

where

k_1 is the rate constant of the singlet oxygen quenching reaction,

$${}^1O_2 \xrightarrow{k_1} {}^3O_2, \tag{39.14}$$

k_2 is the rate constant of the singlet oxygen reaction with a chemical compound A,

$$A + {}^1O_2 \xrightarrow{k_2} AO_2. \tag{39.15}$$

Singlet oxygen reacts with unsaturated bonds in the polymer by the ENE MECHANISM to give allyl hydroperoxide groups with shifting of the double bond,

$$-CH_2-CH{=}CH- + {}^1O_2 \longrightarrow -CH{=}CH-\overset{\overset{\displaystyle OOH}{|}}{C}H-. \tag{39.16}$$

In order to avoid the occurrence of the above reaction it is necessary to add polymer singlet oxygen quenchers, which are different types of organic compounds, e.g. chelates.

Bibliography: 47, 251–253, 619, 720, 721, 740, 985, 986, 1079, 1080, 1089, 1090, 1147, 1263, 1388, 1408, 1437.
References: 2166, 2561, 2718–2720, 3255, 3439, 4413, 4415, 4416, 5235, 5475, 5533, 5936, 5939, 7347.

39.5 OXIDATION OF POLYMERS BY ATOMIC OXYGEN

ATOMIC OXYGEN exists as separate atoms and has an even number of electrons. For that reason it is highly paramagnetic, and can easily be detected by ESR spectroscopy (its spectrum contains six fine-structure lines).

Atomic oxygen can be generated by the following methods:

(i) Electric discharge in molecular oxygen,

$$O_2 + e^- \longrightarrow 2O + e^-. \qquad (39.17)$$

(ii) Mercury-sensitized photodecomposition of dinitrogen oxide (N_2O) by ultraviolet light:

$$N_2O \xrightarrow[H_g]{h\nu} N_2 + O. \qquad (39.18)$$

N_2O under normal atmospheric pressure is passed through a quartz tube which is irradiated by a low-pressure mercury lamp which produces light at 254 nm (Fig. 39.4).

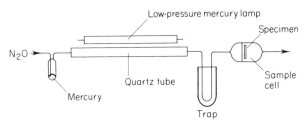

Fig. 39.4 Photoreactor for production of atomic oxygen

Atomic oxygen is very reactive and can easily abstract hydrogen even from the methylene groups present in some types of polymer:

$$-CH_2- + O \longrightarrow -\overset{\bullet}{C}H- + HO^{\bullet}. \qquad (39.19)$$

Bibliography: 566, 567, 632, 1086.
References: 2219, 3042, 3670, 3896, 4039, 4467, 4841, 5004, 5071, 5214.

39.6 OXIDATION OF POLYMERS BY OZONE

OZONE (O_3) is an unstable, blue gas with a characteristic odour. The structure of the ozone molecule is that of an obtuse angle in which a central oxygen atom is attached to two equidistant oxygen atoms. The included angle is about $116°49'$ and the bond length is 1.278 Å.

At $-112\,°C$, ozone condenses to a dark-blue liquid. Liquid ozone is easily exploded. Explosions may be initiated by traces of catalysts, organic matter, shocks, electric sparks, and sudden changes in temperature or pressure.

Ozone has strong absorption bands in the infrared, the visible, and the ultra-violet (maximum about 2540 Å).

Liquid ozone is miscible in all proportions with the following liquids: $CClF_3$, CCl_2F_2, CH_4, CO, F_2, NF_3, and OF_2. Ozone is sparingly soluble in water, especially at low temperatures.

The apparatus designed to produce ozone by electric discharge in an oxygen gas flow is called an OZONATOR or OZONIZER (Fig. 39.5). The discharge

630

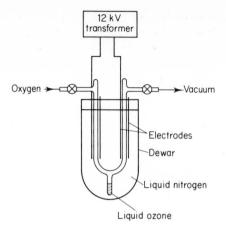

Fig. 39.5   Electric discharge ozonator for
production of ozone

tube is made from Pyrex glass. The cylindrical electrodes are aluminium foil in
close contact with the tube, and the voltage applied is 12 kV. The formation of
ozone in the ozonator is directly proportional to the power dissipated in the
discharge. The discharge tube is placed in a Dewar containing liquid nitrogen.
Ozone (100%) is liquefied and collected in the finger of the discharge tube. It is
recommended that no more than 0.2–0.5 cm^3 of liquid ozone be produced in
one operation. Manipulations of pure ozone should be accomplished behind
a safety shield of Plexiglas.

Ozone concentration can be measured by the iodometric method. Ozone is
absorbed in neutral or alkaline potassium or sodium iodide solution. After
absorption of the ozone the iodide solution is acidified and the liberated iodine
is titrated with standard sodium thiosulphate or determined spectrophotometrically at 360 nm in 1 cm cells.

Bibliography: 40, 162, 354, 866, 1322, 1340, 1455.
References: 3151, 3775, 4300,

Ozone reacts easily with polymers by the following reactions:
(i) Abstraction of hydrogen from methylene groups,

$$-CH_2- + O_3 \longrightarrow \begin{matrix} OO^{\bullet} \\ | \\ -CH- \end{matrix} + HO^{\bullet}. \qquad (39.20)$$

(ii) Reaction with double bonds in unsaturated polymers. During this reaction
OZONIDES are formed as intermediates,

$$\underset{R_1}{\overset{H}{}}C=C\underset{R_2}{\overset{H}{}} + O_3 \rightarrow \underset{O}{\overset{O-O}{C \quad C}} \rightarrow R_1-C\overset{O}{\underset{H}{}} + \underset{O}{\overset{HO}{C-R_2}}. \qquad (39.21)$$

The nature of the solvent determines what ozonolysis products will be obtained. In non-polar hydrocarbon solvents, ozonides will be formed. In more polar solvents, a mixture of ozonides, peroxides, aldehydes, and acids will be produced. In reactive solvents, hydroperoxides are the products from the reaction with the zwitterion,

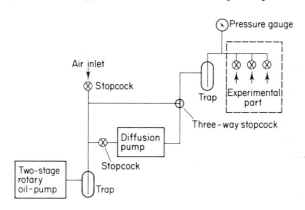

$$\underset{R_2}{\overset{R_1}{\diagup}}C=C\underset{R_4}{\overset{R_3}{\diagdown}} + O_3 \longrightarrow \underset{}{\overset{O-O}{\diagup}}C\underset{O}{\diagdown}C\diagdown \longrightarrow \underset{R_2}{\overset{R_1}{\diagup}}C=O + R_4-\underset{OO^-}{\overset{R_3}{\underset{|}{C^+}}} \qquad (39.22)$$

Zwitterion

Bibliography: 675, 678, 1086.
References: 2961, 3835, 4296, 4297, 5981, 6642–6644, 7206.

39.7 EXPERIMENTAL TECHNIQUES IN VACUUMS

The design and construction of high-vacuum apparatus depend on the specific purpose of its application. A block diagram of a high-vacuum pumping system is shown in Fig. 39.6. The apparatus should be as simple as possible. Too many

Fig. 39.6 Schematic representation of a high-vacuum pumping system

joints, valves, and stopcocks lead to complications in construction and operation, and also result in vacuum leakages. Commonly used oils, greases, and waxes in high-vacuum equipment are shown in Table 39.2.

Silicone greases have very low temperature coefficients and are suitable for the lubrication of taps and joints which operate at low temperatures. They are dissolved in carbon tetrachloride or by warm 10% alcoholic potash. The apiezon greases are soluble in many organic solvents. Solubility can be a disadvantage and limits the application of silicone and apiezon greases in vacuum systems where vapours of organic solvents are present.

Table 39.2   Oils, greases, and waxes commonly used in high-vacuum techniques (from B: 553; reproduced by permission of Marcel Dekker)

Grade	Vapour pressure (CGS)	Recommended use	Maximum temperature in us (°C)
Oils			
Silicon oil 702	5×10^{-6}	Diffusion pumps	
Silicon oil 703	$< 10^{-8}$	For highest vacua	
Silicon oil 704	10^{-8}	Diffusion pumps	
Octoil (2-ethyl hexyl phthalate)	2×10^{-7}	Diffusion pumps	
Octoil S (2-ethyl hexyl sebacate)	5×10^{-8}	Diffusion pumps	
Apiezon oil A	10^{-5}	Diffusion pumps	
Apiezon oil B	10^{-7}	Oil U-tube manometers	
Apiezon oil C		Diffusion pumps	
Apiezon oil G		Diffusion pumps	
Greases			
Silicon D.C.			
High-vacuum grease	$< 10^{-6}$	Taps or joints	-40 to $+150$
Apiezon grease L	10^{-10}	Ground joints and taps	30
Apiezon grease M	10^{-7}	Ground joints	30
Apiezon grease N	10^{-8}	Taps	30
Apiezon grease T	10^{-8}	Heated joints	110
Kel-F		Resistant against corrosive gases	
Waxes			
Apiezon sealing			
Compound Q	10^{-4}	Unground joints	30
Wax W	10^{-6}	Permanent joints	80
Wax W40	10^{-6}	(Semipermanent joints)	30
Wax W100	10^{-6}		50
Khotinsky cement	10^{-3}		< 50
Picein	10^{-6}		< 50
Silver chloride			< 450

Traps placed between the experimental equipment and the rotary oil pump prevent contamination of the latter by vapours and gases.

Bibliography: 188, 387, 553, 801, 848, 914, 1044, 1116, 1142, 1317.

Chapter 40

Units and Constants

40.1 CGS–SI INTERCONVERSIONS

General bibliography: 638, 896, 1004, 1015.

The SI unit system is based on mutally consistent units assigned to the nine physical quantities listed in Table 40.1. This table shows how the conversion

Table 40.1 Basic SI units and their relation to CGS or other common units

Physical quantity	SI unit Name	SI unit Symbol	× Conversion → factor	CGS/common unit Name	CGS/common unit Symbol
Length	metre	m	10^2	centimetre	cm
Mass	kilogram	kg	10^3	gram	g
Time	second	s	1	second	s
Electric current	ampere	A	2.998×10^9	statampere	statamp
Thermodynamic temperature	kelvin	K	1	kelvin	K
Luminous intensity	candela	cd	π	lambert	(cm^2)
Amount of substance	mole	mol	1	mole	mol
Plane angle	radian	rad	$180/\pi$	degree (angle)	°
Solid angle	steradian	sr	1	steradian	sr

SI unit   ← Conversion ÷ CGS/common unit
factor

factors can be used in going from SI to CGS units, and in the reverse direction. From these nine basic quantities numerous other SI units may be derived, e.g.:

(i) physical quantities (Appendix K);
(ii) mechanical quantities (Appendix L).

The following tables are given for rapid unit conversion factors:

(i) force (Table 40.2);
(ii) energy (Table 40.3);
(iii) power (Table 40.4).

The names of multiples and submultiples of SI units may be formed by application of the prefixes (Table 40.5).

Table 40.2   Force unit conversion factors

N	dyn	kp	lbf (pound force)
1	0.1×10^6	0.101 972	0.224 809
10×10^{-6}	1	$1.019\ 72 \times 10^{-6}$	$2.248\ 09 \times 10^{-6}$
9.806 65	$0.980\ 665 \times 10^6$	1	2.204 62
4.448 22	$0.444\ 822 \times 10^6$	0.453 592	1

Table 40.3   Energy unit conversion factors

J, N m, W s	erg	kWh	eV	kpm	kcal	ft-lbf (foot pound force)
1	10×10^6	$0.277\ 778 \times 10^{-6}$	$6.241\ 8 \times 10^{18}$	0.101 972	$0.238\ 846 \times 10^{-3}$	0.737 562
0.1×10^{-6}	1	$27.777\ 8 \times 10^{-15}$	$0.624\ 18 \times 10^{12}$	$10.197\ 2 \times 10^{-9}$	$23.884\ 6 \times 10^{-12}$	73.756×10^{-9}
3.6×10^6	36×10^{12}	1	22.471×10^{24}	$0.367\ 098 \times 10^6$	859.845	$2.655\ 22 \times 10^6$
$0.160\ 21 \times 10^{-18}$	$1.602\ 1 \times 10^{-12}$	44.503×10^{-27}	1	$1.633\ 7 \times 10^{-20}$	38.266×10^{-24}	$0.118\ 21 \times 10^{-18}$
9.806 65	$98.066\ 5 \times 10^6$	$2.724\ 07 \times 10^{-6}$	61.211×10^{18}	1	$2.342\ 28 \times 10^{-3}$	7.233 01
$4.186\ 8 \times 10^3$	41.868×10^9	1.163×10^{-3}	$2.613\ 2 \times 10^{22}$	426.935	1	$3.088\ 03 \times 10^3$
1.355 82	$13.558\ 2 \times 10^6$	$0.376\ 616 \times 10^{-6}$	8.463×10^{18}	0.138 255	$0.323\ 823 \times 10^{-3}$	1

Table 40.4   Power unit conversion factors

W, N m s^{-1}, J s^{-1}	kpm s^{-1}	kcal s^{-1}	kcal h^{-1}	hk	hp (United) Kingdom, USA)	ft-lbf s^{-1}
1	0.101 972	$0.238\ 846 \times 10^{-3}$	0.859 845	$1.359\ 62 \times 10^{-3}$	$1.341\ 02 \times 10^{-3}$	0.737 562
9.806 65	1	$2.342\ 28 \times 10^{-3}$	8.432 20	$1.333\ 33 \times 10^{-2}$	$1.315\ 09 \times 10^{-2}$	7.233 01
$4.186\ 8 \times 10^3$	426.935	1	3.6×10^3	5.692 46	5.614 59	$3.088\ 03 \times 10^3$
1.163	0.118 593	$0.277\ 778 \times 10^{-3}$	1	$1.581\ 24 \times 10^{-3}$	$1.559\ 61 \times 10^{-3}$	0.857 785
735.499	75	0.175 671	632.415	1	0.986 320	542.476
745.700	76.040 2	0.178 107	641.186	1.013 87	1	550
1.355 82	0.138 255	$0.323\ 832 \times 10^{-3}$	1.165 79	$1.843\ 40 \times 10^{-3}$	$1.818\ 18 \times 10^{-3}$	1

Table 40.5   Multiples of units, their names, and symbols

Multiple	Prefix	Symbol
10^{12}	tera	T
10^{9}	giga	G
10^{6}	mega	M
10^{3}	kilo	k
10^{2}	hecto	h
10	deca	da
10^{-1}	deci	d
10^{-2}	centi	c
10^{-3}	milli	m
10^{-6}	micro	μ
10^{-9}	nano	n
10^{-12}	pico	p
10^{-15}	femto	f
10^{-18}	atto	a

40.1.1 Definitions of SI Units

The following material is reproduced from Publication SP-7012 of the Scientific and Technical Information Office of the National Aeronautics and Space Administration (USA):

Ampere (A)   The *ampere* is that constant current which, if maintained in two straight parallel conductors of infinite length, of negligible circular cross-section, and placed 1 metre apart in vacuum, would produce between these conductors a force equal to 2×10^{-7} newton per metre of length.

Candela (cd)   The *candela* is the luminous intensity, in the perpendicular direction, of a surface of 1/600 000 square metres of a black body at the temperature of freezing platinum under a pressure of 101 325 newtons per square metre.

Coulomb (C)   The *coulomb* is the quantity of electricity transported in 1 second by a current of 1 ampere.

Henry (H)   The *henry* is the inductance of a closed circuit in which an electromotive force of 1 volt is produced when the electric current in the circuit varies uniformly at a rate of 1 ampere per second.

Farad (F)   The *farad* is the capacitance of a capacitor between the plates of which there appears a difference of potential of 1 volt when it is charged by a quantity of electricity equal to 1 coulomb.

Joule (J)   The *joule* is the work done when the point of application of 1 newton is displaced a distance of 1 metre in the direction of the force.

Kelvin (K)   The *kelvin*, unit of thermodynamic temperature, is the fraction 1/273.16 of the thermodynamic temperature of the triple point of water.

Kilogram (kg)   The *kilogram* is the unit of mass; it is equal to the mass of the international prototype of the kilogram. (The international prototype of the kilogram is a particular cylinder of platinum–iridium alloy which is preserved in a vault at Sèvres, France, by the International Bureau of Weights and Measures.)

Lumen (lm) The *lumen* is the luminous flux emitted in a solid angle of 1 steradian by a uniform point source having an intensity of 1 candela.

Metre (m) The *metre* is the length equal to 1 650 763.73 wavelengths in vacuum of the radiation corresponding to the transition between the levels $2p_{10}$ and $5d_5$ of the krypton-86 atom.

Mole (mol) The *mole* is the amount of substance of a system which contains as many elementary entities as there are carbon atoms in 0.012 kilograms of carbon 12. The elementary entities must be specified and may be atoms, molecules, ions, electrons, other particles, or specified groups of such particles.

Newton (N) The *newton* is that force which gives to a mass of 1 kilogram an acceleration of 1 metre per second per second.

Ohm (Ω) The *ohm* is the electric resistance between two points of a conductor when a constant difference of potential of 1 volt, applied between these two points, produces in this conductor a current of 1 ampere, this conductor not being the source of any electromotive force.

Radian (rad) The *radian* is the plane angle between two radii of a circle which cut off on the circumference an arc equal in length to the radius.

Second (s) The *second* is the duration of 9 192 631 770 periods of the radiation corresponding to the transition between the two hyperfine levels of the ground state of the caesium-133 atom.

Steradian (sr) The *steradian* is the solid angle which, having its vertex in the centre of a sphere, cuts off an area of the surface of the sphere equal to that of a square with sides of length equal to the radius of the sphere.

Volt (V) The *volt* is the difference of electric potential between two points of a conducting wire carrying a constant current of 1 ampere, when the power dissipated between these points is equal to 1 watt.

Watt (W) The *watt* is the power which gives rise to the production of energy at the rate of 1 joule per second.

Weber (Wb) The *weber* is the magnetic flux which, linking a circuit of one turn, produces in it an electromotive force of 1 volt as it is reduced to zero at a uniform rate in 1 second.

40.1.2 Temperature Scales

The different temperature scales commonly used in the United States and Europe are

    (i) The CELSIUS TEMPERATURE SCALE (sometimes called the CENTI-GRADE SCALE), which employs a degree of the same magnitude as that of the Kelvin scale, but its zero is shifted so that the Celsius temperature of the triple point of water is at 0.01 degrees Celsius (0.01 °C).

Between the Celsius and the Kelvin temperature scales is the following relation:

$$t = T - 273.15 \text{ K} \qquad (40.1)$$

Note: the KELVIN TEMPERATURE SCALE is independent of the properties of any particular substance. The KELVIN is the fraction 1/273.16 of the thermodynamic temperature of the triple point of water.

(ii) The FAHRENHEIT TEMPERATURE, where the ice point ($t = 0$ °C) is 32 °F and the steam point ($t = 100$ °C) is 212 °F. The 100 Celsius or Kelvin degrees between the ice point and the steam point correspond to 180 Fahrenheight degrees.

The different temperature scales are compared in Table 40.6.

Table 40.6 Different temperature scales

Kelvin	Celsius	Fahrenheit
0 K	− 273.15 °C	− 459.67 °F
255.372 2 K	− 17.777 8 °C	0 °F
273.15 K	0 °C	32 °F
273.16 K	0.01 °C	32.018 °F
373.15 K	100 °C	212 °F
1 K	1 °C	1.8 °F
0.555 556 K	0.555 556 °C	1 °F
= 5/9 K	= 5/9 °C	

40.1.3 Pressure Unit Conversion Factors

The following pressure units are commonly used:

Bar 1 bar = 10^5 N m^{-2} = 10^5 Pa.

normal atmosphere 1 atm = 101 325 N m^{-2} = 1.013 25 × 10^5 Pa
= 14.7 lb in^{-2} = 2117 lb ft^{-2}.

Technical atmosphere 1 at = 1 kp cm^{-2} = 98 066.5 N m^{-2} = 9.8066 × 10^4 Pa.

Torr 1 torr = $\frac{1}{760}$ atm = 133.322 N m^{-2} = 1.333 22 × 10^2 Pa.

mm Hg 1 mm Hg = 1 torr.

Pascal (Pa) 1 Pa = 1 N m^{-2} = 1.451 × 10^{-4} lb in^{-2} = 0.209 lb ft^{-2}.

lb in^{-2} 1 lb in^{-2} = 6891 Pa.

lb ft^{-2} 1 lb ft^{-2} = 47.85 Pa.

Table 40.7 is added for rapid location of pressure unit conversion factors.

Table 40.7   Pressure unit conversion factors

$N\ m^{-2}$	bar	$kp\ cm^{-2} = at$	torr = mmHg (0 °C)	atm (normal atmosphere)	$lbf\ in^{-2}$ †
1	10×10^{-6}	$10.197\ 2 \times 10^{-6}$	$7.500\ 62 \times 10^{-3}$	$9.869\ 23 \times 10^{-6}$	$0.145\ 038 \times 10^{-3}$
100×10^3	1	1.019 72	750.062	0.986 923	14.503 8
$98.066\ 5 \times 10^3$	0.980 665	1	735.559	0.967 841	14.223 3
133.322	$1.333\ 22 \times 10^{-3}$	$1.359\ 51 \times 10^{-3}$	1	$1.315\ 79 \times 10^{-3}$	$1.933\ 68 \times 10^{-2}$
101.325×10^3	1.013 25	1.033 23	760	1	14.695 9
$6.894\ 76 \times 10^3$	$6.894\ 76 \times 10^{-2}$	$7.030\ 70 \times 10^{-2}$	51.714 9	$6.804\ 60 \times 10^{-2}$	1

† p.s.i. = *p*ounds per *s*quare inch.

40.2 THE UNIVERSAL GAS CONSTANT

The UNIVERSAL GAS CONSTANT (R) originates from the IDEAL GAS LAW EQUATION, defined as

$$pV = nRT = (m/M)\,RT, \tag{40.2}$$

where

p is the gas pressure,

V is the volume occupied by the definite mass of gas (m),

$$m = nM, \tag{40.3}$$

n is the number of moles,

M is the molecular weight of gas,

R is the universal gas constant,

T is the thermodynamic temperature (in kelvins).

The constant of proportionality (R), called the UNIVERSAL GAS CONSTANT, has the same value for all gases. The numerical value of R depends on the units in which p, V, n, and T are expressed.

In the CGS system, where p is measured in dynes per square centimetre and V is measured in cubic centimetres, the numerical value of R is

$$R = 8.314 \times 10^7\ (dyn\ cm^{-2})\ cm^3\ mol^{-1}\ K^{-1}$$

$$= 8.314 \times 10^7\ erg\ mol^{-1}\ K^{-1}.$$

In the SI system, where p is measured in pascals (newtons per square metre) and V is measured in cubic metres, the numerical value of R is

$$R = 8.314\ (N\ m^{-2})\ m^3\ mol^{-1}\ K^{-1}$$

$$= 8.314\ J\ mol^{-1}\ K^{-1}.$$

Since the units of pressure times volume are the same as units of energy, hence R has units of energy per mole per unit of absolute temperature in all systems of units. In terms of calories,

$$R = 1.99 \text{ cal mol}^{-1} \text{ K}^{-1}.$$

In chemical calculations volumes are commonly expressed in litres, pressure in atmospheres, and temperatures in kelvins. In this system,

$$R = 0.082\,07 \text{ litres atm mol}^{-1} \text{ K}^{-1}.$$

40.3 THE AVOGADRO NUMBER

The AVOGADRO NUMBER (N_A) is the number of molecules in a mole. The most precise value of N_A to date, obtained by using X-rays to measure the distance between the layers of molecules in a crystal, is

$$N_A = 6.022\,17 \times 10^{23} \text{ molecules mol}^{-1}.$$

The MOLECULAR WEIGHT (more correctly the MOLECULAR MASS) (M) equals the mass of one mole, and can be expressed as the product of the Avogadro number (N_A) and the mass (m) of a single molecule,

$$M = N_A m. \tag{40.4}$$

The NUMBER OF MOLES (n) in a given sample of a pure substance equals the total number of molecules (N) divided by the Avogadro number (N_A),

$$n = N/N_A. \tag{40.5}$$

The ideal gas law equation (eq. (40.2)) can be written as

$$pV = nRT = (N/N_A) RT \tag{40.6}$$

or

$$N/V = p(N_A/RT), \tag{40.7}$$

where N/V is the number of molecules per unit volume.

Since N_A and R are universal constants, it follows that at constant temperature T the number of molecules per unit volume is directly proportional to the pressure.

40.4 THE BOLTZMANN CONSTANT

The BOLTZMANN CONSTANT (k) originates from the molecular gas theory and is defined as

$$k = R/N_A, \tag{40.8}$$

where

R is the universal gas constant,
N_A is the Avogadro number.

In the CGS system,

$$k = \frac{R}{N_A} = \frac{8.314 \times 10^7 \text{ erg mol}^{-1} \text{ K}^{-1}}{6.023 \times 10^{23} \text{ molecules mol}^{-1}} = 1.38 \times 10^{-16} \text{ erg molecule}^{-1} \text{ K}^{-1}.$$

(40.9)

In the SI system,

$$k = \frac{R}{N_A} = \frac{8.31 \text{ J mol}^{-1} \text{ K}^{-1}}{6.023 \times 10^{23} \text{ molecules mol}^{-1}} = 1.38 \times 10^{-23} \text{ J molecule}^{-1} \text{ K}^{-1}.$$

(40.10)

Since R and N_A are universal constants, the same is true of k.

40.5 MATHEMATICAL TRANSFORMATIONS

In order to make experimental data more readily understandable, it is recommended that simple mathematical transformations be used. These transformations involve changing the scale of the variables (independent, dependent, or both) to serve one or more of the following purposes:

(i) to make a linear relationship out of a non-linear relationship;
(ii) to make the variation in experimental data more independent of their position on the scale of measurement;
(iii) to make a symmetrical or normal distribution out of skewed data;
(iv) to remove interactions—a measure of the extent to which the effect (on the dependent variable) of one factor depends upon the level of another factor;
(v) to conform to theoretical considerations.

Note: NORMAL DISTRIBUTION is a distribution in which the ordinate is given by

$$y = \left(\frac{1}{\sigma \sqrt{(2\pi)}} \right) \exp \left\{ -\frac{1}{2} \left(\frac{x-m}{\sigma} \right)^2 \right\},$$

(40.11)

where
m is the mean,
σ is the STANDARD DEVIATION, that is the square root of the average of the squares of the deviations of individual observations from their arithmetic mean.

Figure 40.1(A–I) shows how curved lines can be straightened out by the proper choice of transformations. Table 40.8 show basics untransformed relationships and the results of transformations.

641

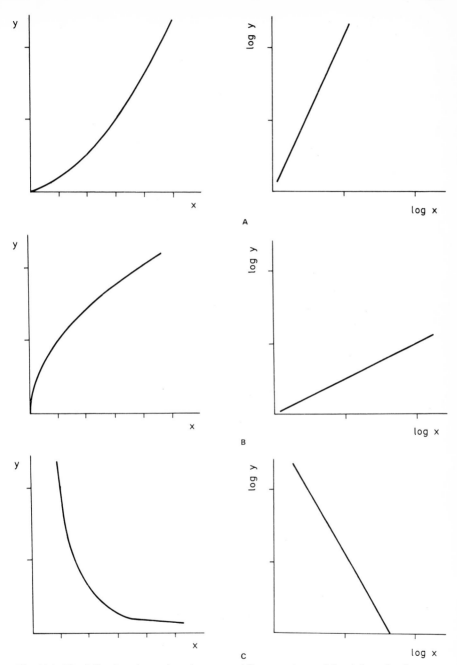

Fig. 40.1. The following charts show how curved lines can be straightened out by the proper choice of mathematical transformations. In these cases, the variation is usually made constant at the same time. Figures on the left-hand side show the untransformed relationship and the figures on the right-hand side show the results of transformations. For the mathematical description of these transformations see Table 40.8.

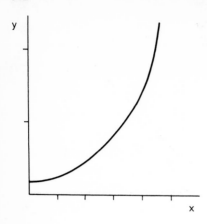

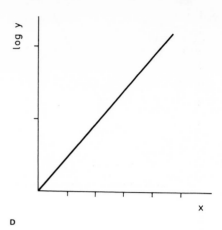

D

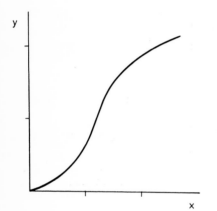

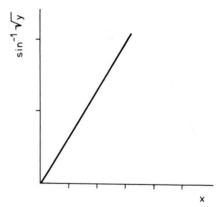

E

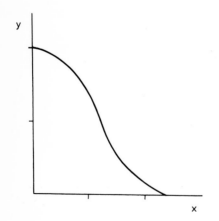

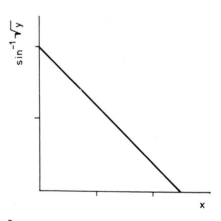

F

Fig. 40.1 (*cont.*)

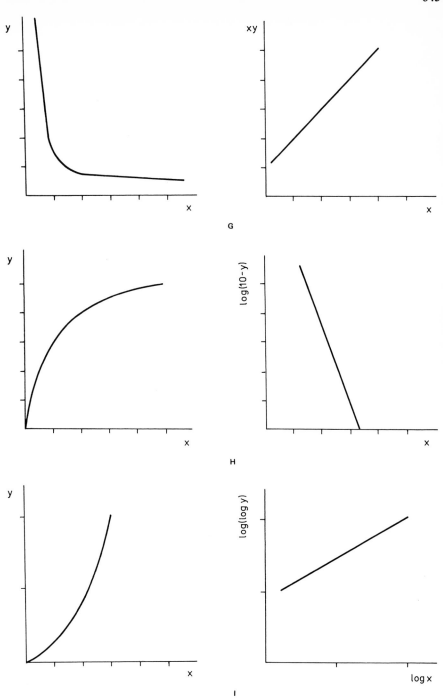

G

H

I

Fig. 40.1 (*cont.*)

Table 40.8. Mathematical transformations

Figure	Basic relationship	Transformations
A	$y = ax^b$	$\log y = \log a + b \log x$
B	$y = ax^{1/b}$	$\log y = \log a + (\log x/b)$
C	$y = ax^{-b}$	$\log y = \log a - b \log x$
D	$y = ab^x$	$\log y = \log a + x \log b$
E	Increase of proportion with time $y = f(x)$	Replace y by $\sin^{-1} \sqrt{y}$ $\sin^{-1} \sqrt{y} = f'(x)$
F	Decrease of proportion with time $y = f(x)$	Replace y by $\sin^{-1} \sqrt{y'}$ $\sin^{-1} \sqrt{y} = f'(x)$
G	$y = a + (b/x)$	$xy = ax + b$
H	$y = a - b\,e^{-x}$	$\ln(a-y) = -x + \ln b$
I	$y = a\,e^{(bx)^n}$	$\ln(\log y/a) = n \ln b + n \ln x$

Appendix A

Wavelength λ (μm)–Wavenumber $\bar{v}$ (cm^{-1}) Conversion Table for the UV–Visible Region (accurate to 1μm)

λ	0	1	2	3	4	5	6	7	8	9
170	58 823	58 479	58 139	57 803	57 471	57 143	56 818	56 497	56 180	55 866
180	55 555	55 249	54 945	54 645	54 348	54 054	53 763	53 476	53 191	52 910
190	52 632	52 356	52 083	51 813	51 546	51 282	51 020	50 761	50 505	50 251
200	50 000	49 751	49 505	49 261	49 020	48 780	48 544	48 309	48 077	47 847
210	47 619	47 393	47 170	46 948	46 729	46 512	46 296	46 083	45 872	45 662
220	45 455	45 249	45 045	44 843	44 648	44 444	44 248	44 053	43 860	43 668
230	43 478	43 290	43 103	42 918	42 735	42 553	42 373	42 194	42 016	41 841
240	41 667	41 494	41 322	41 152	40 984	40 816	40 650	40 486	40 323	40 161
250	40 000	39 841	39 683	39 526	39 370	39 216	39 063	38 911	38 760	38 610
260	38 462	38 314	38 168	38 023	37 879	37 736	37 594	37 453	37 313	37 175
270	37 037	36 900	36 765	36 630	36 496	36 364	36 232	36 101	35 971	35 842
280	35 714	35 587	35 461	35 336	35 211	35 088	34 965	34 843	34 722	34 602
290	34 483	34 364	34 247	34 130	34 014	33 898	33 784	33 670	33 557	33 445
300	33 333	33 223	33 113	33 003	32 895	32 787	32 680	32 573	32 468	32 362
310	32 258	32 154	32 051	31 949	31 847	31 746	31 646	31 546	31 447	31 348
320	31 250	31 153	31 056	30 960	30 864	30 769	30 675	30 581	30 488	30 395
330	30 303	30 211	30 120	30 030	29 940	29 851	29 762	29 674	29 586	29 499
340	29 412	29 326	29 240	29 155	29 070	28 986	28 902	28 818	28 736	28 653
350	28 571	28 490	28 409	28 329	28 249	28 169	28 090	28 011	27 933	27 855
360	27 778	27 701	27 624	27 548	27 473	27 397	27 322	27 248	27 174	27 100
370	27 027	26 954	26 882	26 810	26 738	26 667	26 596	26 525	26 455	26 385
380	26 316	26 247	26 178	26 110	26 042	25 974	25 907	25 840	25 773	25 707
390	25 641	25 575	25 510	25 445	25 381	25 316	25 253	25 189	25 126	25 063
400	25 000	24 938	24 876	24 814	24 752	24 691	24 630	24 570	24 510	24 450
410	24 390	24 331	24 272	24 213	24 155	24 096	24 038	23 981	23 923	23 866
420	23 810	23 753	23 697	23 641	23 585	23 529	23 474	23 419	23 364	23 310
430	23 256	23 202	23 148	23 095	23 041	22 989	22 936	22 883	22 831	22 779
440	22 727	22 676	22 624	22 573	22 523	22 472	22 422	22 371	22 321	22 272
450	22 222	22 173	22 124	22 075	22 026	21 978	21 930	21 882	21 834	21 786
460	21 739	21 692	21 645	21 598	21 552	21 505	21 459	21 413	21 363	21 322
470	21 277	21 231	21 186	21 142	21 097	21 053	21 008	20 964	20 921	20 877
480	20 833	20 790	20 747	20 704	20 661	20 619	20 576	20 534	20 492	20 450
490	20 408	20 367	20 325	20 284	20 243	20 202	20 161	20 121	20 080	20 040
500	20 000	19 960	19 920	19 881	19 841	19 802	19 763	19 724	19 685	19 646

Appendix A (*cont.*)

λ	0	1	2	3	4	5	6	7	8	9
510	19 603	19 569	19 531	19 493	19 455	19 417	19 380	19 342	19 305	19 268
520	19 231	19 194	19 157	19 120	10 084	19 048	19 010	18 975	18 939	18 904
530	18 868	18 832	18 797	18 762	18 727	18 692	18 657	18 622	18 587	18 553
540	18 519	18 484	18 450	18 416	18 382	18 349	18 315	18 282	18 248	18 215
550	18 182	18 149	18 116	18 083	18 051	18 018	17 986	17 953	17 921	17 889
560	17 857	17 825	17 794	17 762	17 731	17 699	17 668	17 637	17 606	17 575
570	17 544	17 513	17 483	17 452	17 422	17 391	17 361	17 331	17 301	17 271
580	17 241	17 212	17 182	17 153	17 123	17 094	17 065	17 035	17 007	16 978
590	16 949	16 920	16 892	16 863	16 835	16 807	16 779	16 750	16 722	16 694
600	16 667	16 639	16 611	16 584	16 556	16 529	16 502	16 474	16 447	16 420
610	16 393	16 367	16 340	16 313	16 287	16 260	16 234	16 207	16 181	16 155
620	16 129	16 103	16 077	16 051	16 026	16 000	15 974	15 949	15 924	15 898
630	15 873	15 848	15 823	15 798	15 773	15 748	15 723	15 699	15 674	15 649
640	15 625	15 601	15 576	15 552	15 528	15 504	15 480	15 456	15 432	15 408
650	15 385	15 361	15 331	15 314	15 291	15 267	15 244	15 221	15 198	15 175
660	15 152	15 129	15 106	15 083	15 060	15 038	15 015	14 993	14 970	14 948
670	14 925	14 903	14 881	14 859	14 837	14 815	14 793	14 771	14 749	14 728
680	14 706	14 684	14 663	14 641	14 620	14 599	14 577	14 556	14 535	14 514
690	14 493	14 472	14 451	14 430	14 409	14 388	14 368	14 347	14 327	14 306
700	14 285	14 265	14 245	14 225	14 205	14 184	14 164	14 144	14 124	14 104
710	14 085	14 065	14 045	14 025	14 006	13 986	13 966	13 947	13 928	13 908
720	13 889	13 870	13 850	13 831	13 812	13 793	13 774	13 755	13 736	13 717
730	13 699	13 680	13 661	13 643	13 624	13 605	13 587	13 569	13 550	13 532
740	13 514	13 495	13 477	13 459	13 441	13 423	13 405	13 387	13 369	13 351
750	13 333	13 316	13 298	13 280	13 263	13 245	13 228	13 210	13 193	13 175
760	13 158	13 141	13 123	13 106	13 089	13 072	13 055	13 038	13 021	13 004
770	12 987	12 970	12 953	12 937	12 920	12 903	12 887	12 870	12 853	12 837
780	12 821	12 804	12 788	12 771	12 755	12 739	12 722	12 706	12 690	12 674
790	12 658	12 642	12 626	12 610	12 594	12 579	12 563	12 547	12 531	12 516
800	12 500	12 484	12 469	12 453	12 438	12 422	12 407	12 392	12 376	12 361
810	12 346	12 330	12 315	12 300	12 285	12 270	12 255	12 240	12 225	12 210
820	12 195	12 180	12 165	12 151	12 136	12 121	12 107	12 092	12 077	12 063
830	12 048	12 034	12 019	12 005	11 990	11 976	11 962	11 947	11 933	11 919
840	11 905	11 891	11 876	11 862	11 848	11 834	11 820	11 806	11 792	11 779
850	11 765	11 751	11 737	11 723	11 710	11 696	11 682	11 669	11 655	11 641
860	11 628	11 614	11 601	11 587	11 574	11 561	11 547	11 534	11 520	11 507
870	11 494	11 481	11 468	11 455	11 442	11 429	11 416	11 403	11 390	11 376
880	11 364	11 351	11 338	11 325	11 312	11 299	11 287	11 274	11 261	11 249
890	11 236	11 223	11 211	11 198	11 186	11 173	11 161	11 148	11 136	11 124
900	11 111	11 099	11 086	11 074	11 062	11 050	11 038	11 025	11 013	11 002
910	10 989	10 977	10 965	10 953	10 941	10 929	10 917	10 905	10 893	10 881
920	10 870	10 858	10 846	10 834	10 823	10 811	10 799	10 787	10 776	10 764
930	10 753	10 741	10 730	10 718	10 707	10 695	10 684	10 672	10 661	10 650
940	10 638	10 627	10 616	10 604	10 593	10 582	10 571	10 560	10 549	10 537
950	10 526	10 515	10 504	10 493	10 482	10 471	10 460	10 449	10 438	10 428

Appendix A (*cont.*)

λ	0	1	2	3	4	5	6	7	8	9
960	10 417	10 406	10 395	10 384	10 373	10 363	10 352	10 341	10 331	10 320
970	10 309	10 299	10 288	10 277	10 267	10 256	10 246	10 235	10 225	10 215
980	10 204	10 194	10 183	10 173	10 163	10 152	10 142	10 132	10 121	10 111
990	10 101	10 091	10 081	10 070	10 060	10 050	10 040	10 030	10 020	10 010
1000	10 000									

Appendix B

Wavenumber $\bar{v}$ (cm^{-1})–Wavelength λ (μm) Conversion Table in the Visible–UV Region (accurate to 100 cm^{-1})

$\bar{v}$	000	100	200	300	400	500	600	700	800	900
10 000	1000.0	990.1	980.4	970.9	961.5	952.4	943.4	934.6	925.9	917.4
11 000	909.1	900.9	892.9	885.0	877.2	869.6	862.1	854.7	847.5	840.3
12 000	833.3	826.4	819.7	813.0	806.5	800.0	793.7	787.4	781.2	775.2
13 000	769.2	763.4	757.6	751.9	746.3	740.7	735.3	729.9	724.6	719.4
14 000	714.3	709.2	704.2	699.3	694.4	689.7	684.9	680.3	675.7	671.1
15 000	666.7	662.3	657.9	653.6	649.4	645.2	641.0	636.9	632.9	628.9
16 000	625.0	621.1	617.3	613.5	609.8	606.1	602.4	598.8	595.2	591.7
17 000	588.2	584.8	581.4	578.0	574.7	571.4	568.2	565.0	561.8	558.7
18 000	555.6	552.5	549.5	546.4	543.5	540.5	537.6	534.8	531.9	529.1
19 000	526.3	523.6	520.8	518.1	515.5	512.8	510.2	507.6	505.1	502.5
20 000	500.0	497.5	495.0	492.6	490.2	487.8	485.4	483.1	480.8	478.5
21 000	476.2	473.9	471.7	469.5	467.3	465.1	463.0	460.8	458.7	456.6
22 000	454.5	425.5	450.5	448.4	446.4	444.4	442.5	440.5	438.6	436.7
23 000	434.8	432.9	431.0	429.2	427.4	425.5	423.7	421.9	420.2	418.4
24 000	416.7	414.9	413.2	411.5	409.8	408.2	406.5	404.9	403.2	401.6
25 000	400.0	398.4	396.8	395.3	393.7	392.2	390.6	389.1	387.6	386.1
26 000	384.6	383.1	381.7	380.2	378.8	377.4	375.9	374.5	373.1	371.7
27 000	370.4	369.0	367.6	366.3	365.0	363.6	362.3	361.0	359.7	358.4
28 000	357.1	355.9	354.6	353.4	352.1	350.9	349.7	348.4	347.2	346.0
29 000	344.8	343.6	342.5	341.3	340.1	339.0	337.8	336.7	335.6	334.4
30 000	333.3	332.2	331.1	330.0	328.9	327.9	326.8	325.7	324.7	323.6
31 000	322.6	321.5	320.5	319.5	318.5	317.5	316.8	315.5	314.5	313.5
32 000	312.5	311.5	310.6	309.6	308.6	307.7	306.7	305.8	304.9	304.0
33 000	303.0	302.1	301.2	300.3	299.4	298.5	297.6	296.7	295.9	295.0
34 000	294.1	293.3	292.4	291.5	290.7	289.9	289.0	288.2	287.4	286.5
35 000	285.7	284.9	284.1	283.3	282.5	281.7	280.9	280.1	279.3	278.6
36 000	277.8	277.0	276.2	275.5	274.7	274.0	273.2	272.5	271.7	271.0
37 000	270.3	269.5	268.8	268.1	267.4	266.7	266.0	265.3	264.6	263.9
38 000	263.2	262.5	261.8	261.1	260.4	259.8	259.1	258.4	257.7	257.1
39 000	256.4	255.8	255.1	254.5	253.8	253.2	252.5	251.9	251.3	250.6
40 000	250 0	249.4	248.8	248.1	247.5	246.9	246.3	245.7	245.1	244.5
41 000	243.9	243.3	242.7	242.1	241.5	241.0	240.4	239.8	239.2	239.7
42 000	238.1	237.5	237.0	236.4	235.8	235.3	234.7	234.2	233.6	233.1
43 000	232.6	232.0	231.5	231.0	230.4	229.9	229.4	228.8	228.3	227.8
44 000	227.3	226.8	226.2	225.7	225.2	224.7	224.2	223.7	223.2	222.7

Appendix B (*cont.*)

$\bar{\nu}$	000	100	200	300	400	500	600	700	800	900
45 000	222.2	221.7	221.2	220.8	220.3	219.8	219.3	218.8	218.3	217.9
46 000	217.4	216.9	216.5	216.0	215.5	215.1	214.6	214.1	213.7	213.2
47 000	212.8	212.3	211.9	211.4	211.0	210.5	210.1	209.6	209.2	208.8
48 000	208.3	207.9	207.5	207.0	206.6	206.2	205.8	205.3	204.9	204.5
49 000	204.1	203.7	203.3	202.8	202.4	202.0	201.6	201.2	200.8	200.4
50 000	200.0									

Appendix C

Transmittance $T(\%)$–Absorbance A Conversion Table (accurate to 0.1% T)

$$A = \log (100/T)$$

$T(\%)$	A	$T(\%)$	A	$T(\%)$	A	$T(\%)$	A
0.1	3.0000	3.6	1.4438	7.1	1.1488	10.6	0.9747
0.2	2.6990	3.7	1.4319	7.2	1.1428	10.7	0.9706
0.3	2.5228	3.8	1.4203	7.3	1.1367	10.8	0.9665
0.4	2.3979	3.9	1.4089	7.4	1.1306	10.9	0.9626
0.5	2.3010	4.0	1.3979	7.5	1.1249	11.0	0.9586
0.6	2.2219	4.1	1.3872	7.6	1.1192	11.1	0.9546
0.7	2.1550	4.2	1.3768	7.7	1.1137	11.2	0.9508
0.8	2.0969	4.3	1.3666	7.8	1.1079	11.3	0.9469
0.9	2.0457	4.4	1.3566	7.9	1.1025	11.4	0.9431
1.0	2.0000	4.5	1.3468	8.0	1.0969	11.5	0.9393
1.1	1.9586	4.6	1.3773	8.1	1.0916	11.6	0.9356
1.2	1.9208	4.7	1.3279	8.2	1.0864	11.7	0.9319
1.3	1.8860	4.8	1.3187	8.3	1.0809	11.8	0.9282
1.4	1.8539	4.9	1.3098	8.4	1.0759	11.9	0.9245
1.5	1.8240	5.0	1.3010	8.5	1.0709	12.0	0.9208
1.6	1.7959	5.1	1.2925	8.6	1.0656	12.1	0.9173
1.7	1.7695	5.2	1.2840	8.7	1.0603	12.2	0.9137
1.8	1.7448	5.3	1.2758	8.8	1.0554	12.3	0.9101
1.9	1.7212	5.4	1.2677	8.9	1.0507	12.4	0.9066
2.0	1.6990	5.5	1.2596	9.0	1.0457	12.5	0.9031
2.1	1.6778	5.6	1.2519	9.1	1.0411	12.6	0.8997
2.2	1.6576	5.7	1.2440	9.2	1.0363	12.7	0.8962
2.3	1.6383	5.8	1.2365	9.3	1.0315	12.8	0.8929
2.4	1.6197	5.9	1.2292	9.4	1.0270	12.9	0.8894
2.5	1.6021	6.0	1.2219	9.5	1.0224	13.0	0.8860
2.6	1.5850	6.1	1.2146	9.6	1.0178	13.1	0.8827
2.7	1.5687	6.2	1.2076	9.7	1.0132	13.2	0.8794
2.8	1.5528	6.3	1.2007	9.8	1.0086	13.3	0.8761
2.9	1.5376	6.4	1.1939	9.9	1.0043	13.4	0.8729
3.0	1.5228	6.5	1.1872	10.0	1.0000	13.5	0.8696
3.1	1.5087	6.6	1.1800	10.1	0.9956	13.6	0.8665
3.2	1.4949	6.7	1.1741	10.2	0.9914	13.7	0.8631
3.3	1.4814	6.8	1.1676	10.3	0.9872	13.8	0.8601
3.4	1.4684	6.9	1.1611	10.4	0.9829	13.9	0.8569
3.5	1.4559	7.0	1.1550	10.5	0.9788	14.0	0.8539

Appendix C (*cont.*)

T (%)	A	T (%)	A	T (%)	A	T (%)	A
14.1	0.8507	18.6	0.7305	23.1	0.6364	27.6	0.5591
14.2	0.8477	18.7	0.7281	23.2	0.6345	27.7	0.5575
14.3	0.8447	18.8	0.7258	23.3	0.6327	27.8	0.5560
14.4	0.8416	18.9	0.7236	23.4	0.6308	27.9	0.5544
14.5	0.8386	19.0	0.7212	23.5	0.6289	28.0	0.5528
14.6	0.8357	19.1	0.7191	23.6	0.6270	28.1	0.5513
14.7	0.8327	19.2	0.7167	23.7	0.6252	28.2	0.5497
14.8	0.8298	19.3	0.7144	23.8	0.6234	28.3	0.5483
14.9	0.8268	19.4	0.7122	23.9	0.6216	28.4	0.5466
15.0	0.8240	19.5	0.7100	24.0	0.6198	28.5	0.5452
15.1	0.8211	19.6	0.7078	24.1	0.6179	28.6	0.5437
15.2	0.8182	19.7	0.7055	24.2	0.6162	28.7	0.5421
15.3	0.8153	19.8	0.7034	24.3	0.6143	28.8	0.5406
15.4	0.8125	19.9	0.7011	24.4	0.6126	28.9	0.5391
15.5	0.8097	20.0	0.6990	24.5	0.6109	29.0	0.5376
15.6	0.8069	20.1	0.6968	24.6	0.6090	29.1	0.5361
15.7	0.8041	20.2	0.6947	24.7	0.6074	29.2	0.5346
15.8	0.8013	20.3	0.6925	24.8	0.6055	29.3	0.5332
15.9	0.7986	20.4	0.6904	24.9	0.6037	29.4	0.5316
16.0	0.7959	20.5	0.6882	25.0	0.6021	29.5	0.5302
16.1	0.7932	20.6	0.6861	25.1	0.6003	29.6	0.5286
16.2	0.7905	20.7	0.6840	25.2	0.5986	29.7	0.5272
16.3	0.7879	20.8	0.6819	25.3	0.5969	29.8	0.5258
16.4	0.7852	20.9	0.6799	25.4	0.5952	29.9	0.5242
16.5	0.7826	21.0	0.6778	25.5	0.5935	30.0	0.5228
16.6	0.7799	21.1	0.6757	25.6	0.5918	30.1	0.5214
16.7	0.7773	21.2	0.6736	25.7	0.5900	30.2	0.5199
16.8	0.7746	21.3	0.6717	25.8	0.5884	30.3	0.5185
16.9	0.7721	21.4	0.6696	25.9	0.5867	30.4	0.5171
17.0	0.7695	21.5	0.6676	26.0	0.5850	30.5	0.5157
17.1	0.7670	21.6	0.6656	26.1	0.5833	30.6	0.5143
17.2	0.7645	21.7	0.6635	26.2	0.5817	30.7	0.5128
17.3	0.7619	21.8	0.6616	26.3	0.5800	30.8	0.5114
17.4	0.7594	21.9	0.6596	26.4	0.5784	30.9	0.5100
17.5	0.7569	22.0	0.6576	26.5	0.5768	31.0	0.5087
17.6	0.7549	22.1	0.6556	26.6	0.5750	31.1	0.5072
17.7	0.7520	22.2	0.6537	26.7	0.5735	31.2	0.5059
17.8	0.7496	22.3	0.6517	26.8	0.5718	31.3	0.5045
17.9	0.7471	22.4	0.6497	26.9	0.5702	31.4	0.5031
18.0	0.7448	22.5	0.6478	27.0	0.5687	31.5	0.5018
18.1	0.7423	22.6	0.6459	27.1	0.5670	31.6	0.5004
18.2	0.7400	22.7	0.6440	27.2	0.5654	31.7	0.4990
18.3	0.7376	22.8	0.6421	27.3	0.5639	31.8	0.4976
18.4	0.7352	22.9	0.6402	27.4	0.5623	31.9	0.4962
18.5	0.7328	23.0	0.6383	27.5	0.5606	32.0	0.4949

652

Appendix C (*cont.*)

T (%)	A	T (%)	A	T (%)	A	T (%)	A
32.1	0.4935	36.6	0.4365	41.1	0.3861	45.6	0.3410
32.2	0.4922	36.7	0.4354	41.2	0.3850	45.7	0.3401
32.3	0.4909	36.8	0.4341	41.3	0.3840	45.8	0.3391
32.4	0.4895	36.9	0.4330	41.4	0.3829	45.9	0.3383
32.5	0.4881	37.0	0.4319	41.5	0.3820	46.0	0.3373
32.6	0.4868	37.1	0.4306	41.6	0.3809	46.1	0.3363
32.7	0.4854	37.2	0.4294	41.7	0.3799	46.2	0.3352
32.8	0.4842	37.3	0.4283	41.8	0.3788	46.3	0.3345
32.9	0.4829	37.4	0.4272	41.9	0.3779	46.4	0.3334
33.0	0.4814	37.5	0.4260	42.0	0.3768	46.5	0.3326
33.1	0.4801	37.6	0.4249	42.1	0.3756	46.6	0.3316
33.2	0.4789	37.7	0.4237	42.2	0.3747	46.7	0.3306
33.3	0.4775	37.8	0.4226	42.3	0.3736	46.8	0.3298
33.4	0.4763	37.9	0.4215	42.4	0.3726	46.9	0.3288
33.5	0.4749	38.0	0.4203	42.5	0.3717	47.0	0.3279
33.6	0.4737	38.1	0.4191	42.6	0.3705	47.1	0.3269
33.7	0.4723	38.2	0.4179	42.7	0.3696	47.2	0.3261
33.8	0.4711	38.3	0.4168	42.8	0.3685	47.3	0.3251
33.9	0.4698	38.4	0.4157	42.9	0.3676	47.4	0.3243
34.0	0.4684	38.5	0.4145	43.0	0.3666	47.5	0.2332
34.1	0.4673	38.6	0.4135	43.1	0.3655	47.6	0.3224
34.2	0.4660	38.7	0.4123	43.2	0.3645	47.7	0.3214
34.3	0.4646	38.8	0.4111	43.3	0.3634	47.8	0.3205
34.4	0.4634	38.9	0.4101	43.1	0.3624	47.9	0.3198
34.5	0.4623	39.0	0.4089	43.5	0.3615	48.0	0.3187
34.6	0.4609	39.1	0.4079	43.6	0.3606	48.1	0.3179
34.7	0.4597	39.2	0.4067	43.7	0.3594	48.2	0.3171
34.8	0.4585	39.3	0.4057	43.8	0.3585	48.3	0.3160
34.9	0.4572	39.4	0.4045	43.9	0.3575	48.4	0.3152
35.0	0.4559	39.5	0.4034	44.0	0.3566	48.5	0.3143
35.1	0.4547	39.6	0.4023	44.1	0.3556	48.6	0.3135
35.2	0.4535	39.7	0.4012	44.2	0.3545	48.7	0.3124
35.3	0.4523	39.8	0.4002	44.3	0.3536	48.8	0.3115
35.4	0.4510	39.9	0.3989	44.4	0.3526	48.9	0.3107
35.5	0.4498	40.0	0.3979	44.5	0.3516	49.0	0.3098
35.6	0.4486	40.1	0.3969	44.6	0.3506	49.1	0.3090
35.7	0.4474	40.2	0.3959	44.7	0.3497	49.2	0.3081
35.8	0.4461	40.3	0.3947	44.8	0.3487	49.3	0.3071
35.9	0.4449	40.4	0.3936	44.9	0.3477	49.4	0.3062
36.0	0.4438	40.5	0.3925	45.0	0.3467	49.5	0.3054
36.1	0.4425	40.6	0.3914	45.1	0.3458	49.6	0.3045
36.2	0.4412	40.7	0.3904	45.2	0.3448	49.7	0.3036
36.3	0.4401	40.8	0.3894	45.3	0.3439	49.8	0.3027
36.4	0.4389	40.9	0.3883	45.4	0.3430	49.9	0.3018
36.5	0.4378	41.0	0.3872	45.5	0.3420	50.0	0.3010

Appendix C (*cont.*)

T (%)	A	T (%)	A	T (%)	A	T (%)	A
50.1	0.3002	54.6	0.2630	59.1	0.2284	63.6	0.1965
50.2	0.2993	54.7	0.2620	59.2	0.2277	63.7	0.1959
50.3	0.2985	54.8	0.2613	59.3	0.2269	63.8	0.1951
50.4	0.2976	54.9	0.2603	59.4	0.2264	63.9	0.1945
50.5	0.2967	55.0	0.2596	59.5	0.2256	64.0	0.1939
50.6	0.2958	55.1	0.2589	59.6	0.2248	64.1	0.1931
50.7	0.2954	55.2	0.2582	59.7	0.2240	64.2	0.1925
50.8	0.2943	55.3	0.2572	59.8	0.2232	64.3	0.1917
50.9	0.2934	55.4	0.2565	59.9	0.2225	64.4	0.1911
51.0	0.2925	55.5	0.2558	60.0	0.2219	64.5	0.1903
51.1	0.2916	55.6	0.2551	60.1	0.2212	64.6	0.1897
51.2	0.2907	55.7	0.2541	60.2	0.2204	64.7	0.1892
51.3	0.2898	55.8	0.2534	60.3	0.2196	64.8	0.1883
51.4	0.2891	55.9	0.2526	60.4	0.2191	64.9	0.1878
51.5	0.2882	56.0	0.2519	60.5	0.2183	65.0	0.1869
51.6	0.2874	56.1	0.2511	60.6	0.2175	65.1	0.1864
51.7	0.2865	56.2	0.2502	60.7	0.2166	65.2	0.1858
51.8	0.2858	56.3	0.2495	60.8	0.2161	65.3	0.1850
51.9	0.2849	56.4	0.2487	60.9	0.2153	65.4	0.1843
52.0	0.2840	56.5	0.2480	61.0	0.2146	65.5	0.1838
52.1	0.2830	56.6	0.2472	61.1	0.2140	65.6	0.1829
52.2	0.2823	56.7	0.2465	61.2	0.2133	65.7	0.1824
52.3	0.2814	56.8	0.2457	61.3	0.2125	65.8	0.1818
52.4	0.2806	56.9	0.2447	61.4	0.2119	65.9	0.1810
52.5	0.2799	57.0	0.2440	61.5	0.2111	66.0	0.1804
52.6	0.2790	57.1	0.2432	61.6	0.2103	66.1	0.1798
52.7	0.2784	57.2	0.2425	61.7	0.2098	66.2	0.1793
52.8	0.2774	57.3	0.2417	61.8	0.2089	66.3	0.1783
52.9	0.2765	57.4	0.2410	61.9	0.2084	66.4	0.1778
53.0	0.2758	57.5	0.2402	62.0	0.2076	66.5	0.1772
53.1	0.2749	57.6	0.2395	62.1	0.2068	66.6	0.1767
53.2	0.2742	57.7	0.2387	62.2	0.2062	66.7	0.1759
53.3	0.2732	57.8	0.2380	62.3	0.2054	66.8	0.1753
53.4	0.2725	57.9	0.2372	62.4	0.2049	66.9	0.1747
53.5	0.2716	58.0	0.2365	62.5	0.2041	67.0	0.1741
53.6	0.2709	58.1	0.2357	62.6	0.2034	67.1	0.1732
53.7	0.2700	58.2	0.2350	62.7	0.2028	67.2	0.1727
53.8	0.2693	58.3	0.2342	62.8	0.2020	67.3	0.1721
53.9	0.2684	58.4	0.2335	62.9	0.2014	67.4	0.1715
54.0	0.2677	58.5	0.2326	63.0	0.2007	67.5	0.1706
54.1	0.2667	58.6	0.2319	63.1	0.2001	67.6	0.1700
54.2	0.2660	58.7	0.2314	63.2	0.1993	67.7	0.1694
54.3	0.2653	58.8	0.2306	63.3	0.1987	67.8	0.1688
54.4	0.2644	58.9	0.2300	63.4	0.1979	67.9	0.1682
54.5	0.2637	59.0	0.2292	63.5	0.1973	68.0	0.1676

Appendix C (*cont.*)

T (%)	A	T (%)	A	T (%)	A	T (%)	A
68.1	0.1668	72.6	0.1390	77.1	0.1130	81.6	0.0885
68.2	0.1662	72.7	0.1386	77.2	0.1123	81.7	0.0878
68.3	0.1656	72.8	0.1380	77.3	0.1120	81.8	0.0871
68.4	0.1650	72.9	0.1373	77.4	0.1113	81.9	0.0867
68.5	0.1641	73.0	0.1367	77.5	0.1106	82.0	0.0864
68.6	0.1638	73.1	0.1361	77.6	0.1103	82.1	0.0856
68.7	0.1632	73.2	0.1354	77.7	0.1096	82.2	0.0852
68.8	0.1623	73.3	0.1348	77.8	0.1089	82.3	0.0845
68.9	0.1617	73.4	0.1341	77.9	0.1086	82.4	0.0842
69.0	0.1611	73.5	0.1338	78.0	0.1079	82.5	0.0835
69.1	0.1605	73.6	0.1332	78.1	0.1072	82.6	0.0831
69.2	0.1599	73.7	0.1326	78.2	0.1069	82 7	0 0823
69.3	0.1593	73.8	0.1319	78.3	0.1062	82.8	0.0820
69.4	0.1587	73.9	0.1313	78.4	0.1059	82.9	0.0813
69.5	0.1580	74.0	0.1306	78.5	0.1052	83.0	0.0809
69.6	0.1574	74.1	0.1303	78.6	0.1045	83.1	0.0802
69.7	0.1568	74.2	0.1297	78.7	0.1041	83.2	0.0799
69.8	0.1562	74.3	0.1290	78.8	0.1035	83.3	0.0792
69.9	0.1556	74.4	0.1284	78.9	0.1028	83.4	0.0789
70.0	0.1550	74.5	0.1277	79.0	0.1025	83.5	0.0785
70.1	0.1544	74.6	0.1271	79.1	0.1018	83.6	0.0778
70.2	0.1538	74.7	0.1268	79.2	0.1014	83.7	0.0074
70.3	0.1529	74.8	0.1262	79.3	0.1007	83.8	0.0766
70.4	0.1523	74.9	0.1255	79.4	0.1000	83.9	0.0763
70.5	0.1516	75.0	0.1249	79.5	0.0997	84.0	0.0755
70.6	0.1510	75.1	0.1245	79.6	0.0990	84.1	0.0753
70.7	0.1504	75.2	0.1239	79.7	0.0986	84.2	0.0749
70.8	0.1498	75.3	0.1232	79.8	0.0979	84.3	0.0742
70.9	0.1492	75.4	0.1225	79.9	0.0976	84.4	0.0738
71.0	0.1485	75.5	0.1222	80.0	0.0969	84.5	0.0730
71.1	0.1479	75.6	0.1216	80.1	0.0962	84.6	0.0727
71.2	0.1473	75.7	0.1209	80.2	0.0958	84.7	0.0723
71.3	0.1470	75.8	0.1202	80.3	0.0951	84.8	0.0716
71.4	0.1464	75.9	0.1199	80.4	0.0948	84.9	0.0712
71.5	0.1459	76.0	0.1189	80.5	0.0941	85.0	0.0705
71.6	0.1453	76.1	0.1186	80.6	0.0937	85.1	0.0701
71.7	0.1446	76.2	0.1179	80.7	0.0930	85.2	0.0697
71.8	0.1440	76.3	0.1176	80.8	0.0927	85.3	0.0690
71.9	0.1433	76.4	0.1168	80.9	0.0920	85.4	0.0686
72.0	0.1428	76.5	0.1162	81.0	0.0916	85.5	0.0682
72.1	0.1422	76.6	0.1155	81.1	0.0909	85.6	0.0675
72.2	0.1415	76.7	0.1152	81.2	0.0906	85.7	0.0672
72.3	0.1409	76.8	0.1145	81.3	0.0899	85.8	0.0668
72.4	0.1402	76.9	0.1139	81.4	0.0895	85.9	0.0660
72.5	0.1396	77.0	0.1137	81.5	0.0888	86.0	0.0656

Appendix C *(cont.)*

T (%)	A	T (%)	A	T (%)	A	T (%)	A
86.1	0.0649	90.6	0.0429	95.1	0.0220	99.6	0.0017
86.2	0.0645	90.7	0.0425	95.2	0.0212	99.7	0.0012
86.3	0.0641	90.8	0.0418	95.3	0.0207	99.8	0.0008
86.4	0.0634	90.9	0.0414	95.4	0.0203	99.9	0.0004
86.5	0.0630	91.0	0.0411	95.5	0.0199		
86.6	0.0626	91.1	0.0407	95.6	0.0195		
86.7	0.0618	91.2	0.0403	95.7	0.0191		
86.8	0.0615	91.3	0.0395	95.8	0.0187		
86.9	0.0611	91.4	0.0391	95.9	0.0182		
87.0	0.0603	91.5	0.0386	96.0	0.0178		
87.1	0.0599	91.6	0.0382	96.1	0.0174		
87.2	0.0596	91.7	0.0378	96.2	0.0170		
87.3	0.0588	91.8	0.0371	96.3	0.0161		
87.4	0.0584	91.9	0.0367	96.4	0.0157		
87.5	0.0580	92.0	0.0363	96.5	0.0153		
87.6	0.0577	92.1	0.0359	96.6	0.0149		
87.7	0.0569	92.2	0.0355	96.7	0.0145		
87.8	0.0565	92.3	0.0346	96.8	0.0140		
87.9	0.0561	92.4	0.0342	96.9	0.0136		
88.0	0.0554	92.5	0.0338	97.0	0.0132		
88.1	0.0550	92.6	0.0334	97.1	0.0128		
88.2	0.0546	92.7	0.0331	97.2	0.0123		
88.3	0.0542	92.8	0.0327	97.3	0.0119		
88.4	0.0535	92.9	0.0319	97.4	0.0115		
88.5	0.0531	93.0	0.0315	97.5	0.0111		
88.6	0.0526	93.1	0.0311	97.6	0.0107		
88.7	0.0519	93.2	0.0306	97.7	0.0103		
88.8	0.0515	93.3	0.0302	97.8	0.0094		
88.9	0.0511	93.4	0.0298	97.9	0.0090		
89.0	0.0507	93.5	0.0294	98.0	0.0086		
89.1	0.0500	93.6	0.0286	98.1	0.0080		
89.2	0.0496	93.7	0.0282	98.2	0.0076		
89.3	0.0492	93.8	0.0278	98.3	0.0072		
89.4	0.0487	93.9	0.0274	98.4	0.0068		
89.5	0.0480	94.0	0.0270	98.5	0.0064		
89.6	0.0476	94.1	0.0265	98.6	0.0060		
89.7	0.0472	94.2	0.0261	98.7	0.0055		
89.8	0.0468	94.3	0.0253	98.8	0.0051		
89.9	0.0461	94.4	0.0249	98.9	0.0047		
90.0	0.0457	94.5	0.0245	99.0	0.0043		
90.1	0.0453	94.6	0.0241	99.1	0.0037		
90.2	0.0448	94.7	0.0237	99.2	0.0033		
90.3	0.0441	94.8	0.0233	99.3	0.0029		
90.4	0.0437	94.9	0.0229	99.4	0.0025		
90.5	0.0433	95.0	0.0224	99.5	0.0021		

Appendix D

Solvents for UV and Visible Spectroscopy

These data are reproduced by permission of BDH, England.

Solvent	UV spectrum	Minimum percentage transmission λ (nm)	T (%)
Acetonitrile CH_3CH			
Molecular weight 41.05		220	80
Melting point $-45.7\,°C$		230	90
Boiling point $81.60\,°C$	(I)	240	93
Density 0.7856 g ml^{-1}		260	96
Density gradient 0.0011		280	98
Refractive index, n_D^{20} 1.3441		300	99
Benzene C_6H_6			
Molecular weight 78.11		280	25
Melting point $5.5\,°C$		290	80
Boiling point $80.1\,°C$	(II)	300	90
Density 0.8790 g ml^{-1}		320	95
Density gradient $0.001\ 05$		350	99
Refractive index, n_D^{20} 1.5011			
Carbon tetrachloride CCl_4			
Molecular weight 153.82		265	25
Melting point $-22.96\,°C$		270	45
Boiling point $77.75\,°C$	(III)	280	78
Density 1.5942 g ml^{-1}		290	91
Density gradient $0.001\ 97$		300	97
Refractive index, n_D^{20} 1.4664			

Appendix D (*cont.*)

Solvent	UV spectrum	Minimum percentage transmission λ (nm)	T (%)

Chloroform
CHCl₃ → $CHCl_3$
Molecular weight 119.38
Melting point −63.5 °C
Boiling point 61.2 °C
Density 1.489 g ml⁻¹
Density gradient 0.001 86
Refractive index, n_D^{20} 1.4460
This material contains about 2% v/v of ethanol as preservative

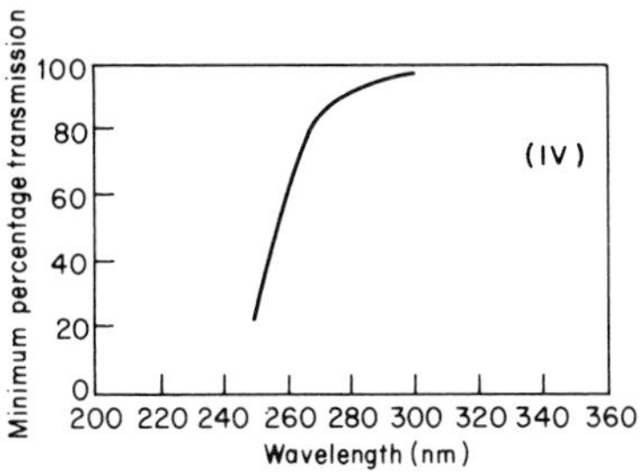

λ (nm)	T (%)
250	20
260	60
270	85
280	92
290	95
300	97

Cyclohexane
$CH_2(CH_2)_4CH_2$
Molecular weight 84.16
Melting point 6.55 °C
Boiling point 80.85 °C
Density 0.778 67 g ml⁻¹
Density gradient 0.000 92
Refractive index, n_D^{20} 1.426 33

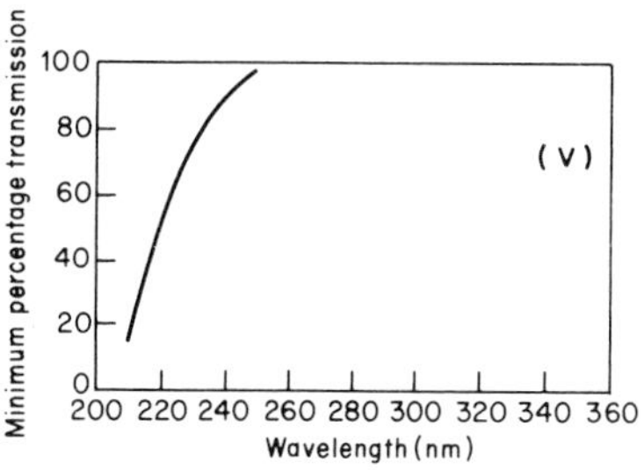

λ (nm)	T (%)
210	15
220	50
230	75
240	90
250	97

Dimethylformamide
$HCON(CH_3)_2$
Molecular weight 73.10
Melting point −61 °C
Boiling point 153 °C
Density 0.9445 g ml⁻¹ at 25 °C
Density gradient 0.000 95
Refractive index, n_D^{20} 1.4269

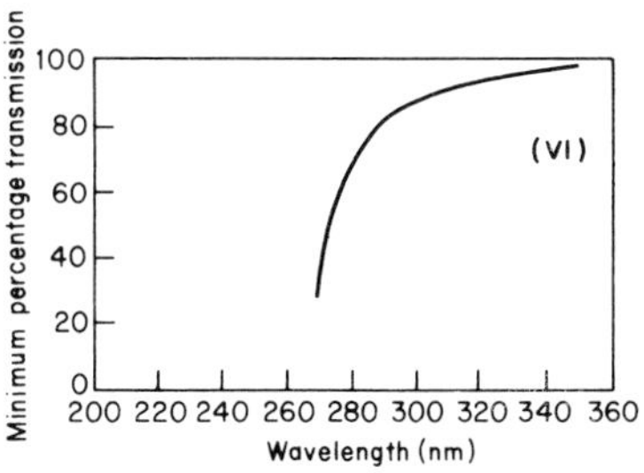

λ (nm)	T (%)
270	25
280	72
290	85
300	90
320	95
350	99

Dimethylsulphoxide
CH_3SOCH_3
Molecular weight 78.13
Melting point 18.45 °C
Boiling point 189 °C
Density 1.1014 g ml⁻¹
Density gradient 0.0008
Refractive index, n_D^{20} 1.4783

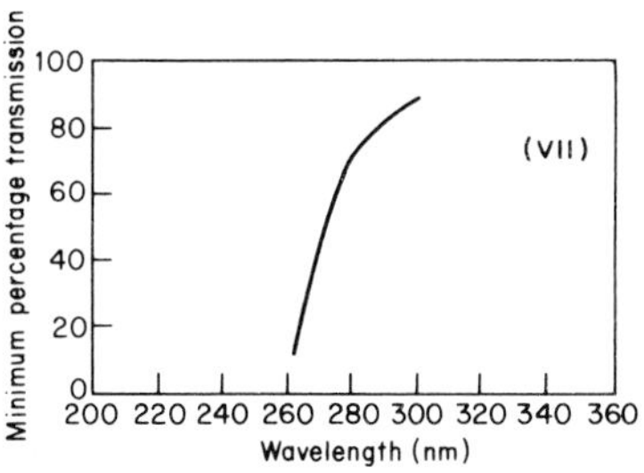

λ (nm)	T (%)
261	10
270	50
280	76
290	84
300	89

Appendix D (*cont.*)

Solvent	UV spectrum	Minimum percentage transmission λ (nm)	T (%)
1,4-Dioxan		220	20
$\overline{CH_2CH_2OCH_22H_2O}$		230	35
Molecular weight 88.11		240	47
Melting point 11.8 °C	(VIII)	250	59
Boiling point 101.3 °C		260	70
Density 1.0337 g ml^{-1}		270	79
Density gradient 0.001 12		280	87
Refractive index, n_D^{20} 1.4224		290	95
Ethanol (absolute)		210	30
C_2H_5OH		220	52
Molecular weight 46.07		230	73
Melting point -117.3 °C	(IX)	240	85
Boiling point 78.32 °C		250	90
Density 0.7893 g ml^{-1}			
Density gradient 0.000 85			
Refractive index, n_D^{20} 1.3610			
Ethyl acetate		255	25
$CH_3COOC_2H_5$		260	40
Molecular weight 88.11		270	67
Melting point -83.6 °C	(X)	280	81
Boiling point 77.11 °C		290	88
Density 0.9009 g ml^{-1}		300	91
Density gradient 0.001 22			
Refractive index, n_D^{20} 1.3724			
1,1,1,3,3,3-Hexafluoropropan-2-ol		195	90
$(CF_3)_2CHOH$		200	93
Molecular weight 168.04		210	95
Melting point -3.4 °C	(XI)	220	96
Boiling point 38.2 °C		230	98
Density 1.59 g ml^{-1} at 25 °C		240	99
Density gradient 0.003		250	99
Refractive index, n_D^{20} 1.2752			

Each UV spectrum plot has y-axis "Minimum percentage transmission" (0–100) and x-axis "Wavelength (nm)" (200 220 240 260 280 300 320 340 360).

Appendix D (*cont.*)

Solvent	UV spectrum	Minimum percentage transmission λ (nm)	T (%)

n-Hexane (fraction from petroleum)
$CH_3\ (CH_2)_4\ CH_3$

Molecular weight	86.17
Melting point	$-94.3\ ^\circ C$
Boiling point	$69.0\ ^\circ C$
Density	$0.6594\ g\ ml^{-1}$
Density gradient	0.000 75
Refractive index, n_D^{20}	1.3753

(XII)

210	10
220	70
230	86
240	92
250	95

Propan-2-ol
$(CH_3)_2CHOH$

Molecular weight	60.10
Melting point	$-89.5\ ^\circ C$
Boiling point	$82.4\ ^\circ C$
Density	$0.785\ g\ ml^{-1}$
Density gradient	0.000 82
Refractive index, n_D^{20}	1.3771

(XIII)

210	30
220	62
230	83
240	94
250	99

Tetrachloroethylene
$CCl_2{=}CCl_2$

Molecular weight	165.83
Melting point	$-22.18\ ^\circ C$
Boiling point	$121.2\ ^\circ C$
Density	$1.6226\ g\ ml^{-1}$
Density gradient	0.001 64
Refractive index, n_D^{20}	1.5053

(XIV)

290	10
300	75
320	85
400	95

Trifluoroacetic acid
CF_3CO_2H

Molecular weight	114.02
Melting point	$-15.25\ ^\circ C$
Boiling point	$72.4\ ^\circ C$
Density	$1.4890\ g\ ml^{-1}$
Density gradient	0.000 81
Refractive index, n_D^{20}	1.2850

(XV)

260	10
270	80
280	90
290	93
300	95
320	95

Appendix D (*cont.*)

Solvent	UV spectrum	Minimum percentage transmission λ (nm)	T (%)
2,2,2-Trifluoroethanol CF_3CH_2OH Molecular weight 86.01 Melting point $-44\ °C$ Boiling point $74–75\ °C$ Density $1.383\ g\ ml^{-1}$ Density gradient 0.0015 Refractive index, n_D^{20} 1.2940		195 200 210 220 230 240 250 260	70 90 92 94 95 96 97 98
2,2,4-Trimethylpentane $(CH_3)_3CCH_2CH(CH_3)_2$ Molecular weight 114.23 Melting point $-107.4\ °C$ Boiling point $99.2\ °C$ Density $0.6918\ g\ ml^{-1}$ Density gradient 0.000 80 Refractive index, n_D^{20} 1.3915		210 220 230 240 250	30 60 85 95 97

Appendix E

Wavelength λ (μm)–Wavenumber $\bar{\nu}$ (cm^{-1}) Conversion Table for the IR Region (accurate to 0·1 μm)

λ	0	1	2	3	4	5	6	7	8	9
1.0	10 000	9090	8333	7692	7143	6667	6250	5882	5555	5263
2.0	5000	4975	4950	4926	4902	4878	4854	4831	4808	4785
2.1	4762	4739	4717	4695	4673	4651	4630	4608	4587	4566
2.2	4545	4525	4505	4484	4464	4444	4425	4405	4386	4367
2.3	4348	4329	4310	4292	4274	4255	4237	4219	4202	4184
2.4	4167	4149	4232	4115	4098	4082	4065	4049	4032	4016
2.5	4000	3984	3968	4953	3937	3922	3006	3891	3876	3861
2.6	3846	3831	3817	3802	3788	3774	3759	3745	3731	3717
2.7	3704	3690	3676	3663	3650	3636	3623	3610	3597	3584
2.8	3571	3559	3546	3534	3521	3509	3497	3484	3472	3460
2.9	3448	3436	3425	3413	3401	3390	3378	3367	3356	3344
3.0	3333	3322	3311	3300	3289	3279	3268	3257	3247	3236
3.1	3226	3215	3205	3195	3185	3175	3165	3155	3145	3135
3.2	3125	3115	3106	3096	3086	3077	3067	3058	3049	3040
3.3	3030	3021	3012	3003	2994	2985	2976	2967	2959	2950
3.4	2941	2933	2924	2915	2907	2899	2890	2882	2874	2865
3.5	2857	2849	2841	2833	2825	2817	2809	2801	2793	2786
3.6	2778	2770	2762	2755	2747	2740	2732	2725	2717	2710
3.7	2703	2695	2688	2681	2674	2667	2660	2653	2646	2639
3.8	2632	2625	2618	2611	2604	2597	2591	2584	2577	2571
3.9	2654	2558	2551	2545	2538	2532	2525	2519	2513	2506
4.0	2500	2494	2488	2481	2475	2469	2463	2457	2451	2445
4.1	2439	2433	2427	2421	2415	2410	2404	2398	2387	2387
4.2	2381	2375	2370	2364	2358	2353	2347	2342	2336	2331
4.3	2326	2320	2315	2309	2304	2299	2294	2288	2283	2278
4.4	2273	2268	2262	2257	2252	2247	2242	2237	2232	2227
4.5	2222	2217	2212	2208	2203	2198	2193	2188	2183	2179
4.6	2174	2169	2165	2160	2155	2151	2146	2141	2137	2132
4.7	2128	2123	2119	2114	2110	2105	2101	2096	2092	2088
4.8	2083	2079	2075	2070	2066	2062	2058	2053	2049	2045
4.9	2041	2037	2033	2028	2024	2020	2016	2012	2008	2004

Appendix E (*cont.*)

λ	0	1	2	3	4	5	6	7	8	9
5.0	2000	1996	1992	1988	1984	1980	1976	1972	1969	1965
5.1	1961	1957	1953	1949	1946	1942	1938	1934	1931	1927
5.2	1923	1919	1916	1912	1908	1905	1901	1898	1894	1890
5.3	1887	1883	1880	1876	1873	1869	1866	1862	1859	1855
5.4	1852	1848	1845	1842	1838	1835	1832	1828	1825	1821
5.5	1818	1815	1812	1808	1805	1802	1799	1795	1792	1788
5.6	1786	1783	1779	1776	1773	1770	1767	1764	1761	1757
5.7	1754	1751	1748	1745	1742	1739	1736	1733	1730	1727
5.8	1724	1721	1718	1715	1712	1709	1706	1704	1701	1698
5.9	1695	1692	1689	1686	1684	1681	1678	1675	1672	1669
6.0	1667	1664	1661	1658	1656	1653	1650	1647	1645	1642
6.1	1639	1637	1634	1631	1629	1626	1623	1621	1618	1616
6.2	1613	1610	1608	1605	1603	1600	1597	1595	1592	1590
6.3	1587	1585	1582	1580	1577	1575	1572	1570	1567	1565
6.4	1563	1560	1558	1555	1553	1550	1548	1546	1543	1541
6.5	1538	1536	1534	1531	1529	1527	1524	1522	1520	1517
6.6	1515	1513	1511	1508	1506	1504	1502	1499	1497	1495
6.7	1493	1490	1488	1486	1484	1481	1479	1477	1475	1473
6.8	1471	1468	1466	1464	1462	1460	1458	1456	1453	1451
6.9	1449	1447	1445	1443	1441	1439	1437	1435	1433	1431
7.0	1429	1427	1425	1422	1420	1418	1416	1414	1412	1410
7.1	1408	1406	1404	1403	1401	1399	1397	1395	1393	1391
7.2	1389	1387	1385	1383	1381	1379	1377	1376	1374	1372
7.3	1370	1368	1366	1364	1362	1361	1359	1357	1355	1353
7.4	1351	1350	1348	1346	1344	1342	1340	1339	1337	1335
7.5	1333	1332	1330	1328	1326	1325	1323	1321	1319	1318
7.6	1316	1314	1312	1311	1309	1307	1305	1304	1302	1300
7.7	1299	1297	1295	1294	1292	1290	1289	1287	1285	1284
7.8	1282	1280	1279	1277	1276	1274	1272	1271	1269	1267
7.9	1266	1264	1263	1261	1259	1258	1256	1255	1253	1252
8.0	1250	1248	1247	1245	1244	1242	1241	1239	1238	1236
8.1	1235	1233	1232	1230	1229	1227	1225	1224	1222	1221
8.2	1220	1218	1217	1215	1214	1212	1211	1209	1208	1206
8.3	1205	1203	1202	1200	1199	1198	1196	1195	1193	1192
8.4	1190	1189	1188	1186	1185	1183	1182	1181	1179	1178
8.5	1176	1175	1174	1172	1171	1170	1168	1167	1166	1164
8.6	1163	1161	1160	1159	1157	1156	1155	1153	1152	1151
8.7	1149	1148	1147	1145	1144	1143	1142	1140	1139	1138
8.8	1136	1135	1134	1133	1131	1130	1129	1127	1126	1125
8.9	1124	1122	1121	1120	1119	1117	1116	1115	1114	1112
9.0	1111	1110	1109	1107	1106	1105	1104	1103	1101	1100
9.1	1099	1098	1096	1095	1094	1093	1092	1091	1089	1088
9.2	1087	1086	1085	1083	1082	1081	1080	1079	1078	1076
9.3	1075	1074	1073	1072	1071	1070	1068	1067	1066	1065
9.4	1064	1063	1062	1060	1059	1058	1057	1056	1055	1054

663

Appendix E (*cont.*)

λ	0	1	2	3	4	5	6	7	8	9
9.5	1053	1052	1050	1049	1048	1047	1046	1045	1044	1043
9.6	1042	1041	1040	1038	1037	1036	1035	1034	1033	1032
9.7	1031	1030	1029	1028	1027	1026	1025	1024	1022	1021
9.8	1020	1019	1018	1017	1016	1015	1014	1013	1012	1011
9.9	1010	1009	1008	1007	1006	1005	1004	1003	1002	1001
10.0	1000	999	998	997	996	995	994	993	992	991
10.1	990	989	988	987	986	985	984	983	982	981
10.2	980	979	978	978	977	976	975	974	973	972
10.3	971	970	969	968	967	966	965	964	963	962
10.4	962	961	960	959	958	957	956	955	954	953
10.5	952	951	951	950	949	948	947	946	945	944
10.6	943	943	942	941	940	939	938	937	936	935
10.7	935	934	933	932	931	930	929	929	928	927
10.8	926	925	924	923	923	922	921	920	919	918
10.9	917	917	916	915	914	913	912	912	911	910
11.0	909	908	907	907	906	905	904	903	903	902
11.1	901	900	899	898	898	897	896	895	894	894
11.2	893	892	891	890	890	889	888	887	887	886
11.3	885	884	883	883	882	881	880	880	879	878
11.4	877	876	876	875	874	873	873	872	871	870
11.5	870	869	868	867	867	866	865	864	864	863
11.6	862	861	861	860	859	858	858	857	856	855
11.7	855	854	853	853	852	851	850	850	849	848
11.8	847	847	846	845	845	844	843	842	842	841
11.9	840	840	839	838	838	837	836	835	835	834
12.0	833	833	832	831	831	830	829	829	828	827
12.1	826	826	825	824	824	823	822	822	821	820
12.2	820	819	818	818	817	816	816	815	814	814
12.3	813	812	812	811	810	810	809	808	808	807
12.4	806	806	805	805	804	803	803	802	801	801
12.5	800	799	799	798	797	797	796	796	795	794
12.6	794	793	792	792	791	791	790	789	789	788
12.7	787	787	786	786	785	784	784	783	782	782
12.8	781	781	780	779	779	778	778	777	776	776
12.9	775	775	774	773	773	772	772	771	770	770
13.0	769	769	768	767	767	766	766	765	765	764
13.1	763	763	762	762	761	760	760	759	759	758
13.2	758	757	756	756	755	755	754	754	753	752
13.3	752	751	751	750	750	749	749	748	747	747
13.4	746	746	745	745	744	743	743	742	742	741
13.5	741	740	740	739	739	738	737	737	736	736
13.6	735	735	734	734	733	733	732	732	731	730
13.7	730	729	729	728	728	727	727	726	726	725
13.8	725	724	724	723	723	722	722	721	720	720
13.9	719	719	718	718	717	717	716	716	715	715

Appendix E (*cont.*)

λ	0	1	2	3	4	5	6	7	8	9
14.0	714	714	713	713	712	712	711	711	710	710
14.1	709	709	708	708	707	707	706	706	705	705
14.2	704	704	703	703	702	702	701	701	700	700
14.3	699	699	698	698	697	697	696	696	695	695
14.4	694	694	693	693	693	692	692	691	691	690
14.5	690	689	689	688	688	687	687	686	686	685
14.6	685	684	684	684	683	683	682	682	681	681
14.7	680	680	679	679	678	678	678	677	677	676
14.8	676	675	675	674	674	673	673	672	672	672
14.9	671	671	670	670	669	669	668	668	668	667
15.0	666.7	666.2	665.8	665.3	664.5	664.5	664.0	663.6	663.1	662.7
15.1	662.3	661.8	661.4	660.9	660.5	660.1	659.6	659.2	658.8	658.3
15.2	*657.9*	*657.5*	*657.0*	*656.6*	*656.2*	*655.7*	*655.2*	*654.9*	*654.5*	*654.0*
15.3	653.6	653.2	652.7	652.3	651.9	651.5	651.0	650.6	650.2	649.8
15.4	649.4	648.9	648.5	648.1	647.7	647.2	646.8	646.4	646.0	645.6
15.5	645.2	644.7	644.3	643.9	643.5	643.1	642.7	642.3	641.8	641.4
15.6	641.0	640.6	640.2	639.8	639.4	639.0	638.6	638.2	637.8	637.3
15.7	*636.9*	*636.5*	*636.1*	*635.7*	*635.3*	*634.9*	*634.5*	*634.1*	*633.7*	*633.3*
15.8	632.9	632.5	632.1	631,7	631.3	630.9	630.5	630.1	629.7	629.3
15.9	628.9	628.5	628.1	627.7	627.4	627.0	626.6	626.2	625.8	625.4
16.0	625.0	624.6	624.2	623.8	623.4	623.1	622.7	622.3	621.9	621.9
16.1	621.1	620.7	620.3	620.0	619.6	619.2	618.8	618.4	618.0	617.7
16.2	*617.3*	*616.9*	*616.5*	*616.1*	*615.8*	*615.4*	*615.0*	*614.6*	*614.3*	*613.9*
16.3	613.5	613.1	612.7	612.4	612.0	611.6	611.2	610.9	610.5	610.1
16.4	609.9	609.4	609.0	608.6	608.3	607.9	607.5	607.2	606.8	606.4
16.5	606.1	605.7	605.3	605.0	604.6	604.2	603.9	603.5	603.1	602.8
16.6	602.4	602.0	601.7	601.3	601.0	600.6	600.2	599.9	599.5	599.2
16.7	*598.8*	*598.4*	*598.1*	*597.7*	*597.4*	*597.0*	*596.7*	*596.3*	*595.9*	*595.6*
16.8	595.2	594.9	594.5	594.2	593.8	593.5	593.1	592.8	592.4	592.1
16.9	591.7	591.4	591.0	590.7	590.3	590.0	589.6	589.3	588.9	588.6
17.0	588.2	587.9	587.5	587.2	586.9	586.5	586.2	585.8	585.5	585.1
17.1	584.8	584.5	584.1	583.8	583.4	583.1	582.8	582.4	582.1	581.7
17.2	*581.4*	*581.1*	*580.7*	*580.4*	*580.0*	*579.7*	*579.4*	*579.0*	*578.7*	*578.4*
17.3	578.0	577.7	577.4	577.0	576.7	576.4	576.0	575.7	575.4	575.0
17.4	574.7	574.4	574.1	573.7	573.4	573.1	572.7	572.4	572.1	571.8
17.5	571.4	571.1	570.8	570.5	570.1	569.8	569.5	569.2	568.8	568.5
17.6	568.2	567.9	567.5	567.2	566.9	566.6	566.3	565.9	565.6	565.3
17.7	*565.0*	*564.7*	*564.3*	*564.0*	*563.7*	*563.4*	*563.1*	*562.7*	*562.4*	*562.1*
17.8	561.8	561.5	561.2	560.9	560.5	560.2	559.9	559.6	559.3	559.0
17.9	558.7	558.3	558.0	557.7	557.4	557.1	556.8	556.5	556.2	555.9
18.0	555.6	555.2	554.9	554.6	554.3	554.0	553.7	553.4	553.1	552.8
18.1	552.5	552.2	551.9	551.6	551.3	551.0	550.7	550.4	550.1	549.8
18.2	*549.5*	*549.1*	*548.8*	*548.5*	*548.2*	*547.9*	*547.6*	*547.3*	*547.0*	*546.7*
18.3	546.4	546.1	545.9	545.6	545.3	545.0	544.7	544.4	544.1	543.8
18.4	543.5	543.2	542.9	542.6	542.3	542.0	541.7	541.4	541.1	540.8

Appendix E (*cont.*)

λ	0	1	2	3	4	5	6	7	8	9
18.5	540.5	540.2	540.0	539.7	539.4	539.1	538.8	538.5	538.2	537.9
18.6	537.6	537.3	537.1	536.8	536.5	536.2	535.9	535.6	535.3	535.0
18.7	*534.8*	*534.5*	*543.2*	*533.9*	*533.6*	*533.3*	*533.0*	*532.8*	*532.5*	*532.2*
18.8	531.9	531.6	531.3	531.1	530.8	530.5	530.2	529.9	529.7	529.4
18.9	529.1	528.8	528.5	528.3	528.0	527.7	527.4	527.1	526.9	526.6
19.0	526.3	526.0	525.8	525.5	525.2	524.9	524.7	524.4	524.1	538.8
19.1	523.6	523.3	523.0	522.7	522.5	522.2	521.9	521.6	521.4	521.1
19.2	*520.8*	*520.6*	*520.3*	*520.0*	*519.8*	*519.5*	*519.2*	*518.9*	*518.7*	*518.4*
19.3	518.1	517.9	517.6	517.3	517.1	516.8	516.5	516.3	516.0	515.7
19.4	515.4	515.2	514.9	514.7	514.4	514.1	513.9	513.6	513.3	513.1
19.5	512.8	512.6	512.3	512.0	511.8	511.5	511.2	511.0	510.7	510.5
19.6	510.2	509.9	509.7	509.4	509.2	508.9	508.6	508.4	508.1	507.9
19.7	*507.6*	*507.4*	*507.1*	*506.8*	*506.6*	*506.3*	*506.1*	*505.8*	*505.6*	*505.3*
19.8	505.1	504.8	504.5	504.3	504.0	503.8	503.5	503.3	503.0	502.8
19.9	502.5	502.3	502.0	501.8	501.5	501.3	501.0	500.8	500.5	500.3
20 0	500.0	499.8	499.5	499.3	499.0	498.8	498.5	498.3	498.0	497.8
20.1	497.5	497.3	497.0	496.8	496.5	496.3	496.0	495.8	495.5	495.3
20.2	*495.0*	*494.8*	*494.6*	*494.3*	*494.1*	*493.8*	*493.6*	*493.3*	*493.1*	*492.9*
20.3	492.6	492.4	492.1	491.9	491.6	491.4	491.2	490.9	490.7	490.4
20.4	490.2	490.0	489.7	489.5	489.2	489.0	488.8	488.5	488.3	488.0
20.5	487.8	487.6	487.3	487.1	486.9	486.6	486.4	486.1	485.9	485.7
20.6	485.4	485.2	485.0	484.7	484.5	484.3	484.0	483.8	483.6	483.3
20.7	*483.1*	*482.9*	*482.6*	*482.4*	*482.2*	*481.9*	*481.7*	*481.5*	*481.2*	*481.0*
20.8	480.8	480.5	480.3	480.1	479.8	479.6	479.4	479.2	478.9	478.7
20.9	478.5	478.2	478.0	477.8	477.6	477.3	477.1	476.9	476.6	476.4
21.0	476.2	476.0	475.7	475.5	475.3	475.1	474.8	474.6	474.4	474.2
21.1	473.9	473.7	473.5	473.3	473.0	472.8	472.6	472.4	472.1	471.9
21.2	*471.7*	*471.5*	*471.8*	*471.0*	*470.8*	*470.6*	*470.4*	*470.1*	*469.9*	*469.7*
21.3	469.5	469.3	469.0	468.8	468.6	468.4	468.2	467.9	467.7	467.5
21.4	467.3	467.1	466.9	466.6	466.4	466.2	466.0	465.8	465.5	465.3
21.5	465.1	464.9	464.7	464.5	464.3	464.0	463.8	463.6	463.4	463.2
21.6	463.0	462.7	462.5	462.3	462.1	461.9	461.7	461.5	461.3	461.0
21.7	*460.8*	*460.6*	*460.4*	*460.2*	*460.0*	*459.8*	*459.6*	*459.3*	*459.1*	*458.9*
21.8	458.7	458.5	458.3	458.1	457.9	457.7	457.5	457.2	457.0	456.8
21.9	456.6	456.4	456.2	456.0	455.8	455.6	455.4	455.2	455.0	454.8
22.0	454.4	454.3	454.1	453.9	453.7	453.5	453.3	453.1	452.9	452.7
22.1	452.5	452.3	452.1	451.9	451.7	451.5	451.3	451.1	450.9	450.7
22.2	*450.5*	*450.2*	*450.0*	*449.8*	*449.6*	*449.4*	*449.2*	*449.0*	*448.8*	*448.6*
22.3	448.4	448.2	448.0	447.8	447.6	447.4	447.2	447.0	446.8	446.6
22.4	446.4	446.2	446.0	445.8	445.6	445.4	445.2	445.0	444.8	444.6
22.5	444.4	444.2	444.0	443.9	443.7	443.5	443.3	443.1	442.9	442.7
22.6	442.5	442.3	442.1	441.9	441.7	441.5	441.3	441.1	440.9	440.7
22.7	*440.5*	*440.3*	*440.1*	*439.9*	*439.8*	*439.6*	*439.4*	*439.2*	*439.0*	*438.8*
22.8	438.6	438.4	438.2	438.0	437.8	437.6	437.4	437.3	437.1	436.9
22.9	436.7	436.5	436.3	436.1	435.9	435.7	435.5	435.4	435.2	435.0

Appendix E (*cont.*)

λ	0	1	2	3	4	5	6	7	8	9
23.0	434.8	434.6	434.4	434.2	434.0	433.8	433.7	433.5	433.3	433.1
23.1	432.9	432.7	432.5	432.3	432.2	432.0	431.8	431.6	431.4	431.2
23.2	*431.0*	*430.8*	*430.7*	*430.5*	*430 3*	*430.1*	*429.9*	*429.7*	*429.6*	*429.4*
23.3	429.2	429.0	428.8	428.6	428.4	428.3	428.1	427.9	427.7	427.5
23.4	427.4	427.2	427.0	426.8	426.6	426.4	426.3	426.1	425.9	425.7
23.5	425.5	425.4	425.2	425.0	424.8	424.6	424.4	424.3	424.1	423.9
23.6	423.7	423.5	423.4	423.2	423.0	422.8	422.7	422.5	422.3	422.1
23.7	*421.9*	*421.8*	*421.6*	*421.4*	*421.2*	*421.1*	*420.9*	*420.7*	*420.5*	*420.3*
23.8	420.2	420.0	419.8	419.6	419.5	419.3	419.1	418.9	418.8	418.6
23.9	418.4	418.2	418.1	417.9	417.7	417.5	417.4	417.2	417.0	416.8
24.0	416.7	416.5	416.3	416.1	416.0	415.8	415.6	514.5	415.3	415.1
24.1	414.9	414.8	414.6	414.4	414.3	414.1	413.9	413.7	413.6	413.4
24.2	*413.2*	*413.1*	*412.9*	*412.7*	*412.5*	*412.4*	*412.2*	*412.0*	*411.9*	*411.7*
24.3	411.5	411.4	411.2	411.0	410.8	410.7	410.5	410.3	410.2	410.0
24.4	409.8	409.7	409.5	409.3	409.2	409.0	408.8	408.7	408.5	408.3
24.5	408.2	408.0	407.8	407.7	407.5	407.3	407.2	407.0	406.8	406.7
24.6	406.5	406.3	406.2	406.0	405.8	405.7	405.5	405.4	405.2	405.0
24.7	*404.9*	*404.7*	*404.5*	*404.4*	*404.2*	*404.0*	*403.9*	*403.7*	*403.6*	*403.4*
24.8	403.2	403.1	402.9	402.7	402.6	402.4	402.3	402.1	401.9	401.8
24.9	401.6	401.4	401.3	401.1	401.0	400.8	400.6	400.5	400.3	400.2

Appendix F

IR Spectra of Some Commercially Produced IR Optical Materials

These data are reproduced with the permission of BDH, England.

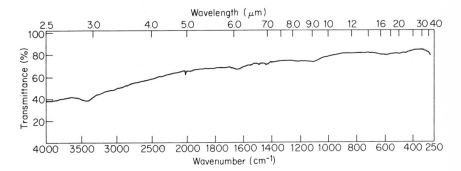

Fig. F1 IR spectrum for caesium bromide. Path length = 1.0 mm; pressure = 40 tons in^{-2} ≃ 620 MPa

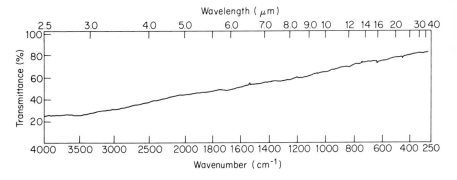

Fig. F2 IR spectrum for caesium iodide. Path length = 1.0 mm; pressure = 40 tons in^{-2} ≃ 620 MPa

668

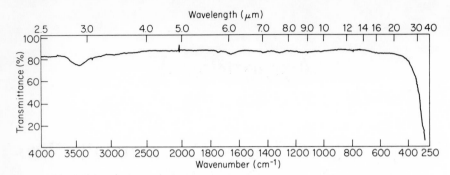

Fig. F3 IR spectrum for potassium bromide. Path length = 1.3 mm; pressure = 40 tons in $^{-2} \simeq 620$ MPa

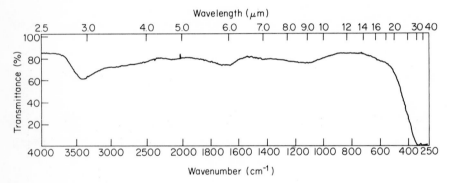

Fig. F4 IR spectrum for potassium chloride. Path length = 2.00 mm; pressure = 40 tons in$^{-2} \simeq 620$ MPa

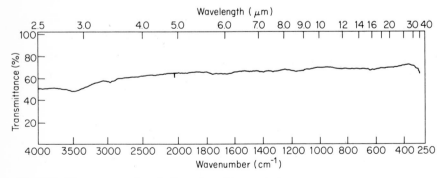

Fig. F5 IR spectrum for thallous bromide. Path length = 0.5 mm; pressure = 40 tons in$^{-2} \simeq 620$ MPa

Appendix G

IR Spectra of Some Commercially Produced Solvents for IR Spectroscopy

These data are reproduced with the permission of BDH, England, with the exceptions of the spectra of n-heptane and methylene dichloride; these two spectra were added by the author.

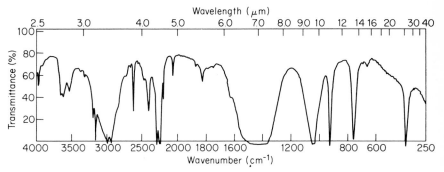

Fig. G1   IR spectrum for acetonitrile. Path length = 0.2 mm

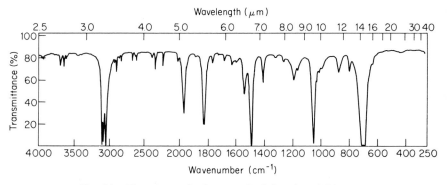

Fig. G2   IR spectrum for benzene. Path length = 0.05 mm

670

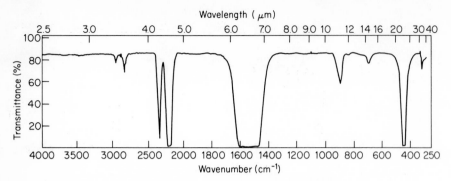

Fig. G3 IR spectrum for carbon disulphide. Path length = 0.2 mm

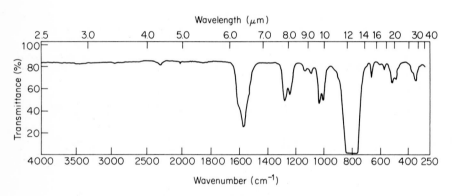

Fig. G4 IR spectrum for carbon tetrachloride. Path length = 0.2 mm

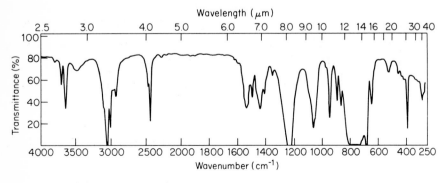

Fig. G5 IR spectrum for chloroform. Path length = 0.2 mm

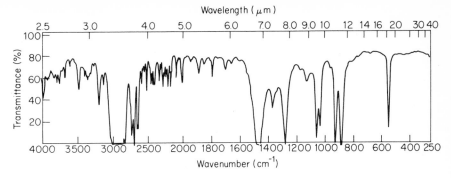

Fig. G6   IR spectrum for cyclohexane. Path length = 0.2 mm

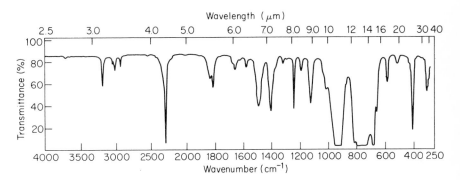

Fig. G7   IR spectrum for deuterochloroform. Path length = 0.2 mm

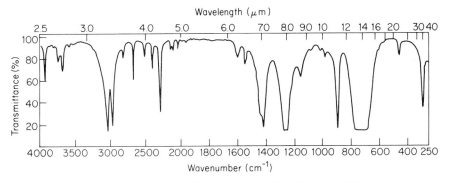

Fig. G8   IR spectrum for dichloromethane. Path length = 0.2 mm

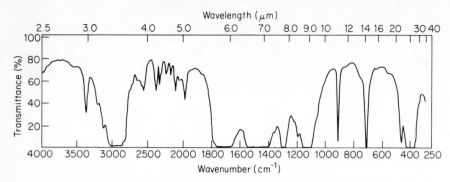

Fig. G9   IR spectrum for dimethylformamide. Path length = 0.05 mm

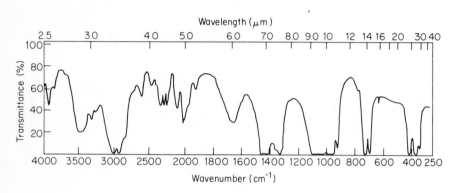

Fig. G10   IR spectrum for dimethylsulphoxide. Path length = 0.05 mm

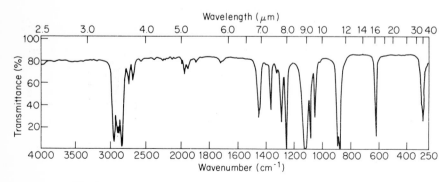

Fig. G11   IR spectrum for 1,4-dioxan. Path length = 0.014 mm

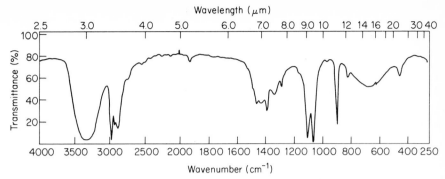

Fig. G12   IR spectrum for absolute ethanol. Path length = 0.014 mm

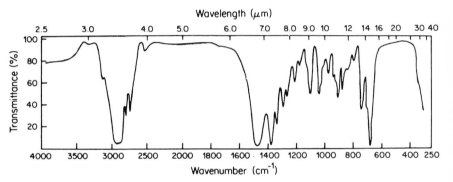

Fig. G13   IR spectrum for *n*-heptane

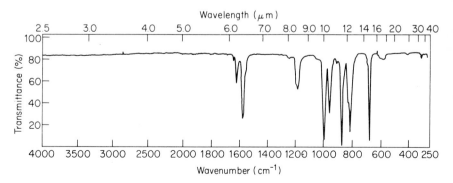

Fig. G14   IR spectrum for hexachlorobuta-1,3-diene. Path length = 0.014 mm

674

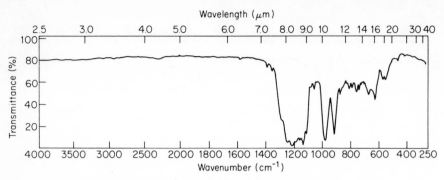

Fig. G15   IR spectrum for 'Kel-F' oil. Path length = 0.014 mm

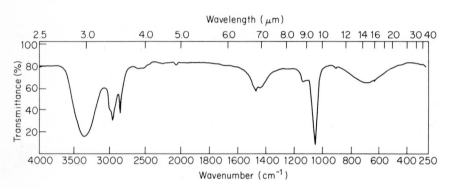

Fig. G16   IR spectrum for methanol. Path length = 0.014 mm

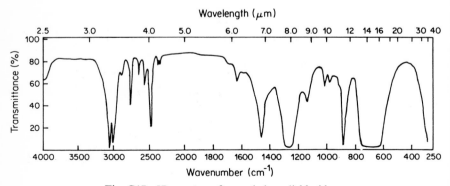

Fig. G17   IR spectrum for methylene dichloride

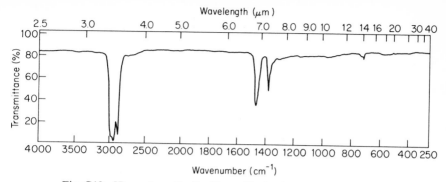

Fig. G18   IR spectrum for liquid paraffin. Path length = 0.014 mm

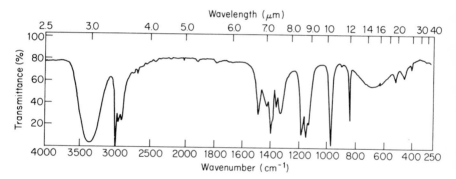

Fig. G19   IR spectrum for propan-2-ol. Path length = 0.014 mm

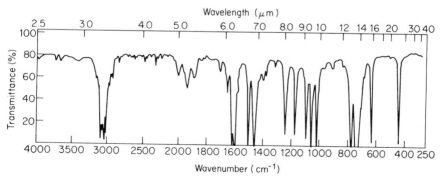

Fig. G20   IR spectrum for pyridine. Path length = 0.05 mm

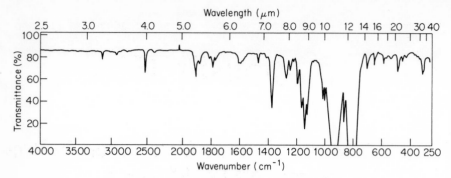

Fig. G21 IR spectrum for tetrachloroethylene. Path length = 0.2 mm

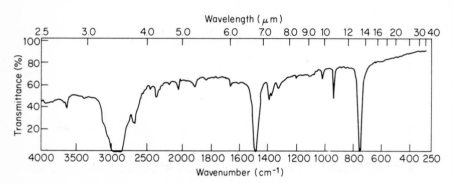

Fig. G22 IR spectra for polyethylene powder. In the upper diagram the path length = 0.16 mm and the pressure = 40 tons in$^{-2} \simeq 620$ MPa; in the lower diagram the path length is 0.5 mm and the pressure is 40 tons in$^{-2} \simeq 620$ MPa

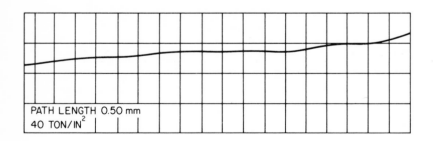

PATH LENGTH 0.50 mm
40 TON/IN^2

Appendix H

NMR Spectra of Some Commercially Produced NMR Solvents

These spectra are reproduced with the permission of BDH, England.

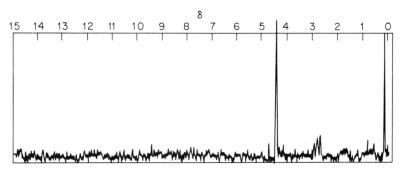

Fig. H1 NMR spectrum for deuterium oxide, D_2O. Molecular weight, 20.30; minimum isotopic purity, 99.7%

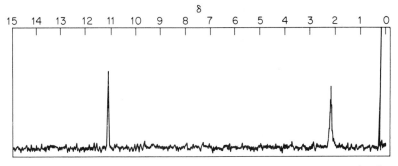

Fig. H2 NMR spectrum for deuteroacetic acid, CD_3COOD. Molecular weight, 64.08; density, 1.120 g ml^{-1}; density gradient 0.0009; refractive index, n_D^{20}, 1.3720; minimum isotopic purity, 99.5%

678

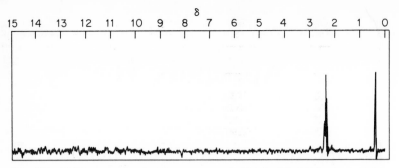

Fig. H3   NMR spectrum for deuteroacetone, $(CD_3)_2$ CO. Molecular weight, 64.12; density, 0.874 g ml^{-1}; density gradient, 0.0010; refractive index, n_D^{20}, 1.3582; minimum isotopic purity, 99.5%

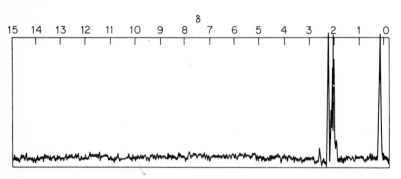

Fig. H4   NMR spectrum for deuteroacetonitrile, CD_3CN. Molecular weight, 44.07; density, 0.839 g ml^{-1}; density gradient, 0.0010; refractive index, n_D^{20}, 1.3438; minimum isotopic purity, 99.0%

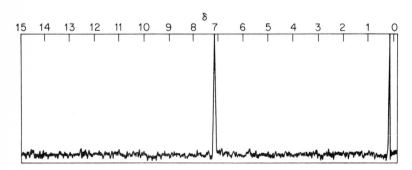

Fig. H5   NMR spectrum for deuterobenzene, C_6D_6. Molecular weight, 84.15; density, 0.948 g ml^{-1}; density gradient, 0.0010; refractive index, n_D^{20}, 1.5001; minimum isotopic purity, 99.5%

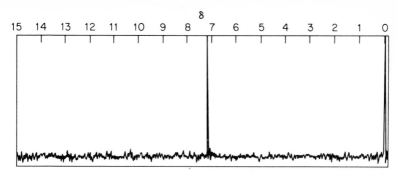

Fig. H6 NMR spectrum for deuterochloroform, $CDCl_3$. Molecular weight, 120.40; density, 1.499 g ml^{-1}; density gradient, 0.0018; refractive index, n_D^{20}, 1.4460; minimum isotopic purity, 99%

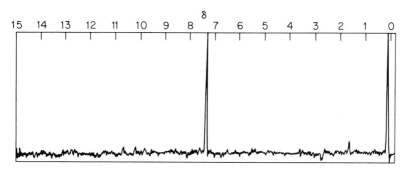

Fig. H7 NMR spectrum for deuterochloroform containing 1% v/v trimethyl silane. Density, ~1.5 g ml^{-1}; minimum isotopic purity of deuterochloroform, 99.0%

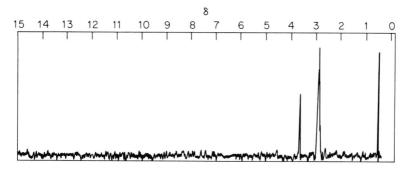

Fig. H8 NMR spectrum for deuterodimethyl sulphoxide, CD_3SOCD_3. Molecular weight, 84.17; density, 1.193 g ml^{-1}; density gradient, 0.0016; refractive index, n_D^{20}, 1.4765; minimum isotopic purity, 99.5%

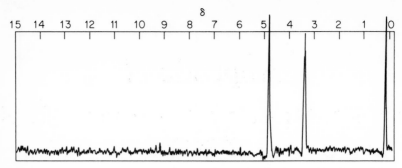

Fig. H9   NMR spectrum for deuteromethanol, CD₃OD. Molecular weight, 36.07; density, 0.892 g ml⁻¹; density gradient, 0.0009; refractive index, n_D^{20}, 1.3318; minimum isotopic purity, 99%

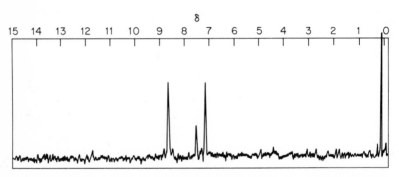

Fig. H10   NMR spectrum for deuteropyridine, C₅D₅N. Molecular weight, 84.13; density, 1.037 g ml⁻¹; density gradient, 0.0008; refractive index, n_D^{20}, 1.5073; minimum isotopic purity, 99%

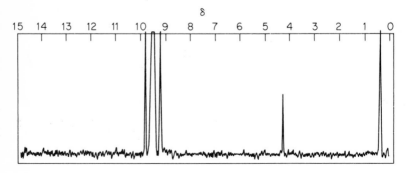

Fig. H11   NMR spectrum for deuterotrifluoroacetic acid, CF₃COOD. Molecular weight, 115.03; density, 1.500 g ml⁻¹; density gradient, 0.0016; minimum isotopic purity, 98%

Appendix I

Chemical Structures of Common Liquid Phases

Reproduced from B: 905 by permission of Varian Associates.

Trade name	Chemical structure
Amine 220	
Carbowax	$OH + CH_2 - CH_2 - O +_n H$
Castorwax	
DEGS	
Dibutyltetra-chloro-phthalate	
Dimethyl-sulpholane	
Epon 1001	
Ethofat	

Trade name	Chemical structure
Flexol 8N8	$CH_3-(CH_2)_3-CH-\overset{\displaystyle O}{\overset{\|}{C}}-N\begin{cases} CH_2CH_2O-\overset{\displaystyle O}{\overset{\|}{C}}-\overset{\displaystyle C_2H_5}{\underset{}{CH}}-(CH_2)_3-CH_3 \\ CH_2CH_2O-\overset{\displaystyle O}{\overset{\|}{C}}-CH-(CH_2)_3-CH_3 \end{cases}$ with C_2H_5 groups
Hallcomid M18	$R-\overset{}{\underset{\|}{\underset{O}{C}}}-N(CH_3)_2$
Hallcomid M 180L	$CH_3(CH_2)_7-CH=CH-(CH_2)_7-\overset{\displaystyle O}{\overset{\|}{C}}-NH_2$
Igepal	$C_9H_{19}-\langle\text{benzene}\rangle-O\text{-}[CH_2-CH_2-O]_n CH_2-CH_2-OH$
Kel F grease	$Cl\text{-}[CF_2CFCl]_n Cl$
Kel F oil No. 10, No. 3	$CF_3\text{-}[CF_2]_n CF_3$
Lexan	$\left[O-\langle\text{benzene}\rangle-\overset{\displaystyle CH_3}{\underset{\displaystyle CH_3}{C}}-\langle\text{benzene}\rangle-O-\overset{\displaystyle O}{\overset{\|}{C}} \right]_n$
Neopentyl glycol succinate	$\left[O-CH_2-\overset{\displaystyle CH_3}{\underset{\displaystyle CH_3}{C}}-CH_2-O-\overset{\displaystyle O}{\overset{\|}{C}}-CH_2-CH_2-\overset{\displaystyle O}{\overset{\|}{C}}-O \right]_n$
QF-1 (FS-1265)	$Si(CH_3)_3\left[O-\overset{\displaystyle CF_3}{\overset{\displaystyle CH_2}{\overset{\displaystyle CH_2}{\underset{\displaystyle CH_3}{Si}}}}\right]_x\left[O-\overset{\displaystyle CH_3}{\underset{\displaystyle CH_3}{Si}}\right]_y O-Si(CH_3)_3$
Quadrol	$\begin{matrix} CH_3-\overset{\displaystyle OH}{\underset{}{CH}}-H_2C & & CH_2-\overset{\displaystyle OH}{\underset{}{CH}}-CH_3 \\ & N-CH_2-CH_2-N & \\ CH_3-\overset{}{\underset{\displaystyle OH}{CH}}-H_2C & & CH_2-\overset{}{\underset{\displaystyle OH}{CH}}-CH_3 \end{matrix}$

Appendix I (*cont.*)

Trade name	Chemical structure
PDEAS	
Porapak	
SAIB	
SE-30 DOW-200 DOW-11 SF-96	

Appendix I (*cont.*)

Trade name	Chemical structure
SE-52 DOW-710 DOW-550	
Squalane	
Tetracyano- ethylpenta- erythritol	
THEED	
Tricresyl- phosphate	
Versamid	
Zonyl E-7	

Appendix I (*cont.*)

Trade name	Chemical structure
XF-1150 XF-1112 XE-60 XF-1125	

$$\text{Si(CH}_3)_3-\text{O}\left[\begin{array}{c}\text{CH}_3\\|\\\text{Si}-\text{O}\\|\\\text{CH}_2\\|\\\text{CH}_2\\|\\\text{C}\equiv\text{N}\end{array}\right]_n\text{Si(CH}_3)_3$$

Appendix J

Recommended Temperatures and Solvents for Liquid Phases*

*(Chemical structure of some Liquid phases is given in the Appendix I)
Reproduced from B: 905 by permission of Varian Associates.

Liquid phase	Recommended maximum temperature (°C)	Polarity†	Solvent‡
Acetonyl acetone (2,5 hexanedione)	25	P, I	1
Adiponitrile	50	I	3, 4
Alkaterge T	75	I	3, 4 h
Amine 220	180	P	3, 4
Apiezon H	275	N	3
Apiezon J	300	N	3, 4
Apiezon L	300	N	3. 4
Apiezon M	275	N	3, 4
Apiezon N	300	N	3, 4
Armeen SD	100	H, P	3, 4
Armeen 12D	100	H, P	3
Armeen 2HT	100	H, P	3
Aroclor 1254 (chlorinated biphenyl)	125	I	3, 4
Asphalt	300	N	3, 4
Bentone 34	200	S	4
7,8-Benzoquinoline	150	I	3, 4
Benzyl cellosolve	50	I	3, 4
Benzyl cyanide (phenyl acetonitrile)	35	I	3, 4
Silver nitrate	25	S	4
Benzyl ether	50	I	3, 4
Bis(2-ethoxyethyl) adipate	150	I	4
Bis(2-ethylhexyl) tetrachlorophthalate	150	I	3, 4
Bis(2-methoxyethyl) adipate	150	I	4
Bis(2(2-methoxyethoxy)ethyl) ether	50	P	3, 4
Butanediol adipate	225	I, P	3, 4
Butanediol succinate	225	I, P	3, 4
Carbowax 20M	250	P	3, 4
Carbowax 20M TPA	250	P	4, 6
Carbowax 300	100	P	3, 4
Carbowax 400	125	P	3, 4
Carbowax 400 monooleate	125	P	3, 4
Carbowax 550	125	P	3, 4
Carbowax 600	125	P	3, 4
Carbowax 600 monostearate	125	P	3 h, 4 h
Carbowax 750	150	P	3, 4
Carbowax 1000	175	P	3, 4
Carbowax 1500	200	P	3, 4
Carbowax 1540	200	P	3, 4

Appendix J (*cont.*)

Liquid phase	Recommended maximum temperature (°C)	Polarity†	Solvent‡
Carbowax 4000	200	P	3, 4
Carbowax 4000 TPA	175	P	4, 6
Carbowax 4000 monostearate	220	P	4
Carbowax 6000	200	P	3, 4
Castorwax	200	P	3, 4
Celanese ester = 9	200	I	3, 4
Chloronaphthalene	75	I	3, 4
Cyanoethylsucrose	175	P	4
Cyclohexane dimethanol succinate	210	I	3
n-Decane	30	N	1, 4
Dibutyl maleate	50	P, I	3, 4
Dibutyl phthalate	100	I	3, 4
Dibutyl tetrachlorophthalate	150	I, P	4
Didecyl phthalate	125	I	3, 4
Diethylene glycol adipate (DEGA)	190	I, P	3, 4
Diethylene glycol glutarate	225	I, P	3
Diethylene glycol sebacate (DEGSE)	190	I, P	3 h, 4 h
Diethylene glycol succinate (DEGS)	190	P	1, 4
Di-(2-ethylhexyl) sebacate	125	I, P	3, 4
Diethyl-D-tartrate	125	P, S	3, 4
Diglycerol	120	H	6, 9
Diisodecyl phthalate	175	I	3, 4
Diisoctyl sebacate	175	I	3, 4
Dimer acid	150	I	3, 4
2,4-Dimethylsulpholane	50	P	?, 4
Dimethylsulphoxide	50	P	3, 4
Dinonyl phthalate	175	I	3, 4
Dinonyl sebacate	125	I	3
Dioctyl phthalate	175	I	3, 4
Dioctyl sebacate	100	I	3, 4
Dowfax 9N9	225	S	1, 3
Dowfax 9N40	225	S	1, 3
EPON resin 1001	225	P	3 , 4 h
Ethofat 60 25	140	I	3 h, 4 h
Ethomeen	75	P	3, 4
Ethylene glycol adipate (EGA)	200	I, P	3 h, 4 h
Ethylene glycol isophthalate (EGIP)	250	P	3 h, 4 h
Ethylene glycol phthalate	250	P	3
Ethylene glycol glutarate	225	I, P	3
Ethylene glycol sebacate (EGSE)	200	I, P	3 h, 4 h
Ethylene glycol succinate (EGS)	200	I, P	3 h, 4 h
Ethylene glycol tetrachlorophthalate	225	P	3
FFAP	275	S, P	4
Flexol plasticizer 8N8	180	P	3, 4
Fluorolube GR 362	100	S	2 h
Glycerol	100	H	6, 9
Hallcomid M-18	150	I	3, 4
Hallcomid M-18-OL	150	I	3, 4
Hercoflex 600	150	P	4
n-Hexadecane	50	N	3, 4
Hexadecene	50	N	3, 4
H.H.K. mix	35	N	4

Appendix J (*cont.*)

Liquid phase	Recommended maximum temperature (°C)	Polarity†	Solvent‡
Hexamethylphosphoramide (HMPA)	50	P	3, 4
2,5-Hexanedione (see acetonyl acetone)			
Hyprose SP-80	190	P	3 h, 4 h
IGEPAL (nonyl phenoxypolyoxyethylene ethanol)	200	I	3 h, 4 h
IGEPAL, CO 990	200	I	3 h, 4 h
Isoquinoline	50	I, P	3, 4
K F grease	200	I	3, 4
Kel F oil = 3	60	I	4
Kel F oil = 10	100	I	3, 4
LAC 1-R-296 (see DEGA—diethylene glycol adipate)			
LAC 2-R-446 (DEGA cross-linked)	190	P	1
LAC 3-R-728 (see DEGS—diethylene glycol succinate)			
LAC 4-R-886 (see EGS—ethylene glycol succinate)			
Lexan (polycarbonate resin)	300	P	hot DMP
Mannitol	200	H	6, 9
Neopentyl glycol adipate (NPGA)	240	I	3, 4
Neopentyl glycol isophthalate (NPGIP)	240	I	3
Neopentyl glycol sebacate (NPGSE)	240	I	3, 4
Neopentyl glycol succinate (NPGS)	240	I	3, 4
Nonyl phenol	125	I	4 h
Nonyl phenoxypolyoxyethylene ethanol (see IGEPAL)			
Nujol (mineral oil)	200	N	3, 4
OV-1 (methyl silicone)	350	N	3
OV-17 (methyl phenyl silicone)	300	I	3
OV-25 (phenyl silicone)	300	I	3
OV-101 (liquid methyl silicone)	300	N	3
OV-210 (trifluoropropyl methyl silicone)	275	I	3
OV-225 (cyanopropylmethyl phenylmethyl silicone)	275	I	3
(β,β′-Oxydiproprionitrile	100	P	4
Phenyl acetonitrile (see benzyl cyanide)			
Phenyl diethanolamine succinate (PDEAS)	225	P	1, 4
Polyethylene glycol (see carbowaxes)			
Polyethylene glycol 600 (Jefferson)	160	P	6
Polyethylene imine	250	P	6 h
Poly-*m*-phenyl ether (5-ring)	250	I	3, 4
Poly-*m*-phenyl ether (6-ring)	300	I	3, 4
Poly-*m*-phenyl ether (high polymer)	400	I	7
Poly-*m*-phenyl ether (6-ring) squalane	100	I	3, 4
Polypropylene glycol	150	H	3, 4
Polypropylene glycol sebacate (Harflex 370)	225	I	3
Polypropylene glycol silver nitrate	75	S	3, 4
Propylene carbonate (1,2-propanediol cyclic carbonate)	60	P	4
Polyvinyl pyrrolidinone	225	P, H	6
Quadrol	150	H	3, 4
Reoplex 400 (polypropylene glycol adipate)	190	I	3, 4

689

Appendix J (*cont.*)

Liquid phase	Recommended maximum temperature (°C)	Polarity†	Solvent‡
Silicone, D.C. 11	300	N	2, 5
Silicone, D.C. 200, 200 CS	250	N	3, 4
Silicone, D.C. 200, 1000 CS	250	N	3, 4
Silicone, D.C. 200, 12 500 CS	250	N	3, 4
Silicone, D.C. 200, 2 500 000 CS	250	N	3, 4
Silicone, D.C. 220	250	I	3, 4
Silicone, D.C. 550	275	I	3, 4
Silicone, D.C. 703	225	I	3, 4
Silicone, D.C. 710	300	I	3, 4
Silicone (fluoro) QF-1 (FS 1265)	250	I	3, 4
Silicone GE SF-96	300	N	3, 4
Silicone GE XE-60 (nitrile gum)	275	I	3, 4
Silicone GE (Versilube F-50)	300	I	3, 4
Silicone, D.C. 560 (F-60)	250	I	1
G.C. grade SE-30 (methyl silicone)	300	N	3 h, 4 h
Silicone gum rubber SE-30 (methyl)	350	N	3 h, 4 h
Silicone gum rubber SE-52 (phenyl)	300	I	3 h, 4 h
Silicone gum rubber SE54 (methyl phenyl vinyl)	300	I	4
Span 80 (Sorbitan monooleate)	150	P	3, 4
Squalane	100	N	3, 4
Squalene	140	I, N	3, 4
STAP (steroid analysis phase)	255	P	1
Sucrose acetate isobutyrate (SAIB)	225	I, P	3 h, 4
Tergitol nonionic NP-35	200	P	3, 4
Tetracyanoethylated pentaerythritol (TCEPE)	180	P	4
Tetraethylene glycol	70	P	4
Tetraethylene glycol dimethyl ether (see *Bis*-(2-methoxyethox)ethyl)ether)			
Tetraethylenepentamine	150	H, P	4
Tetrehydroxyethylethylenediamine (THEED)	150	H	3, 4
β,β′-Thiodipropionitrile	100	P	3, 4
Tributyl phosphate	50	I	4
Tricresyl phosphate (TCP) (tritolyl phosphate)	125	I	3, 4
Triethanolamine	75	H	3, 4
Trimer acid	200	H	3, 4
1,2,3, Tris-(2-cyanoethoxy) propane (TCEP)	180	P	3, 4
Tritolyl phosphate (see tricresyl phosphate)			
Triton X-305	200	P	1
TWEEN 80 (polyoxyethylene sorbitan monooleate)	150	P	3, 4
UCON 50 HB 280X polar	200	P	3, 4
UCON 50 HB 2000 polar	200	P	3, 4
UCON 50 HB 5100	200	P	3, 4
UCON 50 LB 550X	200	P	3, 4
UCON LB 1715 non-polar	200	I	3, 4
Versamid 900	250	P	*
Xylenyl phosphate	175	I	4
Zonyl E-7	200	I	4

* 1:1 butonol and phenol or 87% chloroform and 13% methanol, Versamid is unstable at high temperatures in the presence of oxygen.

690

† Polarity code: N = non-polar; P = polar, I = intermediate; H = hydrogen bonding; S = specific.

‡ Solvent code: 1 = acetone; 2 = benzene; 3 = chloroform; 4 = methylene chloride; 5 = ethyl acetate; 6 = methanol; 7 = toluene; 8 = carbon disulphide; 9 = water; h = hot.

Appendix K

Derived SI Units and Their Relation to CGS or Other Common Units for Physical Quantities

Physical quantity	SI unit	Conversion factor	CGS unit
Acceleration	$m\,s^{-2}$	10^2	$cm\,s^{-2}$
Acceleration, angular	$rad\,s^{-1}$	1	$rad\,s^{-1}$
Area	m^2	10^4	cm^2
		10^{20}	$Å^2$
Capacitance (farad)	$F = m^{-2}\,kg^{-1}\,s^4\,A^2 = C\,V^{-1}$	8.99×10^{11}	statfarad
Charge (coulomb)	$C = A\,s = J\,V^{-1}$	3.00×10^9	statcoulomb (e.s.u.)
Charge density, surface	$C\,m^{-2}$	3.00×10^5	$statcoul\,cm^{-2}$
Charge density, volume	$C\,m^{-3}$	3.00×10^3	$statcoul\,cm^{-3}$
Conductance (siemens)	$S = m^{-2}\,kg^{-1}\,s^3\,A^2 = \Omega^{-1}$	8.99×10^{11}	statmho (statohm^{-1})
Conductivity	$\Omega^{-1}\,m^{-1}$	8.99×10^9	$statmho\,cm^{-1}$
		10^{-2}	$mho\,cm^{-1}$ (ohm^{-1} cm^{-1})
Density	$kg\,m^{-3}$	10^{-3}	$g\,cm^{-3}$
Diffusion coefficient	$m^2\,s^{-1}$	10^4	$cm^2\,s^{-1}$
Dipole moment	$C\,m$	3.00×10^{11}	$statcoul\,cm$
Electric field	$V\,m^{-1}$	3.34×10^{-5}	$statvolt\,cm^{-1}$
		10^{-2}	$V\,cm^{-1}$
Electric potential (volt)	$V = m^2\,kg\,s^{-1}\,A^{-1} = J\,C^{-1}$	3.34×10^{-3}	statvolt
Energy (joule)	$J = m^2\,kg\,s^{-2} = N\,m$	10^7	erg
		0.2390	calorie
Entropy	$J\,K^{-1}$	0.2390	$cal\,K^{-1}$
Force (newton)	$N = m\,kg\,s^{-2}$	10^5	dyne
Frequency (hertz)	$Hz = s^{-1}$	1	s^{-1}
Friction factor	$kg\,s^{-1}$	10^3	$g\,s^{-1}$
Heat capacity	$J\,K^{-1}$	0.2390	$cal\,K^{-1}$
Molarity	$mol\,dm^{-3}$	1	$moles\,litre^{-1}$
Moment, dipole	$C\,m$	3.00×10^{11}	$statcoul\,cm$
Moment, force	$N\,m$	10^7	$dyne\,cm$
Moment, inertia	$kg\,m^2$	10^7	$g\,cm^2$
Momentum	$N\,s$	10^5	$dyne\,s$
Momentum, angular	$J\,s$	10^7	$erg\,s$
Period	s	1	s

Appendix **K** (*cont.*)

Physical quantity	SI unit	Conversion factor	CGS unit
Permittivity	$F\,m^{-1}$	8.99×10^9	statfarad cm^{-1}
Polarization, electric	$C\,m^{-2}$	3.00×10^5	statcoul cm^{-2}
Potential (volt)	V	3.34×10^{-3}	statvolt
Power (watt)	$W = m^2\,kg\,s^{-3} = J\,s^{-1}$	10^7	erg s^{-1}
Pressure (pascal)	$Pa = m^{-1}\,kg\,s^{-2} = N\,m^{-2}$	10	dynes cm^{-2}
		9.87×10^{-6}	atm
Radius of gyration	m	10^2	cm
Resistance (ohm)	$\Omega = m^2\,kg\,s^{-3}\,A^{-2}$ $= V\,A^{-1}$	1.11×10^{-12}	statohm
Specific heat capacity	$J\,kg^{-1}\,K^{-1}$	2.39×10^{-4}	cal $g^{-1}\,K^{-1}$
Stress	$N\,m^{-2}$	10	dynes cm^{-2}
Surface energy	$J\,m^{-2}$	10^3	ergs cm^{-2}
Surface tension	$N\,m^{-1}$	10^3	dynes cm^{-1}
Torque	$N\,m$	10^7	dyne cm
Velocity	$m\,s^{-1}$	10^2	cm s^{-1}
Velocity, angular	$rad\,s^{-1}$	1	rad s^{-1}
Viscosity	$N\,s\,m^{-2}$	10	dyne s cm^{-2} (poise)
Volume	m^3	10^6	cm^3
		10^3	dm^3 (litre)
Wave number	m^{-1}	10^{-2}	cm^{-1}
Weight	N	10^5	dyne

Note: CGS unit = SI unit × Conversion factor
SI unit = CGS unit/Conversion factor

Appendix L

Derived SI Units and Their Relation to CGS or Other Common Units for Mechanical Quantities such as Moduli, Stress, and Viscosity

SI unit	Conversion factor	CGS unit
newtons per square metre (N m^{-2})	10.00	dynes cm^{-2}
dynes cm^{-2}	0.100	newtons per square metre (N m^{-2})
psi	6.895×10^3	N m^{-2}
N m^{-2}	1.405×10^{-4}	psi
dynes cm^{-2}	1.020×10^{-6}	kg cm^{-2}
dynes cm^{-2}	1.020×10^{-8}	kg mm^{-2}
kg mm^{-2}	9.806×10^7	dynes cm^{-2}
dynes cm^{-2}	1.450×10^{-5}	psi
psi	6.895×10^4	dynes cm^{-2}
psi	7.03×10^{-4}	kg mm^{-2}
kg mm^{-2}	1.422×10^3	psi
dynes cm^{-2}	9.869×10^{-7}	atm
atm	1.013×10^6	dynes cm^{-2}
atm	1.013×10^5	N m^{-2}
psi	6.81×10^{-2}	atm
dynes cm^{-2}	1.00×10^{-6}	bars
psi	6.895×10^{-2}	bars
g denier^{-1}	$8.83 \times 10^8 \rho\dagger$	dynes cm^{-2}
g denier^{-1}	$1.28 \times 10^4 \rho\dagger$	psi
bars	1.00×10^5	N m^{-2}
kg cm^{-2}	9.807×10^4	N m^{-2}
poise	1.000×10^{-1}	N S m^{-2}
stokes	1.000×10^{-4}	m^2 S
dynes	1.000×10^{-5}	N

$\rho\dagger$ = density.

BIBLIOGRAPHY

1. Abragam, A., THE PRINCIPLES OF NUCLEAR MAGNETISM, Clarendon Press, Oxford, 1961.
2. Abraham, R. J., THE ANALYSIS OF HIGH RESOLUTION NMR SPECTRA, Elsevier, Amsterdam, 1971.
3. Abu-Shumays, A., and Duffield, J. J., *Anal. Chem.*, **38**, 29A (1966).
4. Ache, H. J., *Angew. Chem.*, **84**, 234 (1972).
5. Ackers, G. K., *Adv. Protein Chem.*, **24**, 343 (1970).
6. Adams, E. T., in Characterization of Macromolecular Structure, Nat. Acad. Sci. U.S.A., Publ. 1573, Washington, D.C., 1968, p. 84.
7. Adams, H. E., in Gel Permeation Chromatography (K. H. Altgelt and L. Segal, eds), Dekker, New York, 1971, p. 391.
8. Adamson, A. W., PHYSICAL CHEMISTRY OF SURFACES, 3rd ed., Wiley–Interscience, New York, 1976.
9. Adler, R. B., Smith, A. C., and Longini, R. K., INTRODUCTION TO SEMICONDUCTOR PHYSICS, Wiley–Interscience, New York, 1964.
10. Aggarwal, S. L., BLOCK COPOLYMERS, Plenum Press, New York, 1970.
11. Aggarwal, S. L., *Polymer*, **17**, 939 (1976).
12. Ahearn, A. J., MASS SPECTROMETRIC ANALYSIS OF SOLIDS, Elsevier, Amsterdam, 1966.
13. Albaugh, E. W., Talarico, P. C., Davis, B. E., and Wirkkala, R. A., in Gel Permeation Chromatography (K. H. Altgelt and L. Segal, eds), Dekker, New York, 1971, p. 569.
14. Alderson, R. H. and Halliday, J. S., in Techniques for Electron Microscopy, 2nd ed. (D. H. Kay, ed.), Blackwell, Oxford, 1965, Chap. 15.
15. Alexander, G., CHROMATOGRAPHY, American Chemical Society, Washington, D.C., 1977.
16. Alexander, L. E., X-RAY DIFFRACTION METHODS IN POLYMER SCIENCE, Wiley–Interscience, New York, 1969.
17. Alger, R. S., ELECTRON PARAMAGNETIC RESONANCE: TECHNIQUES AND APPLICATIONS, Wiley–Interscience, New York, 1968.
18. Allara, D. L., in Characterization of Metal and Polymer Surfaces, Vol. 2 (L. H. Lee, ed.) Academic Press, New York, 1977, p. 193.
19. Allegra, G., and Bassi, I. W., *Fortsch. Hochpolym. Forsch., Adv. Polym. Sci.*, **6**, 549 (1969).
20. Allen, G., *Rev. Pure Appl. Chem.*, **17**, 67 (1967).
21. Allen, G., in Characterization of Macromolecular Structure, Nat. Acad. Sci. U.S.A., Publ. 1573, Washington, D.C., 1968, p. 157.
22. Allen, N. S., and McKellar J. F., *Chem. Soc. Rev.*, **4**, 533 (1975).
23. Allen P. W., TECHNIQUES OF POLYMER CHARACTERIZATION, Butterworths, London, 1959.
24. Allen, P. W., in Techniques of Polymer Characterization (P. W. Allen, ed.), Butterworths, London, 1959, Chap. 1.
25. Alley, C. L., and Atwood, K. W., Jr, SEMICONDUCTOR DEVICES AND CIRCUITS, Wiley–Interscience, New York, 1971.
26. Alliet, D. F., and Pacco, J. M., in Gel Permeation Chromatography (K. H. Altgelt and L. Segal, eds), Dekker, New York, 1971, p. 417.
27. Allport, D. C., and Janes, W. H., BLOCK COPOLYMERS, Halsted, New York, 1973.
28. Altgelt, K. H., *Adv. Chromatog.*, **7**, 3 (1968).
29. Altgelt, K. H., in Gel Permeation Chromatography (K. H. Altgelt and L. Segal, eds), Dekker, New York, 1971, p. 193.
30. Altgelt, K. H., and Moore, J. C., in Polymer Fractionation (K. H. Altgelt, J. C. Moore, and M. J. R. Cantow, eds), Academic Press, New York, 1967, p. 123.

696

31. Altgelt, K. H., Moore, J. C., and Cantow, M. J. R., POLYMER FRACTIONATION, Academic Press, New York, 1967.
32. Altgelt, K. H., and Segal, L., GEL PERMEATION CHROMATOGRAPHY, Dekker, New York, 1971.
33. Altshuler, S. A., and Kozyrev, B. M., ELECTRON PARAMAGNETIC RESONANCE, Academic Press, New York, 1964.
34. Amphlett, C. B., INORGANIC ION EXCHANGERS, Elsevier, London, 1964.
35. Amundson, N. R., and Luss, D., *Rev. Macromol. Sci.*, **3**, 145 (1968).
36. Anderson, A., THE RAMAN EFFECT, Dekker, New York, 1971.
37. Anderson, D. R., *Chem. Rev.*, **66**, 677 (1966).
38. Anderson, H. C., in Techniques and Methods of Polymer Evaluation, Vol. 1, Thermal Analysis (P. E. Slade Jr, ed.), Dekker, New York, 1966, Chap. 3.
39. Andrew, E. R., NUCLEAR MAGNETIC RESONANCE, Cambridge University Press, 1955.
40. Andrews, A. C., *J. Chem. Educ.*, **32**, 154 (1955).
41. Andrews, K. W., Dyson, D. J., and Keown, S. R., INTERPRETATION OF ELECTRON DIFFRACTION PATTERNS, 2nd ed., Hilger, London, 1968.
42. Andrews, R. D., and Hart, T. R., in Characterization of Metal and Polymer Surfaces, Vol. 2 (L. H. Lee, ed.), Academic Press, New York, 1977, p. 207.
43. Angus, J. C., Morrow, D. L., Dunning, J. W., Jr, and French, M. J., *Ind. Eng. Chem.*, **61**, 8 (1969).
44. Arecchi, F. T., Giglio, M., and Tartari, U., *Phys. Rev.*, **163**, 186 (1967).
45. Armstrong, J. L., in Characterization of Macromolecular Structure, Nat. Acad. Sci. U.S.A., Publ. 1573, Washington, D.C., 1968, p. 51.
46. Arnold, P., and Willis, H., in Polymer Science, Vol. 2 (A. J. Jenkins, ed.), North-Holland, Amsterdam, 1972, Chap. 24.
47. Arnold, S. J., Kubo, M., and Ogryzlo, E. A., in Oxidation of Organic Compounds, Advances in Chemistry Series, No. 77, American Chemical Society, Washington, D.C., 1968, p. 133.
48. Arridge, R. G. C., MECHANICS OF POLYMERS, Clarendon Press, Oxford, 1975.
49. Aslin, D., *Intern. Lab.*, July/Aug. 1977, p. 59.
50. Assenheim, H. M., INTRODUCTION TO ELECTRON SPIN RESONANCE, Hilger and Watts, London, 1966.
51. Atherton, N. M., ELECTRON SPIN RESONANCE: THEORY AND APPLICATION, Halsted, New York, 1973.
52. Auerbach, P., MAGNETIC RESONANCE AND RADIOFREQUENCY SPECTROSCOPY, North-Holland, Amsterdam, 1969.
53. Ault, A., PROBLEMS IN ORGANIC STRUCTURE DETERMINATION, McGraw-Hill, New York, 1967.
54. Ault, A., *J. Chem. Educ.*, **47**, 812 (1970).
55. Ault, A., and Dudek, G. O., NUCLEAR MAGNETIC RESONANCE SPECTROSCOPY, Holden-Day, San Francisco, 1976.
56. Avram, M., and Mateescu, G. D., INFRARED SPECTROSCOPY: APPLICATIONS IN ORGANIC CHEMISTRY, Wiley–Interscience, New York, 1972.
57. Azaroff, L. V., and Buerger, M. J., THE POWDER METHOD IN X-RAY CRYSTALLOGRAPHY, McGraw-Hill, New York, 1958.
58. Azzam, R. M. A., and Bashra, N. M., ELLIPSOMETRY AND POLARIZED LIGHT, North-Holland, Amsterdam, 1977.
59. Ayrey, G., *Fortsch. Hochpolym. Forsch., Adv. Polym. Sci.*, **6**, 128 (1969).
60. Ayscough, P. B., ELECTRON SPIN RESONANCE IN CHEMISTRY, Methuen, London, 1967.
61. Axenrod, T., and Webb, G. A., NMR SPECTROSCOPY OF NUCLEI OTHER THAN PROTONS, Wiley–Interscience, New York, 1974.
62. Backon, G. E., NEUTRON DIFFRACTION, Clarendon Press, Oxford, 1975.
63. Bak, B., ELEMENTARY INTRODUCTION TO MOLECULAR SPECTRA, 2nd ed., Elsevier, Amsterdam, 1964.
64. Bak, T. A., PHONONS AND NEUTRON SCATTERING, Benjamin, New York, 1964.

65. Baker, A. J., and Cairns, T., SPECTROSCOPIC TECHNIQUES IN ORGANIC CHEMISTRY, Heyden, London, 1965.
66. Baker, F. L., and Princen, L. H., in Encyclopedia of Polymer Science and Technology, Vol. 15 (H. F. Mark, N. G. Gaylord, and N. M. Bikales, eds), Wiley–Interscience, New York, 1971, p. 498.
67. Baldeschwieler, J. D., and Randall, E. W., *Chem. Rev.*, **63**, 81 (1963).
68. Baldwin, R. L., and Van Holde, K. E., *Fortsch. Hochpolym. Forsch., Adv. Polym. Sci.*, **1**, 451 (1960).
69. Ballinger, H. R., Brenner, M., Gänshirt, H., Mangold, H. K., Seiler, H., Stahl, E., and Waldi, D., THIN LAYER CHROMATOGRAPHY: A LABORATORY HAND-BOOK, Springer, Heidelberg, 1965.
70. Bamford, C. H., Elliot, A., and Hanby, W. E., SYNTHETIC POLYPEPTIDES, Academic Press, New York, 1965, p. 186.
71. Banwell, C. N., FUNDAMENTALS OF MOLECULAR SPECTROSCOPY, McGraw-Hill, New York, 1972.
72. Bareiss, R. E., in Polymer Handbook, 2nd ed. (J. Brandrup, E. H. Immergut, and W. McDowell, eds), Wiley–Interscience, New York, 1975, p. IV-115.
73. Barnes, A. J., in Vibrational Spectroscopy—Modern Trends (A. J. Barnes, and W. J. Orville-Thomas, eds), Elsevier, Amsterdam, 1977, Chap. 6.
74. Barnes, A. J., and Orville-Thomas, W. J., VIBRATIONAL SPECTROSCOPY—MODERN TRENDS, Elsevier, Amsterdam, 1977.
75. Barnes, R. M., EMISSION SPECTROSCOPY, Wiley–Interscience, New York, 1976.
76. Barr, J. K., and Flournoy, P. A., in Physical Methods in Macromolecular Chemistry, Vol. 2 (B. Caroll, ed.), Dekker, New York, 1969, Chap. 2.
77. Barrales-Rienda, J. M., *Rev. Plast. Mod.*, **20**, 609 (1969).
78. Barrall, E. M., II, and Johnson, J. F., in Techniques and Methods of Polymer Evaluation, Vol. i (P. E. Slade, Jr and L. T. Jenkins, eds), Dekker, New York, 1966, Chap. 1.
79. Barral, E. M., II, and Johnson, J. J., in Gel Permeation Chromatography (K. H. Altgelt and L. Segal, eds), Dekker, New York, 1971, p. 25.
80. Barrall, E. M., II, Johnson, J. F., and Cooper, A. R., in Fractionation of Synthetic Polymers: Principles and Applications (L. H. Tung, ed.), Dekker, New York, 1977, Chap. 3.
81. Barrow, G. M., INTRODUCTION TO MOLECULAR SPECTROSCOPY, McGraw-Hill, New York, 1962.
82. Bassett, D. C., *Polymer*, **17**, 460 (1976).
83. Bassett, D. C., *High Temp. High Press.*, **9**, 553 (1977).
84. Battista, O. A., in Polymer Fractionation (M. J. R. Cantow, ed.), Academic Press, New York, 1967, p. 307.
85. Battista, O. A., MICROCRYSTAL POLYMER SCIENCE, McGraw-Hill, New York, 1975.
86. Battista, O. A., Cruz, M. M., and Ferraro, C. F., in Surface and Colloid Science, Vol. 3 (E. Matjevic, ed.), Wiley–Interscience, New York, 1971.
87. Bauer, G., MEASUREMENT OF OPTICAL RADIATIONS, Focal Press, London, 1965.
88. Baumann, F., in Basic Liquid Chromatography (N. Hadden, F. Baumann, F. Mac-Donald, M. Munk, R. Stevenson, D. Gere, F. Zamaroni, and R. Majors, eds), Varian Aerograph, Palo Alto, Calif., 1971.
89. Bauman, R. P., ABSORPTION SPECTROSCOPY, Wiley, New York, 1962.
90. Baxter, R. A., in Thermal Analysis, Vol. 1 (R. F. Schwenker and P. D. Garn, eds), Academic Press, New York, 1969, p. 69.
91. Becker, E. D., HIGH RESOLUTION NMR, Academic Press, New York, 1969.
92. Becker, R. S., THEORY AND INTERPRETATION OF FLUORESCENCE AND PHOSPHORESCENCE, Wiley–Interscience, New York, 1969.
93. Beerbower, A., and Hansen, C. M., in Kirk–Othmer Encyclopedia of Chemical Technology, Suppl. Vol., 2nd ed., Wiley–Interscience, New York, 1971, p. 889.
94. Bekturov, E. A., TERNARY POLYMER SYSTEMS IN SOLUTIONS, Nauka, Alma-Ata, 1975 (in Russian).
95. Bell, R. J., INTRODUCTORY FOURIER TRANSFORM SPECTROSCOPY, Academic Press, New York, 1972.

698

96. Bellamy, L. J., ADVANCES IN INFRARED GROUP FREQUENCIES, Methuen, London, 1968.
97. Bellamy, L. J., THE INFRARED SPECTRA OF COMPLEX MOLECULES, 3rd ed., Halsted-Wiley, New York, 1975.
98. Bello, A., Barrales-Rienda, J. M., and Guzman, G. M., in Polymer Handbook, 2nd ed. (J. Brandrup, E. H. Immergut, and W. McDowell, eds), Wiley, New York, 1975, p. IV-175.
99. Bender, M., J. Chem. Educ., **29**, 15 (1952).
100. Benoit, H., Ber. Bunsenges Phys. Chem., **70**, 286 (1966).
101. Bentley, F., Smithson, L. D., and Rozek, A. L., INFRARED SPECTRA AND CHARACTERISTIC FREQUENCIES (700–300 cm^{-1}), Wiley–Interscience, New York, 1968.
102. Bentley, K. W., TECHNIQUES OF ORGANIC CHEMISTRY, Vol. 4, ELUCIDATION OF STRUCTURE BY PHYSICAL AND CHEMICAL METHODS, Wiley–Interscience, New York, 1972.
103. Benz, W., MASSENSPEKTROMETRIK ORGANISCHER VERBINDUNGEN, Akademie Verlag, Frankfurt/Main, 1968.
104. Berezkin, V. G., Alishovey, V. R., and Nemirovskaya, I. B., GAS CHROMATOGRAPHY OF POLYMERS, Elsevier, Amsterdam, 1977.
105. Berg, E. W., PHYSICAL AND CHEMICAL METHODS OF SEPARATION, McGraw-Hill, New York, 1963.
106. Bergmann, E. D., and Pullman, B., CONFORMATION OF BIOLOGICAL MOLECULES AND POLYMERS, Academic Press, New York, 1973.
107. Berliner, L. J., SPIN LABELING: THEORY AND APPLICATIONS, Academic Press, New York, 1976.
108. Berne, B. J., and Pecora, R., DYNAMIC LIGHT SCATTERING WITH APPLICATIONS TO CHEMISTRY, BIOLOGY AND PHYSICS, Wiley–Interscience, New York, 1976.
109. Berry, G. C., and Cassasa, E. F., Macromol. Rev., **4**, 1 (1970).
110. Berry, G. C., and Fox., T. G., Adv. Polym. Sci., **5**, 261 (1968).
111. Bersohn, M., and Baird, J. C., AN INTRODUCTION TO ELECTRON PARAMAGNETIC RESONANCE, Benjamin, New York, 1966.
112. Berthier, G., and Serre J., in The Chemistry of the Carbonyl Group (S. Patai, ed.), Wiley–Interscience, London, 1966.
113. Bertolaccini, M., Bussolati, C., and Zappa, L., Phys. Rev., **139**, A696 (1965).
114. Bettelheim, F. C., EXPERIMENTAL PHYSICAL CHEMISTRY, Saunders, Philadelphia, 1971.
115. Beychok, S., Science, **154**, 1288 (1966).
116. Beyer, G. L., in Encyclopedia of Industrial Chemical Analyses, Vol. 2 (F. D. Snell, and C. L. Hilton, eds), Wiley–Interscience, New York, 1966, p. 611.
117. Beynon, J. H., MASS SPECTROMETRY AND ITS APPLICATION TO ORGANIC CHEMISTRY, Elsevier, Amsterdam, 1960.
118. Beynon, J. H., Saunders, R. A., and Williams, A. E., THE MASS SPECTRA OF ORGANIC MOLECULES, American Elsevier, New York, 1968.
119. Beynon, J. H., and Williams, A. E., MASS AND ABUNDANCE TABLES FOR USE IN MASS SPECTROMETRY, 2nd ed., Elsevier, Amsterdam, 1970.
120. Bhacca, N. S., and Williams, D. H., APPLICATION OF NMR SPECTROSCOPY IN ORGANIC CHEMISTRY, Holden-Day, San Francisco, 1964.
121. Bible, R. H., INTRODUCTION TO NMR SPECTROSCOPY, Plenum Press, New York, 1965.
122. Bible, R. H. GUIDE TO THE NMR EMPIRICAL METHOD, Plenum Press, New York, 1967.
123. Biemann, K., MASS SPECTROMETRY, APPLICATION TO ORGANIC CHEMISTRY, McGraw-Hill, New York, 1962.
124. Bikales, N. M., CHARACTERIZATION OF POLYMERS, Wiley–Interscience, New York, 1971.
125. Billingham, N. C., MOLAR MASS MEASUREMENTS IN POLYMER SCIENCE, Halsted Press, New York and Kogan Page, London, 1977.
126. Billmeyer, F. W., Jr, in Treatise on Analytical Chemistry, Vol. 5, Part 1 (I. M. Kolthoff and P. J. Eving, eds), Wiley–Interscience, New York, 1964, p. 2839.

699

127. Billmeyer, F. W., Jr, TEXTBOOK OF POLYMER SCIENCE, 2nd ed., Wiley–Interscience, New York, 1971.
128. Billmeyer, F. W., in Characterization of Macromolecular Structure, Nat. Acad. Sci. U.S.A., Publ. 1573, Washington, D.C., 1968, p. 1.
129. Billmeyer, F. W., Jr, *J. Paint Technol.*, **42**, 1 (1969).
130. Billmeyer, F. W. Jr, and Altgert, K. H., in Gel Permeation Chromatography (K. H. Altgert and L. Segal, eds), Dekker, New York, 1971, p. 3.
131. Binsbergen, F. L., *Polym. Symp.*, No. 59, 11 (1977).
132. Birks, J. B., in Progress in Reaction Kinetics, Vol. 4 (G. Porter, ed.), Pergamon Press, New York, 1967, p. 239.
133. Birks, J. B., ORGANIC MOLECULAR PHOTOPHYSICS, Vols 1 and 2, Wiley–Interscience, London, 1973.
134. Birks, J. B., in Organic Molecular Photophysics, Vol. 2 (J. B. Birks, ed.), Wiley–Interscience, London, 1975, Chap. 9.
135. Birshtein, T. M., and Ptitsyn, O. B., CONFORMATION OF MACROMOLECULES, Wiley–Interscience, New York, 1966.
136. Bisi, A., Fasana, A., and Zappa, L., *Phys. Rev.*, **124**, 1487 (1961).
137. Blackadder, D. A., *J. Macromol. Sci. C*, **1**, 297 (1967).
138. Blagrove, R. J., *J. Macromol. Sci. C*, **9**, 71 (1973).
139. Blair, E. J., INTRODUCTION TO CHEMICAL INSTRUMENTATION, ELECTRONIC SIGNALS AND OPERATIONS, McGraw-Hill, New York, 1962.
140. Blair, E. J., in Analysis and Fractionation of Polymers (J. Mitchell Jr and F. W. Billmeyer Jr, eds), Wiley–Interscience, New York, 1965, p. 287.
141. Blanks, R. F., *Polym. Plast. Technol.*, **8**, 13 (1977).
142. Blazek A., THERMAL ANALYSIS, Van Nostrand-Reinhold, London, 1974.
143. Blern, K., and King, G. S., HANDBOOK OF DERIVATIVES FOR CHROMATOGRAPHY, Heyden, London, 1977.
144. Bloch, F., *Phys. Rev.*, **93**, 944 (1954).
145. Bloembergen, N., NUCLEAR MAGNETIC RELAXATION, Benjamin, New York, 1961.
146. Bloembergen, N., Purcell, E. M., and Pound, R. V., *Phys. Rev.*, **73**, 679 (1948).
147. Blumenfeld, L. A., Voyevodskii, V. V., and Siemionov, A. G., APPLICATION OF ELECTRON SPIN RESONANCE IN CHEMISTRY, Izd. Sybirskoyi Akad. Nauk SSSR, 1962 (in Russian).
148. Bly, D. D., *Science*, **168**, 527 (1970).
149. Boerio, F. J., and Koenig, J. L., in Polymer Characterization: An Interdisciplinary Approach (C. D. Craver, ed.), Plenum Press, New York, 1971, p. 1.
150. Boerio, F. J., and Koenig, J. L., *J. Macromol. Sci. C*, **7**, 209 (1972).
151. Boerio, F. J., and Koenig, J. L., *Polym. Symp.*, No. 43, 205 (1973).
152. Böhm, L. L., Chmelir, M., Löhr, G., Schmitt, B. J., and Schultz, G. C., *Fortsch. Hochpolym. Forsch.*, *Adv. Polym. Sci.*, **9**, 1 (1972).
153. Bohn, L., *Kolloid. Z. Z. Polym.*, **213**, 55 (1966).
154. Bohn, L., in Polymer Handbook, 2nd ed. (J. Brandrup, E. H. Immergut, and W. McDowell, eds), Wiley–Interscience, New York, 1975, p. III-211.
155. Bohn, L., in Polymer Handbook, 2nd ed. (J. Brandrup, E. H. Immergut, and W. McDowell, eds), Wiley–Interscience, New York, 1975, p. III-241.
156. Bollinger, J. C., *J. Macromol. Sci. C*, **16**, 23 (1978).
157. Bombaugh, J. J., and Levangie, R. F., in Gel Permeation Chromatography (K. H. Altgelt, and L. Segal, eds), Dekker, New York, 1971, p. 179.
158. Bondarovich, H. A., and Freeman, S. K., in Interpretive Spectroscopy (S. K. Freeman, ed.), Reinhold, New York, 1965.
159. Bonnar, R. V., Dimbat, M., and Stross, F. H., NUMBER AVERAGE MOLECULAR WEIGHT, Wiley–Interscience, New York, 1958.
160. Bonner, D. C., *J. Macromol. Sci. C*, **13**, 263 (1975).
161. Bonner, D. C., and Prausnitz, J. M., *AIChE J.*, **19**, 943 (1973).
162. Bonner, W. A., *J. Chem. Educ.*, **30**, 452 (1953).
163. Booker, G. R., in Modern Diffraction and Imaging Techniques in Materials Science (S. Amelnickx, R. Gevers, G. Remaut, and J. Van Landuyt, eds), Elsevier, New York, 1970, p. 553.

700

164. Booth, C., in Polymer Handbook, 2nd ed. (K. Brandrup, E. H. Immergut, and W. McDowell, eds), Wiley–Interscience, New York, 1975, p. IV-325.
165. Born, M., and Huang, K., DYNAMICAL THEORY OF CRYSTAL LATTICES, Clarendon Press, Oxford, 1954.
166. Born, M., and Wolf, E., PRINCIPLES OF OPTICS, 3rd ed., Pergamon Press, New York, 1965.
167. Böttcher, C. J. F., THEORY OF ELECTRIC POLARIZATION, 2nd ed., Elsevier, Amsterdam, 1973.
168. Boutin, H., and Yip, S., MOLECULAR SPECTROSCOPY WITH NEUTRONS, MIT Press, Cambridge, Mass., 1968.
169. Bovey, F. A., THE EFFECTS OF IONIZING RADIATION ON NATURAL AND SYNTHETIC HIGH POLYMERS, Wiley–Interscience, New York, 1958.
170. Bovey, F. A., Chem. Eng. News, 43 (35), 98 (1965).
171. Bovey, F. A., NMR DATA TABLES FOR ORGANIC COMPOUNDS, Wiley–Interscience, New York, 1967.
172. Bovey, F. A., Polym. Eng. Sci., 7, 128 (1967).
173. Bovey, F. A., Accounts Chem. Res., 1, 175 (1968).
174. Bovey, F. A., in Encyclopedia of Polymer Science and Technology, Vol. 9 (H. F. Mark, N. G. Gaylord, and N. M. Bikales, eds), Wiley–Interscience, New York, 1968, p. 356.
175. Bovey, F. A., POLYMER CONFORMATION AND CONFIGURATION, Academic Press, New York, 1969.
176. Bovey, F. A., NMR SPECTROSCOPY, Academic Press, New York, 1969.
177. Bovey, F. A., HIGH RESOLUTION NMR OF MACROMOLECULES, Academic Press, New York, 1972.
178. Bovey, F. A., and Tiers, G. V. D., Fortsch. Hochpolym. Forsch., Adv. Polym. Sci., 3, 139 (1963).
179. Bowen, E. J., LUMINSCENCE IN CHEMISTRY, Van Nostrand, London, 1968.
180. Bowen, E. J., and Wokes, F., FLUORESCENCE OF SOLUTIONS, Longmans, London, 1953.
181. Bowen, T. J., AN INTRODUCTION TO ULTRACENTRIFUGATION, Wiley–Interscience, London, 1970.
182. Bowler, D. I., in Structure and Properties of Oriented Polymers (I. M. Ward, ed.), Applied Science Publ., London, 1975, Chap. 4.
183. Boyer, R. F., Rubb. Chem. Technol., 36, 1303 (1963).
184. Boyer, R. F., J. Polym. Sci. C, 14, 3 (1966).
185. Boyer, R. F., Polymer, 17, 996 (1976).
186. Boyer, R. F., Macromolecules, 6, 288 (1973).
187. Bracewell, R., THE FOURIER TRANSFORM AND ITS APPLICATIONS, McGraw-Hill, New York, 1965.
188. Braddick, H. J. J., THE PHYSICS OF EXPERIMENTAL METHODS, Reinhold, New York, 1963.
189. Bradley, D. E., in Techniques for Electron Microscopy (D. H. Kay, ed.), Blackwell, Oxford, 1965, Chaps 3, 4 and 5.
190. Brame, E. G., APPLIED SPECTROSCOPY REVIEWS, Dekker, New York, 1969.
191. Brame, E. G., and Grasselli, J. G., INFRARED AND RAMAN SPECTROSCOPY, Vol. 1, Parts A, B, C, Dekker, New York, 1976/1977.
192. Brand, J. C. D., and Eglington, G., APPLICATIONS OF SPECTROSCOPY TO ORGANIC CHEMISTRY, Oldbourne Press, London, 1965.
193. Brandmüller, J., and Moser, H., EINFUHRUNG IN DIE RAMANSPEKTROSKOPIE, Steinkopff, Darmstadt, 1962.
194. Brandt, W., Berko, S., and Walker, W. W., Phys. Rev., 120, 1289 (1960).
195. Brandt, W., and Spirn, I., Phys. Rev., 142, 231 (1966).
196. Brauer, G. M., J. Polym. Sci. C, 8, 3 (1965).
197. Brauer, G. M., in Techniques and Methods of Polymer Evaluation, Vol. 2, Thermal Characterization Techniques (P. E. Slade and L. T. Jenkins, eds), Dekker, New York, 1970, Chap. 2.
198. Braun, J. M., and Guillet, J. E., Fortsch. Hochpolym. Forsch., Adv. Polym. Sci., 21, 107 (1976).
199. Braun, T., and Ghersini, G., EXTRACTION CHROMATOGRAPHY, Elsevier, Amsterdam, 1975.

200. Breitmeier, E., and Bauer., G., ¹³C-NMR SPEKTROSKOPIE, Georg Thieme Verlag, Stuttgart, 1977.
201. Breitmeier, E., and Voelter, W., ¹³C-NMR SPECTROSCOPY, Verlag Chemie, Weinheim, 1974.
202. Brenner, N., Callen, J. E., and Weiss, D. M., GAS CHROMATOGRAPHY, Academic Press, New York, 1962.
203. Bresler, S. E., and Kazbekov, E. N., Fortsch. Hochpolym. Forsch., Adv. Polym, Sci., 3, 688 (1964).
204. Bresler, S. E., and Kazbekov, E. N., in Encyclopedia of Polymer Science and Technology, Vol. 5 (H. F. Mark, N. G. Gaylord, and N. M. Bikales, eds), Wiley–Interscience, New York, p. 669.
205. Brewis, D. M., in Polymer Science, Vol. 2 (A. D. Jenkins, ed.), North-Holland, Amsterdam, 1972, Chap. 13.
206. Briggs, D., in Molecular Spectroscopy (A. R. West, ed.), Heyden, London, 1977, p. 467.
207. Brittain, E., George, W. O., and Wells, C. H. J., INTRODUCTION TO MOLECULAR SPECTROSCOPY THEORY AND EXPERIMENTS, Academic Press, New York, 1970.
208. Brockhouse, B. N., in Phonons and Neutron Scattering (T. A. Bak, ed.), Benjamin, New York, 1964, p. 221.
209. Brotherton, M., MASERS AND LASERS, McGraw-Hill, New York, 1964.
210. Broughton, G., in Techniques of Organic Chemistry, Vol. 8, Part 1 (A. Weissberger, ed.), Wiley–Interscience, New York, 1966, 1966, p. 831.
211. Brown, S. C., and Harvey, A. B., in Infrared and Raman Spectroscopy, Vol. 1, Part C (E. G. Brame, and J. G. Grasselli, eds), Dekker, New York, 1977, Chap. 12.
212. Browning, D., SPECTROSCOPY, McGraw-Hill, New York, 1969.
213. Brugel, W., NMR SPECTRA AND CHEMICAL STRUCTURE, Academic Press, New York, 1967.
214. Brugger, R. M., in Thermal Neutron Scattering (P. A. Egelstaff, ed.), Academic Press, New York, 1965, p. 53.
215. Brumberger, H., SMALL ANGLE X-RAY SCATTERING, Gordon and Breach, New York, 1967.
216. Brzezinski, J., Polimery-Tw. Wielkocz. 14, 421 (1969).
217. Brydson, J. A., FLOW PROPERTIES OF POLYMER MELTS, Van Nostrand-Reinhold, New York, 1971.
218. Brydson, J. A., in Polymer Science, Vol. 1 (A. D. Jenkins, ed.), North-Holland, Amsterdam, 1973, Chap. 3.
219. Buchachenko, A. L., Kovarskii, A. L., and Vasserman, A. M., in Advances in Polymer Science (Z. A. Rogovin, ed.), Wiley–Interscience, New York, 1974, p. 37.
220. Budzikiewicz, H., MASSENSPEKTROSKOPIE, Verlag Chemie, Weinheim, 1972.
221. Budzikiewicz, H., Djerassi, C., and Williams, D. H., MASS SPECTROMETRY OF ORGANIC COMPOUNDS, Holden-Day, San Francisco, 1967.
222. Buerger, M. J., PRECISION METHODS IN X-RAY CRYSTALLOGRAPHY, Wiley, New York, 1964.
223. Bunn, C. W., CHEMICAL CRYSTALLOGRAPHY, 2nd ed., Oxford University Press, London, 1961.
224. Burchard, W., and Cantow, H. J., in Polymer Fractionation (M. J. R. Cantow, ed.), Academic Press, New York, 1967, p. 285.
225. Burke, J. J., and Weiss, V., BLOCK AND GRAFT COPOLYMERS, Syracuse University Press, Syracuse, N.Y., 1973.
226. Burlant, W. J. and Hoffman, A. S., BLOCK AND GRAFT COPOLYMERS, Van Nostrand-Reinhold, Princetown, N.J., 1960
227. Burlingame, A. L., TOPICS IN ORGANIC MASS SPECTROMETRY, Wiley–Interscience, New York, 1970.
228. Burrell, H., Interchem. Rev., 14, 31 (1955).
229. Burrell, H., in Encyclopedia of Polymer Science and Technology, Vol. 12 (H. F. Mark. N. G. Gaylord, and N. M. Bikales, eds), Wiley–Interscience, New York, 1970, p. 618,
230. Burrell, H., in Polymer Handbook, 2nd ed. (K. Brandrup, E. H. Immergut, and W. McDowell, eds), Wiley–Interscience, New York, 1975, p. IV-337.

702

231. Bursey, M. M., and Lehman, T. A., ION CYCLOTRON RESONANCE SPECTRO-SCOPY, Wiley–Interscience, New York, 1978.
232. Bush, C. A., in Physical Techniques in Biochemical Research (G. Oster, ed.), Academic Press, New York, 1971, p. 347.
233. Butyagin, P.Yu., *Pure Appl. Chem.*, **30**, 57 (1972).
234. Butyagin, P.Yu., Dubinskaya, A. M., and Radstig, V. A., *Uspekh. Khim.*, **37**, 539 (1969).
235. Byrne, S. H., Jr, in Modern Practice of Liquid Chromatography (J. J. Kirkland, ed.), Wiley–Interscience, New York, 1971, Chap. 3.
236. Cabannes, J., LA DIFFUSION MOLECULAIRE DE LA LUMIERE, Université de France, Paris, 1929.
237. Cahn, R. W., *Adv. Phys.*, **3**, 363 (1954).
238. Cahn, R. S., Ingold, C., and Prelog, V., *Angew. Chem. Int. Ed.*, **5**, 385 (1966).
239. Cain, D. S., and Stimler, S. S., INFRARED SPECTRA OF PLASTICS AND RESINS, NRL Report 6503, 28 Feb. 1967.
240. Cairns, T., SPECTROSCOPY IN EDUCATION, Vols 1–4, Heyden, London, 1964–1967.
241. Caldwell, D. J., and Eyring, H., THE THEORY OF OPTICAL ACTIVITY, Wiley–Interscience, New York, 1971.
242. Calvert, J. G., and Pitts, J. N., Jr, PHOTOCHEMISTRY, Wiley–Interscience, New York, 1966.
243. Calvet, E., and Prat, H., RECENT PROGRESS IN MICROCALORIMETRY, Pergamon Press, New York, 1963.
244. Cameron, B. F., in Gel Peremeation Chromatography (K. H. Altgert and L. Segal, eds), Dekker, New York, 1971, p. 351.
245. Campbell, D., *Makromol. Rev.*, **4**, 91 (1970).
246. Cannon, C. G., in Physical Methods of Investigating Textiles (S. Meredith and J. W. S. Hearle, eds), Wiley–Interscience, New York, 1959.
247. Cantow, M. J. R., POLYMER FRACTIONATION, Academic Press, New York, 1967.
248. Cantow, M. J. R., in Polymer Fractionation (M. J. R. Cantow, ed.), Academic Press, New York, 1967, Chap. 6.
249. Cantow, M. J. R., and Johnson, J. F., in Encyclopedia of Polymer Science and Technology, Vol. 9 (H. F. Mark, N. G. Gaylord, and N. M. Bikales, eds), Wiley–Interscience, New York, 1968, p. 182.
250. Cantow, M. J. R., Porter, R. S., and Johnson, J. F., *J. Macromol. Sci. C*, **1**, 393 (1966).
251. Carlsson, D. J., and Wiles, D. M., *Nuova Chim.*, **47**, 36 (1971).
252. Carlsson, D. J., and Wiles, D. M., *Rubber Chem. Technol.*, **49**, 991 (1974).
253. Carlsson, D. J., and Wiles, D. M., *J. Macromol. Sci. C*, **14**, 65 (1976).
254. Carpenter, D. K., in Encyclopedia of Polymer Science and Technology, Vol. 12 (H. F. Mark, N. G. Gaylord, and N. M. Bikales, eds), Wiley–Interscience, New York, 1970, p. 627.
255. Carpenter, D. K., and Westerman, L., in Techniques and Methods of Polymer Evaluation, Vol. 4, Polymer Molecular Weights, Part I (P. E. Slade, Jr, ed.), Dekker, New York, 1975, Chap. 7.
256. Carrington, A., and McLachlen, A. D., INTRODUCTION TO MAGNETIC RESONANCE, Harper and Row, New York, 1966.
257. Carrington, R. A. G., COMPUTERS FOR SPECTROSCOPISTS, Hilger, London, 1974.
258. Carter, G. B., and Schenck, V. T. J., in Structure and Properties of Oriented Polymers (I. M. Ward, ed.), Applied Science Publ., London, 1975, Chap. 13.
259. Casale, A., Porter, R. G., and Johnson, J. F., *J. Macromol. Sci. C*, **5**, 387 (1971).
260. Casassa, E. F., in Gel Permeation Chromatography (K. H. Altgelt and L. Segal, eds), Dekker, New York, 1971, p. 119.
261. Casassa, E. F., in Techniques and Methods of Polymer Evaluation, Vol. 4, Polymer Molecular Weights, Part I (P. E. Slade, Jr, ed.), Dekker, New York, 1975, p. 161.
262. Casassa, E. F., in Fractionation of Synthetic Polymers; Principles and Applications (L. H. Tung, ed.), Dekker, New York, 1977, Chap. 1.
263. Casassa, E. F., and Berry, G. C., in Techniques and Methods of Polymer Evaluation, Vol. 4, Polymer Molecular Weights, Part I (P. E. Slade, Jr, ed.), Dekker, New York, Chap. 5.

264. Cassasa, E. F., and Eisenberg, H., *Adv. Protein Chem.*, **19**, 287 (1964).
265. Cassidy, H. G., FUNDAMENTALS OF CHROMATOGRAPHY, Wiley–Interscience, New York, 1957.
266. Causa, A. G., in Characterization of Metal and Polymer Surfaces, Vol. 2 (L. H. Lee, ed.), Academic Press, New York, 1977, p. 267.
267. Cazes, J., *J. Chem. Educ.*, **43**, A567, A625 (1966).
268. Cazes, J., *J. Chem. Educ.*, **47**, A461, A505 (1970).
269. Cazes, J., LIQUID CHROMATOGRAPHY OF POLYMERS AND RELATED MATERIALS, Dekker, New York, 1977.
270. Ceresa, R. J., BLOCK AND GRAFT COPOLYMERS, Butterworth, London, 1962.
271. Ceresa, R. J., in Encyclopedia of Polymer Science and Technology, Vol. 2 (H. F. Mark, N. G. Gaylord, and N. M. Bikales, eds), Wiley–Interscience, New York, 1964, p. 485.
272. Ceresa, R. J., BLOCK AND GRAFT COPOLYMERS, Vols 1 and 2, Wiley, London, 1976.
273. Cerf, R., *Adv. Chem. Phys.*, **33**, 73 (1975).
274. Chang, R., BASIC PRINCIPLE OF SPECTROSCOPY, McGraw-Hill, New York, 1971.
275. Channen, E. W., *Pure Appl. Chem.*, **9**, 225 (1959).
276. Chapiro, A., RADIATION CHEMISTRY OF POLYMER SYSTEMS, Wiley–Interscience, New York, 1962.
277. Chapman, D., and Magnus, P. D., INTRODUCTION TO PRACTICAL HIGH RESOLUTION NMR SPECTROSCOPY, Academic Press, New York, 1966.
278. Charlesby, A., ATOMIC RADIATION AND POLYMERS, Pergamon Press, New York, 1960.
279. Charschah, S. S., LASERS IN INDUSTRY, Van Nostrand-Reinhold, New York, 1972.
280. Chen, H. Y., *Rubber Chem. Technol.*, **41**, 47 (1968).
281. Chervenka, C. H., A MANUAL OF METHODS FOR THE ANALYTICAL ULTRA-CENTRIFUGE, Spinco Division of Beckman Instruments, Inc., Palo Alto, Calif., 1969.
282. Chiang, R., in NEWER METHODS OF POLYMER CHARACTERIZATION (B. Ke, ed.), Wiley–Interscience, New York, 1964, Chap. 12.
283. Chiang, R., in Polymer Handbook, 2nd ed. (J. Brandrup, E. H. Immergut, and W. McDowell, eds), Wiley–Interscience, New York, 1975, p. IV-267.
284. Chiang, R., in Polymer Handbook, 2nd ed. (J. Brandrup, E. H. Immergut, and W. McDowell, eds), Wiley–Interscience, New York, 1975, p. IV-309.
285. Chompff, A. J., in Polymer Networks; Structural and Mechanical Properties (A. J. Chompff and S. Newman, eds), Plenum Press, New York, 1971, p. 145.
286. Chompff, A. J., and Newman, S., POLYMER NETWORKS; STRUCTURAL AND MECHANICAL PROPERTIES, Plenum Press, New York, 1971.
287. Chu, B., *Ann. Rev. Phys. Chem.*, **21**, 145 (1970).
288. Chu, B., LASER LIGHT SCATTERING, Academic Press, New York, 1974.
289. Ciardelli, F., and Salvadori, P., FUNDAMENTAL ASPECTS AND RECENT DEVELOPMENTS IN OPTICAL ROTATORY DISPERSION AND CIRCULAR DICHROISM, Heyden, London, 1973.
290. Clar, E., POLYCYCLIC HYDROCARBONS, Vols 1 and 2, Springer–Academic Press, London, 1965.
291. Clark, D. T., in Electron Emission Spectroscopy (W. Dekeyser, ed.), Reidel Publ., Holland, 1973, p. 373.
292. Clark, D. T., in Structural Studies of Macromolecules by Spectroscopic Methods (K. Ivin, ed.) Wiley, London, 1976, Chap. 9.
293. Clark, D. T., in Molecular Spectroscopy (A. R. West, ed.), Heyden, London, 1977, p. 339.
294. Clark, D. T., *Fortsch. Hochpolym. Forsch., Adv. Polym. Sci.*, **24**, 126 (1977).
295. Clark, D. T., in Characterization of Metal and Polymer Surfaces, Vol. 2 (L. H. Lee, ed.), Academic Press, New York, 1977, p. 5.
296. Clark, D. T., and Dilks, A., in Characterization of Metal and Polymer Surfaces, Vol. 2 (L. H. Lee, ed.), Academic Press, New York, 1977, p. 101.
297. Clark, D. T., and Feast, W. J., *J. Macromol. Sci. C*, **12**, 191 (1975).
298. Clark, G. L. ENCYCLOPEDIA OF X-RAYS AND GAMMA RAYS, Reinhold, New York, 1963.

299. Clerc, J. T., Pretsch, E., and Sternhell S., ¹³C-KERNRESONANTSPEKTROSKOPIE, Akad. Verlagsgesellschaft, Frankfurt, 1973.
300. Cocking, S. J., and Webb, F. J., in Thermal Neutron Scattering (P. A. Egelstaff, ed.), Academic Press, New York, 1965, p. 141.
301. Coleman, B. D., Markovitz, H., and Noll, W., VISCOSIMETRIC FLOW OF NON-NEWTONIAN FLUIDS, Springer, New York, 1966.
302. Coleman, M. M., and Painter, P. C., J. Macromol. Sci. C, 16, 197 (1977–1978).
303. Coleman, H. J., and Weill, G., Polymer, 18, 1235 (1977).
304. Coil, H., in Gel Permeation Chromatography (K. H. Altgelt and L. Segal, eds), Dekker, New York, 1971, p. 135.
305. Coll, H., in Gel Permeation Chromatography (K. H. Altgelt and L. Segal, eds), Dekker, New York, 1971, p. 329.
306. Coll, H., Macromol. Rev., D. 5, 541 (1971).
307. Coll, H., and Stross, F. H., in Characterization of Molecular Structure, Nat. Acad. Sci. U.S.A., Publ. 1537, Washington, D.C., 1968, p. 10.
308. Collier, J. R., Ind. Eng. Chem., 61, 50 (1969).
309. Collins, E. A., Bares, J., and Billmeyer, F. W., Jr, EXPERIMENTS IN POLYMER SCIENCE, Wiley–Interscience, New York, 1973.
310. Collins, E. A., Davidson, J. A., and Jenkins, C. E., in Computer Programs for Plastic Engineers (I. Klein and D. Marchall, eds), Reinhold, New York, 1968, p. 344.
311. Collins, E. A., Haehn, J. B., and Wagner, J., in Computer Programs for Plastic Engineers (I. Klein and D. I. Marchall, eds), Reinhold, New York, 1968, p. 376.
312. Collins, E. A., Haehn, J. B., and Wolff, I., in Computer Programs for Plastic Engineers (I. Klein and D. I. Marchall, eds), Reinhold, New York, 1968, p. 349.
313. Colthrup, N. B., Daly, L. H., and Wiberley, S. E., INTRODUCTION TO INFRARED AND RAMAN SPECTROSCOPY, Academic Press, New York, 1964.
314. Conix, A., and Jeurissen, L., Adv. Chem., 48, 172 (1965).
315. Conley, R. T., THERMAL STABILITY OF POLYMERS, Dekker, New York, 1970.
316. Conley, R. T., INFRARED SPECTROSCOPY, 2nd ed., Allyn and Bacon, Boston, 1972.
317. Conlon, R. D., Anal. Chem., 41, 107A (1969).
318. Connor, T. M., Brit. Polym. J., 1, 116 (1969).
319. Conrad, C. M., Ind. Eng. Chem., 45, 2511 (1953).
320. Conrad, J., in Encyclopedia of Polymer Science and Technology. Vol. 8 (H. F. Mark, N. G. Gaylord, and N. M. Bikales, eds), Wiley–Interscience, New York, 1968, p. 231.
321. Coogan, C. K., Ham, N. S., Stuart, S. N., Pilbrow, J. R., and Wilson, G., MAGNETIC RESONANCE, Plenum Press, New York, 1970.
322. Cooper, A. R., Cain, J. H., Barrall, E. M., II., and Johnson, J. F., in Gel Permation Chromatography (K. H. Altgert and L. Segal, eds), Dekker, New York, 1971, p. 165.
323. Cooper, D. R., in Polymer Handbook, 2nd ed. (K. Brandrup, E. H. Immergut, and W. McDowell, eds), Wiley–Interscience, New York, 1975, p. IV-335.
324. Corio, P. L., STRUCTURE OF HIGH RESOLUTION NMR SPECTRA, Academic Press, New York, 1967.
325. Cornet, C. F., Polymer, 9, 7 (1968).
326. Cornu, A., Massot, R., COMPILATION OF MASS SPECTRAL DATA, Heyden, London, 1966.
327. Corradini, P. in The Stereochemistry of Macromolecules (A. D. Ketley, ed), Dekker, New York, 1968, Chap. 1.
328. Cowan, D. O., and Drisko, R. L., ELEMENTS OF ORGANIC PHOTOCHEMISTRY, Plenum Press, New York, 1976.
329. Crabbé, P., OPTICAL ROTATORY DISPERSION AND CIRCULAR DICHROISM IN ORGANIC CHEMISTRY, Holden-Day, San Francisco, 1965.
330. Crabbé, P., OPTICAL ROTATORY DISPERSION AND CIRCULAR DICHROISM IN CHEMISTRY AND BIOCHEMISTRY: AN INTRODUCTION, Academic Press, New York, 1972.
331. Cragg, L. H., and Hammerschlag, H., Chem. Rev., 39, 79 (1946).
332. Crank J., THE MATHEMATICS OF DIFFUSION, 2nd ed., Clarendon Press, Oxford, 1975.
333. Crank, J., and Parks, G. S., DIFFUSION IN POLYMERS, Academic Press, London, 1968.

334. Craver, C. D., in Infrared and Raman Spectroscopy, Vol. 1, Part C (E. G. Brame and J. G. Grasselli, eds), Dekker, New York, 1977, Chap. 13.
335. Creeth, J. M., and Pain, R. H., *Progr. Biophys. Mol. Biol.*, **17**, 217 (1967).
336. Crescenzi, V., in The Stereoregularity of Macromolecules (A. D. Ketley, ed.), Dekker, New York, 1968, Chap. 6.
337. Cresswell, C. J., Runquist, O., and Campbell, M. M., SPECTRAL ANALYSIS OF ORGANIC COMPOUNDS, 2nd ed., Burgess, Minneapolis, 1972.
338. Crippen, R. C., IDENTIFICATION OF ORGANIC COMPOUNDS WITH THE AID OF GAS CHROMATOGRAPHY, Elsevier, Amsterdam, 1973.
339. Cross, A. D., INTRODUCTION TO PRACTICAL INFRA-RED SPECTROSCOPY, 3rd ed., Butterworths, London, 1969.
340. Cross, L. H., Haw, J., and Shields, D. J., in Molecular Spectroscopy (P. Hepple, ed.), Institute of Petroleum, London, 1968.
341. Cudby, M. E. A., and Willis, H. A., *Ann. Rep. NMR Spectr.*, **4**, 363 (1971).
342. Cullity, B. D., ELEMENTS OF X-RAY DIFFRACTION, Addison-Wesley, Reading, Mass., 1956.
343. Cummins, H. Z., in Photon Correlation and Light Beating Spectroscopy (H. Z. Cummins and E. R. Pike, eds), Plenum Press, New York, 1973.
344. Cummins, H. Z., and Pike, E. R., PHOTON CORRELATION AND LIGHT BEATING SPECTROSCOPY, Plenum Press, New York, 1973.
345. Cummins, H. Z., and Swinney, H. L., *Progr. Opt.*, **8**, 133 (1970).
346. Curling, J., *Amer. Lab.*, **8**(5), 47 (1976).
347. Curling, J., *Intern. Lab.*, May/June 1976, p. 41.
348. Curling, J., *Intern. Lab.*, July/Aug. 1976, p. 37.
349. Dal Nogare, S., and Juvet, R. S., GAS–LIQUID CHROMATOGRAPHY, Wiley–Interscience, New York, 1963.
350. Daniel, V. V., DIELECTRIC RELAXATION, Academic Press, New York, 1967.
351. Daniels, T., THERMAL ANALYSIS, Kogan Page, London, 1977.
352. Danusso, F., *Polymer*, **8**, 281 (1967).
353. Darby, R., VISCOELASTIC FLUIDS: AN INTRODUCTION TO THEIR PROPERTIES AND BEHAVIOUR, Dekker, New York, 1976.
354. Dardin, V. J., Jr, *J. Chem. Educ.*, **43**, 439 (1966).
355. David, C., in Chemical Kinetics, Vol. 14, Degradation of Polymers (C. H. Bamford and C. F. H. Tipper, eds), Elsevier, Amsterdam, 1975, p. 1 and p. 175.
356. David, D. J., GAS CHROMATOGRAPHY DETECTORS, Wiley–Interscience, New York, 1974.
357. David, D. J., and Statley, H. B., ANALYTICAL CHEMISTRY OF POLYURETHANES, Wiley–Interscience, New York, 1959.
358. David, J. J., in Techniques and Methods of Polymer Evaluation, Vol. 1 (P. E. Slade, Jr and L. T. Jenkins, eds), Dekker, New York, 1966, Chap. 2.
359. Davies, M., INFRARED SPECTROSCOPY AND MOLECULAR STRUCTURE, Elsevier, Amsterdam, 1963.
360. Davison, W. H. T., and Bates, G. R., *Rubber Chem. Technol.*, **30**, 771 (1957).
361. Dawson, P. H., QUADRUPOLE MASS SPECTROMETRY AND ITS APPLICATIONS, Elsevier, Amsterdam, 1976.
362. Dawydoff, W. N., BESTIMMUNG DES MOLECULARGEWICHTS VON POLYAMIDEN, VEB Verlag Technik, Berlin, 1954.
363. de Boer, J. H., THE DYNAMICAL CHARACTER OF ADSORPTION, Clarendon Press, Oxford, 1953.
364. Dechant, J., ULTRAROTSPEKTROSKOPISCHE UNTERSUCHUNGEN AN POLYMERER, Akademie-Verlag, Berlin, 1972.
365. Demtröder, W., GRUNDLAGEN UND TECHNIKEN DER LASERSPEKTROSKOPIE, Springer, Berlin, 1977.
366. Denman, H. H., Heller, W., and Pagonis, W. J., ANGULAR SCATTERING FUNCTIONS FOR SPHERES, Wayne State University Press, Detroit, 1966.
367. Derham, K. W., Goldsbrough, J., and Gordon, M., *Pure Appl. Chem.*, **38**, 97 (1974).
368. Desper, C. R., *Crit. Rev. Macromol. Sci.*, **1**, 501 (1972).
369. Desper, C. R., in Characterization of Materials in Research Ceramics and Polymers

706

(J. J. Burke and V. Weiss, eds), Syracuse University Press, Syracuse, N.Y., 1975, Chap. 16.
370. Desroux, V., and Oth A., *Chem. Weekbl.*, **48**, 247 (1952).
371. Determan, H., GEL CHROMATOGRAPHY: GEL FILTRATION, GEL PERMEATION, MOLECULAR SIEVES, 2nd ed., Springer, New York, 1968.
372. Deyl, Z., Macek, K., and Janák, J., LIQUID COLUMN CHROMATOGRAPHY: A SURVEY OF MODERN TECHNIQUES AND APPLICATIONS, Elsevier, Amsterdam, 1975.
373. Diehl, P., Fluck, E., and Kosfeld, R., NMR: BASIC PRINCIPLES AND PROGRESS, Vol. 4, NATURAL AND SYNTHETIC HIGH POLYMERS, Springer, Berlin, 1971.
374. Djerassi, C., OPTICAL ROTATORY DISPERSION: APPLICATION TO ORGANIC CHEMISTRY, McGraw-Hill, New York, 1960.
375. Dole, M., *Fortsch. Hochpolym. Forsch.*, *Adv. Polym. Sci.*, **2**, 221 (1960).
376. Dole, M., THE RADIATION CHEMISTRY OF MACROMOLECULES, Academic Press, New York, 1972.
377. Dollish, F. R., Fateley, W. H., and Bentley, F. F., CHARACTERISTIC RAMAN FREQUENCIES, Wiley–Interscience, New York, 1974.
378. Dolphin, D., and Wick, A. E., TABULATION OF INFRARED SPECTRAL DATA, Wiley–Interscience, New York, 1978.
379. Domsky, I. I., and Perry, J. A., RECENT ADVANCES IN GAS CHROMATOGRAPHY, Dekker, New York, 1971.
380. Done, J. N., Knox, J. H., and Loheac, J., APPLICATIONS OF HIGH SPEED CHROMATOGRAPHY, Wiley–Interscience, New York, 1975.
381. Doppke, H. J., and Heller, W., in Characterization of Macromolecular Structure, Nat. Acad. Sci. U.S.A., Publ. 1573, Washington, D.C., 1968, p. 68.
382. Doty, P., and Edsall, J. T., *Adv. Protein Chem.*, **1**, 90 (1945).
383. Doyle, C. D., in Techniques and Methods of Polymer Evaluation, Vol. 1 (P. E. Slade, Jr, ed.), Dekker, New York, 1966, Chap. 4.
384. Duerksen, J. H., in Gel Permeation Chromatography (K. H. Altgelt and L. Segal, eds), Dekker, New York, 1971, p. 81.
385. Dulog, L., *Fortschr. Hochpolym. Forsch.*, *Adv. Polym. Sci.*, **6**, 427 (1966).
386. Durrans, T. J., SOLVENTS, 8th ed., Chapman and Hall, London, 1971.
387. Dushman, S., SCIENTIFIC FOUNDATION OF VACUUM TECHNIQUE, Wiley–Interscience, New York, 1966.
388. Dwight, D. W., in Characterization of Metal and Polymer Surfaces, Vol. 2 (L. H. Lee, ed.), Academic Press, New York, 1977, p. 313.
389. Dyer, J. R., APPLICATIONS OF ABSORPTION SPECTROSCOPY OF ORGANIC COMPOUNDS, Prentice-Hall, Englewood Cliffs, N.J., 1965.
390. Dyers, J. R. ORGANIC SPECTRAL PROBLEMS, Prentice-Hall, Englewood Cliffs, N.J., 1972.
391. Earnshaw, A., INTRODUCTION TO MAGNETOCHEMISTRY, Academic Press, London, 1968.
392. Egelstaff, P. A., THERMAL NEUTRON SCATTERING, Academic Press, New York, 1965.
393. Ehlers, J., Hepp, K., Weidenmüller, H. A., and Zittartz, J., LASER SPECTROSCOPY, Springer, Berlin, 1975.
394. Eiermann, K., *J. Polym. Sci. C*, **6**, 157 (1963).
395. Eirich, F. R., RHEOLOGY—THEORY AND APPLICATIONS, Academic Press, New York, 1969.
396. Eisenberh, A., *Macromolecules*, **4**, 125 (1971).
397. Eisenberg, A., and Shen, M., *Rubber Chem. Technol.*, **43**, 156 (1970).
398. Elias, H. G., *Chem. Ing. Techn.*, **33**, 359 (1961).
399. Elias, H. G., ULTRACENTRIFUGEN-METHODEN, Beckman Instruments, München, 1961.
400. Elias, H. G., in Characterization of Macromolecular Structure, Nat. Acad. Sci. U.S.A., Publ. 1537, Washington, D.C., 1968, p. 28.
401. Elias, H. G., in Polymer Handbook, 2nd ed. (J. Brandrup, E. H. Immergut, and W. McDowell, eds), Wiley–Interscience, New York, 1975, p. VII-23.

402. Elias, H. G., MACROMOLECULES, Vols 1 and 2, Plenum Press, New York, 1977.
403. Elias, H. G., in Fractionation of Synthetic Polymers: Principles and Practice (L. H. Tung, ed.), Dekker, New York, 1977, Chap. 4.
404. Elias, H. G. and Bührer, H. G., in Polymer Handbook, 2nd ed. (K. Brandrup, E. H. Immergut, and W. McDowell, eds), Wiley–Interscience, New York, 1975, p. IV-57.
405. Elliott, A., INFRARED SPECTRA AND STRUCTURE OF ORGANIC LONG-CHAIN POLYMERS, Arnold, London, 1969, Chap. 3.
406. Elliott, J. H., in Polymer Fractionation (M. J. R. Cantow, ed.), Academic Press, New York, 1967.
407. Elliott, R. M., Craig, R. D., and Errock, G. A., in INSTRUMENTS AND MEASUREMENTS (H. Van Koch and G. Ljungberg, eds), Academic Press, New York, 1961, p. 271.
408. Emery, A. H., Jr, in Polymer Fractionation (M. J. R. Cantow, ed.), Academic Press, 1967, p. 181.
409. Emsley, J. W., Feeney, J., and Sutcliffe, L. H., HIGH RESOLUTION NMR SPECTROSCOPY, Macmillan–Pergamon, New York, 1966.
410. Ende, H. A., and Klärner, P. E. O., in Polymer Handbook, 2nd ed. (K. Brandrup, E. H. Immergut, and W. McDowell, eds), Wiley–Interscience, New York, p. IV-61.
411. Enhelhardt, H., HOCHDRUCK-FLUSSIKGKEITS-CHROMATOGRAPHIE, Springer, Berlin, 1977.
412. Enyedy, G., Jr and Lasch, R., in Computer Programs for Plastic Engineers, Reinhold, New York, 1968, p. 321.
413. Erbeia, A., RESONANCE MAGNETIQUE, Masson, Paris, 1969.
414. Ernst, R. R., and Anderson, W. A., *Rev. Sci. Instrum.*, **37**, 93 (1966).
415. Ernst, R. R., and Anderson, W. A., *Rev. Sci. Instrum.*, **37**, 1323 (1966).
416. Estes, G. M., Cooper, S. L., and Tobolsky, A. V., *J. Macromol. Sci. C*, **4**, 313 (1970).
417. Ettre, L. S., in Gas Chromatography (N. Brenner, J. E. Callen, and M. D. Weiss, eds), Academic Press, New York, 1962, p. 307 and p. 541.
418. Ettre, L. S., OPEN TUBULAR COLUMNS IN GAS CHROMATOGRAPHY, Plenum Press, New York, 1965.
419. Ettre, L. S., PRACTICAL GAS CHROMATOGRAPHY, Perkin-Elmer, 1973.
420. Ettre, L. S., and McFadden, W. H., ANCILLARY TECHNIQUES OF GAS CHROMATOGRAPHY, Wiley–Interscience, New York, 1969.
421. Evans, J. M., in Light Scattering from Polymer Solutions (M. B. Huglin, ed.), Academic Press, New York, 1972, p. 89.
422. Evans, J. M., *Polym. Eng. Sci.*, **13**, 401 (1973).
423. Ezrin, M., in Characterization of Macromolecular Structure, Nat. Acad. Sci. U.S.A., Publ. 1573, Washinton, D.C., 1968, p. 1.
424. Ezrin, M., POLYMER MOLECULAR WEIGHT METHODS, Advances in Chemistry Series, No. 125, American Chemical Society, Washington, D.C., 1973.
425. Fabelinskii, I. L., MOLECULAR SCATTERING BY LIGHT, Plenum Press, New York, 1968.
426. Fairbain, J. W., in Quantitative Paper and Thin Layer Chromatography (E. J. Shellard, ed.), Academic Press, New York, 1968, Chap. 1.
427. Farina, M., and Bressan, G., in the Stereochemistry of Macromolecules (A. D. Ketley, ed.), Dekker, New York, 1968, Chap. 4.
428. Farina, M., Peraldo, M., and Natta, G., *Angew. Chem.*, **77**, 149 (1965).
429. Farrar, T. C., *Anal. Chem.*, **42**, 109A (1970).
430. Farrar, T. C., and Becker, E. D., PULSE AND FOURIER TRANSFORM NMR, Academic Press, New York, 1971.
431. Fateley, W. G., Dollish, F. R., McDevitt, N. T., and Bentley, F. M., INFRARED AND RAMAN SELECTION RULES FOR MOLECULAR AND LATTICE VIBRATIONS: THE CORRELATION METHODS, Wiley–Interscience, New York, 1972.
432. Faucher, J. A., and Reading, F. P., in Crystalline Olefin Polymers (R. A. V. Raff and K. W. Doak, eds), Wiley–Interscience, New York, 1965, p. 681.
433. Fava, R. A., *Brit. Polym. J.*, **1**, 59 (1969).
434. Fava, R. A., *Macromol. Rev.*, *D* **5**, 1 (1971).
435. Feofilov, P. P., PHYSICAL BASIS OF POLARIZED FLUORESCENCE, Consultants Bureau, New York, 1961.

708

436. Ferry, J. D., VISCOELASTIC PROPERTIES OF POLYMERS, 2nd ed., Wiley, Chichester, 1970.
437. Finch, A., Gates, P. N., Radcliffe, K., Dickson, F. N., and Bentley, F. F., CHEMICAL APPLICATIONS OF FAR INFRARED SPECTROSCOPY, Academic Press, New York, 1970.
438. Fischer, E. W., in Newer Methods of Polymer Characterization (B. Ke, ed.), Wiley–Interscience, New York, 1964, Chap. 7.
439. Fischer, E. W., and Goddar, H., in Encyclopedia of Polymer Science and Technology, Vol. 5 (H. F. Mark, N. G. Gaylord, and N. M. Bikales, eds), Wiley–Interscience, New York, 1966, p. 641.
440. Fischer, H., Kunststoffe, 55, 344 (1965).
441. Fischer, H., Chim. Ind. Gen. Chim., 97, 8 (1967).
442. Fischer, H., Proc. Roy. Soc. A 302, 321 (1968).
443. Fischer, H., Fortschr. Hochpolym. Forsch., Adv. Polym. Sci., 5, 463 (1968).
444. Fischer, H., Accounts Chem. Res., 4, 110 (1971).
445. Fischer, H., Nucl. Magn. Reson., 4, 301 (1971).
446. Fischer, H., in Free Radicals, Vol. 2 (J. K. Kochi, ed.), Wiley–Interscience, New York, 1973, p. 435.
447. Fischer, H., in Magnetic Properties of Free Radicals, Landolt-Börnstein, New Series, Group II, Vol. 1, Springer Verlag, Berlin, 1965.
448. Fischer, H., and Hummel, D. O., in Polymer Spectroscopy (D. O. Hummel, ed.), Verlag Chemie, Weinheim, 1974, p. 289.
449. Fischer, L., AN INTRODUCTION TO GEL CHROMATOGRAPHY, 4th ed., Elsevier, Amsterdam, 1974.
450. Fischer, W. G., Glass Instrum. Technol., 11, 562, 775, 1085 (1967).
451. Fitzgerald, J. M., ANALYTICAL PHOTOCHEMISTRY AND PHOTOCHEMICAL ANALYSIS, SOLIDS, SOLUTIONS AND POLYMERS, Dekker, New York, 1971.
452. Fleischer, D., in Polymer Handbook, 2nd ed. (J. Brandrup, E. H. Immergut, and W. McDowell, eds), Wiley–Interscience, New York, 1975, p. VII-25.
453. Flett, M. S. C., CHARACTERISTIC FREQUENCIES OF CHEMICAL GROUPS IN THE INFRARED, American Elsevier, New York, 1969.
454. Flodin, P., DEXTRAN GELS AND THEIR APPLICATIONS IN GEL FILTRATION, Pharmacia, Uppsala, Sweden, 1962.
455. Flory, P. J., Chem. Rev., 35, 57 (1944).
456. Flory, P. J., PRINCIPLES OF POLYMER CHEMISTRY, Cornell University Press, Ithaca, 1953.
457. Flory, P. J., STATISTICAL MECHANICS OF CHAIN MOLECULES, Wiley–Interscience, New York, 1969.
458. Flory, P. J., Ber. Bunsenges Phys. Chem., 81, 885 (1977).
459. Flynn, J. H., in Aspects of Degradation and Stabilization of Polymers (H. H. G. Jellinek, ed.), Elsevier, Amsterdam, 1978, Chap. 12.
460. Fock, J., SOME APPLICATIONS OF THERMAL ANALYSIS, Mettler Instrument Corp., Switz., 1968.
461. Folkes, M. J., and Ward, I. M., in Structure and Properties of Oriented Polymers (I. M. Ward, ed.), Applied Science Publishers, London, 1975, Chap. 6.
462. Folmer, O. F., Jr, in Polymer Characterization: An Interdisciplinary Approach (C. D. Craver, ed.), Plenum Press, New York, 1971, p. 231.
463. Förster, T., FLUORESCENZ ORGANISCHER VERBINDUNGEN, Vandenhöch-Ruprech, Göttingen, 1951.
464. Förster, T., Angew. Chem., 81, 364 (1969).
465. Förster, T., in The Exciplex (M. Gordon and W. R. Ware, eds), Academic Press, New York, 1975.
466. Forziati, A. F., in Analytical Chemistry of Polymers, Part II (G. M. Kline, ed.), Wiley–Interscience, New York, 1962, Chap. 3.
467. Foss, J. G., J. Chem. Educ., 40, 592 (1963).
468. Fowles, G., INTRODUCTION TO MODERN OPTICS, Holt, Rinehart and Winston, New York, 1968.
469. Fox, R. B., and Price, T. R., in Polymer Characterization: An Interdisciplinary Approach (C. D. Craver, ed.), Plenum Press, New York, 1971, p. 259.
470. Franklin, M. L., Horlick, G., and Malmstadt, H. V., Anal. Chem., 41, 2 (1969).

709

471. Fraser, R. D. B., and MacRae, T. P., CONFORMATION IN FIBROUS PROTEINS, Academic Press, New York, 1973.
472. Fredericq, E., and Houssier, C., ELECTRIC DICHROISM AND ELECTRIC BIRE-FRINGENCE, Oxford University Press, London, 1973.
473. Freed, J. H., *Ann. Rev. Phys. Chem.*, **23**, 265 (1972).
474. Freeman, A. J., and Frankel, R. B., HYPERFINE INTERACTIONS, Academic Press, New York, 1967.
475. Freeman, R., and Hill, H., in Molecular Spectroscopy, Institute of Petroleum, London, 1971, p. 105.
476. Freeman, S. K., INTERPRETIVE SPECTROSCOPY, Reinhold, New York, 1965.
477. Freeman, S. K., APPLICATION OF LASER RAMAN SPECTROSCOPY, Wiley, London, 1973.
478. Friedel, R. A., and Orchin, M., ULTRAVIOLET SPECTRA OF AROMATIC COMPOUNDS, Wiley, New York, 1958.
479. Freidman, H. L., in Thermal Analysis of High Polymers (B. Ke, ed.), Wiley–Interscience, New York, 1964, p. 183.
480. Frigerio, A., and Carrington, R., ESSENTIAL ASPECTS OF MASS SPECTRO-METRY, Wiley–Interscience, New York, 1974.
481. Friis, N., and Hamielec, A., *Adv. Chromatogr.*, **13**, 41 (1975).
482. Frisch, H. L., Frisch, K. C., and Klemper, D., *Modern Plast.*, **54**, 76, 84 (1977).
483. Fröhlich, H., THEORY OF DIELECTRICS: DIELECTRIC CONSTANT AND DIELECTRIC LOSS, 2nd ed., Oxford University Press, Oxford, 1958.
484. Frungel, F. B. A., HIGH SPEED PULSE TECHNOLOGY, Vols 1 and 2, Academic Perss, London, 1965.
485. Fry, F. H., in Analytical Photochemistry and Photochemical Analysis (J. M. Fitzgerald, ed.), Dekker, New York, 1971, Chap. 2.
486. Fuchs, O., and Leugering, H. J. in Kunststoffe, Vol. 1 (E. R. Nitsche, ed.), Springer, Berlin 1962, p. 118.
487. Fuchs, O., and Suhr, H. H., in Polymer Handbook, 2nd ed. (J. Brandrup, E. H. Immergut, and W. McDowell, eds), Wiley–Interscience, New York, 1975, p. IV-241.
488. Fujita, H., *Fortsch. Hochpolym. Forsch., Adv. Polym. Sci.*, **3**, 1 (1961).
489. Fujita, H., MATHEMATICAL THEORY OF SEDIMENTATION ANALYSIS, Academic Press, New York, 1962.
490. Fujita, H., in Diffusion of Polymers (J. Crank and G. S. Park, eds), Academic Press, London, 1968, Chap. 3.
491. Fujita, H., FOUNDATIONS OF ULTRACENTRIFUGAL ANALYSIS, Wiley–Interscience, New York, 1975.
492. Gall, M. J., Hendra, P. J., Watson, D. S., and Peacock, C. J., *Appl. Spectr.*, **25**, 423 (1971).
493. Gardiner, K. W., Klaver, R. F., Baumann, F., and Johnson, J. F., in Gas Chromatography (N. Brenner, J. E. Callen, and D. M. Weiss, eds), Academic Press, New York, 1962, Chap. 24.
494. Gardon, J. L., in Encyclopedia of Polymer Science and Technology, Vol. 3 (H. F. Mark, N. G. Gaylord, and N. M. Bikales, eds), Wiley–Interscience, New York, 1966, p. 833.
495. Garmon, R. G., in Techniques and Methods of Polymer Evaluation, Vol. 4, Polymer Molecular Weights, Part I (P. E. Slade, Jr, ed.), Dekker, New York, 1975, Chap. 3.
496. Garn, P. D., THERMOANALYTICAL METHODS OF INVESTIGATION, Academic Press, New York, 1965.
497. Garrett, C. G. B., GAS LASERS, McGraw-Hill, New York, 1967.
498. Gasparic, J., PAPER AND THIN LAYER CHROMATOGRAPHY, Wiley–Interscience, New York, 1978.
499. Gayles, J. N., and Peticolas, W. L., in Light Scattering Spectra of Solids (G. B. Wright, ed.), Springer, New York, 1969, p. 715.
500. Geiduschek, E. P., and Holtzer, A., *Adv. Biol. Med. Phys.*, **6**, 431 (1958).
501. Geil, P. H., POLYMER SINGLE CRYSTALS, Wiley–Interscience, New York, 1963.
502. Geil, P. H., in Encyclopedia of Polymer Science and Technology, Vol. 5 (H. F. Mark, N. G. Gaylord, and N. M. Bikales, eds), Wiley–Interscience, New York, 1966, p. 662.
503. Geil, P. H., Baer, E., and Wada, Y., THE SOLID STATE OF POLYMERS, Dekker, New York, 1974.

504. Gerrard, W., SOLUBILITY OF GASES AND LIQUIDS, Plenum Press, New York, 1976.
505. Gerson, F., HIGH RESOLUTION ELECTRON SPIN SPECTROSCOPY, Wiley–Interscience, New York, 1971.
506. Geschwind, S., ELECTRON PARAMAGNETIC RESONANCE, Plenum Press, New York, 1972.
507. Geuskens, G., in Chemical Kinetics, Vol. 14, Degradation of Polymers (C. H. Bamford and C. F. H. Tipper, eds), Elsevier, Amsterdam, 1975, p. 333.
508. Gibbs, J. H., MODERN ASPECTS OF THE VITREUS STATE, Butterworths, London, 1960.
509. Giddings, J. C., DYNAMICS OF CHROMATOGRAPHY, Part I. PRINCIPLES AND THEORY, Dekker, New York, 1965.
510. Giddings, J. C., J. Chromatogr., 125, 3 (1976).
511. Giddings, J. C., Fisher, S. R., and Myers, M. N., Intern. Lab., May/June 1978, p. 15.
512. Giesekus, H., in Polymer Fractionation (M. J. R. Cantow, ed.), Academic Press, New York, 1967, p. 191.
513. Gillam, A. E., and Stern, E. S., AN INTRODUCTION TO ELECTRONIC ABSORPTION SPECTROSCOPY IN ORGANIC CHEMISTRY, 2nd ed., Arnold, London, 1957.
514. Gilson, T. R., and Hendra, P. J., LASER RAMAN SPECTROSCOPY, Wiley–Interscience, London, 1970.
515. Glasgow, A. R., Jr, in Treatise on Analytical Chemistry, Part I, Theory and Practice (I. M. Kolthoff and P. J. Elving, eds) Wiley–Interscience, New York, 1968, p. 491.
516. Glauert, A. M., and Philips, R., in Techniques for Electron Microscopy (D. H. Kay, ed.), Blackwell, Oxford, 1965, Chap. 8.
517. Glover, C. A., in Advances in Analytical Chemistry and Instrumentation, Vol. 5 (C. N. Reilly and F. W. McLafferty, eds), Wiley–Interscience, New York, 1965, Chap. 1.
518. Glover, C. A., in Techniques and Methods of Polymer Evaluation, Vol. 4, Polymer Molecular Weights, Part I (P. E. Slade, Jr, ed.), Dekker, New York, 1975, p. 79.
519. Goddu, F., in Advances in Analytical Chemistry and Instrumentation, Vol. 1 (C. N. Reilly, ed.), Wiley–Interscience, New York, 1960, p. 347.
520. Goldanskii, V. I., Atomic Energy Rev., 6, 1 (1968).
521. Goldanskii, V. I., and Firsov, V. G., Ann. Rev. Phys. Chem., 22, 209 (1971).
522. Goldfein, S., BREAKDOWN OF PLASTICS, Dekker, New York, 1969.
523. Goldman, L., APPLICATIONS OF THE LASERS, CRS Press Inc., Florida, 1973.
524. Goldstein, M., and Simha, R., THE GLASS TRANSITION AND THE NATURE OF THE GLASSY STATE, New York Academy of Science, New York, 1976.
525. Goodman, G. W., in Quantitative Paper and Thin Layer Chromatography (E. J. Shellard, ed.), Academic Press, New York, 1968, Chap. 7.
526. Goodman, M., and Ueyama, N., in Polymer Handbook, 2nd ed. (K. Brandrup, E. H. Immergut and W. McDowell, eds), Wiley–Interscience, New York, 1975, p. IV-361.
527. Goodrich, F. C., in Polymer Fractionation (M. J. R. Cantow, ed.), Academic Press, New York, 1967, p. 415.
528. Gordon, M., and Ware, W. R., THE EXCIPLEX, Academic Press, New York, 1975.
529. Gosting, L. J., Adv. Protein, Chem., 11, 429 (1956).
530. Gould, R. F., CONTACT ANGLE, WETTABILITY AND ADHESION, Advances in Chemistry Series, No. 43, American Chemical Society, Washington, D.C., 1965.
531. Goynes, W. R., and Carra, J. H., in Characterization of Metal and Polymer Surfaces, Vol. 2 (L. H. Lee, ed.), Academic Press, New York, 1977, p. 251.
532. Graessley, W. W., in Characterization of Macromolecular Structure, Nat. Acad. Sci., U.S.A., Publ. 1573, Washinton, D.C., 1968, p. 371.
533. Graessley, W. W., in Characterization of Materials in Research Ceramics and Polymers (J. J. Burke and V. Weiss, eds), Syracuse University Press, Syracuse, N.Y., 1975, Chap. 15.
534. Graessley, W. W., Accounts Chem. Res., 10, 332 (1977).
535. Grassie, N., CHEMISTRY OF HIGH POLYMER DEGRADATION, Wiley–Interscience, New York, 1956.
536. Grassie, N., DEVELOPMENT IN POLYMER DEGRADATION, Applied Science Publishers, London, 1977.

537. Gray, A. P., in Analytical Calorimetry (R. S. Porter and F. J. Johnson, eds), Plenum Press, New York, 1968, p. 209.
538. Grechanovskii, V. A., *Rubber Chem. Technol.*, **45**, 519 (1972).
539. Gregg, S. J., and Sing, K. S. W., ADSORPTION SURFACE, AREA AND POROSITY, Academic Press, New York, 1967.
540. Greggs, A. R., Dowden, B. F., Barrall, E. M., II, and Horikawa, T. T., in Gel Permeation Chromatography (K. H. Altgelt and L. Segal, eds), Dekker, New York, 1971, p. 529.
541. Griffiths, C. H., in Characterization of Metal and Polymer Surfaces, Vol. 2 (L. H. Lee, ed.), Academic Press, New York, 1977, p. 333.
542. Griffiths, P. R., in Laboratory Methods in Infrared Spectroscopy (R. G. J. Miller and B. C. Stace, eds), Heyden, London, 1972.
543. Griffiths, P. R., CHEMICAL INFRARED FOURIER TRANSFORM SPECTROSCOPY, Wiley–Interscience, New York, 1975.
544. Griffiths, P. R., Foskett, C. T., and Curbelo, R., *Appl. Spectr. Rev.*, **6**, 31 (1972).
545. Gross, R. W. F., and Bott, J. F., HANDBOOK OF CHEMICAL LASERS, Wiley–Interscience, New York, 1976.
546. Grushka, E., BONDED STATIONARY PHASES IN CHROMATOGRAPHY, Wiley–Interscience, New York, 1975.
547. Gudzinowicz, B. J., Gudzinowicz, M. J., and Martin, H. F., FUNDAMENTAL INTEGRATED GS–MS, Dekker, New York, 1977.
548. Guenther, H., NMR SPEKTROSKOPIE, G. Thieme Verlag, Stuttgart, 1973.
549. Guilbaut, G. G., FLUORESCENCE: THEORY, INSTRUMENTAL AND PRACTICE, Dekker, New York, 1967.
550. Guilbaut, G. G., PRACTICAL FLUORESCENCE, Dekker, New York, 1973.
551. Guillet, J. E., in New Developments in Gas Chromatography (J. H. Purnell, ed.), Wiley–Interscience, New York, 1973.
552. Guinier, A., and Fournet, G., SMALL-ANGLE SCATTERING OF X-RAYS, Wiley–Interscience, New York, 1955.
553. Gunning, H. E., and Strausz, O. P., in Creation and Detection of the Excited States, Vol. 1B (A. A. Lamola, ed.), Dekker, New York, 1971, Chap. 12.
554. Gunzler, H., and Böck, H., IR-SPEKTROSKOPIE, Verlag Chemie, Weinheim, 1975.
555. Gupta, V. D., and Beavers, R. B., *Chem. Rev.*, **62**, 665 (1962).
556. Hadden, N., Baumann, F., MacDonald, F., Munk, M., Stevenson, R., Gere, D., Zamaroni, F., and Majors, R., BASIC CHROMATOGRAPHY, Varian Associates, Palo Alto, Calif., 1972.
557. Haine, M. E., and Cosslett, V. E., ELECTRON MICROSCOPY, Wiley–Interscience, New York, 1961.
558. Halbach, K., *Phys. Rev.*, **119**, 1230 (1960).
559. Hall, C. E., INTRODUCTION TO ELECTRON MICROSCOPY, McGraw-Hill, New York, 1966.
560. Hall, R. W., in Techniques of Polymer Characterization (P. W. Allen, ed.), Butterworths, London, 1959, p. 19.
561. Hallpap, P., and Schütz, H., ANWENDUNG DER ^{1}H-NMR SPEKTROSKOPIE, Verlag Chemie, Veinheim, 1975.
562. Hamielec, A. E., Eldrup, M., Mogensen, O., and Jansen, P., *J. Macromol. Sci. C*, **9**, 305 (1973).
563. Hamm, F. A., in Physical Methods of Organic Chemistry, Part 2 (A. Weissberger, ed.), Wiley–Interscience, New York, 1960, p. 1561.
564. Hamming, M. C., and Foster, N. G., INTERPRETATION OF MASS SPECTRA OF ORGANIC COMPOUNDS, Academic Press, New York, 1972.
565. Hampton, R. R., *Rubber Chem. Technol.*, **45**, 546 (1972).
566. Hansen, R. H., in Interface Conversion (P. Weiss and G. D. Cheever, eds), Elsevier, Amsterdam, 1968, p. 287.
567. Hansen, R. H., in Thermal Stability of Polymers (R. T. Conley, ed.), Dekker, New York, 1970, p. 153.
568. Harbourn, C. L. A., GAS CHROMATOGRAPHY, Institute of Petroleum, London, 1969.

712

569. Harmon, D. J., in Gel Permeation Chromatography (K. H. Altgelt and L. Segal, eds), Dekker, New York, 1971, p. 13.
570. Harmon, D. J., in Gel Permeation Chromatography (K. H. Altgelt and L. Segal, eds), Dekker, New York, 1971, p. 39.
571. Harrick, N. J., INTERNAL REFLECTION SPECTROSCOPY, Wiley–Interscience, New York, 1967.
572. Harrick, N. J., in Characterization of Metal and Polymer Surfaces, Vol. 2 (L. H. Lee, ed.), Academic Press, New York, 1977, p. 153.
573. Harriman, J. E., THEORETICAL FOUNDATIONS OF ELECTRON SPIN RESON-ANCE, Academic Press, New York, 1978.
574. Harris, F. W., and Seymour, R. B., STRUCTURE–SOLUBILITY RELATIONSHIP IN POLYMERS, Academic Press, New York, 1977.
575. Hartshorne, N. H., and Stuart, A., CRYSTALS AND THE POLARIZING MICRO-SCOPE, Elsevier, New York, 1970.
576. Harwood, H. J., Angew. Chem. Int. Ed. Engl., 4, 394 (1965).
577. Harwood, H. J., Angew. Chem. Int. Ed. Engl., 4, 1051 (1965).
578. Harwood, H. J., in Characterization of Materials and Research Ceramics and Polymers (J. J. Burke and V. Weiss, eds), Syracuse University Press, Syracuse, N.Y., 1975, Chap. 11.
579. Harwood, H. J., Kodaira, Y., and Newman, D. L., in Computers in Chemistry and Instrumentation, Vol. 6 (J. S. Mattison, H. C. MacDonald and H. B. Mark, Jr, eds), Dekker, New York, 1978.
580. Haslam, J., Willis, H. A., and Squirrell, D. C. M., IDENTIFICATION AND ANALY-SIS OF PLASTICS, 2nd ed., Iliffe, London, 1972.
581. Hassan, A. M., Crit. Rev. Macromol. Sci., 1, 399 (1972).
582. Hausdorff, H., ANALYSIS OF POLYMERS BY INFRARED SPECTROSCOPY, Perkin-Elmer Co., 1951.
583. Hautojärvi, P., and Seeger, A., PROCEEDINGS OF THIRD INTERNATIONAL CONFERENCE ON POSITRON ANNIHILATION, Springer, Berlin, 1975.
584. Havriliak, S., and Negami, S., J. Polym. Sci. C, 14, 99 (1966).
585. Haward, R. N., J. Macromol. Sci. C, 4, 191 (1970).
586. Haward, R. N., PHYSICS OF GLASSY POLYMERS, Applied Science Publ., London, 1973.
587. Haward, R. N., in Molecular Behavior and The Development of Polymeric Materials (A. Ledwith and A. M. North, eds), Chapman and Hall, London, 1975.
588. Hay, J. N., Brit. Polym. J., 3, 74 (1971).
589. Hayat, M. A., PRINCIPLES AND TECHNIQUES OF ELECTRON MICRO-SCOPY—BIOLOGICAL APPLICATIONS, Van Nostrand-Reinhold, New York, 1970.
590. Hayward, D. O., and Trapnell, B. M. W., CHEMISORPTION, Butterworths, London, 1964.
591. Heading, P., DIELECTRIC SPECTROSCOPY OF POLYMERS, Hilger, Bristol, 1977.
592. Hearle, J. W. S., and Simmens, S. C., Polymer, 14, 273 (1973).
593. Hearle, J. W. S., Sparrow, J. T., and Cross, P. M., THE USE OF THE SCANNING ELECTRON MICROSCOPE, Pergamon Press, Oxford, 1972.
594. Hedvig, P., DIELECTRIC SPECTROSCOPY OF POLYMERS, Wiley–Interscience, New York, 1977.
595. Heftmann, E., CHROMATOGRAPHY, Reinhold, New York, 1967.
596. Heidner, R. H., and Gibson, M. E., in Encyclopedia of Industrial Chemical Analysis, Vol. 4 (F. D. Snell and C. L. Hilton, eds), Wiley–Interscience, New York, 1967, p. 219.
597. Heindrich, R. D., FUNDAMENTALS OF TRANSMISSION ELECTRON MICRO-SCOPY, Wiley–Interscience, New York, 1964.
598. Heinrich, K. F. J., Newbury, D. E., and Yakowitz, H., in Characterization of Materials in Research Ceramics and Polymers (J. J. Burke and V. Weiss, eds), Syracuse University Press, Syracuse, N.Y., 1975, Chap. 4.
599. Heitz, W., and Kearn, W., Angew. Chem., 1, 150 (1967).
600. Helfferich, F., INORGANIC ION EXCHANGERS, McGraw-Hill, New York, 1962.
601. Helfferich, F., Adv. Chromatogr., 1, 3 (1966).
602. Hendra, P. J., Adv. Polym. Sci., 6, 151 (1969).
603. Hendra, P. J., in Laboratory Methods in Infrared Spectroscopy, 2nd ed. (R. G. J. Miller and B. C. Stace, eds), Heyden, London, 1972, Chap. 15.

713

604. Hendra, P. J., in Polymer Spectroscopy (D. O. Hummel, ed.) Verlag Chemie, Weinheim, 1974, p. 151.
605. Hendra, P. J., and Stratton, P. M., *Chem. Rev.*, **69**, 325 (1969).
606. Hendra, P. J., and Vear, C. J., *Analyst*, **95**, 321 (1970).
607. Hengstenberg, J., *Z. Elektrochem.*, **60**, 236 (1956).
608. Henniker, J. C., INFRARED SPECTROMETRY OF INDUSTRIAL POLYMERS, Academic Press, New York, 1967.
609. Hennesy, B. J., Mead, J. A., and Stening, T. C., PERMEABILITY OF PLASTIC FILMS, The Plastics Institute, London, 1967.
610. Hercules, D. M., FLUORESCENCE AND PHOSPHORESCENCE ANALYSIS: PRINCIPLES AND APPLICATIONS, Wiley–Interscience, New York, 1966.
611. Herglotz, H. K., in Characterization of Materials in Research Ceramics and Polymers (J. J. Burke and V. Weiss, eds), Syracuse University Press, Syracuse, N.Y., 1975, Chap. 5.
612. Herkstroeter, W. G., in Creation and Detection of the Excited State, Vol. 1, Part A (A. A. Lamola, ed.), Dekker, New York, 1971, Chap. 1.
613. Hermans, J. J., and Ende, H. A., in Newer Methods of Polymer Characterization, (B. Ke, ed.), Wiley–Interscience, New York, 1964, p. 525.
614. Hermans, P. H., *Experientia*, **19**, 553 (1963).
615. Hershenson, H. M., NMR and ESR SPECTRA INDEX, Academic Press, New York, 1965.
616. Hazel, Y., *Intern. Lab.*, May/June 1978, p. 73.
617. Hiemenz, P. C., PRINCIPLES OF COLLOID AND SURFACE CHEMISTRY, Dekker, New York, 1977.
618. Higgins, G. M. C., and Miller, R. G. J., in Laboratory Methods in Infrared Spectroscopy, 2nd ed. (R. G. J. Miller and B. C. Stace, eds), Heyden, London, 1972, Chap. 16.
619. Higgins, R., Foote, C. S., and Cheng, H., in Oxidation of Organic Compounds, Advances in Chemistry Series, No. 77, American Chemical Society, Washington, D.C., 1968, p. 102.
620. Hilado, C. J., PYROLYSIS OF POLYMERS, Technomic Publ. Inc., Westport, 1976.
621. Hildebrand, J. H., VISCOSITY AND DIFFUSITY, Wiley–Interscience, New York, 1977.
622. Hildebrand, J. H., and Scott, R. L., SOLUBILITY OF NONELECTROLYTES, 3rd ed., Reinhold, New York, 1950.
623. Hildebrand, J. H., and Scott, R. L., REGULAR SOLUTIONS, Prentice-Hall, New York, 1962.
624. Hill, H., and Freeman, R., INTRODUCTION TO FOURIER TRANSFORM NMR, Varian Associates, Palo Alto, Calif., 1970.
625. Hill, N. E., Vaugham, W. E., Price, A. H., and Davies, M., DIELECTRIC PROPERTIES AND MOLECULAR BEHAVIOR, Van Nostrand-Reinhold, New York, 1969.
626. Hirayama, K., HANDBOOK OF ULTRAVIOLET AND VISIBLE ABSORPTION SPECTRA OF ORGANIC COMPOUNDS, Plenum Press, New York, 1967.
627. Hoffman, J. P., *SPE Trans.*, **4**, 315 (1964).
628. Hoffman, J. P., Williams, G., and Passaglia, E., *J. Polym. Sci. C.* **14**, 173 (1966).
629. Hoffman, M., *Brit. Polym. J.*, **6**, 243 (1974).
630. Hoffman, M., Krömer, H., and Kuhn, R., POLYMERANALYTIK, G. Thieme Verlag, Stuttgart, 1977.
631. Hoffman, R. A., and Forsen, S., *Progr. NMR Spectr.*, **1**, 15 (1966).
632. Hollahan, J. R., *J. Chem. Educ.*, **43**, A401 (1966).
633. Holliday, L., and Ward, I. M., in Structure and Properties of Oriented Polymers (I. M. Ward, ed.), Applied Science Publ., London, 1975, Chap. 1.
634. Holzmüller, W., *Fortsch. Hochpolym. Forsch., Adv. Polym. Sci.*, **22**, 1 (1978).
635. Hookway, H. T., in Techniques of Polymer Characterization (P. W. Allen, ed.), Butterworths, London, 1959, Chap. 3.
636. Hopfenberg, H. P., THE PERMEABILITY OF PLASTICS FILMS AND COATINGS TO GASES, VAPOURS AND LIQUIDS, Plenum Press, New York, 1975.
637. Hopfinger, A. J., CONFORMATIONAL PROPERTIES OF MACROMOLECULES, Academic Press, New York, 1973.
638. Hopkins, R. A., THE INTERNATIONAL (SI) METRIC SYSTEM AND HOW IT WORKS, Polymetric Services Inc., Reseda, Calif., 1973.
639. Horlick, G., *Appl. Spectr.*, **22**, 617 (1968).

714

640. Hosemann, R., *J. Polym. Sci. C*, **20**, 1 (1967).
641. Hosemann, R., *Crit. Rev. Macromol. Sci.*, **1**, 351 (1972).
642. Hosemann, R., *Makromol. Chem.*, **176**, 559 (1975).
643. Howell, M. G., Kende, A. S., and Webb, J. S., FORMULA INDEX TO NMR LITERATURE DATA, Vols 1 and 2, Plenum Press, New York, 1966.
644. Howsmon, J. A., and Walter, N. M., in Physical Methods in Chemical Analysis, Vol. 1 (W. G. Berl, ed.), Academic Press, New York, p. 154.
645. Huber, J. F. K., and Keulemans, A. I. M., in Gas Chromatography (M. Van Swaay, ed.), Butterworths, Washington, D.C., 1962, p. 26.
646. Hudson, A., and Luckhurst, G. R., *Chem. Rev.*, **69**, 191 (1969).
647. Hudson, R. D., INFRARED SYSTEM ENGINEERING, Wiley–Interscience, New York, 1969.
648. Huggins, M. L., Natta, G., Desreux, V., and Mark, H., *J. Polym. Sci.*, **56**, 153 (1962).
649. Huggins, M. L., and Okamoto, H., in Polymer Fractionation (M. J. R. Cantow, ed.), Academic Press, New York, 1966, p. 1.
650. Huglin, M. B., LIGHT SCATTERING FROM POLYMER SOLUTIONS, Academic Press, London, 1972.
651. Huglin, M. B., in Polymer Handbook, 2nd ed. (K. Brandrup, E. H. Immergut and W. McDowell, eds), Wiley–Interscience, New York, 1975, p. IV-267.
652. Hummel, D. O., KUNSTSTOFF-, LACK-, UND GUMMI-ANALYSE, CHEMIS-CHE UND INFRAROTSPEKTROSKOPISCHE METHODEN, Hanser Verlag, Munchen, 1958.
653. Hummel, D. O., *Rubber Chem. Technol.*, **32**, 854 (1959).
654. Hummel, D. O., INFRARED SPECTRA OF POLYMERS IN THE MEDIUM AND LONG WAVELENGTH REGIONS, Wiley–Interscience, New York, 1966.
655. Hummel, D. O., ATLAS DER KUNSTSTOFF-ANALYSE, Band I, HOCHPOLY-MERE UND HARZE: SPEKTREN UND METHODEN ZUR IDENTIFIZIER-UNG, Hanser Verlag, Munchen and Verlag Chemie, Weinheim, 1968.
656. Hummel, D. O., in Polymer Spectroscopy (D. O. Hummel, ed.), Verlag Chemie, Weinheim, 1974, p. 112.
657. Hummel, D. O., and Scholl, F., IR ANALYSIS OF POLYMERS, RESINS AND ADDITIVES, 2nd ed., Verlag Chemie, Weinheim, 1973.
658. Hummel, D. O., Schüddemage, H. D., and Rübenacker, K., in Polymer Spectroscopy (D. O. Hummel, ed.), Verlag Chemie, Weinheim, 1974, p. 355.
659. Hurley, W. J., *J. Chem. Educ.*, **43**, 237 (1966).
660. Hyde, J. S., EXPERIMENTAL TECHNIQUES IN EPR; 6TH VARIAN NMR-ESR WORKSHOP, Varian Associates, Palo Alto, Calif., 1962.
661. Illers, K. H., *Kolloid Z. Z. Polym.*, **190**, 16 (1963).
662. Illers, K. H., *Ber. Bunsenges Phys. Chem.*, **70**, 353 (1966).
663. Inagaki, H., in Fractionation of Synthetic Polymers: Principles and Applications (L. H. Tung, ed.), Dekker, New York, 1977, Chap. 7.
664. Inczédy, J., ANALYTICAL APPLICATIONS OF ION EXCHANGERS, Pergamon Press, 1966.
665. Ingham, J. D., *J. Macromol. Sci. C*, **2**, 279 (1968).
666. Ingham, J. D., in Reviews in Macromolecular Science, Vol. 3 (G. B. Butler and K. F. O'Driscoll, eds), Dekker, New York, 1968, p. 280.
667. Ingram, P., and Peterlin, A., in Encyclopedia of Polymer Science and Technology, Vol. 9 (H. F. Mark, N. G. Gaylord and N. M. Bikales, eds), Wiley–Interscience, New York, 1968, p. 204.
668. Ishida, H., and Koenig, J. L., *Intern. Lab.*, May/June 1978, p. 49.
669. Ivin, K. J., STRUCTURAL STUDIES OF MACROMOLECULES BY SPECTRO-SCOPIC METHODS, Wiley–Interscience, New York, 1976.
670. Iyengar, P. K., in Thermal Neutron Scattering (P. A. Egelstaff, ed.), Academic Press, New York, 1965, p. 97.
671. Jackman, L. M., APPLICATION OF NMR SPECTROSCOPY IN ORGANIC CHEMISTRY, Macmillan–Pergamon, New York, 1969.
672. Jaffé, H. H., and Orchin, M., THEORY AND APPLICATION OF ULTRAVIOLET SPECTROSCOPY, Wiley–Interscience, New York, 1962.
673. James, J. F., and Sternberg, R. S., THE DESIGN OF OPTICAL SPECTROMETERS, Chapman and Hall, London, 1969.

674. Jellinek, H. H. G., DEGRADATION OF VINYL POLYMERS, Academic Press, New York, 1955.
675. Jellinek, H. H. G., in Fracture Processes in Polymer Solids (B. Rosen, ed.), Wiley–Interscience, New York, 1964, Chap. 4.
676. Jellinek, H. H. G., in Encyclopedia of Polymer Science and Technology, Vol. 4 (H. F. Mark, N. G. Gaylord and N. M. Bikales, eds), Wiley–Interscience, New York, 1966, p. 740.
677. Jellinek, H. H. G., in The Stereochemistry of Macromolecules, Vol. 3 (A. Ketley, ed.), Dekker, New York, 1968, Chap. 10.
678. Jellinek, H. H. G., in Aspects of Degradation and Stabilization of Polymers (H. H. G. Jellinek, ed.), Elsevier, Amsterdam, 1978, p. 431.
679. Jelly, E. E., in Physical Methods of Organic Chemistry, Part 2 (A. Weissberger, ed.), Wiley–Interscience, New York, 1960, p. 1347.
680. Jenkins, A. D., POLYMER SCIENCE, Elsevier, Amsterdam, 1972.
681. Jen-Yuan, C., DETERMINATION OF MOLECULAR WEIGHTS OF HIGH POLYMERS, Izd. Inostrannoy Literatury, Moskov, 1962; translated to English by J. Schmorak, S. Monson Co., Jerusalem, 1963.
682. Jirgensons, B., OPTICAL ACTIVITY OF PROTEINS AND OTHER MACROMOLECULES, 2nd ed., Springer, New York, 1973.
683. Johnsen, U., Ber. Bunsenges Phys. Chem., 70, 320 (1966).
684. Johnson, E., LIQUID CHROMATOGRAPHY BIBLIOGRAPHY, Varian Associates, Palo Alto, Calif., 1977.
685. Johnson, E. G., and Nier, A. O., Phys. Rev., 91, 10 (1953).
686. Johnson, J. F., and Porter, R. S., in The Stereochemistry of Macromolecules, (A. D. Ketley, ed.), Dekker, New York, 1968, Chap. 5.
687. Johnson, J. F., and Porter, R. S., Progr. Polym. Sci., 2, 203 (1970).
688. Johnson, J. F., Porter, R. S., and Cantow, M. J. R., Rev. Macromol. Sci., 1, 393 (1966).
689. Johnson, J. F., Porter, R. S., and Cantow, M. J. R., in Encyclopedia of Polymer Science and Technology, Vol. 7 (H. F. Mark, N. G. Gaylord and N. M. Bikales, eds), Wiley–Interscience, New York, 1967, p. 231.
690. Johnson, L. F., and Jankowski, W. C., CARBON-13 NMR SPECTRA: A COLLECTION OF ASSIGNED, CODED AND INDEXED SPECTRA, Wiley–Interscience, New York, 1972.
691. Johnston, N. W., J. Macromol. Sci. C, 14, 215 (1976).
692. Johnstone, R. A. W., MASS SPECTROMETRY FOR ORGANIC CHEMISTS, Cambridge University Press, Cambridge, 1972.
693. Jones, D. W., INTRODUCTION TO THE SPECTROSCOPY OF BIOLOGICAL POLYMERS, Academic Press, New York, 1976.
694. Jones, R. N., Anal. Chem., 38, 393R (1966).
695. Jones, W. J., Quart. Rev. Chem. Soc., 1, 73 (1969).
696. Jork, H., in Quantitative Paper and Thin Layer Chromatography (E. J. Shellard, ed.), Academic Press, New York, 1968, Chap. 6.
697. Jungnickel, J. L., Anal. Chem., 35, 1985 (1963).
698. Kaelbe, D. H., in Rheology—Theory and Applications, Vol. 5 (F. R. Eirich, ed.), Academic Press, New York, 1969, Chap. 5.
699. Kaelbe, D. H., PHYSICAL CHEMISTRY OF ADHESION, Wiley–Interscience, New York, 1971.
700. Kaelbe, D. H., and Moacanin, J., Polymer, 18, 475 (1977).
701. Kagarise, R. E., and Weinberger, L. A., INFRARED SPECTRA OF PLASTICS AND RESINS, Part 1, Naval Research Laboratory Report 4369, 1954.
702. Kaiser, H., and Menzis, A. C., THE LIMIT OF DETECTION OF A COMPLETE ANALYTICAL PROCEDURE, Hilger, London, 1968.
703. Kaiser, R., GAS PHASE CHROMATOGRAPHY, Vols 1 and 2, Butterworths, Washington, D.C., 1963.
704. Kakudo, M., and Kasai, N., X-RAY DIFFRACTION BY POLYMERS, Kodansha, Tokyo and Elsevier, Amsterdam, 1972.
705. Kallman, H. P., and Spruch, G. M., LUMINESCENCE OF ORGANIC AND INORGANIC MATERIALS, Wiley–Interscience, New York, 1962.

716

706. Kambe, H., in Aspects of Degradation and Stabilization of Polymers (H. H. G. Jellinek, ed.), Elsevier, 1978, p. 393.
707. Kamide, K., in Fractionation of Synthetic Polymers: Principles and Applications (L. H. Tung, ed.), Dekker, New York, 1977, Chap. 2.
708. Kamiya, Y., and Niki, E., in Aspects of Degradation and Stabilization of Polymers (H. H. G. Jellinek, ed.), Elsevier, Amsterdam, 1978, Chap. 3.
709. Kaplany, N. S., FIBER OPTICS: PRINCIPLES AND APPLICATIONS, Academic Press, New York, 1967.
710. Karasz, F. E., DIELECTRIC PROPERTIES OF POLYMERS, Plenum Press, New York, 1972.
711. Kausch-Blecken von Schmeling, H. H., *J. Macromol. Sci. C*, **4**, 243 (1970).
712. Kauzmann, W., *Chem. Rev.*, **43**, 219 (1948).
713. Kavesh, S., and Schultz, J. M., *Polym. Eng. Sci.*, **9**, 452 (1969).
714. Kavesh, S., and Smith, J. M., *Polym. Eng. Sci.*, **9**, 331 (1969).
715. Kay, D. H., TECHNIQUES FOR ELECTRON MICROSCOPY, 2nd ed., Blackwell, Oxford, 1965.
716. Kaye, W., *Spectrochim. Acta*, **6**, 257 (1954).
717. Kaye, W., *Spectrochim. Acta* **7**, 181 (1955).
718. Ke, B., in Newer Methods of Polymer Characterization (B. Ke, ed.), Wiley–Interscience, New York, 1964, Chap. 9.
719. Ke, B., THERMAL ANALYSIS OF HIGH POLYMERS, Wiley–Interscience, New York, 1964.
720. Kearns, D. R., *Chem. Rev.*, **71**, 395 (1971).
721. Kearns, D. R., and Khan, A. U., *Photochem. Photobiol.*, **10**, 193 (1969).
722. Keattch, C. J., and Dollimore, D., INTRODUCTION TO THERMOGRAVIE-METRY, 2nd ed., Heyden, London, 1975.
723. Keiner, J. A., *Polym. Symp.*, No. 42, 1077 (1973).
724. Keller, A., *Kolloid. Z.*, **197**, 98 (1964).
725. Keller, A., *Progr. Rep. Phys.*, **31**, 623 (1968).
726. Keller, A., *Kolloid. Z.*, **231**, 386 (1969).
727. Keller, A., *Polym. Symp.*, No. 59, 1 (1977).
728. Keller, A., and Mackley, M. R., *Pure Appl. Chem.*, **39**, 195 (1974).
729. Kelly, R. N., and Billmeyer, F. W., Jr, in Gel Permeation Chromatography (K. H. Altgelt and L. Segal, eds), Dekker, New York, 1971, p. 47.
730. Kemp, W., ORGANIC SPECTROSCOPY, Macmillan, London, 1973.
731. Kennedy, J. W., Gordon, M., and Alvarez, G. A., *Polimery-Tw. Wielkocz.*, **10**, 1464 (1975).
732. Kenyon, A. S., in Techniques and Methods of Polymer Evaluation, Vol. 1 (P. E. Slade and L. T. Jenkins, eds), Dekker, New York, 1966, Chap. 5.
733. Kerker, M., in Electromagnetic Scattering (R. L. Rowell and R. S. Stein, eds), Gordon and Breach, New York, 1967.
734. Kerker, M., *Ind. Eng. Chem.*, **60**, 31 (1968).
735. Kerker, M., THE SCATTERING OF LIGHT AND OTHER ELECTROMAGNETIC RADIATION, Academic Press, New York, 1969.
736. Kern, H., and Imhof, K., *Intern. Lab.*, Jan./Feb. 1978, p. 65.
737. Ketley, A. D., THE STEREOCHEMISTRY OF MACROMOLECULES, Dekker, New York, 1967/1968.
738. Keulemans, A. I. M., GAS CHROMATOGRAPHY, Reinhold, New York, 1957.
739. Kevan, L., and Kispert, L. D., ELECTRON SPIN DOUBLE RESONANCE SPEC-TROSCOPY, Wiley–Interscience, New York, 1976.
740. Khan, A. U., and Kearns, D. R., in Oxidation of Organic Compounds, Advances in Chemistry Series, No. 77, American Chemical Society, Washington, D.C., 1968, p. 143.
741. Kilian, H. G., and Jenckel, E., in Struktur und Physikalische Verhalten der Kunststoffe (K. A. Wolf, ed.), Springer, Weinheim, 1962, p. 176.
742. Kimel, S., and Speiser, S., *Chem. Rev.*, **77**, 437 (1977).
743. Kinel, P. O., Rånby, B., and Runnström-Reio, V., ESR APPLICATIONS TO POLYMER RESEARCH, Almqvist, Wiksell, Stockholm, 1973.
744. Kingslake, R., APPLIED OPTICS AND OPTICAL ENGINEERING, Vols 1–4, Academic Press, New York, 1965–1969.

717

745. Kinsinger, J. B., in Characterization of Materials in Research Ceramics and Polymers (J. J. Burke and V. Weiss, eds), Syracuse University Press, Syracuse, N.Y., 1975, Chap. 10.
746. Kirkland, J. J., MODERN PRACTICE OF LIQUID CHROMATOGRAPHY, Wiley–Interscience, New York, 1971.
747. Kirshenbaum, G. S., POLYMER SCIENCE STUDY GUIDE, Gordon and Breach, New York, 1973.
748. Kirste, R. G., in Characterization of Macromolecular Structure, Nat. Acad. Sci. U.S.A., Publ. 1573, Washington, D.C., 1968, p. 74.
749. Kiselev, A. V., and Yashin, Ya. I., GAS ABSORPTION CHROMATOGRAPHY, Plenum Press, New York, 1969.
750. Kiser, R. W., INTRODUCTION TO MASS SPECTROSCOPY AND ITS APPLI-CATIONS, Prentice-Hall, Englewood Cliffs, N.J., 1965.
751. Kissinger, H. E., and Newman, S. B., in Analytical Chemistry of Polymers, Part 2, Analysis of Molecular Structure and Chemical Groups (M. Kline, ed.), Wiley–Interscience, New York, 1962, Chap. 4.
752. Kitagawa, T., and Miyazawa, T., *Fortsch. Hochpolym. Forsch., Adv. Polym. Sci.*, **9**, 336 (1972).
753. Kitamaru, R., and Horii, F., *Fortsch. Hochpolym. Forsch., Adv. Polym. Sci.*, **22**, 137 (1978).
754. Klein, I., and Marschall, D. I., COMPUTER PROGRAMS FOR PLASTIC ENGIN-EERS, Reinhold, New York, 1968.
755. Klesper, E., and Sielaff, G., in Polymer Spectroscopy (D. O. Hummel, ed.), Verlag Chemie, Weinheim, 1974, p. 189.
756. Kline, D. E., and Hansen, D., in Techniques and Methods of Polymer Evaluation, Vol. 2, Thermal Characterization Techniques (P. E. Slade, Jr and L. T. Jenkins, eds), Dekker, New York, 1970, Chap. 5.
757. Kline, G. M., ANALYTICAL CHEMISTRY OF POLYMERS, Parts 1 and 2, Wiley–Interscience, New York, 1962.
758. Klug, H. P., and Alexander, L. E., X-RAY DIFFRACTION PROCEDURES, Wiley–Interscience, New York, 1954.
759. Knappe, W., *Fortsch. Hochpolym, Forsch., Adv. Polym. Sci.*, **7**, 477 (1971).
760. Knappe, W., Lohe, P., and Wutschig, R., *Angew. Makromol. Chem.*, **7**, 181 (1969).
761. Koenig, J. L., *Appl. Spectr. Rev.*, **4**, 233 (1971).
762. Koenig, J. L., *Macromol. Rev.*, **6**, 59 (1972).
763. Koningstein, J. A., INTRODUCTION TO THE THEORY OF THE RAMAN EFFECT, Reidel, Dortrecht, Holland, 1972.
764. Koningsveld, R., in Characterization of Macromolecular Structures, Nat. Acad, Sci., U.S.A., Publ. 1573, Washington, D.C., 1968, p. 172.
765. Koningsveld, R., *Fortsch. Hochpolym. Forsch., Adv. Polym. Sci.*, **7**, 1 (1970).
766. Koningsveld, R., in Polymer Science, Vol. 2 (A. D. Jenkins, ed.), North-Holland, Amsterdam, 1972, Chap. 15.
767. Koningsveld, R., *Brit. Polym. J.*, **7**, 435 (1975).
768. Kortüm, G., REFLECTANCE SPECTROSCOPY, Springer Verlag, New York, 1969.
769. Kosler, F., QUANTITATIVE ANALYSIS BY NMR SPECTROSCOPY, Academic Press, London, 1973.
770. Kotera, A., in Polymer Fractionation (M. J. R. Cantow, ed.), Academic Press, New York, 1967, Chap. 2.
771. Kovacs, A. J., *Fortsch. Hochpolym. Forsch., Adv. Polym. Sci.*, **3**, 394 (1963).
772. Kowalski, P., APPLIED PHOTOGRAPHIC THEORY, Wiley–Interscience, New York, 1972.
773. Kratky, O., *Angew. Chem.*, **72**, 467 (1960).
774. Kratohvil, J. P., *Anal. Chem.*, **36**, R458 (1964).
775. Kratohvil, J. P., *Anal. Chem.*, **38**, R517 (1966).
776. Kratohvil, J. P., in Characterization of Macromolecular Structure, Nat. Acad. Sci., U.S.A., Publ. 1537, Washington, D.C., 1968, p. 59.
777. Krause, S., *J. Macromol. Sci. C*, **7**, 251 (1972).
778. Krigbaum, W. R., *Polym. Rev.*, **6**, 1 (1964).
779. Krigbaum, W. R., and Roe, R. J., in Treatise on Analytical Chemistry, Vol. 7, Part 1 (I. M. Kolthoff and P. J. Elving, eds), Wiley–Interscience, New York, 1967, Chap. 79.

718

780. Krimm, S., *Fortsch. Hochpolym. Forsch., Adv. Polym. Sci.*, **2**, 51 (1960).
781. Krugers, J., INSTRUMENTATION IN APPLIED NUCLEAR CHEMISTRY, Plenum Press, New York, 1973.
782. Kruse, P. W., McGlauchlin, L. D., and McQuistan, R. B., ELEMENTS OF INFRA-RED TECHNOLOGY, Wiley–Interscience, New York, 1962.
783. Kryszewski, M., *Plaste Kautschuk*, **16**, 664 (1969).
784. Kumins, C. A., and Kwei, T. K., in Diffusion in Polymers (J. Crank and G. S. Park, eds), Academic Press, London, 1968, Chap. 4.
785. Kupfer, W., *Z. Anal. Chem.*, **192**, 219 (1963).
786. Kurata, M., Tsunashima, Y., Iwama, M., and Kamada, K., in Polymer Handbook, 2nd ed. (K. Brandrup, E. H. Immergut and W. McDowell, eds), Wiley–Interscience, New York, 1975, p. IV-1.
787. Kurata, M., and Stockmayer, W. H., *Fortsch. Hochpolym. Forsch., Adv. Polym. Sci.*, **3**, 196 (1963).
788. Ladd, W. A., and Ladd, M. W., in Encyclopedia of Industrial Chemical Analysis, Vol. 1 (F. D. Snell and C. L. Hilton, eds), Wiley–Interscience, New York, 1966, p. 649.
789. LaMar, G. N., Horrocks, W. D., Jr, and Holm, R. H., CHEMICAL APPLICATION OF NMR IN PARAMAGNETIC MOLECULES: THEORY AND APPLICA-TIONS, Academic Press, New York, 1973.
790. Lambert, A., *Brit. Polym. J.*, **3**, 13 (1971).
791. Lamola, A. A., CREATION AND DETECTION OF THE EXCITED STATE, Vol. 1, Parts A and B, Dekker, New York, 1971.
792. Lang, L., ABSORPTION SPECTRA IN THE ULTRAVIOLET AND VISIBLE REGION, Vols 1–5, Academic Press, New York, 1961–1965.
793. Langer, H. G., and Gohlke, R. S., in Gas Effluent Analysis (W. Lodding, ed.), Dekker, New York, 1967, Chap. 3.
794. Landgraff, W. C., EPR'S ROLE IN FREE RADICAL CHEMISTRY, Varian Associ-ates, Palo Alto, Calif., 1964.
795. Laszlo, P., and Stang, P., ORGANIC SPECTROSCOPY, Harper and Row, New York, 1971.
796. Laurence, D. J. R., in Physical Methods in Macromolecular Chemistry, Vol. 1 (B. Carroll, ed.), Dekker, New York, 1969, Chap. 5.
797. Lawley, K. P., MOLECULAR SCATTERING: PHYSICAL AND CHEMICAL APPLICATIONS, Wiley–Interscience, New York, 1975.
798. Leathard, D. A., and Shurlock, B. C., IDENTIFICATION TECHNIQUES IN GAS CHROMATOGRAPHY, Wiley–Interscience, New York, 1971.
799. Lebedev, Ya. S., and Muromtseev, V. I., EPR AND RELAXATION OF STABILIZED RADICALS, Khimia, Moscow, 1972 (in Russian).
800. Lebovits, A., *Modern Plastics*, **43**, 139 (1966).
801. Leck, J. H., PRESSURE MEASUREMENTS IN VACUUM SYSTEMS, The Institute of Physics, London, 1957.
802. Lederberg, J., COMPUTATION OF MOLECULAR FORMULAS FOR MASS SPECTROMETRY, Elsevier, Amsterdam, Holden-Day, San Francisco, 1964.
803. Lederberg, J., TABLES AND AN ALGORITHM FOR CALCULATING FUNC-TIONAL GROUPS OR ORGANIC MOLECULES IN HIGH RESOLUTION MASS SPECTROMETRY, NASA Sci. Tech. Aerospace Rep. N64-21426 (1964).
804. Ledwith, A., and North, A. M., MOLECULAR BEHAVIOUR AND THE DEVELOP-MENT OF POLYMERIC MATERIALS, Chapman and Hall, London, 1975.
805. Lee, C. J., *J. Macromol. Sci. C*, **16**, 79 (1978).
806. Lee, W. A., and Knight, G. L., *Brit. Polym. J.*, **2**, 73 (1970).
807. Lee, W. A., and Rutherford, R. A., in Polymer Handbook, 2nd ed. (J. Brandrup, E. H. Immergut and W. McDowell, eds), Wiley–Interscience, New York, 1975, p. III-139.
808. LeFèvre, C. G., and LeFèvre, R. J. W., in Techniques of Organic Chemistry, Vol. 1, Part 3 (A. Weissberger, ed.), Wiley–Interscience, New York, 1960, p. 2459.
809. Lehman, T. A., ION CYCLOTRON RESONANCE SPECTROSCOPY, Wiley–Interscience, New York, 1976.
810. Lehrle, R. S., *Progr. High. Polym.*, **1**, 37 (1961).
811. Lengyel, B. A., LASERS—GENERATION OF LIGHT BY STIMULATED EMIS-SION, 2nd ed., Wiley–Interscience, New York, 1971.

812. Levi, D. W., Reich, L., and Lee, H. T., *Polym. Eng. Sci.*, **5**, 135 (1965).
813. Levi, L., APPLIED OPTICS: A GUIDE TO OPTICAL DESIGN, Vol. 1, Wiley–Interscience, New York, 1968.
814. Levy, G. C., TOPICS IN CARBON-13 NMR SPECTROSCOPY, Vols 1 and 2, Wiley–Interscience, New York, 1974.
815. Levy, G. C., and Nelson, G. L., CARBON-13 NUCLEAR MAGNETIC RESONANCE FOR ORGANIC CHEMISTS, Wiley–Interscience, New York, 1972.
816. Levy, R. L., *Chromatogr. Rev.*, **8**, 48 (1966).
817. Levy, R. L., *J. Gas Chromatogr.*, **5**, 107 (1967).
818. Lewis, O. G., PHYSICAL CONSTANTS OF LINEAR HOMOPOLYMERS, Springer Verlag, Heidelberg, 1968.
819. Leyden, D. E., and Cox, R. H., ANALYTICAL APPLICATION OF NMR, Wiley–Interscience, New York, 1977.
820. Liebhafsky, H. A., Pfeiffer, H. G., Winslow, E. H., and Zemany, P. D., X-RAYS, ELECTRONS, AND ANALYTICAL CHEMISTRY, Wiley–Interscience, New York, 1972.
821. Likhtenshtein, G. I., SPIN LABELING METHODS IN MOLECULAR BIOLOGY, Wiley–Interscience, New York, 1976.
822. Lindenmeyer, P. H., *J. Polym. Sci. C*, **1**, 5 (1963).
823. Lindenmeyer, P. H., *J. Polym. Sci. C*, **20**, 145 (1967).
824. Lindman, B., and Forsén, S., CHLORINE, BROMINE AND IODINE NMR, Springer Verlag, Berlin, 1976.
825. Link, G. L., in Polymer Science, Vol. 1 (A. D. Jenkins, ed.), North-Holland, Amsterdam, 1972, Chap. 18.
826. Lipatov, Yu. S., *Fortschr. Hochpolym. Forsch., Adv. Polym. Sci.*, **22**, 63 (1978).
827. Lipatov, Yu. S., and Sergeeva, L. M., ADSORPTION OF POLYMERS, Wiley–Interscience, New York, 1974.
828. Litenau, C., and Gocen, S., GRADIENT LIQUID CHROMATOGRAPHY, Wiley–Interscience, New York, 1974.
829. Little, J. N., Waters, J. L., Bombaugh, K. J., and Pauplis, W. J., in Gel Permeation Chromatography (K. H. Altgelt and L. Segal, eds), Dekker, New York, 1971, p. 203.
830. Littlewood, A. B., GAS CHROMATOGRAPHY, Academic Press, New York, 1970.
831. Liu, K. J., and Anderson, J. E., *J. Macromol. Sci. C*, **5**, 1 (1970).
832. Loader, E. J., BASIC LASER RAMAN SPECTROSCOPY, Heyden–Sadtler, New York, 1970.
833. Lodding, W., GAS EFFLUENT, ANALYSIS, Dekker, New York, 1967.
834. Loewenstein, E. V., *Appl. Opt.*, **5**, 845 (1966).
835. Long, D. A., RAMAN SPECTROSCOPY: AN INTRODUCTORY SURVEY, Elsevier, Amsterdam, 1976.
836. Longworth, J. W., in Creation and Detection of the Excited State, Vol. 1, Part A (A. A. Lamola, ed.), Dekker, New York, 1971, Chap. 7.
837. Longworth, J. W., *Photochem. Photobiol.*, **26**, 665 (1977).
838. Lott, P. F., *J. Chem. Educ.*, **45**, A89, A169, A273 (1968).
839. Loveland, R. P., PHOTOMICROGRAPHY: A COMPREHENSIVE TREATMENT, Wiley–Interscience, New York, 1970.
840. Low, M. J. D., *J. Chem. Educ.*, **43**, 637 (1966).
841. Low, M. J. D., *J. Chem. Educ.*, **47**, A163, 1255 (1970).
842. Low, W., PARAMAGNETIC RESONANCE, Academic Press, New York, 1963.
843. Lowry, G. G., MARKOW CHAINS AND MONTE CARLO CALCULATIONS IN POLYMER SCIENCE, Dekker, New York, 1970.
844. Lowry, T. M., OPTICAL ROTATORY POWER, Dover Publ., New York, 1964.
845. Luff, N., D.M.S. WORKING ATLAS IN INFRARED, Butterworths, London, 1970.
846. Lundberg, J. L., Hardin, I. R., and Tamioka, M. K., in Encyclopedia of Polymer Science and Technology, Vol. 12 (H. F. Mark, N. G. Gaylord and N. M. Bikales, eds), Wiley–Interscience, New York, 1970, p. 355.
847. Lynden-Bell, R., and Harris, R. K., NUCLEAR MAGNETIC RESONANCE SPECTROSCOPY, Appleton-Century-Crofts, New York, 1969.
848. Ma, T. S., and Horak, V., MICROSCALE MANIPULATIONS IN CHEMISTRY, Wiley–Interscience, New York, 1976.

849. MacCordick, J., OPTICAL CIRCULAR DICHROISM: PRINCIPLES, MEASURE-
      MENTS AND APPLICATIONS, Academic Press, New York, 1965.
850. Mackenzie, R. C., DIFFERENTIAL THERMAL ANALYSIS, Vols 1 and 2, Academic
      Press, New York, 1970/1972.
851. Mackenzie, R. C., *J. Polym. Sci. B*, **12**, 523 (1974).
852. Mackenzie, R. C., and Mitchell, B. D., in Differential Thermal Analysis, Vol. 1 (R. C.
      Mackenzie, ed.), Academic Press, New York, 1970, Chap. 3.
853. MacLean, D. L., and White, J. L., *Polymer*, **13**, 124 (1972).
854. Maconnachie, A., and Richards, R. W., *Polymer*, **19**, 739 (1978).
855. Madorski, S. L., THERMAL DEGRADATION OF ORGANIC POLYMERS, Wiley–
      Interscience, New York, 1964.
856. Magill, J. H., in Polymer Handbook, 2nd ed. (J. Brandrup, E. H. Immergut and W.
      McDowell, eds), Wiley–Interscience, New York, 1975, p. III-193.
857. Maitrot, M., Michel, R., and Madru, R., *Rev. Phys. Appl.*, **6**, 369 (1971).
858. Majors, R. E., *Amer. Lab.*, **5** (5), 27 (1972).
859. Makhlis, F. A., RADIATION PHYSICS AND CHEMISTRY OF POLYMERS,
      Wiley–Interscience, New York, 1975.
860. Manche, E. P., and Carroll, B., in Physical Methods in Macromolecular Chemistry,
      Vol. 2 (B. Carroll, ed.), Dekker, New York, 1972, p. 292.
861. Mandelkern L., *Rubber Chem. Technol.*, **32**, 1392 (1959).
862. Mandelkern, L., CRYSTALLIZATION OF POLYMERS, McGraw-Hill, New York,
      1964.
863. Mandelkern, L., *Polym. Eng. Sci.*, **7**, 232 (1967).
864. Mandelkern, L., in Progress in Polymer Science, Vol. 2 (A. D. Jenkins, ed.), Pergamon
      Press, Oxford, 1970, p. 163.
865. Mandelkern, L., in Characterization of Materials and Research Ceramic and Polymers
      (J. J. Burke and V. Weiss, eds), Syracuse University Press, Syracuse, N.Y., 1975,
      Chap. 13.
866. Manley, T. C., and Niegowski, S. J., in Encyclopedia of Chemical Technology, Vol. 14
      (H. F. Mark, J. J. McKetta, Jr, and D. F. Othmer, eds), Wiley–Interscience, New York,
      1967, p. 410.
867. Manley, T. R., in Techniques of Polymer Science, SCI Monograph No. 17, Society for
      the Chemical Industry, London and Gordon and Breach, New York, 1963.
868. Manson, J. A., and Sperling, L. H., POLYMER BLENDS AND COMPOSITES,
      Plenum Press, New York, 1976.
869. Marchessault, R. H., Fisa, B., and Chanzy, H. D., *Crit. Rev. Macromol. Sci.*, **1**, 315
      (1972).
870. Marini-Bettolo, G. B., THIN-LAYER CHROMATOGRAPHY, Elsevier, Amsterdam,
      1964.
871. Mark, H., and Tobolsky, A. V., PHYSICAL CHEMISTRY OF HIGH POLYMERIC
      SYSTEMS, Wiley–Interscience, New York, 1950.
872. Mark, J. E., in Characterization of Materials in Research Ceramics and Polymers
      (J. J. Burke and V. Weiss, eds), Syracuse University Press, Syracuse, N.Y., 1975,
      Chap. 12.
873. Martin, A. E., INFRARED INSTRUMENTATION AND TECHNIQUES, Elsevier,
      New York, 1966.
874. Martin, D. H., SPECTROSCOPIC TECHNIQUES FOR FAR INFRARED, SUB-
      MILIMETRE AND MILIMETRE WAVES, Wiley–Interscience, New York, 1967.
875. Martin, L. C., THE THEORY OF THE MICROSCOPE, Blackie, London, 1966.
876. Mate, R. D., and Ambler, M. R., in Gel Permeation Chromatography (K. H. Altgelt
      and L. Segal, eds), Dekker, New York, 1971, p. 377.
877. Mathes, K. N., in Encyclopedia of Polymer Science and Technology, Vol. 5 (H. F.
      Mark, N. G. Gaylord and N. M. Bikales, eds), Wiley–Interscience, New York, 1966,
      p. 528.
878. Mathieson, D. W., INTERPRETATION OF ORGANIC SPECTRA, Academic Press,
      New York, 1965.
879. Mathieson, D. W., NMR FOR ORGANIC CHEMISTS, Academic Press, New York,
      1967.
880. Mattauch, J., *Phys. Rev.*, **50**, 617 (1936).
881. Mattauch, J., *Phys. Rev.*, **50**, 1089 (1936).

882. Matthews, A. J. W., EPITAXIAL GROWTH, Academic Press, London, 1975.
883. Mattson, J. S., Mark, H. B., Jr, and MacDonald, H. C., Jr, INFRARED CORRE-LATION AND FOURIER TRANSFORM, Dekker, New York, 1977.
884. Mattson, J. S., Mark, H. B., Jr, and MacDonald, H. C., Jr, COMPUTERS IN POLYMER SCIENCE, Dekker, New York, 1977.
885. McBrierty, V. J., Polymer, 15, 503 (1974).
886. McCaffrey, E. L., LABORATORY PREPARATION FOR MACROMOLECULAR CHEMISTRY, McGraw-Hill, New York, 1970.
887. McCall, D. W., and Slichter, W. P., in Newer Methods of Polymer Characterization (B. Ke, ed.), Wiley–Interscience, New York, 1964, Chap. 8.
888. McCall, J. S., and Potter, B. J., ULTRACENTRIFUGATION, Baillière Tindall, London, 1973.
889. McConnell, B. L., Intern. Lab., May/June 1978, p. 89.
890. McCormick, H. W., in Polymer Fractionation (M. J. R. Cantow, ed.), Academic Press, New York, 1967, p. 251.
891. McCown, S. M., and Earnest, C. M., Intern. Lab., May/June 1978, p. 37
892. McCrea, P. F., in New Developments in Gas Chromatography (J. H. Purnell, ed.), Wiley–Interscience, New York, 1973.
893. McCrum, N. G., Read, B. E., and Williams, G., ANELASTIC AND DIELECTRIC EFFECTS IN POLYMERIC SOLIDS, Wiley–Interscience, New York, 1967.
894. McDowell, C. A., MASS SPECTROMETRY, McGraw-Hill, New York, 1963.
895. McFaden, W. H., TECHNIQUES OF COMBINED GAS CHROMATOGRAPHY/MASS SPECTROMETRY, Wiley–Interscience, New York, 1973.
896. McGlashan, M. L., Pure Appl. Chem., 21, 577 (1970).
897. McGlynn, S. P., Azumi, T., and Kinoshita, M., MOLECULAR SPECTROSCOPY OF TRIPLET STATE, Prentice-Hall, Englewood Cliffs, N.J., 1969.
898. McGraw, G. E., in Polymer Characterization: An Interdisciplinary Approach (C. D. Craver, ed.), Plenum Press, New York, 1971, p. 37.
899. McIntyre, D., and Gornick, F., LIGHT SCATTERING FROM DILUTE POLYMER SOLUTIONS, Gordon and Breach, New York, 1964.
900. McLafferty, F. W., MASS SPECTROMETRY OF ORGANIC IONS, Academic Press, New York, 1963.
901. McLafferty, F. W., MASS SPECTRAL CORRELATION, Advances in Chemistry Series, No. 40, American Chemical Society, Washington, D.C., 1963.
902. McLafferty, F. W., INTERPRETATION OF MASS SPECTRA, 2nd ed., Benjamin, London, 1973.
903. McLauchlan, K. A., MAGNETIC RESONANCE, Oxford University Press, Oxford, 1975.
904. McMillan, J. A., ELECTRON PARAMAGNETISM, Reinhold, New York, 1968.
905. McNair, H. M., and Bonelli, E. J., BASIC GAS CHROMATOGRAPHY, 5th ed., Varian Aerograph, Palo Alto, Calif., 1969.
906. McNeil, I. C., in Thermal Analysis (R. F. Schwenker and P. D. Garn, eds), Academic Press, New York, 1969, p. 417.
907. McNeill, I. C., and Neil, D., in Thermal Analysis (R. F. Schwenker and P. D. Garn, eds), Academic Press, New York, 1969, p. 353.
908. Mead, W. L., ADVANCES IN MASS SPECTROMETRY, Elsevier, Amsterdam, 1966.
909. Meares, P., POLYMER STRUCTURE AND BULK PROPERTIES, Van Nostrand, New York, 1965.
910. Meares, P., MEMBRANE SEPARATION PROCESSES, Elsevier, Amsterdam, 1976.
911. Mecke, R., and Langenbucher, J., INFRARED SPECTRA OF SELECTED CHEMICAL COMPOUNDS (1800 SPECTRA), Heyden, London, 1965.
912. Meinecke, E., in Encyclopedia of Polymer Science and Technology, Vol. 9 (H. F. Mark, N. G. Gaylord and N. M. Bikales, eds), Wiley–Interscience, New York, 1968, p. 525
913. Meloan, C. E., ELEMENTARY INFRARED SPECTROSCOPY, Macmillan, New York, 1963.
914. Melville, H., and Gowenlock, B. G., EXPERIMENTAL METHODS IN GAS REACTIONS, Macmillan, London, 1964.
915. Memory, J. D., QUANTUM THEORY OF MAGNETIC RESONANCE PARAMETERS, McGraw-Hill, New York, 1968.

722

916. Merrigan, J. A., Green, J. H., and Tao, S. J., in Physical Methods of Chemistry, Vol. 1, Part 3D (A. Weissberger and B. W. Rossieter, eds), Wiley–Interscience, New York, 1972, p. 501.
917. Merrington, A. L., VISCOSIMETRY, Arnold, London, 1949.
918. Meyer, J. A., Rogers, C. E., Stannet, V., and Szwarc, M., in Permeability of Plastics, Films and Coated Papers to Gases and Vapors, TAPPI Monograph Series, No. 23, The America Polymer and Plastics Institute, New York, 1962, p. 28.
919. Meyer, J. A., Rogers, C. E., Stannet, V., and Szwarc, M., in Permeability of Plastics, Films and Coated Papers to Gases and Vapors, TAPPI Monograph Series, No. 23, The America Polymer and Plastics Institute, New York, 1962, p. 49 and p. 62.
920. Meyerhoff, G., in Gel Permeation Chromatography (K. H. Altgelt and L. Segal, eds), Dekker, New York, 1971, p. 215.
921. Michel, R., Maitrot, M., and Asch, G., Rev. Phys. Appl., 8, 315 (1973).
922. Middleman, S., THE FLOW OF HIGH POLYMERS, Wiley–Interscience, New York, 1968.
923. Mika, J., and Török, T., ANALYTICAL EMISSION SPECTROSCOPY, Butterworths, London, 1974.
924. Mikes, D., HANDBOOK OF CHROMATOGRAPHIC AND SEPARATION METHODS, Wiley–Interscience, New York, 1978.
925. Millane, J. J., and Bryan, S. J., Polym. Plast., 36, 427 (1968).
926. Miller, R. G. J., and Stace, B. C., LABORATORY METHODS IN INFRARED SPECTROSCOPY, 2nd ed., Heyden, London, 1972.
927. Miller, R. L., in Encyclopedia of Polymer Science and Technology, Vol. 4 (H. F. Mark, N. G. Gaylord and N. M. Bikales, eds), Wiley–Interscience, New York, 1966, p. 449.
928. Miller, R. L., in Polymer Handbook, 2nd ed. (J. Brandrup, E. H. Immergut, and W. McDowell, eds), Wiley–Interscience, New York, 1975, p. III-1.
929. Mills, N. J., in Polymer Science, Vol. 1 (A. D. Jenkins, ed.), North-Holland, Amsterdam, 1972, Chap. 7.
930. Milne, G. W. A., MASS SPECTROMETRY: TECHNIQUES AND APPLICATIONS, Wiley–Interscience, New York, 1971.
931. Minslow, K., INTRODUCTION TO STEREOCHEMISTRY, Benjamin, New York, 1965.
932. Mita, I., in Aspects of Degradation and Stabilization of Polymers, (H. H. G. Jellinek, ed.), Elsevier, New York, 1978, p. 247.
933. Moacanin, J., Holden, G., and Tschoegl, N. W., BLOCK COPOLYMERS, Wiley–Interscience, New York, 1969.
934. Mohacsi, E., J. Chem. Educ., 41, 38 (1964).
935. Molau, G. E., in Characterization of Macromolecular Structure, Nat. Acad. Sci., U.S.A., Publ. 1573, Washington, D.C., 1968, p. 245.
936. Molau, G. E., COLLOIDAL AND MORPHOLOGICAL BEHAVIOR OF BLOCK AND GRAFT COPOLYMERS, Plenum Press, New York, 1971.
937. Moll, W. L., Kolloid Z. Z. Polym., 223, 48 (1968).
938. Moller, K. D., and Rotschild, D. W. G., FAR INFRARED SPECTROSCOPY, Wiley–Interscience, New York, 1971.
939. Mooradian, A., Jaeger, T., and Stokseth, P., TUNABLE LASERS AND APPLICATIONS, Springer Verlag, Berlin, 1976.
940. Moore, J. C., in Characterization of Macromolecular Structure, Nat. Acad. Sci. U.S.A., Publ. 1573, Washington, D.C., 1968, p. 273.
941. Moore, J. C., in Gel Permeation Chromatography (K. H. Altgelt and L. Segal, eds), Dekker, New York, 1971, p. 157.
942. Moore, W. R., in Progress in Polymer Science, Vol. 1 (A. D. Jenkins, ed.), Pergamon Press, New York, 1967, Chap. 1.
943. Morawetz, H., MACROMOLECULES IN SOLUTION, 2nd ed., Wiley–Interscience, New York, 1975.
944. Moshay, A., and McGrath, J. E., BLOCK COPOLYMERS: OVERVIEW AND CRITICAL SURVEY, Academic Press, New York, 1977.
945. Moskowitz, A., Adv. Chem. Phys., 4, 67 (1962).
946. Mosley, M. H., in Laboratory Methods in Infrared Spectroscopy (R. G. J. Miller and B. C. Stace, eds), Heyden, London, 1972, Chap. 3.

947. Müllen, K., and Pregosin, P. S., FOURIER TRANSFORM NMR TECHNIQUES: A PRACTICAL APPROACH, Academic Press, London, 1976.
948. Munk, M. N., *J. Chromatogr. Sci.*, **8**, 491 (1970).
949. Munro, I. H., in Luminescence in Chemistry (E. J. Bowen, ed.), Van Nostrand, London, 1968, Chap. 2.
950. Murakami, K., in Aspects of Degradation and Stabilization of Polymers (H. H. G. Jellinek, ed.), Elsevier, New York, 1978, Chap. 7.
951. Murayama, T., DYNAMIC MECHANICAL ANALYSIS OF POLYMERIC MATERIALS, Elsevier, Amsterdam, 1978.
952. Murov, S. L., HANDBOOK OF PHOTOCHEMISTRY, Dekker, New York, 1973.
953. Murphy, C. B., *Anal. Chem.*, **36**, 347R (1964).
954. Murphy, C. B., *Anal. Chem.*, **38**, 443R (1966).
955. Murphy, C. B., *Anal. Chem.*, **40**, 380R (1968).
956. Murphy, C. B., *Anal. Chem.*, **42**, 268R (1970).
957. Murphy, C. B., *Anal. Chem.*, **44**, 513R (1972).
958. Murphy, C. B., in Differential Thermal Analysis (R. C. Mackenzie, ed.), Academic Press, London, 1970, Chap. 23.
959. Muus, L. T., and Atkins, P. W., ELECTRON SPIN RELAXATION IN LIQUIDS, Plenum Press, New York, 1972.
960. Myers, R. J., MOLECULAR MAGNETISM AND MAGNETIC RESONANCE SPECTROSCOPY, Prentice-Hall, Englewood Cliffs, N.J., 1973.
961. Nachod, F. C., and Phillips, W. D., DETERMINATION OF ORGANIC STRUCTURES BY PHYSICAL METHODS, Vols 1 and 2, Academic Press, New York, 1962.
962. Nachod, F. C., and Zuckerman, J. J., DETERMINATION OF ORGANIC STRUCTURES BY PHYSICAL METHODS, Vols 3 and 4, Academic Press, New York, 1971.
963. Nagai, K., in Progress in Polymer Science in Japan, Vol. 1 (M. Imoto and S. Onogi, eds), Kodansha, Tokyo and Wiley, New York, 1971, p. 215.
964. Nakajima, N., in Gel Permeation Chromatography (K. H. Altgelt and L. Segal, eds), Dekker, New York, 1971, p. 356.
965. Nakanishi, K., and Solomon, P. H., INFRARED ABSORPTION SPECTROSCOPY, 3rd ed., Elsevier, Amsterdam, 1977.
966. Natta, G., *Angew. Chem.*, **68**, 393 (1956).
967. Natta, G., *Angew. Chem.*, **76**, 553 (1964).
968. Natta, G., and Danusso, F., STEREOREGULAR POLYMERS AND STEREOSPECIFIC POLYMERIZATIONS, Vols 1 and 2, Pergamon Press, Oxford, 1967.
969. Navratil, J. D., and Sievers, R. E., *Intern. Lab.*, Nov./Dec. 1977, p. 26.
970. Nesterov, A. E., and Lipatov, Yu. S., INVERSE GAS CHROMATOGRAPHY IN THE THERMODYNAMICS OF POLYMERS, Naukova Dumka, Kiev, 1976 (in Russian).
971. Newman, S. B., in Analytical Chemistry of Polymers, Part 3 (G. M. Kline, ed.), Wiley–Interscience, New York, 1962, Chap. 3.
972. Niederwieser, A., PROGRESS IN THIN-LAYER CHROMATOGRAPHY AND RELATED METHODS, Vols 1–3, Wiley–Interscience, New York, 1970–1972.
973. Nielsen, L. E., MECHANICAL PROPERTIES OF POLYMERS, Reinhold, New York, 1962.
974. Nishijima, Y., *J. Polym. Sci. C*, **31**, 353 (1970).
975. Noggle, J. H., and Schrimer, R. E., THE NUCLEAR OVERHAUSER EFFECT, CHEMICAL APPLICATIONS, Academic Press, New York, 1971.
976. Nordhaus, D. E., and Kinsinger, J. B., *J. Polym. Sci. C*, **43**, 251 (1973).
977. North, A. M., *Brit. Polym. J.*, **7**, 119 (1975).
978. Noshay, A., and McGrath, J. E., BLOCK COPOLYMERS: OVERVIEW AND CRITICAL SURVEY, Academic Press, New York, 1977.
979. Nuffield, E. W., X-RAY DIFFRACTION METHODS, Wiley–Interscience, New York, 1966.
980. Nussbaum, A., ELECTROMAGNETIC AND QUANTUM PROPERTIES OF MATERIALS, Prentice-Hall, Englewood Cliffs, N.J., 1966.
981. Nussbaum, A., and Phillips, R. A., CONTEMPORARY OPTICS FOR SCIENTISTS AND ENGINEERS, Prentice-Hall, Englewood Cliffs, N.J., 1976.

724

982. Nyquist, R. A., INFRARED SPECTRA OF PLASTICS AND RESINS, 2nd ed., Dow Chemical Corps., Willmington, Mich., 1961.

983. Oatley, C. W., THE SCANNING ELECTRON MICROSCOPE, Cambridge University Press, New York, 1972.

984. Ogata, K., and Haykawa, T., RECENT DEVELOPMENTS IN MASS SPECTROMETRY, University of Tokyo Press, Tokyo, 1970.

985. Ogryzlo, E. A., *Photophysiol.*, **5**, 35 (1970).

986. Ogryzlo, E. A., in Singlet Oxygen: Reactions with Organic Compounds and Polymers (B. Rånby and J. F. Rabek, eds), Wiley, Chichester, 1978, p. 4 and p. 17.

987. Ohnishi, S., Ikeda, Y., Kashiwagi, M., and Nitta, I., *Polymer*, **2**, 119 (1961).

988. Ohnishi, S., and Morokuma, K., ELECTRON SPIN RESONANCE—ITS APPLICATION TO CHEMISTRY, Kagakudojin, Kyoto, 1964.

989. O'Konski, C. T., in Encyclopedia of Polymer Science and Technology, Vol. 9 (N. M. Mark, N. G. Gaylord and N. M. Bikales, eds), Wiley–Interscience, New York, 1968, p. 551.

990. O'Konski, C. T., MOLECULAR ELECTRO-OPTICS, Part 1, THEORY AND METHODS, Dekker, New York, 1976.

991. Olsen, E. D., MODERN OPTICAL METHODS OF ANALYSIS, McGraw-Hill, New York, 1975.

992. Onyon, P. F., in Techniques of Polymer Characterization (P. W. Allen, ed.), Butterworths, London, 1959, p. 171.

993. O'Reilly, J. M., and Karasz, F. E., *J. Polym. Sci. C*, **14**, 49 (1966).

994. Oster, G., *Chem. Rev.*, **43**, 319 (1948).

995. Oster, G., in Techniques of Organic Chemistry, Vol. 1 (A. Weissberger, ed.), Wiley–Interscience, New York, 1960, Chap. 32.

996. Oster, G., and Nishijima, Y., *Fortsch. Hochpolym. Forsch., Adv. Polym. Sci.*, **3**, 313, 1964.

997. Oster, G., and Nishijima, Y., in Newer Methods of Polymer Characterization (B. Ke, ed.), Wiley–Interscience, New York, 1964, Chap. 5.

998. Otocka, E. P., *Accounts Chem. Res.*, **6**, 348 (1973).

999. Ouano, A. C., *J. Macromol. Sci. C*, **9**, 123 (1973).

1000. Ouano, A. C., Barrall, E. M., II, and Johnson, J. F., in Techniques and Methods of Polymer Evaluation, Vol. 4, Polymer Molecular Weight, Part 2 (P. E. Slade, Jr, ed.), Dekker, New York, 1975, p. 287

1001. Overhauser, A. W., *Phys. Rev.*, **89**, 689 (1953).

1002. Overhauser, A. W., *Phys. Rev.*, **92**, 477 (1956).

1003. Overton, J. R., in Determination of Thermodynamic and Surface Properties, Vol. 1, Part 5 (A. Weissberger and B. W. Rossitier, eds), Wiley–Interscience, New York, 1971 p. 309.

1004. Page, C. H., and Vigoreux, P., THE INTERNATIONAL SYSTEM OF UNITS (SI), National Bureau of Standards Special Publication 330, National Bureau of Standards, Washinton, D.C., 1974.

1005. Pake, G. E., PARAMAGNETIC RESONANCE, Benjamin, New York, 1962.

1006. Palit, S. R., and Mandal, B. M., *J. Macromol. Sci. C*, **2**, 225 (1968).

1007. Park, G. S., in Diffusion in Polymers (J. Crank and G. S. Park, eds), Academic Press, 1968, Chap. 5.

1008. Parker, C. A., *Adv. Photochem.*, **2**, 305 (1964).

1009. Parker, C. A., PHOTOLUMINESCENCE OF SOLUTION, Elsevier, Amsterdam, 1968.

1010. Parker, T. G., in Polymer Science, Vol. 2 (A. D. Jenkins, ed.), North-Holland, Amsterdam, 1972, Chap. 19.

1011. Parris, N. A., INSTRUMENTAL LIQUID CHROMATOGRAPHY, Elsevier, Amsterdam, 1976.

1012. Pasto, D., and Johnson, C., ORGANIC STRUCTURE DETERMINATIONS, Prentice-Hall, Englewood Cliffs, N.J., 1969.

1013. Patai, S., THE CHEMISTRY OF THE CARBONYL GROUP, Wiley–Interscience, London, 1966.

1014. Patrick, R. L., TREATISE ON ADHESION AND ADHESIVES, Vol. 1, Dekker, New York, 1967.

1015. Paul, M. A., *J. Chem. Educ.*, **48**, 569 (1971).
1016. Parker, F. W., in Techniques of Polymer Characterization (P. W. Allen, ed.), Butterworths, London, 1959, Chap. 5.
1017. Peaker, F. W., *Analyst*, **85**, 235 (1960).
1018. Pease, D. C., HISTOLOGICAL TECHNIQUES FOR ELECTRON MICROSCOPY, Academic Press, New York, 1960.
1019. Pecora, B., in Photochemistry of Macromolecules (R. F. Reinish, ed.), Plenum Press, New York, 1970, p. 145.
1020. Pecora, B., *Nature*, **231**, 73 (1971).
1021. Pecora, R., *Ann. Rev. Biophys. Bioenerg.*, **1**, 257 (1972).
1022. Peebles, L. H., Jr, MOLECULAR WEIGHT DISTRIBUTIONS IN POLYMERS, Wiley–Interscience, New York, 1971.
1023. Peiser, H. S., Rooksby, H. P., and Wilson, A. J. C., X-RAY DIFFRACTION BY POLYCRYSTALLINE MATERIALS, Institute of Physics, London, 1955.
1024. Pennings, A. J., in Characterization of Macromolecular Structure, Nat. Acad. Sci. U.S.A., Publ. 1573, Washington, D.C., 1968, p. 214.
1025. Pennings, A. J., *Polym. Symp.* No. 59, 55 (1977).
1026. Perone, S. P., and Drew, H. D., in Analytical Photochemistry and Photochemical Analysis (J. M. Fitzgerald, ed.), Dekker, New York, 1971, Chap. 7.
1027. Perry, S. G., *J. Gas Chromatogr.*, **2**, 54 (1964).
1028. Perry, S. G., *J. Gas Chromatogr*, **5**, 77 (1967).
1029. Perry, S. G., Amos, R., and Brewer, P. I., PRACTICAL LIQUID CHROMATOGRAPHY, Plenum Press, New York, 1972.
1030. Peterlin, A., *J. Mater. Sci.*, **6**, 490 (1971).
1031. Peterlin, A., *J. Macromol. Sci. B*, **11**, 57 (1975).
1032. Peterlin, A., in Structure and Properties of Oriented Polymers (I. M. Ward, ed.), Applied Science Publ., London, 1975, Chap. 2.
1033. Peticolas, W. L., *Fortsch. Hochpolym. Forsch., Adv. Polym. Sci.*, **9**, 285 (1972).
1034. Peticolas, W. L., Small, E. W., and Fanconi, B., in Polymer Characterization: An Interdisciplinary Approach (C. D. Craver, ed.), Plenum Press, New York, 1971, p. 47.
1035. Phifer, L. H., and Dyer, J., in Gel Permeation Chromatography (K. H. Altgelt and L. Segal, eds), Dekker, New York, 1971, p. 465.
1036. Phillips, D. W., and Pethrick, R. A., *J. Macromol. Sci. C*, **16**, 1 (1978).
1037. Phillips, J. P., SPECTRA STRUCTURE CORRELATION, Academic Press, New York, 1964.
1038. Phillips, R. A., and Gherz, R. D., *Amer. J. Phys.*, **38**, 429 (1970).
1039. Pidcock, A., in Polymer Science, Vol. 2 (A. D. Jenkins, ed.), North-Holland, Amsterdam, 1972, Chap. 21.
1040. Pinner, S. H., A PRACTICAL COURSE IN POLYMER CHEMISTRY, Pergamon Press, New York, 1961.
1041. Pino, P., *Fortsch. Hochpolym. Forsch., Adv. Polym. Sci.*, **4**, 236 (1965).
1042. Pino, P., Ciardelli, F., and Zandomeneghi, M., *Ann. Rev. Phys. Chem.*, **21**, 561 (1970).
1043. Pinsker, Z. G., ELECTRON DIFFRACTION, Butterworth, London, 1953.
1044. Pirani, M., and Yarwood, J., PRINCIPLES OF VACUUM ENGINEERING, Reinhold, London, 1961.
1045. Placzek, G., RAYLEIGH AND RAMAN SCATTERING, U.S. Atomic Energy Commission Translation UCRL-TRANS-526 (L) (1962).
1046. Platzer, N. A. J., COPOLYMERS, POLYBLENDS AND COMPOSITIONS, Advances in Chemistry Series, No. 142, American Chemical Society, Washington, D.C., 1975.
1047. Point, J. J., *Polym. Symp.*, No. 59, 87 (1977).
1048. Poland, D., and Scheraga, H. A., THEORY OF HELIX–COIL TRANSITIONS IN BIOPOLYMERS: STATISTICAL MECHANICAL THEORY OF ORDER–DISORDER TRANSITIONS IN BIOLOGICAL MACROMOLECULES, Academic Press, New York, 1970.
1049. Ponec, V., Knor, Z., and Cerny, S., ADSORPTION ON SOLIDS, Butterworths, London, 1974.
1050. Poole, C. P., Jr, ELECTRON SPIN RESONANCE: A COMPREHENSIVE TREATISE ON EXPERIMENTAL TECHNIQUE, Wiley–Interscience, New York, 1967.

726

1051. Poole, C. P., Jr and Farrah, H. A., THE THEORY OF MAGNETIC RESONANCE, Wiley–Interscience, New York, 1972.
1052. Pople, J. A., Schneider, W. G., and Bernstein, H. J., HIGH RESOLUTION NMR, McGraw-Hill, New York, 1959.
1053. Porod, G., Fortsch. Hochpolym. Forsch., Adv. Polym. Sci., **2**, 363 (1961).
1054. Porter, G., in Techniques of Organic Chemistry, Vol. 8, Part 2, Investigations of Rates and Mechanisms of Reactions (S. L. Fries, E. S. Lewis and A. Weissberger, eds), Wiley–Interscience, New York, 1963, p. 1055.
1055. Porter, Q. N., and Baldas, J., BIOCHEMICAL APPLICATIONS OF MASS SPECTROMETRY, Wiley–Interscience, New York, 1971.
1056. Porter, R. S., and Johnson, J. F., in Polymer Fractionation (M. J. R. Cantow, ed.), Academic Press, New York, 1967, p. 95.
1057. Porter, R. S., and Johnson, J. F., in Characterization of Macromolecular Structure, Nat. Acad. Sci., U.S.A., Publ. 1573, Washington, D.C., 1968, p. 297.
1058. Porter, R. S., and Johnson, J. F., ANALYTICAL CALORIMETRY, Vols 1 and 2, Plenum Press, New York, 1968/1970.
1059. Potts, J. E., in Aspects of Degradation and Stabilization of Polymers (H. H. G. Jellinek, ed.), Elsevier, Amsterdam, 1978, p. 617.
1060. Potts, W. J., Jr, CHEMICAL INFRARED SPECTROSCOPY, Vol. 1, TECHNIQUES, Wiley–Interscience, New York, 1963.
1061. Pouchert, C. J., ALDRICH LIBRARY OF INFRARED SPECTRA, Aldrich Chemical Co., Milwaukee, 1971.
1062. Powell, R. W., in Thermal Conductivity, Vol. 2 (R. P. Tye, ed.), Academic Press, New York, 1969, Chap. 6.
1063. Powles, J. G., Polymer, **1**, 219 (1960).
1064. Pressley, R. J., LASERS WITH SELECTED DATA ON OPTICAL TECHNOLOGY, CRC Press Inc., Florida, 1971.
1065. Price, G. F., in Techniques of Polymer Characterization (P. W. Allen, ed.), Butterworths, London, 1959, Chap. 7.
1066. Price, W. J., in Laboratory Methods in Infrared Spectroscopy (R. G. J. Miller and B. C. Stace, eds), Heyden, London, 1972, Chap. 8.
1067. Prigogine, I., THE MOLECULAR THEORY OF SOLUTIONS, North-Holland, Amsterdam, 1957.
1068. Princen, L. H., SCANNING ELECTRON MICROSCOPY OF POLYMERS AND COATINGS, Appl. Polym. Symp., No. 16 (1971).
1069. Provder, T., and Rosen, E. M., in Gel Permeation Chromatography (K. H. Altgelt and L. Segal, eds), Dekker, New York, 1971, p. 243.
1070. Provder, T., Woodbrey, J. C., and Clark, J. H., in Gel Permeation Chromatography (K. H. Altgelt and L. Segal, eds), Dekker, New York, 1971, p. 493.
1071. Pshezhetskii, S. Ya., Kotov, A. G., Millinchuk, V. K., Roginskii, V. A., and Tupikov, V. I., EPR OF FREE RADICALS IN RADIATION CHEMISTRY, Halsted, New York, 1974.
1072. Purnell, J. H., GAS CHROMATOGRAPHY, Wiley, New York, 1962.
1073. Purnell, J. H., NEW DEVELOPMENT IN GAS CHROMATOGRAPHY, Wiley–Interscience, New York, 1973.
1074. Pusey, P. N., in Photon Correlation and Light Beating Spectroscopy (H. Z. Cummins and E. R. Pike, eds), Plenum Press, New York, 1973.
1075. Quayle, O. R., Chem. Rev., **53**, 439 (1953).
1076. Rabek, J. F., Pure Appl. Chem., **8**, 29 (1971).
1077. Rabek, J. F., in Chemical Kinetics, Vol. 14, Degradation of Polymers (C. H. Bamford and C. F. H. Tipper, eds), Elsevier, Amsterdam, 1975, p. 425.
1078. Rabek, J. F., PRINCIPLES OF PHYSICAL CHEMISTRY OF POLYMERS, Technical University Press, Wroclaw, Poland, 1977 (in Polish).
1079. Rabek, J. F., and Rånby, B., Polym. Eng. Sci., **15**, 40 (1975).
1080. Rabek, J. F., and Rånby, B., Photochem. Photobiol., **24**, 557 (1978).
1081. Raff, R. A. V., and Doak, K. W., CRYSTALLINE OLEFIN POLYMERS, Wiley–Interscience, New York, 1965.
1082. Rafikov, S. R., Pavlova, S. A., and Tverdochlebova, I. I., DETERMINATION OF

727

MOLECULAR WEIGHTS AND POLYDISPERSITY OF HIGH POLYMERS; translated into English by J. Eliass, S. Monson Co., Jerusalem, 1964.
1083. Ramachandran, G. N., CONFORMATION OF BIOPOLYMERS, 2nd ed., Academic Press, London, 1967.
1084. Ramey, K. C., and Brey, W. S., Jr, *J. Macromol. Sci. C*, **1**, 263 (1967).
1085. Ramsey, N. F., *Phys. Rev.*, **78**, 699 (1950).
1086. Rånby, B., and Rabek, J. F., PHOTODEGRADATION, PHOTOOXIDATION AND PHOTOSTABILIZATION OF POLYMERS, Wiley, London, 1975.
1087. Rånby, B., and Rabek, J. F., in Ultraviolet Light Induced Reactions in Polymers (S. S. Labana, ed.), ACS Symposium Series, No. 25, American Chemical Society, Washington, D.C., 1976, p. 391.
1088. Rånby, B., and Rabek, J. F., ESR SPECTROSCOPY IN POLYMER RESEARCH, Springer Verlag, Berlin, 1977.
1089. Rånby, B., and Rabek, J. F., SINGLET OXYGEN: REACTIONS WITH ORGANIC COMPOUNDS AND POLYMERS, Wiley, Chichester, 1978.
1090. Rånby, B., and Rabek, J. F., in Singlet Oxygen: Reactions with Organic Compounds and Polymers (B. Rånby and J. F. Rabek, eds), Wiley, Chichester, 1978, p. 211.
1091. Randall, J. C., POLYMER SEQUENCE DETERMINATION BY CARBON-13 NMR METHOD, Academic Press, New York, 1977.
1092. Randerath, K., THIN-LAYER CHROMATOGRAPHY, Academic Press, New York, 1968.
1093. Rao, C. N. R., ULTRAVIOLET AND VISIBLE SPECTROSCOPY, 2nd ed., Plenum Press, New York, 1967.
1094. Read, B. E., in Structure and Properties of Oriented Polymers (I. M. Ward, ed.), Applied Science Publishers, London, 1975, Chap. 4.
1095. Redfield, A. G., *Adv. Mag. Reson.*, **1**, 1 (1966).
1096. Reed, R. I., ION PRODUCTION BY ELECTRON IMPACT, Academic Press, London, 1962.
1097. Reed, R. I., APPLICATION OF MASS SPECTROMETRY TO ORGANIC CHEMISTRY, Academic Press, New York, 1966.
1098. Ress, D. V., and Bassett, D. C., *J. Mater. Sci.*, **6**, 1021 (1971).
1099. Rehage, G., and Borchard, W., in The Physics of Glassy Polymers (R. N. Haward, ed.), Applied Science Publishers, London, 1973.
1100. Reich, L., *Macromol. Rev.*, **3**, 49 (1968).
1101. Reich, L., and Levi, D. W., *Macromol. Rev.*, **1**, 173 (1967).
1102. Reich, L., and Levi, D. W., in Encyclopedia of Polymer Science and Technology, Vol. 14 (H. F. Mark, N. G. Gaylord and N. M. Bikales, eds), Wiley–Interscience, New York, 1971, p. 1.
1103. Reich, L. and Stivala, S. S., AUTOOXIDATION OF HYDROCARBONS AND POLYMERS, Dekker, New York, 1969.
1104. Reich, L., and Stivala, S. S., ELEMENTS OF POLYMER DEGRADATION, McGraw-Hill, New York, 1971.
1105. Reilley, C. N., and Sawyer, D. T., EXPERIMENTS FOR INSTRUMENTAL METHODS, Elsevier, Amsterdam, 1961.
1106. Reimer, L., ELEKTRONENMIKROSKOPISCHE UNTERSUCHUNGEN UND PRÄPARATIONSMETHODEN, Springer Verlag, Göttingen, 1959.
1107. Rempp, P., Herz, J. E., and Borchard, W., *Fortsch. Hochpolym. Forsch., Adv. Polym. Sci.*, **22**, 105 (1978).
1108. Richardson, J. H., OPTICAL MICROSCOPY IN MATERIALS SCIENCE, Dekker, New York, 1971.
1109. Richardson, M. J., *Brit. Polym. J.*, **1**, 132 (1969).
1110. Riddle, E. H., MONOMERIC ACRYLIC ESTERS, Reinhold, New York, 1954, p. 59.
1111. Riess, G., and Callot, P., in Fractionation of Synthetic Polymers, Principles and Applications (L. H. Tung, ed.), Dekker, New York, 1977, Chap. 5.
1112. Robb, J. C., and Peaker, F. W., FRACTIONATION OF HIGH POLYMERS, Heywood, London, 1961.
1113. Robb, I. D., and Tiddy, G. J. T., *Nucl. Mag. Reson.*, **3**, 279 (1974).

728

1114. Roberts, J. D., NUCLEAR MAGNETIC RESONANCE, APPLICATIONS TO ORGANIC CHEMISTRY, McGraw-Hill, New York, 1959.
1115. Roberts, J. D., INTRODUCTION TO SPIN–SPIN SPLITTING IN HIGH RESO-LUTION NMR, Benjamin, New York, 1961.
1116. Roberts, R. W., and Vanderlice, T. A., ULTRAHIGH VACUUM AND ITS APPLI-CATION, Prentice-Hall, Englewood Cliffs, N.J., 1963.
1117. Robertson, R. E., *Ann. Rev. Mater. Sci.*, **5**, 73 (1975).
1118. Robinson, C. F., Perkins, G. D., and Bell, N. W., in Instruments and Measurements (H. Van Hock and G. Ljunberg, eds), Academic Press, New York, 1961, p. 261.
1119. Roboz, J., INTRODUCTION TO MASS SPECTROMETRY, Wiley–Interscience, New York, 1968.
1120. Rodriguez, F., PRINCIPLES OF POLYMER SYSTEMS, Elsevier, Amsterdam, 1970.
1121. Roff, W. J., and Scott, J. R., FIBRES, FILMS AND RUBBERS, HANDBOOK OF COMMON PROPERTIES, Butterworths, London, 1971.
1122. Rogers, C. E., in Engineering Design For Plastics (E. Baer, ed.), Reinhold, New York, 1964, Chap. 9.
1123. Rogers, C. E., in Physics and Chemistry of Organic Solid State (D. Fox, M. N. Labes and A. Weissberger, eds), Wiley–Interscience, New York, 1965, Chap. 6.
1124. Rogers, C. E., and Machin, D., *Crit. Rev. Macromol. Sci.*, **1**, 245 (1972).
1125. Rogers, C. E., Meyer, J. A., Stannet, V., and Szwarc, M., in Permeability of Plastic Films and Coated Papers to Gases and Vapors, TAPPI Monograph No. 23, The American Polymer and Plastics Institute, New York, 1962, pp. 1, 12, and 78.
1126. Rosen, E. M., and Provder, T., in Gel Permeation Chromatography (K. H. Altgelt and L. Segal, eds), Dekker, New York, 1971, p. 291.
1127. Ross, S., in Encyclopedia of Chemical Technology, Vol. 1 (H. F. Mark, J. J. McKetta and D. F. Othmer, eds.), Wiley–Interscience, New York, 1963, p. 421.
1128. Rosen, S. L., *Polym. Eng. Sci.*, **7**, 115 (1967).
1129. Ross, W. D., in Separation and Purification Methods, Vol. 3 (E. S. Perry, C. J. Vaoss and E. Grushka, eds), Dekker, New York, 1974, p. 1.
1130. Rozantsev, E. G., FREE NITROXYL RADICALS, Plenum Press, New York, 1970.
1131. Rudman, R., *J. Chem. Educ.*, **44**, A7, A99, A187, A289, A399, A499 (1967).
1132. Runyon, J. R., in Gel Permeation Chromatography (K. H. Altgelt and L. Segal, eds), New York, 1971, p. 407.
1133. Rushman, D. F., in Techniques of Polymer Characterization (P. W. Allen, ed.), Butterworths, London, 1959.
1134. Sadron, C., DYNAMIC ASPECTS OF CONFORMATION CHANGES IN BIO-LOGICAL MACROMOLECULES, Reidel, Dordrecht, 1973.
1135. Safford, G. J., and Neuman, A. W., *Fortsch. Hochpolym. Forsch., Adv. Polym. Sci.*, **5**, 1 (1967).
1136. Sakurada, I., and Kai, K., *J. Polym. Sci. C*, **31**, 57 (1970).
1137. Samsonov, G. V., *Uspekh. Khim.*, **30**, 1410 (1961).
1138. Samuels, R. J., in The Science and Technology of Polymer Films, Vol. 1 (O. J. Sweeting, ed.), Wiley–Interscience, New York, 1968, p. 255.
1139. Samuels, R. J., STRUCTURED POLYMER PROPERTIES: THE IDENTIFI-CATION, INTERPRETATION AND APPLICATION OF CRYSTALLINE POLY-MER STRUCTURE, Wiley–Interscience, New York, 1974.
1140. Sanchez, I. C., *J. Macromol. C*, **11**, 113 (1974).
1141. Sanchez, I. C., *Polym. Symp.*, No. 59, 109 (1977).
1142. Sanderson, R. T., VACUUM MANIPULATION OF VOLATILE COMPOUNDS, Wiley–Interscience, New York, 1948.
1143. Saratovkin, D. D., DENDRIC CRYSTALLIZATION, Consultants Bureau, New York, 1959.
1144. Sauer, J. A., Richardson, G. C., and Morrow, D. R., *J. Macromol. Sci. C*, **9**, 149 (1973).
1145. Sauer, J. A., and Woodward, A. E., *Rev. Modern Phys.*, **32** 88 (1960).
1146. Scatchard, G., *Chem. Rev.*, **8**, 321 (1931).
1147. Schaap, A. P., SINGLET OXYGEN, Wiley–Interscience, New York, 1976.
1148. Schachman, H. K., ULTRACENTRIFUGATION IN BIOCHEMISTRY, Academic Press, New York, 1962.

1149. Schaefer, J., Stejskal, E. O., and Buchdahl, R., *Macromolecules*, **8**, 291 (1975).
1150. Schaefer, J., Stejskal, E. O. and Buchdahl, R., *Macromolecules*, **10**, 384 (1977).
1151. Schaeffer, H. F., MICROSCOPY FOR CHEMISTS, Van Nostrand, New York, 1953.
1152. Schäfer, F. P., DYE LASERS, 2nd ed., Springer Verlag, Berlin, 1977.
1153. Schaufele, R. F., *Trans. N.Y. Acad. Sci.*, **30**, 69 (1967).
1154. Schaufele, R. F., *Macromol. Rev.*, **4**, 67 (1970).
1155. Schaufele, R. F., in Encyclopedia of Polymer Science and Technology, Vol. 12 (H. F. Mark, N. G. Gaylord, and N. M. Bikales, eds), Wiley–Interscience, New York, 1970, p. 397.
1156. Schawlow, A. L., *Scient. Amer.*, **204**, 52 (1961).
1157. Scheffler, L., and Stegman, H. B., ELEKTRONENSPINRESONANZ, Springer Verlag, Berlin, 1970.
1158. Scheinmann, F., AN INTRODUCTION TO SPECTROSCOPIC METHODS FOR THE IDENTIFICATION OF ORGANIC COMPOUNDS, Vol. 1, NUCLEAR MAGNETIC RESONANCE AND INFRARED SPECTROSCOPY, Pergamon Press, Oxford, 1970.
1159. Schellman, J. A., *Chem. Rev.*, **75**, 323 (1975).
1160. Schlag, E. W., Schneider, S., and Fischer, S. F., *Ann. Rev. Phys. Chem.*, **22**, 465 (1971).
1161. Schnabel, W., in Aspects of Degradation and Stabilization of Polymers (H. H. G. Jellinek, ed.), Elsevier, New York, 1978, Chap. 4.
1162. Schnabel, W., and Kiwi, J., in Aspects of Degradation and Stabilization of Polymers (H. H. G. Jellinek, ed.), Elsevier, New York, 1978, Chap. 5.
1163. Schneider, N. S., *Anal. Chem.*, **33**, 1829 (1961).
1164. Schneider, N. S., *J. Polym. Sci. C*, **8**, 179 (1965).
1165. Schnell, G., *Ber. Bunsenges Phys. Chem.*, **70**, 297 (1966).
1166. Schoffa, G., ELEKTRONSPINRESONANZ IN DER BIOLOGIE, Braun, Karlsruhe, 1963.
1167. Scholte, T. H., in Techniques and Methods of Polymer Evaluation, Vol. 4, Polymer Molecular Weights, Part 2 (P. E. Slade, Jr, ed.), Dekker, New York, 1975, Chap. 8.
1168. Schomburg, G., GASCHROMATOGRAPHIE, Verlag Chemie, Weinheim, 1977.
1169. Schoon, T. G. F., *Brit. Polym. J.*, **2**, 86 (1970).
1170. Schulman, S. G., FLUORESCENCE AND PHOSPHORESCENCE: PHYSICO-CHEMICAL PRINCIPLES AND PRACTICE, Pergamon Press, New York, 1977.
1171. Schulz, G. V., in Die Physik der Hochpolymern, Vol. 2 (H. A. Stuart, ed.), Springer Verlag, Berlin, 1953, p. 748.
1172. Schulz, G. V., in Copolymers, Polyblends and Composites (N. A. Platzer, ed.), Advances in Chemistry Series, No. 142, American Chemical Society, Washington, D.C., 1975, p. 14.
1173. Schulz, R. C., *Fortsch. Hochpolym. Forsch.*, *Adv. Polym. Sci.*, **4**, 235 (1965).
1174. Schulze, D., DIFFERENTIALTHERMOANALYSE, Verlag Chemie, Weinheim, 1969.
1175. Schumacher, R. T., INTRODUCTION TO MAGNETIC RESONANCE: PRINCIPLES AND APPLICATIONS, Benjamin, New York, 1970.
1176. Schupp, O. E., TECHNIQUES OF ORGANIC CHEMISTRY, Vol. 13, GAS CHROMATOGRAPHY (E. S. Perry and A. Weissberger, eds), Wiley–Interscience, New York, 1968.
1177. Schurz, J., VISCOSITY MEASUREMENTS OF HIGH POLYMERS, Berliner Union Verlag, Stuttgart, 1972.
1178. Schurz, J., PHYSIKALISCHE CHEMIE DER HOCHPOLYMEREN, Springer Verlag, Berlin, 1974.
1179. Schwarz, J. C. P., PHYSICAL METHODS IN ORGANIC CHEMISTRY, Oliver and Boyd, Edinburgh, 1964.
1180. Schwenker, R. F., THERMOANALYSIS OF FIBERS AND FIBER-FORMING POLYMERS, Wiley–Interscience, New York, 1966.
1181. Schwenker, R. F., and Garn, P. D., THERMAL ANALYSIS, Vol. 1. INTRUMENT-ATION, ORGANIC MATERIALS AND POLYMERS, Academic Press, New York, 1969.
1182. Scott, A. I. INTERPRETATION OF THE ULTRAVIOLET SPECTRA OF NAT-URAL PRODUCTS, Pergamon Press, New York, 1964.

730

1183. Scott, R. P. W., TECHNIQUES OF ORGANIC CHEMISTRY, Vol. 11, CONTEM-PORARY LIQUID CHROMATOGRAPHY (E. S. Perry and A. Weissberger, eds), Wiley–Interscience, New York, 1976.
1184. Scott, R. P. W., LIQUID CHROMATOGRAPHY DETECTORS, Elsevier, Amsterdam, 1977.
1185. Scott-Blair, G. W., ELEMENTARY RHEOLOGY, Academic Press, London, 1969.
1186. Screaton, R. M., in Newer Methods of Polymer Characterization (B. Ke, ed.), Wiley–Interscience, New York, 1964.
1187. Searle, N. Z., in Analytical Photochemistry and Photochemical Analysis (J. M. Fitzgerald, ed.), Dekker, New York, 1971, Chap. 9.
1188. Seibl, J., MASSENSPEKTROMETRIE, Akademischer Verlag, Frankfurt/Main, 1970.
1189. Ševčik, J., DETECTORS IN GAS CHROMATOGRAPHY, Elsevier, Amsterdam, 1976.
1190. Sewell, P. R., Ann. Rev. NMR Spectr., 1, 165 (1968).
1191. Shafrin, E. G., in Polymer Handbook, 2nd ed. (J. Brandrup, E. H. Immergut, and W. McDowell, eds), Wiley–Interscience, New York, 1975, p. III-221.
1192. Shapiro, R. H., SPECIAL EXERCISES IN STRUCTURE DETERMINATIONS OF ORGANIC COMPOUNDS, Rinehart and Winston, New York, 1969.
1193. Shapkin, V. V., ELECTRON PARAMAGNETIC RESONANCE, Gosud. Pedagog. Inst., Leningrad, 1974 (in Russian).
1194. Sharples, A., INTRODUCTION TO POLYMER CRYSTALLIZATION, Arnold, London, 1966.
1195. Sharples, A., in Polymer Science, Vol. 1 (A. D. Jenkins, ed.), North-Holland, Amsterdam, Chap. 4.
1196. Shellard, E. J., QUANTITATIVE PAPER AND THIN LAYER CHROMATO-GRAPHY, Academic Press, New York, 1968.
1197. Shen, M. C., and Eisenberg, A., Rubber Chem. Technol., 43, 95 (1970).
1198. Shepherd, I. W., Rep. Progr. Phys., 38, 565 (1975).
1199. Shimoda, K., HIGH RESOLUTION LASER SPECTROSCOPY, Springer Verlag, Berlin, 1977.
1200. Shitrin, K. S., and Zelmanovich, I. L., TABLES OF LIGHT SCATTERING, Part 1, TABLES OF ANGULAR FUNCTIONS, Hydrometerological Publ. House, Leningrad, 1966.
1201. Shlyapintokh, V. Y., Karpukhin, O. N., Postnikov, L. M., Tsepalov, V. F., Vichutienskii, V. F., and Zakharov, I. V., CHEMILUMINESCENCE TECHNIQUES IN CHEMICAL REACTIONS, Consultants Bureau, Plenum Press, New York, 1968.
1202. Shrader, S. R., INTRODUCTION TO MASS SPECTROMETRY, Allyn and Bacon, Boston, 1971.
1203. Shultz, A. R., in Characterization of Macromolecular Structures, Nat. Acad. Sci. U.S.A., Publ. 1573, Washington, D.C., 1968, p. 407.
1204. Shur, H., ANWENDUNGEN DER KERNMAGNETISCHEN RESONANZ IN ORGANISCHEN CHEMIE, Springer Verlag, New York, 1965.
1205. Shurcliff, W. A., POLARIZED LIGHT, PRODUCTION AND USE, Harvard University Press, Cambridge, Mass., 1962.
1206. Shurcliff, W. and Ballard, S., POLARIZED LIGHT, Van Nostrand, New York, 1964.
1207. Sieghban, K., J. Electron Spectr. Rel. Phenom., 5, 3 (1974).
1208. Sieghban, K., in Molecular Spectroscopy (A. R. West, ed.), Heyden, London, 1977, p. 227.
1209. Sieghban, K., Nordling, C., Johansson, G., Hedman, J., Héden, P. F., Hamrin, K., Gelius, U., Bergmark, T., Werme, L. O., Manne, R., and Bear, Y., ESCA APPLIED TO FREE MOLECULES, North-Holland, Amsterdam, 1969.
1210. Siegman, A. E., AN INTRODUCTION TO LASERS AND MASERS, McGraw-Hill, New York, 1971.
1211. Sievers, R. E., NUCLEAR MAGNETIC RESONANCE SHIFT REAGENTS, Academic Press, New York, 1973.
1212. Siggia, S., QUANTITATIVE ORGANIC ANALYSIS VIA FUNCTIONAL GROUPS, 3rd ed., Wiley–Interscience, New York, 1963.
1213. Silverstein, R. M., Bassler, G. C., and Moril, T. C., SPECTROMETRIC IDENTIFI-CATION OF ORGANIC COMPOUNDS, 2nd ed., Wiley–Interscience, New York, 1974.

1214. Simha, R., and Utracki, L. A., in Block Copolymers (S. L. Aggarwal, ed.), Plenum Press, New York, 1970, p. 107.
1215. Simon, I., INFRARED RADIATION, Van Nostrand, Princeton, N.J., 1966.
1216. Simpson, C., GAS CHROMATOGRAPHY, Kogan Page, London, 1977.
1217. Sinnott, K. M., *J. Polym. Sci. C*, **14**, 141 (1966).
1218. Sjöstrand, F. S., ELECTRON MICROSCOPY OF CELLS AND TISSUES, Vol. 1, INSTRUMENTATION AND TECHNIQUES, Academic Press, New York, 1967.
1219. Slade, P. E., Jr, in Techniques and Methods of Polymer Evaluation, Vol. 4, Polymer Molecular Weights, Part 1 (P. E. Slade, Jr and L. T. Jenkins, eds), Dekker, New York, 1975, Chap. 1.
1220. Sladkova, J., INTERFERENCE OF LIGHT, Illiffe Books, London, 1968.
1221. Slichter, C. P., PRINCIPLES OF MAGNETIC RESONANCE, Harper and Row, New York, 1963.
1222. Slichter, W. P., *Fortsch. Hochpolym. Forsch., Adv. Polym. Sci.*, **1**, 35 (1958).
1223. Slichter, W. P., *J. Polym. Sci. C*, **14**, 33 (1966).
1224. Sloane, H. J., in Polymer Characterization: An Interdisciplinary Approach (C. D. Craver, ed.), Plenum Press, New York, 1971, p. 15.
1225. Slonim, I. Ya., Lyubimov, A. N., THE NMR OF POLYMERS, Plenum Press, New York, 1970.
1226. Small, P. A., *Fortsch. Hochpolym. Forsch., Adv. Polym. Sci.*, **18**, 1 (1975).
1227. Smith, E. J., in *Analytical Photochemistry and Photochemical Analysis* (J. M. Fitzgerald, ed.), Dekker, New York, 1971, Chap. 1.
1228. Smith, H. F., in Polymer Characterization: An Interdisciplinary Approach (C. D. Craver, ed), Plenum Press, New York, 1971, p. 249.
1229. Smith, R. A., Jones, F. E., and Chasmar, R. P., THE DETECTION AND MEASUREMENTS OF INFRARED RADIATION, Oxford University Press, Oxford, 1957.
1230. Smith, W. J., MODERN OPTICAL ENGINEERING, THE DESIGN OF OPTICAL SYSTEMS, McGraw-Hill, New York, 1966.
1231. Smith, W. V., *Rubber Chem. Technol.*, **45**, 667 (1972).
1232. Smith, W. V., THE LASER, McGraw-Hill, New York, 1966.
1233. Smothers, W. J., and Chiang, Y., HANDBOOK OF DIFFERENTIAL THERMAL ANALYSIS, Chemical Publ. Co., New York, 1966.
1234. Smyth, C. P., DIELECTRIC BEHAVIOR AND STRUCTURE, McGraw-Hill, New York, 1955.
1235. Snavely, B. B., in Organic Molecular Photophysics, Vol. 1 (J. B. Birks, ed.), Wiley–Interscience, London, 1973, Chap. 5.
1236. Snyder, L. R., *Adv. Chromatogr.*, **4**, 3 (1967).
1237. Snyder, L. R., in Chromatography (E. Heftman, ed.), Reinhold, New York, 1967, p. 93.
1238. Snyder, L. R., PRINCIPLES OF ADSORPTION CHROMATOGRAPHY, Dekker, New York, 1968.
1239. Snyder, L. R., and Kirkland, J. J., INTRODUCTION TO MODERN LIQUID CHROMATOGRAPHY, Wiley–Interscience, New York, 1974.
1240. Soignet, D. M., in Characterization of Metal and Polymers Surfaces, Vol. 2 (L. H. Lee, ed.), Academic Press, New York, 1977, p. 73.
1241. Solomon, I., *Phys. Rev.*, **99**, 559 (1955).
1242. Sommersall, A. C., and Guillet, J. E., *J. Macromol. Sci. C*, **13**, 135 (1975).
1243. Southall, J. P. C., MIRRORS, PRISMS AND LENSES, Dover, New York, 1964.
1244. Sperling, L. H., RECENT ADVANCES IN POLYMER BLENDS, GRAFTS AND BLOCKS, Plenum Press, New York, 1974.
1245. Sperling, L. H., in Encyclopedia of Polymer Science and Technology, Suppl. No. 1 (N. Bikales, ed.), Wiley–Interscience, New York, 1976, p. 288.
1246. Sperling, L. H., *Makromol. Rev.*, **12**, 141 (1977).
1247. Spiteller, G., MASSENSPEKTROMETRISCHE STRUKTURANALYSE ORGANISCHER VERBINDUNGEN, Verlag Chemie, Weinheim, 1966.
1248. Squires, T. L., AN INTRODUCTION TO ELECTRON SPIN RESONANCE, Academic Press, New York, 1964.
1249. Stace, B. C., in Laboratory Methods in Infrared Spectroscopy, 2nd ed. (R. G. J. Miller and B. C. Stace, eds), Heyden, London, 1972, Chap. 17.

732

1250. Stacey, K. A., LIGHT SCATTERING IN PHYSICAL CHEMISTRY, Butterworths, London, 1956.
1251. Stahl, E., THIN LAYER CHROMATOGRAPHY, Springer Verlag, Berlin, 1969.
1252. Stannett, V., in Diffusion of Polymers (J. Crank and G. S. Park, eds), Academic Press, London, 1968, Chap. 2.
1253. Stannett, V., and Yasuda, H., in Testing of Polymers, Vol. 1 (V. Schmitz, ed.), Wiley–Interscience, New York, 1965, Chap. 13.
1254. Statton, W. O., in Newer Methods of Polymer Characterization (B. Ke, ed.), Wiley–Interscience, New York, 1964, Chap. 6.
1255. Statton, W. O., in Handbook of X-Rays in Research and Analysis (M. Kaeble, ed.), McGraw-Hill, New York, 1967, Chap. 21.
1256. Staudinger, H., DIE HOCHMOLEKULAREN ORGANISCHER VERBINDUN-GEN, Springer Verlag, 1960.
1257. Steel, W. H., INTERFEROMETRY, Cambridge University Press, New York, 1967.
1258. Stein, R. S., in Newer Methods of Polymer Characterization (B. Ke, ed.), Wiley–Interscience, New York, 1964, Chap. 4.
1259. Stein, R. S., in Polymer Science and Technology, Vol. 1, Structure and Properties of Polymer Films (R. W. Lentz and R. S. Stein, eds), Plenum Press, New York, 1973, p. 1.
1260. Stein, R. S., and Wilkes, G. L., in Structure and Properties of Oriented Polymers (I. M. Ward, ed.), Applied Science Publishers, London, 1975, Chap. 3.
1261. Stenhagen, E., Abrahamson, S., and McLafferty, F. W., REGISTRY OF MASS SPECTRAL DATA, Vols 1–4, Wiley–Interscience, New York, 1974.
1262. Stern, E. S., and Timmons, T. C. J., ELECTRONIC ABSORPTION SPECTRO-SCOPY IN ORGANIC CHEMISTRY, St Martin's Press, New York, 1971.
1263. Stevens, B., in Singlet Oxygen: Reactions with Organic Compounds and Polymers (B. Rånby and J. F. Rabek, eds), Wiley, Chichester, 1978, p. 54.
1264. Stevens, M. P., TECHNIQUES AND METHODS OF POLYMER EVALUATION, Vol. 3, CHARACTERIZATION AND ANALYSIS OF POLYMERS BY GAS CHROMATOGRAPHY (P. E. Slade and L. T. Jenkins, eds), Dekker, New York, 1969.
1265. Stevenson, R., and Johnson, R., BASIC LIQUID CHROMATOGRAPHY, Varian Associates, Palo Alto, Calif., 1977.
1266. Stewart, A. T., and Roellig, L., POSITRON ANNIHILATION, Academic Press, New York, 1967.
1267. Stewart, J. E., in Interpretive Spectroscopy (S. K. Freeman, ed.), Reinhold, New York, 1965.
1268. Stimler, S. S., and Kagarise, R. E., INFRARED SPECTRA OF PLASTICS AND RESINS, Part 2, NRL Report 6392, May 23, 1966.
1269. Stivala, S. S., and Gabbay, S. M., in Aspects of Degradation and Stabilization of Polymers (H. H. G. Jellinek, ed.), Elsevier, Amsterdam, 1978, Chap. 13.
1270. Stothers, J. B., Quart. Rev. (London), 19, 144 (1965).
1271. Stothers, J. B., CARBON-13 NMR SPECTROSCOPY, Academic Press, New York, 1972.
1272. Stout, G. H., and Jensen, L. H., X-RAY STRUCTURE DETERMINATION, Macmillan, New York, 1968.
1273. Strauss, E. L., in Aspects of Degradation and Stabilization of Polymers (H. H. G. Jellinek, ed.), Elsevier, Amsterdam, 1978, Chap. 11.
1274. Strazielle, C., and Benoit, H., Pure Appl. Chem., 26, 451 (1971).
1275. Strobel, H. A., CHEMICAL INSTRUMENTATION: A SYSTEMATIC APPROACH, Addison-Wesley, Reading, Mass., 1973.
1276. Sundelöf, L. O., Arkiv Kemi, 25, 1 (1965).
1277. Sütterlin, N., in Polymer Handbook, 2nd ed. (K. Bradrup, E. H. Immergut, and W. McDowell, eds), Wiley–Interscience, New York, 1975, p. IV-135.
1278. Svedberg, T., and Kedersen, K. C., ULTRACENTRIFUGATION, Clarendon Press, Oxford, 1940.
1279. Swartz, H. M., Bolton, J. R., and Borg, D. C., BIOLOGICAL APPLICATIONS OF ELECTRON SPIN RESONANCE, Wiley–Interscience, New York, 1972.

733

1280. Swift, H. F., ELECTRON MICROSCOPY, Kogan Page, London, 1977.
1281. Szwarc, M., *Polym. Eng. Sci.*, **13**, 1 (1973).
1282. Szymanski, H. A., IR THEORY AND PRACTICE OF INFRARED SPECTRO-SCOPY, Plenum Press, New York, 1964.
1283. Szymanski, H. A., INTERPRETED INFRARED SPECTRA, Vols 1–3, Plenum Press, New York, 1964–1967.
1284. Szymanski, H. A., CORRELATION OF INFRARED AND RAMAN SPECTRA OF ORGANIC COMPOUNDS, Hertillon Press, Cambridge Springs, Pa. 1969.
1285. Szymanski, H. A., INFRARED BAND HANDBOOK, Vols 1 and 2, Plenum Press, New York, 1970.
1286. Szymanski, H. A., RAMAN SPECTROSCOPY: THEORY AND PRACTICE, Vols 1 and 2, Plenum Press, New York, 1967/1970.
1287. Tadekoro, H., and Kobyashi, M., in Polymer Spectroscopy (D. O. Hummel, ed.), Verlag Chemie, Weinheim, 1974, p. 3.
1288. Tager, A. A., and Dreval, V. E., *Uspekh. Khim.*, **36**, 888 (1967).
1289. Tandford, C., PHYSICAL CHEMISTRY OF MACROMOLECULES, 3rd ed., Wiley–Interscience, New York, 1965.
1290. Tasumi, M., in Vibrational Spectroscopy—Modern Trends (A. J. Barnes, and W. J. Orville-Thomas, eds), Elsevier, Amsterdam, 1977, Chap. 23.
1291. Taylor, H. A., in Analytical Photochemistry and Photochemical Analysis (J. M. Fitzgerald, ed.), Dekker, New York, 1971, Chap. 3.
1292. Taylor, R. J., THE PHYSICS OF CHEMICAL STRUCTURE, Unilever, London, 1969.
1293. Thomas, W., Jr, SPSE HANDBOOK OF PHOTOGRAHPIC SCIENCE AND ENGINEERING, Wiley–Interscience, New York, 1973.
1294. Thornton, P. R., SCANNING ELECTRON MICROSCOPY, Chapman and Hall, London, 1968.
1295. Tinoco, I., Jr, *Adv. Chem. Phys.*, **4**, 113 (1962)
1296. Tobias, R. S., *J. Chem. Educ.*, **44**, 70 (1967).
1297. Tobin, M. C., LASER RAMAN SPECTROSCOPY, Wiley–Interscience, Chichester, 1971.
1298. Tobolsky, A. V., PROPERTIES AND STRUCTURE OF POLYMERS, Wiley–Interscience, New York, 1960.
1299. Tobolsky, A. V., POLYMER SCIENCE AND MATERIALS, Wiley–Interscience, New York, 1971.
1300. Todd, H. N., PHOTOGRAPHIC SENSITOMETRY, Wiley–Interscience, New York, 1976.
1301. Tombs, M. P., and Peacock, A. R., THE OSMOTIC PRESSURE OF BIOLOGICAL MACROMOLECULES, Clarendon Press, Oxford, 1975.
1302. Tompa, H., POLYMER SOLUTIONS, Butterworths, London, 1956.
1303. Touchstone, J. C., QUANTITATIVE THIN LAYER CHROMATOGRAPHY, Wiley–Interscience, New York, 1973.
1304. Touchstone, J. C., and Dobbins, M. F., PRACTICE OF THIN LAYER CHROMATO-GRAPHY, Wiley–Interscience, New York, 1978.
1305. Trost, B., PROBLEMS IN SPECTROSCOPY, Benjamin, New York, 1967.
1306. Tsuji, K., *Fortsch. Hochpolym. Forsch.*, *Adv. Polym. Sci.*, **12**, 131 (1973).
1307. Tsvetkov, V. N., Eskin, V. Yu., and Frenkel, S. Ya., STRUCTURE OF MACRO-MOLECULES IN SOLUTION, Butterworths, London, 1970.
1308. Tung, L. H., in Encyclopedia of Industrial Chemical Analysis, Vol. 2 (F. D. Snell and C. L. Hilton, eds), Wiley–Interscience, New York, 1966, p. 412.
1309. Tung, L. H., in Polymer Fractionation (M. J. R. Cantow, eds), Academic Press, New York, 1966, p. 379.
1310. Tung, L. H., in Characterization of Macromolecular Structure, Nat. Acad. Sci., U.S.A., Publ. 1573, Washington, D.C., 1968, p. 273.
1311. Tung, L. H., *J. Macromol. Sci. C*, **6**, 51 (1971).
1312. Tung, L. H., in Gel Permeation Chromatography (K. H. Altgelt and L. Segal, eds), Dekker, New York, 1971, p. 73
1313. Tung, L. H., in Gel Permeation Chromatography (K. H. Altgelt and L. Segal, eds), Dekker, New York, 1971, p. 145.

734

1314. Tung, L. H., FRACTIONATION OF SYNTHETIC POLYMERS: PRINCIPLES AND PRACTICE, Dekker, New York, 1977.
1315. Tung, L. H. and Moore, J. C., in Fractionation of Synthetic Polymers: Principles and Practice (L. H. Tung, ed.), Dekker, New York, 1977, Chap. 6.
1316. Tunnicliff, D. D., Weadsworth, P. A., and Schissler, D. O., MASS AND ABUNDANCE TABLES, Shell Development Co., Emeryville, Calif., 1965.
1317. Turnbull, H., Barton, R. S., and Rieviere, J. C., AN INTRODUCTION TO VACUUM TECHNIQUE, Wiley–Interscience, New York, 1962
1318. Turner, D. T., *Polymer*, **19**, 789 (1978).
1319. Turner, J. J., in Vibrational Spectroscopy—Modern Trends (A. J. Barnes and W. J. Orville-Thomas, eds), Elsevier, Amsterdam, 1977, p. 7.
1320. Tye, R. P., THERMAL CONDUCTIVITY, Academic Press, New York, 1969.
1321. Ueberreiter, K., in Diffusion in Polymers (J. Crank and G. S. Park, eds), Academic Press, London, 1968, Chap. 7.
1322. Uhrig, K., *J. Chem. Educ.*, **22**, 582 (1945).
1323. Ulrich, R. D., in Techniques and Methods of Polymer Evaluation, Vol. 4, Polymer Molecular Weights, Part I (P. E. Slade, Jr, ed.), Dekker, New York, 1975, Chap. 2.
1324. Urbanski, J., Czerwinski, W., and Zowall, H., HANDBOOK OF ANALYSIS OF SYNTHETIC POLYMERS, Halstead–Wiley, New York, 1977.
1325. Urwin, J. R., in Light Scattering From Polymer Solution (M. B. Huglin, ed.), Academic Press, New York, 1972, p. 789.
1326. Urusu, I., ELECTRON SPIN RESONANCE, Dunod, Paris, 1968 (in French).
1327. Vainshtein, B. K., STRUCTURE ANALYSIS BY ELECTRON DIFFRACTION, Pergamon Press, Oxford, 1964.
1328. Vainshtein, B. K., DIFFRACTION OF X-RAYS BY CHAIN MOLECULES, Elsevier, Amsterdam, 1966.
1329. Van Amerongen, G. J., *Rubber Chem. Technol.*, **37**, 1065 (1964).
1330. Vanasse, G. A., and Sakai, H., in Progress in Optics, Vol. 6 (E. Wolf, ed.), North-Holland, Amsterdam, 1967, p. 261.
1331. Van den Hulst, H. C., LIGHT SCATTERING BY SMALL PARTICLES, Wiley–Interscience, New York, 1957.
1332. Van der Maas, J. M., BASIC INFRARED SPECTROSCOPY, Heyden, London, 1969.
1333. Van der Vegt, A. K., and Smith, P. P. A., *Fortsch. Hochpolym. Forsch.*, *Adv. Polym. Sci.*, **2**, 313 (1967).
1334. Van der Ziel, A., SOLID STATE PHYSICAL ELECTRONICS, 2nd ed., Prentice-Hall, New York, 1968.
1335. Van Dam, J., in Characterization of Macromolecular Structure, Nat. Acad. Sci., U.S.A., Publ. 1573, Washington, D.C., 1968, p. 336.
1336. Van Krevelen, D. W., and Hoftyzer, P. J., PROPERTIES OF POLYMERS: CORRELATIONS WITH CHEMICAL STRUCTURE, Elsevier, Amsterdam, 1972.
1337. Van Oene, H., in Characterization of Macromolecular Structure, Nat. Acad. Sci., U.S.A., Publ. 1573, Washington, D.C., 1968, p. 353.
1338. Van Turnhout, J., THERMALLY STIMULATED DISCHARGE OF POLYMER ELECTRETS, Elsevier, Amsterdam, 1975.
1339. Van Wazer, J. R., Lyons, J. W., Kim, K. V., and Colwell, R. E., VISCOSITY AND FLOW MEASUREMENTS—A LABORATORY HANDBOOK OF RHEOLOGY, Wiley–Interscience, New York, 1963.
1340. Vassy, A., ATMOSPHERIC OZONE, Academic Press, New York, 1965.
1341. Vaugham, W. E., Smith, C. P., and Powels, J. G., in Physical Methods of Organic Chemistry, Vol. 5 (A. Weissberger, ed.), Wiley–Interscience, New York, 1972, Chap. 5.
1342. Veening, H., *J. Chem. Educ.*, **47**, A549, A675, A749 (1970).
1343. Velluz, L., Legrand, M., and Grosjean, M., OPTICAL CIRCULAR DICHROISM, Academic Press, New York, 1965.
1344. Vinograd, J., and Hearst, J. E., *Fortsch. Chem. Organ. Naturstoffe*, **20**, 372 (1962).
1345. Vinogradov, G., and Malkin, A. Ya., RHEOLOGY OF POLYMERS, Khimya, Moscow, 1977 (in Russian).
1346. Vitek, R. K., and Kent, D. M., *Intern. Lab.*, March/Apr. 1978, p. 73.

1347. Vleck, J. H., *Phys. Rev.*, **74**, 1169 (1948).
1348. Voight, J., and Fischer, W. G., *Chemiker Ztg*, **88**, 919 (1964).
1349. Volkenstein, M. V., CONFIGURATION STATISTICS OF POLYMER CHAINS, Wiley–Interscience, New York, 1963.
1350. Vollmert, B., POLYMER CHEMISTRY, Springer Verlag, Berlin, 1973.
1351. Von Hippel, A. R., DIELECTRIC MATERIALS AND APPLICATIONS, Wiley–Interscience, New York, 1954.
1352. Voorn, M. J., *Fortsch. Hochpolym. Forsch.*, *Adv. Polym. Sci.*, **1**, 192 (1959).
1353. Wahlstrom, E. E., OPTICAL CRYSTALLOGRAPHY, Wiley–Interscience, New York, 1969.
1354. Waldron, J. D., ADVANCES IN MASS SPECTROMETRY, Pergamon Press, London, 1959.
1355. Wales, M., in Characterization of Macromolecular Structure, Nat. Acad. Sci., U.S.A., Publ. 1573, Washington, D.C., 1968, p. 343.
1356. Wall, L. A., in Analytical Chemistry of Polymers, Vol. 2 (G. M. Kline, ed.), Wiley–Interscience, New York, 1962, p. 249.
1357. Walters, K., RHEOMETRY, Halsted–Wiley, New York, 1975.
1358. Walther, H., LASER SPECTROSCOPY, Springer Verlag, Berlin, 1975.
1359. Walton, H. F., ION EXCHANGE CHROMATOGRAPHY, Wiley–Interscience, New York, 1976.
1360. Ward, I. M., STRUCTURE AND PROPERTIES OF ORIENTED POLYMERS, Applied Science Publisher, London, 1975.
1361. Ware, W. R., in Creation and Detection of the Excited State, Vol. 1, Part A (A. A. Lamola, ed.), Dekker, New York, 1971, Chap. 5.
1362. Wayne, R. P., PHOTOCHEMISTRY, Butterworths, London, 1970.
1363. Weeks, N. E., Kohlmayer, G. M., and Otocka, E. P., in Characterization of Metal and Polymer Surfaces, Vol. 2 (L. H. Lee, ed.), Academic Press, New York, 1977, p. 289.
1364. Weiss, P., ADHESION AND COHESION, Elsevier, Amsterdam, 1962.
1365. Weissberg, S. G., and Brown, J. E., in Encyclopedia of Polymer Science and Technology, Vol. 9 (H. F. Mark, N. G. Gaylord, and N. M. Bikales, eds), Wiley–Interscience, New York, 1968, p. 659.
1366. Weissberg, S. G., Rothman, S., and Wiles, M., in Analytical Chemistry of Polymers, Vol. 2 (G. M. Kline, ed.), Wiley–Interscience, New York, 1962, p. 1.
1367. Wells, C. H. J., INTRODUCTION TO MOLECULAR PHOTOCHEMISTRY, Chapman and Hall, London, 1972.
1368. Wendlandt, W. W., in Chemical Analysis, Vol. 19 (P. J. Elving and I. M. Kolthoff, eds), Wiley–Interscience, New York, 1964.
1369. Wendlandt, W. W., MODERN ASPECTS OF REFLECTANCE SPECTROSCOPY, Plenum Press, New York, 1968.
1370. Wendlandt, W. W., *J. Chem. Educ.*, **49**, A624, A671 (1972).
1371. Wendlandt, W. W., HANDBOOK OF COMMERCIAL SCIENTIFIC INSTRUMENTS, Vol. 2, THERMOANALYTICAL TECHNIQUES, Dekker, New York, 1974.
1372. Wendlandt, W. W., THERMAL METHODS OF ANALYSIS, 2nd ed., Wiley–Interscience, New York, 1974.
1373. Wendlandt, W. W., and Collins, L. W., THERMAL ANALYSIS, Wiley–Interscience, New York, 1977.
1374. Wendlandt, W. W., and Hecht, H. G., REFLECTANCE SPECTROSCOPY, Wiley–Interscience, New York, 1966.
1375. Wertz, J. E., and Bolton, J. R., ELECTRON SPIN RESONANCE: ELEMENTARY THEORY AND PRACTICAL APPLICATIONS, McGraw-Hill, New York, 1972.
1376. West, M. A., LASERS IN CHEMISTRY, Elsevier, Amsterdam, 1977.
1377. West, R. N., *Adv. Phys.*, **22**, 263 (1973).
1378. Wetton, R. E., and Moneypenny, H. G., *Brit. Polym. J.*, **7**, 51 (1975).
1379. Whiffen, D. H., and Fleming, I., SPECTROSCOPY, Wiley–Interscience, New York, 1967.
1380. White, C. E., and Argauer, R. J., FLUORESCENCE ANALYSIS: A PRACTICAL APPROACH, Dekker, New York, 1970.

1381. White, F. A., MASS SPECTROMETRY IN SCIENCE AND TECHNOLOGY, Wiley–Interscience, New York, 1968.
1382. White, J. W., in Polymer Science, Vol. 2 (A. D. Jenkins, ed.), North-Holland, Amsterdam, 1972, Chap. 27.
1383. Wiberg, K. B., and Nist, B. J., THE INTERPRETATION OF NMR SPECTRA, Benjamin, New York, 1962.
1384. Wienefordner, J. D., Schulman, S. G., and O'Haver, T. C., LUMINESCENCE SPECTROMETRY IN ANALYTICAL CHEMISTRY, Wiley–Interscience, New York, 1972.
1385. Wijga, P. W. O., Z. Anal. Chem. 205, 342 (1964).
1386. Wilde, P. F., THIN LAYER CHROMATOGRAPHY, United Trades Press, London, 1964.
1387. Wilkes, G. L., Fortschr. Hochpolym. Forsch., Adv. Polym. Sci., 8, 91 (1971).
1388. Wilkinson, F., in Singlet Oxygen: Reactions with Organic Compounds and Polymers (B. Rånby and J. F. Rabek, eds), Wiley, Chichester, 1978, p. 27.
1389. Wilks, P. A., Jr, in Laboratory Methods in Infrared Spectroscopy (R. G. J. Miller and B. C. Stace, eds), Heyden, London, 1972, Chap. 14.
1390. Wilks, P. A., Jr, Hirschfeld, T., and Brame, E. G., APPLIED SPECTROSCOPY REVIEWS, Vol. 1, Arnold, New York, 1968.
1391. Willardson, R. K., and Beer, A. C., SEMICONDUCTORS AND SEMIMETALS, Vol. 5, INFRARED DETECTORS, Academic Press, New York, 1970.
1392. Willems, J., Experientia, 23, 409 (1967).
1393. Williams, D., and Fleming, I., SPECTROSCOPIC METHODS IN ORGANIC CHEMISTRY, 2nd ed., Elsevier, Amsterdam, 1973.
1394. Williams, D. E., and Davis, L. E., in Characterization of Metal and Polymer Surfaces, Vol. 2 (L. H. Lee, ed.), Academic Press, New York, 1977, p. 53.
1395. Williams, D. H., MASS SPECTROMETRY, The Chemical Society, London, 1971.
1396. Williams, D. J., POLYMERS SCIENCE AND ENGINEERING, Prentice-Hall, Englewood Cliffs, N.J., 1971.
1397. Williams, J. H., Van Holde, K. E., Baldwin, R. L., and Fujita, H., Chem. Rev., 58, 715 (1958).
1398. Williams, J. W., ULTRACENTRIFUGAL ANALYSIS IN THEORY AND EXPERIMENT, Academic Press, New York, 1963.
1399. Williams, J. W., ULTRACENTRIFUGATION OF MACROMOLECULES, Academic Press, New York, 1972.
1400. Williams, R. C., and Yphantis, D. A., in Encyclopedia of Polymer Science and Technology, Vol. 14 (H. F. Mark, N. G. Gaylord and N. M. Bikales, eds), Wiley–Interscience, New York, 1971, p. 97.
1401. Williams, T., J. Mater. Sci., 5, 811 (1970).
1402. Willis, H. A., and Cudby, M. E. A., Appl. Spectr. Rev., 1, 237 (1968).
1403. Willis, H. A., and Cudby, M. E. A., Anal. Chem., 263, 291 (1973).
1404. Wilski, H., in Polymer Handbook, 2nd ed. (J. Brandrup, E. H. Immergut and W. McDowell, eds), Wiley–Interscience, New York, 1975, p. III-215.
1405. Wilson, D. E., and Hamaker, F. M., in Thermal Analysis, Vol. 1 (R. F. Schwenker and P. D. Garn, eds), Academic Press, New York, 1969, p. 217.
1406. Wilson, J. E., RADIATION CHEMISTRY OF MONOMERS, POLYMERS AND PLASTICS, Dekker, New York, 1974.
1407. Wilson, R. K., Johnson, P. O., and Stump, R., Phys. Rev., 129, 2091 (1963).
1408. Wilson, T., and Hastings, J. W., Photophysiol., 5, 49 (1970).
1409. Winslow, F. H., Pure Appl. Chem., 49, 495 (1977).
1410. Winstead, M. B., ORGANIC CHEMISTRY STRUCTURAL PROBLEMS, Sadtler Research Laboratories, Philadelphia, 1968.
1411. Winzor, D. J., in Physical Principles and Techniques of Protein Chemistry, Part A (S. J. Lech, ed.), Academic Press, New York, 1969, p. 451.
1412. Winchnitzer, S., INTRODUCTION TO ELECTRON MICROSCOPY, Pergamon Press, New York, 1962.
1413. Wolf, B. A., in Polymer Handbook, 2nd ed. (K. Brandrup, E. H. Immergut, and W. McDowell, eds), Wiley–Interscience, New York, 1975, p. IV-131.
1414. Wolf, H. F., SEMICONDUCTORS, Wiley–Interscience, New York, 1971.
1415. Wood, E., CRYSTALS AND LIGHT, Van Nostrand, New York, 1964.

1416. Woodbrey, J. C., in The Stereochmistry of Macromolecules (A. D. Ketley, ed.), Dekker, New York, 1968, Chap. 2.

1417. Woodward, A. E., in Ordered Fluids and Liquid Crystals, Advances in Chemistry Series No. 63, American Chemical Society, Washington, D.C., 1967, p. 298.

1418. Woody, R. W., *Macromol. Rev.*, **12**, 181 (1977).

1419. Wrasidlo, W., THERMAL ANALYSIS OF POLYMERS, Advances in Polymer Science Series, Vol. 13, Springer Verlag, Berlin, 1974.

1420. Wright, G. B., LIGHT SCATTERING SPECTRA OF SOLIDS, Springer, New York, 1969.

1421. Wright, W. W., in Thermal Degradation of Polymers, SCI Monograph No. 13, Society for Chemical Industry, London, 1961, p. 248.

1422. Wunderlich, B., *J. Polym. Sci. C*, **1**, 41 (1963).

1423. Wunderlich, B., HEAT CAPACITY OF LINEAR HIGH POLYMERS, Rensselaer Polytechnic Institute, Department of Chemistry, Troy, N.Y., Nov. 1968.

1424. Wunderlich, B., CRYSTALLINE HIGH POLYMERS: NOMENCLATURE, STRUCTURE AND THERMODYNAMICS, American Chemical Society, Washington D.C., 1969.

1245. Wunderlich, B., in Physical Methods of Chemistry, Vol. 1 (A. Wissberger and B. W. Rossiter, eds), Wiley–Interscience, New York, 1971, Chap. 8.

1426. Wunderlich, B., *Polym. Symp.*, No. 43, 29 (1973).

1427. Wunderlich, B., MOLECULAR PHYSICS, Vols 1 and 2, Academic Press, New York, 1973/1976.

1428. Wunderlich, B., and Bodily, D. M., *J. Polym. Sci. C*, **6**, 137 (1964).

1429. Yamakawa, H., MODERN THEORY OF POLYMER SOLUTIONS, Harper and Row, New York, 1971.

1430. Yamakawa, H., *Ann. Rev. Phys. Chem.*, **25**, 179 (1974).

1431. Yamashita, Y., *Fortsch. Hochpolym. Forsch., Adv. Polym. Sci.*, **28**, 1 (1978).

1432. Yasuda, H., and Stannett, V., in Polymer Handbook, 2nd ed. (J. Brandrup, E. H. Immergut, and W. McDowell, eds), Wiley–Interscience, New York, 1975, p. III-229.

1433. Yau, W. W., Malone, C. P., and Suchan, H. L., in Gel Permeation Chromatography, (K. H. Altgelt and L. Segal), Dekker, New York, 1971, p. 105.

1434. Yeh, G. S. Y., *Crit. Rev. Macromol. Sci.*, **1**, 173 (1972).

1435. Yguerabide, J., *Rev. Sci. Instrum.*, **36**, 1734 (1965).

1436. Young, C. L., *Chromatogr. Rev.*, **10**, 129 (1968).

1437. Young, R. H., and Brewer, D. R., in Singlet Oxygen: Reactions with Organic Compounds and Polymers (B. Rånby and J. F. Rabek, eds), Wiley, Chichester, 1978, p. 36.

1438. Young, R. P., in Infrared and Raman Spectroscopy Vol. 1, Part B (E. G. Brame and J. G. Grasselli, eds), Dekker, New York, 1977, Chap. 5.

1439. Zand, R., in Encyclopedia of Polymer Science and Technology, Vol. 9 (H. F. Mark, N. G. Gaylord, and N. M. Bikales, eds), Wiley–Interscience, New York, 1968, p. 610.

1440. Zander, M., PHOSPHORIMETRY, Academic Press, New York, 1968.

1441. Zbinden, R., INFRARED SPECTROSCOPY OF HIGH POLYMERS, Academic Press, New York, 1964.

1442. Zemach, A. C., and Glauber, R. J., *Phys. Rev.*, **101**, 118 (1956).

1443. Zerbi, G., in Vibrational Spectroscopy—Modern Trends (A. J. Barnes and W. J. Orville-Thomas, eds), Elsevier, Amsterdam, 1977, Chap. 24.

1444. Zettlemoyer, A. C., NUCLEATION, Dekker, New York, 1969.

1445. Zhbankov, R. G., INFRARED SPECTRA OF CELLULOSE AND ITS DERIVATIVES, Consultants Bureau, New York, 1966.

1446. Zichy, E. L., TECHNIQUES OF POLYMER SCIENCE, SCI Monograph No. 17, Society for Chemical Industry, London and Gordon and Breach, New York, 1966, p. 122.

1447. Zichy, V. J. I., in Laboratory Methods in Infrared Spectroscopy, 2nd ed. (R. G. J. Miller and B. C. Stace, eds), Heyden, London, 1972, Chap. 5.

1448. Zisman, W. A., in Contact Angle, Wettability and Adhesion (R. F. Gould, ed.), Advances in Chemistry Series No. 43, American Chemical Society, Washington, D.C., 1964, Chap. 1.

1449. Zlatkis, A., and Kaiser, R. E., HPTLC HIGH PERFORMANCE THIN-LAYER CHROMATOGRAPHY, Elsevier, Amsterdam, 1977.
1450. Zlatkis, A., and Pretorious, V., PREPARATIVE GAS CHROMATOGRAPHY, Wiley–Interscience, New York, 1971.
1451. Zweig, G., CHROMATOGRAPHY, Vols 1 and 2, CRC Press, Florida, 1972.
1452. *J. Polym. Sci.*, **8**, 257 (1952).
1453. *Macromolecules*, **1**, 193 (1968).
1454. IUPAC Information Bulletin, Appendix No. 29, 1972.
1455. Basic Manual of Applications and Laboratory Ozonization Techniques, Welsbach Corp., Philadelphia, 1967.
1456. A User's Guide to Chromatography: Gas Liquid, TLC, Regis Innovative Products, Morton Grove, Illinois, 1976.
1457. *Perkin-Elmer Thermal Analysis Newsletter*, No. 10, Feb. 1972.
1458. Price List U-Ion Exchange Gel Filtration Adsorption, Bio-Rad Laboratory, Richmond, Calif., 1969.
1459. RAPRA Data Handbook Series on Transition Temperatures. Rubber and Plastics Research Association, Shawbury, England, 1973.
1460. Sephadex-Gel Filtration in Theory and Practice. Pharmacia Fine Chemicals, N.J., no date.
1461. ASTM AMD 11: Index of Mass Spectral Data, American Society for Testing and Materials, Philadelphia.
1462. ASTM D276-61T: Identification of Fibers in Textiles, American Society for Testing and Materials, Philadelphia.
1463. ASTM STP-356: Index of Mass Spectral Data, American Society for Testing and Materials, Philadelphia.
1464. ASTM D1525: Standard Method of Test for Vicat Softening Point of Plastics, American Society for Testing and Materials, Philadelphia.
1465. ASTM D1434: Standard Methods of Test for Gas Transmission Rate of Plastic Film and Sheeting, American Society for Testing and Materials, Philadelphia.
1466. ASTM D2857-69: Standard Methods of Test for Dilute Solution Viscosity of Polymers, American Society for Testing and Materials, Philadelphia.
1467. ASTM C177: Standard Method of Test for Thermal Conductivity of Materials by Means of the Guarded Hot Plate, American Society for Testing and Materials, Philadelphia.
1468. ASTM A131: Wanadotte Index, Molecular Formula List of Compounds, Name and References to Published Infrared Spectra, American Society for Testing and Materials, Philadelphia.
1469. ALLTECH CATALOGUE: EVERYTHING FOR CHROMATOGRAPHY, Alltech Assoc., 1977.
1470. BECKMAN CATALOGUE: INFRARED SPECTROMETER ACCESSORIES, Bulletin L8612, Beckman, Geneva, 1975.
1471. BDH MATERIALS FOR SPECTROSCOPY, BDH Chemicals Ltd, Poole, England, 1978.
1472. DMS UV ATLAS OF ORGANIC COMPOUNDS, Verlag Chemie, Weinheim and Butterworths, London, 1956–.
1473. HELLMA KUVETTEN KATALOGE, Hellma, Mullheim, 1976.
1474. LADD CATALOG OF SCIENTIFIC EQUIPMENT, Ladd Research Inc., Burlington, Vt, 1976.
1475. ORIEL INSTRUMENTS AND ACCESSORIES FOR SPECTROSCOPY, Oriel Optic GmbH, Darmstadt, 1973.
1476. SADTLER STANDARD ULTRAVIOLET SPECTRA, Sadtler Research Labs, Philadelphia and Heyden, London 1972.
1477. SADTLER CATALOG OF INFRARED SPECTROGRAMS (Specifinder), Sadtler Research Labs, Philadelphia, 1972.
1478. SADTLER STANDARD INFRARED GRATING SPECTRA, Sadtler Research Labs, Philadelphia, 1972.
1479. SADTLER STANDARD INFRARED PRISM SPECTRA, Sadtler Research Labs, Philadelphia, 1972.
1480. SADTLER STANDARD NMR SPECTRA, Sadtler Research Labs, Philadelphia, 1972.

1481. VARIAN HIGH RESOLUTION NMR SPECTRA CATALOG, Vols 1 and 2, Varian Associates, Palo Alto, Calif., 1962/1963.
1482. WATERS ASSOCIATES, Manual Instruction No. CU84037, March 1976.

REFERENCES

2001. Abbâs, K. B., Bovey, F. A., and Schilling, F. C., *Makromol. Chem.*, Suppl. 1, 227 (1975).
2002. Abe, A., *Macromolecules*, 10, 34 (1977).
2003. Abe, A., and Goodman, M., *J. Polym. Sci. A*, 1, 2193 (1963).
2004. Abe, A., Jernigen, R. L., and Flory, P. J., *J. Amer. Chem. Soc.*, 88, 631 (1966).
2025. Aharoni, S. M., *J. Appl. Polym. Sci.*, 21, 1323 (1977).
2026. Aharoni, S. M., *Macromolecules*, 10, 1408 (1977).
2027. Aharoni, S. M., Reimschuessel, A. C., and Turi, E. A., *J. Polym. Sci. A2*, 14, 2109 (1976).
2008. Abraham, K. P., and Ehrlich, P., *Macromolecules*, 8, 945 (1975).
2009. Abraham, R. J., and Whiffen, D. H., *Trans. Faraday Soc.*, 54, 1291 (1948).
2010. Abu-Isa, I., and Dole, M., *J. Phys. Chem.*, 69, 2668 (1965).
2011. Acierno, D., La Mantia, F. P., Polizzotti, G., Alfonso, G. C., and Ciferri, A., *J. Polym. Sci., B*, 15, 323 (1977).
2012. Acierno, D., Titomanlio, G., and Marruchi, G., *J. Polym. Sci. A2*, 12, 2177 (1974).
2013. Adachi, K., Fujihara, I., and Ishida, Y., *J. Polym. Sci. A2*, 13, 2155 (1975).
2014. Adachi, K., Hattori, M., and Ishida, Y., *J. Polym. Sci. A2*, 15, 693 (1977).
2015. Adachi, K., and Ishida, Y., *J. Polym. Sci. A2*, 14, 2219 (1976).
2016. Adam, G., and Müller, F. H., *Kolloid Z. Z. Polym.*, 192, 29 (1963).
2017. Adam, M., and Delsanti, M., *Macromolecules*, 10, 1229 (1977).
2018. Adams, D., Clark, D., Diks, A., Peeling, J., and Thomas, H. R., *Makromol. Chem.*, 177, 2139 (1976).
2019. Addelman, R. L., and Zichy, V. J. I., *Polymer*, 13, 391 (1972).
2020. Adelman, S. A., and Deutch, J. M., *Macromolecules*, 8, 58 (1975).
2021. Aden, A. L., and Kerker, M., *J. Appl. Phys.*, 22, 1242 (1951).
2022. Adicoff, A., and Murbach, W. J., *Anal. Chem.*, 39, 302 (1967).
2023. Afifi-Effat, A. M., Hay, J. N., and Wiles, M., *J. Polym. Sci. B*, 11, 87 (1973).
2024. Aggarwal, S. L., and Tilley, G. P., *J. Polym. Sci.*, 18, 17 (1955).
2025. Aharoni, S. M., *J. Appl. Polym. Sci.*, 21, 1323 (1977).
2026. Aharoni, S. M., *Macromolecules*, 10, 1408 (1977).
2027. Aharoni, S. M., Reimschuessel, A. C., and Turi, E. A., *J. Polym. Sci. A2*, 14, 2109 (1976).
2028. Ahmed, A. U., Ahmed, N., Aslam, J., Butt, N. M., Khan, Q. H., and Atta, M. A., *J. Polym. Sci. B*, 14, 561 (1976).
2029. Ainsworth, S., and Winter, E., *Appl. Opt.*, 3, 371 (1964).
2030. Aitken, D., Glotin, M., Hendra, P. J., Jobic, H., and Marsden, E., *J. Polym. Sci. B*, 14, 619 (1976).
2031. Ajo, D., Granozzi, G., and Zannetti, R., *Makromol. Chem.*, 178, 2471 (1977).
2032. Akana, Y., and Stein, R. S., *J. Polym. Sci. A2*, 13, 2195 (1975).
2033. Albert, A. C., *J. Chem. Phys.*, 27, 1002, 1014 (1957).
2034. Albert, R., and Malone, W. M., *J. Makromol. Sc. A*, 6, 347 (1972).
2035. Aldersley, J. W., Bertram, V. M. R., Harper, G. R., and Stark, B. P., *Brit. Polym. J.*, 1, 101 (1969).
2036. Aleman, J. V., *J. Polym. Sci. A1*, 9, 3501 (1971).
2037. Alexander, L. E., Ohlberg, S., and Taylor, G. R., *J. Appl. Phys.*, 26, 1608 (1955).
2038. Alexander, P., and Stacey, K. A., *Trans. Faraday Soc.*, 51, 299 (1955).
2039. Alexandrowicz, Z., *Macromolecules*, 6, 255 (1973).
2040. Alexandrowicz, Z., and Accad, Y., *Macromolecules*, 6, 251 (1973).
2041. Alexandru, L., and Somersall, A. C., *J. Polym. Sci. A1*, 15, 2013 (1977).
2042. Alfonso, G. C., Fiorina, L., Martuscelli, E., Pedemonte, E., and Russo, S., *Polymer*, 14, 373 (1973).
2043. Alfonso, G. C., Olivero, L., Turturro, A., and Pedemonte, E., *Brit. Polym. J.*, 5, 141 (1973).

2044. Alfonso, G. C., Valenti, B., and Pedemonte, E., *Makromol. Chem.*, **175**, 1917 (1974).
2045. Alford, S., and Dole, M., *J. Amer. Chem. Soc.*, **77**, 4774 (1955).
2046. Alfrey, T., Jr, Gurnee, E. F., and Lloyd, W. G., *J. Polym. Sci. C*, **12**, 249 (1966).
2047. Ali, M. S., and Sheldon, R. P., *J. Appl. Polym. Sci.*, **14**, 2619 (1970).
2048. Ali, M. S., and Sheldon, R. P., *J. Polym. Sci. C*, **38**, 97 (1972).
2049. Alishov, V. R., Berezkin, V. G., and Melnikova, Yu. V., *Zhur. Fiz. Khim.*, **39**, 105 (1965).
2050. Allan, G. C., and Neogi, A. N., *J. Appl. Polym. Sci.*, **14**, 999 (1970).
2051. Allegra, G., Benedetti, E., and Pedone, C., *Macromolecules*, **3**, 727 (1970).
2052. Allegra, G., and Brückner, S., *Macromolecules*, **10**, 106 (1977).
2053. Allegra, G., Calligaris, M., and Randaccio, L., *Macromolecules*, **6**, 390 (1973).
2054. Allegra, G., Calligaris, M., Rondaccio, L., and Morgalio, G., *Macromolecules*, **6**, 397 (1973).
2055. Allen, G., *Proc. Roy. Soc., A*, **351**, 381 (1976).
2056. Allen, G., and Barker, C. H., *Polymer*, **6**, 181 (1965).
2057. Allen, G., Booth, C., Gee, G., and Jones, M. N., *Polymer*, **5**, 367 (1964).
2058. Allen, G., Booth, C., and Hurst, S. J., *Polymer*, **8**, 385 (1967).
2059. Allen, G., Booth, C., and Jones, M. N., *Polymer*, **5**, 257 (1964).
2060. Allen, G., Booth, C., and Price, C., *Polymer*, **8**, 397 (1967).
2061. Allen, G., Egerton, P. L., and Walsh, D. J., *Polymer*, **17**, 65 (1976).
2062. Allen, G., Gee, G., Mangaraj, D., Sims, D., and Wilson, G. J., *Polmer*, **1**, 467 (1960).
2063. Allen, G., Gee, G., and Wilson, G. J., *Polymer*, **1**, 456 (1960).
2064. Allen, G., and Higgins, J. S., *Macromolecules*, **10**, 1006 (1977).
2065. Allen, G., McAinsh, J., and Jeffs, G. M., *Polymer*, **12**, 85 (1971).
2066. Allen, G., and Tanaka, T., *Polymer*, **19**, 271 (1978).
2067. Allen, N. S., Cundall, R. B., Jones, M. W., and McKellar, J. F., *Chem. Ind. (London)*, **1976**, 110.
2068. Allen, N. S., Homer, J., and McKellar, J. F., *Chem. Ind. (London)*, **1975**, 533.
2069. Allen, N. S., Homer, J., and McKellar, J. F., *J. Appl. Polym. Sci.*, **20**, 2553 (1976).
2070. Allen, N. S., Homer, J., and McKellar, J. F., *J. Appl. Polym. Sci.*, **21**, 2261 (1977).
2071. Allen, N. S., Homer, J., and McKellar, J. F., *Makromol. Chem.*, **179**, 1575 (1978).
2072. Allen, N. S., Homer, J., McKellar, J. F., and Phillips, G. O., *Brit. Polym. J.*, **7**, 11 (1975).
2073. Allen, N. S., Homer, J., McKellar, J. F., and Wood, D. G. M., *J. Appl. Polym. Sci.*, **21**, 3147 (1977).
2074. Allen, N. S., and McKellar, J. F., *Polymer*, **18**, 986 (1977).
2075. Allen, N. S., and McKellar, J. F., *J. Appl. Polym. Sci.*, **21**, 1129 (1977).
2076. Allen, N. S., and McKellar, J. F., *Makromol. Chem.*, **179**, 523 (1978).
2077. Allen, N. S., and McKellar, J. F., *J. Appl. Polym. Sci.*, **22**, 625 (1978).
2078. Allen, N. S., and McKellar, J. F., *J. Appl. Polym. Sci.*, **22**, 2085 (1978).
2079. Allen, N. S., McKellar, J. F., and Phillips, G. O., *Chem. Ind. (London)*, **1974**, 300.
2080. Allen, N. S., McKellar, J. F., and Phillips, G. O., *J. Polym. Sci. B*, **12**, 253 (1974).
2081. Allen, N. S., McKellar, J. F., Phillips, G. O., and Wood, D. G. M., *J. Polym. Sci. A1*, **12**, 2647 (1974).
2082. Allen, N. S., McKellar, J. F., and Wilson, D., *J. Polym. Sci. A1*, **15**, 2793 (1977).
2083. Allen, P. E. M., Downer, J. M., Hastings, G. W., Melville, H. W., Molyneux, P., and Urwin, J., *Nature*, **177**, 910 (1956).
2084. Allen, P. E. M., Hardy, R., Majer, J. R., and Molyneux, P., *Makromol. Chem.*, **39**, 52 (1960).
2085. Allen, P. W., and Place, M. A., *J. Polym. Sci.*, **26**, 386 (1957).
2086. Allendoerfer, R. D., *J. Chem. Phys.*, **55**, 3615 (1971).
2087. Allerhand, A., and Hailstone, R. K., *J. Chem. Phys.*, **56**, 3718 (1972).
2088. Alliet, D. F., and Pacco, J. M., *J. Polym. Sci. C*, **21**, 199 (1968).
2089. Alon, Y., *Rev. Sci. Instrum.*, **40**, 20 (1969).
2090. Alper, T., Barlow, A. J., and Gray, R. W., *Polymer*, **17**, 665 (1976).
2091. Altares, T. A., Jr, Wyman, D. P., and Aller, V. R., *J. Polym. Sci. A*, **2**, 4533 (1964).
2092. Alter, H., and Hsiao, H. Y., *J. Polym. Sci. B*, **6**, 363 (1968).
2093. Alvång, F., and Samuelson, O., *J. Polym. Sci.*, **24**, 353 (1957).
2094. Amano, T., Fischer, E. W., and Hinrichsen, G., *J. Macromol. Sci. B*, **3**, 209 (1969).
2095. Amaral, L. Q., Vinhas, L. A., and Herdade, S. B., *J. Polym. Sci. A2*, **14**, 1077 (1976).
2096. Amaya, K., and Fujishiro, R., *Bull. Chem. Soc. Japan*, **29**, 270 (1956).

2097. Ambler, M. R., *J. Polym. Sci. A1*, **11**, 191 (1973).
2098. Ambler, M. R., *J. Polym. Sci. B*, **14**, 683 (1976).
2099. Ambler, M. R., *J. Appl. Polym. Sci.*, **21**, 1655 (1977).
2100. Ambler, M. R., Fetters, L. J., and Kesten, Y., *J. Appl. Polym. Sci.*, **21**, 2439 (1977).
2101. Ambler, M. R., and McIntyre, D., *J. Polym. Sci. B*, **13**, 589 (1975).
2102. Ambler, M. R., McIntyre, D., and Fetters, L. J., *Macromolecules*, **11**, 300 (1978).
2103. Ambler, M. R., Mate, R. D., and Purdon, J. R., *J. Polym. Sci. A1*, **12**, 1759, 1771 (1964).
2104. Ambler, M. R., and McIntyre, D., *J. Appl. Polym. Sci.*, **21**, 2269 (1977).
2105. Ambrose, E. J., *J. Sci. Instrum.*, **25**, 134 (1948).
2106. Ambrose, E. J., and Elliot, A., *Proc. Roy. Soc. A*, **206**, 206 (1951).
2107. Ammon, G., Funke, W., and Pechhold, W., *Kolloid Z. Z. Polym.*, **206**, 9 (1965).
2108. Amrhein, E., and Frischorn, H., *Kolloid Z. Z. Polym.*, **251**, 369 (1973).
2109. Anagnostopoulos, C. E., Coran, A. Y., and Gamrath, H. R., *J. Appl. Polym. Sci.*, **4**, 181 (1960).
2110. Anand, N., Murthy, N. S. R. K., Naider, F., and Goodman, M., *Macromolecules*, **4**, 565 (1971).
2111. Ananthanarayanan, V. S., Leroy, E., and Scheraga, H. A., *Macromolecules*, **6**, 553 (1973).
2112. Anderegg, J. W., Beeman, W. W., Sullivan, S., and Kaesburg, P., *J. Amer. Chem. Soc.*, **77**, 2927 (1955).
2113. Andersen, H. C., and Pecora, R., *J. Chem. Phys.*, **54**, 2584 (1971).
2114. Andersen, M., and Nir, S., *Polymer*, **18**, 867 (1977).
2115. Anderson, D. A., and Freeman, E. S., *J. Polym. Sci.*, **54**, 253 (1961).
2116. Anderson, D. H., and Covert, G. L., *Analyt. Chem.*, **39**, 1288 (1967).
2117. Anderson, F. R., *J. Appl. Phys.*, **35**, 64 (1964).
2118. Anderson, H. C., *J. Polym. Sci. B*, **2**, 115 (1964).
2119. Anderson, J. E., and Liu, K. J., *J. Chem. Phys.*, **49**, 2850 (1968).
2120. Anderson, J. E., and Liu, K. J., *Macromolecules*, **4**, 260 (1971).
2121. Anderson, J. E., Sillescu, H., and Subramanian, S., *Macromolecules*, **10**, 375 (1977).
2122. Anderson, J. S., and Vaughan, W. E., *Macromolecules*, **8**, 454 (1975).
2123. Andersson, P., *Makromol. Chem.*, **177**, 271 (1976).
2124. Ando, I., Kato, Y., and Nishioka, A., *Makromol. Chem.*, **177**, 1759 (1976).
2125. Ando, I., Nishioka, A., and Asakura, T., *Makromol. Chem.*, **176**, 411 (1975).
2126. Andrews, E. H., Bennett, M. W., and Markham, A., *J. Polym. Sci. A2*, **5**, 1235 (1967).
2127. Andrews, P., *Biochem. J.*, **96**, 595 (1965).
2128. Andrews, R. D., *ACS Polymer Preprints*, **6**, 717 (1965).
2129. Andrews, R. D., *J. Polym. Sci. C*, **14**, 261 (1966).
2130. Angood, A. C., and Koenig, J. L., *Macromolecules*, **2**, 37 (1969).
2131. Angulo-Sanchez, J. L., Gallegos, A., Ponce-Velez, M. A., and Campos-Lopez, E., *Polymer*, **18**, 922 (1977).
2132. Anton, A., *J. Appl. Polym. Sci.*, **12**, 2117 (1968).
2133. Aoki, Y., and Brittain, J. O., *J. Polym. Sci. A2*, **14**, 1297 (1976).
2134. Aoki, Y., Chiba, A., and Kaneko, M., *J. Phys. Soc. Japan*, **27**, 1579 (1969).
2135. Aoki, Y., Nobuta, A., Chiba, A., and Kaneko, M., *Polymer J.*, **2**, 502 (1971).
2136. Arakawa, T., and Wunderlich, B., *J. Polym. Sci. A2*, **4**, 53 (1966).
2137. Arai, K., Komine, S., and Negishi, M., *J. Polym. Sci. A1*, **8**, 917 (1970)
2138. Archibald, W. J., *Ann. N.Y. Acad. Sci.*, **43**, 211 (1942).
2139. Archibald, W. J., *J. Phys. Colloid Chem.*, **51**, 1204 (1947).
2140. Ariyama, T., Nakayama, T., and Inoue, N., *J. Polym. Sci. B*, **15**, 427 (1977).
2141. Arlie, J. P., and Skoulios, A. E., *Makromol. Chem.*, **99**, 160 (1966).
2142. Arlie, J. P., Spegt, P., and Skoulios, A. E., *Makromol. Chem.*, **104**, 212 (1967).
2143. Arnett, R. L., Smith, M. E., and Buell, B. O., *J. Polym. Sci. A*, **1**, 2753 (1963).
2144. Arnold, N. D., and Guenther, A. H., *J. Appl. Polym. Sci.*, **10**, 731 (1966).
2145. Arond, L. H., and Frank, H. P., *J. Phys. Chem.*, **58**, 953 (1954).
2146. Aroney, M., LeFèvre, R. J. W., and Parkins, G. M., *J. Chem. Soc.*, **1960**, 2890 (1960).
2147. Aronowitz, S., and Eichinger, B. E., *J. Polym. Sci. A2*, **13**, 1655 (1975).
2148. Aronowitz, S., and Eichinger, B. E., *Macromolecules*, **9**, 377 (1976).
2149. Arslanov, V. V., Ivanova, T. I., and Ogarov, V. A., *Dokl. Akad. Nauk SSSR*, **198**, 1113 (1971).
2150. Arthur, J. C., and Demint, R. J., *J. Phys. Chem.*, **64**, 1332 (1960).

2151. Asakura, T., Ando, I., and Nishioka, A., *Makromol. Chem.*, **177**, 523 (1976).
2152. Asakura, T., Ando, I., Nishioka, A., Doi, Y., and Keii, T., *Makromol. Chem.*, **178**, 791 (1977).
2153. Asano, T., and Fujiwara, Y., *Polymer*, **19**, 99 (1978).
2154. Ashby, G. E., *J. Polym. Sci.*, **50**, 99 (1961).
2155. Ashby, C. E., Reitenour, J. S., and Hammer, C. F., *J. Amer. Chem. Soc.*, **79**, 5806 (1957).
2156. Ashcraft, C. R., and Boyd, R. H., *J. Polym. Sci. A2*, **14**, 2153 (1976).
2157. Ashida, M., Ueda, Y., and Watanabe, T., *J. Polym. Sci. A2*, **16**, 179 (1978).
2158. Ashman, P. C., and Booth, C., *Polmer*, **12**, 459 (1972).
2159. Ashman, P. C., and Booth, C., *Polymer*, **16**, 889 (1975).
2160. Ashman, P. C., and Booth, C., *Polymer*, **17**, 105 (1976).
2161. Ashman, P. C., Booth, C., Cooper, D. R., and Price, C., *Polymer*, **16**, 897 (1975).
2162. Asmussen, F., and Ueberreiter, K., *Kolloid. Z.*, **185**, 1 (1962).
2163. Asmussen, F., and Ueberreiter, K., *Makromol. Chem.*, **52**, 164 (1962).
2164. Asmussen, F., and Ueberreiter, K., *J. Polym. Sci.*, **57**, 187 (1962).
2165. Asmussen, F., and Ueberreiter, K., *Kolloid. Z.*, **223**, 6 (1968).
2166. Aspler, J., Carlsson, D. J., and Wiles, D. M., *Macromolecules*, **9**, 691 (1976).
2167. Assink, R. A., *J. Polym. Sci. A2*, **12**, 2281 (1974).
2168. Assink, R. A., *J. Polym. Sci. A2*, **15**, 59 (1977).
2169. Assink, R. A., *J. Polym. Sci. A2*, **15**, 227 (1977).
2170. Astarita, G., Greco, G., and Nicodemo, L., *AIChE J.*, **15**, 4 (1969).
2171. Atkins, E. D. T., Isaac, D. H., Keller, A., and Miyasaka, K., *J. Polym. Sci. A2*, **15**, 211 (1977).
2172. Atkinson, E. B., *J. Polym. Sci. A2*, **15**, 795 (1977).
2173. Atkinson, C. M. L., and Dietz, R., *Makromol. Chem.*, **177**, 213 (1976).
2174. Attenburrow, G. E., and Bassett, D. C., *J. Mater. Sci.*, **12**, 192 (1977).
2175. Audebert, R., and Aubieneau, C., *Europ. Polym, J.*, **6**, 965 (1970).
2176. Audisio, G., and Bajo, G., *Makromol. Chem.*, **176**, 991 (1975).
2177. Auerbach, I., *Polymer*, **7**, 283 (1966).
2178. Auerbach, I., Miller, W. R., Kuryla, W. C., and Gehman, S. D., *J. Polym. Sci.*, **28**, 129 (1958).
2179. Aughey, W. H., and Baum, F. J., *J. Opt. Soc. Amer.*, **44**, 833 (1954).
2180. Avitabile, G., Napolitano, R., Pirozzi, B., Rouse, K. D., Thomas, M. W., and Willis, B. T. M., *J. Polym. Sci. B*, **13**, 351 (1975).
2181. Avrami, M., *J. Chem. Phys.*, **7**, 1103 (1939).
2182. Avrami, M., *J. Chem. Phys.*, **8**, 212 (1940).
2183. Ayscough, P. B., Ivin, K. J., and O'Donell, J. H., *Proc. Chem. Soc.*, **1961**, 71.
2184. Azori, M., Tüdös, F., Rockenbauer, A., and Simon, P., *Europ. Polym. J.*, **14**, 173 (1978).
2185. Axelson, D. E., and Mandelkern, L., *J. Polym. Sci. A2*, **16**, 1135 (1978).
2186. Axelson, D. E., Mandelkern, L., and Levy, G. C., *Macromolecules*, **10**, 557 (1977).

2187. Baba, Y., Fujimoto, K., Kagemoto, A., and Fujishiro, R., *Makromol. Chem.*, **178**, 1439 (1977)
2188. Baba, Y., Katayama, H., and Kagemoto, A., *Makromol. Chem.*, **175**, 209 (1974).
2189. Bagley, E. B., and Long, F. A., *J. Amer. Chem. Soc.*, **77**, 2172 (1955).
2190. Bagley, E. B., and Duffey, H. J., *Trans. Soc. Rheol.*, **14**, 545 (1970).
2191. Bahl, S. K., Cornell, D. D., Boerio, F. J., and McGraw, G. E., *J. Polym. Sci. B*, **12** (1974).
2192. Baier, R. E., and Zisman, W. A., *Macromolecules*, **3**, 462 (1970).
2193. Baijal, M. D., Diller, R. M., and Pool, R. F., *Macromolecules*, **2**, 679 (1969).
2194. Bailey, E. A., and Rollefson, G. K., *J. Chem. Phys.*, **21**, 1315 (1953).
2195. Bailey, J., Block, H., Cowden, D. R., and Walker, S. M., *Polymer*, **14**, 45 (1973).
2196. Bailey, J., and Walker, S. M., *Polymer*, **13**, 561 (1972).
2197. Bailey, G. W., *J. Polym. Sci.*, **62**, 241 (1962).
2198. Bailey, R. T., Hyde, A. J., and Kim, J. J., *Spectrochim. Acta*, **30**, 91 (1974).
2199. Bair, H. E., *Polym. Eng. Sci.*, **10**, 247 (1970).
2200. Bair, H. E., and Salovey, R., *Macromolecules*, **3**, 677 (1970).
2201. Bair, H. E., Salovey, R., and Huseby, T. W., *Polymer*, **8**, 9 (1967).
2202. Baird, M. E., and Houston, E., *Polymer*, **16**, 308 (1975).

744

2203. Baird, M. E., and Sengupta, C. R., *Polymer*, **15**, 608 (1974).
2204. Baker, C., and Maddams, W. F., *Makromol. Chem.*, **177**, 437 (1976).
2205. Baker, C., Maddams, W. F., and Preedy, J. E., *J. Polym. Sci. A2*, **15**, 1041 (1977).
2206. Baker, C. A., and Williams, R. J. P., *J. Chem. Soc.*, **1956**, 2352.
2207. Backer, C. H., and Mandelkern, L., *Polymer*, **7**, 7 (1966).
2208. Baker, R. W., *J. Appl. Polym. Sci.*, **13**, 369 (1969).
2209. Baldwin, J. P., Bradbury, E. M., McLuckie, I. F., and Stephens, R. M., *Macromolecules*, **6**, 83 (1973).
2210. Baldwin, R. L., *J. Phys. Chem.*, **58**, 1081 (1954).
2211. Balke, S. T., and Hamielec, A. E., *J. Appl. Polym. Sci.*, **13**, 1381 (1969).
2212. Balke, S. T., Hamielec, A. E., Leclair, B. P., and Pearce, S. L., *Ind. Eng. Chem., Prod. Res. Develop.*, **8**, 54 (1969).
2213. Balard, H., Fritz, H., and Meybeck, J., *Makromol. Chem.*, **178**, 2393 (1977).
2214. Ballard, D. G. H., Cheshire, P., Longman, G. W., and Schelten, J., *Polymer*, **19**, 379 (1978).
2215. Ballard, D. G. H., Cunningham, A., and Schelten, J., *Polymer*, **18**, 259 (1977).
2216. Ballard, D. G. H., Rayner, M. G., and Schelten, J., *Polymer*, **17**, 640 (1976).
2217. Ballard, D. G. H., Rayner, M. G., and Schelten, J., *Polymer*, **17**, 349 (1976).
2218. Ballenger, T. F., Chen, I. J., Crowder, J. W., Hagler, G. E., Bogue, D.C., and White, J. L., *Trans. Soc. Rheol.*, **15**, 195 (1971).
2219. Bamford, C. H., Crighton, J. S., and Ward, J. C., Society for Chemical Industry Monograph No. 17, Society for Chemical Industry, London and Gordon and Breach, New York, 1963, p. 284.
2220. Bamford, C. H., and Dewar, M. J. S., *Proc. Roy. Soc. A*, **197**, 356 (1949).
2221. Bamford, C. H., Eastmond, G. C., and Whittle, D., *Polymer*, **16**, 377 (1975).
2222. Bamford, C. H., and Tompa, H., *J. Polym. Sci.*, **10**, 345 (1953).
2223. Banerjeee, D. K., and Budke, C. C., *Anal. Chem.*, **36**, 792 (1964).
2224. Bank, M., Leffingwell, J., and Thies, C., *Macromolecules*, **4**, 43 (1971).
2225. Bank, M., Leffingwell, J., and Thies, C., *J. Polym. Sci. A2*, **10**, 1097 (1972).
2226. Bank, M. I., and Krimm, S., *J. Appl. Polym. Sci.*, **39**, 4951 (1968).
2227. Bank, M. I., and Krimm, S., *J. Polym. Sci. A2*, **7**, 1785 (1969).
2228. Bank, M. I., and Krimm, S., *J. Appl. Phys.*, **40**, 4248 (1969).
2229. Bank, M. I., and Krimm, S., *J. Polym. Sci. B*, **8**, 143 (1970).
2230. Banks, W., Gordon, M., Roe, R. J., and Sharples, A., *Polymer*, **4**, 61 (1963).
2231. Banks, W., and Greenwood, C. T., *Europ. Polym. J.*, **4**, 249 (1968).
2232. Banks, W., Sharples, A., and Hay, J. N., *J. Polym. Sci. A*, **2**, 4059 (1964).
2233. Banthia, A. K., Mandal, B. M., and Palit, S. R., *J. Polym. Sci. A1*, **15**, 845 (1977).
2234. Baranov, V. G., Frenkel, S. Ya., Gromov, V. I., Volkov, T. I., and Zurabian, R. S., *J. Polym. Sci. C*, **38**, 61 (1972).
2235. Barb, W. G., *J. Polym. Sci.*, **37**, 515 (1959).
2236. Barbour, W. M., *J. Gas Chromatogr.*, **3**, 228 (1965).
2237. Bareiss, R. E., and Bellido, J., *Makromol. Chem.*, **177**, 3571 (1976).
2238. Bares, J., *Macromolecules*, **8**, 244 (1975).
2239. Bares, V., and Wunderlich, B., *J. Polym. Sci. A2*, **11**, 861 (1973).
2240. Bargon, J., Hellwege, K. H., and Johnsen, U., *Makromol. Chem.*, **95**, 187 (1966).
2241. Barham, P. J., Atkins, E. D. T., and Nieduszynski, I. A., *Polymer*, **15**, 763 (1974).
2242. Barham, P. J., and Keller, A., *J. Polym. Sci. B*, **13**, 197 (1975).
2243. Barker, C. H., Brown, W. B., Gee, G., Rowlinson, J. S., Stabley, D., and Yeadon, R. R., *Polymer*, **3**, 215 (1965).
2244. Barker, R. E., Daane, J. H., and Rentzepis, P. M., *J. Polym. Sci. A*, **3**, 2033 (1965).
2245. Barlow, A., Lehrle, R. S., and Robb, J. C., *Polymer*, **2**, 27 (1961).
2246. Barlow, A., Lehrle, R. S., and Robb, J. C., *Makromol. Chem.*, **54**, 230 (1962).
2247. Barlow, A., Lehrle, R. S., Robb, J. C., and Sunderland, D., *Polymer*, **8**, 523 (1967).
2248. Barlow, A., Lehrle, R. S., Robb, J. C., and Sunderland, D., *Polymer*, **8**, 537 (1967).
2249. Barlow, A., Wild, L., and Ranganath, R., *J. Appl. Polym. Sci.*, **21**, 3319 (1977).
2250. Barlow, A., Wild, L., and Roberts, T., *J. Chromatogr.*, **55**, 155 (1971).
2251. Barnes, W. J., and Price, F. P., *Polymer*, **5**, 283 (1964).
2252. Barnikol, I., Barnikol, W. K. R., Beck, A., Campagnari-Terbojevic, M., Jovanovic, N., and Schulz, G. V., *Makromol. Chem.*, **137**, 111 (1970).

2253. Barnikol, I., Barnikol, W. K. R., Jovanovic, M., and Schulz, G. V., *Makromol. Chem.*, **137,** 123 (1970).
2254. Barnikol, W. K. R., and Schulz, G. V., *Makromol. Chem.*, **145,** 299 (1971).
2255. Barrales-Rienda, J. M., and Fatou, J. M. G., *Polymer*, **13,** 407 (1972).
2256. Barrall, E. M., II, and Cain, J. H., *J. Polym. Sci. C*, **21,** 253 (1968).
2257. Barrall, E. M., II, Cantow, M. J. R., and Johnson, J. F., *J. Appl. Polym. Sci.*, **12,** 1373 (1968).
2258. Barrall, E. M., II, Clarke, H. T. C., and Gregges, A. R., *J. Polym. Sci. A2*, **16,** 1355 (1978).
2259. Barrall, E. M., II, and Rogers, L. B., *Anal. Chem.*, **34,** 1106 (1962)
2260. Barriault, R. J., and Gronholz, L. F., *J. Polym. Sci.*, **18,** 393 (1955).
2261. Barson, C. A., and Robb, J. C., *Brit. Polym. J.*, **3,** 53 (1971).
2262. Barton, J. M., *Makromol. Chem.*, **171,** 247 (1973).
2263. Bartosiewicz, R. L., Booth, C., and Marschall, A., *Europ. Polym. J.*, **10,** 783 (1974).
2264. Bartosiewicz, R. L., and Mencik, Z., *J. Polym. Sci. A2*, **12,** 1163 (1974).
2265. Baruya, A., Booth, A. D., Maddams, W. F., Grasselli, J. G., and Hazle, M. A. S., *J. Polym. Sci. B*, **14,** 329 (1976).
2266. Basedow, A. M., Ebert, K. H., Ederer, H., and Hunger, H., *Makromol. Chem.*, **177,** 1501 (1976).
2267. Bassett, D. C., *Phil. Mag.*, **10,** 595 (1964).
2268. Bassett, D. C., Block, S., and Piermarini, G. J., *J. Appl. Phys.*, **45,** 4146 (1974).
2269. Bassett, D. C., and Carder, D. R., *Polymer*, **14,** 387 (1973).
2270. Bassett, D. C., Damont, F. R., and Salovey, R., *Polymer*, **5,** 579 (1964).
2271. Bassett, D. C., and Davitt, R., *Polymer*, **15,** 721 (1974).
2272. Bassett, D. C., and Hodge, A. M., *Polymer*, **19,** 469 (1978).
2273. Bassett, D. C., and Hodge, A. M., *Proc. Roy. Soc. A*, **359,** 121 (1978).
2274. Bassett, D. C., and Keller, A., *Phil. Mag.*, **101,** 817 (1964).
2275. Bassett, D. C., Keller, A., and Mitsuhashi, S., *J. Polym. Sci. A*, **1,** 763 (1963).
2276. Bassett, D. C., and Khalifa, B. A., *Polymer*, **14,** 390 (1973).
2277. Bassett, D. C., and Khalifa, B. A., *Polymer*, **17,** 275 (1976).
2278. Bassett, D. C., Khalifa, B. A., and Olley, R. H., *Polymer*, **17,** 284 (1976).
2279. Bassett, D. C., Khalifa, B. A., and Olley, R. H., *J. Polym. Sci. A2*, **15,** 995 (1977).
2280. Bassett, D. C., Khalifa, B. A., and Turner, B., *Nature Phys. Sci.*, **239,** 106 (1972).
2281. Bassett, D. C., and Turner, B., *Nature Phys. Sci.*, **240,** 146 (1972).
2282. Bassi, I. W., Allegra, G., and Scordamaglia, R., *Macromolecules*, **4,** 575 (1971).
2283. Bassi, I. W., and Scordamaglia, R., *Makromol. Chem.*, **176,** 1503 (1975).
2284. Basu, S., *J. Polym. Sci.*, **5,** 735 (1950).
2285. Bateman, J., Richards, R. E., Farrow, G., and Ward, I. M., *Polymer*, **1,** 63 (1960).
2286. Bates, T. W., *Trans. Faraday Soc.*, **63,** 1825 (1967).
2287. Bates, T. W., and Stockmayer, W. H., *Macromolecules*, **1,** 12 (1968).
2288. Bates, T. W., and Stockmayer, W. H., *Macromolecules*, **1,** 17 (1968).
2289. Batra, S. K., and Syed, N., *J. Polym. Sci. A2*, **13,** 369 (1975).
2290. Battaerd, H. A., Jr, *J. Polym. Sci. C*, **49,** 149 (1975).
2291. Batzer, H., *Makromol. Chem.*, **10,** 13 (1953).
2292. Baudisch, J., Dechant, J., van Nghi, D., Phillip, D., and Ruscher, C., *Faserforsch. Textiltech.*, **19,** 62 (1968).
2293. Bauer, D. R., Brauman, J. I., and Pecora, R., *Macromolecules*, **8,** 443 (1975).
2294. Baughman, R. H., *J. Polym. Sci. A2*, **12,** 1511 (1974).
2295. Baughman, R. H., and Chance, R. H., *J. Polym. Sci. A2*, **14,** 2037 (1976).
2296. Baughman, R. H., Gleiter, H., and Sendfeld, N., *J. Polym. Sci. A2*, **13,** 1871 (1975).
2297. Baum, F. J., and Billmeyer, F. W., Jr, *J. Opt. Soc. Amer.*, **51,** 452 (1962).
2298. Baumann, H., *J. Polym. Sci. B*, **3,** 1069 (1965).
2299. Baumann, U., Schreiber, H., and Tessmar, K., *Makromol. Chem.*, **36,** 81 (1959).
2300. Bawn, C. E. H., Freeman, R. F., and Kamaliddin, A. R., *Trans. Faraday Soc.*, **46,** 677 (1950).
2301. Bawn, C. E. H., Freeman, R. F. J., and Kamaliddin, A. R., *Trans. Faraday Soc.*, **46,** 862 (1950).
2302. Bawn, C. E. H., and Patel, R. D., *Trans. Faraday Soc.*, **52,** 1664 (1956).
2303. Bawn, C. E. H., and Wajid, M. J., *Trans. Faraday Soc.*, **52,** 1658 (1956).

2304. Baysal, B., and Tobolsky, A. V., *J. Polym. Sci.*, **9**, 171 (1952).
2305. Beahan, P., Bevis, M., and Hull, D., *Polymer*, **14**, 97 (1973).
2306. Beall, G., *J. Polym. Sci.*, **4**, 483 (1949).
2307. Beaman, R. G., *J. Polym. Sci.*, **9**, 470 (1952).
2308. Beams, J. W., Boyle, R. D., and Hexner, P. E., *J. Polym. Sci.*, **57**, 161 (1962).
2309. Bearman, R. J., *J. Phys. Chem.*, **65**, 1961 (1961).
2310. Beattie, W. H., *J. Polym. Sci. A*, **3**, 527 (1965).
2311. Beattie, W. H., and Booth, C., *J. Polym. Sci.*, **44**, 81 (1960)
2312. Beattie, W. H., and Booth, C., *J. Chem. Phys.*, **64**, 696 (1960).
2313. Beattie, W. H., and Booth, C., *J. Appl. Polym. Sci.*, **7**, 507 (1963).
2314. Beatty, K. O., Armstrong, A. A., Jr, and Schoenborn, E. M., *Ind. Eng. Chem.*, **42**, 1527 (1950).
2315. Beaty, C. L., and Karasz, F. E., *J. Polym. Sci. A2*, **13**, 971 (1975).
2316. Beaumont, R. H., Clegg, B., Gee, G., Herbert, J. B. M., Marks, D. J., Robert, R. C., and Sims, D., *Polymer*, **7**, 401 (1966).
2317. Beck, G., Dobrowolski, G., Kiwi, J., and Schnabel, W., *Macromolecules*, **8**, 9 (1975).
2318. Beck, H. N., *J. Appl. Polym. Sci.*, **11**, 673 (1967).
2319. Beck, H. N., and Ledbetter, H. D., *J. Appl. Polym. Sci.*, **9**, 2131 (1965).
2320. Bedborough, D. S., and Jackson, D. A., *Polymer*, **17**, 573 (1976).
2321. Beebe, D. H., *Polymer*, **19**, 231 (1978).
2322. Beech, D. R., and Booth, C., *Polymer*, **13**, 355 (1972).
2323. Beech, D. R., Booth, C., Dodgson, D. V., Sharpe, R. R., and Waring, J. R. S., *Polymer*, **13**, 73 (1972).
2324. Beech, D. R., Booth, C., Pickles, C. J., Sharpe, R. R., and Waring, J. R. S., *Polymer*, **13**, 246 (1972).
2325. Beer, M., *Proc. Roy. Soc. A*, **236**, 136 (1956).
2326. Beevers, M., Crossley, J., Garrington, D. C., and Williams, G., *J. Chem. Soc. Faraday Trans.*, **72**, 1482 (1976).
2327. Beevers, R. B., *Trans. Faraday Soc.*, **58**, 1465 (1962).
2328. Beevers, R. B., *Lab. Prac.*, **22**, 272 (1973).
2329. Beevers, R. B., *J. Polym. Sci. A2*, **12**, 1407 (1974).
2330. Beevers, R. B., and White, E. F. T., *Trans. Faraday Soc.*, **56**, 744 (1960).
2331. Beevers, R. B., and White, E. F. T., *Trans. Faraday Soc.*, **56**, 1529 (1960).
2332. Beigen, J. R., and Czanderna, A. W., *J. Therm. Anal.*, **4**, 39 (1972).
2333. Bekkedahl, N., *J. Res. Nat. Bur. Stand.*, **43**, 145 (1969).
2334. Belanskii, V. G., and Gankina, E. S., *Dokl. Akad. Nauk SSSR*, **186**, 857 (1969).
2335. Belanskii, V. G., and Gankina, E. S., *J. Chromatogr.*, **53**, 3 (1970).
2336. Belisle, J., *Anal. Chim. Acta*, **43**, 515 (1968).
2337. Bellemans, A., and Janssens, M., *Macromolecules*, **7**, 809 (1974).
2338. Bender, G. W., and Gaines, G. L., Jr, *Macromolecules*, **3**, 128 (1970).
2339. Bendler, J., Solc, K., and Gobush, W., *Macromolecules*, **10**, 635 (1977).
2340. Benedetti, E., Pedone, C., and Allegra, G., *Macromolecules*, **3**, 16 (1970).
2341. Bennett, C. E., DiCave, L. W., Jr, Paul, D. G., Wegener, J. A., and Levase, L. J., *Amer. Lab.*, **3**, 67 (1971).
2342. Bennett, R. G., *Rev. Sci. Instrum.*, **31**, 1275 (1960).
2343. Benoit, H., *Ann. Phys.*, **6**, 561 (1951).
2344. Benoit, H., *J. Polym. Sci.*, **11**, 507 (1953).
2345. Benoit, H., Cotton, J. P., Dekker, D., Farnoux, B., Higgins, J. S., Jannik, G., Ober, R., and Picot, C., *Nature*, **245**, 13 (1973).
2346. Benoit, H., Decker, D., Duplessix, R., Picot, C., Rempp, P., Cotton, J. P., Farnoux, B., Jannik, G., and Ober, R., *J. Polym. Sci. A2*, **14**, 2119 (1976).
2347. Benoit, H., Duplessix, R., Ober, R., Daoud, M., Cotton, J. P., Farnoux, B., and Jannik, G., *Macromolecules*, **8**, 451 (1975).
2348. Benoit, H., Grubisic, Z., and Rempp, P., *Rubber Chem. Technol.*, **42**, 636 (1969).
2349. Benoit, H., Grubisic, Z., Rempp, P., Dekker, D., and Zilliox, J. G., *J. Chim. Phys.*, **63**, 1507 (1966).
2350. Benoit, H., Holtzer, A. M., and Doty, P. M., *J. Phys. Chem.*, **58**, 635 (1954).
2351. Benoit, H., and Horn, P., *J. Polym. Sci.*, **10**, 29 (1953).
2352. Benoit, H., Rempp, P., and Franta, E., *C.R. Acad. Sci.*, **257**, 1288 (1963).
2353. Benz, F. W., Feeney, J., and Roberts, G. C. K., *J. Mag. Res.*, **8**, 114 (1972).

2354. Berek, D., Bakoš, D., Bleha, T., and Šoltes, L., *Makromol. Chem.*, **176**, 391 (1975).
2355. Berek, D., Bleha, T., and Pevna, Z., *J. Polym. Sci. B*, **14**, 323 (1976).
2356. Berens, A. R., *Polymer*, **18**, 697 (1977).
2357. Beresniewicz, A., *J. Polym. Sci.*, **35**, 321 (1959).
2358. Beret, S., and Prausnitz, J. M., *Macromolecules*, **7**, 536 (1974)
2359. Beret, S., and Prausnitz, J. M., *Macromolecules*, **8**, 878 (1975).
2360. Berger, K. C., *Makromol. Chem.*, **175**, 2121 (1974).
2361. Berger, K. C., *Makromol. Chem.*, **176**, 399 (1975).
2362. Berger, K. C., *Makromol. Chem.*, **179**, 719 (1978).
2363. Berghmans, H., Groeninckx, G., and Hautecler, S., *J. Polym. Sci. A2*, **13**, 151 (1975).
2364. Berghmans, H., and Overbergh, N., *J. Polym. Sci. A2*, **15**, 1757 (1977).
2365. Berglund, C. A., *J. Polym. Sci. A2*, **15**, 2037 (1977).
2366. Berglund, C. A., and Taylor, C. R., *J. Polym. Sci. A2*, **14**, 2129 (1976).
2367. Bergman, R., and Sundelöf, L. O., *Europ. Polym. J.*, **13**, 881 (1977).
2368. Bergmann, K., *J. Polym. Sci. A2*, **16**, 1611 (1978).
2369. Bergmann, K., and Nawotky, K., *Kolloid Z. Z. Polym.*, **219**, 132 (1967).
2370. Bergmann, K., and Nawotky, K., *Ber. Bunsenges Phys. Chem.*, **74**, 912 (1970).
2371. Bergmann, K., and O'Konski, C. T., *J. Phys. Chem.*, **67**, 2169 (1963).
2372. Berndt, H. J., and Bossmann, A., *Polymer*, **17**, 241 (1976).
2373. Bernet, M. K., *Macromolecules*, **7**, 917 (1974).
2374. Bernstein, R. E., Cruz, C. A., Paul, D. R., and Barlow, J. W., *Macromolecules*, **10**, 681 (1977).
2375. Berry, G. C., *J. Chem. Phys.*, **44**, 4350 (1966).
2376. Berry, G. C., *J. Chem. Phys.*, **46**, 1338 (1967).
2377. Berry, G. C., *J. Polym. Sci. A2*, **9**, 687 (1971).
2378. Berry, G. C., and Eisman, P. R., *J. Polym. Sci. A2*, **12**, 2253 (1974).
2379. Berry, G. C., and Fox, T. G., *J. Amer. Chem. Soc.*, **86**, 3540 (1964).
2380. Berry, G. C., and Liwak, S. M., *J. Polym. Sci. A2*, **14**, 1717 (1976).
2381. Berry, G. C., and Wong, C. P., *J. Polym. Sci. A2*, **13**, 1761 (1975).
2382. Bersted, B. H., *J. Appl. Polym. Sci.*, **17**, 1415 (1973).
2383. Bersted, B. H., *J. Appl. Polym. Sci.*, **18**, 2399 (1974).
2384. Berticat, P., Ai, B., Giam, H. T., Chatain, D., and Lacabanne, C., *Makromol. Chem.*, **177**, 1583 (1976).
2385. Bertie, J. E., and Whalley, E., *J. Chem. Phys.*, **41**, 575 (1964).
2386. Bevington, J. C., *Makromol. Chem.*, **34**, 153 (1959).
2387. Beynon, J. H., and Fontaine, A. E., *Instrum. Rev.*, **14**, 470 (1967).
2388. Beynon, J. H., and Fontaine, A. E., *Instrum. Rev.*, **14**, 501 (1967).
2389. Beynon, J. H., and Fontaine, A. E., *Instrum. Rev.*, **15**, 34 (1968).
2390. Bianchi, E., Ciferri, A., Tealdi, A., Torre, R., and Valenti, B., *Macromolecules*, **7**, 495 (1974).
2391. Bianchi, U., Bruzzone, M., and Mormino, M., *J. Polym. Sci. B*, **15**, 345 (1977).
2392. Bianchi, U., and Magnasco, V., *J. Polym. Sci.*, **4**, 177 (1959).
2393. Bianchi, U., and Peterlin, A., *J. Polym. Sci. A1*, **6**, 1759 (1968).
2394. Biangardi, H. J., and Zachmann, H. G., *Makromol. Chem.*, **177**, 1173 (1976).
2395. Bilen, C. S., and Morantz, D. J., *Polymer*, **17**, 1091 (1976).
2396. Biernacki, P., Chrzczonowicz, C., and Włodarczyk, M., *Europ. Polym. J.*, **7**, 739 (1971).
2397. Biesenberger, J. A., Tan, M., and Duvdevani, I., *J. Appl. Polym. Sci.*, **15**, 1549 (1971).
2398. Biesenberger, J. A., Tan, M., Duvdevani, I., and Maurer, T., *J. Polym. Sci. B*, **9**, 353 (1971).
2399. Billick, I. H., *J. Polym. Sci.*, **62**, 167 (1962).
2400. Billmeyer, F. W., Jr, *J. Polym. Sci.*, **4**, 83 (1949).
2401. Billmeyer, F. W., Jr, and de Than, C. B., *J. Amer. Chem. Soc.*, **77**, 4763 (1955).
2402. Billmeyer, F. W., Jr, Johnson, G. W., and Kelly, R. N., *J. Chromatogr.*, **34**, 316 (1968).
2403. Billmeyer, F. W., Jr, and Katz, I., *Macromolecules*, **2**, 105 (1969).
2404. Billmeyer, F. W., Jr, and Kelly, R. N., *J. Chromatogr.*, **34**, 322 (1968).
2405. Billmeyer, F. W., Jr, and Levine, H. I., *J. Colloid Interface Sci.*, **35**, 204 (1971).
2406. Billmeyer, F. W., Jr, and Stockmayer, W. H., *J. Polym. Sci.*, **5**, 121 (1950).
2407. Binder, J. L., *Anal. Chem.*, **26**, 1877 (1954).
2408. Binder, J. L., *J. Polym. Sci. A*, **1**, 37 (1963).
2409. Binder, J. L., *J. Polym. Sci. A*, **1**, 47 (1963).

748

2410. Binder, J. L., and Ransaw, H. C., *Anal. Chem.*, **29**, 503 (1957).
2411. Binsbergen, F. L., and DeLange, B. G. M., *Polymer*, **9**, 23 (1968).
2412. Binsbergen, F. L., and DeLange, G. G. M., *Polymer*, **11**, 309 (1970).
2413. Birkley, A. W., Dawkins, J. V., and Kyriacos, D., *Polymer*, **19**, 350 (1978).
2414. Birks, J. B., *Nature*, **214**, 1187 (1967).
2415. Birks, J. B., and Dyson, D. J., *J. Sci. Instrum.*, **38**, 282 (1961).
2416. Birks, J. B., Dyson, D. J., and Munro, I. H., *Proc. Roy. Soc. A*, **275**, 575 (1963).
2417. Birks, L. S., and Brooks, E. J., *Anal. Chem.*, **22**, 1017 (1950).
2418. Biroš, J., Zeman, L., and Patterson, D., *Macromolecules*, **4**, 30 (1971).
2419. Birshtein, T. M., Skvortsov, A. M., and Sariban, A. A., *Macromolecules*, **9**, 892 (1976).
2420. Bischoff, J., and Desreux, V., *Bull. Soc. Chim. Belges*, **60**, 137 (1951).
2421. Bishop, J. H., and Silva, S. R., *Appl. Polym. Symp.*, No. 16, 195 (1971).
2422. Biskup, U., and Cantow, H. J., *Makromol. Chem.*, **148**, 31 (1971).
2423. Biskup, U., and Cantow, H. J., *Macromolecules*, **5**, 546 (1972).
2424. Biskup, U., and Cantow, H. J., *Makromol. Chem.*, **168**, 315, 329 (1973).
2425. Black, J. T., *Appl. Polym. Symp.*, No. 16, 105 (1971).
2426. Blackadder, D. A., and Keniry, J. S., *J. Appl. Polym. Sci.*, **17**, 351 (1973).
2427. Blackadder, D. A., Keniry, J. S., and Richardson, M. J., *Polymer*, **13**, 584 (1972).
2428. Blackadder, D. A., and LePoidevin, G. J., *Polymer*, **17**, 387 (1976).
2429. Blackadder, D. A., and LePoidevin, G. J., *Polymer*, **17**, 769 (1976).
2430. Blackadder, D. A., and Lewell, P. A., *Polymer*, **9**, 249 (1968).
2431. Blackadder, D. A., and Roberts, T. L., *Makromol. Chem.*, **126**, 116 (1969).
2432. Blackmore, W. R., *Can. J. Chem.*, **37**, 1508 (1959).
2433. Blackmore, W. R., *Can. J. Chem.*, **37**, 1517 (1959).
2434. Blackmore, W. R., *Can. J. Chem.*, **38**, 565 (1960).
2435. Blackmore, W. R., *Rev. Sci. Instrum.*, **31**, 317 (1960).
2436. Blair, J. E., and Williams, J. W., *J. Phys. Chem.*, **68**, 161 (1964).
2437. Blais, P., and Manley, R. S. J., *Science*, **153**, 539 (1966).
2438. Blais, P., and Manley, R. S. J., *J. Polym. Sci. A1*, **6**, 291 (1968).
2439. Blanchard, L. P., and Baijal, M. D., *J. Polym. Sci. B*, **4**, 837 (1966).
2440. Blanks, R. F., and Prausnitz, J. M., *Ind. Eng. Chem. Fundam.*, **3**, 1 (1964).
2441. Blaum, G., and Wolf, B. A., *Macromolecules*, **9**, 579 (1976).
2442. Blazo, M., and Verhegyi, G., *Europ. Polym. J.*, **14**, 625 (1978).
2443. Bleha, T., Bakoš, D., and Berek, D., *Polymer*, **18**, 897 (1977).
2444. Bleha, T., and Valko, L., *Polymer*, **17**, 298 (1976).
2445. Block, H., Collinson, M. E., and Walker, S. M., *Polymer*, **14**, 68 (1973).
2446. Block, H., Ions, W. D., and Walker, S. M., *J. Polym. Sci. A2*, **16**, 989 (1978).
2447. Block, H., Lord, P. W., and Walker, S. M., *Polymer*, **16**, 739 (1975).
2448. Blondin, D., Regis, J., and Prud'homme, J., *Macromolecules*, **7**, 187 (1974).
2449. Bloomfield, V. A., and Benbasat, J. A., *Macromolecules*, **4**, 609 (1971).
2450. Bloomfield, V. A., and Sharp, P. A., *Macromolecules*, **1**, 380 (1968).
2451. Blout, E. R., Bird, G. R., and Grey, D. S., *J. Opt. Soc. Amer.*, **40**, 306 (1950).
2452. Bluestone, S., Mark, J. E., and Flory, P. J., *Macromolecules*, **7**, 325 (1974).
2453. Blum, L., *Macromolecules*, **8**, 457 (1975).
2454. Blum, L., and Frisch, H. L., *J. Polym. Sci. A2*, **14**, 1743 (1976).
2455. Blundell, D. J., and Keller, A., *J. Macromol. Sci. B*, **2**, 227 (1968).
2456. Blundell, D. J., Keller, A., and Connor, T. M., *J. Polym. Sci. A2*, **5**, 991 (1967).
2457. Blundell, D. J., Keller, A., and Kovacs, A. J., *J. Polym. Sci. B*, **4**, 481 (1966).
2458. Blundell, D. J., Longman, G. W., Wignall, G. D., and Bowden, M. J., *Polymer*, **15**, 33 (1974).
2459. Bly, D. D., *J. Polym. Sci. A2*, **4**, 731 (1966).
2460. Bly, D. D., *J. Polym. Sci. C*, **21**, 13 (1968).
2461. Böckman, O. C., *J. Polym. Sci. A1*, **3**, 3399 (1965).
2462. Boerio, F. J., and Bahl, S. K., *J. Polym. Sci. A2*, **14**, 1029 (1976).
2463. Boerio, F. J., and Koenig, J. L., *J. Chem. Phys.*, **52**, 3425 (1970).
2464. Boerio, F. J., and Koenig, J. L., *J. Chem. Phys.*, **52**, 4826 (1970).
2465. Boerio, F. J., Schoenlein, L. H., and Greivenkamp, J. E., *J. Appl. Polym. Sci.*, **22**, 203 (1978).
2466. Boerio, F. J., and Yuann, J. K., *J. Polym. Sci. A2*, **11**, 1841 (1973).
2467. Boersma, S. L., *J. Amer. Ceramic Soc.*, **38**, 281 (1955).

2468. Bohdanecký, M., *Macromolecules*, **10**, 971 (1977).
2469. Bohlin, L., and Kubat, J., *J. Polym. Sci. A2*, **14**, 1169 (1976).
2470. Böhm, L. L., Casper, R. H., and Schulz, G. V., *J. Polym. Sci. A2*, **12**, 239 (1974).
2471. Bohon, R. L., *Anal. Chem.*, **35**, 1845 (1963).
2472. Boisserie, C., and Marchessault, R. H., *J. Polym. Sci. B*, **14**, 293 (1976).
2473. Boisserie, C., and Marchessault, R. H., *J. Polym. Sci. A2*, **15**, 1211 (1977).
2474. Bolland, J. L., *Proc. Roy. Soc. A*, **186**, 218 (1946).
2475. Bollinger, L. M., and Thomas, G. E., *Rev. Sci. Instrum.*, **32**, 1044 (1961).
2476. Bombaugh, K. J., *J. Chromatogr.*, **53**, 27 (1970).
2477. Bombaugh, K. J., Cook, C. E., and Clampitt, B. H., *Anal. Chem.*, **35**, 1834 (1963).
2478. Bombough, K. J., Dark, W. A., and King, R. N., *J. Polym. Sci. C*, **21**, 131 (1968).
2479. Bombaugh, K. J., Dark, W. A., and Levangie, R. F., *Z. Anal. Chem.*, **236**, 443 (1968).
2480. Bombaugh, K. J., Dark, W. A., and Levangie, R. F., *Separation Sci.*, **3**, 375 (1968).
2481. Bombaugh, K. J., Dark, W. A., and Levangie, R. F., *J. Chromatogr. Sci.*, **7**, 42 (1969).
2482. Bombaugh, K. J., and Levangie, R. F., *Anal. Chem.*, **41**, 1337 (1969).
2483. Bondi, A., *J. Polym. Sci. A*, **2**, 3159 (1964).
2484. Bondi, A., and Simkin, D. J., *AIChE J*, **3**, 473 (1957).
2485. Boni, K. A., Sliemers, F. A., and Stickney, P. B., *J. Polym. Sci. A2*, **6**, 1579 (1968).
2486. Bonner, D. C., and Cheng, Y. L., *J. Polymer Sci. B*, **13**, 259 (1975).
2487. Bonner, D. C., and Prausnitz, J. M., *J. Polym. Sci. A2*, **12**, 51 (1974).
2488. Bonse, U., and Hart, M., *Z. Phys.*, **189**, 151 (1966).
2489. Booth, A., and Hay, J. N., *Polymer*, **12**, 365 (1971).
2490. Booth, A., and Hay, J. N., *Brit. Polym. J.*, **4**, 9 (1972).
2491. Booth, C., and Beason, L. R., *J. Polym. Sci.*, **42**, 81 (1960).
2492. Booth, C., and Beason, L. R., *J. Polym. Sci.*, **42**, 93 (1960).
2493. Booth, C., Gee, G., Holden, G., and Williamson, G., *Polymer*, **5**, 343 (1954).
2494. Booth, C., Gee, G., and Taylor, W. D., *Polymer*, **5**, 353 (1964).
2495. Booth, C., Higginson, W. C. E., and Powell, E., *Polymer*, **5**, 479 (1964).
2496. Booth, C., Jones, M. N., and Powell, E., *Nature*, **196**, 772 (1962).
2497. Booth, C., and Pickles, C. J., *J. Polym. Sci. A2*, **11**, 595 (1973).
2498. Booth, C., and Price, C., *Polymer*, **7**, 85 (1966).
2499. Boots, H., and Deutch, J. M., *Macromolecules*, **10**, 1163 (1977).
2500. Borchard, W., and Rehage, G., Multicomponent Polymer Systems, Advances in Chemistry Series, No. 99, American Chemical Society, Washington, D.C., 1971, p. 42.
2501. Borchardt, H. J., and Daniels, F., *J. Amer. Chem. Soc.*, **79**, 41 (1957).
2502. Borisova, N. P., and Birshtein, T. M., *Vysokomol. Soedin.*, **5**, 279 (1963).
2503. Borisova, N. P., and Birshtein, T. M., *Vysokomol. Soedin.*, **6**, 1234 (1964).
2504. Borman, W. F. H., *J. Appl. Polym. Sci.*, **11**, 2119 (1978).
2505. Borsig, E., Lazár, M., Čapla, M., and Florián, Š., *Angew. Makromol. Chem.*, **9**, 89 (1969).
2506. Boss, B. D., Stejskal, E. O., and Ferry, J. D., *J. Phys. Chem.*, **71**, 1501 (1967).
2507. Bosworth, P., Masson, C. R., Melville, H. W., and Peaker, F. W., *J. Polym. Sci.*, **9**, 565 (1952).
2508. Böttcher, C. J. F., and Scholte, T. G., *Rev. Trav. Chim.*, **70**, 209 (1951).
2509. Boucher, E. A., and Hines, P. M., *J. Polym. Sci. A2*, **14**, 2241 (1976).
2510. Bourdariat, J., Berton, A., Chaussy, J., Isnard, R., and Odin, J., *Polymer*, **14**, 167 (1973).
2511. Bourdariat, J., Isnard, R., and Odin, J., *J. Polym. Sci. A2*, **11**, 1817 (1973).
2512. Boutros, S., and Hanna, A. A., *J. Polym. Sci A1*, **16**, 89 (1978).
2513. Bovey, F. A., *Pure Appl. Chem.*, **12**, 525 (1966).
2514. Bovey, F. A., Anderson, E. W., Douglass, D. C., and Manson, J. A., *J. Chem. Phys.*, **39**, 1199 (1963).
2515. Bovey, F. A., Hood, F. P., Anderson, E. W., and Kornegay, R. D., *J. Phys. Chem.*, **71**, 312 (1967).
2516. Bovey, F. A., Hood, F. P., Anderson, E. W., and Snyder, L. C., *J. Chem. Phys.*, **42**, 3900 (1975).
2517. Bovey, F. A., Sacchi, M. C., and Zambelli, A., *Macromolecules*, **7**, 752 (1974).
2518. Bovey, F. A., Schilling, F. C., Kwei, T. K., and Frisch, H. L., *Macromolecules*, **10**, 559 (1977).
2519. Bovey, F. A., Schilling, F. C., McCrackin, F. L., and Wagner, H. L., *Macromolecules*, **9**, 76 (1976).

2520. Bovey, F. A., and Tiers, G. V. D., *J. Polym. Sci.*, **44**, 173 (1960).
2521. Bovey, F. A., and Tiers, G. V. D., *Chem. Ind. (London)*, **1962**, 1826.
2522. Bovey, F. A., Tiers, G. V. D., and Filipovich, G., *J. Polym. Sci.*, **38**, 73 (1959).
2523. Bower, D. I., *J. Polym. Sci. A2*, **10**, 2135 (1972).
2524. Boyarchuk, Yu. M., Rappoport, L. Ya., Nikitin, V. N., and Apukhtina, N. P., *Polym. Sci., USSR*, **7**, 859 (1965).
2525. Boyd, R. H., and Breitling, S. M., *Macromolecules*, **5**, 1 (1972).
2526. Boyd, R. H., and Breitling, S. M., *Macromolecules*, **5**, 279 (1972).
2527. Boyd, R. H., and Breitling, S. M., *Macromolecules*, **7**, 855 (1974).
2528. Boyd, R. H., and Rahalkar, R. R., *J. Polym. Sci. B*, **15**, 515 (1977).
2529. Boyer, R. F., *J. Polym. Sci.*, **9**, 197 (1952).
2530. Boyer, R. F., *J. Appl. Phys.*, **25**, 825 (1954).
2531. Boyer, R. F., *J. Polym. Sci. C*, **14**, 267 (1966).
2532. Boyer, R. F., *Macromolecules*, **7**, 142 (1974).
2533. Boyer, R. F., and Miller, R. L., *Macromolecules*, **10**, 1167 (1977).
2534. Boyer, R. F., and Snyder, R. G., *J. Polym. Sci. B*, **15**, 315 (1977).
2535. Boyer, R. F., and Spencer, R. S., *J. Polym. Sci.*, **3**, 97 (1948).
2536. Brader, J. J., *J. Appl. Polym. Sci.*, **3**, 370 (1960).
2537. Braddy, D. G., *J. Appl. Polym. Sci.*, **20**, 2541 (1976).
2538. Brady, G. W., and Frisch, H. L., *J. Chem. Phys.*, **35**, 2234 (1961).
2539. Brame, E. G., Jr, *Anal. Chem.*, **37**, 1183 (1965).
2540. Branauer, S., Emmett, P. H., and Teller, E., *J. Amer. Chem. Soc.*, **60**, 309 (1938).
2541. Brandrup, J., and Peebles, L. H., *Macromolecules*, **1**, 64 (1968).
2542. Brant, D. A., and Min, B. K., *Macromolecules*, **2**, 1 (1969).
2543. Braud, C., and Vert, M., *Polymer*, **16**, 115 (1975).
2544. Braun, D., Törmälä, P., and Weber, G., *Polymer*, **19**, 598 (1978).
2545. Braun, D., and Törmälä, P., *Makromol. Chem.*, **179**, 1025 (1978).
2546. Braun, G., *J. Appl. Polym. Sci.*, **15**, 2321 (1971).
2547. Braun, J. M., and Guillet, J. E., *Macromolecules*, **8**, 557 (1975).
2548. Braun, J. M., and Guillet, J. E., *Macromolecules*, **8**, 882 (1975).
2549. Braun, J. M., and Guillet, J. E., *J. Polym. Sci. A1*, **13**, 1119 (1975).
2550. Braun, J. M., and Guillet, J. E., *Macromolecules*, **9**, 340 (1976).
2551. Braun, J. M., and Guillet, J. E., *Macromolecules*, **9**, 617 (1976).
2552. Braun, J. M., and Guillet, J. E., *J. Polym. Sci. A1*, **14**, 1073 (1976).
2553. Braun, J. M., and Guillet, J. E., *Macromolecules*, **10**, 101 (1977).
2554. Braun, J. M., Cutajar, M., Guillet, J. E., Schreiber, H. P., and Patterson, D., *Macromolecules*, **10**, 864 (1977).
2555. Braun, J. M., Lavoie, A., and Guillet, J. E., *Macromolecules*, **8**, 311 (1975).
2556. Braun, J. M., Poos, S., and Guillet, J. E., *J. Polym. Sci. B*, **14**, 257 (1976).
2557. Braun, T., and Farag, A. B., *Talanta*, **22**, 699 (1975).
2558. Braun, W., Hellwege, K. H., and Knappe, W., *Kolloid Z. Z. Polym.*, **215**, 10 (1967).
2559. Bray, J. C., and Merrill, E. W., *J. Appl. Polym. Sci.*, **17**, 3779 (1973).
2560. Brazier, D. W., and Schwartz, N. V., *J. Appl. Polym. Sci.*, **22**, 113 (1978).
2561. Breck, A. K., Taylor, C. L., Russell, K. E., and Wan, J. K. S., *J. Polym. Sci. A1*, **12**, 1505 (1974).
2562. Brehm, B., Demtröder, W., and Osberghaus, O., *Z. Naturforsch.*, **16A**, 843 (1961).
2563. Brehm, G. A., and Bloomfield, V. A., *Macromolecules*, **8**, 663 (1975).
2564. Breitschwerdt, K., and Weller, A., *Z. Phys. Chem. (Frankfurt)*, **20**, 353 (1959).
2565. Brenner, M., and Niederwieser, A., *Experientia*, **17**, 237 (1961).
2566. Brenner, S., and Horne, R. W., *Biochim. Biophys. Acta*, **34**, 103 (1959).
2567. Brereton, M. G., and Davies, G. R., *Polymer*, **18**, 764 (1977).
2568. Brereton, M. G., Davies, G. R., Rushworth, A., and Spence, J., *J. Polym. Sci. A2*, **15**, 583 (1977).
2569. Breuer, H., and Rehage, G., *Kolloid Z. Z. Polym.*, **216**, 159 (1967).
2570. Brewer, L., Berg, R. A., and Rosenblatt, G. M., *J. Chem. Phys.*, **38**, 1381 (1963).
2571. Brewer, P. I., *Polymer*, **9**, 545 (1968).
2572. Brice, B. A., and Halwer, M., *J. Opt. Soc. Amer.*, **41**, 1033 (1951).
2573. Brice, B. A., Halwer, M., and Speiser, R., *J. Opt. Soc. Amer.*, **40**, 768 (1950).
2574. Brice, B. A., and Speiser, R., *J. Opt. Soc. Amer.*, **38**, 363 (1946).
2575. Brigodiot, M., Cheradame, H., Fontanille, M., and Vairon, J., *Polymer*, **17**, 254 (1976).

2576. Brill, O. L., Weill, C. G., and Schmidt, P. W., *J. Colloid Interface Sci.*, **27**, 479 (1968).
2577. Bristow, G. M., and Watson, W. F., *Trans. Faraday Soc.*, **54**, 1731, 1742 (1958).
2578. Bristow, G. M., and Watson, W. F., *Trans. Faraday Soc.*, **54**, 1567 (1958).
2579. Brochard, F., and de Gennes, P. G., *Macromolecules*, **10**, 1157 (1977).
2580. Brockmeier, N. F., McCoy, R. W., and Meyer, J. A., *Macromolecules*, **5**, 464 (1972).
2581. Brockmeier, N. F., McCoy, R. W., and Meyer, J. A., *Macromolecules*, **6**, 176 (1973).
2582. Broda, A., *J. Polym. Sci.*, **25**, 117 (1957).
2583. Broda, A., Niwinska, T., and Polowinski, S., *J. Polym. Sci.*, **29**, 183 (1958).
2584. Broda, A., Niwinska, T., and Polowinski, S., *J. Polym. Sci.*, **32**, 343 (1958).
2585. Brody, E. M., Lubell, C. J., and Beatty, C. L., *J. Polym. Sci. A2*, **13**, 295 (1975).
2586. Brody, S. S., *Rev. Sci. Instrum.*, **28**, 1021 (1957).
2587. Brooks, M. C., and Badger, R. M., *J. Amer. Chem. Soc.*, **72**, 1705 (1950).
2588. Broser, W., *Makromol. Chem.*, **2**, 248 (1948).
2589. Brostow, W., *Macromolecules*, **4**, 142 (1971).
2590. Brown, B. L., and Jones, G. P., *J. Polym. Sci. A2*, **13**, 599 (1975).
2591. Brown, D. S., Fulcher, K. U., and Wetteon, R. E., *Polymer*, **14**, 379 (1973).
2592. Brown, D. W., and Lowry, R. E., *J. Polym. Sci. A1*, **15**, 2623 (1977).
2593. Brown, D. W., Lowry, R. E., and Well, L. A., *J. Polym. Sci. A2*, **12**, 1303 (1974).
2594. Brown, D. W., and Wall, L. A., *J. Polym. Sci. A2*, **7**, 601 (1967).
2595. Brown, D. S., Warner, F. P., and Wetton, R. E., *Polymer*, **13**, 575 (1972).
2596. Brown, I. G., Wetton, R. E., Richardson, M. J., and Savill, N. G., *Polymer*, **19**, 659 (1978).
2597. Brown, R. G., *J. Chem. Phys.*, **38**, 221 (1963).
2598. Brown, R. G., *J. Chem. Phys.*, **40**, 2900 (1964).
2599. Browning, H. L., Jr, Ackerman, H. D., and Patton, H. W., *J. Polym. Sci. A1*, **4**, 1433 (1966).
2600. Brownstein, S., Bywater, S., and Cowie, J. M. G., *Trans. Faraday Soc.*, **65**, 2480 (1969).
2601. Brownstein, S., Bywater, S., and Worsfold, D. J., *J. Phys. Chem.*, **66**, 2067 (1962).
2602. Brownstein, S. M., and Wiles, D. M., *J. Polym. Sci. A1*, 1901 (1964).
2603. Brubaker, W. M., *Rev. Sci. Instrum.*, **35**, 1007 (1964).
2604. Bruck, S. D., *J. Polym. Sci. A1*, **5**, 2458 (1967).
2605. Brügel, W., *Kunststoffe*, **46**, 47 (1956).
2606. Brumberger, H., and Deslatts, R., *J. Res. Nat. Bur. Std. C*, **68**, 173 (1964).
2607. Bruno, G., Foti, S., Maravigna, P., and Montaudo, G., *Polymer*, **18**, 1149 (1977).
2608. Bruno, G., and Freed, J. H., *Chem. Phys. Lett.*, **25**, 328 (1974).
2609. Bruss, D. B., and Stross, F. H., *Anal. Chem.*, **32**, 1456 (1960).
2610. Bruss, D. B., and Stross, F. H., *J. Polym. Sci. A*, **1**, 2439 (1963).
2611. Brüssau, R. J., *Makromol. Chem.*, **175**, 691 (1974).
2612. Bryce, W. A. J., *Polymer*, **10**, 803 (1969).
2613. Bryngdahl, O., *Acta Chem. Scand.*, **12**, 684 (1958).
2614. Brzezinski, J., Glowala, H., and Kornas-Calka, A., *Europ. Polym. J.*, **9**, 1251 (1973).
2615. Bucci, G., and Simonazzi, T., *J. Polym. Sci. C*, **7**, 203 (1964).
2616. Bucario, J. A., and Litovitz, T. A., *J. Chem. Phys.*, **54**, 3846 (1971).
2617. Buckley, D. J., Berger, M., and Poller, D., *J. Polym. Sci.*, **56**, 163 (1962).
2618. Buckley, D. J., and Berger, M., *J. Polym. Sci.*, **57**, 175 (1962).
2619. Bucknall, C. B., Drinkwater, I. C., and Keast, W. E., *Polymer*, **13**, 115 (1972).
2620. Budd, M. S., *Anal. Chem.*, **34**, 1147 (1962).
2621. Bueche, F., *J. Chem. Phys.*, **48**, 1410 (1968).
2622. Bueche, F., Cashin, W. M., and Debye, P., *J. Chem. Phys.*, **20**, 1956 (1952).
2623. Buerger, M. J., *J. Appl. Phys.*, **16**, 501 (1945).
2624. Büld, G., and Meyerhoff, G., *Makromol. Chem.*, Suppl. 1, 359 (1975).
2625. Buleon, A., and Chanzy, H., *J. Polym. Sci. A2*, **16**, 833 (1978).
2626. Buleon, A., Chanzy, H., and Roche, E., *J. Polym. Sci. A2*, **14**, 1913 (1976).
2627. Buleon, A., Chanzy, H., and Roche, E., *J. Polym. Sci. B*, **15**, 265 (1977).
2628. Bullock, A. T., Butterworth, J. H., and Cameron, G. G., *Europ. Polym. J.*, **7**, 445 (1971).
2629. Bullock, A. T., Cameron, G. G., and Elsom, J. M., *Polymer*, **15**, 74 (1974).
2630. Bullock, A. T., Cameron, G. G., and Elsom, J. M., *Polymer*, **18**, 931 (1977).
2631. Bullock, A. T., Cameron, G. G., and Smith, P. M., *Polymer*, **13**, 89 (1972).
2632. Bullock, A. T., Cameron, G. G., and Smith, P. M., *Polymer*, **14**, 525 (1973).
2633. Bullock, A. T., Cameron, G. G., and Smith, P. M., *J. Phys. Chem.*, **77**, 1635 (1973).

752

2634. Bullock, A. T., Cameron, G. G., and Smith, P. M., *Makromol. Chem.*, **176**, 2153 (1975).
2635. Bullock, A. T., Cameron, G. G., and Smith, P. M., *Europ. Polym. J.*, **11**, 617 (1975).
2636. Bullock, A. T., Cameron, G. G., and Smith, P. M., *Macromolecules*, **9**, 650 (1976).
2637. Buniyat-Azde, A. A., Osipov, Ya. E., and Azimova, A. B., *Vysokomol. Soedin. A*, **14**, 722 (1972).
2638. Bunk, A. J. H., Regtui, H. G., van der Berg, J. W. A., Gayamns, R. J., and Schuyer, J., *Makromol. Chem.*, **176**, 3511 (1975).
2639. Bunn, A., and Cudby, M. E. A., *Polymer*, **17**, 548 (1976).
2640. Bunn, C. W., *Proc. Roy. Soc. A*, **180**, 67 (1942).
2641. Bunn, C. W., *J. Chem. Soc.*, **1947**, 297.
2642. Bunn, C. W., *J. Polym. Sci.*, **16**, 323 (1955).
2643. Bunn, C. W., *J. Polym. Sci. B*, **1**, 209 (1963).
2644. Bunn, C. W., Cobbold, A. J., and Palmer, R. P., *J. Polym. Sci.*, **28**, 365 (1958).
2645. Bunn, C. W., and Holmes, D. R., *Discuss. Faraday Soc.*, **25**, 95 (1958).
2646. Bunn, C. W., and Howells, E. R., *Nature*, **174**, 549 (1954).
2647. Bunn, C. W., and Howells, E. R., *J. Polym. Sci.*, **18**, 307 (1955).
2648. Buntjakov, A. S., and Averyanova, V. M., *J. Polym. Sci. C*, **38**, 109 (1972).
2649. Bur, A. J., and Fetters, L. J., *Macromolecules*, **6**, 874 (1973).
2650. Burchard, W., *Macromol. Chem.*, **50**, 20 (1961).
2651. Burchard, W., *Brit. Polym. J.*, **3**, 209, 214 (1971).
2652. Burchard, W., *Macromolecules*, **5**, 604 (1972).
2653. Burchard, W., *Macromolecules*, **7**, 835, 841 (1974).
2654. Burchard, W., *Macromolecules*, **10**, 919 (1977).
2655. Burchard, W., *Macromolecules*, **11**, 455 (1978).
2656. Burchard, W., Kajiwara, K., Gordon, M., Kálal, J., and Kennedy, J. W., *Macromolecules*, **6**, 642 (1973).
2657. Burge, D. E., and Bruss, D. B., *J. Polym. Sci. A*, **1**, 1927 (1963).
2658. Burille, P., Bert, M., Michel, A., and Guyot, A., *J. Polym. Sci. B*, **16**, 181 (1978).
2659. Burkhardt, R. D., *Macromolecules*, **9**, 234 (1976).
2660. Burnett, B. B., and McDevitt, W. F., *J. Appl. Phys.*, **28**, 1101 (1957).
2661. Burnett, L. J., Rottler, C. L., and Laughon, D. H., *J. Polym. Sci. A2*, **16**, 341 (1978).
2662. Burns, E. A., and Muraca, R. F., *Anal. Chem.*, **31**, 397 (1959).
2663. Burrell, H., *J. Paint Technol.*, **40**, 197 (1968).
2664. Bush, D. G., Kunzelsauer, L. J., and Merrill, S. H., *Anal. Chem.*, **35**, 1250 (1963).
2665. Bushuk, W., and Benoit, H., *Can. J. Chem.*, **36**, 1616 (1958).
2666. Busse, W. F., *J. Polym. Sci. C*, **14**, 15 (1966).
2667. Buter, R., Tan, Y. Y., and Challa, G., *Polymer*, **14**, 171 (1973).
2668. Caccamese, S., Foti, S., Maravigna, P., Montaudo, G., Recca, A., Luderwald, I., and Przybylski, M., *J. Polym. Sci. A1*, **15**, 5 (1977).
2669. Caironi, E., *Intern. Lab.*, March/April 1978, p. 55.
2670. Cais, R. E., and Bovey, F. A., *Macromolecules*, **10**, 169 (1977).
2671. Cais, R. E., and Bovey, F. A., *Macromolecules*, **10**, 752 (1977).
2672. Cais, R. E., and Bovey, F. A., *Macromolecules*, **10**, 757 (1977).
2673. Cais, R. E., and O'Donnell, J. H., *Macromolecules*, **9**, 279 (1976).
2674. Cais, R. E., and Stuk, G. J., *Polymer*, **19**, 179 (1978).
2675. Calleja, F. J. B., Ortega, J. C. G., and Salazar, J. M., *Polymer*, **19**, 1094 (1978)
2676. Călugăru, E. M., and Schneider, I. A., *Europ. Polym. J.*, **10**, 729 (1974).
2677. Calvert, P. D., *J. Polym. Sci. A2*, **14**, 2211 (1976).
2678. Calvert, P. D., and Uhlmann, D. R., *J. Polym. Sci. A2*, **10**, 1811 (1972).
2679. Calvert, P. D., and Uhlmann, D. R., *J. Polym. Sci. A2*, **11**, 457 (1973).
2680. Cameron, G. G., and Fortune, J. D., *Europ. Polym. J.*, **4**, 333 (1968).
2681. Cameron, G. G., Kerr, G. P., and Gourlay, A. R., *J. Macromol. Sci. A*, **2**, 761 (1968).
2682. Campbell, D., *J. Polym. Sci. B*, **8**, 313 (1970).
2683. Campbell, D. R., and Warner, W. C., *Anal. Chem.*, **37**, 276 (1965).
2684. Campbell, G. C., Schurr, G. G., Slawikowski, E. D., and Spense, J. W., *J. Paint. Technol.*, **46**, 59 (1974).
2685. Campbell, H., Kane, P. O., and Ottewill, I. G., *J. Polym. Sci.*, **12**, 611 (1954).
2686. Campbell, J. R., *Aldrichimica Acta*, **4**, 55 (1971).
2687. Camposa, A., and Figueruelo, J. E., *Polymer*, **18**, 1296 (1977).

2688. Campos, A., and Figueruelo, J. E., *Makromol. Chem.*, **178**, 3249 (1977).
2689. Campos-Lopez, E., and Angulo-Sanchez, J. L., *J. Polym. Sci. B*, **14**, 649 (1976).
2690. Canbäck, G., and Rånby, B., *Macromolecules*, **10**, 797 (1977).
2691. Candau, F., Dufour, C., and Francois, J., *Makromol. Chem.*, **177**, 3359 (1976).
2692. Candau, F., Francois, J., and Benoit, H., *Polymer*, **15**, 626 (1974).
2693. Candau, F., Rempp, P., and Benoit, H., *Macromolecules*, **5**, 627 (1972).
2694. Cane, F., and Capaccioli, T., *Europ. Polym. J.*, **14**, 185 (1978).
2695. Cannon, C. G., and Harris, P. H., *J. Macromol. Sci. B*, **3**, 357 (1969).
2696. Cantow, H. J., *Angew. Chem.*, **70**, 318 (1958).
2697. Cantow, H. J., *Makromol. Chem.*, **30**, 169 (1959).
2698. Cantow, H. J., Siefert, E., and Kuhn, R., *Chem. Eng. Technol.*, **38**, 1032 (1966).
2699. Cantow, M. J. R., and Johnson, J. F., *J. Appl. Polym. Sci.*, **11**, 1851 (1967).
2700. Cantow, M. J. R., and Johnson, J. F., *J. Polym. Sci. A1*, **5**, 2835 (1967).
2701. Cantow, M. J. R., Meyerhoff, G., and Schulz, G. V., *Makromol. Chem.*, **49**, 1 (1961).
2702. Cantow, M. J. R., Porter, R. S., and Johnson, J. F., *Nature*, **192**, 752 (1961).
2703. Cantow, M. J. R., Porter, R. S., and Johnson, J. F., *J. Polym. Sci, C*, **1**, 187 (1963).
2704. Cantow, M. J. R., Porter, R. S., and Johnson, J. F., *J. Polym. Sci. A*, **2**, 2547 (1964).
2705. Cantow, M. J. R., Porter, R. S., and Johnson, J. F., *J. Appl. Polym. Sci.* **8**, 2963 (1964).
2706. Cantow, M. J. R., Porter, R. S., and Johnson, J. F., *J. Polym. Sci. B*, **4**, 707 (1966).
2707. Cantow, M. J. R., Porter, R. S., and Johnson, J. F., *J. Polym. Sci. A1*, **5**, 987 (1967).
2708. Cantow, M. J. R., Porter, R. S., and Johnson, J. F., *J. Polym. Sci. A1*, **5**, 1391 (1967).
2709. Cantow, M. J. R., Porter, R. S., and Johnson, J. F., *Nature*, **192**, 752 (1969).
2710. Cappacio, G., and Ward, I. M., *Nature*, **243**, 143 (1973).
2711. Cappacio, G., and Ward, I. M., *Polymer*, **16**, 239 (1975).
2712. Caplan, S. R., *J. Sci. Instrum.*, **31**, 295 (1954).
2713. Caplan, S. R., *J. Polym. Sci.*, **35**, 409 (1959).
2714. Capolla, G., Fillippini, R., and Pallesi, B., *Polymer*, **16**, 546 (1975).
2715. Cardenas, J. N., and O'Driscoll, K. F., *J. Polym. Sci. B*, **13**, 657 (1975).
2716. Carlini, C., Ciardelli, F., Lardicci, L., and Menicagli, R., *Makromol. Chem.*, **174**, 27 (1973).
2717. Carlson, F. D., and Fraser, A., *J. Mol. Biol.*, **89**, 273 (1974).
2718. Carlsson, D. J., Suprunchuk, T., and Wiles, D. M., *J. Polym. Sci.*, **14**, 493 (1976).
2719. Carlsson, D. J., and Wiles, D. M., *J. Polym. Sci. B*, **11**, 759 (1973).
2720. Carlsson, D. J., and Wiles, D. M., *Macromolecules*, **7**, 259 (1974).
2721. Carman, C. J., *Macromolecules*, **6**, 725 (1973).
2722. Carman, C. J., *Macromolecules*, **7**, 789 (1974).
2723. Carman, C. J., Harrington, R. A., and Wilkes, C. E., *Macromolecules*, **10**, 536 (1977).
2724. Carman, C. J., Tarpley, A. R., Jr, and Goldstein, J. H., *Macromolecules*, **4**, 445 (1971).
2725. Carman, C. J., Tarpley, A. R., Jr, and Goldstein, J. H., *J. Amer. Chem. Soc.*, **93**, 2864 (1971).
2726. Carman, C. J., and Wilkes, C. E., *Macromolecules*, **7**, 40 (1967).
2727. Carman, C. J., and Wilkes, C. E., *Rubber Chem. Technol.*, **44**, 781 (1971).
2728. Carmichael, J. B., *ACS Polym. Prepr.*, **9**, 572 (1968).
2729. Carmichael, J. B., *J. Polym. Sci. A2*, **6**, 517 (1968).
2730. Carmichael, J. B., *IUPAC Intern. Symp. Macromol. Chem. (Toronto)* **11**, A137 (1968).
2731. Carmichael, J. B., *Macromolecules*, **1**, 526 (1969).
2732. Carpenter, D. K., *J. Polym. Sci. A2*, **4**, 923 (1966).
2733. Carr, C. I., and Zimm, B. H., *J. Chem. Phys.*, **18**, 1616 (1950).
2734. Carr, S. H., Keller, A., and Baer, E., *J. Polym. Sci. A2*, **8**, 1467 (1970).
2735. Carrazzolo, G., and Mammi, M., *J. Polym. Sci. B*, **2**, 1053 (1964).
2736. Carstensen, P., *Makromol. Chem.*, **135**, 219 (1970).
2737. Carstensen, P., *Makromol. Chem.*, **142**, 131 (1971).
2738. Carter, S. R., and Record, B. R., *J. Chem. Soc.*, **1939**, 664.
2739. Carter, V. B., *J. Mol. Spectr.*, **34**, 356 (1970).
2740. Casassa, E. F., *J. Chem. Phys.*, **37**, 2178 (1962).
2741. Casassa, E. F., *J. Chem. Phys.*, **41**, 3213 (1964).
2742. Casassa, E. F., *J. Polym. Sci. B*, **5**, 773 (1967).
2743. Casassa, E. F., in Characterization of Macromolecular Structure, Nat. Acad. Sci., U.S.A., Publ. No. 1573, Washington, D.C., 1968, p. 285.
2744. Casassa, E. F., *J. Polym. Sci. A2*, **8**, 1651 (1970).

754

2745. Casassa, E. F., *ACS Div. Petrol. Chem. Preprints*, **15**, A75 (1970).
2746. Casassa, E. F., *J. Phys. Chem.*, **75**, 3929 (1971).
2747. Casassa, E. F., *Macromolecules*, **8**, 242 (1975).
2748. Casassa, E. F., *Macromolecules*, **9**, 182 (1976).
2749. Casassa, E. F., and Katz, S., *J. Polym. Sci.*, **14**, 385 (1954).
2750. Casassa, E. F., and Markovitz, H., *J. Chem. Phys.*, **29**, 493 (1958).
2751. Casassa, E. F., and Markovitz, H., *J. Chem. Phys.*, **31**, 800 (1959).
2752. Casassa, E. F., and Tagami, Y., *J. Polym. Sci. A2*, **6**, 63 (1968).
2753. Casassa, E. F., and Tagami, Y., *Macromolecules*, **2**, 14 (1969).
2754. Case, L. C., *J. Phys. Chem.*, **62**, 895 (1958).
2755. Case, L. C., *Makromol. Chem.*, **41**, 61 (1960).
2756. Casey, M., *Polymer*, **18**, 1219 (1977).
2757. Casper, R. H., Biskup, U., Lange, H., and Pohl, U., *Makromol. Chem.*, **177**, 1111 (1976).
2758. Casper, R. H., and Schulz, G. V., *Separation Sci.*, **6**, 321 (1971).
2759. Casteleijn, G., Bosh, J. J., and Smidt, J., *J. Appl. Phys.*, **39**, 4375 (1968).
2760. Casiff, E. H., and Hewett, W. A., *J. Appl. Polym. Sci.*, **6**, S30 (1962).
2761. Cavalli, L., Borsini, G. C., Carraro, G., and Canfalonieri, G., *J. Polym. Sci. A1*, **8**, 801 (1970).
2762. Cazes, J., and Gaskill, D. R., *Separation Sci.*, **2**, 421 (1967).
2763. Ceccorulli, G., and Menescalchi, F., *Makromol. Chem.*, **168**, 303 (1973).
2764. Ceccorulli, G., Menescalchi, F., and Pizzoli, M., *Makromol. Chem.*, **176**, 1163 (1975).
2765. Černes, F., Osredkar, U., Može, A., Vizovišek, I., and Lapanje, S., *Makromol. Chem.*, **178**, 2197 (1977).
2766. Cessac, G. L., and Curro, J. G., *J. Polym. Sci. A2*, **12**, 695 (1974).
2767. Cha, C., *J. Polym. Sci. B*, **2**, 1069 (1964).
2768. Chabert, B., Chauchard, J., Edel, G., and Soulier, J. P., *Europ. Polym. J.*, **9**, 993 (1973).
2769. Chachaty, C., Forchini, A., and Ronfard-Haret, J. C., *Makromol. Chem.*, **173**, 213 (1973).
2770. Challa, G., *Makromol. Chem.*, **38**, 105 (1960).
2771. Challa, G., Hermans, O. H., and Weidinger, A., *Makromol. Chem.*, **56**, 169 (1962).
2772. Chalmers, J. M., *Polymer*, **18**, 681 (1977).
2773. Chalykh, A. Y., and Vasenin, R. M., *Vysokomol. Soedin.*, **7**, 586 (1965).
2774. Chambers, R. D., Clark, D. T., Kilcast, D., and Partington, S., *J. Polym. Sci. A1*, **12**, 1647 (1974).
2775. Champion, J. V., Dandridge, A., Downer, D., McGrath, J. C., and Meeten, G. H., *Polymer*, **17**, 511 (1976).
2776. Champion, J. V., Meeten, G. H., and Southwell, G. W., *J. Chem. Soc., Faraday Trans.*, **71**, 225 (1975).
2777. Champion, J. V., Meeten, G. H., and Southwell, G. W., *Polymer*, **17**, 651 (1976).
2778. Champion, J. V., Meeten, G. H., and Whittle, C. D., *J. Chim. Phys.*, **67**, 1864 (1970).
2779. Champion, J. V., Meeten, G. H., and Whittle, C. D., *Trans. Faraday Soc.*, **66**, 2671 (1970).
2780. Chan, M. G., and Hawkins, W. L., *ACS Polymer Preprints*, **9**, 1638 (1968).
2781. Chandra, G., Kulkarni, V. G., Lagu, R. G. and Thosar, B. V., *Phys. Lett.*, **19**, 201 (1965).
2782. Chang, C., Pfeiffer, D., and Stein, R. S., *J. Polym. Sci. A2*, **12**, 1441 (1974).
2783. Chang, E. P., and Chien, J. C. W., *J. Polym. Sci A2*, **11**, 737 (1973).
2784. Chang, M., Pound, T. C., and Manley, R. S. J., *J. Polym. Sci. A2*, **11**, 399 (1973).
2785. Chang, P. I., and Slagowski, E. L., *J. Appl. Polym. Sci.*, **22**, 769 (1978).
2786. Chang, S. S., and Bestul, A. B., *J. Polym. Sci. A2*, **6**, 849 (1968).
2787. Chang, S. S., and Bestul, A. B., *J. Res. Nat. Bur. Std. A*, **75**, 113 (1971).
2788. Chang, T. L., and Mead, T. E., *Anal. Chem.*, **43**, 534 (1971).
2789. Chang, Y. J., Chen, C. T., and Tobolsky, A. V., *J. Polym. Sci. A2*, **12**, 1 (1974).
2790. Chang, Y. I., and Summerfield, G. C., *J. Polym. Sci. A2*, **7**, 405 (1969).
2791. Chantry, G. W., Fleming, J. W., Nicol, E. A., Willis, H. A., and Cudby, M. E. A., *Chem. Phys. Lett.*, **16**, 141 (1972).
2792. Chantry, G. W., Fleming, J. W., Nicol, E. A., Willis, H. A., Cudby, M. E. A., and Boerio, F. J., *J. Polym. Sci.*, **15**, 69 (1974).

2793. Chantry, G. W., Fleming, J. W., Pardoe, G. W. F., Reddish, W., and Willis, H. A., *Infrared Phys.*, **11**, 109 (1971).
2794. Chanzy, H. D., and Roche, E. J., *J. Polym. Sci. A2*, **12**, 1117 (1974).
2795. Chanzy, H. D., and Roche, E. J., *J. Polym. Sci. A2*, **12**, 2583 (1974).
2796. Chanzy, H. D., Roche, E. J., and Vuong, R. K., *J. Polym. Sci. A2*, **11**, 1859 (1973).
2797. Chapiro, A., Cordier, P., Jozefowicz, J., and Sebban-Danon, J., *J. Polym. Sci. C*, **4**, 491 (1963).
2798. Chappel, F. P., *Polymer*, **1**, 409 (1960).
2799. Charch, W. H., and Mosley, W. W., Jr, *Text. Res. J.*, **29**, 525 (1959).
2800. Charlesby, A., and Callanghan, L., *Phys. Chem. Solid.*, **4**, 227 (1958).
2801. Charlesby, A., Libby, D., and Ormerod, M. G., *Proc. Roy. Soc. A*, **262**, 207 (1961).
2802. Charlesby, A., and Partridge, R. H., *Proc. Roy. Soc., A*, **271**, 170 (1963).
2803. Charlesby, A., and Partridge, R. H., *Proc. Roy. Soc. A*, **283**, 329 (1965).
2804. Charlson, R. J., Harrison, H., and Hardwick, R., *Rev. Sci. Instrum.*, **31**, 46 (1960).
2805. Chatain, D., Gautier, P., and Lacabanne, C., *J. Polym. Sci. A2*, **11**, 1631 (1973).
2806. Chau, T. C., and Rudin, A., *Polymer*, **15**, 593 (1974).
2807. Chaurasia, N. M., Kumar, A., and Gupta, S. K., *Polymer*, **17**, 570 (1976).
2808. Che, M., and Tench, A. J., *J. Polym. Sci. B*, **13**, 345 (1975).
2809. Chen, A. K., and Woody, R. W., *J. Amer. Chem. Soc.*, **93**, 29 (1970).
2810. Chen, D. T. L., and Morawetz, H., *Macromolecules*, **9**, 463 (1976).
2811. Chen, F. S., Geusic, J. E., Kurtz, S. K., Skinner, J. G., and Wemple, S. H., *J. Appl. Phys.*, **37**, 388 (1966).
2812. Chen, S. A., *J. Appl. Polym. Sci.*, **15**, 1247 (1971).
2813. Chen, S. P., and Ferry, J. D., *Macromolecules*, **1**, 270 (1968).
2814. Chen, S. P., and Ferry, J. D., *Macromolecules*, **1**, 2699 (1968).
2815. Chen, V. Y., Allegra, G., Corradini, P., and Goodman, M., *Macromolecules*, **3**, 274 (1970).
2816. Cheng, H. N., Schilling, F. C., and Bovey, F. A., *Macromolecules*, **9**, 365 (1976).
2817. Cheng, Y. L., and Bonner, D. C., *Macromolecules*, **7**, 687 (1974).
2818. Chermin, H. A. G., and Kennedy, J. W., *Macromolecules*, **5**, 655 (1972).
2819. Chermin, H. A. G., and Koningsveld, R., *Proc. Roy. Soc. A*, **319**, 3311 (1970).
2820. Cherry, W. W., and Holmes, C. M., *J. Colloid Interface Sci.*, **29**, 174 (1969).
2821. Cherwenka, C. H., *Anal. Chem.*, **38**, 356 (1966).
2822. Chiang, R., *J. Polym. Sci.*, **36**, 91 (1959).
2823. Chiang, R., and Flory, P. J., *J. Amer. Chem. Soc.*, **83**, 2857 (1961).
2824. Chien, J. C. W., and Chang, E. P., *Macromolecules*, **5**, 610 (1972).
2825. Chikahisa, Y., and Tanaka, Y., *Polymer Symp.*, No. 30, 105 (1970).
2826. Chiklis, C. K., and Grosshoff, J. M., *J. Polym. Sci. A2*, **8**, 1617 (1970).
2827. Childers, C. W., *J. Amer. Chem. Soc.*, **85**, 229 (1963).
2828. Chinai, S. N., *J. Polym. Sci.*, **25**, 413 (1957).
2829. Chinai, S. N., and Guzzi, R. A., *J. Polym. Sci.*, **21**, 417 (1956).
2830. Chinai, S. N., and Samuels, R. J., *J. Polym. Sci.*, **19**, 463 (1956).
2831. Chiu, D. S., Takahashi, Y., and Mark, J. E., *Polymer*, **17**, 670 (1976).
2832. Choji, N., and Karasawa, M., *J. Polym. Sci. A2*, **15**, 1309 (1977).
2833. Choy, C. L., *J. Polym. Sci. A2*, **13**, 1263 (1975).
2834. Choy, C. L., Tse, Y. K., Tsui, S. M., and Hsu, B. S., *Polymer*, **16**, 501 (1975).
2835. Choy, C. L., and Young, K., *Polymer*, **19**, 1001 (1978).
2836. Christensen, A. K., *J. Cellular Biol.*, **51**, 772 (1971).
2837. Christopher, P. C., *J. Appl. Polym. Sci.*, **20**, 2989 (1976).
2838. Chu, B., *Rev. Sci. Instrum.*, **35**, 1201 (1964).
2839. Chu, B., Pallesen, M., Kao, W. P., Andrews, D. E., and Schmidt, P. W., *J. Chem. Phys.*, **43**, 2950 (1965).
2840. Chu, B., and Schoenes, F. J., *J. Colloid Interface Sci.*, **27**, 424 (1968).
2841. Chu, S. G., and Munk, P., *J. Polym. Sci. A2*, **15**, 1163 (1977).
2842. Chu, S. G., and Munk, P., *Macromolecules*, **11**, 101 (1978).
2843. Chu, S. P., and Ferry, J. D., *Macromolecules*, **1**, 270 (1968).
2844. Chu, W. H., Gipstein, E., and Ouano, A. C., *J. Appl. Polym. Sci.*, **21**, 1045 (1977).
2845. Chu, W. H., and Horne, D. E., *J. Polym. Sci. A2*, **15**, 303 (1977).
2846. Chu, W. H., and Stein, R. S., *J. Polym. Sci. A2*, **8**, 489 (1970).

756

2847. Chu, W. H., and Yoon, D. Y., *J. Polym. Sci. C*, **61**, 17 (1977).
2848. Chuang, J. C., and Morawetz, H., *Macromolecules*, **6**, 43 (1973).
2849. Chuang, J. Y., and Johnson, J. F., *J. Appl. Polym. Sci.*, **17**, 2123 (1973).
2850. Chuang, S. Y., and Tao, S. J., *J. Appl. Phys.*, **44**, 5171 (1973).
2851. Chuang, S. Y., Tao, S. J., and Wang, T. T., *Macromolecules*, **10**, 713 (1977).
2852. Chuang, S. Y., Tao, S. J., and Wilkenfeld, J. M., *J. Appl. Phys.*, **43**, 737 (1972).
2853. Chûjô, R., Maeda, K., Okuda, K., Murauama, N., and Hoshino, K., *Makromol. Chem.*, **176**, 213 (1975).
2854. Chûjô, R., Satoh, S., Ozeki, T., and Nagai, E., *J. Polym. Sci.*, **61**, 512 (1962).
2855. Chung, F. H., and Scott, R. W., *J. Appl. Cryst.*, **6**, 225 (1973).
2856. Chung, Y. J., Yamakawa, S., and Stannett, V., *Macromolecules*, **7**, 204 (1974).
2857. Ciardelli, F., Lanzillo, S., and Pieroni, O., *Macromolecules*, **7**, 174 (1974).
2858. Cieplinski, E. W., Ettre, L. S., Kolb, B., and Kemner, G., *Fresenius' Z. Anal. Chem.*, **205**, 357 (1964).
2859. Cieplinski, E. W., Ettre, L. S., Kolb, B., and Kemner, G., *Fresenius' Z. Anal. Chem.*, **209**, 302 (1965).
2860. Cincu, C., and Uglea, C. V., *Makromol. Chem.*, **177**, 2457 (1976).
2861. Claesson, S., *Nature*, **158**, 834 (1946).
2862. Claesson, S., *Arkiv Kemi*, **1**, 81 (1949).
2863. Claesson, S., *J. Polym. Sci.*, **16**, 193 (1955).
2864. Claesson, S., *Polymer*, **3**, 471 (1962).
2865. Claesson, S., Boman, N., and Gellerstedt, N., *Makromol. Chem.*, **92**, 51 (1966).
2866. Claesson, S., Finnström, B., and Hunt, J. E., *Chem. Scripta*, **8**, 197 (1975).
2867. Claesson, S., Gehatia, M., and Sundelöf, L. O., *J. Chim. Phys.*, **54**, 894 (1957).
2868. Claesson, S., and Hayward, L. D., *Chem. Scripta*, **9**, 18 (1976).
2869. Claesson, S., and Jacobson, G., *Acta Chem. Scand.*, **8**, 1835 (1954).
2870. Claesson, S., and Lindqvist, L., *Arkiv Kemi*, **11**, 535 (1957).
2871. Claesson, S., and Lindqvist, L., *Arkiv Kemi*, **12**, 1 (1957).
2872. Claesson, S., Lindqvist, L., and Strong, R. L., *Arkiv Kemi*, **22**, 245 (1964).
2873. Claesson, S., and Lohmander, U., *Makromol. Chem.*, **18/19**, 310 (1956).
2874. Claesson, S., Malmrud, S., and Lundgren, B., *Trans. Faraday Soc.*, **66**, 3048 (1970).
2875. Claesson, S., Matsuda, H., and Sundelöf, L. O., *Chem. Scripta*, **6**, 94 (1974).
2876. Claesson, S., and Mörning-Claesson, I., *Anal. Methods Protein Chem.*, **3**, 121 (1961).
2877. Claesson, S., and Öhman, J., *Arkiv Kemi*, **23**, 69 (1964).
2878. Claesson, S., and Svenson, S. A., *Exp. Cell Res.*, **11**, 105 (1956).
2879. Claesson, S., and Wettermark, G., *Arkiv Kemi*, **11**, 561 (1957).
2880. Clague, A. D. H., van Broehoven, J. A. M., and Blaauw, L. P., *Macromolecules*, **7**, 348 (1974).
2881. Clague, A. D. H., van Broehoven, J. A. M., and de Haan, J. W., *J. Polym. Sci. B*, **11**, 299 (1973).
2882. Clampitt, B. H., and Hughes, R. H., *J. Polym. Sci. C*, **6**, 43 (1964).
2883. Clampitt, B. H., and Hughes, R. H., *Anal. Chem.*, **40**, 449 (1969).
2884. Clark, D. T., Adams, D. B., Dilks, A., Peeling, J., and Thomas, H. R., *J. Electr. Spectr. Rel. Phenom.*, **8**, 51 (1976).
2885. Clark, D. T., and Dilks, A., *J. Polym. Sci. A1*, **14**, 533 (1976).
2886. Clark, D. T., and Dilks, A., *J. Electr. Spectr. Rel. Phenom.*, **11**, 225 (1977).
2887. Clark, D. T., and Dilks, A., *J. Polym. Sci. A1*, **15**, 15 (1977).
2888. Clark, D. T., and Dilks, A., *J. Polym. Sci. A1*, **15**, 2321 (1977).
2889. Clark, D. T., Dilks, A., Peeling, J., and Thomas, H. R., *Trans. Faraday Soc., Faraday Discuss.*, **60**, 183 (1975).
2890. Clark, D. T., Feast, W. J., Musgrave, W. K. R., and Ritchie, I., *J. Polym. Sci. A1*, **13**, 857 (1975).
2891. Clark, D. T., Feast, W. J., Ritchie, I., Musgrave, W. K. R., Modena, M., and Ragazzini, M., *J. Polym. Sci. A1*, **12**, 1049 (1974).
2892. Clark, D. T., Kilcast, D., Feast, W. J., and Musgrave, W. K. R., *J. Polym. Sci. A1*, **10**, 1637 (1972).
2893. Clark, D. T., Kilcast, D., Feast, W. J., and Musgrave, W. K. R., *J. Polym. Sci. A1*, **11**, 389 (1973).
2894. Clark, D. T., and Peeling, J., *J. Polym. Sci. A1*, **14**, 543 (1976).
2895. Clark, D. T., and Peeling, J., *J. Polym. Sci. A1*, **14**, 2941 (1976).

2896. Clark, D. T., Peeling, J., and O'Malley, J. M., *J. Polym. Sci. A1*, **14**, 543 (1976).
2897. Clark, D. T., and Thomas, H. R., *J. Polym. Sci. A1*, **14**, 1701 (1976).
2898. Clark, D. T., and Thomas, H. R., *J. Polym. Sci. A1*, **14**, 1671 (1976).
2899. Clark, D. T., and Thomas, H. R., *J. Polym. Sci. A1*, **15**, 2843 (1977).
2900. Clark, D. T., and Thomas, H. R., *J. Polym. Sci. A1*, **16**, 791 (1978).
2901. Clark, D. T., Thomas, H. R., Dilks, A., and Shuttleworth, D., *J. Electr. Spectr. Rel. Phenom.*, **10**, 455 (1977).
2902. Clark, E. S., *SPE J.*, **23**, 46 (1967).
2903. Clark, E. S., Amer. Crystl. Assoc. Meeting, Bozeman, Montana, 26–31 July 1964, Abstract F-12.
2904. Clark, E. S., and Muus, L. T., *Z. Krist.*, **117**, 108, 109 (1962).
2905. Clark, H. G., *Makromol. Chem.*, **86**, 107 (1965).
2906. Clark, N. A., Lunacek, J. H., and Benedek, G. B., *Amer. J. Phys.*, **38**, 575 (1970).
2907. Clark-Monks, C., and Ellis, B., *J. Colloid Interface Sci.*, **44**, 37 (1973).
2908. Clegg, G. A., Gee, D. R., Melia, T. P., and Tyson, A., *Polymer*, **9**, 501 (1968).
2909. Clegg, G. A., and Melia, T. P., *Polymer*, **11**, 245 (1970).
2910. Clemens, J., Jakeways, R., and Ward, I. M., *Polymer*, **19**, 639 (1978).
2911. Cleverdon, D., and Laker, D., *J. Appl. Chem.*, **1**, 2 (1951).
2912. Climie, I. E., and White, E. F. T., *J. Polym. Sci.*, **47**, 149 (1960).
2913. Clough, S., van Aartesen, J. J., and Stein, R. S., *J. Appl. Phys.*, **31**, 1873 (1960).
2914. Clough, S., Rhodes, M. B., and Stein, R. S., *J. Polym. Sci. C*, **18**, 1 (1967).
2915. Cluett, M. L., *Anal. Chem.*, **36**, 2199 (1964).
2916. Coats, A. W., and Redfern, J. P., *Nature*, **201**, 68 (1964).
2917. Cochran, W., Crock, F. H. C., and Vand, V., *Acta Cryst.*, **5**, 581 (1952).
2918. Cohen, C., *J. Polym. Sci. A2*, **15**, 291 (1977).
2919. Cohen, H. H., and Turnbull, D., *J. Chem. Phys.*, **31**, 1164 (1959).
2920. Cohen-Addad, J. P., and Roby, C., *Macromolecules*, **10**, 738 (1977).
2921. Cojazzi, G., Fichera, A., Garbuglio, C., Malta, V., and Zannetti, R., *Makromol. Chem.*, **168**, 289 (1973).
2922. Cole, H. M., Petterson, D. L., Siljaka, V. A., and Smith, D. S., *Rubber Chem. Technol.*, **39**, 259 (1966).
2923. Coleman, B. D., *J. Polym. Sci.*, **31**, 155 (1958).
2924. Coleman, B. D., and Fox, T. G., *J. Polym. Sci. A*, **1**, 3183 (1963).
2925. Coleman, M. M., Painter, P. C., Tabb, D. L., and Koenig, J. L., *J. Polym. Sci. B*, **12**, 577 (1974).
2926. Coleman, M. M., and Petcavich, R. J., *J. Polym. Sci. A2*, **16**, 1821 (1978).
2927. Coleman, M. M., Tabb, D. L., Farmer, D. L., and Koenig, J. L., *J. Polym. Sci. A2*, **12**, 445 (1974).
2928. Coleman, M. M., Zarian, J., Varnell, D. F., and Painter, P. C., *J. Polym. Sci. B*, **15**, 745 (1977).
2929. Coles, H. J., *Polymer*, **18**, 554 (1977).
2930. Coles, H. J., Gupta, A. K., and Marchal, E., *Macromolecules*, **10**, 182 (1977).
2931. Coll, H., *J. Polym. Sci. A2*, **4**, 659 (1966).
2932. Coll, H., *Makromol. Chem.*, **109**, 38 (1967).
2933. Coll, H., *Separation Sci.*, **5**, 273 (1970).
2934. Coll, H., and Gilding, D. K., *J. Polym. Sci. B*, **6**, 621 (1968).
2935. Coll, H., and Gilding, D. K., *J. Polym. Sci. A2*, **8**, 89 (1970).
2936. Coll, H., and Prusinowski, L. R., *J. Polym. Sci B*, **5**, 1153 (1967).
2937. Collins, E. A., and Chandler, L. A., *Rubber Chem. Technol.*, **39**, 193 (1966).
2938. Collins, R. L., *J. Polym. Sci.*, **27**, 75 (1958).
2939. Colonna-Cesari, F., and Premilat, S., *Polymer*, **17**, 267 (1976).
2940. Colthup, N. B., *J. Opt. Soc. Amer.*, **40**, 397 (1950).
2941. Conciatori, A. B., Chenevey, E. C., Bohrer, T. C., and Prince, A. E., *J. Polym. Sci. C*, **19**, 49 (1967).
2942. Conio, G., Patrone, E., and Salaris, F., *Macromolecules*, **4**, 283 (1971).
2943. Conio, G., Trefiletti, V., and Patrone, E., *Makromol. Chem.*, **154**, 311 (1972).
2944. Conix, A., *Makromol. Chem.*, **26**, 226 (1958).
2945. Connor, T. M., and McLauchlan, K. A., *J. Phys. Chem.*, **69**, 1888 (1965).
2946. Constantin, D., *Europ. Polym. J.*, **13**, 907 (1977).
2947. Conte, G., D'Illario, L., Pavel, N. V., and Giglio, E., *J. Polym. Sci. A2*, **14**, 1553 (1976).

758

2948. Conti, F., Acquaviva, L., Chiellini, E., Ciardelli, P., Delfini, M., and Segre, A. L., *Polymer*, **17**, 901 (1976).
2949. Conti, F., Delfini, M., Segre, A. L., Pini, D., and Porri, L., *Polymer*, **15**, 816 (1974).
2950. Conti, F., Segre, A., Pini, P., and Porri, L., *Polymer*, **15**, 5 (1974).
2951. Cook, M., Williams, G., and Jones, T. T., *Polymer*, **16**, 835 (1975).
2952. Cook, W. D., and Delatycki, O., *J. Polym. Sci. A2*, **12**, 1925 (1974).
2953. Cooke, D., and Kerker, M., *J. Opt. Soc. Amer.*, **59**, 43 (1969).
2954. Cooper, A. R., *Brit. Polym. J.*, **5**, 109 (1973).
2955. Cooper, A. R., *J. Polym. Sci. A2*, **12**, 1969 (1974).
2956. Cooper, A. R., and Barrall, E. M., II, *J. Appl. Polym. Sci.*, **17**, 1253 (1973).
2957. Cooper, A. R., Bruzzone, A. R., Cain, J. H., and Barrall, E. M., II, *J. Appl. Polym. Sci.*, **15**, 571 (1971).
2958. Cooper, A. R., Cain, J. H., Barrall, E. M., II, and Johnson, J. F., *Separation Sci.*, **5**, 787 (1970).
2959. Cooper, A. R., and Johnson, J. J., *J. Appl. Polym. Sci.*, **13**, 1487 (1969).
2960. Cooper, A. R., and Kiss, I., Jr, *Brit. Polym. J.*, **5**, 433 (1973).
2961. Cooper, G. D., and Prober, M., *J. Polym. Sci.*, **44**, 397 (1960).
2962. Cooper, M., and Manley, R. S. J., *Macromolecules*, **8**, 219 (1975).
2963. Cooper, W., Eaves, D. E., and Vaugham, G., *Makromol. Chem.*, **67**, 229 (1963).
2964. Cooper, W., Vaugham, G., and Madden, R. W., *J. Appl. Polym. Sci.*, **1**, 329 (1959).
2965. Cooper, W., Vaugham, G., and Yardley, J., *J. Polym. Sci.*, **59**, S2 (1962).
2966. Coover, H. W., McConnell, R. J., Joyner, F. B., Slonaker, D. F., and Combs, R. L., *J. Polym. Sci. A1*, **4**, 2563 (1966).
2967. Coppick, S., Battista, O. A., and Lytton, M. R., *Ind. Eng. Chem.*, **42**, 2533 (1950).
2968. Coran, A. Y., and Anagnostopoulos, C. E., *J. Polym. Sci.*, **57**, 1 (1962).
2969. Corish, P. J., *J. Appl. Polym. Sci.*, **4**, 86 (1960).
2970. Corish, P. J., *J. Appl. Polym. Sci.*, **5**, 53 (1961).
2971. Corish, P. J., *J. Appl. Polym. Sci.*, **7**, 727 (1963).
2972. Cornel, S. W., and Koenig, J. L., *J. Appl. Polym. Sci.*, **39**, 4883 (1968).
2973. Cornet, C. F., and van Ballegooijen, H., *Polymer*, **7**, 293 (1966).
2974. Cornibert, J., Marchessault, R. H., Benoit, H., and Weill, G., *Macromolecules*, **3**, 741 (1970).
2975. Cornell, S. W., and Koenig, J. L., *Macromolecules*, **2**, 540 1969).
2976. Cornell, S. W., and Koenig, J. L., *Macromolecules*, **2**, 546 (1969).
2977. Corradini, P., and Guerra, G., *Macromolecules*, **10**, 1410 (1977).
2978. Corradini, P., Martuscelli, E., and Martynov, M. A., *Makromol. Chem.*, **108**, 285 (1967).
2979. Corradini, P., Petroccone, V., and Allegra, G., *Macromolecules*, **4**, 771 (1971).
2980. Cosani, A., Terbojevich, M., Palumbo, M., Peggione, E., and Scoffone, E., *Macromolecules,*, **6**, 571 (1973).
2981. Cosgrove, T., and Warren, R. F., *Polymer*, **18**, 255 (1977).
2982. Costagliola, M., Greco, R., and Martuscelli, E., *Polymer*, **19**, 860 (1978).
2983. Cote, J. A., and Shida, M., *J. Polym. Sci. A2*, **9**, 421 (1971).
2984. Cotton, J. P., Decker, D., Benoit, H., Farnoux, B., Higgins, J., Jannik, G., Ober, R., Picot, C., and des Cloizeaux, J., *Macromolecules*, **7**, 863 (1974).
2985. Cotton, J. P., Nierlich, M., Boue, F., Donald, M., Farnoux, B., Jannik, G., Duplessix, R., and Picot, C., *J. Chem. Phys.*, **65**, 1101 (1976).
2986. Couchman, P. R., and Karasz, F. E., *J. Polym. Sci. A2*, **15**, 1037 (1977).
2987. Couchman, P. R., and Karasz, F. E., *Macromolecules*, **11**, 117 (1978).
2988. Coulson, C. A., Cox, J. T., Ogston, A. G., and Philpot, S. T., *Proc. Roy. Soc. A*, **192**, 382 (1948).
2989. Coumou, D. J., *J. Colloid Sci.*, **15**, 408 (1960).
2990. Cournoyer, R. F., Rhodes, M. B., and Siggia, S., *J. Polym. Sci. A2*, **13**, 1023 (1975).
2991. Courval, G., and Gray, D. G., *Macromolecules*, **8**, 326 (1975).
2992. Courval, G. J., and Gray, D. G., *Macromolecules*, **8**, 916 (1975).
2993. Courval, G. J., Gray, D. G., and Goring, D. A. I., *J. Polym. Sci. B*, **14**, 231 (1976).
2994. Cowie, J. M. G., *Polymer*, **7**, 487 (1966).
2995. Cowie, J. M. G., *Europ. Polym. J.*, **9**, 1041 (1973).
2996. Cowie, J. M. G., Macconnachie, A., and Ranson, R. J., *Macromolecules*, **4**, 57 (1971).
2997. Cowie, J. M. G., and McEwen, I. J., *Polymer*, **14**, 423 (1973).

759

2998. Cowie, J. M. G., and McEwen, I. J., *J. Chem. Soc. Faraday Trans.*, **70**, 171 (1974).
2999. Cowie, J. M. G., and McEwen, I. J., *J. Polym. Sci. A2*, **12**, 443 (1974).
3000. Cowie, J. M. G., and McEwen, I. J., *Macromolecules*, **7**, 291 (1974).
3001. Cowie, J. M. G., and McEwen, I. J., *Polymer*, **16**, 144 (1975).
3002. Cowie, J. M. G., and McEwen, I. J., *Macromolecules*, **10**, 1124 (1977).
3003. Cowie, J. M. G., Ranson, R. J., and Burchard, W., *Brit. Polym. J.*, **1**, 187 (1969).
3004. Cowie, J. M. G., and Toporowski, P. M., *Europ. Polym. J.*, **4**, 621 (1968).
3005. Cox, B. C., and Ellis, B., *Anal. Chem.*, **36**, 90 (1964).
3006. Cox, R. A., Wilkinson, M. C., Creasey, J. M., Goodall, A. R., and Hearn, J., *J. Polym. Sci. A1*, **15**, 2311 (1977).
3007. Craig, L. H., and Biegelow, C. C., *J. Polym. Sci.*, **16**, 177 (1955).
3008. Crain, W. O., Zambelli, A., and Roberts, J. D., *Macromolecules*, **4**, 330 (1971).
3009. Crank, J., *Trans. Faraday Soc.*, **47**, 450 (1951).
3010. Crank, J., and Henry, M. E., *Trans. Faraday Soc.*, **45**, 636, 1119 (1949).
3011. Crank, J., and Park, G. S., *Trans. Faraday Soc.*, **47**, 1072 (1951).
3012. Crank, J., and Robinson, C., *Proc. Roy. Soc. A*, **204**, 549 (1951).
3013. Craubner, H., *Ber. Bunsenges. Phys. Chem.*, **74**, 1262 (1970).
3014. Craubner, H., *J. Polym. Sci. C*, **42**, 889 (1973).
3015. Craubner, H., *Makromol. Chem.*, **175**, 2171 (1974).
3016. Craubner, H., *Makromol. Chem.*, **175**, 2461 (1974).
3017. Cravit, S., and van Duuren, B. L., *Chem. Instrum.*, **1**, 71 (1969).
3018. Creeth, J. M., *J. Amer. Chem. Soc.*, **77**, 6428 (1955).
3019. Crellin, R. A., and Ledwith, A., *Macromolecules*, **8**, 93 (1975).
3020. Crissman, J. M., *J. Polym. Sci. A2*, **13**, 1407 (1975).
3021. Crist, B., *J. Polym. Sci. A2*, **11**, 635 (1973).
3022. Crist, B., and Morosoff, N., *J. Polym. Sci. A2*, **11**, 1023 (1973).
3023. Crooks, J. E., Gererd, D. L., and Maddams, W. F., *Anal. Chem.*, **45**, 11 (1973).
3024. Crowley, J. D., Teague, G. S., Jr, and Lowe, J. W., Jr, *J. Paint. Technol.*, **38**, 269 (1966).
3025. Crowley, J. D., Teague, G. S., Jr, Lowe, J. W., Jr, *J. Paint. Technol.*, **39**, 19 (1967).
3026. Crummett, W. B., *Anal. Chem.*, **36**, 126 (1964).
3027. Crystal, R. C., and Hansen, H., *J. Appl. Phys.*, **38**, 3103 (1967).
3028. Crystal, R. C., and Hansen, H., *J. Polym. Sci. A2*, **6**, 981 (1968).
3029. Cudby, M. E. A., and Bunn, A., *Polymer*, **17**, 345 (1976).
3030. Cudby, M. E. A., Willis, H. A., Hendra, P. J., and Peacock, C. J., *Chem. Ind. (London)*, 1971, 531.
3031. Cumberbirch, R. J. E., and Spedding, H., *J. Appl. Chem.*, **12**, 83 (1962).
3032. Cummins, H. Z., Knable, N., and Yeh, Y., *Phys. Rev. Lett.*, **12**, 150 (1964).
3033. Cuniberti, C., and Bianchi, U., *Polymer*, **15**, 346 (1974).
3034. Cuniberti, C., and Ferrendo, R., *Polymer*, **13**, 379 (1972).
3035. Cuniberti, C., and Perico, A., *Europ. Polym. J.*, **13**, 369 (1977).
3036. Cunneen, J. I., Higgins, G. M., and Watson, W. F., *J. Polym. Sci.*, **40**, 1 (1959).
3037. Cunningham, A., Davis, G. R., and Ward, I. M., *Polymer*, **15**, 743 (1974).
3038. Cunningham, A., Manuel, A. J., and Ward, I. M., *Polymer*, **17**, 125 (1976).
3039. Cunningham, A., Ward, I. M., Willis, H. A., and Zichy, V., *Polymer*, **15**, 749 (1974).
3040. Curry, J. E., and McKinley, M. D., *J. Polym. Sci. A2*, **11**, 2209 (1973).
3041. Cutnell, J. D., and Glasel, J. A., *Macromolecules*, **9**, 71 (1976).
3042. Cvetanovič, R. J., *J. Chem. Phys.*, **23**, 1203 (1955).
3043. Czornyj, G., and Wunderlich, B., *Makromol. Chem.*, **178**, 843 (1971).
3044. Czornyj, G., and Wunderlich, B., *J. Polym. Sci. A2*, **15**, 1905 (1977).

3045. Dainton, F. S., Evans, D. M., Hoare, F. E., and Melia, T. P., *Polymer*, **3**, 263 (1962).
3046. Dainton, F. S., and Sutherland, G. B. B. M., *J. Polym. Sci.*, **4**, 37 (1949).
3047. D'Alò, B., Coppola, G., and Pallesi, B., *Polymer*, **15**, 130 (1974).
3048. Dan, E., Sommersall, A. C., and Guillet, J. E., *Macromolecules*, **6**, 228 (1973).
3049. Dandliker, W. B., and Kraut, J., *J. Amer. Chem. Soc.*, **78**, 2380 (1956).
3050. Daňhelka, J., and Kössler, I., *J. Polym. Sci. A1*, **14**, 287 (1976).
3051. Daniels, T., and Lehrle, R. S., *J. Polym. Sci. C*, **16**, 4533 (1969).
3052. Danon, J., and Jozefowicz, J., *Europ. Polym. J.*, **5**, 405 (1969).
3053. Danusso, F., and Moraglio, G., *J. Polym. Sci.*, **24**, 161 (1957).
3054. Daoud, M., *J. Polym. Sci. C*, **61**, 305 (1977).

3055. Daoud, M., Cotton, J. P., Farnoux, B., Jannik, G., Sarma, G., Benoit, H., Duplessix, R., Picot, C., and de Gennes, P. G., *Macromolecules*, **8**, 804 (1975).
3056. Dark, W. A., Limpart, R. J., and Carter, J. D., *Polym. Eng. Sci.*, **15**, 831 (1975).
3057. Darlington, M. W., and Smith, G. R., *Polymer*, **16**, 459 (1975).
3058. Darr, W. C., Gemeinhardt, P. G., and Saunders, J. H., *J. Cell Plastics*, **2**, 266 (1966).
3059. Datta, P. K., and Pethrick, R. A., *Polymer*, **18**, 919 (1977).
3060. Datta, P. K., and Pethrick, R. A., *Polymer*, **19**, 145 (1978).
3061. Daune, M., and Benoit, H., *J. Chim. Phys.*, **51**, 233 (1954).
3062. Daune, M., and Freund, L., *J. Polym. Sci.*, **23**, 115 (1957).
3063. Dautzenberg, H., *J. Polym. Sci. C*, **61**, 83 (1977).
3064. Davenport, J. A., Rowlinson, J. S., and Saville, G., *Trans. Faraday Soc.*, **62**, 322 (1966).
3065. Davenport, R. A., and Manuel, A. J., *Polymer*, **18**, 557 (1977).
3066. David, C., Demerteau, W., and Geuskens, G., *Europ. Polym. J.*, **6**, 1397 (1970).
3067. David, C., Lempereur, M., and Geuskens, G., *Europ. Polym. J.*, **8**, 417 (1972).
3068. David, C., Lempereur, M., and Geuskens, G., *Europ. Polym. J.*, **8**, 1315 (1973).
3069. David, C., Piens, M., and Geuskens, G., *Europ. Polym. J.*, **8**, 1019 (1972).
3070. David, C., Piret, W., Sakaguchi, M. and Geuskens, G., *Makromol. Chem.*, **179**, 181 (1978).
3071. David, D. J., *J. Therm. Anal.*, **3**, 247 (1971).
3072. Davidson, T., and Wunderlich, B., *J. Polym. Sci. A2*, **7**, 377 (1969).
3073. Davis, G. J., and Heigh, J., *Infrared Phys.*, **14**, 183 (1974).
3074. Davis, H. A., *J. Polym. Sci. A2*, **4**, 1009 (1966).
3075. Davis, R. E., and McCrea, J. M., *Anal. Chem.*, **29**, 1114 (1957).
3076. Davis, T. E., and Tobias, R. L., *J. Polym. Sci.*, **50**, 227 (1961).
3077. Davison, W. H. T., *Chem. Ind. (London)*, **1957**, 131.
3078. Davison, W. H. T., Slaney, S., and Wragg, A. L., *Chem. Ind. (London)*, **1954**, 1356.
3079. Dawis, D. D., and Slichter, W. P., *Macromolecules*, **6**, 729 (1973).
3080. Dawkins, J. V., *J. Macromol. Sci. Phys.*, **B2**, 623 (1968).
3081. Dawkins, J. V., *Europ. Polym. J.*, **6**, 831 (1970).
3082. Dawkins, J. V., *Chem. Ind. (London)*, **1970**, 118.
3083. Dawkins, J. V., *Brit. Polym. J.*, **4**, 87 (1972).
3084. Dawkins, J. V., *Polymer*, **19**, 705 (1978).
3085. Dawkins, J. V., Denyer, R., and Maddock, J. W., *Polymer*, **10**, 154 (1969).
3086. Dawkins, J. V., and Hemming, M., *Polymer*, **13**, 553 (1972).
3087. Dawkins, J. V., and Hemming, M., *Polymer*, **16**, 554 (1975).
3088. Dawkins, J. V., and Hemming, M., *Makromol. Chem.*, **176**, 1777, 17795, 1815 (1975).
3089. Dawkins, J. V., and Hemming, M., *J. Appl. Polym. Sci.*, **19**, 3107 (1975).
3090. Dawkins, J. V., Holdsworth, P. J., and Keller, A., *Makromol. Chem.*, **118**, 361 (1968).
3091. Dawkins, J. V., and Peaker, F. W., *Europ. Polym. J.*, **6**, 209 (1970).
3092. Dawkins, J, V., Stone, T., and Yeadon, G., *Polymer*, **18**, 1179 (1977).
3093. Dawkins, J. V., and Taylor, G., *Polymer*, **15**, 687 (1974).
3094. Dawkins, J. V., and Taylor, G., *J. Polym. Sci. B*, **13**, 29 (1975).
3095. Dayantis, J., *J. Phys. Chem.*, **76**, 400 (1972).
3096. Dayantis, J., *J. Phys. Chem.*, **77**, 2977 (1973).
3097. Dayantis, J., *Makromol. Chem.*, **176**, 2437 (1975).
3098. Dayantis, J., and Benoit, H., *J. Chem. Phys.*, **61**, 773 781 (1964).
3099. Dean, G. D., and Martin, D. H., *Chem. Phys. Lett.*, **1**, 415 (1967).
3100. Deb, P. C., Prased, J., and Chatterjee, S. R., *Makromol. Chem.*, **178**, 1455 (1977).
3101. de Boer, A., Alberda van Ekenstein, G. O. R., and Challa, G., *Polymer*, **16**, 930 (1975).
3102. Debska-Kotlowska, M., and Kielich, S., *Polymer*, **17**, 1039 (1976).
3103. Debska-Kotlowska, M., and Kielich, S., *J. Polym. Sci., C*, **61**, 101 (1977).
3104. Debye, P., *J. Appl. Polym.*, **15**, 338 (1944).
3105. Debye, P., *J. Appl. Phys.*, **17**, 392 (1946).
3106. Debye, P., *J. Phys. Chem.*, **51**, 18 (1947).
3107. Debye, P., Bashaw, J., Chu, B., and Tancreti, D. M., *J. Chem. Phys.*, **44**, 4302 (1966).
3108. Debye, P., and Bueche, F., *J. Chem. Phys.*, **19**, 589 (1951).
3109. Debye, P., Coll, H., and Woermann, D., *J. Chem. Phys.*, **33**, 1746 (1960).
3110. Debye, P., Woermann, D., and Chu, B., *J. Polym. Sci. A*, **1**, 225 (1963).
3111. De Candia, F., Romano, G., and Vittoria, V., *J. Polym. Sci. A2*, **11**, 2291 (1973).
3112. Dechant, J., *Faserforsch. Textiltech.*, **25**, 24 (1974).

3113. deCindio, B., Migliaresi, C., and Acierno, D., *Polymer*, **19**, 526 (1978).
3114. De Gennes, P. G., *Phys. Lett. A*, **38**, 339 (1972).
3115. De Kock, R. J., and Hol, P. A. H. M., *Rec. Trav. Chem.*, **85**, 102 (1966).
3116. DeLaney, D. E., and Krause, S., *Macromolecules*, **9**, 455 (1976).
3117. Delf, B. W., and MacKnight, W. J., *Macromolecules*, **2**, 309 (1969).
3118. Dellinger, J. A., George, G. D., and Roberts, C. W., *J. Polym. Sci. B*, **16**, 357 (1978).
3119. Dellinger, J. A., and Roberts, C. W., *J. Polym. Sci. B*, **14**, 167 (1976).
3120. Delmas, G., Daviet, V., and Filiatrault, D., *J. Polym. Sci. A2*, **14**, 1629 (1976).
3121. Delmas, G., and De Saint-Romain, P., *Europ. Polym. J.*, **10**, 1133 (1974).
3122. Delmas, G., and Patterson, D., *Polymer*, **7**, 513 (1966)
3123. Delmas, G., and Patterson, D., *J. Polym. Sci. C*, **30**, 1 (1970).
3124. Delmas, G., Patterson, D., and Somcynski, T., *J. Polym. Sci.*, **57**, 79 (1962).
3125. Delsarte, J., and Weill, G., *Macromolecules*, **7**, 450 (1974).
3126. De Member, J. R., Haas, H. C., and MacDonald, R. L., *J. Polym. Sci. B*, **10**, 385 (1972).
3127. De Member, J. R., Taylor, L. D., Trummer, S., Rubin, L. E., and Chilkis, C. K., *J. Appl. Polym. Sci.*, **21**, 621 (1977).
3128. Denecke, M., and Broecker, H. C., *Makromol. Chem.*, **176**, 1471 (1975).
3129. Denes, F., Asandei, N. N., and Simionescu, C. I., *Anal. Chem.*, **40**, 629 (1968).
3130. Dentini, M., De Santis, P., Savino, M., Sforza, M. P., and Verdini, A. S., *Makromol. Chem.*, **179**, 1041 (1978).
3131. Derby, J. V., and Freedman, R. W., *Intern. Lab.*, July/Aug. 1974, p. 11.
2132. Derge, K., *Glas Instrum. Techn.*, **11**, 164 (1967).
3133. Derham, K. W., Goldsbrough, J., Gordon, M., Koningsveld, R., and Kleintjens, L.A., *Makromol. Chem.*, Suppl. 1, 401 (1975).
3134. De Santis, P., Morosetti, S., and Rizzo, R., *Macromolecules*, **7**, 52 (1974).
3135. Des Cloizeaux, J., *J. Phys. (Paris)*, **31**, 715 (1970).
3136. Desphande, D. D., Patterson, D., Schreiber, H. P., and Su, C. S., *Macromolecules*, **7**, 530 (1974).
3137. Desphande, D. D., and Prabhu, C. S., *Macromolecules*, **10**, 433 (1977).
3138. Desphande, D. D., and Tyagi, O. S., *Macromolecules*, **11**, 746 (1978).
3139. Desper, R. J., *J. Macromol. Sci. B*, **7**, 105 (1973).
3140. Desper, C. R., and Stein, R. S., *J. Polym. Sci. B*, **5**, 893 (1967).
3141. D'Esposito, L., and Koenig, J. L., *J. Polym. Sci. A2*, **14**, 1731 (1976).
3142. Desreux, V., and Spiegels, M. C., *Bull. Soc. Chim. Belges*, **59**, 476 (1950).
3143. Destor, C., Langevin, D., and Rondelez, F., *J. Polym. Sci. B*, **16**, 229 (1978).
3144. Deur-Šiftar, D., Bistrički, T., and Tandi, T., *J. Chromatogr.*, **24**, 404 (1966).
3145. Dover, G. R., Karasz, F. E., MacKnight, W. J., and Lenz, R. W., *Macromolecules*, **8**, 439 (1975).
3146. DeVoe, H., *J. Chem. Phys.*, **43**, 3199 (1965).
3147. De Vos, E., and Bellemans, A., *Macromolecules*, **7**, 812 (1974).
3148. De Vos, E., and Bellemens, A., *Macromolecules*, **8**, 651 (1975).
3149. De Vries, A. J., Le Page, M., Beau, R., and Guillemin, C. L., *Anal. Chem.*, **39**, 935 (1968).
3150. De Zeeuw, R. A., *Anal. Chem.*, **40**, 915 (1968).
3151. Diaper, W. E. J., *Chem. Technol.*, **2**, 368 (1972).
3152. Dietl, J. J., *Kunststoffe* **59**, 792 (1969).
3153. Dietz, W. A., *J. Gas Chromatogr.*, **5**, 68 (1967).
3154. Dietz, W. A., *J. Chromatogr. Sci.*, **10**, 423 (1972).
3155. DiMarzio, E. A., *J. Res. Nat. Bur. Stand. A*, **68**, 611 (1964).
3156. DiMarzio, E. A., *J. Appl. Phys.*, **45**, 4143 (1974).
3157. DiMarzio, E. A., *Macromolecules*, **10**, 1407 (1977).
3158. DiMarzio, E. A., and Gibbs, J. H., *J. Chem. Phys.*, **28**, 373 (1958).
3159. DiMarzio, E. A., Gibbs, J. H., Fleming, P. D., and Sanchez, I. C., *Macromolecules*, **9**, 763 (1976).
3160. DiMarzio, E. A., and Guttman, C. M., *J. Polym. Sci. B*, **7**, 267 (1969).
3161. DiMarzio, E. A., and Guttman, C. M., *Macromolecules*, **3**, 131 (1970).
3162. Dimbat, M., and Eggertsen, F. T., *Microchem. J.*, **9**, 500 (1965).
3163. Dimbat, M., and Stross, F. H., *Anal. Chem.*, **29**, 1517 (1957).
3164. DiMeo, A., Maglio, G., Martuscelli, E., and Palumbo, R., *Polymer*, **17**, 802 (1976).
3165. Dincer, S., and Bonner, D. C., *Macromolecules*, **11**, 107 (1978).

3166. Dinse, H., and Praeger, K., *Faserforsch. Textiltech.*, **21**, 305 (1970).
3167. DiPaola-Baranyi, G., Braun, J. M., and Guillet, J. E., *Macromolecules*, **11**, 224 (1978).
3168. DiPaola-Baranyi, G., and Guillet, J. E., *Macromolecules*, **11**, 228 (1978).
3169. Discher, C. A., Smith, P. F., Lippman, I., and Turse, R., *J. Phys. Chem.*, **67**, 2501 (1963).
3170. Dismore, P. F., and Statton, W. O., *J. Polym. Sci. B*, **2**, 1113 (1964).
3171. Dismore, P. F., and Statton, W. O., *J. Polym. Sci. C*, **13**, 133 (1966).
3172. Djadoun, S., Goldberg, R. N., and Morawetz, H., *Macromolecules*, **10**, 1015 (1977).
3173. Dlugosz, J., Fraser, G. V., Grubb, D., Keller, A., and Odell, J. A., *Polymer*, **17**, 471 (1976).
3174. Dlugosz, J., Keller, A., and Pedemonte, E., *Kolloid Z. Z. Polym.*, **242**, 1125 (1970).
3175. Dobinson, F., and Preston, J., *J. Polym. Sci. A1*, **4**, 2093 (1966).
3176. Dobry, A., *J. Polym. Sci.*, **2**, 623 (1947).
3177. Dobry, A., and Boyer-Kawenoki, F., *J. Polym. Sci.*, **2**, 90 (1947).
3178. Doddi, N., Forsman, W. C., and Price, C. C., *Macromolecules*, **4**, 649 (1971).
3179. Doddi, N., Forsman, W. C., and Price, C. C., *J. Polym. Sci. A2*, **12**, 1395 (1974).
3180. Dodge, D. W., and Metzner, A. B., *AIChE J.*, **5**, 189 (1959).
3181. Dohner, R. E., Wachter, A. H., and Simon, W., *Helv. Chim. Acta*, **50**, 2193 (1967).
3182. Doi, Y., and Asakura, T., *Makromol. Chem.*, **176**, 507 (1975).
3183. Dole, M., *J. Polym. Sci. C*, **18**, 57 (1967).
3184. Dole, M., Hettinger, W. P., Jr, Larsen, N. R., and Wethington, J. A., Jr, *J. Phys. Chem.*, **20**, 781 (1952).
3185. Dole, M., Mack, L. L., Hines, R. L., Mobley, R. C., Ferguson, L. D., and Alice, M. B., *J. Chem. Phys.*, **49**, 2240 (1968).
3186. Dolinskaya, E. R., Khatchaturov, A. S., Poletayeva, I. A., and Kormer, V. A., *Makromol. Chem.*, **179**, 409 (1978).
3187. Domb, C., *Polymer*, **15**, 259 (1974).
3188. Domb, C., and Barrett, A. J., *Polymer*, **17**, 179 (1976).
3189. Dombroski, J. R., Sarko, A., and Schuerch, C., *Macromolecules*, **4**, 93 (1971).
3190. Donatelli, A. A., Sperling, L. H., and Thomas, D. A., *J. Appl. Polym. Sci.*, **21**, 1189 (1977).
3191. Dondos, A., *Makromol. Chem.*, **135**, 181 (1970).
3192. Dondos, A., *Polymer*, **18**, 1250 (1977).
3193. Dondos, A., *Makromol. Chem.*, **178**, 2421 (1977).
3194. Dondos, A., and Benoit, H., *Europ. Polym. J.*, **4**, 561 (1968).
3195. Dondos, A., and Benoit, H., *J. Polym. Sci. B*, **7**, 335 (1969).
3196. Dondos, A., and Benoit, H., *Europ. Polym. J.*, **6**, 1439 (1970).
3197. Dondos, A., and Benoit, H., *Macromolecules*, **4**, 279 (1971).
3198. Dondos, A., and Benoit, H., *Macromolecules*, **6**, 242 (1973).
3199. Dondos, A., and Benoit, H., *Makromol. Chem.*, **176**, 3441 (1975).
3200. Dondos, A., and Benoit, H., *J. Polym. Sci. A2*, **15**, 137 (1977).
3201. Dondos, A., and Benoit, H., *Polymer*, **18**, 1161 (1977).
3202. Dondos, A., and Benoit, H., *Polymer*, **19**, 523 (1978).
3203. Dondos, A., and Benoit, H., *Makromol. Chem.*, **179**, 1051 (1978).
3204. Dondos, A., Havredaki, V., and Mitsou, A., *Makromol. Chem.*, **176**, 1481 (1975).
3205. Dondos, A., and Paterson, D., *J. Polym. Sci. A2*, **7**, 209 (1969).
3206. Dondos, A., Rempp, P., and Benoit, H., *Makromol. Chem.*, **171**, 135 (1973).
3207. Dondos, A., Rempp, P., and Benoit, H., *Makromol. Chem.*, **175**, 1659 (1974).
3208. Donkai, N., and Inagaki, H., *J. Chromatogr.*, **71**, 473 (1972).
3209. Donkai, N., Murayama, N., Miyamoto T., and Inagaki, H., *Makromol. Chem.*, **175**, 187 (1974).
3210. Donnelly, T. H., *J. Phys. Chem.*, **70**, 1862 (1966).
3211. Doolittle, A. K., *J. Appl. Phys.*, **22**, 1031, 1471 (1951).
3212. Doolittle, A. K., *J. Appl. Phys.*, **23**, 236 (1952).
3213. Doolittle, A. K., and Doolittle, D. B., *J. Appl. Phys.*, **26**, 901 (1957).
3214. Dorio, M. M., and Chien, J. C. W., *Macromolecules*, **8**, 734 (1975).
3215. Dorio, M. M., and Chien, J. C. W., *J. Chem. Phys.*, **62**, 3963 (1975).
3216. Dorman, D. E., Otocka, E. P., and Bovey, F. A., *Macromolecules*, **5**, 574 (1972).
3217. Dosch, E. I., and Wendlandt, W. W., *Thermochim. Acta*, **1**, 181 (1970).
3218. Dosière, M., and Point, J. J., *J. Polym. Sci. A2*, **15**, 1655 (1977).

3219. Doskočilová, D., and Schneider, B., *Macromolecules*, **5**, 125 (1972).
3220. Doskočilová, D., and Schneider, B., *Macromolecules*, **6**, 76 (1973).
3221. Doty, P., *J. Polym. Sci.*, **3**, 750 (1948).
3222. Doty, P. T., and Steiner, R. F., *J. Chem. Phys.*, **18**, 1221 (1950).
3223. Doty, P. T., and Zable, H. S., *J. Polym. Sci.*, **1**, 90 (1946).
3224. Douglas, D. C., and McBrierty, V. J., *Macromolecules*, **11**, 766 (1978).
3225. Douglas, D. C., and McCall, D. W., *J. Phys. Chem.*, **62**, 1102 (1958).
3226. Douglas, T. B., and Harman, A. W., *J. Res. Nat. Bur. Stand. A*, **69**, 149 (1965).
3227. Dows, D. A., *J. Chem. Phys.*, **41**, 2656 (1964).
3228. Doyle, C. D., *Anal. Chem.*, **33**, 77 (1961).
3229. Doyle, C. D., *J. Appl. Polym. Sci.*, **5**, 285 (1961).
3230. Doyle, C. D., *Makromol. Chem.*, **80**, 220 (1964).
3231. Dreval, V. E., Malkin, A. Ya., and Botvinnik, G. O., *J. Polym. Sci. A2*, **11**, 1055 (1973).
3232. Dröscher, M., Hertwig, K., Reimann, H., and Wegner, G., *Makromol. Chem.*, **177**, 1695 (1976).
3233. Dröscher, M., Lieser, G., Reimann, H., and Wegner, G., *Polymer*, **16**, 497 (1975).
3234. Drott, E. E., and Mendelson, R. A., *J. Polym. Sci. A2*, **8**, 1361, 1373 (1970).
3235. Drushel, H. V., Ellerbe, J. S., Cox, R. C., and Lane, L. H., *Anal. Chem.*, **40**, 370 (1968).
3236. Dubin, P. L., Koontz, S., and Wright, K. L., *J. Polym. Sci. A1*, **15**, 2047 (1977).
3237. Dubin, S. B., Lunacek, J. H., and Benedek, G. B., *Proc. Nat. Acad. Sci. U.S.A.*, **57**, 1164 (1967).
3238. Duch, E., and Küchler, L., *Z. Elektrochem.*, **60**, 218 (1956).
3239. Duch, M. W., and Grant, D. M., *Macromolecules*, **3**, 165 (1970).
3240. Duckett, R. A., Goswami, B. C., and Ward, I. M., *J. Polym. Sci. A2*, **15**, 333 (1977).
3241. Duda, J. L., Ni, Y. C., and Vrentas, J. S., *J. Appl. Polym. Sci.*, **22**, 689 (1978).
3242. Duerksen, J. H., and Hamielec, A. E., *J. Polym. Sci. C*, **21**, 83 (1968).
3243. Duiser, J. A., and Keijzers, A. E. M., *Polymer*, **19**, 889 (1978).
3244. Duke, R. W., and DuPré, D. B., *Macromolecules*, **7**, 374 (1974).
3245. Dunning, J. W., Jr, and Angus, J. C., *J. Appl. Polym. Sci.*, **39**, 2479 (1968).
3246. Dupont, Y., Gabriel, A., Chabre, M., Bulik-Krzywicki, T., and Schecter, E., *Nature*, **238**, 331 (1972).
3247. Durbetaki, A. J., and Miles, C. M., *Anal. Chem.*, **37**, 1231 (1965).
3248. Dušek, K., *Collect. Czech. Chem. Commun.*, **33**, 1100 (1968).
3249. Dušek, K., *Brit. Polym. J.*, **2**, 257 (1970).
3250. Dušek, K., Gordon, M., and Ross-Murphy, B., *Macromolecules*, **11**, 236 (1978).
3251. Düssel, H. J., Rosen, H., and Hummel, D. O., *Makromol. Chem.*, **177**, 2343 (1976).
3252. Duval, M., Duplissex, R., Picot, C., Decker, D., Rempp, P., Benoit, H., Cotton, J. P., Jannik, G., Farnoux, B., and Ober, R., *J. Polym. Sci. B*, **14**, 585 (1976).
3253. Duvdevani, I., Biesenberger, J. A., and Tan, M., *J. Polym. Sci. B*, **9**, 429 (1971).
3254. Duveau, N., and Piguet, A., *J. Polym. Sci.*, **57**, 357 (1962).
3255. Duynstee, E. E. J., and Mevis, M. E. A. H., *Europ. Polym. J.*, **8**, 1375 (1972).
3256. Dybowski, C. R., and Vaugham, R. W., *Macromolecules*, **8**, 50 (1975).

3257. Early, D. S., *Anal. Chem.*, **40**, 894 (1968).
3258. Eastmond, G. C., and Smith, E. G., *Polymer*, **14**, 509 (1973).
3259. Eastmond, G. C., and Smith, E. G., *Polymer*, **17**, 367 (1976).
3260. Eastmond, G. C., and Smith, E. G., *Polymer*, **18**, 244 (1977).
3261. Ebdon, J. R., *Polymer*, **15**, 782 (1974).
3262. Ebdon, J. R., *Polymer*, **19**, 1232 (1978).
3263. Ebdon, J. R., and Huckerby, T. N., *Polymer*, **17**, 170 (1976).
3264. Edel, G., and Chabert, B., *C. R. Acad. Sci. C*, **267**, 54 (1968).
3265. Edel, G., and Chabert, B., *C. R., Acad. Sci. C*, **268**, 226 (1969).
3266. Edsall, J. T., Rich, A., and Goldstein, M., *Rev. Sci. Instrum.*, **23**, 695 (1952).
3267. Edward, J. M., Mulley, R. D., Pape, G. A., Booth, C., and Shepherd, I. W., *Polymer*, **18**, 1190 (1977).
3268. Edwards, B. C., *J. Polym. Sci. A2*, **13**, 1387 (1976).
3269. Edwards, B. C., and Phillips, P. J., *Polymer*, **15**, 491 (1974).
3270. Edwards, B. C., and Phillips, P. J., *J. Polym. Sci. A2*, **13**, 2117 (1975).
3271. Edwards, B. C., and Phillips, P. J., *J. Polym. Sci. A1*, **14**, 391 (1976).
3272. Edwards, G. D., and Ng, Q. Y., *J. Polym. Sci. C*, **21**, 105 (1968).

764

3273. Edwards, T. J., and Newman, J., *Macromolecules*, **10**, 609 (1977).
3274. Eggertsen, F. T., and Stross, F. H., *J. Appl. Polym. Sci.*, **10**, 1171 (1966).
3275. Eggertsen, F. T., and Stross, F. H., *Thermochim. Acta*, **1**, 451 (1970).
3276. Ehrlich, P., *J. Polym. Sci. A1*, **3**, 131 (1965).
3277. Eichhoff, U., and Zachmann, H. G., *Ber. Bunsenges. Phys. Chem.*, **74**, 919 (1970).
3278. Eichhoff, U., and Zachmann, H. G., *Kolloid Z. Z. Polym.*, **241**, 928 (1970).
3279. Eichhoff, U., and Zachmann, H. G., *Faserforsch. Textiltechn.*, **22**, 395 (1971).
3280. Eichinger, B. E., and Flory, P. J., *Macromolecules*, **1**, 285 (1968).
3281. Eichinger, B. E., and Flory, P. J., *Trans. Faraday Soc.*, **64**, 2035, 2053, 2061, 2066 (1968).
3282. Eiermann, K., and Hellwege, K. H., *J. Polym. Sci.*, **57**, 99 (1962).
3283. Eiermann, K., Hellwege, K. H., and Knappe, W., *Kolloid Z.*, **174**, 134 (1961).
3284. Eisenberg, A., and Rovira, E., *J. Polym. Sci. B*, **2**, 269 (1964).
3285. Eisenberg, A., and Tobolsky, A. V., *J. Polym. Sci.*, **61**, 483 (1962).
3286. Eisinger, J., *Photochem. Photobiol.*, **9**, 247 (1969).
3287. Eissler, R. L., and Baker, F. L., *Appl. Polym. Symp.*, No. 16, 171 (1971).
3288. Eldib, I. A., *Ind. Eng. Chem. Process Design Develop.*, **1**, 2 (1962).
3289. Elgert, K. F., Quack, G., and Stützel, B., *Polymer*, **15**, 612 (1974).
3290. Elgert, K. F., Quack, G., and Stützel, B., *Makromol. Chem.*, **175** 1955 (1974).
3291. Elgert, K. F., Quack, G., and Stützel, B., *Makromol. Chem.*, **176**, 759 (1975).
3292. Elgert, K. F., Quack, G., and Stützel, B., *Polymer*, **16**, 154 (1975).
3293. Elgert, K. F., and Ritter, W., *Makromol. Chem.*, **177**, 2021 (1976).
3294. Elgert, K. F., and Ritter, W., *Makromol. Chem.*, **177**, 2781 (1976).
3295. Elgert, K. F., and Ritter, W., *Makromol. Chem.*, **178**, 2827 (1977).
3296. Elgert, K. F., and Stützel, B., *Polymer*, **16**, 758 (1975).
3297. Elgert, K. F., Stützel, B., and Forgo, I., *Polymer*, **16**, 761 (1975).
3298. Elgert, K. F., Stützel, B., Frenzel, P., Cantow, H. J., and Streck, R., *Makromol. Chem.*, **170**, 257 (1973).
3299. Elgert, K. F., Wicke, R., Stützel, B., and Ritter, W., *Polymer*, **16**, 465 (1975).
3300. Elias, H. G., *J. Polym. Sci.*, **28**, 648 (1958).
3301. Elias, H. G., *Makromol. Chem.*, **33**, 140 (1959).
3302. Elias, H. G., Bellido, J., and Bareiss, R. E., *Makromol. Chem.*, **176**, 439 (1975).
3303. Elias, H. G., and Gruber, U., *J. Polym. Sci. B*, **1**, 337 (1963).
3304. Elias, H. G., and Gruber, U., *Makromol. Chem.*, **78**, 72 (1964).
3305. Elias, H. G., Ritscher, T. A., and Patat, F., *Makromol. Chem.*, **27**, 1 (1958).
3306. Elias, H. G., and Šolc, K., *Makromol. Chem.*, **176**, 365 (1975).
3307. Ellerstein, S. M., *J. Phys. Chem.*, **69**, 2471 (1965).
3308. Ellerstein, S. M., *Appl. Polym. Symp.*, No. 2, 111 (1965).
3309. Elliot, A., *J. Sci. Instrum.*, **42**, 312 (1965).
3310. Elliot, A., Ambrose, E. J., and Temple, R. B., *J. Sci. Instrum.*, **27**, 21 (1950).
3311. Ellis, G. W., *Science*, **154**, 1195 (1966).
3312. Ellis, K. C., and Warwicker, J. O., *J. Polym. Sci. A*, **1**, 1185 (1963).
3313. Ende, H. A., and Hermans, J. J., *J. Colloid Sci.*, **17**, 601 (1962).
3314. Ende, H. A., and Hermans, J. J., *J. Polym. Sci. A*, **2**, 4053 (1964).
3315. Ende, H. A., and Stannett, V., *J. Polym. Sci. A*, **2**, 4047 (1964).
3316. Enoksson, B., *J. Polym. Sci.*, **3**, 314 (1948).
3317. Enoksson, B., *J. Polym. Sci.*, **6**, 575 (1951).
3318. Enomoto, S., *J. Polym. Sci.*, **55**, 95 (1961).
3319. Enomoto, S., Asahina, M., and Satoh, S., *J. Polym. Sci. A1*, **3**, 1373 (1966).
3320. Enscore, D. J., Hopfenberg, H. B., and Stannett, V. T., *Polymer*, **18**, 793 (1977).
3321. Enscore, D. J., Hopfenberg, H. B., Stannett, V. T., and Berens, A. R., *Polymer*, **18**, 1105 (1977).
3322. Eppe, R., Fischer, E. W., and Stuart, H. A., *J. Polym. Sci.*, **34**, 721 (1959).
3323. Erä, V. A., *Makromol. Chem.*, **175**, 2191 (1974).
3324. Ergoz, E., Fatou, J. G., and Mandelkern, L., *Macromolecules*, **5**, 147 (1972).
3325. Erhardt, P. F., and Stein, R. S., *J. Polym. Sci. B*, **3**, 553 (1965).
3326. Erhardt, P. F., and Stein, R. S., *Appl. Polym. Symp.*, **5**, 113 (1967).
3327. Esposito, G. G., *Anal. Chem.*, **36**, 2183 (1964).
3328. Estes, G. M., Seymour, R. W., and Cooper, S. L., *Macromolecules*, **4**, 452 (1971).
3329. Ettre, L. S., *J. Chromatogr.*, **8**, 525 (1962).

3330. Ettre, L. S., and Varadi, P. F., *Anal. Chem.*, **35**, 69 (1963).
3331. Eustache, H., Robin, N., Daniel, J. C., and Carrega, M., *Europ. Polym. J.*, **14**, 239 (1978).
3332. Evans, H., and Woodward, A. E., *Macromolecules*, **9**, 88 (1976).
3333. Evans, J. M., Huglin, M. B., and Lindley, J., *J. Appl. Polym. Sci.*, **11**, 2159 (1967).
3334. Evans, L. J., Bucknall, C. B., and Hall, M. M., *Plastics Polymers*, **39**, 118 (1971).
3335. Evans, R. A., and Hallam, H. E., *Polymer*, **17**, 839 (1976).
3336. Evans, W. J., McCourtney, E. J., and Carbey, W. B., *Anal. Chem.*, **40**, 262 (1968).
3337. Ewald, P. P., *Z. Krystal.*, **56**, 129 (1921).
3338. Ezrin, M., and Claver, G. C., *ACS Polym. Preprints*, **3**, 308 (1962).
3339. Fairbairn, J. W., and Relph, S. J., *J. Chromatogr.*, **33**, 494 (1968).
3340. Fakirov, S., Fischer, E. W., Hoffman, R., and Schmidt, G. F., *Polymer*, **18**, 1121 (1977).
3341. Fakirov, S., Fischer, E. W., and Schmidt, G. F., *Makromol. Chem.*, **176**, 2459 (1975).
3342. Falgoux, D., Mangin, P., Engel, J., and Cranger, C., *Anal. Chem.*, **236**, 228 (1968).
3343. Fallgatter, M. B., and Dole, M., *J. Phys. Chem.*, **68**, 1988 (1964).
3344. Fanconi, B., and Crissman, J., *J. Polym. Sci. B*, **13**, 421 (1975).
3345. Farmer, J., and Little, K., *Brit. Polym. J.*, **1**, 259 (1969).
3346. Farone, W. A., and Overfeld, C. W., *J. Opt. Soc. Amer.*, **56**, 476 (1966).
3347. Farré-Rius, F., and Guiochon, G., *Anal. Chem.*, **40**, 998 (1968).
3348. Farritor, R. E., and Tao, L. C., *Thermochim. Acta*, **1**, 297 (1970).
3349. Farrow, G., Ravens, D. A. S., and Ward, I. M., *Polymer*, **3**, 17 (1962).
3350. Farrow, G., and Ward, I. M., *Brit. J. Appl. Phys.*, **11**, 543 (1960).
3351. Farrow, G., and Ward, I. M., *Polymer*, **1**, 330 (1960).
3352. Fatou, J. G., and Mandelkern, L., *J. Phys. Chem.*, **69**, 417 (1965).
3353. Fatou, J. G., Riande, E., and Valdecasas, R. G., *J. Polym. Sci. A2*, **13**, 2103 (1975).
3354. Faucher, J. A., *J. Polym. Sci. B*, **3**, 143 (1965).
3355. Faucher, J. A., Koleske, J. V., Santee, E. R., Stratta, J. J., and Wilson, C. W., *J. Appl. Phys.*, **37**, 3962 (1966).
3356. Faure, J., *Pure Appl. Chem.*, **49**, 487 (1977).
3357. Faure, J., Fouassier, J. P., and Lougnot, D. J., *J. Photochem.*, **5**, 13 (1976).
3358. Fawcett, A. H., Heatley, F., Ivin, K. J., Steward, C. D., and Watt, P., *Macromolecules*, **10**, 765 (1977).
3359. Fawcett, A. H., and Ivin, K. J., *Polymer*, **16**, 569, 573 (1975).
3360. Fawcett, J. S., and Morris, C. J. O. R., *Separation Sci.*, **1**, 9 (1966).
3361. Fake, G. T., and Prins, W., *Macromolecules*, **7**, 527 (1974).
3362. Fendler, H. G., Husman, W., and Stuart, H. A., *Z. Naturforsch.*, **9a**, 522 (1954).
3363. Ferguson, R. C., *J. Polym. Sci. A*, **2**, 4735 (1964).
3364. Ferguson, R. C., *Anal. Chem.*, **36**, 2204 (1964).
3365. Ferguson, R. C., *Kautschuk, Gummi. Kunststoffe*, **18**, 723 (1965).
3366. Ferguson, R. C., *Trans. N.Y. Acad. Sci.*, **29**, 495 (1967).
3367. Ferguson, R. C., *Macromolecules*, **2**, 237 (1969).
3368. Ferguson, R. C., *Macromolecules*, **4**, 324 (1971).
3369. Ferguson, R. C., and Phillips, W. D., *Science*, **157**, 257 (1967).
3370. Ferington, T. E., and Tobolsky, A. V., *J. Polym. Sci.*, **31**, 25 (1958).
3371. Fetters, L. J., *J. Appl. Polym. Sci.*, **20**, 3437 (1976).
3372. Fetters, L. J., and McIntyre, D., *Polymer*, **19**, 463 (1978).
3373. Fichera, A., and Zannetti, R., *Makromol. Chem.*, **176**, 1885 (1975).
3374. Fijolka, P., Lenz, I., and Runge, F., *Makromol. Chem.*, **23**, 60 (1957).
3375. Findeis, A. F., Rosinski, K. D. W., Petro, P. P., and Earp, R. E. W., *Thermochim. Acta*, **1**, 383 (1970).
3376. Fiorentini, A., *Phys. Lett. A*, **25**, 401 (1967).
3377. Fischer, E. J., *J. Macromol. Sci.*, **A2**, 1285 (1968).
3378. Fischer, D. G., Mann, J., *J. Polym. Sci.*, **42**, 189 (1960).
3379. Fischer, E. W., *Z. Naturforsch.*, **12a**, 753 (1957).
3380. Fischer, E. W., *Discuss. Faraday Soc.*, **25**, 204 (1957).
3381. Fischer, E. W., *Kolloid Z.*, **159**, 108 (1958).
3382. Fischer, E. W., Goddar, H., and Pisczek, W., *J. Polym. Sci. C*, **32**, 149 (1971).
3383. Fischer, E. W., Goddar, H., and Schmidt, G. F., *J. Polym. Sci. B*, **5**, 619 (1967).

766

3384. Fischer, E. W., Herchenröder, P., Manley, R. S., and Stamm, M., *Macromolecules*, **11**, 213 (1978).
3385. Fischer, E. W., Kloos, F., and Lieser, G., *J. Polym. Sci. B*, **7**, 845 (1969).
3386. Fischer, E. W., and Lorenz, R., *Kolloid Z. Z. Polym.*, **189**, 97 (1963).
3387. Fischer, H., and Hellwege, K. H., *J. Polym. Sci.*, **56**, 33 (1963).
3388. Fischer, H., Hellwege, K. H., and Johnsen, U., *Kolloid. Z.*, **170**, 107 (1960).
3389. Fischer, P., and Röhl, P., *J. Polym. Sci. A2*, **14**, 531 (1976).
3390. Fischer, L., *Photochem. Photobiol.*, **2**, 411 (1963).
3391. Fitchmun, D. R., and Mencik, Z., *J. Polym. Sci. A2*, **11**, 951 (1973).
3392. Fite, W., *J. Vac. Sci. Technol.*, **11**, 351 (1974).
3393. Fleming, J. W., Chantry, G. W., Turner, P. A., Nicol, E. A., Willis, H. A., and Cudby, M. E. A., *Chem. Phys. Lett.*, **17**, 84 (1972).
3394. Flett, M. S. C., and Plesch, P. H., *J. Chem. Soc.*, **1952**, 3355.
3395. Flocke, H. A., *Kolloid Z.*, **180**, 118 (1962).
3396. Flodin, P., *J. Chromatogr.*, **5**, 103 (1961).
3397. Flodin, P., *Anal. Chim. Acta*, **38**, 89 (1967).
3398. Flodin, P., and Killander, J., *Biochem. Biophys. Acta*, **63**, 403 (1962).
3399. Florin, R. E., Wall, L. A., and Brown, D. W., *J. Polym. Sci. A*, **1**, 1521 (1963).
3400. Flory, P. J., *J. Chem. Phys.*, **10**, 51 (1942).
3401. Flory, P. J., *J. Amer. Chem. Soc.*, **65**, 372 (1943).
3402. Flory, P. J., *J. Chem. Phys.*, **17**, 303 (1949).
3403. Flory, P. J., *J. Chem. Phys.*, **18**, 108 (1950).
3404. Flory, P. J., *Proc. Roy. Soc. A*, **234**, 73 (1956).
3405. Flory, P. J., *J. Polym. Sci.*, **49**, 105 (1961).
3406. Flory, P. J., *J. Amer. Chem. Soc.*, **84**, 2857 (1962).
3407. Flory, P. J., *J. Amer. Chem. Soc.*, **87**, 1833 (1965).
3408. Flory, P. J., *Discuss. Faraday Soc.*, **49**, 7 (1970).
3409. Flory, P. J., *Macromolecules*, **3**, 613 (1970).
3410. Flory, P. J., *J. Polym. Sci. A2*, **11**, 621 (1973).
3411. Flory, P. J., and Abe, Y., *Macromolecules*, **2**, 335 (1969).
3412. Flory, P. J., and Bueche, A. M., *J. Polym. Sci.*, **27**, 219 (1958).
3413. Flory, P. J., Eichinger, B. E., and Orwell, R. A., *Macromolecules*, **1**, 287 (1968).
3414. Flory, P. J., Ellson, J. L., and Eichinger, B. E., *Macromolecules*, **1**, 279 (1968).
3415. Flory, P. J., and Fox, T. G., Jr, *J. Polym. Sci.*, **5**, 745 (1950).
3416. Flory, P. J., and Fox, T. G., Jr, *J. Amer. Chem. Soc.*, **73**, 1904 1915 (1951).
3417. Flory, P. J., and Fujiwara, Y., *Macromolecules*, **2**, 315 (1969).
3418. Flory, P. J., and Fujiwara, Y., *Macromolecules*, **2**, 327 (1969).
3419. Flory, P. J., and Höcker, H., *Trans. Faraday Soc.*, **67**, 2258 (1971).
3420. Flory, P. J., and Krigbaum, W. R., *J. Chem. Phys.*, **18**, 1086 (1950).
3421. Flory, P. J., Mandelkern, L., and Hall, H. K., *J. Amer. Chem. Soc.*, **73**, 2532 (1951).
3422. Flory, P. J., Mark, J. E., and Abe, A., *J. Amer. Chem. Soc.*, **88**, 639 (1966).
3423. Flory, P. J., Orwoll, R. A., and Wrij, A., *J. Amer. Chem. Soc.*, **86**, 3507, 3515 (1964).
3424. Flory, P. J., and Rehner, J., *J. Chem. Phys.*, **11**, 512 (1943).
3425. Flory, P. J., and Shih, H., *Macromolecules*, **5**, 761 (1972).
3426. Flory, P. J., and Wrij, A., *J. Amer. Chem. Soc.*, **85**, 3548 (1963).
3427. Flossdorf, J., Schillig, H., and Schindler, K., *Makromol. Chem.*, **179**, 1617 (1978).
3428. Flowers, D. L., Hewett, W. A., and Mullineaux, R. D., *J. Polym. Sci. A*, **2**, 2305 (1964).
3429. Fluornoy, P. A., and Schaffers, W. J., *Spectrochim. Acta*, **22**, 5 (1966).
3430. Flynn, J. H., and Wall, L. A., *J. Res. Nat. Bur. Stand. A*, **70**, 487 (1966).
3431. Flynn, J. H., and Wall, L. A., *J. Polym. Sci. B*, **4**, 323 (1966).
3432. Fok, J. S., Robson, J. W., and Youngken, F. C., *Anal. Chem.*, **43**, 38 (1971).
3433. Folkes, M. J., and Keller, A., *J. Polym. Sci. A2*, **14**, 833 (1976).
3434. Folkes, M. J., Keller, A., and Odell, J. A., *J. Polym. Sci.*, *A2*, **14**, 847 (1976).
3435. Folland, R., and Charlesby, A., *J. Polym. Sci. B*, **16**, 339 (1978).
3436. Folland, R., Steven, J. H., and Charlesby, A., *J. Polym. Sci. A2*, **16**, 1041 (1978).
3437. Folmer, O. F., Jr, and Azarraga, L. V., *J. Chromatogr. Sci.*, **7**, 665 (1969).
3438. Folt, V. L., Shipman, J. J., and Krimm, S., *J. Polym. Sci.*, **61**, S17 (1962).
3439. Foote, C. S., and Denny. R. W., *J. Amer. Chem. Soc.*, **90**, 6233 (1958).
3440. Foote, C. S., and Wexler, S., *J. Amer. Chem. Soc.*, **86**, 3879 (1964).
3441. Foote, C. S., Wexler, S., and Ando, W., *Tetrahedron Lett.*, **46**, 4111 (1965).

3442. Foote, C. S., Wexler, S., Ando, W., and Higgins, R., *J. Amer. Chem. Soc.*, **90**, 975 (1968).
3443. Ford, N. C., Jakeman, E., Oliver, C. J., Pike, E. R., Blakgrove, R. J., Wood, R. J., and Peacocke, A. R., *Nature*, **227**, 242 (1970).
3444. Ford, N. C., Karasz, F. E., and Owen, J. E. M., *Discuss. Faraday Soc.*, **49**, 228 (1970).
3445. Ford, N. C., Lee, W., and Karasz, F. E., *J. Chem. Phys.*, **50**, 3098 (1969).
3446. Ford, R. W., and Ilavsky, J. D., *J. Appl. Polym. Sci.*, **12**, 2299 (1968).
3447. Forelich, D., Strazielle, C., Bernardi, G., and Benoit, H., *Biophys.*, **3**, 115 (1963).
3448. Forman, J. W., Jr, George, E. W., and Lewis, R. D., *Appl. Phys. Lett.*, **7**, 77 (1965).
3449. Formoshino, S. J., Porter, G., and West, M. A., *Chem. Phys. Lett.*, **6**, 7 (1970).
3450. Forrestal, L. J., and Hodgson, W. G., *J. Polym. Sci. A*, **2**, 1275 (1964).
3451. Förster, T., and Kasper, K., *Z. Elektrochem.*, **59**, 976 (1955).
3452. Fort, R. J., and Polyzoidis, T. M., *Makromol. Chem.*, **178**, 3229 (1977).
3453. Fortelny, I., *J. Polym. Sci. A2*, **12**, 2319 (1974).
3454. Fortelny, I., *J. Polym. Sci. A2*, **13**, 1011 (1975).
3455. Fortelny, I., *J. Polym. Sci. A2*, **14**, 1931 (1976).
3456. Forziati, F. H., and Rowden, J. W., *J. Res. Nat. Bur. Stand.*, **46**, 38 (1951).
3457. Foster, G. N., *J. Appl. Polym. Sci.*, **8**, 1357 (1964).
3458. Fouassier, J. P., Lougnot, D. J., and Faure, J., *Makromol. Chem.*, **179**, 437 (1978).
3459. Fourche, G., and Bothorel, P., *J. Chim. Phys.*, **66**, 54 (1969).
3460. Fourche, G., and Jacq, M. T., *Polym. J.*, **4**, 465 (1973).
3461. Fourche, G., and Lemaire, B., *Polym. J.*, **4**, 476 (1973).
3462. Foweraker, A. R., and Jennings, B. R., *Polymer*, **17**, 508 (1976).
3463. Fowkes, F. M., *J. Phys. Chem.*, **66**, 382 (1962).
3464. Fox, H. W., and Zisman, W. A., *J. Colloid Sci.*, **5**, 514 (1950).
3465. Fox, R. B., Price, T. R., Cozzens, R. F., and Echols, W. H., *Macromolecules*, **7**, 937 (1974).
3466. Fox, R. B., Price, T. R., Cozzens, R. F., and McDonald, J. R., *J. Chem. Phys.*, **57**, 534 (1972).
3467. Fox, T. G., *Bull. Amer. Phys. Soc.*, **1**, 123 (1956).
3468. Fox, T. G., and Flory, P. J., *J. Appl. Phys.*, **21**, 581 (1950).
3469. Fox, T. G., and Flory, P. J., *J. Amer. Chem. Soc.*, **73**, 1909 (1951).
3470. Fox, T. G., and Flory, P. J., *J. Polym. Sci.*, **14**, 315 (1954).
3471. Fox, T. G., and Loshaek, S., *J. Polym. Sci.*, **15**, 371 (1955).
3472. Fraga, D. W., *J. Polym. Sci.*, **41**, 522 (1959).
3473. Fraga, D. W., *Rubber Chem. Technol.*, **33**, 982 (1960).
3474. Francis, P. S., Cooke, R. C., Jr, and Elliott, J. H., *J. Polym. Sci.*, **31**, 453 (1958).
3475. Francis, S. A., and Ellison, A. H., *J. Opt. Soc. Amer.*, **49**, 131 (1959).
3476. Francois, J., and Candau, F., *Europ. Polym. J.*, **9**, 1355 (1973).
3477. Francois, J., Candau, F., and Benoit, H., *Polymer*, **15**, 618 (1974).
3478. Francois, J., Jacob, M., Grubisic-Gallot, Z., and Benoit, H., *J. Appl. Polym. Sci.*, **22**, 1159 (1978).
3479. Frank, C. W., *Macromolecules*, **8**, 305 (1975).
3480. Frank, F. C., *Proc. Roy. Soc. A*, **201**, 586 (1950).
3481. Frank, F. C., *Proc. Roy. Soc. A*, **282**, 9 (1964).
3482. Frank, F. C., Keller, A., and O'Connor, A., *Phil. Mag.*, **4**, 200 (1959).
3483. Frank, F. C., and Tosi, M., *Proc. Roy Soc. A*, **263**, 323 (1960).
3484. Frank, F. C., Ward, I. M., and Williams, T., *J. Polym. Sci. A2*, **6**, 1357 (1968).
3485. Frank, W., *J. Polym. Sci. B*, **15**, 679 (1977).
3486. Frank, W., and Rabus, G., *Kolloid Z. Z. Polym.*, **252**, 1003 (1974).
3487. Frank, W., Schmidt, H., and Wulff, W., *J. Polym. Sci. C*, **16**, 317 (1977).
3488. Frank, W., and Wulff, W., *Kolloid Z. Z. Polym.*, **254**, 534 (1976).
3489. Franken, I., and Burchard, W., *Macromolecules*, **6**, 849 (1973).
3490. Fraser, G. V., Hendra, P. J., Cudby, M. E. A., and Willis, H. A., *Chem. Commun.*, **1973**, 16.
3491. Fraser, G. V., Hendra, P. J., Chalmers, J. M., Cudby, M. E. A., and Willis, H. A., *Makromol. Chem.*, **173**, 195 (1973).
3492. Fraser, G. V., Hendra, P. J., Walker, J. H., Cudby, M. E., and Willis, H. A., *Makromol. Chem.*, **173**, 205 (1973).

768

3493. Fraser, G. V., Hendra, P. J., Watson, D. S., Gall, M. J., Willis, H. A., and Cudby, M. E. A., *Spectrochim. Acta*, **29**, 1525 (1973).
3494. Fraser, G. V., Keller, A., and Pope, D. P., *J. Polym. Sci. B*, **13**, 341 (1975).
3495. Fraser, R. D. B., *J. Chem. Phys.*, **21**, 1511 (1953).
3496. Fraser, R. D. B., and Suzuki, E., *Spectrochim. Acta*, **26A**, 423 (1970).
3497. Frederick, J. E., Reed, T. F., and Kramer, O., *Macromolecules*, **4**, 242 (1971).
3498. Freeberg, F. F., and Alleman, T. G., *Anal. Chem.*, **38**, 1806 (1966).
3499. Freed, K., *Adv. Chem. Phys.*, **22**, 1 (1972).
3500. Freed, K. F., and Edwards, S. F., *J. Chem. Phys.*, **62**, 4032 (1975).
3501. Freed, J. H., and Fraenkel, G. K., *J. Chem. Phys.*, **39**, 326 (1963).
3502. Freed, J. H., Leniart, D. S., and Connor, H. D., *J. Chem. Phys.*, **58**, 3089 (1973).
3503. Freeman, E. S., and Caroll, B., *J. Phys. Chem.*, **62**, 394 (1958).
3504. Freeman, P. I., and Rowlinson, J. S., *Polymer*, **1**, 20 (1960).
3505. French, D. M., and Ewart, R. H., *Anal. Chem.*, **19**, 165 (1947).
3506. French, M. J., Angus, J. C., and Walton, A. G., *Science*, **163**, 345 (1969).
3507. Frenkel, S., *Pure Appl. Chem.*, **38**, 117 (1974).
3508. Freund, L., and Daune, M., *J. Polym. Sci.*, **29**, 161 (1958).
3509. Friedman, E. A., and Andrews, R. D., *J. Appl. Phys.*, **40**, 4243 (1969).
3510. Friedman, E. A., Ritger, A. J., and Andrews, R. D., *J. Appl. Phys.*, **40**, 4243 (1969).
3511. Friedman, E. A., Ritger, A. J., and Huang, Y. Y., *Bull. Amer. Phys. Soc.*, **15**, 282 (1970).
3512. Friedman, H. L., *J. Polym. Sci. C*, **6**, 183 (1965).
3513. Friedman, H. L., *Thermochim. Acta*, **1**, 199 (1970).
3514. Frisch, H. L., *J. Polym. Sci. B*, **3**, 13 (1965).
3515. Frisch, H. L., *J. Polym. Sci. A2*, **7**, 879 (1969).
3516. Frisch, H. L., *Macromolecules*, **6**, 758 (1973).
3517. Frisch, H. L., *J. Polym. Sci. A2*, **16**, 1651 (1978).
3518. Frisch, H. L., Klempner, D., and Kwei, T. K., *Macromolecules*, **4**, 237 (1971).
3519. Frisch, H. L., and Lundberg, J. L., *J. Polym. Sci.*, **37**, 123 (1959).
3520. Frisch, H. L. Mallows, C. L., and Bovey, F. A., *J. Chem. Phys.*, **45**, 1565 (1966).
3521. Frisch, H. L., Mallows, C. L., Heatley, F., and Bovey, F. A., *Macromolecules*, **1**, 533 (1968).
3522. Frisch, H. L., Wang, T. T., and Kwei, T. K., *J. Polym. Sci. A2*, **7**, 879 (1969).
3523. Froehling, P. E., Koenhen, D. M., Bantjes, A., and Smolders, C. A., *Polymer*, **17**, 835 (1976).
3524. Froix, M. F., Beatty, C. L., Pochan, J. M., and Hinman, D. D., *J. Polym. Sci. A2*, **13**, 1269 (1975).
3525. Froix, M. F., and Goedde, A. O., *Polymer*, **17**, 758 (1976).
3526. Froix, M. F., and Goedde, A. O., *Macromolecules*, **9**, 428 (1976).
3527. Froix, M. F., and Nelson, R., *Macromolecules*, **8**, 726 (1975).
3528. Froix, M. F., and Pochan, J. M., *J. Polym. Sci. A2*, **14**, 1047 (1976).
3529. Froix, M. F., Williams, D. J., and Goedde, A. O., *Macromolecules*, **9**, 81 (1976).
3530. Froix, M. F., Williams, D. J., and Goedde, A. O., *Macromolecules*, **9**, 354 (1976).
3531. Frosini, V., Magagnini, P. L., and Newman, B. A., *J. Polym. Sci. A2*, **12**, 23 (1974).
3532. Frost, R. A., and Caroline, D., *Macromolecules*, **10**, 616 (1977).
3533. Frost, R. S., Barker, R. E., Jr, and Chen, R. Y. S., *J. Polym. Sci. A2*, **16**, 689 (1978).
3534. Fuchs, O., *Makromol. Chem.*, **5**, 245 (1950).
3535. Fuchs, O., *Makromol. Chem.*, **7**, 259 (1952).
3536. Fuchs, O., *Z. Elektrochem.*, **60**, 229 (1956).
3537. Fujime, S., *J. Phys., Soc., Japan*, **31**, 1805 (1971).
3538. Fujime, S., and Maruyama, M., *Macromolecules*, **6**, 237 (1973).
3539. Fujimura, T., Hayakawa, N., and Kuriyama, I., *J. Polym. Sci. A2*, **16**, 945 (1978).
3540. Fujimura, T., Hayakawa, N., and Kuriyama, I., *Polymer*, **19**, 1031 (1978).
3541. Fujita, H., *J. Chem. Phys.*, **24**, 1084 (1956).
3542. Fujita, H., *J. Amer. Chem. Soc.*, **78**, 3598 (1956).
3543. Fujita, H., *J. Phys. Chem.*, **63**, 194 (1959).
3544. Fujita, H., *J. Chem. Phys.*, **32**, 1739 (1960).
2545. Fujita, H., and Kishimota, A., *J. Chem. Phys.*, **34**, 393 (1961).
3546. Fujita, H., Linklater, A. M., and Williams, J. W., *J. Amer. Chem. Soc.*, **82**, 379 (1960).
3547. Fujita, H., Taki, H., Norisuye, T., and Sotobayashi, H., *J. Polym. Sci. A2*, **15**, 2255 (1977).
3548. Fujita, S., and Work, R. A., *J. Chem. Phys.*, **44**, 3779 (1966).

3549. Fujiwara, Y., and Flory, P. J., *Macromolecules*, **3**, 43 (1970).
3550. Fujiwara, Y., and Flory, P. J., *Macromolecules*, **3**, 280 (1970).
3551. Fujiwara, Y., and Flory, P. J., *Macromolecules*, **3**, 288 (1970).
3552. Funke, W., Hellweg, G. E. H., and Neumann, A. W., *Angew. Makromol. Chem.*, **8**, 185 (1969).
3553. Fuoss, R. M., and Mead, D. J., *J. Phys. Chem.*, **47**, 59 (1943).
3554. Fuoss, R. M., Salyer, I. O., and Wilson, H. S., *J. Polym. Sci. A*, **2**, 3147 (1964).
3555. Funt, B. L., and Hornof, V., *J. Appl. Polym. Sci.*, **15**, 2439 (1971).
3556. Funt, J. M., and Magill, J. J., *J. Polym. Sci. A2*, **12**, 217 (1974).
3557. Furukawa, G. T., McCoskey, R. E., and King, G. J., *J. Res. Nat. Bur. Stand.*, **49**, 273 (1952).
3558. Furukawa, G. T., McCoskey, R. E., and King, G. J., *J. Res. Nat. Bur. Stand.*, **50**, 357 (1953).
3559. Furukawa, G. T., McCoskey, R. E., and Reilly, M. L., *J. Res. Nat. Bur. Stand.*, **55**, 127 (1955).
3560. Furukawa, G. T., and Reilly, M. L., *J. Res. Nat. Bur. Stand.*, **56**, 285 (1956).
3561. Furukawa, J., Kobyashi, E., Katsuki, N., and Kawagoe, T., *Makromol. Chem.*, **175**, 237 (1974).
3562. Furukawa, J., Kobyashi, E., and Kawage, T., *Polym. J.*, **5**, 231 (1973).
3563. Furukawa, J., Kobyashi, E., and Kawage, T., *J. Polym. Sci. B*, **11**, 239 (1973).
3564. Furukawa, T., and Fukuda, E., *J. Polym. Sci. A2*, **14**, 1979 (1976).

3565. Gabbay, S. M., and Stivala, S. S., *Polymer*, **17**, 61 (1976).
3566. Gabbay, S. M., and Stivala, S. S., *Polymer*, **17**, 121 (1976).
3567. Gabbay, S. M., Stivala, S. S., and Reed, P. R., *Anal. Chem. Acta*, **78**, 359 (1975).
3568. Gabbay, S. M., Stivala, S. S., and Reich, L., *J. Appl. Polym. Sci.*, **19**, 2391 (1975).
3569. Gabriel, A., and Dupont, Y., *Rev. Sci. Instrum.*, **43**, 1600 (1972).
3570. Gaeckle, D., Kao, W. P., and Patterson, D., *J. Chem. Soc., Faraday Trans.*, **69**, 1849 (1973).
3571. Gaecke, D., and Patterson, D., *Macromolecules*, **5**, 136 (1972).
3572. Galassi, S., and Audisio, G., *Makromol. Chem.*, **175**, 2975 (1974).
3573. Galembeck, F., *J. Polym. Sci. B*, **15**, 107 (1977).
3574. Galeski, A., and Kryszewski, M., *J. Polym. Sci. A2*, **12**, 455 (1974).
3575. Galeski, A., and Kryszewski, M., *J. Polym. Sci. A2*, **12**, 471 (1974).
3576. Galeski, A., Pakula, T., and Kryszewski, M., *J. Polym. Sci. C*, **61**, 35 (1977).
3577. Galin, M., *J. Macromol. Sci. A*, **7**, 873 (1973).
3578. Galin, M., *Macromolecules*, **10**, 1239 (1977).
3579. Galin, M., and Guillet, J. E., *J. Polym. Sci. B*, **11**, 233 (1973).
3580. Galin, M., and Rupprecht, M. C., *Polymer*, **19**, 506 (1978).
3581. Gall, M. J., *Polymer*, **15**, 272 (1974).
3582. Gall, M. J., Hendra, P. J., Peacock, C. J., Cudby, M. E. A., and Willis, H. A., *Polymer*, **13**, 104 (1972).
3583. Gallot, Y., Franta, E., Rempp, P., and Benoit, H., *J. Polym. Sci. C*, **4**, 473 (1963).
3584. Gamble, L. W., Wipke, W. T., and Lane, T., *J. Appl. Polym. Sci.*, **9**, 1503 (1965).
3585. Gandica, A., and Magill, J. H., *Polymer*, **13**, 595 (1972).
3586. Garber, C. A., and Clark, E. S., *J. Macromol. Sci. B*, **4**, 499 (1970).
3587. Garber, C. A., and Geil, P. H., *J. Appl. Phys.*, **37**, 4034 (1966).
3588. Garber, C. A., and Geil, P. H., *Makromol. Chem.*, **113**, 251 (1968).
3589. Garber, C. A., and Geil, P. H., *Kolloid. Z.*, **229**, 140 (1969).
3590. Garbuglio, C., Mula, A., and Chinellato, L., *J. Polym. Sci. C*, **16**, 1529 (1967).
3591. Gardon, J. L., *J. Paint. Technol.*, **38**, 43 (1966).
3592. Garg, S. K., and Stivala, S. S., *J. Polym. Sci. A2*, **16**, 1419 (1978).
3593. Gargallo, L., and Russo, M., *Makromol. Chem.*, **176**, 2735 (1975).
3594. Garmon, R. G., and Gibson, M. E., *Anal. Chem.*, **37**, 1309 (1965).
3595. Garn, P. D., *Talanta*, **11**, 1417 (1964).
3596. Garner, H. R., and Packer, H., *Appl. Spectr.*, **22**, 122 (1968).
3597. Garrett, R. R., Hargreaves, C. A., and Robinson, D. N., *J. Macromol. Sci. A*, **4**, 1967 (1970).
3598. Garton, A., Stepaniak, R. F., Carlsson, D. J., and Wiles, D. M., *J. Polym. Sci. A2*, **16**, 599 (1978).

3599. Gatti, G., and Carbonaro, A., *Makromol. Chem.*, **175**, 1627 (1974).
3600. Gaylord, R. J., *J. Polym. Sci. A2*, **14**, 1827 (1976).
3601. Geacintov, C., Miles, R. B., and Schuurmans, H. J. L., *J. Polym. Sci. A1*, **4**, 431 (1966).
3602. Geacintov, C., Miles, R. B., and Schuurmans, H. J. L., *J. Polym. Sci. C*, **14**, 283 (1966).
3603. Gee, G., *Polymer*, **7**, 179 (1966).
3604. Gee, G., *Contemp. Phys.*, **10**, 353 (1970).
3605. Gee, G., Hartley, P. N., Herbert, T. B. M., and Lancely, H. A., *Polymer*, **1**, 365 (1960).
3606. Gee, D. R., and Melia, T. P., *Makromol. Chem.*, **116**, 122 (1968).
3607. Gee, D. R., and Melia, T. P., *Polymer*, **10**, 239 (1969).
3608. Gee, G., *Trans. Faraday Soc.*, **40**, 463 (1944).
3609. Gee, G., and Orr, W. J. C., *Trans. Faraday Soc.*, **42**, 507 (1946).
3610. Geil, P. H., *J. Polym. Sci.*, **44**, 449 (1960).
3611. Geil, P. H., *J. Polym. Sci.*, **47**, 291 (1960).
3612. Geil, P. H., *J. Appl. Phys.*, **33**, 642 (1962).
3613. Geil, P. H., *J. Polym. Sci. A2*, **2**, 3813 (1964).
3614. Geil, P. H., *J. Polym. Sci. A2*, **2**, 3835 (1964).
3615. Geil, P. H., and Reneker, D. H., *J. Polym. Sci.*, **51**, 569 (1961).
3616. Geissler, E., *J. Polym. Sci. A2*, **13**, 1301 (1975).
3617. Gelote, B., and Emnéus, A., *Chem. Ing. Techn.*, **38**, 445 (1966).
3618. Gent, A. N., *Macromolecules*, **2**, 262 (1969).
3619. Gény, F., and Monnerie, L., *J. Polym. Sci. A2*, **15**, 1 (1977).
3620. Gény, F., and Monnerie, L., *Macromolecules*, **10**, 1003 (1977).
3621. Georgiadis, T., and Manley, R. S. T., *J. Polym. Sci. B*, **9**, 297 (1971).
3622. Georgiadis, T., and Manley, R. S. T., *Polymer*, **13**, 567 (1972).
3623. Gerasimov, V. I., Genin, Ya. V., and Tsvankin, D. Ya., *J. Polym. Sci. A2*, **12**, 2035 (1974).
3624. Germar, H., *Makromol. Chem.*, **84**, 36 (1965).
3625. Gernert, J. F., Cantow, M. J. R., Porter, R. S., and Johnson, J. F., *J. Polym. Sci. C*, **1**, 195 (1963).
3626. Gerrard, D. L., and Maddams, W. F., *Macromolecules*, **8**, 54 (1975).
3627. Gerrard, D. L., and Maddams, W. F., *Macromolecules*, **10**, 1221 (1977).
3628. Ghesquiere, D., Ban, B., and Chachaty, C., *Macromolecules*, **10**, 743 (1977).
3629. Ghesquiere, D., and Chacharty, C., *Macromolecules*, **11**, 246 (1978).
3630. Ghiggino, K. P., Wright, R. D., and Phillips, D., *J. Polym. Sci. A2*, **16**, 1499 (1978).
3631. Ghosh, P., Chadha, S. C., and Palit, S. R., *J. Polym. Sci. A1*, **2**, 4433 (1964).
3632. Ghosh, P., Chadha, S. C., and Palit, S. R., *J. Polym. Sci. A1*, **2**, 4441 (1964).
3633. Ghotra, J. S., Stevens, G. C., and Bloor, D., *J. Polym. Sci. A1*, **15**, 1155 (1977).
3634. Gibbs, J. H., *J. Chem. Phys.*, **25**, 185 (1956).
3635. Gibbs, J. H., and DiMarzio, E. A., *J. Chem. Phys.*, **28**, 373 (1958).
3636. Giddings, J. C., *Separation Sci.*, **1**, 123 (1966).
3637. Giddings, J. C., *J. Chem. Phys.*, **49**, 1 (1968).
3638. Giddings, J. C., *Separation Sci.*, **8**, 567 (1973).
3639. Giddings, J. C., *J. Chem. Educ.*, **50**, 667 (1973).
3640. Giddings, J. C., Bowman, L. M., and Myers, M. N., *Macromolecules*, **10**, 443 (1977).
3641. Giddings, J. C., Caldwell, K. D., and Myers, M. N., *Macromolecules*, **9**, 106 (1976).
3642. Giddings, J. C., Hovingh, M. E., and Thompson, G. H., *J. Phys. Chem.*, **74**, 4291 (1970).
3643. Giddings, J. C., Kucera, E., Russell, C. P., and Myers, M. N., *J. Phys. Chem.*, **72**, 4397 (1968).
3644. Giddings, J. C., Lin, G. C., and Myers, M. N., *J. Liquid Chromatogr.*, **1**, 1 (1978).
3645. Giddings, J. C., and Mallik, K. L., *Anal. Chem.*, **38**, 997 (1966).
3646. Giddings, J. C., Meyers, M. N., and Moellmer, J. F., *J. Chromatogr.*, **149**, 501 (1978).
3647. Giddings, J. C., Yang, F. J. F., and Myers, M. N., *Anal. Chem.*, **46**, 1917 (1974).
3648. Giddings, J. C., Yoon, Y. H., and Myers, M. N., *Anal. Chem.*, **47**, 126 (1975).
3649. Gieniewski, C., and Moore, R. S., *Macromolecules*, **2**, 385 (1969).
3650. Giesekus, H., *Kolloid Z.*, **158**, 35 (1958).
3651. Gilbert, G. A., Graff-Baker, C., and Greenwood, C. T., *J. Polym. Sci.*, **6**, 585 (1951).
3652. Gilbert, M., and Hybart, F. J., *Polymer*, **13**, 327 (1972).
3653. Gilbert, M., and Hybart, F. J., *Polymer*, **15**, 407 (1974).
3654. Gilbert, M., and Vyvoda, J. C., *Polymer*, **19**, 863 (1978).

3655. Giles, C. H., and Easton, I. A., *Adv. Chromatogr.*, **3**, 70 (1966).
3656. Giles, W. H., Brandkamp, W., Farace, L., and Bergvist, J., *ACS Polym. Preprints*, **8**, 522 (1967).
3657. Gilespie, T., *J. Polym. Sc.*, **46**, 383 (1966).
3658. Gills, J., and Kedem, O., *J. Polym. Sci.*, **11**, 545 (1953).
3659. Gilmour, I. W., and Hay, J. N., *Polymer*, **18**, 281 (1977).
3660. Ginnings, D. C., and Furukawa, G. T., *J. Amer. Chem. Soc.*, **75**, 522 (1953).
3661. Girifalco, L. A., and Good, R. J., *J. Phys. Chem.*, **61**, 904 (1957).
3662. Girifalco, L. A., and Good, R. J., *J. Phys. Chem.*, **64**, 561 (1960).
3663. Girolamo, M., Keller, A., and Stejny, J., *Makromol. Chem.*, **176**, 1489 (1975).
3664. Girolamo, M., Keller, A., Miyasaka, K., and Overbergh, N., *J. Polym. Sci. A2*, **14**, 39 (1976).
3665. Gladkih, I., Kunchenko, A. B., Ostanevich, Yu. M., and Csér, L., *J. Polym. Sci. C*, **61**, 359 (1977).
3666. Gladney, H. M., *J. Comp. Phys.*, **2**, 255 (1968).
3667. Gladney, H. M., Dowden, B. F., and Swalen, J. D., *Anal. Chem.*, **41**, 883 (1969).
3668. Glasgow, A. R., Jr, Krouskop, N. C., and Rossini, F. D., *Anal. Chem.*, **22**, 1521 (1950).
3669. Glasgow, A. R., Streiff, A. J., and Rossini, F. D., *J. Res. Nat. Bur. Stand.*, **35**, 355 (1945).
3670. Gleit, C. E., and Holland, W. D., *Anal. Chem.*, **34**, 1454 (1962).
3671. Glenz, W., and Peterlin, A., *J. Macromol. Sci. B*, **4**, 473 (1970).
3672. Glockner, G., and Müller, K. D., *Plaste Kautsch.*, **15**, 812 (1968).
3673. Glover, C. A., *Adv. Anal. Chem. Instrum.*, **5**, 1 (1966).
3674. Glover, C. A., *ACS Polym. Preprints*, **12**, 763 (1971).
3675. Glover, C. A., and Kirn, J. E., *J. Polym. Sci. B*, **3**, 27 (1965).
3676. Glover, C. A., and Stanley, R. R., *Anal. Chem.*, **33**, 447 (1961).
3677. Go, N., and Okuyama, K., *Macromolecules*, **9**, 867 (1976).
3678. Go, S., Mandelkern, L., Prud'homme, R., and Stein, R., *J. Polym. Sci. A2*, **12**, 1485 (1974).
3679. Godovskii, Yu. K., Levin, V. Yu., Slonimski, G. L., Shadanov, A. A., and Andrianov, K. A., *Vysokomol. Soedin. A*, **11**, 2444 (1969).
3680. Godovskii, Yu. K., and Lipatov, Yu. S., *Vysokomol. Soedin. A*, **10**, 32 (1968).
3681. Godovskii, Yu. K., and Slonimsky, G. L., *J. Polym. Sci. A2*, **12**, 1053 (1974).
3682. Godovskii, Yu. K., Slonimsky, G. L., and Garbar, N. M., *J. Polym. Sci. C*, **28**, 1 (1972).
3683. Godwin, R. W., and Kunkel, R. K., *Appl. Polym. Symp.*, No. 16, 165 (1971).
3684. Goebel, K. D., and Berry, G. C., *J. Polym. Sci. A2*, **15**, 555 (1972).
3685. Goedhart, D., and Opschoor, A., *J. Polym. Sci. A2*, **8**, 1227 (1970).
3686. Goffin, A., Dosiere, M., Point, J. J., and Gillot, M., *J. Polym. Sci. C*, **38**, 135 (1972).
3687. Gogolewski, S., *Kolloid Z. Z. Polym.*, **256**, 323 (1978).
3688. Gogolewski, S., and Pennings, A. J., *Polymer*, **16**, 673 (1975).
3689. Gohil, R. M., Patel, K. C., and Patel, R. D., *Makromol. Chem.*, **169**, 291 (1973).
3690. Gohil, R. M., Patel, K. C., and Patel, R. D., *Polymer*, **15**, 403 (1974).
3691. Goldberg, R. J., *J. Phys. Chem.*, **57**, 194 (1953).
3692. Goldfarb, L., Garret, T. B., Rittenhouse, J. R., and Messersmith, D. C., *Macromol. Chem.*, **175**, 2483 (1974).
3693. Goldfein, S., and Calderon, J., *J. Appl. Polym. Sci.*, **9**, 2985 (1965).
3694. Goldstein, M., *J. Appl. Polym. Sci.*, **33**, 3377 (1962).
3695. Goldstein, M., *J. Chem. Phys.*, **21**, 1255 (1953).
3696. Goldstein, M., *Macromolecules*, **10**, 1407 (1977).
3697. Goldstein, M., Seeley, M. E., Willis, H. A., and Zichy, V. J., *Polymer*, **14**, 530 (1973).
3698. Goldstein, M., Seely, M. E., Willis, H. A., and Zichy, V. J., *Polymer*, **19**, 1118 (1978).
3699. Golub, M. A., *J. Polym. Sci.*, **11**, 281 (1953).
3700. Golub, M. A., *J. Polym. Sci.*, **18**, 27 (1955).
3701. Golub, M. A., *J. Polym. Sci.*, **36**, 523 (1959).
3702. Golub, M. A., and Shipman, J. J., *Spectrochim. Acta*, **16**, 1165 (1960).
3703. Gölz, W. L. F., and Zachmann, H. G., *Kolloid Z. Z. Polym.*, **247**, 814 (1971).
3704. Gonzales, J. J., and Hemmer, P. C., *J. Polym. Sci. B*, **14**, 645 (1976).
3705. Gooberman, G., *J. Polym. Sci.*, **40**, 469 (1959).
3706. Good, R. J., Kvikstad, J. A., and Bailey, W. O., *Appl. Polym. Symp.*, No. 16, 153 (1971).

772

3707. Goode, W. E., Owen, F. H., Fellman, R. P., Snyder, W. H., and Moore, J. E., *J. Polym. Sci.*, **46**, 317 (1960).
3708. Goodman, M., and Abe, A., *J. Polym. Sci.*, **59**, 537 (1962).
3709. Goodman, M., and Brandrup, J., *J. Polym. Sci. A1*, **3**, 327 (1965).
3710. Goodman, M., and Chen, S. C., *Macromolecules*, **3**, 398 (1970).
3711. Goodman, M., and Chen, S. C., *Macromolecules*, **4**, 625 (1971).
3712. Goodman, M., Clark, K. J., Stake, M., and Abe, A., *Makromol. Chem.*, **72**, 131 (1964).
3713. Goodman, M., and D'Alagni, M., *J. Polym. Sci. B*, **5**, 575 (1967).
3714. Goodrich, F. C., and Cantow, M. J. R., *J. Polym. Sci. C*, **8**, 269 (1965).
3715. Gordienko, A., Griehl, W., and Seiber, H., *Faserforsch. Textiltech.*, **6**, 105 (1955).
3716. Gordon, G. A., *J. Polym. Sci. A2*, **9**, 1693 (1971).
3717. Gordon, J. L., and Mason, S. G., *Can. J. Chem.*, **33**, 1477 (1955).
3718. Gordon, M., *Macromolecules*, **8**, 247 (1975).
3719. Gordon, M., Chermin, H. A. G., and Koningsveld, R., *Macromolecules*, **2**, 207 (1969).
3720. Gordon, M., Irvine, P., and Kennedy, J. W., *J. Polym. Sci. C*, **61**, 199 (1977).
3721. Gordon, M., Kajiwara, K., Peniche-Covas, C. A. L., and Ross-Murphy, S. B., *Makromol. Chem.*, **176**, 2413 (1975).
3722. Gordon, M., and Polowinski, S., *Brit. Polym. J.*, **2**, 183 (1970).
3723. Gordon, M., and Taylor, J. S., *J. Appl. Chem.*, **2**, 493 (1952).
3724. Gorning, D. A. I., *Can. J. Chem.*, **31**, 1078 (1953).
3725. Gornick, F., *J. Polym. Sci. C*, **25**, 131 (1968).
3726. Gosting, L. J., *J. Amer. Chem. Soc.*, **74**, 1548 (1952).
3727. Goth, R., Takenaka, T., and Hayama, N., *Kolloid. Z.*, **205**, 18 (1965).
3728. Gotto, K., and Nishioka, A., *Bull. Chem. Soc. Japan*, **44**, 877 (1971).
3729. Gouinlock, E. V., *J. Polym. Sci. A2*, **13**, 961 (1975).
3730. Gouinlock, E. V., *J. Polym. Sci. A2*, **13**, 1533 (1975).
3731. Gouw, T. H., and Jentoft, R. E., *J. Chromatogr.*, **66**, 303 (1972).
3732. Gouw, T. H., and Jentoft, R. E., *Adv. Chromatogr.*, **13**, 1 (1975).
3733. Grabec, I., *J. Polym. Sci. B*, **12**, 573 (1974).
3734. Grabec, I., and Peterlin, A., *J. Polym. Sci. A2*, **14**, 651 (1976).
3735. Graessley, W. W., *Macromolecules*, **8**, 186 (1975).
3736. Graessley, W. W., *Macromolecules*, **8**, 865 (1975).
3737. Graessley, W. W., Glasscock, S. D., and Crawley, R. L., *Trans. Soc. Rheol.*, **14**, 519 (1970).
3738. Graessley, W. W., Masuda, T., Roovers, J. E. L., and Hadjichristidis, N., *Macromolecules*, **9**, 127 (1976).
3739. Graessley, W. W., and Pennline, H. W., *J. Polym. Sci. A2*, **12**, 2347 (1974).
3740. Graessley, W. W., and Segal, L., *Macromolecules*, **2**, 49 (1969).
3741. Graessley, W. W., and Shinbach, E. S., *J. Polym. Sci. A2*, **12**, 2047 (1974).
3742. Gramain, P., and Libeyre, R., *J. Appl. Polym. Sci.*, **14**, 383 (1970).
3743. Gramberg, G., *Angew. Chem.*, **73**, 117 (1961).
3744. Granath, K. A., and Flodin, P., *Makromol. Chem.*, **48**, 160 (1961).
3745. Grandjean, J., Sillescu, H., and Willenberg, B., *Makromol. Chem.*, **178**, 1445 (1977).
3746. Grant, I. J., and Ward, I. M., *Polymer*, **6**, 223 (1965).
3747. Grant, P. M., *IBM J. Res. Develop.*, **13**, 15 (1969).
3748. Grassie, N., Scotney, A., and Mackinnon, L., *J. Polym. Sci. A1*, **15**, 251 (1977).
3749. Grassie, N., and Torrance, B. J. D., *J. Polym. Sci. A1*, **6**, 3315 (1968).
3750. Grassie, N., and Weir, N. A., *J. Appl. Polym. Sci.*, **9**, 963, 975 (1965).
3751. Grates, J. A., Thomas, D. A., Hickey, E. C., and Sperling, L. H., *J. Appl. Polym. Sci.*, **15**, 1731 (1975).
3752. Grattan, D. W., Carlsson, D. J., and Wiles, D. M., N.R.C.C. Report No. 15995, National Research Council of Canada, Ottawa.
3753. Gray, D. G., and Guillet, J. E., *Macromolecules*, **4**, 129 (1971).
3754. Gray, D. G., and Guillet, J. E., *Macromolecules*, **5**, 316 (1972).
3755. Gray, D. G., and Guillet, J. E., *Macromolecules*, **6**, 223 (1973).
3756. Gray, D. G., and Guillet, J. E., *Macromolecules*, **7**, 244 (1974).
3757. Gray, D. G., and Guillet, J. E., *J. Polym. Sci. B*, **12**, 831 (1974).
3758. Gray, D. O., *J. Chromatogr.*, **37**, 320 (1968).
3759. Grebenova, B., and Hloušek, M., *J. Polym. Sci. C*, **61**, 139 (1977).
3760. Greco, R., Nicodemo, L., and Nicolais, L., *Macromolecules*, **9**, 686 (1976).

3761. Greco, R., and Nicolais, L., *Polymer*, **17**, 1049 (1976).
3762. Green, J. H. S., *Nature*, **183**, 818 (1959).
3763. Green, J. H. S., and Vaugham, M. F., *Chem. Ind.* (*London*), **29**, 829 (1958).
3764. Greenler, R. G., *J. Chem. Phys.*, **44**, 310 (1966).
3765. Greenler, R. G., *J. Chem. Phys.*, **50**, 1963 (1969).
3766. Greenler, R. G., Rahn, R. R., and Schwartz, J. P., *J. Catal.*, **23**, 42 (1971).
3767. Greenwood, C. T., and Hourston, D. J., *Polymer*, **16**, 474 (1975).
3768. Greth, G. G., Smith, R. G., and Rudkin, G. O., *J. Cell. Plastics*, **1**, 159 (1965).
3769. Grewer, T., and Wilski, H., *Kolloid Z. Z. Polym.*, **226**, 46 (1968).
3770. Grehl, W., and Neue, S., *Faserforsch. Textiltechn.*, **5**, 423 (1954).
3771. Griffin, G. J. L., *Appl. Polym. Symp.*, No. 16, 67 (1971).
3772. Griffith, J. H., and Rånby, B., *J. Polym. Sci.*, **44**, 369 (1960).
3773. Griffiths, C. H., *J. Polym. Sci. A2*, **13**, 1167 (1975).
3774. Griffiths, M. D., and Maisey, L. J., *Polymer*, **17**, 869 (1976).
3775. Griggs, M., and Kaye, S., *Rev. Sci. Instrum.*, **39**, 1685 (1968).
3776. Grime, D., and Ward, I. M., *Trans. Faraday Soc.*, **54**, 959 (1959).
3777. Grisenthwaite, R. J., and Hunter, R. F., *Chem. Ind.* (*London*), **1958**, 719.
3778. Griskey, R. G., and Hubbel, D. O., *J. Appl. Polym. Sci.*, **12**, 853 (1968).
3779. Gritter, R. J., Seeger, M., and Gipstein, E., *J. Polym. Sci. A1*, **16**, 353 (1978).
3780. Gritter, R. J., Seeger, M., and Johnson, D. E., *J. Polym. Sci. A1*, **16**, 169 (1978).
3781. Griva, A. P., and Denisov, E. T., *J. Polym. Sci. A1*, **14**, 1051 (1976).
3782. Grob, R. L., Mercer, D., Gribben, T., and Wells, J., *J. Chromatogr.*, **3**, 545 (1960).
3783. Groenickx, G., Berghmans, H., Overbergh, N., and Smets, G., *J. Polym. Sci. A2*, **12**, 303 (1974).
3784. Grohn, H., and Fredrich, H., *J. Polym. Sci. C*, **16**, 3737 (1968).
3785. Grohn, H., and Huu-Binh, H., *Plaste Kautschuk*, **8**, 63 (1961).
3786. Gronski, W., *Makromol. Chem.*, **178**, 2949 (1977).
3787. Gronski, W., and Murayama, N., *Makromol. Chem.*, **177**, 3017 (1976).
3788. Gronski, W., and Murayama, N., *Makromol. Chem.*, **179**, 1509, 1521 (1978).
3789. Gronski, W., Murayama, N., and Cantow, H. J., *Polymer*, **17**, 358 (1976).
3790. Gronski, W., Murayama, N., Mannewitz, C., and Cantow, H. J., *Makromol. Chem.*, Suppl. 1, 485 (1975).
3791. Groseclose, B. C., and Loper, G. D., *Phys. Rev.*, **137**, A939 (1965).
3792. Gross, S. C., *J. Polym. Sci. A1*, **9**, 3327 (1971).
3793. Grossman, H. P., and Frank, W., *Polymer*, **18**, 341 (1977).
3794. Groten, B., *Anal. Chem.*, **36**, 1206 (1964).
3795. Groten, B., *Rubber Chem. Technol.*, **39**, 248 (1966).
3796. Gruber, U., and Elias, H. G., *Makromol. Chem.*, **78**, 58 (1964).
3797. Gruber, E., Sezen, M. C., and Schurz, J., *Makromol. Chem.*, Suppl. 1, 387 (1975).
3798. Gruber, U., and Elias, H. G., *Makromol. Chem.*, **86**, 168 (1965).
3799. Grubisic, Z., and Benoit, H., *C. R. Acad. Sci. C*, **266**, 1275 (1968).
3800. Grubisic, Z., Reibel, L., and Spach, G., *C. R. Acad. Sci. C*, **264**, 1690 (1967).
3801. Grubisic, Z., Rempp, P., and Benoit, H., *J. Polym. Sci. B*, **5**, 753 (1967).
3802. Grubisic-Gallot, Z., Marias, L., and Benoit, H., *J. Polym. Sci. A2*, **14**, 959 (1976).
3803. Grubisic-Gallot, Z., Picot, M., Gramain, P., and Benoit, H., *J. Appl. Polym. Sci.*, **16**, 2931 (1972).
3804. Gruner, C. L., Wunderlich, B., and Bopp, R. C., *J. Polym. Sci. A2*, **7**, 2099 (1969).
3805. Guaita, M., and Chiantore, O., *Makromol. Chem.*, **176**, 185 (1975).
3806. Gubler, M., and Kovacs, J., *J. Polym. Sci.*, **34**, 551 (1959).
3807. Gucker, F. T., Chiu, G., Osborne, E. C., and Tuma, J., *J. Colloid Interface Sci.*, **27**, 395 (1968).
3808. Gudzinowicz, B. J., and Alden, K., *J. Chromatogr. Sci.*, **9**, 65 (1971).
3809. Guenet, J. M., Gallot, Z., Picot, C., and Benoit, H., *J. Appl. Polym. Sci.*, **21**, 2181 (1977).
3810. Guhaniyogi, S., Ibin, K. J., and Lillie, E. D., *Polymer*, **18**, 345 (1977).
3811. Guilbaut, L. J., and Harwood, H. J., *J. Polym. Sci. A1*, **12**, 1461 (1974).
3812. Guillet, J. E., Combs, R. L., Slonaker, D. F., and Coover, H. W., Jr, *J. Polym. Sci.*, **47**, 307 (1960).
3813. Guillet, J. E., Combs, R. L., Slonaker, D. F., Summers, J. T., and Coover, H. W., Jr, *SPE Trans.* **2**, 164 (1964).

774

3814. Guillet, J. E., Dan, E., Mitchell, R. S., and Valleau, J. P., *Nature*, **234**, 135 (1971).
3815. Guillet, J., Seytre, G., Chatain, D., Lacabanne, C., and Monpagnes, J. C., *J. Polym. Sci. A2*, **15**, 541 (1977).
3816. Guillet, J. E., and Stein, A. N., *Macromolecules*, **3**, 102 (1970).
3817. Guillet, J. E., Wooten, W. C., and Combe, R. L., *J. Appl. Polym. Sci.*, **3**, 61 (1960).
3818. Gupta, D., and Forsman, W. C., *Macromolecules*, **2**, 304 (1969).
3819. Gupta, S. K., and Forsman, W. C., *Macromolecules*, **5**, 779 (1972).
3820. Gupta, S. K., and Forsman, W. C., *Macromolecules*, **6**, 285 (1973).
3821. Gupta, S. K., and Forsman, W. C., *Macromolecules*, **7**, 853 (1974).
3822. Gupta, S. K., Kumar, A., and Deo, S. R., *Polymer*, **19**, 895 (1978).
3823. Gupta, S. K., Marchal, E., and Burchard, W., *Macromolecules*, **8**, 843 (1975).
3824. Gupta, V. D., Treviono, S., and Boutin, H., *J. Chem. Phys.*, **48**, 3008 (1968).
3825. Gurato, G., Fichera, A., Grandi, F. Z., Zannetti, R., and Canal, P., *Makromol. Chem.*, **175**, 953 (1974).
3826. Gurato, G., Gaidano, D., and Zannetti, R., *Makromol. Chem.*, **179**, 231 (1978).
3827. Gurnee, E. G., *J. Polym. Sci. A2*, **5**, 799 (1967).
3828. Gurnee, E. F., *J. Appl. Phys.*, **25**, 1232 (1954).
3829. Gurran, B. T., O'Brian, R. J., and Anderson, D. H., *Anal. Chem.*, **42**, 115 (1970).
3830. Gutowsky, H. S., and Pake, G. E., *J. Chem. Phys.*, **18**, 162 (1950).
3831. Guttman, C. M., and DiMarzio, E. A., *Macromolecules*, **3**, 681 (1970).
3832. Guzmán, G. M., *Progr. High Polym.*, **1**, 113 (1961).
3833. Guzmán, G. M., and Fatou, J. M. G., *Anales Real. Soc. Esp. Fis. Quim.*, **54B**, 263 (1958).
3834. Guzman, G. M., and Fatou, J. M. G., *Anales Real. Soc. Esp. Fis. Quim.*, **54B**, 609 (1958).
3835. Haagen-Smith, A. J., Brunelle, M. F., and Haagen-Smith, J. W., *Rubber Chem. Technol.*, **32**, 1134 (1959).
3836. Haberland, G. G., and Carmichael, J. B., *J. Polym. Sci. C*, **14**, 291 (1966).
3837. Hackathron, M. J., and Brock, M. J., *J. Polym. Sci. B*, **8**, 617 (1970).
3838. Hackathorn, M. J., and Brock, M. J., *ACS Polym. Preprints*, **14**, 42 (1973).
3839. Haddon, W. F., Jr, Porter, R. S., and Johnson, J. F., *J. Appl. Polym. Sci.*, **8**, 1371 (1964).
3840. Hadjichristidis, N., *Polymer*, **16**, 848 (1975).
3841. Hadjichristidis, N., *Makromol. Chem.*, **178**, 1463 (1977).
3842. Hadjichristidis, N., and Roovers, J. E. L., *J. Polym. Sci. A2*, **12**, 2521 (1974).
3843. Hadjichristidis, N., and Roovers, J. E. L., *J. Polym. Sci. A2*, **12**, 2551 (1974).
3844. Hadjichristidis, N., and Roovers, J., *J. Polym. Sci. A2*, **16**, 851 (1978).
3845. Haeringer, A., Hild, G., Rempp, P., and Benoit, H., *Makromol. Chem.*, **169**, 249 (1973).
3846. Hager, N. E., Jr, *Rev. Sci. Instrum.*, **35**, 618 (1964).
3847. Hager, N. E., Jr, *Rev. Sci. Instrum.*, **42**, 678 (1971).
3848. Hager, N. E., Jr, *J. Polym. Sci. C*, **43**, 77 (1973).
3849. Heigh, J., Ali, A. S. M., and Davies, G. J., *Polymer*, **16**, 714 (1975).
3850. Haken, J. K., and Werner, R. L., *Brit. Polym. J.*, **3**, 263 (1971).
3851. Haken, J. K., and Werner, R. L., *Brit. Polym. J.*, **4**, 147 (1972).
3852. Haken, J. K., and Werner, R. L., *Brit. Polym. J.*, **5**, 451 (1973).
3853. Halasz, I., and Walking, P., *J. Chromatogr. Sci.*, **7**, 129 (1969).
3854. Hall, I. H., and Pass, M. G., *Polymer*, **17**, 807 (1976).
3855. Haller, W. J., *Nature*, **206**, 693 (1965).
3856. Haller, W. J., *Chem. Phys.*, **42**, 686 (1965).
3857. Halpaap, H., and Klatyk, K., *J. Chromatogr.*, **33**, 80 (1968).
3858. Hama, Y., Hosono, K., Furui, Y., and Shinohara, K., *J. Polym. Sci. A1*, **9**, 1411 (1971).
3859. Hama, Y., Nishi, K., Watanabe, K., and Shinohara, K., *J. Polym. Sci. A2*, **12**, 1109 (1974).
3860. Hama, Y., Ooi, T., Shiotsubo, M., and Shinohara, K., *Polymer*, **15**, 787 (1974).
3861. Hamada, F., Fujisawa, K., and Nakajima, A., *Polym. J.*, **4**, 316 (1973).
3862. Hamada, F., Wunderlich, B., Sumida, T., Hayashi, S., and Nakajima, A., *J. Phys. Chem.*, **72**, 178 (1968).
3863. Hamazaki, T., Kanchiku, Y., Handa, R., and Izumi, M., *J. Appl. Polym. Sci.*, **21**, 1569 (1977).

3864. Hamburger, J. W., *Text. Res. J.*, **18**, 705 (1948).
3865. Hamermesh, C. L., and Dynes, P. J., *J. Polym. Sci. B*, **13**, 663 (1975).
3866. Hamielec, A. E., and Ray, W. H., *J. Appl. Polym. Sci.*, **13**, 1319 (1969).
3867. Hamilton, J. G., Ivin, K. J., Kuan-Essig, L. C., and Watt, P., *Polymer*, **16**, 763 (1975).
3868. Hamilton, J. G., Ivin, K. J., Kuan-Essig, L. C., and Watt, P., *Macromolecules*, **9**, 67 (1976).
3869. Hammel, J. J., Mickey, J., and Golub, H. R., *J. Colloid Interface Sci.*, **27**, 329 (1968).
3870. Hammers, W. E., Bos, B. C., Vaas, L. H., Loomans, Y. J. W. A., and Deligny, G. L., *J. Polym. Sci. A2*, **13**, 401 (1975).
3871. Hammers, W. E., and De Ligny, C. L., *J. Polym. Sci. A2*, **12**, 2065 (1974).
3872. Hammers, W. E., De Ligny, C. L., and Vaas, L. H., *J. Polym. Sci. A2*, **11**, 499 (1973).
3873. Hammon, H. G., Ernst, K., and Newton, J. C., *J. Appl. Polym. Sci.*, **21**, 1989 (1977).
3874. Han, C. D., *J. Appl. Polym. Sci.*, **15**, 2567, 2579, 2591 (1971).
3875. Han, C. D., and Charles, M., *Polym. Eng. Sci.*, **10**, 148 (1970).
3876. Han, C. D., Charles, M., and Philippoff, W., *Trans. Soc. Rheol.*, **14**, 393 (1970).
3877. Han, C. D., and Kim, K. U., *Polym. Eng. Sci.*, **11**, 395 (1971).
3878. Han, C. D., Kim, K. U., Siskovic, N., and Huang, C. R., *J. Appl. Polym. Sci.*, **17**, 95 (1973).
3879. Han, C. D., and Mozer, B., *Macromolecules*, **10**, 44 (1977).
3880. Han, C. D., and Villamizar, C. A., *J. Appl. Polym. Sci.*, **22**, 1677 (1978).
3881. Han, C. D., Villamizar, C. A., Kim, Y. W., and Chen, S. J., *J. Appl. Polym. Sci.*, **21**, 353 (1977).
3882. Han, C. D., Yu, T. C., and Kim, K. U., *J. Appl. Polym. Sci.*, **15**, 1149, 1163 (1971).
3883. Hann, N. D., *J. Polym. Sci. A1*, **15**, 1331 (1977).
3884. Hanna, F. F., and Abou-Bakr, A. F., *Brit. Polym. J.*, **5**, 49 (1973).
3885. Hanna, J. G., and Siggia, S., *J. Polym. Sci.*, **56**, 297 (1962).
3886. Hannon, M. J., Boerio, F. J., and Koenig, J. L., *J. Chem. Phys.*, **50**, 2829 (1969).
3887. Hannon, M. J., and Koenig, J. L., *J. Polym. Sci. A2*, **7**, 1085 (1969).
3888. Hansen, C. M., *J. Paint. Technol.*, **39**, 104 (1967).
3889. Hansen, C. M., *Ind. Eng. Chem. Prod. Res. Dev.*, **8**, 2 (1969).
3890. Hansen, C. M., and Sather, H. A., *J. Appl. Polym. Sci.*, **8**, 2479 (1964).
3891. Hansen, C. M., and Skarrup, K., *J. Paint. Technol.*, **39**, 511 (1967).
3892. Hansen, D., and Bernier, G. A., *Polym. Eng. Sci.*, **12**, 204 (1972).
3893. Hansen, D., and Ho, C. C., *J. Polym. Sci. A*, **3**, 659 (1965).
3894. Hansen, D., Kantayya, R. C., and Ho, C. C., *Polym. Eng. Sci.*, **6**, 260 (1966).
3895. Hansen, L. C., and Sievers, R. E., *J. Chromatogr.*, **99**, 123 (1974).
3896. Hansen, R. H., Pascale, J. V., DeBenedictis, T., and Rentzepis, P. M., *J. Polym. Sci. A*, **3**, 2205 (1965).
3897. Hazell, J. E., Prince, L. A., and Stapelfeldt, H. E., *J. Polym. Sci. C*, **21**, 43 (1968).
3898. Happ, G. P., and Maier, D. P., *Anal. Chem.*, **36**, 1678 (1964).
3899. Happey, F., and Keighley, J. H., *J. Sci. Instrum.*, **35**, 116 (1958).
3900. Hara, T., and Seto, T., *Rep. Progr. Polym. Phys. Japan*, **12**, 189 (1969).
3901. Harada, A., Furue, M., and Nozakura, S., *Macromolecules*, **10**, 676 (1977).
3902. Harland, W. G., *J. Textile Inst. Trans.*, **46**, 483 (1955).
3903. Harland, W. G., Khadr, M. M., and Peters, R. H., *Polymer*, **13**, 13 (1972).
3904. Harmathy, T. Z., *J. Appl. Polym. Sci.*, **35**, 1190 (1964).
3905. Harmon, D. J., *J. Polym. Sci. C*, **8**, 243 (1965).
3906. Harmon, D. J., *ACS Div. Petrol. Chem. Preprints*, **15**, A29 (1970).
3907. Harper, B. G., and Moore, J. C., *Ind. Eng. Chem.*, **49**, 411 (1957).
3908. Harpst, J. A., Krasna, A. I., and Zimm, B. H., *Biopolymers*, **6**, 585 (1968).
3909. Harpst, J. A., Krasna, A. I., and Zimm, B. H., *Biopolymers*, **6**, 595 (1968).
3910. Harrah, L. A., *J. Chem. Phys.*, **56**, 385 (1972).
3911. Harrington, R. E., and Pecoraro, P. G., *J. Polym. Sci. A1*, **4**, 475 (1966).
3912. Harrington, R. E., and Zimm, B. H., *ACS Polym. Preprints*, **6**, 346 (1965).
3913. Harris, I., and Miller, R. G. J., *J. Polym. Sci.*, **7**, 377 (1951).
3914. Harrison, I. R., *J. Polym. Sci. A2*, **11**, 991 (1973).
3915. Hart, V. E., *J. Polym. Sci.*, **17**, 207 (1955).
3916. Hart, W. W., Painter, P. C., Koenig, J. L., and Coleman, M. M., *J. Appl. Spectrosc.*, **31**, 220 (1977).
3917. Hartley, A. J., Leung, Y. K., Booth, C., and Shepherd, I. W., *Polymer*, **17**, 354 (1976).

3918. Hartley, A. J., Leung, Y. K., McMahon, J., Booth, C., and Shepherd, I. W., *Polymer*, **18**, 337 (1977).
3919. Hartley, A. J., and Shepherd, I. W., *J. Polym. Sci. A2*, **14**, 643 (1976).
3920. Hartley, F. D., *J. Polym. Sci.*, **34**, 397 (1959).
3921. Hartmann, H. A., *Kolloid Z.*, **192**, 1 (1964).
3922. Hartmann, W., and Klesper, E., *J. Polym. Sci. B*, **15**, 713 (1977).
3923. Harvey, M. R., Stewart, J. E., and Achhammer, B. G., *J. Res. Nat. Bur. Stand.*, **56**, 225 (1956).
3924. Harwood, H. J., *J. Polym. Sci. C*, **25**, 37 (1968).
3925. Harwood, C., Wostenholm, G. H., Yates, B., and Badami, D. V., *J. Polym. Sci. A2*, **16**, 759 (1978).
3926. Hase, Y., and Geil, P. H., *Polymer J.*, **2**, 560 (1971).
3927. Haseley, E. A., *J. Polym. Sci.*, **35**, 309 (1959).
3928. Hashimoto, T., Kawasaki, H., and Kawai, H., *J. Polym. Sci. A2*, **16**, 271 (1978).
3929. Hashimoto, T., Murakami, Y., Haysahi, N., and Kawai, H., *Polymer J.*, **6**, 132 (1974).
3930. Hashimoto, T., Murakami, Y., and Kawai, H., *J. Polym. Sci. A2*, **13**, 1613 (1975).
3931. Hashimoto, T., Nagatoshi, K., Toda, A., Hasegawa, H., and Kawai, H., *Macromolecules*, **7**, 364 (1974).
3932. Hashimoto, T., Nagatoshi, K., Toda, A., and Kawai, H., *Polymer*, **17**, 1063 (1976).
3933. Hashimoto, T., Nagatoshi, K., Toda, A., and Kawai, H., *Polymer*, **17**, 1075 (1976).
3934. Hashimoto, T., Prud'homme, R. E., Keedy, D. A., and Stein, R. S., *J. Polym. Sci. A2*, **11**, 693 (1973).
3935. Hashimoto, T., Prud'homme, R. E., and Stein, R. S., *J. Polym. Sci. A2*, **11**, 709 (1973).
3936. Hashimoto, T., and Stein, R. S., *J. Polym. Sci. A2*, **8**, 1503 (1970).
3937. Hashimoto, T., Todo, A., Murakami, Y., and Kawai, H., *J. Polym. Sci. A2*, **15**, 501 (1977).
3938. Hashiyama, M., Gaylord, R., and Stein, R. S., *Makromol. Chem.*, Suppl. 1, 579 (1975).
3939. Hashiyama, M., and Stein, R. S., *J. Polym. Sci. A2*, **16**, 29 (1978).
3940. Hassan, A. M., *J. Polym. Sci. A2*, **12**, 655 (1974).
3941. Hastings, G. W., and Peaker, F. W., *J Polym. Sci.*, **36**, 351 (1959).
3942. Hatada, K., Ishikawa, H., Kitayama, T., and Yuki, H., *Makromol. Chem.*, **178**, 2753 (1977).
3943. Hatada, K., Kitayama, T., Okamoto, Y., Ohta, K., Umemura, Y., and Yuki, H., *Makromol. Chem.*, **179**, 485 (1978).
3944. Hatada, K., Ohta, K., Okamoto, Y., Kitayama, Y., Umemura, Y., and Yuki, H., *J. Polym. Sci. B*, **14**, 531 (1976).
3945. Hatada, K., Ohta, K., and Yuki, H., *J. Polym. Sci. B*, **5**, 225 (1967).
3946. Hatada, K., Tanaka, Y., Terawaki, Y., and Okuda, H., *Polymer J.*, **5**, 327 (1973).
3947. Hatada, K., Terawaki, Y., Okuda, H., Tanaka, Y., and Sato, H., *J. Polym. Sci. B*, **12**, 305 (1974).
3948. Hatchard, C. G., and Parker, C. A., *Proc. Roy. Soc. A*, **235**, 518 (1956).
3949. Hathaway, C. E., and Nielsen, J. R., *J. Chem. Phys.*, **41**, 2203 (1964).
3950. Haugen, G. R., and Marcus, R. J., *Appl. Opt.*, **3**, 1049 (1964).
3951. Havrilak, S., *Polymer*, **9**, 289 (1968).
3952. Hay, J. N., *J. Polym. Sci. A1*, **8**, 1201 (1970).
3953. Hay, J. N., *J. Polym. Sci. B*, **14**, 543 (1976).
3954. Hay, J. N., *Makromol. Chem.*, **177**, 2559 (1976).
3955. Hay, J. N., *Polymer*, **19**, 1224 (1978).
3956. Hay, J. N., and Booth, A., *Brit. Polym. J.*, **4**, 18 (1972).
3957. Hay, J. N., and Booth, A., *Brit. Polym. J.*, **4**, 27 (1972).
3958. Hay, J. N., Fitzgerald, P. A., and Wiles, M., *Polymer*, **17**, 1015 (1976).
3959. Hay, J. N., and Przekop, Z. J., *J. Polym. Sci. A2*, **16**, 81 (1978).
3960. Hayakawa, R., Tanabe, Y., and Wada, Y., *J. Macromol. Sci. B*, **8**, 445 (1973).
3961. Hayashi, H., Hamada, F., and Nakajima, A., *Macromolecules*, **7**, 960 (1974).
3962. Hayashi, H., Hamada, F., and Nakajima, A., *Macromolecules*, **9**, 543 (1976).
3963. Hayashi, H., Hamada, F., and Nakajima, A., *Makromol. Chem.*, **178**, 827 (1977).
3964. Hayduk, W., and Bromfield, H. A., *J. Appl. Polym. Sci.*, **22**, 149 (1978).
3965. Hayes, M. J., and Park, G. S., *Trans. Faraday Soc.*, **51**, 1134 (1955).
3966. Hayes, M. J., and Park, G. S., *Trans. Faraday Soc.*, **52**, 949 (1956).
3967. Hearle, J. W. S., and Lomas, B., *J. Appl. Polym. Sci.*, **21**, 1103 (1977).

3968. Hearst, J. E., and Vinograd, J., *Proc. Nat. Acad. Sci., U.S.A.*, **47**, 999 (1961).
3969. Heatley, F., *Polymer*, **13**, 218 (1972).
3970. Heatley, F., *Polymer*, **16**, 493 (1975).
3971. Heatley, F., and Begum, A., *Polymer*, **17**, 399 (1976).
3972. Heatley, F., and Bovey, F. A., *Macromolecules*, **1**, 301 (1968).
3973. Heatley, F., and Bovey, F. A., *Macromolecules*, **2**, 241 (1969).
3974. Heatley, F., and Cox, M. K., *Polymer*, **18**, 225 (1977).
3975. Heatley, F., Salovey, R., and Bovey, F. A., *Macromolecules*, **2**, 619 (1969).
3976. Heatley, F., and Scrivens, J. H., *Polymer*, **16**, 489 (1975).
3977. Heatley, F., and Zambelli, A., *Macromolecules*, **2**, 618 (1969).
3978. Hebert, J. J., Carra, J. H., Esposito, C. R., and Rollins, M. L., *Text. Res.*, **43**, 260 (1973).
3979. Hebert, J. J., and Muller, L., *J. Appl. Sci.*, **18**, 3373 (1974).
3980. Hedvig, P., *J. Polym. Sci. A1*, **7**, 1145 (1969).
3981. Heffelfinger, C. J., and Lippert, E. L., *J. Appl. Polym. Sci.*, **15**, 2699 (1971).
3982. Heikens, D., and Barentsen, W., *Polymer*, **18**, 69 (1977).
3983. Heine, S., Kratky, O., Porod, G., and Schmitz, P. J., *Makromol. Chem.*, **44/46**, 682 (1961).
3984. Heine, S., Kratky, O., and Rappert, J., *Makromol. Chem.*, **56**, 150 (1952).
3985. Heitz, W., *Makromol. Chem.*, **127**, 113 (1969).
3986. Heitz, W., and Bier, P., *Makromol. Chem.*, **176**, 657 (1975).
3987. Heitz, W., Bömer, B., and Ullner, H., *Makromol. Chem.*, **121**, 102 (1969).
3988. Heitz, F., Cary, P. D., and Crane-Robinson, C., *Macromolecules*, **10**, 526 (1977).
3989. Hefele, J. R., Lundberg, J. L., and Salovey, R., *Rev. Sci. Instrum.*, **33**, 1256 (1962).
3990. Helfand, E., *Macromolecules*, **11**, 682 (1978).
3991. Helfand, E., and Lauritzen, J. I., Jr, *Macromolecules*, **6**, 631 (1973).
3992. Helfand, E., and Tonelli, A. E., *Macromolecules*, **7**, 832 (1974).
3993. Heller, J., and Lyman, D. J., *J. Polym. Sci. B*, **1**, 317 (1963).
3994. Heller, J., Tiezen, D. O., and Parkinson, D. B., *J. Polym. Sci. A1*, 125 (1963).
3995. Heller, W., *J. Phys. Chem.* **69**, 1123 (1965).
3996. Heller, W., *J. Polym. Sci. A2*, **4**, 209 (1966).
3997. Heller, W., and Pugh, T. L., *J. Polym. Sci.*, **47**, 203 (1960).
3998. Hellfritz, H., *Makromol. Chem.*, **7**, 184 (1951).
3999. Hellfritz, H., *Kunststoffe*, **50**, 502 (1960).
4000. Hellfritz, H., Krämer, H., and Schmieder, W., *Kunststoffe*, **49**, 391 (1959).
4001. Hellmuth, E., and Wunderlich, B., *J. Appl. Phys.*, **36**, 3039 (1965).
4002. Hellwege, K. H., Henning, J., and Knappe, W., *Kolloid Z. Z. Polym.*, **188**, 121 (1963).
4003. Hellwege, K. H., Knappe, W., and Wetzel, W., *Kolloid Z.*, **180**, 126 (1962).
4004. Hemming, D. F., Collins, J. D., and Datyner, A., *J. Appl. Polym. Sci.*, **20**, 597 (1976).
4005. Henderson, J. F., and Hulme, J. M., *J. Appl. Polym. Sci.*, **11**, 2349 (1967).
4006. Hendra, P. J., *J. Mol. Spectr.*, **28**, 119 (1968).
4007. Hendra, P. J., Jobic, H. P., and Holland-Moritz, K., *J. Polym. Sci. B*, **13**, 365 (1975).
4008. Hendra, P. J., and Marsden, E. P., *J. Polym. Sci. B*, **15**, 259 (1977).
4009. Hendra, P. J., Marsden, E. P., Cudby, E. A., and Willis, H. A., *Makromol. Chem.*, **176**, 2443 (1975).
4010. Hendra, P. J., Peacock, C. J., Watson, D. S., and Gall, M. J., *Appl. Spectr.*, **25**, 423 (1971).
4011. Hendrickson, J. G., *Anal. Chem.*, **36**, 126 (1964).
4012. Hendrickson, J. G., *Anal. Chem.*, **40**, 49 (1968).
4013. Hendrickson, J. G., *J. Polym. Sci. A2*, **6**, 1903 (1968).
4014. Hendrickson, J. G., and More, J. C., *J. Polym. Sci. A1*, **4**, 167 (1966).
4015. Hendrickson, J. R., and Bray, P. J., *J. Magnet. Reson.*, **9**, 341 (1972).
4016. Hendrik, C., Whitting, D. A., and Woodward, A. E., *Macromolecules*, **4**, 571 (1971).
4017. Hendrix, J., Kugler, M., Gnädig, K., and De Maeyer, L., to be published.
4018. Hendrix, J., Saleh, B., Gnädig, K., and De Maeyer, L., *Polymer*, **18**, 10 (1977).
4019. Hendus, H., and Illers, K. H., *Kunststoffe*, **57**, 193 (1967).
4020. Hendus, H., and Schnell, G., *Kunststoffe*, **51**, 69 (1961).
4021. Henry, A. W., and Safford, G. J., *J. Polym. Sci. A2*, **7**, 433 (1969).
4022. Henry, P. M., *J. Polym. Sci.*, **36**, 3 (1959).
4023. Herdan, G., *J. Polym. Sci.*, **10**, 1 (1953).

4024. Herkstroeter, W. G., Lamola, A. A., and Hammond, G. S., *J. Amer. Chem. Soc.*, **86**, 4537 (1964).
4025. Hermann, G., and Weil, G., *Macromolecules*, **8**, 171 (1975).
4026. Hermans, J. J., *J. Polym. Sci. A2*, **6**, 1217 (1968).
4027. Hermans, J. J., and Ende, H. A., *J. Polym. Sci. C*, **1**, 161 (1963).
4028. Hermans, J. J., and Ende, H. A., *J. Polym. Sci. C*, **4**, 519 (1963).
4029. Hermans, J. J., Hermans, P. H., Vermaas, D., and Weidinger, A., *Rec. Trav. Chim.*, **65**, 427 (1946).
4030. Hermans, J. J., and Levinson, S., *J. Opt. Soc. Amer.*, **41**, 460 (1951).
4031. Hermans, P. H., Heikens, D., and Weidinger, A., *J. Polym. Sci.*, **35**, 145 (1959).
4032. Hermans, P. H., and Platzek, P., *Kolloid Z.*, **88**, 68 (1939).
4033. Hermans, P. H., and Weidinger, A., *J. Appl. Phys.*, **19**, 491 (1948).
4034. Hermans, P. H., and Weidinger, A., *J. Polym. Sci.*, **4**, 135 (1949).
4035. Hermans, P. H., and Weidinger, A., *J. Polym. Sci.*, **4**, 709 (1949).
4036. Hermans, P. H., and Weidinger, A., *J. Polym. Sci.*, **5**, 565 (1950).
4037. Hermans, P. H., and Weidinger, A., *Makromol. Chem.*, **44/46**, 24 (1961).
4038. Hermans, P. H., and Weidinger, A., *Makromol. Chem.*, **50**, 98 (1961).
4039. Herron, J. T., and Schiff, H. I., *Can. J. Chem.*, **36**, 1159 (1958).
4040. Hershey, H. C., and Zakin, J. L., *Chem. Eng. Sci.*, **22**, 1847 (1967).
4041. Hert, M., and Strazielle, C., *Europ. Polym. J.*, **9**, 543 (1973).
4042. Hert, M., and Strazielle, C., *Makromol. Chem.*, **175**, 2149 (1974).
4043. Hert, M., and Strazielle, C., *Makromol. Chem.*, **176**, 1849 (1975).
4044. Hert, M., Strazielle, C., and Benoit, H., *Makromol. Chem.*, **172**, 169 (1973).
4045. Hess, M., and Kratz, R. F., *J. Polym. Sci. A2*, **4**, 731 (1966).
4046. Hettinger, J., and Hubbard, J., *Amer. Lab.*, **6** (2), 99 (1974).
4047. Heuvel, H. M., Huisman, R., and Lind, K. C. J. B., *J. Polym. Sci. A2*, **14**, 921 (1976).
4048. Heymann, D., and Keur, E., *J. Sci. Instrum.*, **42**, 121 (1965).
4049. Hibi, S., Maeda, M., Kubota, H., and Miura, T., *Polymer*, **18**, 137 (1977).
4050. Hibi, S., Maeda, M., Kubota, H., and Miura, T., *Polymer*, **18**, 143 (1977).
4051. Hickman, J. J., and Ikeda, R. M., *J. Polym. Sci. A2*, **11**, 1713 (1973).
4052. Higgins, J. S., Allen, G., and Brier, P. N., *Polymer*, **13**, 157 (1972).
4053. Higgs, P. W., *Proc. Roy. Soc. A*, **220**, 472 (1953).
4054. Higgs, P. W., *J. Chem. Phys.*, **23**, 1450 (1955).
4055. Hildebrand, J. H., *J. Chem. Phys.*, **15**, 225 (1947).
4056. Hileman, F. D., Sievers, R. E., Hess, G. G., and Ross, W. D., *Anal. Chem.*, **45**, 1126 (1973).
4057. Hill, F. N., and Brown, A., *Anal. Chem.*, **22**, 562 (1950).
4058. Hiller, I. H., *J. Polym. Sci. A1*, **3**, 3067 (1965).
4059. Hindeleh, A. M., and Johnson, D. J., *Polymer*, **11**, 666 (1970).
4060. Hindeleh, A. M., and Johnson, D. J., *Polymer*, **13**, 27 (1972).
4061. Hindeleh, A. M., and Johnson, D. J., *Polymer*, **15**, 697 (1974).
4062. Hindeleh, A. M., and Johnson, D. J., *Polymer*, **19**, 27 (1978).
4063. Hino, T., *J. Appl. Phys.*, **46**, 1956 (1975).
4064. Hirata, E., Ijitsu, T., Soen, T., Hashimoto, T., and Kawai, H., *Polymer*, **16**, 249 (1975).
4065. Hirayama, F., *J. Chem. Phys.*, **42**, 3163 (1965).
4066. Hiroka, H., Kanda, H., and Nakaguchi, K., *J. Polym. Sci. B*, **1**, 701 (1963).
4067. Hiroka, H., and Lee, W. Y., *Macromolecules*, **11**, 624 (1978).
4068. Hirota, S., Saito, S., and Nakajima, T., *Kolloid Z.*, **213**, 109 (1966).
4069. Hlavacek, B., Rollin, L. A., and Schreiber, H. P., *Polymer*, **17**, 81 (1976).
4070. Ho, F. F. L., *J. Polym. Sci. B*, **9**, 491 (1969).
4071. Hoashi, K., and Andrews, R. D., *J. Polym. Sci. C*, **38**, 387 (1972).
4072. Hobbs, S. Y., *J. Polym. Sci.*, *A2*, **16**, 1321 (1978).
4073. Hobbs, S. Y., and Pratt, C. F., *Polymer*, **16**, 462 (1975).
4074. Hoch, M. J. R., Bovey, F. A., Davis, D. D., Douglass, D. C., Falcone, D. R., McCall, D. W., and Slichter, W. P., *Macromolecules*, **4**, 712 (1971).
4075. Höcker, H., and Flory, P. J., *Trans. Faraday Soc.*, **67**, 2270 (1971).
4076. Höcker, H., and Musch, R., *Makromol. Chem.*, **175**, 1395 (1974).
4077. Höcker, H., Shih, H., and Flory, P. J., *Trans. Faraday Soc.*, **67**, 2275 (1971).
4078. Hodge, I. M., and Eisenberg, A., *Macromolecules*, **11**, 283, 289 (1978).
4079. Hoff, E. A. W., Robinson, D. W., and Willbourn, A. H., *J. Polym. Sci.*, **18**, 161 (1955).

4080. Hoffman, J. D., *J. Chem. Phys.*, **28**, 1192 (1958).
4081. Hoffman, J. D., Frolen, L. J., Ross, G. S., and Lauritzen, J. I., Jr, *J. Res. Natl. Bur. Stand. A*, **79**, 671 (1975).
4082. Hoffman, J. D., Lauritzen, J. I., Jr, Passaglia, E., Ross, G. S., Frolen, L. J., and Weeks, J. J., *Kolloid Z. Z. Polym.*, **231**, 567 (1969).
4083. Hoffman, J. D., and Zimm, B. H., *J. Polym. Sci.*, **15**, 405 (1955).
4084. Hoffman, M., *Makromol. Chem.*, **57**, 96 (1962).
4085. Hoffman, M., *Makromol. Chem.*, **174**, 167 (1973).
4086. Hoffman, M., *Makromol. Chem.*, **175**, 613 (1974).
4087. Hoffman, M., and Kuhn, R., *Makromol. Chem.*, **174**, 149 (1973).
4088. Hoffman, M., and Unbehend, M., *Makromol. Chem.*, **88**, 256 (1965).
4089. Hoffman, M., and Urban, H., *Makromol. Chem.*, **178**, 1683 (1977).
4090. Hoffman, M., and Urban, H., *Makromol. Chem.*, **178**, 2661 (1977).
4091. Hoffman, R., and Knappe, W., *Kolloid Z. Z. Polym.*, **240**, 784 (1970).
4092. Hoffman, R., and Knappe, W., *Kolloid Z. Z. Polym.*, **247**, 763 (1971).
4093. Holdworth, P. J., and Keller, A., *J. Polym. Sci. B*, **5**, 605 (1967).
4094. Holleman, T., *Rheol. Acta*, **10**, 194 (1971).
4095. Holleran, P. M., and Billmeyer, F. W., Jr, *J. Polym. Sci. B*, **6**, 137 (1968).
4096. Holmes, D. R., Bunn, C. W., and Smith, D. J., *J. Polym. Sci.*, **17**, 159 (1955).
4097. Holoubek, J., *J. Polym. Sci. A2*, **11**, 683 (1973).
4098. Holoubek, J., *J. Polym. Sci. A2*, **16**, 1665 (1978).
4099. Holt, C., and Sellen, D. B., *J. Polym. Sci. A2*, **13**, 1 (1975).
4100. Holtrup, W., *Makromol. Chem.*, **178**, 2335 (1977).
4101. Holtzer, A., and Rice, S. A., *J. Amer., Chem. Soc.*, **79**, 4847 (1957).
4102. Holtzmann, G., *Science*, **111**, 550 (1950).
4103. Homma, T., Kawahara, K., Fujita, H., and Heda, M., *Makromol. Chem.*, **67**, 132 (1963).
4104. Hommel, H., Facchini, L., Legrand, A. P., and Lecourtier, J., *Europ. Polym. J.*, **14**, 803 (1978).
4105. Hon, N. S., *J. Polym. Sci. A1*, **13**, 955, 1347, 1933, 2363, 2641, 2653 (1975).
4106. Hon, N. S., *J. Polym. Sci. B*, **14**, 225 (1976).
4107. Hon, N. S., *J. Appl. Polym. Sci.*, **19**, 2789 (1976).
4108. Hon, N. S., *J. Polym. Sci. A1*, **14**, 2497, 2513 (1976).
4109. Hon, N. S., *J. Polym. Sci. A1*, **15**, 725 (1977).
4110. Honaker, C. B., and Horton, A. D., *J. Gas Chromatogr.*, **3**, 396 (1965).
4111. Hong, S. D., Chang, C., and Stein, R. S., *J. Polym. Sci. A2*, **13**, 1447 (1975).
4112. Honjo, G., and Watanabe, M., *Nature*, **181**, 326 (1958).
4113. Hoogervorst, C. J. P., Smith, J. A. M., and Staverman, A., *J. Polym. Sci. A2*, **16**, 297 (1978).
4114. Hookway, H. T., and Townsend, R., *J. Chem. Soc.*, **1952**, 3190 (1952).
4115. Hope, P., Anderson, R., and Bloss, A. S., *Brit. Polym. J.*, **5**, 67 (1973).
4116. Hope, P., Stark, B. P., and Zahir, S. A., *Brit. Polym. J.*, **5**, 363 (1973).
4117. Hopfenberg, H. B., and Frisch, H. L., *J. Polym. Sci. B*, **7**, 405 (1969).
4118. Hopfenberg, H. B., Holley, R. H., and Stannett, V., *Polym. Eng. Sci.*, **9**, 242 (1969).
4119. Hopfenberg, H. B., Stannett, V. T., and Folk, G. M., *Polym. Eng. Sci.*, **15**, 261 (1975).
4120. Hopfinger, A. J., *Macromolecules*, **4**, 731 (1971).
4121. Hopkins, R. C., *J. Polym. Sci. A2*, **7**, 1907 (1969).
4122. Horaček, H., *Makromol. Chem.*, **179**, 1291 (1978).
4123. Hori, Y., Shimada, S., and Kashiwabara, H., *Polymer*, **18**, 567 (1977).
4124. Hori, Y., Shimada, S., and Kashiwabara, H., *Polymer*, **18**, 1143 (1977).
4125. Horii, F., and Ikada, Y., *J. Polym. Sci. B*, **12**, 27 (1974).
4126. Horii, F., Ikada, Y., and Sakurada, I., *J. Polym. Sci. A1*, **13**, 755 (1975).
4127. Horii, F., and Kitamaru, R., *J. Polym. Sci. A2*, **16**, 265 (1978).
4128. Horii, F., Kitamaru, R., and Suzuki, T., *J. Polym. Sci. B*, **15**, 65 (1977).
4129. Horlicek, G., *Anal. Chem.*, **43**, 61A (1971).
4130. Horn, P., Benoit, H., and Oster, G., *J. Chem. Phys.*, **48**, 530 (1951).
4131. Horvath, C., and Lipsky, S. R., *Anal. Chem.*, **41**, 1227 (1969).
4132. Hosemann, R., and Schramek, W., *J. Polym. Sci.*, **59**, 29 (1962).
4133. Hoskins, R. H., and Pastor, R. C., *J. Appl. Phys.*, **31**, 1506 (1960).
4134. Hostetler, R. E., and Swanson, J. W., *J. Polym. Sci. A1*, **12**, 29 (1974).

4135. Houriet, J. P., Ghiste, P., and Stoeckli, H. F., *Helv. Chim. Acta*, **57**, 851 (1974).
4136. Hovingh, M. E., Thompson, G. H., and Giddings, J. C., *Anal. Chem.*, **42**, 195 (1970).
4137. Howard, G. J., *J. Polym. Sci.*, **59**, S4 (1962).
4138. Howard, G. J., *J. Polym. Sci. A1*, 2667 (1963).
4139. Howell, J. A., *Macromolecules*, **2**, 301 (1969).
4140. Hoy, K. L., *J. Paint Technol.*, **42**, 76 (1970).
4141. Hoyle, C. E., and Guillet, J. E., *J. Polym. Sci. B*, **16**, 185 (1978).
4142. Hoyle, C. E., Nemzek, T. L., Mar, A., and Guillet, J. E., *Makromolecules*, **11**, 429 (1978).
4143. Hsieh, H. W. S., Post, B., and Morawetz, H., *J. Polym. Sci. A2*, 1241 (1976).
4144. Hsu, B. S., and Kwan, S. H., *J. Polym. Sci. A2*, **14**, 1591 (1976).
4145. Hsu, B. S., Kwan, S. H., and Wong, L. W., *J. Polym. Sci. A2*, **13**, 2079 (1975).
4146. Hsu, C. C., and Prausnitz, J. M., *Macromolecules*, **7**, 320 (1974).
4147. Hu, W. K. H., and Zisman, W. A., *Macromolecules*, **4**, 689 (1971).
4148. Huang, R. Y. M., and Jenkins, R. G., *TAPPI*, **52**, 1503 (1969).
4149. Huang, W. N., and Frederick, J. E., *J. Chem. Phys.*, **58**, 4022 (1973).
4150. Huang, W. N., and Frederick, J. E., *Macromolecules*, **7**, 34 (1974).
4151. Huang, W. N., Vrancken, E., and Frederick, J. E., *Macromolecules*, **6**, 58 (1973).
4152. Huang, Y. C., and Eichinger, B. E., *Polymer*, **18**, 55 (1977).
4153. Huang, Y. S., and Koenig, J. L., *J. Appl. Polym. Sci.*, **15**, 1237 (1971).
4154. Hubbel, D. S., and Cooper, S. L., *J. Polym. Sci. A2*, **15**, 1143 (1977).
4155. Hubbell, W. H., Jr, Brandt, H., and Munir, Z. A., *J. Polym. Sci. A2*, **13**, 493 (1975).
4156. Huber, J. F. K., *J. Chromatogr. Sci.*, **7**, 85 (1969).
4157. Huber, J. F. K., *J. Chromatogr. Sci.*, **7**, 172 (1969).
4158. Huber, J. F. K., and Gerritse, R. G., *J. Chromatogr.*, **58**, 137 (1971).
4159. Huber, J. F. K., and Hulsman, J. A. R. J., *Anal. Chim. Acta*, **38**, 305 (1967).
4160. Hudec, P., *Makromol. Chem.*, **178**, 1187 (1977).
4161. Huelck, V., Thomas, D. A., and Sperling, L. H., *Macromolecules*, **5**, 340 (1972).
4162. Huggins, M. L., *Ann. N.Y. Acad. Sci.*, **43**, 1 (1942).
4163. Huggins, M. L., *J. Amer. Chem. Soc.*, **64**, 2716 (1942).
4164. Huggins, M. L., *Makromol. Chem.*, **92**, 260 (1966).
4165. Huggins, M. L., *J. Phys. Chem.*, **74**, 371 (1970).
4166. Huggins, M. L., *Macromolecules*, **4**, 274 (1971).
4167. Huggins, M. L., *J. Phys. Chem.*, **75**, 1255 (1971).
4168. Huggins, M. L., *J. Phys. Chem.*, **80**, 1317 (1976).
4169. Huggins, M. L., Natta, G., Desreux, V., and Mark, H., *J. Polym. Sci.*, **56**, 153 (1962).
4170. Hughes, J., and Rhoden, F., *J. Sci. Instrum.*, **2**, 1134 (1969).
4171. Hughes, R. E., and Lauer, J. L., *J. Chem. Phys.*, **30**, 1165 (1959).
4172. Huglin, M. E., *J. Appl. Polym. Sci.*, **9**, 3963 (1965).
4173. Huglin, M. B., *J. Appl. Polym. Sci.*, **9**, 4003 (1965).
4174. Huglin, M. B., and Richards, R. W., *Polymer*, **17**, 587 (1976).
4175. Huguet, J., and Vert, M., *J. Polym. Sci. A1*, **14**, 1257 (1976).
4176. Huisman, R., and Heuvel, H. M., *J. Polym. Sci. A2*, **14**, 941 (1976).
4177. Hulme, J. M., and McLeod, L. A., *Polymer*, **3**, 153 (1962).
4178. Hummel, D. O., *Kunststoffe*, **55**, 102 (1965).
4179. Hummel, D. O., and Düssel, H. J., *Makromol. Chem.*, **175**, 655 (1974).
4180. Hummel, D. O., Düssel, H. J., Rosen, H., and Rübenacker, K., *Makromol. Chem.*, Suppl. 1, 471 (1975).
4181. Hummel, D. O., Düssel, H. J., and Rübenacker, K., *Makromol. Chem.*, **145**, 267 (1971).
4182. Hummel, K., and Ast, W., *Makromol. Chem.*, **166**, 30 (1973).
4183. Hummel, K., Wewerka, D., Lorber, F., and Zeplichal, G., *Makromol. Chem.*, **166**, 45 (1973).
4184. Hundley, L., Coburn, T., Garwin, E., and Stryer, L., *Rev. Sci. Instrum.*, **38**, 488 (1967).
4185. Hunter, E., and Oakes, W. G., *Trans. Faraday Soc.*, **41**, 56 (1945).
4186. Husby, T. W., and Bair, H. E., *J. Polym. Sci. B*, **5**, 265 (1967).
4187. Huston, D. L., *J. Polym. Sci. A1*, **14**, 713 (1976).
4188. Hutchinson, J. M., and Kovacs, A. J., *J. Polym. Sci. A2*, **14**, 1575 (1976).
4189. Hyde, J. S., *J. Chem., Phys.*, **43**, 1806 (1965).
4190. Hyde, J. S., Chien, J. C. W., and Freed, J. H., *J. Chem. Phys.*, **48**, 4211 (1968).

4191. Hyde, J. S., Smigel, M. D., Dalton, L. R., and Dalton, L. A., *J. Chem. Phys.*, **62**, 1655 (1975).
4192. Hyman, A. S., *Macromolecules*, **8**, 849 (1975).
4193. Ibel, K., *J. Appl. Crystallogr.*, **9**, 296 (1976).
4194. Ibrahim, F., and Elias, H. G., *Makromol. Chem.*, **76**, 1 (1964).
4195. Ichihara, S., Komatsu, A., and Hata, T., *Polymer, J.*, **2**, 650 (1971).
4196. Ichihara, S., Komatsu, A., Tsujita, Y., Nose, T., and Hata, T., *Polym. J.*, **2**, 530 (1971).
4197. Ifft, J. B., Voet, D. H., and Vinograd, J., *J. Phys. Chem.*, **65**, 1138 (1961).
4198. Iguchi, M., *Brit. Polym. J.*, **5**, 195 (1973).
4199. Iguchi, M., Kanetsuma, H., and Kawai, T., *Brit. Polym. J.*, **3**, 177 (1971).
4200. Iguchi, M., and Murase, I., *J. Cryst. Growth*, **24/25**, 596 (1974).
4201. Iguchi, M., and Murase, I., *J. Polym. Sci. A2*, **13**, 1461 (1975).
4202. Iguchi, M., and Murase, I., *Makromol. Chem.*, **176**, 2113 (1975).
4203. Iguchi, M., Murase, I., and Watanabe, K., *Brit. Polym. J.*, **6**, 61 (1974).
4204. Iguchi, M., and Watanabe, Y., *Polymer*, **18**, 265 (1977).
4205. Iida, Y., Yano, S., Aoki, K., and Ohnuma, H., *J. Polym. Sci. B*, **14**, 23 (1976).
4206. Ikada, E., Sugimura, T., Aoyama, T., and Watanabe, T., *Polymer*, **16**, 101 (1975).
4207. Ikada, E., Sugimura, T., and Watanabe, T., *J. Polym. Sci. A2*, **16**, 907 (1978).
4208. Ikeda, M., Suga, H., and Seki, S., *Polymer*, **16**, 634 (1975).
4209. Illers, K. H., *Makromol. Chem.*, **118**, 88 (1968).
4210. Imai, S., *Brit. Polym. J.*, **1**, 161 (1969).
4211. Imhof, L. G., *J. Appl. Polym. Sci.*, **10**, 1137 (1966).
4212. Immergut, E. H., Kollmann, G., and Malatesta, A., *J. Polym. Sci.*, **51**, S57 (1961).
4213. Immergut, E. H., Rollin, S., Salkind, A., and Mark, H. F., *J. Polym. Sci.*, **12**, 439 (1954).
4214. Inagaki, H., and Kamiyama, F., *Macromolecules*, **6**, 107 (1973).
4215. Inagaki, H., Kamiyama, F., and Yagi, T., *Macromolecules*, **4**, 133 (1971).
4216. Inagaki, H., Matsuda, H., and Kamiyama, F., *Macromolecules*, **1**, 520 (1968).
4217. Inagaki, H., Miyamoto, T., and Kamiyama, F., *J. Polym. Sci. B*, **7**, 329 (1969).
4218. Inagaki, H., Miyamoto, T., and Kamiyama, F., *Polym. J.*, **1**, 46 (1970).
4219. Inagaki, H., Suzuki, H., Fuji, M., and Matsuo, T., *J. Phys. Chem.*, **70**, 1718 (1966).
4220. Inagaki, H., Suzuki, H., and Kurata, M., *J. Polym. Sci. C*, **15**, 409 (1966).
4221. Ingalls, R. B., and Wall, L. A., *J. Chem. Phys.*, **35**, 370 (1961).
4222. Ingelstam, E., *Arkiv Fysik*, **9**, 197 (1955).
4223. Ingersoll, K. A., *Appl. Opt.*, **10**, 2781 (1971).
4224. Ingham, J. D., and Lawson, D. D., *J. Polym. Sci. A1*, **3**, 2707 (1965).
4225. Inoue, K., and Hoshino, S., *J. Polym. Sci. A2*, **11**, 1077 (1973).
4226. Inoue, K., and Hoshino, S., *J. Polym. Sci. A2*, **14**, 1513 (1976).
4227. Inoue, K., and Hoshino, S., *J. Polym. Sci. A2*, **15**, 1363 (1977).
4228. Inoue, Y., Ando, I., and Nishioka, A., *Polym. J.*, **3**, 246 (1972).
4229. Inoue, Y., and Konno, T., *Makromol. Chem.*, **179**, 1311 (1978).
4230. Inoue, Y., Koyama, K., Chûjô, R., and Nishioka, A., *J. Polym. Sci. B*, **11**, 55 (1973).
4231. Inoue, Y., Koyama, K., Chûjô, R., and Nishioka, A., *Makromol. Chem.*, **175**, 277 (1974).
4232. Inoue, Y., Nishioka, A., and Chûjô, R., *Makromol. Chem.*, **152**, 15 (1972).
4233. Inoue, Y., Nishioka, A., and Chûjô, R., *Makromol. Chem.*, **156**, 207 (1972).
4234. Inoue, Y., Nishioka, A., and Chûjô, R., *Makromol. Chem.*, **168**, 163 (1973).
4235. Inoue, Y., Nishioka, A., and Chûjô, R., *J. Polym. Sci. A2*, **11**, 2237 (1973).
4236. Inoue, T., Moritani, M., Hashimoto, T., and Kawai, H., *Macromolecules*, **4**, 500 (1971).
4237. Inverarity, G., *Brit. Polym. J.*, **1**, 245 (1969).
4238. Irie, S., Irie, M., Yamamoto, Y., and Hayashi, K., *Macromolecules*, **8**, 424 (1975).
4239. Ishida, Y., *J. Polym. Sci. A2*, **7**, 1835 (1969).
4240. Ishida, Y., Togami, S., and Yamafuji, K., *Kolloid Z.*, **222**, 18 (1968).
4241. Ishida, Y., Yoshino, M., Takayanagi, M., and Irie, F., *J. Appl. Polym. Sci.*, **1**, 227 (1969).
4242. Ishida, Y., Watanabe, M., and Yamafuji, K., *Kolloid Z.*, **200**, 48 (1964).
4243. Ishigure, K., Tabata, Y., and Oshima, K., *J. Polym. Sci B*, **4**, 669 (1966).
4244. Ishii, T., Matsunaga, S., and Handa, T., *Makromol. Chem.*, **177**, 283 (1976).

782

4245. Ishii, T., Handa, T., and Matsunaga, S., *Makromol. Chem.*, **178**, 2351 (1971).
4246. Ishii, T., Handa, T., and Matsunaga, S., *Macromolecules*, **11**, 40 (1978).
4247. Ishikawa, S., *J. Polym. Sci.*, **3**, 4075 (1965).
4248. Isobst, S. A., and Manson, J. A., *ACS Polym. Preprints*, **11**, 765 (1970).
4249. Ito, E., *J. Polym. Sci. A2*, **12**, 1477 (1974).
4250. Ito, K., Sakakura, H., and Yamashita, Y., *J. Polym. Sci. B*, **15**, 755 (1977).
4251. Ito, M., Kanamoto, T., and Tanaka, K., *J. Polym. Sci. B*, **14**, 189 (1971).
4252. Ito, M., Nakatani, S., Gokan, A., and Tanaka, K., *J. Polym. Sci. A2*, **15**, 605 (1977).
4253. Ito, Y., Katsura, S., and Tabata, Y., *J. Polym. Sci. A2*, **9**, 1525 (1971).
4254. Ito, Y., Okuda, K., and Tabata, Y., *Bull. Chem. Soc. Japan*, **44**, 1764 (1971).
4255. Ito, Y., and Shishido, S., *J. Polym. Sci. A2*, **11**, 2283 (1973).
4256. Ito, Y., and Shishido, S., *J. Polym. Sci. A2*, **16**, 725 (1978).
4257. Ivin, K. J., Ende, H. A., and Meyerhoff, G., *Polymer*, **3**, 129 (1962).
4258. Ivin, K. J., Kuan–Essig, L. C., Lillie, E. D., and Watt, P., *Polymer*, **17**, 656 (1976).
4259. Ivin, K. J., and Navratil, M., *J. Polym. Sci. B*, **8**, 51 (1970).
4260. Iwamoto, R., Bopp, R. C., and Wunderlich, B., *J. Polym. Sci. A2*, **13**, 1925 (1975).
4261. Iwamoto, R., and Wunderlich, B., *J. Polym. Sci. A2*, **11**, 2403 (1973).
4262. Iwanow, N., and Schneider, R., *Bull. Inst. Textile France*, **1958**, 55 (1958).
4263. Iwasa, Y., and Chiba, A., *J. Polym. Sci. A2*, **15**, 881 (1977).
4264. Iwasa, Y., Inoue, S., Mashino, S., and Chiba, A., *Rep. Progr. Polym. Phys. Japan*, **16**, 431 (1973).
4265. Izumi, Z., and Rånby, B., *Macromolecules*, **8**, 151 (1975).
4266. Izydroczyk, J., Podkowka, J., Slawinski, J., and Grzywa, Z., *J. Appl. Polym. Sci.*, **21**, 1835 (1977).

4267. Jachowicz, J., and Kryszewski, M., *Polymer*, **19**, 93 (1978).
4268. Jackson, D. A., Pentecost, H. T. A., and Powles, J. G., *Mol. Phys.*, **23**, 425 (1972).
4269. Jackson, J. F., and Mandelkern, L., *Macromolecules*, **1**, 546 (1968).
4270. Jackson, J. F., Mandelkern, L., and Long, O. C., *Macromolecules*, **1**, 218 (1968).
4271. Jackson, M. T., and Henry, R. A., *Intern. Lab.*, Nov./Dec. 1974, p. 57.
4272. Jackson, R. A., Oldland, S. R. D., and Pajaczkowski, A., *J. Appl. Polym. Sci.*, **12**, 1297 (1968).
4273. Jacob, M., Freund, L., and Daune, M., *J. Chim. Phys.*, **58**, 521 (1961).
4274. Jacob, M., Varoqui, R., Klenine, S., and Daune, M., *J. Chim. Phys.*, **59**, 865 (1962).
4275. Jacobi, E., Lüderwald, I., and Schulz, R. C., *Makromol. Chem.*, **179**, 277 (1978).
4276. Jacobi, E., Lüderwald, I., and Schulz, R. C., *Makromol. Chem.*, **179**, 429 (1978).
4277. Jacobs, H., and Jenckel, E., *Makromol. Chem.*, **47**, 72 (1961).
4278. Jacobson, H., *Macromolecules*, **2**, 650 (1969).
4279. Jacques, C. H. M., Hopfenberg, H. B., and Stannett, V., *Polym. Eng. Sci.*, **13**, 81 (1973).
4280. Jacques, C. H. M., and Hopfenberg, H. B., *Polym. Eng. Sci.*, **14**, 441, 449 (1974).
4281. Jaffe, J. H., and Jaffe, H., *J. Opt. Soc. Amer.*, **40**, 53 (1950).
4282. Jagodic, F., Borstnik, B., and Azman, A., *Makromol. Chem.*, **173**, 221 (1973).
4283. Jain, P. C., Bhatnagar, S., and Gupta, A., *J. Phys. C*, **5**, 2156 (1972).
4284. Jakeman, E., Oliver, C. J., and Pike, E. R., *J. Phys. A*, **1**, 406 (1968).
4285. James, A. T., Ravenhill, J. R., and Scott, R. P. W., *Chem. Ind. (London)*, **1964**, 746 (1964).
4286. James, D. I., *Brit. Polym. J.*, **1**, 205 (1969).
4287. Jamieson, A. M., and Presley, C. T., *Macromolecules*, **6**, 358 (1973).
4288. Jamieson, G. R., *J. Chromatogr.*, **3**, 464, 494 (1960).
4289. Jamieson, G. R., *J. Chromatogr.*, **4**, 420 (1960).
4290. Jamroz, M., Kozlowski, K., Sieniakowski, M., and Jachym, B., *J. Polym. Sci. A1*, **15**, 1359 (1977).
4291. Janacek, J., *J. Polym. Sci. B*, **13**, 401, 409 (1975).
4292. Janaczek, H., Turska, E., Szekely, T., Lengyel, M., and Till, E., *Polymer*, **19**, 85 (1978).
4293. Janneret, C., and Stoeckli, H. F., *Helv. Chim. Acta*, **57**, 851 (1974).
4294. Jansson, J. F., and Yannas, I. V., *J. Polym. Sci. A2*, **15**, 2103 (1977).
4295. Janssens, M., and Bellemans, A., *Macromolecules*, **9**, 303 (1976).
4296. Jellinek, H. H. G., and Burkhardt, A., *Angew. Makromol. Chem.*, **7**, 101 (1969).
4297. Jellinek, H. H. G. and Kryman, F., *Environ. Sci. Technol.*, **1**, 658 (1967).
4298. Jenckel, E., and Heusch, R., *Kolloid Z.*, **130**, 89 (1953).

4299. Jenckel, E., and Keller, G., *Z. Naturforsch.*, **5a**, 317 (1950).
4300. Jenkins, A. C., and Birdsall, C. M., *J. Chem., Phys.*, **20**, 1158 (1952).
4301. Jennings, B. R., *Brit. Polym. J.*, **1**, 70 (1969).
4302. Jennings, B. R., *Brit. Polym. J.*, **1**, 252 (1969).
4303. Jennings, B. R., and Brown, B. L., *Europ. Polym. J.*, **7**, 805 (1971).
4304. Jennings, B. R., and Jerrard, H. G., *J. Phys. Chem.*, **69**, 2817 (1965).
4305. Jennings, B. R., and Jerrard, H. G., *J. Chem. Phys.*, **44**, 1291 (1966).
4306. Jentoft, R. E., and Gouw, T. H., *Anal. Chem.*, **38**, 949 (1966).
4307. Jentoft, R. E., and Gouw, T. H., *Anal. Chem.*, **40**, 923 (1968).
4308. Jentoft, R. E., and Gouw, T. H., *Anal. Chem.*, **40**, 1787 (1968).
4309. Jentoft, R. E., and Gouw, T. H., *J. Polym. Sci. B*, **7**, 811 (1969).
4310. Jentoft, R. E., and Gouw, T. H., *J. Chromatogr. Sci.*, **8**, 138 (1970).
4311. Jernejcic, M., and Premru, L., *Rubber Chem. Technol.*, **41**, 411 (1968).
4312. Jerrard, H. G., and Sellen, D. B., *Appl. Opt.*, **1**, 243 (1962).
4313. Jessup, R. S., *J. Res. Nat. Bur. Stand.*, **60**, 47 (1959).
4314. Jeziorny, A., *Polymer*, **19**, 1142 (1978).
4315. Johansen, R. M., *Chem. Scripta*, **2**, 31 (1972).
4316. Johari, O., and Hill, V. L., *Appl. Polym. Symp.*, No. 16, 183 (1971).
4317. Johnsen, A., Klesper, E., and Wirthlin, T., *Makromol. Chem*, **177**, 2397 (1976).
4318. Johnsen, U., *J. Polym. Sci.*, **54**, 56 (1961).
4319. Johnsen, U., and Kolbe, K., *Kolloid Z.*, **221**, 64 (1967).
4320. Johnson, D. R., Cassels, J. W., Brame, E. G., and Westneat, D. F., *Anal. Chem.*, **34**, 1610 (1962).
4321. Johnson, D. R., Wen, W. Y., and Dole, M., *J. Phys. Chem.*, **77**, 2174 (1973).
4322. Johnson, H. W., Campanile, V. A., and Lefebre, H. A., *Anal. Chem.*, **39**, 33 (1967).
4323. Johnson, H. W., Seibert, E. E., and Stross, F. H., *Anal. Chem.*, **40**, 403 (1968).
4324. Johnson, J. F., and Cantow, M. J. R., *J. Chromatogr.*, **28**, 128 (1967).
4325. Johnson, J. F., Macphail, M. G., Cooper, A. R., and Bruzzone, A. R., *Separation Sci.*, **8**, 577 (1973).
4326. Johnson, L. F., Heatley, F., and Bovey, F. A., *Macromolecules*, **3**, 175 (1970).
4327. Johnson, R. G. L., Delf, B. W., and MacKnight, W. J., *J. Polym. Sci. A2*, **11**, 571 (1953).
4328. Johnston, N. W., *Macromolecules*, **6**, 453 (1973).
4329. Johnston, N. W., *ACS Polym. Preprints*, **10**, 609 (1969).
4330. Johnston, N. W., and Harwood, H. J., *Macromolecules*, **2**, 221 (1969).
4331. Johnston, N. W., and Harwood, H. J., *J. Polym. Sci. C*, **22**, 591 (1969).
4332. Johnston, N. W., and Harwood, H. J., *Macromolecules*, **3**, 20 (1970).
4333. Jones, A. A., *J. Polym. Sci. A2*, **15**, 863 (1977).
4334. Jones, A. A., Lubianez, R. P., Hanson, M. A., and Shostak, S. L., *J. Polym. Sci. A2*, **16**, 1685 (1978).
4335. Jones, A. A., and Stockmayer, W. H., *J. Polym. Sci. A2*, **15**, 847 (1977).
4336. Jones, C. E. R., and Reynolds, G. E. J., *J. Gas Chromatogr.*, **5**, 25 (1967).
4337. Jones, C. E. R., and Reynolds, G. E. J., *Brit. Polym. J.*, **1**, 197 (1969).
4338. Jones, D. H., Latham, A. J., Keller, A., and Girolamo, M., *J. Polym. Sci. A2*, **11**, 1759 (1973).
4339. Jones, P., *J. Chem. Phys.*, **49**, 3730 (1968).
4340. Jones, R. B., Felderhof, B. U., and Deutch, J. M., *Macromolecules*, **8**, 680 (1975).
4341. Jones, R. G., Nicol, E. A., Birch, J. R., Chantry, G. W., Fleming, J. W., Willis, H. A., and Cudby, M. E. A., *Polymer*, **17**, 153 (1976).
4342. Jones, R. N., *Appl. Opt.*, **8**, 597 (1969).
4343. Jordan, E. F., Jr, Artymyshin, B., Riser, G. R., and Wrigley, A. N., *J. Appl. Polym. Sci.*, **20**, 2715 (1976).
4344. Jordan, E. F., Jr, Smith, S., Jr, Koos, R. E., Parker, W. F., Artymyshin, B., and Wrigley, A. N., *J. Appl. Polym. Sci.*, **22**, 1509 (1978).
4345. Jordan, E. F., Jr, Smith, S., Jr, Zabarsky, R. D., Austin, R., and Wrighley, A. N., *J. Appl. Polym. Sci.*, **22**, 1529 (1978).
4346. Jordan, E. F., Jr, Smith, S., Jr, Zabarsky, R. D., and Wrighley, A. N., *J. Appl. Polym. Sci.*, **22**, 1547 (1978).
4347. Juhasz, A. A., Omar Doali, J., and Rocchio, J. J., *Intern. Lab.*, July/Aug. 1974, p. 21.

784

4348. Julemont, M., Walckiers, E., Warin, R., and Teyssie, P., *Makromol. Chem.*, **175**, 1673 (1974).
4349. Jungnickel, J. L., and Weiss, F. T., *J. Polym. Sci.*, **49**, 437 (1961).
4350. Jupille, T. H., and Curtice, A. B., *Chromatographia*, **8**, 193 (1975).
4351. Jupille, T. H., and Perry, J. A., *J. Chromatogr.*, **99**, 231 (1974).
4352. Jupille, T. H., and Perry, J. A., *J. Chromatogr. Sci.*, **13**, 163 (1975).
4353. Jurs, C., Kowalski, B. R., Isenhour, T. L., and Reilley, C. N., *Anal. Chem.*, **41**, 1949 (1969).
4354. Jyo, Y., Nozaki, C., and Matsuo, M., *Macromolecules*, **4**, 517 (1971).
4355. Kabayama, M. A., and Daoust, H., *J. Phys. Chem.*, **62**, 1127 (1968).
4356. Kabot, F. J., and Ettre, L. S., *J. Gas Chromatogr.*, **1**, 7 (1963).
4357. Kabin, S. P., Makevich, S. G., and Mikhilov, G. P., *Vysokomol. Soedin.*, **3**, 618 (1961).
4358. Kagan, D. F., and Popova, L. A., *Vyskomol. Soedin. A*, **12**, 2774 (1970).
4359. Kagemoto, A., and Fujishiro, R., *Makromol. Chem.*, **114**, 139 (1968).
4360. Kagemoto, A., and Fujishiro, R., *Bull. Chem. Soc. Japan*, **41**, 2201 (1968).
4361. Kagemoto, A., Itoi, Y., Baba, Y., and Fujishiro, R., *Makromol. Chem.*, **150**, 255 (1971).
4362. Kagemoto, A., Katayama, H., and Baba, Y., *Polym. J.*, **6**, 230 (1974).
4363. Kagemoto, A., Murakami, S., and Fujishiro, R., *Bull. Chem. Soc. Japan*, **39**, 15 (1966).
4364. Kagemoto, A., Murakami, S., and Fujishiro, R., *Bull. Chem. Soc. Japan*, **40**, 11 (1967).
4365. Kagemoto, A., Murakami, S., and Fujishiro, R., *Makromol. Chem.*, **105**, 154 (1967).
4366. Kahar, N., Duckett, R. A., and Ward, J. M., *Polymer*, **19**, 137 (1978).
4367. Kahn, D. S., and Polson, A., *J. Phys. Chem.*, **51**, 816 (1947).
4368. Kaji, K., *Makromol. Chem.*, **175**, 311 (1974).
4369. Kaji, K., and Sakurada, I., *Makromol. Chem.*, **179**, 209 (1978).
4370. Kajiwara, K., Burchard, W., and Gordon, M., *Brit. Polym. J.*, **2**, 110 (1970).
4371. Kajiwara, K., and Ribeiro, C. A. M., *Macromolecules*, **7**, 121 (1974).
4372. Kajiwara, K., and Ross-Murphy, S. B., *Europ. Polym. J.*, **11**, 365 (1975).
4373. Kajiyama, T., Oda, T., Stein, R. S., and MacKnight, W. J., *Macromolecules*, **4**, 198 (1971).
4374. Kakudo, M., and Ullman, R., *J. Polym. Sci.*, **45**, 91 (1960).
4375. Kalkal, J., Morousek, U., and Svec, F., *Angew. Makromol. Chem.*, **38**, 45 (1974).
4376. Kalfus, M., and Dudek, B., *Makromol. Chem.*, **178**, 1609 (1977).
4377. Kam, Z., *Macromolecules*, **10**, 927 (1977).
4378. Kamath, P. M., and Wild, L., *Polym. Eng. Sci.*, **6**, 213 (1966).
4379. Kambour, R. P., Gruner, C. L., and Romagosa, E. E., *Macromolecules*, **7**, 248 (1974).
4380. Kamel, I., Kusy, R. P., and Corneliussen, R. D., *Macromolecules*, **6**, 53 (1973).
4381. Kamide, K., Manabe, S., and Osafune, E., *Makromol. Chem.*, **168**, 173 (1973).
4382. Kamide, K., and Miyakawa, Y., *Makromol. Chem.*, **179**, 359 (1978).
4383. Kamide, K., and Miyazaki, Y., *Makromol. Chem.*, **176**, 1029, 1051, 1427, 1447, 2393, 3453 (1975).
4384. Kamide, K., Miyazaki, Y., and Abe, T., *Makromol. Chem.*, **177**, 485 (1976).
4385. Kamide, K., Miyazaki, Y., and Sugamiya, K., *Makromol. Chem.*, **173**, 113 (1973).
4386. Kamide, K., Miyazaki, Y., and Yamaguchi, K., *Makromol. Chem.*, **173**, 157, 175 (1973).
4387. Kamide, K., and Moore, W. R., *J. Polym. Sci. B*, **2**, 809 (1964).
4388. Kamide, K., and Nagayama, C., *Makromol. Chem.*, **129**, 280 (1969).
4389. Kamide, K., and Ogata, T., *Makromol. Chem.*, **168**, 195 (1973).
4390. Kamide, K., Ogawa, T., and Nakayama, C., *Makromol. Chem.*, **132**, 65 (1970).
4391. Kamide, K., and Sugamiya, K., *Makromol. Chem.*, **139**, 197 (1970).
4392. Kamide, K., and Sugamiya, K., *Makromol. Chem.*, **156**, 259 (1972).
4393. Kamide, K., Sugamiya, K., and Nakayama, C., *Makromol. Chem.*, **132**, 75 (1970).
4394. Kamide, K., Sugamiya, K., and Nakayama, C., *Makromol. Chem.*, **133**, 101 (1970).
4395. Kamide, K., Sugamiya, K., and Nakayama, C., *Makromol. Chem.*, **135**, 9 (1970).
4396. Kamide, K., Sugamiya, K., Ogawa, T., Nakayama, C., and Baba, N., *Makromol. Chem.*, **135**, 23 (1970).
4397. Kamide, K., Terakawa, T., and Uchiki, H., *Makromol. Chem.*, **177**, 1447 (1976).
4398. Kamide, K., and Yamaguchi, K., *Makromol. Chem.*, **167**, 287 (1973).
4399. Kamide, K., Yamaguchi, K., and Miyazaki, Y., *Makromol. Chem.*, **173**, 133 (1973).
4400. Kaminow, I. P., and Turner, E. H., *Appl. Opt.*, **5**, 1612 (1966).
4401. Kamiya, Y., and Takahashi, F., *J. Appl. Polym. Sci.*, **21**, 1945 (1977).

4402. Kamiyama, F., Inagaki, H., and Kotaka, T., *Polym. J.*, **3**, 470 (1972).
4403. Kamiyama, F., Matsuda, H., and Inagaki, H., *Makromol. Chem.*, **125**, 286 (1969).
4404. Kamiyama, F., Matsuda, H., and Inagaki, H., *Polymer J.*, **1**, 518 (1970).
4405. Kammerer, H., Rocaboy, F., and Kern, W., *Makromol. Chem.*, **51**, 222 (1962).
4406. Kammerer, H., Rocaboy, F., Steinfort, K. G., and Kern, W., *Makromol. Chem.*, **53**, 80 (1962).
4407. Kammerer, H., Sextro, G., and Seyed-Mozaffari, A., *J. Polym. Sci. A1*, **14**, 609 (1976).
4408. Kämpf, G., Hoffman, M., and Krömer, H., *Ber. Bunsenges. Phys. Chem.*, **74**, 851 (1970).
4409. Kanamoto, T., *J. Polym. Sci. A2*, **12**, 2535 (1974).
4410. Kanig, G., *Kolloid Z.*, **190**, 1 (1963).
4411. Kanjilal, C., Mitra, B. C., and Palit, S. R., *Makromol. Chem.*, **178**, 1707 (1977).
4412. Kaplan, D. S., *J. Appl. Polym. Sci.*, **20**, 2615 (1976).
4413. Kaplan, M. L., and Kelleher, P. G., *Science*, **169**, 1206 (1970).
4414. Kaplan, M. L., and Kelleher, P. G., *J. Polym. Sci. A1*, **8**, 3163 (1970).
4415. Kaplan, M. L., and Kelleher, P. G., *J. Polym. Sci. B*, **8**, 565 (1971).
4416. Kaplan, M. L., and Kelleher, P. G., *Rubber Chem. Technol.*, **45**, 423 (1972).
4417. Kapur, R., Shelton, J. R., and Koenig, J. L., *Rubber Chem. Technol.*, **48**, 348 (1976).
4418. Karasev, A. N., *Plast. Massy*, **1967**, 52 (1967).
4419. Karasz, F. E., Bair, H. E., and O'Reilly, J. M., *J. Phys. Chem.*, **69**, 2657 (1965).
4420. Karasz, F. E., Bair, H. E., and O'Reilly, J. M., *Polymer*, **8**, 547 (1967).
4421. Karasz, F. E., Bair, H. E., and O'Reilly, J. M., *J. Polym. Sci. A2*, **6**, 1141 (1968).
4422. Karasz, F. E., Couchman, P. R., and Klemper, D., *Macromolecules*, **10**, 88 (1977).
4423. Karasz, F. E., and MacKnight, W. J., *Macromolecules*, **1**, 537 (1968).
4424. Karayannidis, P., and Dondos, A., *Makromol. Chem.*, **147**, 135 (1971).
4425. Kardos, J. L., Christiansen, A. W., and Baer, E., *J. Polym. Sci. A2*, **4**, 777 (1966).
4426. Kardos, J. L., Li, H. M., and Huckshold, K. A., *J. Polym. Sci. A2*, **9**, 2061 (1971).
4427. Karger, B. L., *Anal. Chem.*, **39**, 24A (1967).
4428. Karim, K. A., and Bonner, D. C., *J. Appl. Polym. Sci.*, **22**, 1277 (1978).
4429. Karl, V. H., Asmussen, F., and Ueberreiter, K., *Makromol. Chem.*, **178**, 1649 (1977).
4430. Karl, V. H., Asmussen, F., and Ueberreiter, K., *Makromol. Chem.*, **179**, 1601 (1978).
4431. Karmen, A., *Anal. Chem.*, **38**, 286 (1966).
4432. Karoll, G. W., and Singh, A., *Photochem. Photobiol.*, **28**, 611 (1978).
4433. Kasai, N., and Kukudo, M., *Rept. Progr. Polym. Phys. Japan*, **6**, 319 (1963).
4434. Kasai, N., and Kukudo, M., *J. Polym. Sci. A*, **2**, 1955 (1964).
4435. Kasai, N., and Kukudo, M., *Rept. Progr. Polym. Phys. Japan*, **9**, 243 (1966).
4436. Kasai, N., Kukudo, M., and Watase, T., *Rept. Progr. Polym. Phys. Japan*, **9**, 239 (1966).
4437. Kashiwabara, H., *J. Phys. Soc. Japan*, **16**, 2493 (1961).
4438. Kashiwabara, H., *J. Phys. Soc. Japan*, **17**, 567 (1962).
4439. Kashiwabara, H., and Shinohara, K., *J. Phys. Soc. Japan*, **15**, 1129 (1960).
4440. Kashiwagi, M., Cunningham, A., Manuel, A. J., and Ward, I. M., *Polymer*, **14**, 111 (1973).
4441. Kashiwagi, M., Folkes, M. J., and Ward, I. M., *Polymer*, **12**, 697 (1971).
4442. Kashiwagi, M., and Ward, I. M., *Polymer*, **13**, 145 (1972).
4443. Kashmiri, M. I., and Sheldon, R. P., *J. Polym. Sci. B*, **7**, 51 (1969).
4444. Kashmiri, M. I., and Sheldon, R. P., *Brit. Polym. J.*, **1**, 65 (1969).
4445. Kashyap, A. K., Kalpagam, V., and Reddy, C. R., *Polymer*, **18**, 978 (1977).
4446. Kassenbeck, P., *J. Polym. Sci. C*, **20**, 49 (1967).
4447. Katayama, H., Baba, Y., Kagemoto, A., and Fujishiro, R., *Makromol. Chem.*, **175**, 209 (1974).
4448. Katchalski, A., and Eisenberg, H., *J. Polym. Sci.*, **6**, 145 (1951).
4449. Katime, I., and Strazielle, C., *Makromol. Chem.*, **178**, 2295 (1977).
4450. Kato, K., *Electron Microsc.*, **14**, 220 (1965).
4451. Kato, K., *J. Polym. Sci. B*, **4**, 35 (1966).
4452. Kato, K., *Polym. Eng. Sci.*, **7**, 38 (1967).
4453. Kato, K., *Kolloid Z. Z. Polym.*, **220**, 24 (1967).
4454. Kato, T., *Rep. Progr. Polym. Phys. Japan*, **12**, 1 (1969).
4455. Kato, Y., Kido, S., Yamamoto, M., and Hashimoto, T., *J. Polym. Sci. A2*, **12**, 1339 (1974).
4456. Kato, Y., and Hashimoto, T., *J. Polym. Sci. A2*, **12**, 813 (1974).

786

4457. Kato, T., Hashimoto, T., Fujimoto, T., and Nagasawa, M., *J. Polym. Sci. A2*, **13**, 1849 (1975).
4458. Kato, T., Itsubo, A., Yamamoto, Y., Fujimoto, T., and Nagasawa, M., *Polymer J.*, **7**, 123 (1975).
4459. Kato, Y., Kametani, T., Furukawa, K., and Hashimoto, T., *J. Polym. Sci. A2*, **13**, 1695 (1975).
4460. Kato, Y., Kametani, T., and Hashimoto, T., *J. Polym. Sci. A2*, **14**, 2105 (1976).
4461. Kato, Y., Kido, S., and Hashimoto, T., *J. Polym. Sci. A2*, **11**, 2329 (1973).
4462. Kato, Y., Kido, S., Yamamoto, M., and Hashimoto, T., *J. Polym. Sci. A2*, **12**, 1339 (1974).
4463. Kato, T., Miyaso, K., Noda, I., Fujimoto, T., and Nagasawa, M., *Macromolecules*, **3**, 777 (1970).
4464. Kato, Y., and Nishioka, A., *Bull. Chem. Soc. Japan*, **37**, 1622 (1964).
4465. Kato, Y., Takamatsu, T., Fukutomi, M., Fukuda, M., and Hashimoto, T., *J. Appl. Polym. Sci.*, **21**, 577 (1977).
4466. Katzer, H., and Heusinger, H., *Makromol. Chem.*, **163**, 195 (1973).
4467. Kaufman, F., and Kelso, J. R., *Chem. Phys.*, **32**, 301 (1960).
4468. Kaufman, H. S., and Solomon, E., *Ind. Eng. Chem.*, **45**, 1779 (1953).
4469. Kavesh, S., and Schultz, J. M., *J. Polym. Sci. A2*, **9**, 85 (1971).
4470. Kawaguchi, T., Ito, T., Kawai, H., Keedy, D., and Stein, R. S., *Macromolecules*, **1**, 126 (1968).
4471. Kawahara, K., and Okada, R., *J. Polym. Sci.*, **56**, S7 (1962).
4472. Kawai, T., *J. Polym. Sci.*, **32**, 425 (1958).
4473. Kawai, T., *Makromol. Chem.*, **102**, 125 (1967).
4474. Kawai, T., *Kolloid Z. Z. Polym.*, **229**, 116 (1969).
4475. Kawai, T., Goto, T., and Maeda, H., *Kolloid Z.*, **223**, 117 (1968).
4476. Kawai, T., Kato, M., Matsumoto, T., and Maeda, H., *Kolloid Z. Z. Polym.*, **222**, 1 (1968).
4477. Kawai, T., and Keller, A., *J. Polym. Sci. B*, **2**, 333 (1964).
4478. Kawai, T., Shiozaki, S., Sonoda, S., Nakagawa, H., Matsumoto, T., and Maeda, H., *Makromol. Chem.*, **128**, 252 (1969).
4479. Kawai, T., and Ueyama, T., *J. Appl. Polym. Sci.*, **3**, 227 (1960).
4480. Kawamura, T., and Matsuzaki, K., *Makromol. Chem.*, **179**, 1003 (1978).
4481. Kazhdan, M. V., Bakeyev, N. F., and Berestneva, Z. Ya., *J. Polym. Sci. C*, **38**, 443 (1972).
4482. Kaye, A., Lodge, A. S., and Vale, D. G., *Rheol. Acta*, **7**, 368 (1968).
4483. Kaye, H., and Chou, H. J., *J. Polym. Sci. A2*, **13**, 477 (1975).
4484. Kaye, H. F., and Newman, B. A., *J. Appl. Polym. Sci.*, **38**, 4105 (1968).
4485. Kaye, S., *Anal. Chem.*, **24**, 1038 (1952).
4486. Kaye, W., *Anal. Chem.*, **45**, 221A (1973).
4487. Kaye, W., and Havlik, A., *J. Appl. Opt.*, **12**, 541 (1973).
4488. Kaye, W., Havlik, A. J., and McDaniel, J. B., *J. Polym. Sci. B*, **9**, 695 (1971).
4489. Ke, B., *J. Polym. Sci.*, **42**, 15 (1960).
4490. Keeling, C. D., and Dole, M., *J. Polym. Sci.*, **14**, 105 (1954).
4491. Keith, H. D., *J. Polym. Sci. A2*, **2**, 4339 (1964).
4492. Keith, H. D., *J. Appl. Phys.*, **35**, 3115 (1964).
4493. Keith, H. D., and Padden, F. J., Jr, *J. Polym. Sci.*, **39**, 101, 123 (1959).
4494. Keith, H. D., and Padden, F. J., Jr, *J. Appl. Phys.*, **34**, 2409 (1963).
4495. Keith, H. D., and Padden, F. J., Jr, *J. Appl. Phys.*, **35**, 1270, 1286 (1964).
4496. Keith, H. D., Padden, F. J., Jr, and Vadimsky, R. G., *Science*, **150**, 1026 (1965).
4497. Keith, F. H., Padden, F. J., Jr, Vadimsky, R. G., *J. Polym. Sci. A2*, **4**, 267 (1966).
4498. Keith, H. D., Padden, F. J., Jr, and Vadimsky, R. G., *J. Appl. Phys.*, **37**, 4027 (1966).
4499. Keith, H. D., Padden, F. J., Jr, and Vadimsky, R. G., *J. Appl. Phys.*, **42**, 4585 (1971).
4500. Keith, H. D., Padden, F. J., Jr, Walter, N. M., and Wyckoff, H. W., *J. Appl. Phys.*, **30**, 1485 (1959).
4501. Keith, H. D., Vadimsky, R. G., and Padden, F. J., Jr, *J. Polym. Sci. A2*, **8**, 1687 (1970).
4502. Keizers, A. E. M., van Aartsen, J. J., and Prins, W., *J. Appl. Phys.*, **36**, 2874 (1965).
4503. Keizers, A. E. M., van Aartsen, J. J., and Prins, W., *J. Amer. Chem. Soc.*, **90**, 3167 (1968).
4504. Kelker, J. V., Mashelkar, R. A., and Ulbrecht, J., *J. Appl. Polym. Sci.*, **17**, 3069 (1973).

4505. Keller, A., *J. Polym. Sci.*, **15**, 31 (1955).
4506. Keller, A., *J. Polym. Sci.*, **17**, 291, 351, 447 (1955).
4507. Keller, A., *Phil. Mag.*, **2**, 1171 (1957).
4508. Keller, A., *Makromol. Chem.*, **34**, 1 (1959).
4509. Keller, A., *J. Polym. Sci.*, **36**, 361 (1959).
4510. Keller, A., *J. Polym. Sci.*, **39**, 151 (1959).
4511. Keller, A., *Polymer*, **3**, 393 (1962).
4512. Keller, A., *J. Polym. Sci.*, **17**, 291 (1965).
4513. Keller, A., and Machin, M. J., *J. Macromol. Sci. B*, **1**, 41 (1967).
4514. Keller, A., and Martuscelli, E., *Makromol. Chem.*, **141**, 189 (1971).
4515. Keller, A., Martuscelli, E., Priest, D. J., and Udagawa, Y., *J. Polym. Sci. A2*, **9**, 1807 (1971).
4516. Keller, A., and O'Connor, A., *Discuss. Faraday Soc.*, **25**, 114 (1958).
4517. Keller, A., and O'Connor, A., *Polymer*, **1**, 163 (1960).
4518. Keller, A., and Priest, D. J., *J. Macromol. Sci. B*, **2**, 479 (1968).
4519. Keller, A., and Sawada, S., *Makromol. Chem.*, **74**, 190 (1964).
4520. Keller, A., and Snyder, L. R., *J. Chromatogr. Sci.*, **9**, 346 (1971).
4521. Keller, A., and Udagawa, Y., *J. Polym. Sci. A2*, **9**, 1793 (1971).
4522. Keller, A., and Willmouth, F. M., *J. Polym. Sci. A2*, **8**, 1443, (1970).
4523. Keller, A., and Willmouth, F. M., *J. Macromol. Sci. B*, **6**, 493 (1972).
4524. Keller, A., and Willmouth, F. M., *J. Macromol. Sci. B*, **6**, 539 (1972).
4525. Kelly, R. N., and Billmeyer, F. W., Jr, *Anal. Chem.*, **41**, 874 (1969).
4526. Kelly, R. N., and Billmeyer, F. W., Jr, *Separation Sci.*, **5**, 291 (1970).
4527. Kelly, R. N., and Billmeyer, F. W., Jr, *Anal. Chem.*, **42**, 399 (1970).
4528. Kelly, K. M., Patterson, G. D., and Tonelli, A. E., *Macromolecules*, **10**, 859 (1977).
4529. Kemp, D. R., and Paul, D. R., *J. Polym. Sci. A2*, **12**, 485 (1974).
4530. Kendrich, E., *Anal. Chem.*, **35**, 2146 (1963).
4531. Kenyon, A. S., and Salyer, I. O., *J. Polym. Sci.*, **43**, 427 (1960).
4532. Kenyon, A. S., Salyer, I. O., Kurz, J. E., and Brown, D. R., *J. Polym. Sci C*, **8**, 205 (1965).
4533. Kerker, M., and Farone, W. A., *J. Opt. Soc. Amer.*, **56**, 481 (1966).
4534. Kerker, M., Kratochvil, J. P., and Matijevic, E., *J. Opt. Soc. Amer.*, **52**, 551 (1962).
4535. Kerker, M., and Matijevic, E., *J. Opt. Soc. Amer.*, **51**, 506 (1961).
4536. Kern, W., Achon, M. A., and Schulz, R., *Makromol. Chem.*, **15**, 161 (1955).
4537. Kern, W., Munk, R., Sabel, A., and Schmidt, K. H., *Makromol. Chem.*, **17**, 201 (1956).
4538. Kern, W., Munk, R., and Schmidt, K. R., *Makromol. Chem.*, **17**, 201 (1956).
4539. Keskkula, H., and Traylor, P. A., *J. Appl. Polym. Sci.*, **11**, 2361 (1967).
4540. Khalifa, B. A., and Bassett, D. C., *Polymer*, **17**, 291 (1976).
4541. Khambatta, F. B., Warner, F., Russell, T., and Stein, R. S., *J. Polym. Sci. A2*, **14**, 1391 (1976).
4542. Khan, M. N. G. A., *J. Chem. Phys.*, **50**, 3639 (1969).
4543. Khan, M. N. G. A., *J. Phys. D*, **3**, 663 (1970).
4544. Khan, M. N. G. A., Carswell, D. J., and Bell, J., *Phys. Lett. A*, **29**, 237 (1969).
4545. Khatchaturov, A. S., Dolinskaya, E. R., Prozenko, L. K., Abramenko, E. L., and Kormer, V. A., *Polymer*, **18**, 871 (1977).
4546. Khoury, F., *J. Res. Nat. Bur. Stand. A*, **70**, 29 (1966).
4547. Kieslig, H., *Kolloid Z.*, **152**, 62 (1957).
4548. Kiho, H., Peterlin, A., and Geil, P. H., *J. Appl. Phys.*, **35**, 1599 (1964).
4549. Kijima, T., Imamura, M., and Kusumoto, N., *Polymer*, **17**, 249 (1976).
4550. Killgoar, P. C., and Dickie, R. A., *J. Appl. Polym. Sci.*, **21**, 1813 (1977).
4551. Kilp, T., and Guillet, J. E., *Macromolecules*, **10**, 90 (1977).
4552. Kilp, T., Houvenaghel-Defoort, B., Panning, W., and Guillet, J. E., *Rev. Sci. Instrum.*, **47**, 1496 (1976).
4553. Kim, H. G., *Macromolecules*, **5**, 594 (1972).
4554. Kim, H. G., *J. Appl. Polym. Sci.*, **22**, 889 (1978).
4555. Kim, S. C., Klempner, D., Frisch, K. C., Frisch, H. L., and Ghiradella, H., *Polym. Eng. Sci.*, **15**, 339 (1975).
4556. Kimmich, R., *Polymer*, **16**, 851 (1975).
4557. Kimmich, R., and Schmauder, K., *Polymer*, **18**, 239 (1977).
4558. Kimoto, S., and Russ, J. C., *Amer. Sci.*, **57**, 112 (1969).

4559. Kimura, I., Ishihara, H., Ono, H., Yoshihara, N., Nomura, S., and Kawai, H., *Macromolecules*, **7**, 355 (1974).
4560. Kimura, I., Kagiyama, M., Nomura, S., and Kwei, H., *J. Polym. Sci. A2*, **7**, 709 (1969).
4561. Kinell, P. O., *Arkiv Kemi*, **14**, 337 (1959).
4562. King, M. V., *Acta. Crystal.*, **21**, 629 (1966).
4563. King, T. A., Knox, A., and McAdam, J. D. G., *Polymer*, **14**, 293 (1973).
4564. Kinkelaar, E. W., Rozsa, J. T., and Vavniska, L. J., *J. Paint Technol.*, **46**, 63 (1974).
4565. Kinoshita, Y., *Makromol. Chem.*, **33**, 21 (1959).
4566. Kiran, E., and Gillham, J. K., *J. Macromol. Sci. A*, **8**, 211 (1974).
4567. Kiran, E., and Gillham, J. K., *J. Appl. Polym. Sci.*, **20**, 931, 2045 (1976).
4568. Kiran, E., Gillham, J. K., and Gipstein, E., *J. Appl. Polym. Sci.*, **21**, 1159 (1977).
4569. Kirby, J. R., and Baldwin, A. J., *Anal. Chem.*, **40**, 689 (1968).
4570. Kireenko, O. Ph., Marchin, V. A., Miasnikova, L. P., and Regel, V. R., *J. Polym. Sci. C*, **38**, 363 (1972).
4571. Kirkwood, J. G., Baldwin, R. L., Dunlop, P. J., Gosting, L. J., and Kegles, G., *J. Chem. Phys.*, **33**, 1505 (1960).
4572. Kirkwood, J. G., and Reisman, J., *J. Chem. Phys.*, **16**, 565 (1948).
4573. Kirste, R. G., Kruse, W. A., and Ibel, K., *Polymer*, **16**, 120 (1975).
4574. Kirste, R. G., Kruse, W. A., and Schelten, J., *Makromol. Chem.*, **162**, 299 (1973).
4575. Kirste, R. G., and Lehnen, B. R., *Makromol. Chem.*, **177**, 1137 (1976).
4576. Kirste, R. G., and Wild, G., *Makromol. Chem.*, **121**, 174 (1969).
4577. Kirste, R. G., and Wunderlich, W., *J. Phys. Chem.*, **58**, 133 (1968).
4578. Kishore, K., Paiverneker, V. R., and Nair, M. N. R., *J. Appl. Polym. Sci.*, **20**, 2355 (1976).
4579. Kinsinger, J. B., and Ballard, L. E., *J. Polym. Sci. B*, **2**, 879 (1964).
4580. Kiss, K., Carr, S. H., Walton, A. G., and Baer, E., *J. Polym. Sci. B*, **5**, 1087 (1967).
4581. Kitamaru, R., Chu, H. D., and Hyon, S. H., *Macromolecules*, **6**, 337 (1973).
4582. Kitamaru, R., Horii, F., and Hyon, A. S. H., *J. Polym. Sci A2*, **15**, 821 (1977).
4583. Kitamaru, R., and Hyon, S. H., *Makromol. Chem.*, **175**, 255 (1974).
4584. Kitchen, R. G., Preston, B. N., and Wells, J. D., *J. Polym. Sci. C*, **55**, 39 (1976).
4585. Kitson, R. E., Oemler, E. N., and Mitchell, J., *Anal. Chem.*, **21**, 404 (1949).
4586. Kitagawa, T., *J. Chem. Phys.*, **47**, 337 (1967).
4587. Kitagawa, T., *Rept. Progr. Polym. Phys. Japan*, **10**, 185 (1967).
4588. Kitagawa, T., *Rept. Progr. Polym. Phys. Japan*, **11**, 219 (1968).
4589. Kitagawa, T., *J. Polym. Sci. B*, **6**, 83 (1968).
4590. Kitamaru, R., and Hyon, S. H., *J. Polym. Sci. A2*, **13**, 1085 (1975).
4591. Kitao, T., Furukawa, J., and Yamashita, S., *J. Polym. Sci. A2*, **11**, 1091 (1973).
4592. Kiwi, J., and Schnabel, W., *Macromolecules*, **8**, 430 (1975).
4593. Kiwi, J., and Schnabel, W., *Macromolecules*, **9**, 468 (1976).
4594. Klein, E., Möbius, K., and Winterhoff, H., *Z. Naturforsch.*, **22a**, 1707 (1967).
4595. Klein, J., *J. Polym. Sci. A2*, **15**, 2057 (1977).
4596. Klein, J., and Briscoe, B. J., *Nature*, **257**, 386 (1975).
4597. Klein, J., and Briscoe, B. J., *Polymer*, **17**, 481 (1976).
4598. Klein, J., and Briscoe, B. J., *J. Polym. Sci. A2*, **15**, 2065 (1977).
4599. Klein, J., and Conrad, J. D., *Makromol. Chem.*, **179**, 1635 (1978).
4600. Klein, J., and Werner, M., *Makromol. Chem.*, **179**, 475 (1978).
4601. Klein, J., and Werner, M., *Makromol. Chem.*, **179**, 991 (1978).
4602. Kleinberg, G. A., Geiger, D. L., and Gormley, W. T., *Makromol. Chem.*, **175**, 483 (1974).
4603. Kleine, L. W., Radloff, M. R., Schultz, J. M., and Chou, T. W., *J. Polym. Sci. A2*, **12**, 819 (1974).
4604. Klement, W., Jr, *J. Polym. Sci. A2*, **12**, 815 (1974).
4605. Klesper, E., and Hartman, W., *J. Polym. Sci. B*, **15**, 9 (1977).
4606. Klesper, E., and Hartman, W., *J. Polym. Sci. B*, **15**, 707 (1977).
4607. Klesper, E., Johnsen, A., and Gronski, W., *J. Polym. Sci. B*, **8**, 369 (1970).
4608. Klesper, E., Johnsen, A., Gronski, W., and Wehrli, F. W., *Makromol. Chem.*, **176**, 1071 (1975).
4609. Kline, D. E., *J. Polym. Sci.*, **50**, 441 (1961).
4610. Kloos, F., Go, S., and Mandelkern, L., *J. Polym. Sci. A2*, **12**, 1145 (1974).
4611. Klöpffer, W., *J. Chem. Phys.*, **50**, 2337 (1969).

4612. Klöpffer, W., *Ber. Bunsenges. Phys. Chem.*, **73**, 864 (1969).
4613. Klug, A., Crick, F. H. C., and Wyckoff, H. W., *Acta Cryst.*, **11**, 199 (1958).
4614. Knapstein, P., and Touchstone, J. C., *J. Chromatogr.*, **37**, 83 (1968).
4615. Kneübuhl, F. K., *J. Chem. Phys.*, **33**, 1074 (1960).
4616. Knox, B. H., *J. Appl. Polym. Sci.*, **21**, 225, 249, 267 (1977).
4617. Knox, J. H., and Saleem, H., *J. Chromatogr. Sci.*, **7**, 614 (1969).
4618. Ko, J. H., and Mark, J. E., *Macromolecules*, **8**, 869 (1975).
4619. Ko, H. W., and Eitel, M. J., *Macromolecules*, **1**, 364 (1968).
4620. Kobayashi, E., Okamura, S., and Singer, R., *J. Appl. Polym. Sci.*, **12**, 1661 (1968).
4621. Kobayashi, H., *J. Polym. Sci.*, **26**, 230 (1957).
4622. Kobayashi, H., and Fujisaki, Y., *J. Polym. Sci. B*, **1**, 15 (1963).
4623. Kobayashi, K., and Nagasawa, T., *J. Macromol. Sci. B*, **4**, 351 (1970).
4624. Kobayashi, M., Tashiro, K., and Tadeokoro, H., *Macromolecules*, **8**, 158 (1975).
4625. Kobayashi, S., Murakami, K., Chatani, Y., and Tadeokoro, H., *J. Polym. Sci. B*, **14**, 591 (1976).
4626. Kobayashi, T., Chitale, A., and Frank, H. P., *J. Polym. Sci.*, **24**, 156 (1957).
4627. Koda, S., Nomura, H., and Miyahara, Y., *Macromolecules*, **11**, 604 (1978).
4628. Kodaira, Y., and Harwood, H. J., *ACS Polym. Preprints*, **14**, 323 (1973).
4629. Koenhen, D. M., and Smolders, C. A., *J. Polym. Sci. A2*, **15**, 155 (1977).
4630. Koenhen, D. M., and Smolders, C. A., *J. Polym. Sci. A2*, **15**, 167 (1977).
4631. Koenhen, D. M., Smolders, C. A., and Gordon, M., *J. Polym. Sci. C*, **61**, 93 (1977).
4632. Koenig, J. L., *Appl. Spectr.*, **29**, 293 (1975).
4633. Koenig, J. L., and Agboatwalla, M. C., *J. Macromol. Sci. B*, **2**, 391 (1968).
4634. Koenig, J. L., and Antoon, M. K., *J. Polym. Sci. A2*, **15**, 1379 (1977).
4635. Koenig, J. L., and Boerio, F. J., *J. Chem. Phys.*, **50**, 2823 (1969).
4636. Koenig, J. L., and Cornell, S. W., *J. Macromol. Sci. B*, **1**, 279 (1967).
4637. Koenig, J. L., Cornell, S. W., and Witenhafer, D. E., *J. Polym. Sci. A2*, **5**, 301 (1967).
4638. Koenig, J. L., and Druesdow, D., *J. Polym. Sci. A2*, **7**, 1075 (1969).
4639. Koenig, J. L., and Hannon, M. J., *J. Macromol. Sci. B*, **1**, 119 (1967).
4640. Koenig, J. L., Wolfram, L., and Grasselli, J., *ACS Polym. Preprints*, **10**, 959 (1969).
4641. Koepp, H. M., and Werner, H., *Makromol. Chem.*, **32**, 79 (1959).
4642. Kohler, A., Zilliox, J. G., Rempp, P., Polacek, J., and Koessler, I., *Europ. Polym. J.*, **8**, 627 (1972).
4643. Kojima, M., Magill, J. H., and Merker, R. L., *J. Polym. Sci. A2*, **12**, 317 (1974).
4644. Kojima, T., and Morishita, F., *J. Chromatogr. Sci.*, **8**, 471 (1970).
4645. Kolarik, J., and Janacek, J., *J. Polym. Sci. C*, **16**, 441 (1967).
4646. Kolb, B., and Kaiser, K. H., *J. Gas Chromatogr.*, **2**, 233 (1964).
4647. Kolb, H. J., and Izard, E. F., *J. Appl. Phys.*, **20**, 564 (1949).
4648. Kolesov, V. P., Paukov, I. E., and Skuratov, S. M., *Zh. Fiz. Khim.*, **36**, 770 (1962).
4649. Kolpak, F. J., and Blackwell, J., *Macromolecules*, **9**, 273 (1976).
4650. Komoroski, R. A., and Mandelkern, L., *J. Polym. Sci. B*, **14**, 253 (1976).
4651. Komoroski, R. A., Maxfield, J., and Mandelkern, L., *Macromolecules*, **10**, 545 (1977).
4652. Komoroski, R. A., Maxfield, J., Sakaguchi, F., and Mandelkern, L., *Macromolecules*, **10**, 550 (1977).
4653. Konda, A., Nose, K., and Ishikawa, H., *J. Polym. Sci. A2*, **14**, 1495 (1976).
4654. Kondo, Y., Lizuka, E., Oka, A., and Hayakawa, T., *Polymer*, **18**, 111 (1977).
4655. Kong, J. M., and Hawkes, S. J., *Macromolecules*, **8**, 685 (1975).
4656. Koningsveld, R., *Discuss. Faraday Soc.*, **49**, 144 (1970).
4657. Koningsveld, R., and Kleintjens, L. A., *Macromolecules*, **4**, 637 (1971).
4658. Koningsveld, R., and Kleintjens, L. A., *J. Polym. Sci. C*, **21**, 221 (1977).
4659. Koningsveld, R., Kleintjens, L. A., and Markert, G., *Macromolecules*, **10**, 1105 (1977).
4660. Koningsveld, R., Kleintjens, L. A., and Shultz, A. R., *J. Polym. Sci. A2*, **8**, 1261 (1970).
4661. Koningsveld, R., and Pennings, A. J., *Rec. Trav. Chim.*, **83**, 552 (1964).
4662. Koningsveld, R., and Staverman, A. J., *J. Polym. Sci. A2*, **6**, 305, 325 (1968).
4663. Koningsveld, R., and Staverman, A. J., *J. Polym. Sci. A2*, **6**, 367, 383 (1968).
4664. Koningsveld, R., and Stepto, R. F. T., *Macromolecules*, **10**, 1166 (1977).
4665. Koningsveld, R., Stockmayer, W. H., Kennedy, J. W., and Kleintjens, L. A., *Macromolecules*, **7**, 73 (1974).
4666. Koningsveld, R., and Tuijnman, C. A. F., *J. Polym. Sci.*, **39**, 445 (1959).

4667. Konno, S., Saeki, S., Kuwahara, N., Nakata, M., and Kaneko, M., *Macromolecules*, **8**, 799 (1975).
4668. Koppel, D. E., *J. Appl. Phys.*, **42**, 3216 (1971).
4669. Koppel, D. E., *J. Chem. Phys.*, **57**, 4814 (1972).
4670. Koros, W. J., and Paul, D. R., *J. Polym. Sci. A2*, **14**, 1903 (1976).
4671. Koros, W. J., Paul, D. R., Fuji, M., Hopfenberg, H. B., and Stannett, V., *J. Appl. Polym. Sci.*, **21**, 2899 (1977).
4672. Koros, W. J., Paul, D. R., and Rocha, A. A., *J. Polym. Sci. A2*, **14**, 687 (1976).
4673. Korshak, V. V., Paulova, S. A., Boiko, L. V., Babchinster, T. M., Vinogradova, S. V., Vygodskii, Ya. S., and Gulobeva, N. A., *Vysokomol. Soedin. A*, **12**, 56 (1970).
4674. Korte, F., and Glet, W., *J. Polym. Sci. B*, **4**, 685 (1966).
4675. Kosfeld, R., and von Mylius, V., *Kolloid Z. Z. Polym.*, **250**, 1088 (1972).
4676. Kössler, I., and Krejsa, J., *J. Polym. Sci.*, **35**, 308 (1959).
4677. Kössler, I., and Krejsa, J., *J. Polym. Sci.*, **57**, 509 (1962).
4678. Kössler, I., and Vodehnal, J., *J. Polym. Sci B*, **1**, 415 (1963).
4679. Koszewski, J., and Grabowski, Z. R., *Bull. Acad. Polon. Sci., Ser. Sci. Chim.*, **11**, 165 (1963).
4680. Kosztersznitz, G., Barnikol, W. K. R., and Schulz, G. V., *Makromol. Chem.*, **178**, 1133 (1977).
4681. Kosztersznitz, G., Greschner, G. S., and Schulz, G. V., *Makromol. Chem.*, **178**, 1169 (1977).
4682. Kosztersznitz, G., and Schulz, G. V., *Makromol. Chem.*, **178**, 1149 (1977).
4683. Kosztersznitz, G., and Schulz, G. V., *Makromol. Chem.*, **178**, 2437 (1977).
4684. Kotaka, T., *J. Appl. Polym. Sci.*, **21**, 501 (1977).
4685. Kotaka, T., and Donkai, N., *J. Polym. Sci. A2*, **6**, 1457 (1968).
4686. Kotaka, T., and White, J. L., *Macromolecules*, **7**, 106 (1974).
4687. Kotera, A., Saito, T., Takamisawa, K., Miyazawa, Y., Nomura, H., Kamata, T., Yamaguchi, K., and Kawaguchi, H., *Rept. Progr. Polym. Phys. Japan*, **3**, 58 (1960).
4688. Koutsky, J. A., Walton, A. G., and Baer, E., *J. Polym. Sci. A2*, **4**, 611 (1966).
4689. Koutsky, J. A., Walton, A. G., and Baer, E., *J. Polym. Sci. B*, **5**, 177 (1967).
4690. Koutsky, J. A., Walton, A. G., and Baer, E., *J. Polym. Sci. B*, **5**, 185 (1967).
4691. Kovacs, A. J., *Rheol. Acta*, **5**, 262 (1966).
4692. Kovacs, A. J., *Rubber Chem. Technol.*, **41**, 555 (1968).
4693. Kovacs, A. J., and Hobbs, S. Y., *J. Appl. Polym. Sci.*, **16**, 301 (1972).
4694. Kovacs, A. J., Straupe, C., and Gonthier, A., *J. Polym. Sci. C*, **59**, 31 (1977).
4695. Kovar, J., Fortelny, I., and Bohdanecky, M., *Makromol. Chem.*, **178**, 2375 (1977).
4696. Kovar, J., Mrkvičková, L., and Bohdanecký, M., *Makromol. Chem.*, **176**, 1829 (1975).
4697. Kovarskaya, B. M., Golubenkova, L. I., Akutin, M. S., and Levantovskaya, I. I., *Vysokomol. Soedin.*, **1**, 1042 (1959).
4698. Kovarskii, A. L., Placek, J., and Szöcs, F., *Polymer*, **19**, 1137 (1978).
4699. Kowalski, B. R., Jurs, P. C., Isenhour, T. L., and Reilley, C. N., *Anal. Chem.*, **41**, 1945 (1969).
4700. Kozlowski, W., *J. Polym. Sci. C*, **38**, 47 (1972).
4701. Kraemer, E. O., *Ind. Eng. Chem.*, **30**, 1200 (1938).
4702. Kramer, O., and Ferry, J. D., *J. Polym. Sci. A2*, **15**, 761 (1977).
4703. Kramer, O., and Frederick, J. E., *Macromolecules*, **4**, 613 (1971).
4704. Kramer, O., and Frederick, J. E., *Macromolecules*, **5**, 69 (1972).
4705. Kratky, O., *Z. Elektrochem.*, **58**, 49 (1954).
4706. Kratky, O., *Kolloid Z.*, **144**, 110 (1955).
4707. Kratky, O., *Z. Elektrochem.*, **62**, 66 (1958).
4708. Kratky, O., Leopold, H., and Stalinger, H., *Angew. Phys.*, **27**, 273 (1969).
4709. Kratky, O., and Porod, G., *J. Colloid Sci.*, **4**, 35 (1949).
4710. Kratky, O., Porod, G., Paletta, B., and Sekora, A., *J. Polym. Sci.*, **16**, 163 (1955).
4711. Kratky, O., and Sekora, A., *Monatsh. Chem.*, **85**, 660 (1954).
4712. Kratky, O., and Skala, Z., *Z. Elektrochem.*, **62**, 73 (1958).
4713. Kratohvil, J. P., *Anal. Chem.*, **38**, 517R (1966).
4714. Kratohvil, J. P., and Vorliček, J., *J. Polym. Sci. A2*, **14**, 1561 (1976).
4715. Kraus, G., and Stacy, C. J., *J. Polym. Sci. A2*, **10**, 657 (1972).
4716. Krause, S., and O'Konski, C. T., *Biopolymers*, **1**, 503 (1963).

4717. Krause, S., and Stroud, D. E., *J. Polym. Sci. A2*, **11**, 2253 (1973).
4718. Krebs, K. F., and Wunderlich, W., *Angew. Makromol. Chem.*, **20**, 203 (1971).
4719. Krejsa, J., *Makromol. Chem.*, **33**, 244 (1959).
4720. Kremen, J., and Shapiro, J. J., *J. Opt. Soc.*, **44**, 500 (1954).
4721. Kreshkov, A. P., Shevetsova, L. N., and Emelin, E. A., *Plast. Massy*, **10**, 52 (1968).
4722. Kricheldorf, H. R., Leppert, E., and Schilling, G., *Makromol. Chem.*, **176**, 1629 (1975).
4723. Kricheldorf, H. R., and Luderwald, I., *Makromol. Chem.*, **179**, 421 (1978).
4724. Krigbaum, W. R., *J. Polym. Sci.*, **18**, 315 (1955).
4725. Krigbaum, W. R., Balta, Y. I., and Via, G. H., *Polymer*, **7**, 61 (1966).
4726. Krigbaum, W. R., and Flory, P. J., *J. Polym. Sci.*, **11**, 37 (1953).
4727. Krigbaum, W. R., and Geymer, D. O., *J. Amer. Chem. Soc.*, **81**, 1859 (1959).
4728. Krigbaum, W. R., and Kurz, J. E., *J. Polym. Sci.*, **41**, 275 (1962).
4729. Krigbaum, W. R., Kurz, J. E., and Smith, P., *J. Phys. Chem.*, **65**, 1984 (1961).
4730. Krigbaum, W. R., and Salaria, F., *J. Polym. Sci. A2*, **16**, 883 (1978).
4731. Krigbaum, W. R., and Wall, F. T., *J. Polym. Sci.*, **5**, 505 (1950).
4732. Krigbaum, W. R., Yazgan, S., and Tolbert, W. R., *J. Polym. Sci. A2*, **11**, 511 (1973).
4733. Krimm, S., *Pure Appl. Chem.*, **16**, 369 (1968).
4734. Krimm, S., and Bank, M. I., *J. Chem. Phys.*, **42**, 4059 (1965).
4735. Krimm, S., and Bank, M. I., *J. Polym. Sci. A2*, **7**, 1785 (1969).
4736. Krimm, S., Berens, A. R., Folt, V. L., and Shipman, J. J., *Chem. Ind. (London)*, **1958**, 1512 (1958).
4737. Krimm, S., Berens, A. R., Folt, V. L., and Shipman, J. J., *Chem. Ind. (London)*, **1959**, 433 (1959).
4738. Krimm, S., and Enomoto, S., *J. Polym. Sci. A2*, **2**, 669 (1964).
4739. Krimm, S., Folt, V. L., Shipman, J. J., and Berens, A. R., *J. Polym. Sci. A1*, **1**, 2621 (1963).
4740. Krimm, S., and Liang, C. Y., *J. Polym. Sci.*, **22**, 95 (1956).
4741. Krimm, S., Liang, C. Y., and Sutherland, G. B. B. M., *J. Polym. Sci.*, **22**, 227 (1956).
4742. Krimm, S., Liang, C. Y., and Sutherland, G. B. B. M., *J. Chem. Phys.*, **25**, 549 (1956).
4743. Krimm, S., and Stein, R. S., *Rev. Sci. Instrum.*, **22**, 920 (1951).
4744. Krimm, S., and Tobolsky, A. V., *J. Polym. Sci.*, **7**, 57 (1951).
4745. Krishnamurthy, S., McIntyre, D., Santee, E. R., Jr, and Wilson, C. W., *J. Polym. Sci. A2*, **11**, 427 (1972).
4746. Krozer, S., *Makromol. Chem.*, **175**, 1905 (1974).
4747. Krozer, S., Carstensen, P., Bodelsen, M., and Thirsig, A., *Makromol. Chem.*, **175**, 1893 (1974).
4748. Krozer, S., Wainryb, M., and Silina, L., *Vysokomol. Soedin.*, **2**, 1876 (1960).
4749. Krška, F., and Dušek, K., *J. Polym. Sci. C*, **38**, 121 (1972).
4750. Krueger, D., and Yeh, G. S. Y., *J. Macromol. Sci. B*, **6**, 431 (1972).
4751. Kruse, W. A., Kirste, R. G., Haas, J., Schmitt, B. J., and Stein, D. J., *Makromol. Chem.*, **177**, 1145 (1976).
4752. Kryszewski, M., Galeski, A., and Milczarek, P., *Plaste u. Kautschuk*, **9**, 656 (1974).
4753. Kryszewski, M., Galeski, A., Pakula, T., and Grebowicz, J., *J. Colloid Interface Sci.*, **44**, 85 (1973).
4754. Kryszewski, M., Zielinski, M., and Sapieha, S., *Polymer*, **17**, 212 (1976).
4755. Kubata, K., *J. Polym. Sci. B*, **3**, 403 (1965).
4756. Kubitz, K. A., *Anal. Chem.*, **29**, 814 (1957).
4757. Kubo, S., and Dole, M., *Macromolecules*, **6**, 774 (1973).
4758. Kubo, S., and Dole, M., *Macromolecules*, **7**, 190 (1974).
4759. Kubota, K., *J. Polym. Sci. B*, **3**, 545 (1965).
4760. Kuhlmann, K. F., Grant, D. M., and Harris, R. K., *J. Chem. Phys.*, **52**, 3439 (1970).
4761. Kuhn, R., *Makromol. Chem.*, **177**, 1525 (1976).
4762. Kuhn, W. H., and Schulz, G. V., *Makromol. Chem.*, **50**, 52 (1961).
4763. Kulshreshta, A. K., Khan, A. H., and Madan, G. L., *Polymer*, **19**, 819 (1978).
4764. Kumbar, M., and Windwer, S., *Macromolecules*, **7**, 624 (1974).
4765. Kumler, P. L., and Boyer, R. F., *Macromolecules*, **9**, 903 (1976).
4766. Kupke, D. W., *Adv. Protein Chem.*, **15**, 57 (1960).
4767. Kurata, M., *J. Polym. Sci. A2*, **6**, 1607 (1968).
4768. Kurata, M., Abe, M., Iwama, M., and Matsushima, M., *Polymer J.*, **3**, 729 (1972).

4769. Kurata, M., and Fukatsu, M., *J. Chem. Phys.*, **41**, 2934 (1964).
4770. Kurata, M., Fukatsu, M., Sotobayashi, H., and Yamakawa, H., *J. Chem. Phys.*, **41**, 139 (1964).
4771. Kurata, M., and Stockmayer, W. H., *Adv. Polym. Sci.*, **3**, 196 (1963).
4772. Kurata, M., Stockmayer, W. H., and Roig, A., *J. Chem. Phys.*, **33**, 151 (1960).
4773. Kurata, M., and Yamakawa, H., *J. Chem. Phys.*, **29**, 311 (1959).
4774. Kusakov, M. M., Koshevnik, A. Yu., Nekrasov, D. N., Chirkova, V. F., and Sholpina, L. M., *Vysokomol. Soedin.*, **8**, 1040 (1966).
4775. Kusanagi, H., Tadekoro, H., and Chatani, Y., *Macromolecules*, **9**, 531 (1976).
4776. Kusanagi, H., Tadekoro, H., Chatani, Y., and Suehiro, K., *Macromolecules*, **10**, 405 (1977).
4777. Kusanagi, H., Takase, M., Chatani, Y., and Tadekoro, H., *J. Polym. Sci. A2*, **16**, 131 (1978).
4778. Kushner, L. M., *J. Opt. Soc.*, **44**, 155 (1954).
4779. Kusumoto, N., and Mukoyama, H., *Rep. Progr. Polym. Phys. Japan*, **15**, 581 (1972).
4780. Kusumoto, N., and Mukoyama, H., *Rep. Progr. Polym. Phys. Japan*, **16**, 551, 555 (1973).
4781. Kusumoto, N., Sano, S., Zaitsu, N., and Motozato, Y., *Polymer*, **17**, 448 (1976).
4782. Kusumoto, N., Yamaoka, T., and Takayanagi, M., *J. Polym. Sci. A2*, **9**, 1173 (1971).
4783. Kusumoto, N., Yonezawa, M., and Motozato, Y., *Polymer*, **15**, 793 (1974).
4784. Kusy, R. P., *J. Polym. Sci. A1*, **14**, 1527 (1976).
4785. Kusy, R. P., and Turner, D. T., *Macromolecules*, **8**, 235 (1975).
4786. Kusy, R. P., and Turner, D. T., *Polymer*, **18**, 391 (1977).
4787. Kusy, R. P., and Turner, D. T., *Macromolecules*, **10**, 493 (1977).
4788. Kuwahara, N., *J. Polym. Sci. A*, **1**, 2395 (1963).
4789. Kuwahara, N., Kojima, J., and Kaneko, M., *J. Polym. Sci. A2*, **11**, 2307 (1973).
4790. Kuwahara, N., Nakata, M., and Kaneko, M., *Polymer*, **14**, 415 (1973).
4791. Kuwahara, N., Saeki, S., Chiba, T., and Kaneko, M., *Polymer*, **15**, 777 (1974).
4792. Kuwahara, N., Saeki, S., Konno, S., and Kaneko, M., *Polymer*, **15**, 66 (1974).
4793. Kwei, K. P. S., *J. Polym. Sci. A*, **1**, 2309 (1963).
4794. Kwei, T. K., Nishi, T., and Roberts, R. F., *Macromolecules*, **7**, 667 (1974).
4795. Kwei, T. K., and Wang, T. T., *Macromolecules*, **5**, 128 (1972).
4796. Kwei, T. K., Wang, T. T., and Zupko, H. M., *Macromolecules*, **1**, 65 (1969).
4797. Kwei, T. K., and Zupko, H. M., *J. Polym. Sci. A2*, **7**, 867 (1969).
4798. Kyotani, M., and Kanetsuna, H., *J. Polym. Sci. A2*, **12**, 2331 (1974).
4799. Labsky, J., Pilar, J., and Kalal, J., *Macromolecules*, **10**, 1153 (1977).
4800. Lacabanne, C., Chatain, D., Guillet, J., Seytre, G., and May, J. F., *J. Polym. Sci. A2*, **13**, 445 (1975).
4801. Lacharojana, S., and Caroline, D., *Macromolecules*, **10**, 365 (1977).
4802. Lacey, P. M. C., *Chem. Eng. Sci.*, **29**, 1495 (1974).
4803. Lai, J. H., *Macromolecules*, **10**, 1253 (1977).
4804. Lai, J. H., *J. Appl. Polym. Sci.*, **20**, 847 (1977).
4805. Lal, M., and Spencer, D., *Mol. Phys.*, **22**, 649 (1971).
4806. La Mantia, F. P., and Acierno, D., *Polymer*, **19**, 851 (1978).
4807. Lambert, A., *Polymer*, **10**, 319 (1969).
4808. Lang, M. C., Nöel, C., and Legrand, A. P., *J. Polym. Sci. A2*, **15**, 1319 (1977).
4809. Lang, M. C., Nöel, C., and Legrand, A. P., *J. Polym. Sci. A2*, **15**, 1329 (1977).
4810. Langelaar, J., de Vries, G. A., and Bebelaar, D., *J. Sci. Instrum.*, **2**, 149 (1969).
4811. Langevin, D., and Rondelez, F., *Polymer*, **19**, 975 (1978).
4812. Langhammer, G., *Svensk. Kem. Tidskr.*, **69**, 328 (1957).
4813. Langhammer, G., *J. Polym. Sci.*, **29**, 505 (1959).
4814. Langhammer, G., Pfennig, H., and Quitzsch, K., *Z. Elektrochem.*, **62**, 458 (1958).
4815. Langhammer, G., and Quitzsch, K., *Makromol. Chem.*, **17**, 74 (1955).
4816. Lanikova, J., and Hlousek, M., *Europ. Polym. J.*, **6**, 25 (1970).
4817. Lansing, W. D., and Kraemer, E. O., *J. Amer. Chem. Soc.*, **57**, 1369 (1935).
4818. Lapčik, K., Panak, J., Kellö, V., and Polavka, J., *J. Polym. Sci. A2*, **14**, 981 (1976).
4819. Lapčik, L., and Sundelöf, L. O., *Chem. Scripta*, **2**, 41 (1972).
4820. Larkins, J. H., Perrings, J. D., Shepherd, G. R., and Noland, B. J., *J. Chem. Educ.*, **42**, 555 (1965).

4821. Larsen, D. W., and Strange, J. H., *J. Polym. Sci. A2*, **11**, 65 (1973).
4822. Larsen, D. W., and Strange, J. H., *J. Polym. Sci. A2*, **11**, 449 (1973).
4823. Latimer, P., and Tully, B., *J. Colloid Interface Sci.*, **27**, 475 (1968).
4824. Lau, W. W. Y., and Burns, C. M., *Surface Sci.*, **30**, 478 (1972).
4825. Lau, W. W. Y., and Burns, C. M., *Surface Sci.*, **30**, 497 (1972).
4826. Lau, W. W. Y., and Burns, C. M., *J. Colloid Interface Sci.*, **45**, 295 (1973).
4827. Lau, W. W. Y., and Burns, C. M., *J. Polym. Sci. A2*, **12**, 431 (1974).
4828. Lau, K. H., and Young, K., *Polymer*, **16**, 477 (1975).
4829. Lau, K. H., and Young, K., *Polymer*, **17**, 7 (1976).
4830. Laupretre, F., and Geny, F., *Europ. Polym. J.*, **14**, 401 (1978).
4831. Laupretre, F., Nöel, C., and Monnerie, L., *J. Polym. Sci. A2*, **15**, 2127 (1977).
4832. Laurent, T. C., and Killander, J., *J. Chromatogr.*, **14**, 317 (1964).
4833. Lauritzen, J. I., and Hoffman, J. D., *J. Res. Nat. Bur. Stand. A*, **64**, 73 (1960).
4834. Laustriat, G., Coche, A., Lami, H., and Pfeffer, G., *Compt. Rend*, **257**, 434 (1963).
4835. Laven, J., van den Esker, M. W. J., and Vrij, A., *J. Polym. Sci. A2*, **13**, 443 (1975).
4836. Lavoie, A., and Guillet, J. E., *Macromolecules*, **2**, 443 (1969).
4837. Law, R. D., *J. Polym. Sci. C*, **21**, 225 (1968).
4838. Law, R. D., *J. Polym. Sci. A2*, **7**, 2097 (1969).
4839. Lawrence, K. G., *Chem. Ind. (London)*, **1966**, 1338 (1966).
4840. Lawson, K. D., and Flautt, T. J., *J. Phys. Chem.*, **69**, 4256 (1965).
4841. Lawton, E. J., *J. Polym. Sci. A1*, **10**, 1857 (1972).
4842. Lawton, E. J., and Balwit, J. S., *J. Phys. Chem.*, **65**, 815 (1961).
4843. Lawton, E. J., Balwit, J. S., and Powell, R. S., *J. Chem. Phys.*, **33**, 395, 405 (1960).
4844. Lax, M., and Gillis, J., *Macromolecules*, **10**, 334 (1977).
4845. Lebedev, B. V., Rabinovich, I. B., and Budarina, V. A., *Vysokomol. Soedin. A*, **9**, 488 (1967).
4846. Lebedev, B. V., Rabinovich, I. B., and Martynenko, L. Ya., *Vysokomol. Soedin. A*, **9**, 1640 (1967).
4847. Leblanc, J. L., *Polymer*, **17**, 235 (1976).
4848. Lechner, M. D., *J. Polym. Sci. C*, **61**, 63 (1977).
4849. Lechner, M. D., *Europ. Polym. J.*, **14**, 51 (1978).
4850. Lechner, M. D., and Schulz, G. V., *Makromol. Chem.*, **172**, 161 (1973).
4851. Lecomte, L., and Desreux, V., *Polymer*, **16**, 765 (1975).
4852. Lednicky, F., and Pelzbauer, Z., *J. Polym. Sci. C*, **38**, 375 (1972).
4853. Lee, A. K., and Sedgwick, R. D., *J. Polym. Sci. A1*, **16**, 685 (1978).
4854. Lee, H., Swartz, M. L., and Stoffey, D. G., *Appl. Polym. Symp.*, No. 16, **1** (1971).
4855. Lee, J. L., and Wunderlich, B., *J. Polym. Sci. A2*, **13**, 607 (1975).
4856. Lee, L. H., *J. Polym. Sci. A2*, **5**, 1103 (1967).
4857. Lee, L. H., *J. Appl. Polym. Sci.*, **12**, 719 (1968).
4858. Lee, S., and Simha, R., *Macromolecules*, **7**, 909 (1974).
4859. Lee, W. A., *J. Polym. Sci. A2*, **8**, 555 (1970)
4860. Lee, W. K., and Choy, C. L., *J. Polym. Sci. A2*, **13**, 619 (1975).
4861. Lee, W. K., Lau, P. C., and Choy, C. L., *Polymer*, **15**, 487 (1974).
4862. LeFèvre, R. J. W., and Sundaraman, K. M. S., *J. Chem. Soc.*, **1962**, 1494, 4003 (1962).
4863. LeFèvre, R. J. W., and Sundaraman, K. M. S., *J. Chem. Soc.*, **1963**, 1880, 3188, 3547 (1963).
4864. LeFèvre, R. J. W., and Sundaraman, K. M. S., *J. Chem. Soc.*, **1964**, 556, 3518 (1964).
4865. LeFèvre, R. J. W., Sundaraman, A., and Sundaraman, K. M. S., *J. Chem. Soc.*, **1963**, 3180 (1963).
4866. Leffingwell, J., Thies, C., and Gertzman, H., *ACS Polym. Preprints*, **14**, 596 (1973).
4867. LeGrand, D. G., *J. Polym. Sci. A2*, **8**, 1937 (1970).
4868. LeGrand, D. G., *Macromolecules*, **3**, 764 (1970).
4869. LeGrand, D. G., and Haaf, W. R., *J. Polym. Sci. C*, **5**, 153 (1964).
4870. Lehmann, F. A., and Brauer, G. M., *Anal. Chem.*, **33**, 673 (1961).
4871. Lehrle, R. S., and Majury, T. G., *J. Polym. Sci.*, **29**, 219 (1958).
4872. Lehrle, R. S., and Robb, J. C., *J. Gas Chromatogr.*, **5**, 89 (1967).
4873. Leighton, W. G., and Forbes, G. S., *J. Amer. Chem. Soc.*, **52**, 3139 (1930).
4874. Leite, R. C. C., Moore, R. S., and Porto, S. P. S., *J. Chem. Phys.*, **40**, 3741 (1964).
4875. Lekkerkerker, H. N. W., and Laidlaw, W. G., *Phys. Chem. Liquids*, **3**, 175 (1972).
4876. Lelievre, J., *Polymer*, **17**, 854 (1976).

794

4877. Lemaire, B., and Fourche, G., *J. Polym. Sci. A2*, **12**, 1137 (1974).
4878. Lemaire, B., Fourche, G., and Sanchez, E., *J. Polym. Sci. A2*, **12**,, 417 (1974).
4879. Lemstra, P. J., and Challa, G., *J. Polym. Sci. A2*, **13**, 1809 (1975).
4880. Lemstra, P. J., Postma, J., and Challa, G., *Polymer*, **15**, 757 (1974).
4881. Lemstra, P. J., Schouten, A. J., and Challa, G., *J. Polym. Sci. A2*, **12**, 1565 (1874).
4882. Lenz, R. W., Lüderwald, I., Montaudo, G., Przybylski, M., and Ringsdorf, H., *Makromol. Chem.*, **175**, 2441 (1974).
4883. Lenz, R. W., Regel, W., and Westfelt, L., *Makromol. Chem.*, **176**, 781 (1975).
4884. Leonis, C. G., Suzuki, H., and Gordon, M., *Makromol. Chem.*, **178**, 2867 (1977).
4885. Le Page, M., Beau, R., and De Vries, A. J., *J. Polym. Sci. C*, **21**, 119 (1968).
4886. Lepori, L., and Mollica, V., *J. Polym. Sci. A2*, **16**, 1123 (1978).
4887. Leray, J., and Gramain, P., *J. Chim. Phys.*, **62**, 976 (1965).
4888. Lerner, N. R., *J. Chem. Phys.*, **50**, 2902 (1969).
4889. Lesiak, T., Hetper, J., Pielichowski, J., Prewesz-Kwinto, A., and Marzec, K., *Polymer*, **17**, 1110 (1976).
4890. Leugering, H. J., *Makromol. Chem.*, **109**, 204 (1967).
4891. Leung Yu-Kwan, *Polymer*, **17**, 374 (1976).
4892. Leung Yu-Kwan and Eichinger, B. E., *J. Phys. Chem.*, **78**, 60 (1974).
4893. Leung Yu-Kwan and Eichinger, B. E., *Macromolecules*, **7**, 685 (1974).
4894. Leung Yu-Kwan and Eichinger, B. E., *Rubber Chem. Technol.*, **48**, 108, 119 (1975).
4895. Leute, U., *J. Polym. Sci. A2*, **15**, 715 (1977).
4896. Leute, U., and Smith, T. L., *Macromolecules*, **11**, 707 (1978).
4897. Levene, A., Pullen, W. J., and Roberts, J., *J. Polym. Sci. A1*, **3**, 697 (1965).
4898. Levesque, D., and Prud'homme, R. E., *J. Polym. Sci. A2*, **15**, 1613 (1977).
4899. Levy, A. C., *J. Polym. Sci. B*, **15**, 219 (1977).
4900. Levy, P. F., Nieuweboer, G., and Szymanski, L. C., *Thermochim. Acta*, **3**, 259 (1972).
4901. Lewis, G., and Johanson, A. F., *J. Chem. Soc. A*, **12**, 1816 (1969).
4902. Lewis, I. C., and Petro, B. A., *J. Polym. Sci. A1*, **14**, 1975 (1976).
4903. Lewis, P. R., and Price, C., *Polymer*, **12**, 258 (1971).
4904. Li, H. M., and Magill, J. H., *Polymer*, **19**, 829 (1978).
4905. Li, L. S., and Kargin, V. A., *Vysokomol. Soedin.*, **3**, 1102 (1961).
4906. Liang, C. Y., and Krimm, S., *J. Chem. Phys.*, **25**, 563 (1956).
4907. Liang, C. Y., and Krimm, S., *J. Polym. Sci.*, **27**, 241 (1958).
4908. Liang, C. Y., and Krimm, S., *J. Polym. Sci.*, **31**, 513 (1958).
4909. Liang, C. Y., and Krimm, S., *J. Mol. Spectr.*, **3**, 554 (1959).
4910. Liang, C. Y., and Lytton, M. R., *J. Polym. Sci.*, **61**, S45 (1962).
4911. Liang, C. Y., Lytton, M. R., and Boone, C. J., *J. Polym. Sci.*, **44**, 549 (1960).
4912. Liang, C. Y., Lytton, M. R., and Boone, C. J., *J. Polym. Sci.*, **54**, 523 (1961).
4913. Liang, C. Y., and Pearson, F. G., *J. Polym. Sci.*, **35**, 303 (1959).
4914. Liang, C. Y., and Pearson, F. G., *J. Mol. Spectr.*, **5**, 290 (1960).
4915. Liang, C. Y., Pearson, F. G., and Marchessault, R. H., *Spectrochim. Acta*, **17**, 568 (1961).
4916. Liang, C. Y., and Watt, W. R., *J. Polym. Sci.*, **51**, S14 (1961).
4917. Liaw, Gin-Chain, Zakin, J. L., and Patterson, G. K., *AIChE J.*, **17**, 391 (1971).
4918. Libby, D., Ormerod, M. G., and Charlesby, A., *Polymer*, **1**, 212 (1960).
4919. Liberman, M. H., Debolt, L. C., and Flory, P. J., *J. Polym. Sci. A2*, **12**, 187 (1974).
4920. Licht, W. R., and Kline, D. E., *J. Polym. Sci. A2*, **4**, 313 (1966).
4921. Lichtenthalter, R. N., Liu, D. D., and Prausnitz, J. M., *Macromolecules*, **7**, 565 (1974).
4922. Lichtenthalter, R. N., Newman, R. D., and Prausnitz, J. M., *Macromolecules*, **6**, 650 (1973).
4923. Liddell, A. H., and Swinton, F. L., *Discuss. Faraday Soc.*, **49**, 115 (1970).
4924. Liebman, S. A., Ahlstrom, D. H., and Foltz, C. R., *J. Chromatogr.*, **67**, 153 (1972).
4925. Liebman, S. A., Foltz, C. R., Reuwer, J. F., and Obremski, R. J., *Macromolecules*, **4**, 134 (1971).
4926. Liebman, S. A., Reuwer, J. F., Gollatz, K. A., and Nauman, C. D., *J. Polym. Sci. A1*, **9**, 1823 (1971).
4927. Lieser, G., Fischer, E. W., and Ibel, K., *J. Polym. Sci. B*, **13**, 39 (1975).
4928. Light, T. S. Fitzpatrick, L. F., and Phaneuf, J. P., *Anal. Chem.*, **37**, 79 (1965).
4929. Limpert, R. J., Cotter, R. L., and Dark, W. A., *Amer. Lab.*, May 1974, p. 63.
4930. Lin, L. S., Chou, Y. T., and Hu, H., *J. Polym. Sci. A2*, **13**, 1659 (1975).

795

4931. Lin, O. C. C., *Macromolecules*, **3**, 80 (1970).
4932. Lin, T. P., and Koenig, J. L., *J. Mol. Spectr.*, **9**, 228 (1963).
4933. Lindenmeyer, P. H., *J. Chem. Phys.*, **46**, 1902 (1967).
4934. Linder, W. L., *Polymer*, **14**, 9 (1973).
4935. Lindsay, S. M., Hartley, A. J., and Shepherd, I. W., *Polymer*, **17**, 501 (1976).
4936. Lindström, T., Söremark, C., and Westman, L., *J. Appl. Polym. Sci.*, **21**, 2873 (1977).
4937. Linkens, A., and Vanderschueren, J., *J. Polym. Sci.*, **15**, 41 (1977).
4938. Lipatov, Yu. S., *J. Polym. Sci. C*, **61**, 369 (1977).
4939. Lipatov, Yu. S., Chramova, T. S., Sergeeva, L. M., and Karabanova, L. V., *J. Polym. Sci. A1*, **15**, 427 (1977).
4940. Liaptov, Yu. S., Lobodina, A. P., Maistruk, V. K., Privalko, V. P., and Fainerman, A. E., *Macromolecules*, **7**, 257 (1974).
4941. Lipatov, Yu. S., and Nesterov, A. E., *Macromolecules*, **8**, 889 (1975).
4942. Lipson, H., Nelson, J. B., Riley, B., *J. Sci. Instrum.*, **22**, 184 (1945).
4943. Liquori, A. M., *Acta Crystal.*, **8**, 345 (1955).
4944. Little, J. N., Waters, J. L., Bombaugh, K. J., and Pauplis, W. J., *ACS Polym. Preprints*, **10**, 326 (1969).
4945. Little, J. N., Waters, J. L., Bombaugh, K. J., and Pauplis, W. J., *J. Polym. Sci. A2*, **7**, 1775 (1969).
4946. Little, J. N., Waters, J. L., Bombaugh, K. J., and Pauplis, W. J., *Separation Sci.*, **5**, 765 (1970).
4947. Little, J. N., Waters, J. L., Bombaugh, K. J., and Pauplis, W. J., *J. Chromatogr.*, **9**, 341 (1971).
4948. Liu, D. D., and Prausnitz, J. M., *J. Polym. Sci. A2*, **15**, 145 (1977).
4949. Liu, H. Z., and Liu, K. J., *Macromolecules*, **1**, 157 (1968).
4950. Liu, J. J., and Anderson, J. E., *Macromolecules*, **3**, 163 (1970).
4951. Liu, K. J., *J. Polym. Sci.*, *A2*, **5**, 697 (1967).
4952. Liu, K. J., and Liganowski, S. J., *J. Polym. Sci. B*, **6**, 191 (1968).
4953. Liu, K. J., and Ullman, R., *Macromolecules*, **2**, 525 (1969).
4954. Livingston, H. K., *Macromolecules*, **2**, 98 (1969).
4955. Lloyd, D. R., and Burns, C. M., *J. Appl. Polym. Sci.*, **22**, 593 (1978).
4956. Lobach, M. I., Poletayeva, I. A., Khatchaturov, A. S., Druz, N. N., and Kormer, V. A., *Polymer*, **18**, 1196 (1977).
4957. Loboda-Cačković, J., Hosemann, R., Cačković, H., Ferrerp, F., and Ferrecini, E., *Polymer*, **17**, 303, (1976).
4958. Loboda-Cačković, J., Hosemann, R., and Wilke, W., *Kolloid Z. Z. Polym.*, **235**, 1253 (1969).
4959. Loconti, J. D., and Cahill, J. W., *J. Polym. Sci. A*, **1**, 3163 (1963).
4960. Löfgren, B., Spiliä, R., and Heleskivi, J., *J. Polym. Sci. A2*, **12**, 1547 (1974).
4961. Lombardi, E., Segre, A., Zambelli, A., and Marinangeli, A., *J. Polym. Sci. C*, **16**, 2539 (1967).
4962. Longman, G. W., Wignall, G. D., and Sheldon, R. P., *Polymer*, **17**, 485 (1976).
4963. Longsworth, L. G., *J. Amer. Chem. Soc.*, **74**, 4155 (1952).
4964. Longsworth, L. G., *J. Amer. Chem. Soc.*, **75**, 5705 (1953).
4965. Longworth, J. W., *Photochem. Photobiol.*, **8**, 589 (1968).
4966. Loper, G. D., Wayne, J. P., and Giles, J. W., Jr, *Phys. Lett. A.*, **30**, 403 (1969).
4967. Lorimer, J. W., *Polymer*, **13**, 46, 274 (1972).
4968. Lorimer, J. W., and Jones, D. E. G., *Polymer*, **13**, 52 (1972).
4969. Lotz, B., Kovacs, A. J., Bassett, G. A., and Keller, A., *Kolloid. Z. Z. Polym.*, **209**, 115 (1966).
4970. Lotz, B., Kovacs, A. J., and Wittmann, J. C., *J. Polym. Sci. A2*, **13**, 909 (1975).
4971. Lougnot, D. J., Fouassier, J. P., and Faure, J., *J. Chim. Phys.*, **72**, 125 (1975).
4972. Lovell, R., and Windle, A. H., *Polymer*, **17**, 408 (1976).
4973. Lovinger, A. J., Chua, J. O., and Gryte, C. C., *J. Polym. Sci. A2*, **15**, 641 (1977).
4974. Lovinger, A. J., and Gryte, C. C., *Macromolecules*, **9**, 247 (1976).
4975. Lovinger, A. J., Lau Chi-Ming, and Gryte, C. C., *Polymer*, **17**, 581 (1976).
4976. Lovric, I. L., *J. Polym. Sci. A2*, **7**, 1357 (1969).
4977. Lovric, I. L., *J. Polym. Sci. A2*, **8**, 807 (1970).
4978. Lowry, G. G., *J. Polym. Sci. B*, **1**, 489 (1963).
4979. Loy, B. R., *J. Polym. Sci.*, **50**, 145 (1961).

796

4980. Loy, B. R., *J. Polym. Sci. A*, **1**, 2251 (1963).
4981. Lucas, H. C., Jackson, D. A., Powels, J. G., and Simic-Glavski, B., *Mol. Phys.*, **18**, 505 (1970).
4982. Luderwald, I., Montaudo, G., Przybylski, M., and Ringdorf, H., *Makromol. Chem.*, **175**, 2423 (1974).
4983. Luderwald, I., and Urrutia, H., *Makromol. Chem.*, **177**, 2079, 2093 (1976).
4984. Lue, P. C., Smyth, C. P., and Tobolsky, A. V., *Macromolecules*, **2**, 446 (1969).
4985. Luft, N. W., *J. Chem. Phys.*, **22**, 1814 (1954).
4986. Luft, N. W., *Z. Elektrochem.*, **59**, 46 (1955).
4987. Lukovin, G. M., Komarova, O. P., Torchilin, V. P., and Kirsh, Yu. E., *Vysokomol. Soedin.*, **15**, 443 (1973).
4988. Lumry, R., Müller, A., and Kokubun, H., *Rev. Sci. Instrum.*, **36**, 1214 (1965).
4989. Lundberg, J. L., *J. Colloid Interface Sci.*, **29**, 565 (1969).
4990. Lundberg, J. L., Hellman, M. V., and Frish, H. L., *J. Polym. Sci.*, **46**, 3 (1960).
4991. Lundberg, J. L., Mooney, E. J., and Gardner, K. R., *Science*, **145**, 1308 (1964).
4992. Lundh, S., *J. Polym. Sci. A2*, **15**, 733 (1977).
4993. Lundin, R. E., Elsken, R. H., Flath, R. A., and Terenishi, R., *Appl. Spectr. Rev.*, **1**, 131 (1967).
4994. Lundtrom, J. E., and Bearman, R. J., *J. Polym. Sci. A2*, **12**, 97 (1974).
4995. Luongo, J. P., *J. Appl. Polym, Sci.*, **3**, 302 (1960).
4996. Luongo, J. P., and Salovey, R., *J. Polym. Sci. B*, **3**, 513 (1965).
4997. Luston, J., Manasek, Z., and Kosik, M., *J. Appl. Polym. Sci.*, **21**, 915 (1977).
4998. Lütje, H., *Makromol. Chem.*, **133**, 295 (1970).
4999. Lütje, H., *Makromol. Chem.*, **142**, 81 (1971).
5000. Lyerla, J. R., McIntyre, H. M., and Torchia, D. A., *Macromolecules*, **7**, 11 (1974).
5001. Lynch, J. E., Jr, Summerfield, G. C., Feldkamp, L. A., and King, J. S., *J. Chem. Phys.*, **48**, 912 (1968).
5002. Lynn, T. R., Rushneck, D. R., and Cooper, A. R., *J. Chromatogr. Sci.*, **12**, 76 (1974).
5003. Lyons, B. J., and Fox, A. S., *J. Polym. Sci. C.*, **21**, 159 (1968).

5004. MacCallum, J. R., and Rankin, C. T., *Makromol. Chem.*, **175**, 2477 (1974).
5005. MacCallum, J. R., and Tanner, J., *Europ. Polym. J.*, **6**, 1033 (1970).
5006. Macchi, E. M., Morosoff, N., and Morawetz, H., *J. Polym. Sci. A1*, **6**, 2033 (1968).
5007. Machek, A. L., *J. Elastoplastics*, **1**, 213 (1969).
5008. Machin, D., and Rogers, C. E., *J. Polym. Sci. A2*, **10**, 887 (1972).
5009. Machin, D., and Rogers, C. E., *J. Polym. Sci. A2*, **11**, 1535, 1555 (1973).
5010. Mächtle, W., and Klodwig, U., *Makromol. Chem.*, **177**, 1607 (1976).
5011. Mack, L. L., Kralik, P., Rheude, A., and Dole, M., *J. Chem. Phys.*, **52**, 4977 (1970).
5012. MacKenzie, I. K., *Phys. Lett. A*, **30**, 115 (1969).
5013. MacKenzie, I. K., Eady, J. A., and Gingerich, R. R., *Phys. Lett. A*, **33**, 279 (1970).
5014. Mackenzie, R. C., *Talanta*, **16**, 1227 (1969).
5015. Mackenzie, R. C., Keattch, C. J., Dollimore, D., Forrester, J. A., Hodgson, A. A., and Redfern, J. P., *Thermochim. Acta*, **5**, 71 (1972).
5016. Mackenzie, R. C., Keattch, C. J., Hodgson, A. A., and Redfern, J. P., *Chem. Ind. (London)*, **1970**, 272 (1970).
5017. Mackey, R. C., Pollack, S. A., and White, R. S., *Rev. Sci. Intrum.*, **36**, 1715 (1965).
5018. Mackie, J. S., and Rudin, A., *J. Polym. Sci.*, **49**, 407 (1961).
5019. Mackley, M. R. and Keller, A., *Polymer*, **14**, 16 (1973).
5020. Maclaine, J. Q. G., Ashman, P. C., and Booth, C., *Polymer*, **17**, 109 (1976).
5021. Maclaine, J. Q. G., and Booth, C., *Polymer*, **16**, 191 (1975).
5022. Maclaine, J. Q. G., and Booth, C., *Polymer*, **16**, 680 (1975).
5023. Macosco, C. W., and Miller, D. R., *Macromolecules*, **9**, 199 (1976).
5024. Madorsky, S. L., *J. Polym. Sci.*, **9**, 133 (1952).
5025. Madorsky, S. L., and Strauss, S., *J. Res. Nat. Bur. Stand.*, **55**, 223 (1955).
5026. Maeda, M., Hibi, S., Itoh, F., Namura, S., Kawaguchi, T., and Kawai, H., *J. Poly. Sci. A2*, **8**, 1303 (1970).
5027. Maeda, Y., and Kanetsuna, H., *J. Polym. Sci. A2*, **12**, 2551 (1974).
5028. Maeda, Y., and Kanetsuna, H., *J. Polym. Sci., A2*, **14**, 2057 (1976).
5029. Magagnini, P. L., *Chem. Ind.*, **49**, 1041 (1967).
5030. Magill, J. H., *J. Polym. Sci. A2*, **4**, 243 (1966).

5031. Magill, J. H., and Li, H. M., *Polymer*, **19**, 416 (1978).
5032. Maglio, G., Martuscelli, E., Palumbo, R., and Soldati, I., *Polymer*, **17**, 185 (1976).
5033. Mahabadi, H. K., and O'Driscoll, K. F., *J. Appl. Polym. Sci.*, **21**, 1283 (1977).
5034. Mahl, H., and Weitsch, W., *Z. Naturforsch.*, **15a**, 1051, (1960).
5035. Mair, R. D., and Graupner, A. J., *Anal. Chem.*, **36**, 194 (1964).
5036. Majewska, J., and Warzywoda, J., *Meliand Textilber.*, **43**, 480 (1962).
5037. Majors, R. E., *Intern. Lab.*, Nov./Dec. 1975, p. 11.
5038. Majors, R. E., *J. Chromatogr. Sci.*, **15**, 333 (1977).
5039. Makarewicz, P. J., and Wilkes, G. L., *J. Polym. Sci. A2*, **16**, 1529 (1978).
5040. Makarewicz, P. J., and Wilkes, G. L., *J. Polym. Sci. A2*, **16**, 1559 (1978).
5041. Makaya, G., *Polymer*, **16**, 852 (1975).
5042. Malcolm, G. N., and Rowlinson, J. S., *Trans. Faraday Soc.*, **59**, 921 (1957).
5043. Maley, L. E., *J. Polym. Sci. C*, **8**, 253 (1965).
5044. Malhotra, S. L., Hess, J., and Blanchard, L. P., *Polymer*, **16**, 81 (1975).
5045. Malmberg, J. H., *Rev. Sci. Instrum.*, **28**, 1027 (1957).
5046. Malmstadt, H. V., Barnes, R. M., and Rodgriguez, P. A., *J. Chem. Educ.*, **41**, 263 (1964).
5047. Malone, C. P., Suchan, H. L., and Yau, W. W., *J. Polym. Sci. B*, **7**, 781 (1969).
5048. Manaresi, P., Parrini, P., Semeghini, G. L., and de Fornasari, E., *Polymer*, **17**, 595 (1976).
5049. Manatt, S. L., Lawson, D. D., Ingham, J. D., Rapp, N. S., and Hardy, J. P., *Anal. Chem.*, **38**, 1063 (1966).
5050. Mancarella, C., Martuscelli, E., and Pracella, M., *Polymer*, **17**, 541 (1976).
5051. Mandelkern, L., *J. Appl. Phys.*, **26**, 443 (1955).
5052. Mandelkern, L., Allou, A. L., and Goplan, M., *J. Phys. Chem.*, **72**, 309 (1968).
5053. Mandelkern, L., Fatou, G., Denison, R., and Justin, J., *J. Polym. Sci. B*, **3**, 803 (1965).
5054. Mandelkern, L., Fatou, J. G., and Howard, C., *J. Phys. Chem.*, **68**, 3386 (1964).
5055. Mandelkern, L., Fatou, J. G., and Howard, C., *J. Phys. Chem.*, **69**, 956 (1965).
5056. Mandelkern, L., and Flory, P. J., *J. Amer. Chem. Soc.*, **73**, 3206 (1951).
5057. Mandelkern, L., and Flory, P. J., *J. Chem. Phys.*, **20**, 212 (1952).
5058. Mandelkern, L., Go, S., Peiffer, D., and Stein, R. S., *J. Polym. Sci. A2*, **15**, 1189 (1977).
5059. Mandelkern, L., Jain, N. L., and Kim, H., *J. Polym. Sci. A2*, **6**, 165 (1968).
5060. Mandelkern, L., Krigbaum, W. R., Scheraga, H. A., and Flory, P. J., *J. Chem. Phys.*, **20**, 1392 (1952).
5061. Mandelkern, L., and Quinn, F. A., *J. Appl. Phys.*, **26**, 443 (1955).
5062. Mandelkern, L., Quinn, F. A., and Flory, P. J., *J. Appl. Phys.*, **25**, 830 (1954).
5063. Mandelkern, L., Sharma, R. K., and Jackson, J. F., *Macromolecules*, **2**, 644 (1969).
5064. Mandema, W., and Zeldenrust, H., *Polymer*, **18**, 835 (1977).
5065. Manescalchi, F., Pizzoli, M., Drusiani, A., and Zanetti, F., *Makromol. Chem.*, **178**, 863 (1977).
5066. Mangaraj, D., Bhatnagar, S. K., and Rath, S. B., *Makromol. Chem.*, **67**, 75 (1963).
5067. Manley, R. S. J., *J. Polym. Sci. A2*, **11**, 2303 (1973).
5068. Manely, R. S. J., *J. Polym. Sci. A2*, **12**, 1347 (1974).
5069. Mann, J., and Roldan-Gonzales, L., *J. Polym. Sci.*, **60**, 1 (1962).
5070. Mansfield, M., and Boyd, R. H., *J. Polym. Sci. A2*, **16**, 1227 (1978).
5071. Mantell, R. M., and Ormand, W. L., *I&EC Product Res. Develop.*, **3**, 300 (1964).
5072. Manziani, G., Crescenzi, V., and Furlanetto, R., *Macromolecules*, **8**, 198 (1975).
5073. Manzione, L., Jameel, H., and Wilkes, G. L., *J. Polym. Sci. B*, **16**, 237 (1978).
5074. Marais, L., Gallot, Z., and Benoit, H., *J. Appl. Polym. Sci.*, **21**, 1955 (1977).
5075. Marchall, C. A., and Mock, R. A., *J. Polym. Sci.*, **17**, 591 (1955).
5076. Marchessault, R. H., Okamura, K., and Su, C. J., *Macromolecules*, **3**, 735 (1970).
5077. Marchetti, A., and Martuscelli, E., *J. Polym. Sci. A2*, **12**, 1649 (1974).
5078. Marchin, V. A., Miasnikova, L. P., Sutchkov, V. A., Tuchvatullina, M.Sh., and Novak, I. I., *J. Polym. Sci. C*, **38**, 195 (1972).
5079. Marcinčin, K., and Romanov, A., *Polymer*, **16**, 173 (1975).
5080. Marcinčin, K., and Romanov, A., *Polymer*, **16**, 177 (1975).
5081. Marco, C., Bello, A., and Fatou, J. G., *Makromol. Chem.*, **179**, 1333 (1978).
5082. Margerison, D., Bain, D. R., and Kiely, B., *Polymer*, **14**, 133 (1973).
5083. Mark, H., and Howink, R., *J. Prakt. Chem.*, **157**, 15 (1941).
5084. Mark, H., and Simha, R., *Trans. Faraday Soc.*, **36**, 611 (1940).
5085. Mark, J. E., Chiu, D. S., and Su, T. K., *Polymer*, **19**, 407 (1978).

5086. Mark, J. E., and Flory, P. J., *Macromolecules*, **6**, 300 (1973).
5087. Mark, J. E., and Ko, J. H., *Macromolecules*, **8**, 874 (1975).
5088. Mark, J. E., and Ko, J. H., *J. Polym. Sci. A2*, **13**, 2221 (1975).
5089. Mark, J. E., and Krimm, S., *Macromolecules*, **2**, 175 (1969).
5090. Marker, L., Early, R., and Aggarwal, S. L., *J. Polym. Sci.*, **38**, 369 (1959).
5091. Maron, S. H., and Lou, R. L. H., *J. Phys. Chem.*, **59**, 231 (1955).
5092. Marrinan, H. J., *J. Polym. Sci.*, **39**, 461 (1959).
5093. Marrucci, G., and Ciferri, A., *J. Polym. Sci. B.*, **15**, 643 (1977).
5094. Marsh, D. G., Yanus, J. F., and Pearson, J. M., *Macromolecules*, **8**, 427 (1975).
5095. Martic, P. A., Daly, R. C., Williams, J. L. R., and Farid, S., *J. Polym. Sci. B*, **15**, 295 (1977).
5096. Martin, G. M., and Passaglia, E., *J. Res. Nat. Bur. Stand. A*, **70**, 221 (1966).
5097. Martiscelli, E., and Mancarella, C., *Polymer*, **14**, 71 (1973).
5098. Martuscelli, E., and Martynov, M. A., *Makromol. Chem.*, **111**, 50 (1968).
5099. Marupov, R., Kolontarov, I. Ya., Asrorov, Y., Mavlyanov, A. M., and Niyazi, F. F., *J. Polym. Sci.*, *A1*, **15**, 2835 (1977).
5100. Marx, P., and Dole, M., *J. Amer. Chem. Soc.*, **77**, 4771 (1955).
5101. Marx, P., Smith, C. W., Worthington, A. E., and Dole, M., *J. Phys. Chem.*, **59**, 1015 (1955).
5102. Masetti, G., Cabassi, F., Marelli, G., and Zerbi, G. *Macromolecules*, **6**, 700 (1973).
5103. Mashimo, S., *Macromolecules*, **9**, 91 (1976).
5104. Mashino, S., and Chiba, A., *Polym. J.*, **5**, 41 (1973).
5105. Mason, P., *Trans. Faraday Soc.*, **55**, 1461 (1959).
5106. Masson, C. R., and Melville, H. W., *J. Polym. Sci.*, **4**, 323 (1949).
5107. Masson, C. R., and Melville, H. W., *J. Polym. Sci.*, **6**, 21 (1951).
5108. Matheson, I. B. C., and Lee, J., *J. Amer. Chem. Soc.*, **94**, 3310 (1972).
5109. Matheson, I. B. C., Lee, J., Yamanashi, E. S., and Wolbarsht, M. L., *J. Amer. Chem. Soc.*, **96**, 3343 (1974).
5110. Mathieson, A. R., *J. Colloid Sci.*, **15**, 387 (1960).
5111. Matsubara, I., and Magill, J. H., *J. Polym. Sci. A2*, **11**, 1173 (1973).
5112. Matsuda, H., Aonuma, H., and Kuroiwa, S., *J. Appl. Polym. Sci.*, **14**, 335 (1970).
5113. Matsui, Y., Kubota, T., Tadekoro, H., and Yashihara, T., *J. Polym. Sci. A*, **3**, 3375 (1965).
5114. Matsumoto, T., and Bogue, D. C., *J. Polym. Sci. A2*, **15**, 1663 (1977).
5115. Matsuo, M., *Polym. Eng. Sci.*, **7**, 38 (1967).
5116. Matsuo, M., Nomura, S., Hashimoto, T., and Kawai, H., *Polymer J.*, **6**, 151 (1974).
5117. Matsuo, M., Sagae, S., and Asai, H., *Polymer*, **10**, 79 (1969).
5118. Matsuo, K., and Stockmayer, W. H., *Macromolecules*, **8**, 661 (1975).
5119. Matsuo, S., Tohara, A., Iwakura, Y., and Iwata, K., *Makromol. Chem.*, **168**, 241 (1973).
5120. Matsuoka, S., and Ishida, Y., *J. Polym. Sci. C*, **14**, 247 (1966).
5121. Matsushige, K., and Takemura, T., *J. Polym. Sci. A2*, **16**, 921 (1978).
5122. Matsuura, H., and Miyazawa, T., *Bull. Chem. Soc. Japan*, **41**, 1798 (1968).
5123. Matsuura, H., and Miyazawa, T., *Bull. Chem. Soc, Japan*, **42**, 372 (1972).
5124. Matsuura, H., and Miyazawa, T., *J. Chem. Phys.*, **50**, 915 (1969).
5125. Matsuzaki, K., and Ito, H., *J. Polym. Sci. A2*, **12**, 2507 (1974).
5126. Matsuzaki, K., and Ito, H., *J. Polym. Sci. A1*, **15**, 647 (1977).
5127. Matsuzaki, K., Ishida, A., Osawa, Z. and Sano, K., *J. Chem. Soc. Japan*, **68**, 855 (1965).
5128. Matsuzaki, K., Kanai, T., Matsubara, T., Matsumoto, S., *J. Polym. Sci. A1*, **14**, 1475 (1976).
5129. Matsuzaki, K., and Ohshima, O., *Makromol. Chem.*, **164**, 265 (1973).
5130. Matsuzaki, K., and Uryu, T., *J. Polym. Sci.*, **4**, 255 (1966).
5131. Matsuzaki, K., Uryu, T., Osada, K., and Kawamura, T., *Macromolecules*, **5**, 816 (1972).
5132. Matsuzaki, K., Uryu, T., Seki, T., Osada, K., and Kawamura, T., *Makromol. Chem.*, **176**, 3051 (1975).
5133. Matsuzaki, K., Uryu, T., Tameda, K., and Takeuchi, M., *J. Chem. Soc. Japan*, **68**, 1366 (1965).
5134. Matsuzaki, K., Uryu, T., and Takeuchi, M., *J. Polym. Sci. B*, **3**, 835 (1965).
5135. Mattauch, J., and Herzog, R. F. K., *Z. Phys.*, **89**, 786 (1934).
5136. Matthews, J. L., Peiser, H. S., and Richards, R. B. *Acta Crystal.*, **2**, 85 (1949)
5137. Mattice, W. L., *Macromolecules*, **8**, 644 (1975).

5138. Mattice, W. L., *Macromolecules*, **10**, 1171 (1977).
5139. Mattice, W. L., *Macromolecules*, **10**, 1177 (1977).
5140. Mattice, W. L., *Macromolecules*, **10**, 1182 (1977).
5141. Mattice, W. L., *Macromolecules*, **11**, 517 (1978).
5142. Mattice, W. L., and Carpenter, D. K., *Macromolecules*, **9**, 53 (1976).
5143. Mattson, J. S., *Ind. Res.*, **1975**, 57 (1975).
5144. Matyi, R. J., and Crist, B., Jr, *J. Polym. Sci. A2*, **16**, 1329 (1978).
5145. Maurer, J. J., *Rubber Chem. Technol.*, **42**, 110 (1969).
5146. Maurer, J. J., and Tsien, H. C., *J. Appl. Polym. Sci.*, **8**, 1719 (1964).
5147. Maurice, M. J., *Anal. Chim. Acta*, **26**, 406 (1962).
5148. Maurice, M. J., and Huizinga, F., *Anal. Chim, Acta*, **22**, 363 (1960).
5149. Mauritz, K. A., Baer, E., and Hopfinger, A. J., *J. Polym. Sci. A2*, **11**, 2185 (1973).
5150. Mauritz, K. A., and Hopfinger, A. J., *J. Polym. Sci. A2*, **13**, 787 (1975).
5151. Mauzac, M., Vairon, J. P., and Sigwalt, P., *Polymer*, **18**, 1193 (1977).
5152. Mazur, J., Guttman, C. M., and McCrankin, F. L., *Macromolecules*, **6**, 872 (1973).
5153. Mazur, J., and McCrankin, F., *Macromolecules*, **10**, 326 (1977).
5154. Mazur, J., and McIntyre, D., *Macromolecules*, **8**, 464 (1975).
5155. May, J. A., Jr, and Knight, G. W., *J. Chromatogr.*, **55**, 11 (1971).
5156. Maynard, J. T., and Moohel, W. E., *J. Polym. Sci.*, **13**, 251 (1954).
5157. Maxfield, J., and Mandelkern, L., *Macromolecules*, **10**, 1141 (1977).
5158. Maxfield, J., and Shepherd, I. W., *Chem. Phys.*, **2**, 433 (1973).
5159. Maxfield, J., and Shepherd, I. W., *Chem. Phys. Lett.*, **19**, 541 (1973).
5160. Maxfield, J., and Shepherd, I. W., *Polymer*, **16**, 227 (1975).
5161. Maxfield, J., and Shepherd, I. W., *Polymer*, **16**, 505 (1975).
5162. Maxfield, J., Stein, R. S., and Chen, M. C., *J. Polym. Sci. A2*, **16**, 37 (1978).
5163. McAdam, J. D. G., King, T. A., and Knox, A., *Chem. Phys. Lett.*, **26**, 6 (1974).
5164. McAdie, H. G., *Anal. Chem.*, **39**, 543 (1967).
5165. McAdie, H. G., *Anal. Chem.*, **44**, 640 (1972).
5166. McAllister, P. B., Carter, T. J., and Hinde, R. M., *J. Polym. Sci. A2*, **16**, 49 (1978).
5167. McBain, J. W., and Working, E. B., *J. Phys. Chem.*, **51**, 974 (1947).
5168. McBrierty, V. J., and Douglass, D. C., *Macromolecules*, **10**, 855 (1977).
5169. McBrierty, V. J., Douglass, D. C., and Weber, T. A., *J. Polym. Sci. A2*, **14**, 1271 (1976).
5170. McBrierty, V. J., McCall, D. W., Douglass, D. C., and Falcone, D. R., *Macromolecules*, **4**, 584 (1971).
5171. McBrierty, V. J., and McDonald, I. R., *Polymer*, **16**, 125 (1975).
5172. McCall, D. W., Douglass, D. C., and Anderson, E. W., *J. Chem. Phys.*, **30**, 771 (1959).
5173. McCall, D. W., Douglass, D. C., and Anderson, E. W., *J. Polym. Sci. A*, **1**, 1709 (1963).
5174. McCall, D. W., and Falcone, D. R., *Trans. Faraday Soc.*, **66**, 262 (1970).
5175. McCall, D. W., and Slichter, W. P., *J. Polym. Sci.*, **26**, 171 (1957).
5176. McCally, R. L., and Farrell, R. A., *Polymer*, **18**, 445 (1977).
5177. McCammon, R. D., Saba, R. G., and Work, R. N., *J. Polymer. Sci. A2*, **7**, 1721 (1969).
5178. McCarthy, D. E., *Appl. Opt.*, **2**, 591, 596 (1963).
5179. McCleod, L. A., and McIntosh, R., *Can. J. Res.*, **29**, 1104 (1951).
5180. McCormick, H., *J. Chromatogr.*, **40**, 1 (1969).
5181. McCrackin, F. L., *J. Appl. Polym. Sci.*, **21**, 191 (1977).
5182. McCrackin, F. L., Mazur, J., and Guttman, C. M., *Macromolecules*, **6**, 859 (1973).
5183. McCullough, R. L., *ACS Polym. Preprints*, **3**, 53 (1962).
5184. McCullough, R. L., *J. Polym. Sci. A2*, **15**, 1805 (1977).
5185. McCullough, R. L., Eisenstein, A. J., and Weikart, D. F., *J. Polym. Sci. A2*, **15**, 1837 (1977).
5186. McDonald, C. J., and Claesson, S., *Chem. Scripta*, **9**, 36 (1976).
5187. McDonald, M. P., and Ward, I. M., *Polymer*, **2**, 341 (1961).
5188. McDonald, M. P., and Ward, I. M., *Chem. Ind. (London)*, **1961**, 631 (1961).
5189. McDonnell, M. E., and Jamieson, A. M., *J. Appl. Polym. Sci.*, **21**, 3261 (1977).
5190. McElwain, J. W., *Anal. Chem.*, **21**, 194 (1949).
5191. McGuchan, R., and McNeill, I. C., *J. Polym. Sci. A1*, **4**, 2051 (1966).
5192. McHugh, A. J., *J. Appl. Polym. Sci.*, **19**, 125 (1975).
5193. McHugh, A. J., and Forrest, E. H., *J. Polym. Sci. A2*, **13**, 1643 (1975).
5194. McHugh, A. J., and Schultz, J. M., *Kolloid Z. Z. Polym.*, **251**, 13 (1973).
5195. McInally, I., Soutar, I., and Steedman, W., *J. Polym. Sci. A1*, **15**, 2511 (1977).

800

5196. McIntyre, D., and Doderer, G. C., *J. Res. Nat. Bur. Stand.*, **62**, 153 (1959).
5197. McIntyre, P. R., and Campos-Lopez, E., *Macromolecules*, **3**, 321 (1970).
5198. McKague, E. L., Reynolds, J. D., and Halkias, J. E., *J. Appl. Polym. Sci.*, **22**, 1643 (1978).
5199. McKinney, J. E., and Simha, R., *Macromolecules*, **9**, 430 (1976).
5200. McKinney, P. J., and Foltz, C. R., *J. Appl. Polym. Sci.*, **11**, 1189 (1967).
5201. McKinney, R. W., *J. Gas Chromatogr.*, **2**, 432 (1964).
5202. McMaster, L. P., *Macromolecules*, **6**, 760 (1973).
5203. McMaster, L. P., Advances in Chemistry Series No. 142, American Chemical Society, Washington, D.C., 1975, p. 1975.
5204. McNair, H. M., and Chandler, C. D., *J. Chromatogr. Sci.*, **14**, 477 (1976).
5205. McNeill, I. C., *J. Polym. Sci. A1*, **4**, 2479 (1966).
5206. McNeill, I. C., *Europ. Polym. J.*, **3**, 409 (1967).
5207. McNeill, I. C., *Europ. Polym. J.*, **6**, 373 (1970).
5208. McNeill, I. C., Ackerman, L., Gupta, S. N., and Zulfiqar, M., *J. Polym. Sci. A1*, **15**, 2381 (1977).
5209. McRae, M. A., and Maddams, W. F., *Makromol. Chem.*, **177**, 449, 461 (1976).
5210. McRae, M. A., and Maddams, W. F., *Makromol. Chem.*, **177**, 473 (1976).
5211. McRae, M. A., and Maddams, W. F., *Polymer*, **18**, 525 (1977).
5212. Mead, W. T., Porter, R. S., and Reed, P. E., *Macromolecules*, **11**, 56 (1978).
5213. Meares, P., *Trans. Faraday Soc.*, **53**, 31 (1957).
5214. Mearns, A. M., and Morris, A. J., *Nature*, **225**, 59 (1970).
5215. Meehan, E. J., *J. Colloid Interface Sci.*, **27**, 388 (1968).
5216. Meeks, A. C., and Goldfarb, I. J., *Anal. Chem.*, **39**, 908 (1967).
5217. Meffroy-Biget, A. M., *J. Polym. Sci., B*, **14**, 11 (1976).
5218. Mehta, A., Gaur, U., and Wunderlich, B., *J. Polym. Sci. A2*, **16**, 289 (1978).
5219. Mehta, A., and Wunderlich, B., *Makromol. Chem.*, **175**, 977 (1974).
5220. Meinel, G., and Peterlin, A., *J. Polym. Sci. B*, **5**, 197 (1967).
5221. Mele, D., *Europ. Polym. J.*, **14**, 623 (1978).
5222. Melia, T., *Polymer*, **3**, 317 (1962).
5223. Melia, T., and Tyson, A., *Makromol. Chem.*, **109**, 87 (1967).
5224. Melillo, L., and Wunderlich, B., *Kolloid Z. Z. Polym.*, **250**, 417 (1972).
5225. Melveger, A. J., and Baughman, R. H., *J. Polym. Sci. A2*, **11**, 603 (1973).
5226. Melville, H. W., and Stead, B. D., *J. Polym. Sci.*, **16**, 505 (1955).
5227. Menčik, Z., *J. Polm. Sci.*, **17**, 147 (1955).
5228. Menčik, Z., *J. Polym. Sci. A2*, **11**, 1585 (1973).
5229. Menčik, Z., *J. Polym. Sci. A2*, **13**, 2173 (1975).
5230. Menčik, Z., and Fitchmun, D. R., *J. Polym. Sci. A2*, **11**, 973 (1973).
5231. Mendelson, R. A., *J. Polym. Sci. A1*, 2361 (1963).
5232. Mendelson, R. M., and Droth, E. E., *J. Polym. Sci. B*, **6**, 795 (1968).
5233. Mendenhall, G. D., Hassell, J. A., and Nathan, R. A., *J. Polym. Sci. A1*, **15**, 99 (1977).
5234. Menin, J. P., and Roux, R., *J. Polym. Sci. A1*, **10**, 855 (1972).
5235. Merkel, P. B., and Kearns, D. R., *J. Amer. Chem. Soc.*, **94**, 7244 (1972).
5236. Merle-Aubry, L., Merle, Y., and Selegny, E., *Makromol. Chem.*, **176**, 709 (1975).
5237. Merrill, E. W., Salzman, E. W., Wong, P. S. L., and Silliman, J. L., *ACS Polym. Preprints*, **13**, 511 (1973).
5238. Merritt, C., and Walsh, J. T., *Anal. Chem.*, **35**, 110 (1963).
5239. Merritt, C., and Robertson, D. H., *J. Gas Chromatogr.*, **5**, 96 (1967).
5240. Merz, E. H., and Raetz, R. W., *J. Polym. Sci.*, **5**, 587 (1950).
5241. Meselson, M., Stahl, F. W., and Vinograd, J., *Proc. Nat. Acad. Sci., U.S.A.*, **43**, 581 (1957).
5242. Messner, A. E., Rosie, D. M., and Argabright, P. A., *Anal. Chem.*, **31**, 230 (1959).
5243. Metcalf, W. S., Natusch, D. F. S., Page, S. G., Shipley, E. D., and Wiggins, P. M., *J. Sci. Instrum.*, **42**, 603 (1965).
5244. Metzer, A. B., Houghton, W. T., Sailor, R. A., and White, J. L., *Trans. Soc. Rheol.*, **5**, 131 (1961).
5245. Meyer, R., *Polymer*, **15**, 137 (1974).
5246. Meyerhoff, G., *Makromol. Chem.*, **12**, 45 (1954).
5247. Meyerhoff, G., *Z. Naturforsch.*, **11b**, 302 (1956).
5248. Meyerhoff, G., *Makromol. Chem.*, **22**, 237 (1957).
5249. Meyerhoff, G., *Z. Electrochem.*, **61**, 325 (1957).

5250. Meyerhoff, G., *Z. Electrochem.*, **61**, 1249 (1957).
5251. Meyerhoff, G., *Adv. Polym. Sci.*, **3**, 196 (1963).
5252. Meyerhoff, G., *J. Polym. Sci. C*, **21**, 31 (1968).
5253. Meyerhoff, G., *Makromol. Chem.*, **118**, 265 (1968).
5254. Meyerhoff, G., *Separation Sci.*, **6**, 239 (1971).
5255. Meyerhoff, G., and Buldt, G., *Makromol. Chem.*, **175**, 675 (1974).
5256. Meyerhoff, G., Hack, H., and Raczek, J. *Polym. Sci. C*, **61**, 169 (1977).
5257. Meyerhoff, G., and Romatowski, J., *Makromol. Chem.*, **74**, 222 (1964).
5258. Meyerhoff, G., and Shimotsuma, S., *Makromol. Chem.*, **109**, 263 (1967).
5259. Meyerson, S., *Anal. Chem.*, **25**, 338 (1953).
5260. Michaels, A. S., and Bixler, H. J., *J. Polym. Sci.*, **50**, 393, 413 (1961).
5261. Michaels, A. S., Bixler, H. J., and Hopfenberg, H. B., *J. Appl. Polym. Sci.*, **12**, 991 (1968).
5262. Michaels, A. S., and Parker, R. B., *J. Polym. Sci.*, **41**, 53 (1959).
5263. Michaels, A. S., Vieth, W. R., and Barrie, J. A., *J. Appl. Phys.*, **34**, 13 (1963).
5264. Michajlov, L., Zugenmaier, P., and Cantow, H. J., *Polymer*, **9**, 325 (1968).
5265. Michel, R. E., *J. Polym. Sci. A2*, **10**, 1841 (1972).
5266. Michel, R., Seytre, G., and Maitrot, M., *J. Polym. Sci. A2*, **13**, 1333 (1975).
5267. Mijnlieff, P. F., and Comou, D. J., *J. Colloid Interface Sci.*, **27**, 533 (1968).
5268. Minjlieff, P. F., and Wiegel, F. W., *J. Polym. Sci. A2*, **16**, 245 (1978).
5269. Miles, M., and Gleiter, H., *J. Polym. Sci. A2*, **16**, 171 (1978).
5270. Millen, W., and Hawkes, S. J., *J. Polym., Sci. B*, **15**, 463 (1977).
5271. Miller, D. R., and Macosco, C. W., *Macromolecules*, **9**, 206 (1976).
5272. Miller, G. A., *J. Phys. Chem.*, **71**, 2305 (1967).
5273. Miller, G. A., *Macromolecules*, **3**, 125 (1970).
5274. Miller, G. A., *Macromolecules*, **3**, 674 (1970).
5275. Miller, G. A., San Filippo, F. I., and Carpenter, D. K., *Macromolecules*, **3**, 125 (1970).
5276. Miller, G. J. and Willis, H. A., *J. Appl. Chem.*, **6**, 385 (1956).
5277. Miller, G. W., *J. Appl. Polym. Sci.*, **15**, 2335 (1971).
5278. Miller, G. W., *J. Polym., Sci. A2*, **13**, 1831 (1975).
5279. Miller, L. E., and Hamm, F. A., *J. Phys. Chem.*, **57**, 110 (1953).
5280. Miller, P. J., Jackson, J. F., and Porter, R. S., *J. Polym. Sci. A2*, **11**, 2001 (1973).
5281. Miller, R. G., and Willis, H. A., *J. Polym. Sci.*, **19**, 485 (1956).
5282. Miller, R. G. J., *Chem. Ind. (London)*, **1957**, 190 (1957).
5283. Miller, R. L., *J. Polym. Sci.*, **56**, 375 (1962).
5284. Miller, R. L., *J. Polym. Sci.*, **57**, 975 (1962).
5285. Miller, R. L., and Nielsen, L. E., *J. Polym. Sci.*, **44**, 391 (1960).
5286. Miller, R. L., and Nielsen, L. E., *J. Polym. Sci.*, **46**, 303 (1960).
5287. Miller, R. L., and Nielsen, L. E., *J. Polym. Sci.*, **55**, 643 (1961).
5288. Miller, R. L., and Raisoni, J., *J. Polym. Sci. A2*, **14**, 2325 (1976).
5289. Miller, W., and Stepto, R. F. T., *Europ. Polym. J.*, **7**, 65 (1971).
5290. Miller, W. G., *Macromolecules*, **6**, 100 (1973).
5291. Milstein, J. B., and Charney, E., *Macromolecules*, **2**, 678 (1969).
5292. Mikhailov, G. P., Labanov, A. M. and Platonov, M. P., *Vysokomol Soedin. A*, **9**, 2267 (1967).
5293. Min, K. W., *J. Appl. Polym. Sci.*, **22**, 589 (1978).
5294. Mirabella, F. M., Jr, Barrall, E. M., II, and Johnson, J. F., *J. Appl. Polym. Sci.*, **19**, 2131 (1975).
5295. Mirabella, F. M., Jr, Barrall, E. M., II, and Johnson, J. F., *Polymer*, **17**, 17 (1976).
5296. Mirabella, F. M., Jr., Barrall, E. M., II, and Johnson, J. F., *J. Appl. Polym. Sci.*, **20**, 765 (1976).
5297. Mirabella, F. M., Jr, Johnson, J. F., and Barrall, E. M., II, *Intern. Lab.* Nov./Dec. 1975, p. 37.
5298. Mirkamilov, D. M., and Platonov, M. P., *Vysokomol. Soedin. A*, **11**, 1017 (1969).
5299. Misra, A., Prud'homme, R. E., and Stein, R. S., *J. Polym. Sci. A2*, **12**, 1235 (1974).
5300. Misra, A., Stein, R. S., Chu, C., Wilkes, G. L., and Desai, A. B., *J. Polym. Sci. B*, **13**, 303 (1975).
5301. Mita, I., Imai, I., and Kambe, H., *Thermochim. Acta*, **2**, 337 (1971).
5302. Mitchell, J. Jr, and Perkins, L. R., *Appl. Polym. Symp.*, No. 4, 167 (1967).
5303. Mitchel, R. L., *Ind. Eng. Chem.*, **45**, 2526 (1953).
5304. Mitchel, R. S., and Guillet, J. E., *J. Polym. Sci. A2*, **12**, 713 (1974).

802

5305. Mitomo, H., Nakazato, K., and Kuriyama, I., *J. Polym. Sci. A2*, **15**, 915 (1977).
5306. Mitzner, B. M., *J. Polym. Sci.*, **21**, 323 (1956).
5307. Mizushima, S., Shimanouchi, T., Nakomura, K., Haysahi, M., and Tsuchiga, S., *J. Chem. Phys.*, **26**, 970 (1957).
5308. Miyagi, A., and Wunderlich, B., *J. Polym. Sci. A2*, **10**, 2073 (1972).
5309. Miyake, A., *J. Polym. Sci.*, **38**, 479, 496 (1959).
5310. Miyake, A., *J. Polym. Sci.*, **45**, 232 (1960).
5311. Miyamoto, T., and Inagaki, H., *Macromolecules*, **2**, 554 (1969).
5312. Miyamoto, T., and Inagaki, H., *Polym. J.*, **1**, 46 (1970).
5313. Miyamoto, Y., Nakafuku, C., and Takemura, T., *Polym. J.*, **3**, 120 (1972).
5314. Miyatake, R., and Kumanotani, J., *Makromol. Chem.*, **177**, 2749 (1976).
5315. Miyazawa, T., *J. Polym. Sci.*, **55**, 215 (1961).
5316. Miyazawa, T., *J. Chem. Phys.*, **35**, 693 (1961).
5317. Miyazawa, T., *J. Polym. Sci. B*, **2**, 847 (1964).
5318. Miyazawa, T., Fukushima, K., and Ideguchi, Y., *J. Chem. Phys.*, **12**, 2764 (1962).
5319. Miyazawa, T., and Ideguchi, Y., *Bull. Chem. Soc. Japan*, **37**, 1065 (1964).
5320. Miyazawa, T., Ideguchi, Y., and Fukushima, K., *J. Chem. Phys.*, **38**, 2709 (1963).
5321. Moacanin, J., Felicetta, V. F., and McCarthy, J. L., *J. Amer. Chem. Soc.*, **81**, 2052 (1959).
5322. Moacanin, J., Nelson, H., Back, E., Felicetta, V. F., and McCarthy, J. L., *J. Amer. Chem. Soc.*, **81**, 2054 (1959).
5323. Mochel, V. D., *J. Polym. Sci. A1*, **10**, 1009 (1972).
5324. Mochizuki, T., *J. Polym. Sci. B*, **14**, 623 (1976).
5325. Mohadger, Y., and Wilkes, G. L., *J. Polym. Sci. A2*, **14**, 963 (1976).
5326. Mohlin, U. B., and Gray, D. G., *J. Colloid Interface Sci.*, **47**, 747 (1974).
5327. Mokhtari-Nejad, E., Berger, K. C., Hammel, R., and Lederer, K., *Makromol. Chem.*, **179**, 159 (1978).
5328. Molyneux, P., *Makromol. Chem.*, **43**, 31 (1961).
5329. Molyneux, P., *Kolloid Z. Z. Polym.*, **226**, 15 (1968).
5330. Moneva, I. T., *J. Polym. Sci. A2*, **15**, 1501 (1977).
5331. Montagne, P. G., and Peaker, F. W., *J. Polym. Sci. C*, **43**, 277 (1973).
5332. Montague, P. G., Peaker, F. W., Bosworth, P., and Lemon, P., *Brit. Polym. J.*, **3**, 93 (1971).
5333. Montaudo, G., Przybylski, M., and Ringsdorf, H., *Makromol. Chem.*, **176**, 1753, 1763 (1975).
5334. Montfort, J. P., Marin, G., Arman, J., and Monge, P., *Polymer*, **19**, 277 (1978).
5335. Montoya, E. F., and Leung, A. T., *Intern. Lab.*, July/Aug. 1976, p. 45.
5336. Moore, J. C., *J. Polym. Sci. A*, **2**, 835 (1964).
5337. Moore, J. C., *J. Polym. Sci. C*, **21**, 1 (1968).
5338. Moore, J. C., *Separation Sci.*, **5**, 723 (1970).
5339. Moore, J. C., and Hendrickson, J. G., *J. Polym. Sci. C*, **8**, 233 (1965).
5340. Moore, L. D., Jr, and Adock, J. I., in Characterization of Macromolecular Structure, Nat. Acad. Sci. U.S.A., Publ. 1537, Washington, D.C., 1968, p. 289.
5341. Moore, L. D., Jr, Greear, A., and Aharp, J. O., *J. Polym. Sci.*, **59**, 339 (1962).
5342. Moore, L. D., Jr, and Overton, J. R., *J. Chromatogr.*, **55**, 137 (1971).
5343. Moore, R. S., and Ferry, J. D., *J. Phys. Chem.*, **66**, 2699 (1962).
5344. Moore, W. H., and Krimm, S., *Makromol. Chem.*, Suppl. 1, 491 (1975).
5345. Moore, W. R. A. D., and Millns, W., *Brit. Polym. J.*, **1**, 81 (1969).
5346. Morantz, D. J., and Bilen, C. S., *Polymer*, **16**, 745 (1975).
5347. Morawetz, H., Cho, J. R., and Gans, P. J., *Macromolecules*, **6**, 624 (1973).
5348. Morese-Séguéla, B., St. Jacques, M., Renauld, J. M., and Prud'homme, J., *Macromolecules*, **10**, 431 (1977).
5349. Moretti, P., and Mesnard, G., *J. Polym. Sci. A2*, **15**, 1989 (1977).
5350. Morey, D. R., *J. Colloid Sci.*, **6**, 407 (1951).
5351. Morgan, P. W., *Macromolecules*, **10**, 1381 (1977).
5352. Morgan, P. W., and Kwolek, S. L., *J. Polym. Sci. A*, **1**, 1147 (1963).
5353. Morgan, R. J., and O'Neal, J. E., *J. Polym. Sci. A2*, **14**, 1053 (1976).
5354. Mori, S., *J. Appl. Polym. Sci.*, **21**, 1921 (1977).
5355. Mori, S., Mochizuki, K., Watanabe, M., and Saito, M., *Intern. Lab.*, Nov./Dec. 1977, p. 49.

5356. Mori, T., Ogawa, K., and Tanaka, T., *J. Appl. Polym. Sci.*, **21**, 3381 (1977).
5357. Mori, Y., Ueda, A., Tanazawa, H., Matsuzaki, K., and Kobayashi, H., *Makromol. Chem.*, **176**, 699 (1975).
5358. Morie, G. P., Powers, T. A., and Glover, C. A., *Thermochim. Acta*, **1**, 429 (1970).
5359. Morimoto, H., and Ueda, R., *Acta Crystal*, **16**, 1107 (1963).
5360. Morishima, Y., Iimuro, H., Irie, Y., and Nozakura, S., *J. Polym. Sci. B*, **13**, 157 (1975).
5361. Morita, S., Shen, M., Sawa, G., and Ieda, M., *J. Polym. Sci. A2*, **14**, 1917 (1976).
5362. Moritani, T., and Fujiwara, Y., *Macromolecules*, **10**, 532 (1977).
5363. Morosoff, N., Morawetz, H., and Post, B., *J. Amer. Chem. Soc.*, **87**, 3035 (1965).
5364. Morris, C. E. M., *J. Appl. Polym. Sci.*, **21**, 435 (1977).
5365. Morris, R. B., amd Bassett, D. C., *J. Polym. Sci. A2*, **13**, 1501 (1975).
5366. Morrow, D. R., *Trans. N.Y. Acad. Sci.*, **30**, 1130 (1968).
5367. Moseley, M. E., and Stilbs, P., *Polymer*, **19**, 1133 (1978).
5368. Moseley, W. W., Jr, *J. Appl. Polym. Sci.*, **3**, 266 (1960).
5369. Moser, M., and Boudeulle, M., *J. Polym. Sci. A2*, **16**, 971 (1978).
5370. Moser, P., Squire, P. G., and O'Konski, C. T., *J. Phys. Chem.*, **70**, 744 (1966).
5371. Motozato, Y., Kusumoto, N., Hirayama, C., Murakami, R., and Isozaki, H., *Polymer*, **16**, 321 (1975).
5372. Mottley, C., Chang, K., and Kispert, L. D., *J. Mag. Reson.*, **19**, 130 (1975).
5373. Mountain, R. D., *J. Res. Nat. Bur. Stand. A*, **70**, 207 (1966).
5374. Mountain, R. D., and Deutch, J. M., *J. Chem. Phys.*, **50**, 1103 (1969).
5375. Mozisek, M., *Intern. J. Appl. Radiation Isotopes*, **21**, 11 (1970).
5376. Mozisek, M., and Klimanek, L., *Plaste Kaustchuk*, **15**, 99 (1968).
5377. Moyles, A. F., and Reynolds, G. E. J., *Brit. Polym. J.*, **1**, 180 (1969).
5378. Moynihan, R. E., *J. Amer. Chem. Soc.*, **81**, 1045 (1959).
5379. Muenker, A. H., and Hudson, B. E., *J. Macromol. Sci. A*, **3**, 1465 (1969).
5380. Mula, A., and Chinellato, L., *Europ. Polym. J.*, **6**, 1 (1970).
5381. Müller, R. H., and Solten, H. J., *Anal. Chem.*, **25**, 1103 (1953).
5382. Muller, T. E., and Alexander, W. J., *J. Polym. Sci. C*, **21**, 283 (1968).
5383. Mullins, L., *J. Polym. Sci.*, **19**, 225 (1956).
5384. Munch, J. P., and Candau, S., *J. Polym. Sci. A2*, **15**, 11 (1972).
5385. Munch, J. P., Candau, S., Duplessix, R., Picot, C., and Benoit, H., *J. Phys. (L).*, **35**, 239 (1976).
5386. Munch, J. P., Candau, S., Duplessix, R., Picot, C., Hertz, H., and Benoit, H., *J. Polym. Sci. A2*, **14**, 1097 (1976).
5387. Munk, P., *Macromolecules*, **11**, 387 (1978).
5388. Munk, P., Abijaoude, M. T., and Halbrook, M. E., *J. Polym. Sci. A2*, **16**, 105 (1978).
5389. Munk, P., and Halbrook, M. E., *Macromolecules*, **9**, 441 (1976).
5390. Munk, P., and Halbrook, M. E., *Macromolecules*, **9**, 568 (1976).
5391. Münster, A., *J. Polym. Sci.*, **5**, 333 (1950).
5392. Murakami, S., Kimura, F., and Fujishiro, R., *Makromol. Chem.*, **176**, 3425 (1975).
5393. Murano, M., and Harwood, H. J., *Macromolecules*, **3**, 605 (1970).
5394. Murano, M., and Yamadera, R., *J. Polym. Sci. A1*, **5**, 843 (1968).
5395. Murphy, E. B., and O'Neil, W. A., *SPE Journal*, **18**, 191 (1962).
5396. Murray, R. W., and Kaplan, M. L., *J. Amer. Chem. Soc.*, **80**, 4161 (1968).
5397. Murray, R. W., and Kaplan, M. L., *J. Amer. Chem. Soc.*, **91**, 5358 (1969).
5398. Mussa, C., *J. Polym. Sci.*, **26**, 67 (1957).
5399. Muthukumar, M., and Freed, K. F., *Macromolecules*, **10**, 899 (1977).
5400. Myers, M. N., Caldwell, K. D., and Giddings, J. C., *Separation Sci.*, **9**, 47 (1974).
5401. Myers, W., Donovan, J. L., and Kings, J. S., *J. Chem. Phys.*, **42**, 4299 (1965).
5402. Myers, W., Summerfield, G. C. and King, J. S., *J. Chem. Phys.*, **44**, 184 (1966).

5403. Naar, R. Z., Zabusky, H. H., and Heitmiller, R. F., *J. Appl. Polym. Sci.*, **7**, S30 (1963).
5404. Nagai, H., *J. Appl. Polym. Sci.*, **7**, 1597 (1963).
5405. Nagai, K., *J. Chem. Phys.*, **47**, 4690 (1967).
5406. Nagai, K., and Ishikawa, T., *J. Chem. Phys.*, **43**, 4508 (1965).
5407. Nagai, K., and Kobyashi, M., *J. Chem. Phys.*, **36**, 1268 (1961).
5408. Nagai, E., Kuribyashi, S., Shiraki, M., and Ukita, M., *J. Polym. Sci.*, **35**, 295 (1959).
5409. Nagamura, T., Fukitani, K., and Takayanagi, M., *J. Polym. Sci. A2*, **13**, 1515 (1975).
5410. Nagamura, T., Kusumoto, N., and Takayanagi, M., *J. Polym. Sci. A2*, **11**, 2357 (1973).

5411. Nagamura, T., and Takayanagi, M., *J. Polym. Sci. A2*, **12**, 2019 (1974).
5412. Nagamura, T., and Takayanagi, M., *J. Polym. Sci. A2*, **13**, 567 (1975).
5413. Nagamura, T., and Woodward, A. E., *J. Polym. Sci. A2*, **14**, 275 (1976).
5414. Nagasawa, T., and Shimomura, Y., *J. Polym. Sci. A2*, **12**, 2291 (1974).
5415. Nagasawa, T., Shimomura, Y., and Nishihara, Y., *Bull. Inst. Chem. Res., Kyoto Univ.*, **55**, 168 (1977).
5416. Nakagawa, K., and Ishida, Y., *J. Polym. Sci. A2*, **11**, 1503 (1973).
5417. Nakagawa, K., and Ishida, Y., *J. Polym. Sci. A2*, **11**, 2153 (1973).
5418. Nakajima, N., *J. Polym. Sci. A2*, **5**, 101 (1966).
5419. Nakajima, N., *J. Polym. Sci. C*, **21**, 153 (1968).
5420. Nakajima, A., and Fujiwara, H., *Bull. Chem. Soc. Japan*, **37**, 909 (1964).
5421. Nakajima, A., Fujiwara, H., and Hamada, F., *J. Polym. Sci. A2*, **4**, 507 (1966).
5422. Nakajima, A., Hamada, F., and Hayashi, S., *J. Polym. Sci. C*, **15**, 285 (1966).
5423. Nakajima, A., Hamada, F., Yasue, K., Fujisawa, K., and Shiomi, T., *Makromol. Chem.*, **175**, 197 (1974).
5424. Nakajima, A., Sakurada, I., and Yamakawa, H., *J. Polym. Sci.*, **35**, 489, 497 (1959).
5425. Nakamoto, K., Suga, H., Seki, S., Teramoto, A., Norisuye, T., and Fujita, H., *Macromolecules*, **7**, 784 (1974).
5426. Nakamura, K., Endo, R., and Takeda, M., *J. Polym. Sci. A2*, **15**, 2095 (1977).
5427. Nakamura, S., Sasaki, T., and Matsuzaki, K., *Makromol. Chem.*, **176**, 3471 (1975).
5428. Nakamura, S., Shindo, S., and Matsuzaki, K., *J. Polym. Sci. B*, **9**, 591 (1971).
5429. Nakamura, S., and Tobolsky, A. V., *Rept. Progr. Polym. Phys. Japan*, **12**, 303 (1969).
5430. Nakano, S., *J. Polym. Sci. A2*, **12**, 1499 (1974).
5431. Nakano, S., and Goto, Y., *J. Appl. Polym. Sci.*, **20**, 3313 (1976).
5432. Naoki, M., and Nose, T., *J. Polym. Sci. A2*, **13**, 1747 (1975).
5433. Naoki, M., and Nose, T., *J. Polym. Sci. A2*, **13**, 1893 (1975).
5434. Naoki, M., Motomura, M., Nose, T., and Hata, T., *J. Polym. Sci. A2*, **13**, 1737 (1975)
5435. Napper, D. H., *Polymer*, **10**, 181 (1969).
5436. Nara, S., Kashiwabara, H., and Sohma, J., *J. Polym. Sci. A2*, **5**, 929 (1967).
5437. Nara, S., Shimada, S., Kashiwabara, H., and Sohma, J., *J. Polym. Sci. A2*, **6**, 1435 (1968).
5438. Narisawa, I., *J. Polym. Sci. A2*, **10**, 1789 (1972).
5439. Narita, S., Ichinohe, S., and Enomoto, S., *J. Polym. Sci.*, **36**, 389 (1959).
5440. Narita, S., Ichinohe, S., and Enomoto, S., *J. Polym. Sci.*, **37**, 251 (1959).
5441. Narita, S., Ichinohe, S., and Enomoto, S., *J. Polym. Sci.*, **37**, 273 (1959).
5442. Nash, D. W., and Pepper, D. C., *Polymer*, **16**, 105 (1975).
5443. Nasini, A., and Mussa, C., *Makromol. Chem.*, **22**, 59 (1957).
5444. Natarajan, R. T., Prud'homme, R. E., Bourland, L., and Stein, R. S., *J. Polym. Sci A2*, **14**, 1541 (1976).
5445. Natta, G., *Angew. Makromol. Chem.*, **35**, 94 (1960).
5446. Natta, G., Allegra, G., Bassi, I. W., Carlini, C., Chielini, E., and Montagnoli, G., *Macromolecules*, **2**, 311 (1969).
5447. Natta, G., and Corradini, P., *J. Polym. Sci.*, **20**, 251 (1956).
5448. Natta, G., and Corradini, P., *J. Polym. Sci.*, **39**, 29 (1959).
5449. Natta, G., and Corradini, P., *Nuovo Cimento*, Suppl. 15, 9 (1960).
5450. Natta, G., and Corradini, P., *Nuovo Cimento*, Suppl. 15, 40 (1960).
5451. Natta, G., Corradini, P., and Bassi, I. W., *Nuovo Cimento*, Suppl. *15*, 52, 83, 111 (1960).
5452. Natta, G., Corradini, P., and Bassi, I. W., *Makromol. Chem.*, **33**, 247 (1959).
5453. Natta, G., Corradini, P., and Bassi, I. W., *J. Polym. Sci.*, **51**, 505 (1961).
5454. Natta, G., Corradini, P., and Ganis, P., *Makromol. Chem.*, **39**, 238 (1960).
5455. Natta, G., Corradini, P., and Ganis, P., *J. Polym. Sci.*, **58**, 1191 (1962).
5456. Natta, G., Dall'Asta, G., Mazzanti, G., Pasquon, I., Valvassori, A., and Zambelli, A., *Makromol. Chem.*, **54**, 95 (1962).
5457. Natta, G., Danusso, F., and Morgalio, G., *Makromol. Chem.*, **20**, 37 (1956).
5458. Natta, G., Farina, M., and Peraldo, M., *Makromol. Chem.*, **38**, 13 (1960).
5459. Natta, G., Mazzanti, G., Corradini, P., and Bassi, I. W., *Makromol. Chem.*, **37**, 156 (1960).
5460. Natta, G., Peraldo, M., and Allegra, G., *Makromol. Chem.*, **75**, 215 (1964).
5461. Natta, G., Porri, L., Carbonaro, A., and Greco, A., *Makromol. Chem.*, **71**, 207 (1964).
5462. Neilson, R. D. M., Soutar, I., and Steedman, W., *Macromolecules*, **10**, 1193 (1977).

5463. Nelson, D. F., Yee, J. Y., and Kirk, P. L., *Microchem. J.*, **6**, 225 (1962).
5464. Nemoto, N., Mitsuda, Y., Schrag, J. L., and Ferry, J. D., *Macromolecules*, **7**, 253 (1974).
5465. Nencini, G., Giuliano, G., and Salvatori, T., *J. Polym. Sci. B*, **3**, 483 (1965).
5466. Netwitt, E. J., and Kokle, V., *J. Polym. Sci. A2*, **4**, 705 (1966).
5467. Neves, D. E., and Scott, R. A., *Macromolecules*, **10**, 339 (1977).
5468. Newing, M. J., *Trans. Faraday Soc.*, **46**, 613 (1950).
5469. Newman, B. A., Frosini, V., and Magagnini, P. L., *J. Polym. Sci. A2*, **13**, 87 (1975).
5470. Newman, R. D., and Prausnitz, J. M., *J. Paint Technol.*, **45**, 33 (1973).
5471. Newman, S., *J. Polym. Sci.*, **47**, 111 (1960).
5472. Newman, S., *J. Colloid. Interface Sci.*, **26**, 207 (1968).
5473. Newman, S., and Cox, W. P., *J. Polym. Sci.*, **46**, 29 (1960).
5474. Newmayer, J. J., *Anal. Chem. Acta*, **20**, 519 (1959).
5475. Ng, H. C., and Guillet, J. E., *Photochem. Photobiol.*, **28**, 571 (1978).
5476. Nguyen, U. L., Berticat, P., May, J. F., and Vallet, G., *Makromol. Chem.*, **171**, 229 (1973).
5477. Nichols, J. B., *J. Appl. Polym. Sci.*, **25**, 840 (1954).
5478. Nicodemo, L., Nicolas, L., Romeo, G., and Scafora, E., *Polymer*, **19**, 230 (1978).
5479. Nicolais, L., Drioli, E., Hopfenberg, H. B., and Tidone, D., *Polymer*, **18**, 1137 (1977).
5480. Niederwieser, A., *Chromatography*, **2**, 23, 362 (1969).
5481. Niederwieser, A., and Brenner, M., *Experientia*, **21**, 50 (1965).
5482. Niederwieser, A., and Honegger, C. G., *Adv. Chromatogr.*, **2**, 123 (1966).
5483. Niegisch, W., *J. Appl. Polym. Sci.*, **37**, 4041 (1966).
5484. Nielsen, J. R., and Holland, R. F., *J. Mol. Spectr.*, **4**, 488 (1960).
5485. Nielsen, J. R., and Holland, R. F., *J. Mol. Spectr.*, **6**, 394 (1961).
5486. Nielsen, J. R., and Woollett, A. H., *J. Chem. Phys.*, **26**, 1391 (1957).
5487. Nielsen, L. E., *J. Appl. Phys.*, **25**, 1209 (1954).
5488. Nielsen, L. E., Dahm, D. J., Berger, P. A., Murty, V. S., and Kardos, J. L., *J. Polym. Sci. A2*, **12**, 1239 (1974).
5489. Niesel, W., Lubbers, D. W., Schnee, W. D., Richter, J., and Botticher, W., *Rev. Sci. Instrum*, **35**, 578 (1964).
5490. Nieuwenhuysen, P. C., and Clauwaert, J., *J. Polym. Sci. C*, **61**, 163 (1977).
5491. Niezette, J., Hadjichristidis, N., and Desreux, V., *Makromol. Chem.*, **177**, 2069 (1976).
5492. Nikitin, V., Mihailova, N., and Volkova, L., *Vyskomol. Soedin. A*, **7**, 1235 (1965).
5493. Nilsson, R., and Sundelöf, L. O., *Makromol. Chem.*, **66**, 11 (1963).
5494. Nishi, T., *J. Polym. Sci. A2*, **12**, 685 (1974).
5495. Nishi, T., and Kwei, T. K., *Polymer*, **16**, 285 (1975).
5496. Nishi, T., and Wang, T. T., *Macromolecules*, **8**, 909 (1975).
5497. Nishi, T., Wang, T. T., and Kwei, T. K., *Macromolecules*, **8**, 227 (1975).
5498. Nishijima, Y., Fujimoto, T., and Onogi, Y., *Rept. Progr. Polym. Phys. Japan*, **9**, 457 (1966).
5499. Nishijima, Y., Midorikawa, T., and Taki, F., *Rept. Progr. Polym. Phys. Japan*, **9**, 497 (1966).
5500. Nishijima, Y., and Mito, Y., *Rep. Progr. Polym. Phys. Japan*, **11**, 425 (1968).
5501. Nishijima, Y., Onogi, Y. and Asai, T., *J. Polym. Sci., C*, **15**, 237 (1966).
5502. Nishijima, Y., and Onogi, Y., *Rep. Progr. Polym. Phys. Japan*, **11**, 411 (1968).
5503. Nishijima, Y., Onogi, Y., and Asai, T., *Rep. Progr. Polym. Phys. Japan*, **11**, 391 (1968).
5504. Nishijima, Y., Onogi, Y., Yamazaki, R., and Kawakami, K., *Rep. Progr. Polym. Phys. Japan*, **11**, 407 (1968).
5505. Nishijima, Y., and Oster, G., *J. Polym. Sci.*, **19**, 337 (1956).
5506. Nishijima, Y., Takadono, S., and Oku, S., *Rep. Progr. Polym. Phys. Japan*, **9**, 497 (1966).
5507. Nishijima, Y., Teramoto, A., Yamamoto, M., and Hiratsuka, S., *J. Polym. Sci., A2*, **5**, 23 (1967).
5508. Nishijima, Y., Yamamoto, M., Oku, S., and Umegae, M., *Rep. Progr. Polym. Phys. Japan*, **9**, 501 (1966).
5509. Nishimo, Y., *Analyst*, **4**, 174 (1955).
5510. Nishioka, M., Kikuchi, K., and Yoshika, K., *Polymer*, **16**, 791 (1975).
5511. Nobbs, J. H., Bower, D. I., and Ward, I. M., *Polymer*, **17**, 25 (1976).
5512. Nobbs, J. H., Bower, D. I., Ward, I. M., and Patterson, D., *Polymer*, **14**, 287 (1974).
5513. Noda, I., Horikawa, T., Kato, T., Fujimoto, T., and Nagasawa, M., *Macromolecules*, **3**, 795 (1970).

806

5514. Noda, I., Kitano, T., and Nagasawa, M., *J. Polym. Sci. A2*, **15**, 1129 (1977).
5515. Noda, I., Mizutani, K., and Kato, T., *Macromolecules*, **10**, 618 (1977).
5516. Noda, I., Mizutani, K., Kato, T., Fujimoto, T., and Nagasawa, M., *Macromolecules*, **3**, 787 (1970).
5517. Noether, H. D., and Whitney, W., *Kolloid. Z. Z. Polym.*, **251**, 991 (1973).
5518. Noffz, D., Benz, W., and Pfab, W., *Fresenius' Z. Anal. Chem.*, **235**, 121 (1968).
5519. Nojiri, A., and Okamoto, S., *Polymer J.*, **2**, 689 (1971).
5520. Nomura, S., Kawai, H., Kimura, I., and Kagiyama, M., *J. Polym. Sci. A2*, **5**, 479 (1967).
5521. Nomura, S., Matsuo, M., and Kawai, H., *J. Polym. Sci. A2*, **12**, 1371 (1974).
5522. Noordermeer, J. W. M., and Ferry, J. D., *J. Polym. Sci. A1*, **14**, 509 (1976).
5523. Norberg, P. H., and Sundelöf, L. O., *Makromol. Chem.*, **77**, 77 (1964).
5524. Norisuye, T., and Fujita, H., *J. Polym. Sci. A2*, **16**, 999 (1978).
5525. North, A. M., Pethrick, R. A., and Phillips, D. W., *Polymer*, **18**, 324 (1977).
5526. North, A. M., Pethrick, R. A., and Wilson, A. D., *Polymer*, **19**, 913, 923 (1978).
5527. North, A. M., and Phillips, P. J., *Brit. Polym. J.*, **1**, 76 (1969).
5528. Northolt, M. G., *J. Polym. Sci. C*, **38**, 205 (1972).
5529. Northolt, M. G., and Stuut, H. A., *J. Polym. Sci. A2*, **16**, 939 (1978).
5530. Nose, T., *Polym. J.*, **2**, 124, 428, 437, 445 (1971).
5531. Nose, T., *Polymer J.*, **4**, 217 (1973).
5532. Nose, T., and Tan, T. V., *J. Polym. Sci. B*, **14**, 705 (1976).
5533. Nowakowska, M., Najbar, J., and Waligora, B., *Europ. Polym. J.*, **12**, 387 (1976).
5534. Nozakura, S. I., Morishima, Y., Iimuro, H., and Irie, Y., *J. Polym. Sci.*, **14**, 759 (1976).
5535. Nozakura, S. I., Takeuchi, S., Yuki, H., and Murahasji, S., *Bull. Chem. Soc. Japan*, **34**, 1673 (1961).
5536. Nukushina, Y., Itoh, Y., and Fischer, E. W., *J. Polym. Sci. B*, **3**, 383 (1965).
5537. Nyström, B., and Bergman, R., *Europ. Polym. J.*, **14**, 431 (1978).
5538. Ober, R., Cotton, J. P., Farnoux, B., and Higgins, J. S., *Macromolecules*, **7**, 634 (1974).
5539. Oberholtzer, J. E., and Rogers, L. B., *Anal. Chem.*, **41**, 1234 (1969).
5540. Oberthur, R. C., *Makromol. Chem.*, **176**, 3593 (1975).
5541. O'Brien, J., Cashell, E., Wardell, G. E., and McBrierty, V. J., *Macromolecules*, **9**, 653 (1976).
5542. O'Connor, C. L., and Schlupf, J. P., *J. Acoust. Soc. Amer.*, **40**, 663 (1966).
5543. Odajima, A., Woodward, A. E., and Sauer, J. A., *J. Polym. Sci.*, **55**, 181 (1961).
5544. Odell, J. A., Dlugosz, J., and Keller, A., *J. Polym. Sci. A2*, **14**, 861 (1976).
5545. Odell, J. A., Grubb, D. T., and Keller, A., *Polymer*, **19**, 617 (1978).
5546. Oeder, D., and Schneider, G. M., *Ber. Bunsenges. Phys. Chem.*, **73**, 229 (1969).
5547. Odijk, T., and Houwaart, A. C., *J. Polym. Sci. A2*, **16**, 627 (1978).
5548. Ogasawara, K., Yuasa, K., and Matsuzawa, S., *Makromol. Chem.*, **177**, 3403 (1976).
5549. Ogata, A., Tabata, Y., and Hamaguschi, H., *Bull. Chem. Soc. Japan*, **40**, 2205 (1967).
5550. Ogata, A., and Tao, S. J., *J. Appl. Polym. Sci.*, **41**, 4261 (1970).
5551. Ogawa, T., and Inaba, T., *J. Polym. Sci. A2*, **12**, 785 (1974).
5552. Ogawa, T., and Inaba, T., *J. Appl. Polym. Sci.*, **21**, 2979 (1977).
5553. Ogawa, T., Suzuki, Y., and Inaba, T., *J. Polym. Sci. A1*, **10**, 737 (1972).
5554. Ogawa, T., Tanaka, S., and Inaba, T., *J. Appl. Polym. Sci.*, **17**, 779 (1973).
5555. Oguchi, M., *Makromol. Chem.*, **177**, 549 (1976).
5556. Oguni, N., Lee, K., and Tani, H., *Macromolecules*, **5**, 819 (1972).
5557. Oguni, N., Maeda, S., and Tani, H., *Macromolecules*, **6**, 459 (1973).
5558. Oguni, N., Watanabe, S., Maki, M., and Tani, H., *Macromolecules*, **6**, 195 (1973).
5559. Ogura, K., *Brit. Polym. J.*, **7**, 221 (1975).
5560. Ogura, K., Kawamura, S., and Sobue, H., *Macromolecules*, **4**, 79 (1971).
5561. Ogura, K., Miyachi, Y., Sobue, H., and Nakamura, S., *Makromol. Chem.*, **176**, 1173 (1975).
5562. Ohmine, I., Silbey, R., and Deutch, J. M., *Macromolecules*, **10**, 862 (1977).
5563. Ohnishi, S., *Bull. Chem. Soc. Japan*, **35**, 254 (1962).
5564. Ohnishi, S., Ikeda, Y., Sugimoto, S., and Nitta, I., *J. Polym. Sci.*, **47**, 503 (1960).
5565. Ohnishi, S., Ikeda, Y., Kashiwagi, M., and Nitta, I., *Polymer*, **2**, 119 (1961).
5566. Ohnishi, S., Nakajima, Y., and Nitta, I., *J. Appl. Polym. Sci.*, **6**, 629 (1962).
5567. Ohnishi, S., and Nukada, K., *J. Polym. Sci. B*, **3**, 179, 1001 (1965).
5568. Ohnishi, S., Sugimoto, S., and Nitta, I., *J. Chem. Phys.*, **37**, 1283 (1962).
5569. Ohnishi, S., Sugimoto, S., and Nitta, I., *J. Polym. Sci. A*, **1**, 605 (1963).

807

5570. Ohnishi, S., Sugimoto, S., and Nitta, I., *J. Chem. Phys.*, **39**, 2647 (1963).
5571. Onishi, T., and Krimm, S., *J. Appl. Phys.*, **32**, 2320 (1961).
5572. Okada, M., Yamashita, Y., and Ishii, J., *Makromol. Chem.*, **94**, 181 (1966).
5573. Okada, T., and Ikushige, T., *J. Polym. Sci. A1*, **14**, 2059 (1976).
5574. Okada, T., and Mandelkern, L., *J. Polym. Sci. A2*, **5**, 239 (1967).
5575. Okamoto, H., *J. Polym. Sci.*, **37**, 173 (1959).
5576. Okamoto, H., *J. Polym. Sci. A*, **2**, 3451 (1964).
5577. Okamoto, Y., and Overberger, C. G., *Macromolecules*, **7**, 614 (1974).
5578. Okazawa, T., *Macromolecules*, **8**, 371 (1975).
5579. Okazawa, T., and Kaneko, M., *Polym. J.*, **2**, 747 (1971).
5580. O'Konski, C. T., and Bergmann, K., *J. Chem. Phys.*, **37**, 1573 (1962).
5581. O'Konski, C. T., and Haltner, A. J., *J. Amer. Chem. Soc.*, **78**, 3804 (1956).
5582. Okui, N., Li, H. M., and Magill, J. H., *Polymer*, **19**, 411 (1978).
5583. Okumoto, T., Takeuchi, T., and Tsuge, S., *Macromolecules*, **6**, 922 (1973).
5584. Okumoto, T., Tsuge, S., Yamamoto, Y., and Takeuchi, T., *Macromolecules*, **7**, 376 (1974).
5585. Olabisi, O., *Macromolecules*, **8**, 316 (1975).
5586. Olaj, O. F., *Makromol. Chem.*, **177**, 3427 (1976).
5587. Olaj, O. F., and Pelinka, K. H., *Makromol. Chem.*, **177**, 3413 (1976).
5588. Olaj, O. F., and Pelinka, K. H., *Makromol. Chem.*, **177**, 3447 (1976).
5589. Olcese, T., Ackerman, J., Radici, P., and Bianchi, U., *Makromol. Chem.*, **174**, 137 (1973).
5590. O'Leary, K., and Geil, P. H., *J. Macromol. Sci. B*, **1**, 147 (1967).
5591. Olf, H. G., and Peterlin, A., *Kolloid Z. Z. Polym.*, **215**, 97 (1967).
5592. Olf, H. G., Peterlin, A., and Peticolas, W. L., *J. Polym. Sci. A2*, **12**, 359 (1974).
5593. Olley, R. H., and Bassett, D. C., *J. Polym. Sci. A2*, **15**, 1011 (1977).
5594. O'Malley, J. J., *J. Polym. Sci. A2*, **13**, 1353 (1975).
5595. O'Mara, J. H., and McIntyre, D., *J. Phys. Chem.*, **63**, 1435 (1959).
5596. Onder, K., Peters, R. H., and Spark, L. C., *Polymer*, **13**, 133 (1972).
5597. O'Neill, M. J., *Anal. Chem.*, **36**, 1238 (1964).
5598. Ong, C. S. M., and Stein, R. S., *J. Polym. Sci. A2*, **12**, 1599 (1974).
5599. Ong., C. S. M., and Stein, R. S., *J. Polym. Sci. A2*, **12**, 1899 (1974).
5600. Ong, C. S. M., Yoon, D. Y., and Stein, R. S., *J. Polym. Sci. A2*, **12**, 1319 (1974).
5601. Onodera, M., *J. Polym. Sci. C*, **61**, 271 (1977).
5602. Onu, A., Legras, R., and Mercier, J. P., *J. Polym. Sci. A2*, **14**, 1187 (1976).
5603. Ooi, T., Shiotsubo, M., Hama, Y., and Shinohara, K., *Polymer*, **16**, 510 (1975).
5604. Opaskar, C. G., and Krimm, S., *Spectrochim. Acta A*, **23**, 2261 (1967).
5605. O'Reilly, J. M., Karasz, F. E., and Bair, H. E., *J. Polym. Sci. C*, 109 (1963).
5606. Orofino, T. A., *Polymer*, **2**, 305 (1961).
5607. Orofino, T. A., *J. Polym. Sci. A2*, **6**, 575 (1968).
5608. Orofino, T. A., and Flory, P. J., *J. Chem. Phys.*, **26**, 1067 (1957).
5609. Orofino, T. A., and Wenger, F., *J. Phys. Chem.*, **67**, 566 (1963).
5610. Orth, H., and Fischer, E. W., *Makromol. Chem.*, **88**, 188 (1965).
5611. Orttung, W. H., *J. Amer. Chem. Soc.*, **87**, 924 (1965).
5612. Orwoll, R. A., and Flory, P. J., *J. Amer. Chem. Soc.*, **89**, 6814, 6822 (1967).
5613. Orwoll, R. A., and Small, J. A., *Macromolecules*, **6**, 755 (1973).
5614. Osaki, S., and Ishida, Y., *J. Polym. Sci. A2*, **12**, 1727 (1974).
5615. Osaki, S., and Ishida, Y., *J. Polym. Sci. A2*, **13**, 1071 (1975).
5616. Osaki, S., and Schrag, J. L., *J. Polym. Sci. A2*, **11**, 549 (1973).
5617. Oshumi, Y., Higashimura, T., Okamura, S., Chûjô, R., and Kuroda, T., *J. Polym. Sci. A1*, **6**, 3015 (1968).
5618. Oster, G., *Anal. Chem.*, **25**, 1165 (1963).
5619. Oster, G., and Nishijima, Y., *J Amer. Chem. Soc.*, **78**, 1581 (1956).
5620. Oster, G., and Riley, D. P., *Acta Crystalogr.*, **5**, 272 (1952).
5621. Osterhoudt, H. W., and Ray, L. N., Jr, *J. Polym. Sci. A2*, **5**, 569 (1967).
5622. Osterhoudt, H. W., and Ray, L. N., Jr, *J. Polym. Sci. C*, **21**, 5 (1968).
5623. Osterhoudt, H. W., and Williams, J. W., *J. Phys. Chem.*, **69**, 1050 (1965).
5624. Oth, A., and Desreux, V., *Bull. Soc. Chim. Belges*, **63**, 137 (1951).
5625. Otocka, E. P., *Macromolecules*, **3**, 691 (1970).
5626. Otocka, E. P., *ACS Polym. Preprints*, **12**, 645 (1971).

808

5627. Otocka, E. P., and Hellman, M. Y., *Macromolecules*, **3**, 362 (1970).
5628. Otocka, E. P., and Hellman, M. Y., *Macromolecules*, **3**, 392 (1970).
5629. Otocka, E. P., and Hellman, M. Y., *J. Polym. Sci. B*, **12**, 331 (1974).
5630. Otocka, E. P., Hellman, M. Y., and Muglia, P. M., *Macromolecules*, **5**, 227 (1972).
5631. Otocka, E. P., Muglia, P. M., and Frisch, H. L., *Macromolecules*, **4**, 512 (1971).
5632. Otocka, E. P., Roe, R. J., Hellman, M. Y., and Muglia, P. M., *Macromolecules*, **4**, 507 (1971).
5633. Ottenill, R. H., and Parreira, H. C., *J. Phys. Chem.*, **62**, 912 (1958).
5634. Ouano, A. C., *J. Polym. Sci. A1*, **9**, 2179 (1971).
3635. Ouano, A. C., *J. Polym. Sci. A1*, **10**, 2169 (1972).
5636. Ouano, A. C., *J. Chromatogr.*, **118**, 303 (1976).
5637. Ouano, A. C., and Kaye, W., *J. Polym. Sci. A1*, **12**, 1151 (1974).
5638. Ouano, A. C., and Mercier, P. L., *J. Polym. Sci. C*, **21**, 309 (1968).
5639. Ouchi, I., Hosoi, M., and Shimotsuma, S., *J. Appl. Polym. Sci.*, **21**, 3445 (1977).
5640. Outer, P., Carr, C. I., and Zimm, B. H., *J. Chem. Phys.*, **18**, 830 (1950).
5641. Ovchinnikov, Y. K., Antipov, E. M., Markova, G. S., and Bakeev, N. F., *Makromol. Chem.*, **177**, 1567 (1976).
5642. Ovenhall, O. W., and Peaker, F. W., *Makromol. Chem.*, **33**, 222 (1959).
5643. Overberger, C. G., Jarovitzky, P. A., and Mukamal, H., *ACS Polym. Preprints*, **8**, 401 (1967).
5644. Overberger, C. G., Ozaki, S., and Braunstein, D. M., *Makromol. Chem.*, **93**, 13 (1966).
5645. Overberger, C. G., and Takehoshi, T., *Macromolecules*, **1**, 7 (1968).
5646. Overbergh, N., Berghmans, H., and Smets, G., *J. Polym. Sci. C*, **38**, 237 (1972).
5647. Overbergh, N., Girolamo, M., and Keller, A., *J. Polym. Sci. A2*, **15**, 1475 (1977).
5648. Overbergh, N., Sadler, D. M., and Keller, A., *J. Polym. Sci. A2*, **15**, 1485 (1977).
5649. Overbergh, N., Berghmans, H., and Reynaers, H., *J. Polym. Sci. A2*, **14**, 1177 (1976).
5650. Owens, D. K., and Wendt, R. C., *J. Appl. Polym. Sci.*, **13**, 1741 (1969).
5651. Ozawa, T., *Bull. Chem. Soc. Japan*, **38**, 1881 (1965).
5652. Ozawa, T., *Bull. Chem. Soc. Japan*, **39**, 2071 (1966).
5653. Ozawa, T., *Therm. Anal.*, **7**, 601 (1975).
5654. Oyda, O., and Kuriyama, I., *J. Polym. Sci. A2*, **15**, 773 (1977).

5655. Pacor, P., *Anal. Chim. Acta*, **37**, 200 (1967).
5656. Padden, F. J., Jr, and Keith, H. D., *J. Appl. Phys.*, **30**, 1479 (1959).
5657. Pae, K. D., Morrow, D. R., and Sauer, J. A., *Nature*, **211**, 514 (1966).
5658. Page, T. F., and Bresler, W. E., *Anal. Chem.*, **36**, 1981 (1964).
5659. Paglini, S., *Anal. Biochem.*, **23**, 247 (1968).
5660. Painter, P. C., Havens, J., Hart, W. W., and Koenig, J. L., *J. Polym. Sci. A2*, **15**, 1237 (1977).
5661. Painter, P. C., and Koenig, J. L., *J. Polym. Sci. A2*, **15**, 1885 (1977).
5662. Painter, P. C., Runt, J., Coleman, M. M., and Harrison, I. R., *J. Polym. Sci. A2*, **15**, 1647 (1977).
5663. Painter, P. C., Watzek, M., and Koenig, J. L., *Polymer*, **18**, 1169 (1977).
5664. Pakula, T., and Kryszewski, M., *J. Polym. Sci. C*, **38**, 87 (1972).
5665. Pakula, T., and Soukup, Z., *J. Polym. Sci. A2*, **12**, 2437 (1974).
5666. Palit, S. R., *Makromol. Chem.*, **36**, 89 (1959).
5667. Palit, S. R., *Kunststoffe*, **50**, 513 (1960).
5668. Palit, S. R., *Makromol. Chem.*, **38**, 96 (1960).
5669. Palit, S. R., and Ghosh, P., *J. Polym. Sci.*, **58**, 1225 (1962).
5670. Palit, S. R., Mukherjee, B. N., and Konar, R. S., *J. Polym. Sci.*, **50**, 45 (1961).
5671. Palit, S. R., and Saha, M. K., *J. Polym. Sci.*, **58**, 1233 (1962).
5672. Palit, S. R., and Singh, U. N., *J. Indian Chem. Soc.*, **33**, 507 (1956).
5673. Palm, A., *J. Phys. Chem.*, **55**, 1320 (1951).
5674. Palmer, G. L., *Instrum. News*, **21**, 15 (1971).
5675. Palmer, R. P., and Cobbold, A. J., *Makromol. Chem.*, **74**, 174 (1964).
5676. Pan, C. Y., Jensen, C. D., Bielech, C., and Habgood, H. W., *J. Appl. Polym. Sci.*, **22**, 2307 (1978).
5677. Pannell, J., *Polymer*, **13**, 2 (1972).
5678. Pannell, J., *Polymer*, **13**, 277 (1972).
5679. Panton, C. J., Plesch, P. H., and Rutherford, P. P., *J. Chem. Soc.*, **1964**, 2586.

5680. Paoletti, S., Van der Touw, F., and Mandel, M., *J. Polym. Sci. A2*, **16**, 641 (1978).
5681. Pappalardo, L. T., *J. Appl. Polym. Sci.*, **21**, 809 (1977).
5682. Paralikar, K. M., and Betrabet, S. M., *J. Appl. Polym. Sci.*, **21**, 899 (1977).
5683. Pardue, H. L., and Rodriguez, P. A., *Anal. Chem.*, **39**, 901 (1967).
5684. Park, W. S., and Graessley, W. W., *J. Polym. Sci. A2*, **15**, 71 (1977).
5685. Park, W. S., and Graessley, W. W., *J. Polym. Sci. A2*, **15**, 85 (1977).
5686. Parrini, P., Sebastiano, F., and Messina, G., *Makromol. Chem.*, **38**, 27 (1960).
5687. Persons, W. F., Faust, M. A., and Brady, L. E., *J. Polym. Sci. A2*, **16**, 775 (1978).
5688. Pasch, N. F., and Webber, S. E., *Macromolecules*, **11**, 727 (1978).
5689. Paschke, E. E., Bildingmeyer, B. A., and Bergmann, J. G., *J. Polym. Sci. A1*, **15**, 983 (1977).
5690. Pasika, W. M., and West, A. C., *J. Polym. Sci. A2*, **13**, 2237 (1975).
5691. Passaglia, E., and Kevorkian, H. K., *J. Appl. Phys.*, **34**, 90 (1963).
5692. Passaglia, E., and Kevorkian, H. K., *J. Appl. Polym. Sci.*, **7**, 119 (1963).
5693. Pasternak, R. A., Brady, P., Ehrmantraut, H. C., *Dachema Monograph*, No. 44, 205 (1962).
5694. Pasternak, R. A., Burns, G. L., and Heller, J., *Macromolecules*, **4**, 470 (1971).
5695. Pasternak, R. A., Christensen, M. V., and Heller, J., *Macromolecules*, **3**, 366 (1970).
5696. Patat, F., *Z. Elektrochem.*, **60**, 208 (1956).
5697. Patat, F., and Klein, J., *J. Polym. Sci. B*, **3**, 615 (1965).
5698. Patel, G. N., and Patel, R. D., *J. Polym. Sci. A2*, **8**, 47 (1970).
5699. Patel, K. S., Patel, C. K., and Patel, R. D., *Polymer*, **18**, 275 (1977).
5700. Patrick, R. L., Brown, J. A., Cameron, N. M., and Gehman, W. G., *Appl. Polym. Symp.*, No. 16, 87 (1971).
5701. Patt, S. L., and Sykes, B. D., *J. Chem. Phys.*, **56**, 3182 (1972).
5702. Patterson, D., *Rubber Chem. Technol.*, **40**, 1 (1967).
5703. Patterson, D., *J. Polym. Sci. C*, **16**, 3379 (1968).
5704. Patterson, D., *Macromolecules*, **2**, 672 (1969).
5705. Patterson, D., *Macromolecules*, **4**, 30 (1971).
5706. Patterson, D., Delmas, G., and Somcynsky, T., *Polymer*, **8**, 503 (1967).
5707. Patterson, D., and Robard, A., *Macromolecules*, **11**, 690 (1978).
5708. Patterson, D., Tewari, Y. B., Schreiber, H. P., and Guillet, J. E., *Macromolecules*, **4**, 356 (1971).
5709. Patterson, D., and Ward, I. M., *Trans. Faraday Soc.*, **53**, 291 (1957).
5710. Patterson, D., and Ward, I. M., *Trans. Faraday Soc.*, **53**, 1516 (1957).
5711. Patterson, G. D., *Macromolecules*, **7**, 220 (1974).
5712. Patterson, G. D., *J. Polym. Sci. B*, **13**, 415 (1975).
5713. Patterson, G. D., *J. Polym. Sci. A2*, **14**, 143 (1976).
5714. Patterson, G. D., *J. Polym. Sci. A2*, **14**, 741 (1976).
5715. Patterson, G. D., *J. Polym. Sci. A2*, **14**, 1909 (1976).
5716. Patterson, G. D., *J. Polym. Sci. A2*, **15**, 455 (1977).
5717. Patterson, G. D., and Alms, G. R., *Macromolecules*, **10**, 1237 (1977).
5718. Patterson, G. D., Kennedy, A. P., and Latham, J. P., *Macromolecules*, **10**, 667 (1977).
5719. Patterson, G. D., and Latham, J. P., *Macromolecules*, **10**, 736 (1977).
5720. Patterson, G. D., Nishi, T., and Wang, T. T., *Macromolecules*, **9**, 603 (1976).
5721. Patzold, W., *J. Polym. Sci. B*, **1**, 269 (1963).
5722. Paul, D. G., and Umbreit, G. E., *Res. Dev. Ind.*, **21**, 18 (1970).
5723. Paul, D. R., *J. Polym. Sci. A2*, **12**, 1221 (1974).
5724. Paul, D. R., and DiBenedetto, A. T., *J. Polym. Sci. C*, **10**, 17 (1965).
5725. Paul, D. R., Garcin, M., and Garmon, W. E., *J. Appl. Polym. Sci.*, **20**, 609 (1976).
5726. Paul, D. R., and Koros, W. J., *J. Polym. Sci. A2*, **14**, 675 (1976).
5727. Paul, D. R., Mavichak, V., and Kemp, D. R., *J. Appl. Polym. Sci.*, **15**, 1533 (1971).
5728. Paul, W., and Steinwedel, H., *Z. Naturforsch.*, **8a**, 448 (1953).
5729. Pavlinov, L. I., Rabinovich, I. B., Okladnov, N. K., and Arzhakov, S. A., *Vysokomol. Soedin. A*, **9**, 483 (1967).
5730. Pazonyi, T., and Dimitrov, M., *Rubber Chem. Technol.*, **40**, 1119 (1967).
5731. Peacock, C. J., Hendra, P. J., Willis, H. A., and Cudby, M. E. A., *J. Chem. Soc. A*, **1970**, 2943 (1970).
5732. Peaker, F. W., *J. Polym. Sci.*, **22**, 25 (1956).
5733. Peaker, F. W., and Rayner, M. G., *Europ. Polym. J.*, **6**, 107 (1970).

5734. Peaker, F. W., and Robb, J. D., *Nature*, **182**, 1591 (1958).
5735. Peaker, F. W., and Tweedale, C. R., *Nature*, **216**, 75 (1967).
5736. Pearson, D. S., and Graessley, W. W., *Macromolecules*, **11**, 528 (1978).
5737. Pecora, R., *J. Chem. Phys.*, **40**, 1604 (1964).
5738. Pecora, R., *Macromolecules*, **2**, 31 (1969).
5739. Pecora, R., *Discuss. Faraday Soc.*, **49**, 222 (1970).
5740. Pecora, R., and Tagami, Y., *J. Chem. Phys.*, **51**, 3298 (1969).
5741. Pecsok, R. L., Painter, P. C., Shelton, J. R., and Koenig, J. L., *Rubber Chem. Technol.*, **49**, 1010 (1976).
5742. Pedersen, H. L., and Lyngae-Jörgensen, J., *Brit. Polym. J.*, **1**, 138 (1969).
5743. Peebles, L. H., Jr, *Macromolecules*,, **7**, 872 (1974).
5744. Peeters, F. A. H., and Staverman, A. J., *Macromolecules*, **10**, 1164 (1977).
5745. Peifer, J. J., *Michrochem. Acta*, **1962**, 529.
5746. Peitscher, G., and Holtrup, D. I., *Angew. Macromol. Chem.*, **47**, 111 (1975).
5747. Pelzbauer, Z., and Galeski, A., *J. Polym. Sci. C*, **38**, 23 (1972).
5748. Pelzbauer, Z., and Lednicky, F., *J. Polym. Sci. A2*, **12**, 2173 (1974).
5749. Peled, S., El-Hanny, U., and Yatsiv, S., *Rev. Sci. Instrum.*, **37**, 1649 (1966).
5750. Pennings, A. J., *J. Polym. Sci. C*, **16**, 1799 (1967).
5751. Pennings, A. J., *J. Polym. Sci. C*, **20**, 145 (1967).
5752. Pennings, A. J., and Kiel, A. M., *Kolloid Z. Z. Polym.*, **206**, 160 (1965).
5753. Pennings, A. J., Kiel, A. M., and Van der Mark, J. M. A. A., *Kolloid Z. Z. Polym.*, **237**, 336 (1970).
5754. Pennings, A. J., Schouteten, C. J. H., and Kiel, A. M., *J. Polym. Sci. C*, **38**, 167 (1972).
5755. Pennings, A. J., and Smith, P., *Brit. Polym. J.*, **7**, 460 (1975).
5756. Pennings, A. J., Van der Mark, J. M. A. A., and Booij, H. C., *Kolloid Z. Z. Polym.*, **236**, 99 (1970).
5757. Pennings, A. J., Van der Mark, J. M. A. A., and Kiel, A. M., *Kolloid Z. Z. Polym.*, **237**, 336 (1970).
5758. Pennings, A. J., Zwijnenburg, A., and Lageveen, R., *Kolloid Z. Z. Polym.*, **251**, 500 (1973).
5759. Penwell, R. C., and Graessley, W. W., *J. Polym. Sci. A2*, **12**, 213 (1974).
5760. Penwell, R. C., Graessley, W. W., and Kovacs, A., *J. Polym. Sci. A2*, **12**, 1771 (1974).
5761. Penzel, E., Pohlemann, H., and Swoboda, J., *Makromol. Chem.*, Suppl. 1, 367 (1975).
5762. Peppas, N. A., *Makromol. Chem.*, **176**, 3433 (1975).
5763. Peppas, N. A., and Merrill, E. W., *J. Polym. Sci. A1*, **14**, 459 (1976).
5764. Peppas, N. A., and Merrill, E. W., *J. Appl. Polym. Sci.*, **21**, 1763 (1977).
5765. Pepper, D. C., and Railley, P. J., *Proc. Chem. Soc.*, **1961**, 460.
5766. Paper, D. C., and Rutherford, P. P., *J. Appl. Polym. Sci.*, **2**, 100 (1959).
5767. Peraldo, M., and Cambini, M., *Spectrochim. Acta*, **21**, 1509 (1965).
5768. Percival, D. F., and Stevens, M. P., *Anal. Chem.*, **36**, 1574 (1964).
5769. Periard, J., Banderet, A., and Reiss, G., *Angew. Makromol. Chem.*, **15**, 37 (1971).
5770. Perico, A., and Cuniberti, C., *Macromolecules*, **8**, 828 (1975).
5771. Perico, A., and Rossi, C., *J. Chem. Phys.*, **53**, 1217 (1970).
5772. Perrault, G., Tremblay, M., Bedard, M., Duschesne, G., and Voyzelle, R., *Europ. Polym., J.*, **10**, 143 (1974).
5773. Perry, J. A., *J. Chromatogr.*, **113**, 267 (1975).
5774. Perry, J. A., *J. Chromatogr.*, **110**, 27 (1975).
5775. Perry, J. A., and Glunz, L. J., *J. Assoc. Off. Anal. Chem.*, **57**, 832 (1974).
5776. Perry, J. A., Haag, K. W., and Glunz, L. J., *J. Chromatogr. Sci.*, **11**, 447 (1973).
5777. Perry, J. A., Jupille, T. H., and Curtice, A. B., *Separation Sci.*, **10**, 571 (1973).
5778. Perry, J. A., Jupille, T. H., and Glunz, L. J., *Anal. Chem.*, **47**, 65A (1975).
5779. Perry, J. A., Jupille, T. H., and Glunz, L. J., *Separation Purification Methods*, **4**, 97 (1975).
5780. Peter, S., *Chem. Ing. Technol.*, **32**, 437 (1960).
5781. Peterlin, A., *Makromol. Chem.*, **9**, 244 (1952).
5782. Peterlin, A., *J. Polym. Sci.*, **10**, 425 (1953).
5783. Peterlin, A., *J. Polym. Sci.*, **47**, 403 (1960).
5784. Peterlin, A., *J. Polym. Sci. B*, **3**, 1083 (1965).
5785. Peterlin, A., *Makromol. Chem.*, **124**, 136 (1969).
5786. Peterlin, A., *Text. Res. J.*, **42**, 20 (1972).

5787. Peterlin, A., and Holbrook, J. D., *Kolloid Z.*, **203**, 68 (1965).
5788. Peterlin, A., Kiho, H., and Geil, P. H., *J. Poly. Sci. B*, **3**, 151 (1965).
5789. Peterlin, A., and Meinel, G., *J. Polym. Sci. B*, **2**, 751 (1964).
5790. Peterlin, A., and Meinel, G., *J. Appl. Phys.*, **35**, 3221 (1964).
5791. Peterlin, A., and Meinel, G., *J. Polym. Sci. B*, **3**, 783 (1965).
5792. Peterlin, A., and Meinel, G., *J. Polym. Sci. B*, **3**, 1059 (1965).
5793. Peterlin, A., and Meinel, G., *Appl. Polym. Symp.*, **2**, 85 (1966).
5794. Peterlin, A., Meinel, G., Olf, H. G., *J. Polym. Sci. B*, **4**, 399 (1966).
5795. Peterlin, A., and Yasuda, H., *J. Polym. Sci. A2*, **12**, 1215 (1974).
5796. Peterlin-Neumaier, T., *J. Polym. Sci. A2*, **14**, 1351 (1976).
5797. Petermann, J., and Gleiter, H., *J. Polym. Sci. A2*, **13**, Z1939 (1975).
5798. Petermann, J., and Gleiter, H., *J. Polym. Sci. B*, **15**, 649 (1977).
5799. Petersen, R. J., Corneliussen, R. D., and Rozelle, L. T., *ACS Polym. Preprints*, **10**, 385 (1969).
5800. Peterson, C. M., *J. Appl. Polym. Sci.*, **12**, 2669 (1968).
5801. Peterson, G. V., and Poole, J. S., *Intern. Lab.*, July/Aug. 1974, p. 63.
5802. Peterson, J. M., and Fixman, M., *J. Chem. Phys.*, **39**, 2516 (1963).
5803. Pétiaud, R., and Pham, Q. T., *Makromol. Chem.*, **178**, 741 (1977).
5804. Peticolas, W. L., Stegeman, G. I. A., and Stoicheff, B. P., *Phys. Rev. Lett.*, **18**, 1130 (1967).
5805. Petraccone, V., Allegra, G., and Corradini, P., *J. Polym. Sci. C*, **38**, 419 (1972).
5806. Petraccone, V., Sanchez, I. C., and Stein, R. S., *J. Polym. Sci. A2*, **13**, 1991 (1975).
5807. Petrie, S. E. B., *J. Polym. Sci. A2*, **10**, 1255 (1972).
5808. Pezzin, G., Ceccorulli, G., Pizzoli, M., and Peggion, E., *Macromolecules*, **8**, 762 (1975).
5809. Pezzin, G., Zilio-Grandi, F., and Sanmartin, P., *Europ. Polym. J.*, **6**, 1053 (1970).
5810. Peyrouset, A., and Panaris, R., *J. Appl. Polym. Sci.*, **16**, 315 (1972).
5811. Peyrouset, A., Prechner, R., Panaris, R., and Benoit, H., *J. Appl. Polym. Sci.*, **19**, 1363 (1975).
5812. Peyser, P., and Little, R. C., *J. Appl. Polym. Sci.*, **15**, 2623 (1971).
5813. Pham, Q. T., Millan, J. L., and Madruga, E. L., *Makromol. Chem.*, **175**, 945 (1974).
5814. Phaovibul, O., Čačković, H., Loboda-Čačković, J., Hosemann, R., *J. Polym. Sci. A2*, **11**, 2377 (1973).
5815. Philipp, H. J., *J. Polym. Sci.*, **6**, 371 (1951).
5816. Philipp, H. J., and Bjork, C. F., *J. Polym. Sci.*, **6**, 383 (1951).
5817. Phillips, D., Anissimov, V., Karpukhin, O., and Shlyapintokh, V. Ya., *Nature*, **215**, 1163 (1967).
5818. Phillips, P. J., and Edwards, B. C., *J. Polym. Sci. A1*, **13**, 1819 (1975).
5819. Phillips, P. J., and Edwards, B. C., *J. Polym. Sci. A1*, **14**, 377 (1976).
5820. Phillips, P. J., and Edwards, B. C., *J. Polym. Sci. B*, **14**, 449 (1976).
5821. Phillips, P. J., Emerson, F. A., and MacKnight, W. J., *Macromolecules*, **3**, 771 (1970).
5822. Phillips, P. J., and Singh, G., *J. Polym. Sci. A2*, **13**, 1377 (1975).
5823. Phillips, P. J., Wilkes, G. L., Delf, B. W., and Stein, R. S., *J. Polym. Sci. A2*, **9**, 499 (1971).
5824. Pickett, H. E., Cantow, M. J. R., and Johnson, J. F., *J. Appl. Polym. Sci.*, **10**, 917 (1966).
5825. Pickett, H. E., Cantow, M. J. R., and Johnson, J. F., *J. Polym. Sci. C*, **21**, 67 (1968).
5826. Picot, C., Duplessix, R., Decker, D., Benoit, H., Boue, F., Cotton, J. P., Daoud, M., Farnoux, B., Jannik, G., Nierlich, M., deVries, A. J., and Pincus, P., *Macromolecules*, **10**, 436 (1977).
5827. Picot, C., Fukuda, M., Ong, C., and Stein, R. S., *J. Macromol. Sci. B*, **6**, 263 (1972).
5828. Picot, C., Stein, R. S., Marchessault, R. H., Borch, J., and Sarko, A., *Macromolecules*, **4**, 467 (1971).
5829. Picot, C., and Weill, G., *J. Polym. Sci. A2*, **12**, 1733 (1974).
5830. Picot, C., Weill, G., and Benoit, H. B., *J. Polym. Sci. C*, **16**, 3973 (1968).
5831. Pierce, P. E., and Armonas, J. E., *J. Polym. Sci. C*, **21**, 23 (1968).
5832. Pierre, J., and Desreaux, V., *Polymer*, **15**, 685 (1974).
5833. Pierre, L. E. St, and Price, C. C., *J. Amer. Chem. Soc.*, **78**, 3432 (1956).
5834. Pilar, J., and Ulbert, K., *Polymer*, **16**, 730 (1975).
5835. Pillai, P. K. C., and Ahuja, R. C., *Polymer*, **17**, 192 (1976).
5836. Pincus, P., *Macromolecules*, **9**, 386 (1976).

812

5837. Pineri, M., Meyer, C., and Bourret, A., *J. Polym. Sci. A2*, **13**, 1881 (1975).
5838. Pines, E., and Prins, W., *Macromolecules*, **6**, 888 (1973).
5839. Pino, P., Ciardelli, C., Lorenzi, G. P., and Montagnoli, G., *Makromol. Chem.*, **61**, 207 (1963).
5840. Pino, P., and Lorenzi, G. P., *J. Amer. Chem. Soc.*, **82**, 4747 (1960).
5841. Piseri, L., and Zerbi, G., *J. Chem. Phys.*, **48**, 3561 (1968).
5842. Pittman, R. A., and Tripp, V. W., *J. Polym. Sci.*, **8**, 969 (1970).
5843. Pitts, J. N., Jr, Margerum, J. D., Taylor, R. P., and Brim, W., *J. Amer. Chem. Soc.*, **77**, 5499 (1955).
5844. Plaza, A., Norris, F. H., and Stein, R. S., *J. Polym. Sci. C*, **24**, 455 (1957).
5845. Pleštil, J., and Baldrian, J., *Makromol. Chem.*, **174**, 183 (1973).
5846. Pochan, J. M., Beatty, C. L., Hinman, D. D., and Karasz, F. E., *J. Polym. Sci. A2*, **13**, 977 (1975).
5847. Pochan, J. M., and Hinman, D. F., *J. Polym. Sci. A2*, **13**, 1365 (1975).
5848. Pochan, J. M., and Hinman, D. F., *J. Polym. Sci. A2*, **14**, 1871 (1976).
5849. Pochan, J. M., Hinman, D. F., and Froix, M. F., *Macromolecules*, **9**, 611 (1976).
5850. Pochan, J. M., Hinman, D. F., Froix, M. F., and Davidson, T., *Macromolecules*, **10**, 113 (1977).
5851. Pochan, J. M., Pacansky, T. J., and Hinman, D. F., *Polymer*, **19**, 431 (1978).
5852. Podesva, J., Bohdanecky, M., Kratochvil, P., and Samay, G., *J. Polym. Sci. A2*, **15**, 1521 (1977).
5853. Pogany, G. A., *Polymer*, **17**, 690 (1976).
5854. Pohl, H. A., *Anal. Chem.*, **26**, 1614 (1954).
5855. Pohl, H. A., Bacskai, R., and Purcell, W. P., *J. Phys. Chem.* **64**, 1701 (1960).
5856. Point, J. J., Gillot, M., Dosiere, M., and Goffin, A., *J. Polym. Sci. C*, **38**, 261 (1972).
5857. Point, J. J., and Goffin, A., *J. Polym. Sci. B*, **13**, 249 (1975).
5858. Poláček, J., *Europ. Polym. J.*, **6**, 81 (1970).
5859. Pollack, S. S., Robinson, W. H., Chiang, R., and Flory, P. J., *J. Appl. Phys.*, **33**, 237 (1962).
5860. Poltoratzkiy, B. F., and Sachkov, K. N., *J. Polym. Sci. C*, **61**, 73 (1977).
5861. Ponevski, C., and Solunov, S., *J. Polym. Sci. A2*, **13**, 1467 (1975).
5862. Pope, D. P., *J.. Polym. Sci. A2*, **14**, 811 (1976).
5863. Pope, D. P., and Keller, A., *J. Polym. Sci. A2*, **13**, 533 (1975).
5864. Pope, D. P., and Keller, A., *J. Polym. Sci. A2*, **14**, 821 (1976).
5865. Pope, N. T., Weakley, T. J. R., and Williams, R. J. P., *J. Chem. Soc.*, **1959**, 3442. (1959).
5866. Porath, J., *Pure Appl. Chem.*, **6**, 233 (1963).
5867. Porath, J., and Flodin, P., *Nature*, **183**, 1657 (1959).
5868. Porsch, B., and Kubin, M., *Coll. Czech. Chem. Commun.*, **36**, 4046 (1971).
5869. Porter, C. H., and Boyd, R. H., *Macromolecules*, **4**, 589 (1971).
5870. Porter, C. H., Lawer, J. H. L., and Boyd, R. H., *Macromolecules*, **3**, 308 (1970).
5871. Porter, G., *Proc. Roy. Soc. A*, **200**, 284 (1950).
5872. Porter, G., and Topp, M. R., *Nature* **220**, 1228 (1968).
5873. Porter, G., and Topp, M. R., *Proc. Roy. Soc. A*, **315**, 163 (1970).
5874. Porter, K., and Volman, D. H., *J. Amer. Chem. Soc.*, **84**, 2011 (1962).
5875. Porter, K., and Volman, D. H., *Anal. Chem.*, **34**, 748 (1962).
5876. Porter, R. S., Cantow, M. J. R., and Johnson, J. F., *Makromol. Chem.*, **94**, 143 (1966).
5877. Porter, R. S., Hoffman, A. S., and Johnson, J. F., *Anal. Chem.*, **34**, 1179 (1962).
5878. Porto, S. P. S., *J. Opt. Soc. Amer.*, **56**, 1585 (1966).
5879. Portzehl, H., *Z. Naturforsch.*, **56**, 75 (1965).
5880. Pochlý, J., Máša, Z., Biroš, J., Zivný, A., and Trekoval, J., *J. Polym. Sci. A2*, **13**, 1511 (1975).
5881. Pouchlý, J., and Patterson, D., *Macromolecules*, **6**, 465 (1973).
5882. Pouchlý, J., and Patterson, D., *Macromolecules*, **9**, 574 (1976).
5883. Pouyet, G., Candau, F., and Dauantis, J., *Makromol. Chem.*, **177**, 2973 (1976).
5884. Powell, E., *Polymer*, **8**, 211 (1967).
5885. Powels, J. G., *Proc. Phys. Soc. (London) B*, **69**, 281 (1956).
5886. Powers, J. C., *J. Amer. Chem. Soc.*, **88**, 3679 (1966).
5887. Powers, J. C., *J. Amer. Chem. Soc.*, **89**, 1780 (1967).
5888. Powers, J. C., and Peticolas, W. L., *J. Phys. Chem.*, **71**, 3191 (1967).

5889. Powles, J. G., Hunt, B. I., and Sandiford, D. J. H., *Polymer*, **5**, 585 (1964).
5890. Powles, J. G., Strange, J. H., and Sandiford, D. J. H., *Polymer*, **4**, 401 (1963).
5891. Prager, S., Bagley, E. B., and Long, F. A., *J. Amer. Chem. Soc.*, **75**, 2742 (1953).
5892. Pratt, C. F., and Hobbs, S. Y., *Polymer*, **17**, 12 (1976).
5893. Prausnitz, J. A., and Anderson, R., *AIChE J.*, **7**, 96 (1961).
5894. Pravikova, N. A., Ryabova, L. G., and Vyrskii, Yu. A., *Vysokomol. Soedin.*, **5**, 1165 (1963).
5895. Preedy, J. E., *Brit. Polym. J.*, **5**, 13 (1973).
5896. Preedy, J. E., Wallis, S., and Wheeler, E. J., *Macromol. Chem.*, **179**, 1325 (1978).
5897. Preston, J., Smith, R. W., and Stehman, C. J., *J. Polym. Sci. C*, **19**, 7 (1967).
5898. Price, C., *Polymer*, **16**, 585 (1975).
5899. Price, C., Evans, K. A., and Booth, C., *Polymer*, **16**, 196 (1975).
5900. Price, F. P., *J. Chem. Phys.*, **31**, 1679 (1959).
5901. Price, F. P., *J. Polym. Sci.*, **39**, 139 (1959).
5902. Priest, D. J., *J. Polym. Sci. A2*, **9**, 1777 (1971).
5903. Prime, R. B., and Wunderlich, B., *J. Polym. Sci. A2*, **7**, 2061 (1969).
5904. Prime, R. B., and Wunderlich, B., *J. Polym. Sci. A2*, **7**, 2073 (1969).
5905. Prime, R. B., Wunderlich, B., and Lelillo, L., *J. Polym. Sci. A2*, **7**, 2091 (1969).
5906. Princen, L. H., *Appl. Polym. Symp.*, No. 16, 209 (1971).
5907. Prins, W., Rimai, L., and Chompff, A. J., *Macromolecules*, **5**, 104 (1972).
5908. Privalko, V. P., *Macromolecules*, **6**, 111 (1973).
5909. Privalko, V. P., and Lipatov, Y. S., *Makromol. Chem.*, **175**, 641 (1974).
5910. Prosser, R. A., *Anal. Chem.*, **39**, 1125 (1965).
5911. Provder, T., and Rosen, E. M., *Separation Sci.*, **5**, 437 (1970).
5912. Provencher, S. W., and Gobush, W., in Characterization of Macromolecular structure, Nat. Acad. Sci. U.S.A. Publ. 1573, Washington, D.C., 1968, p. 143.
5913. Prud'homme, J., and Bywater, S., *ACS Polym. Preprints*, **10**, 518 (1969).
5914. Prud'homme, J., and Bywater, S., *Macromolecules*, **4**, 543 (1971).
5915. Prud'homme, J., Goursot, P., and Laramée, A., *Makromol. Chem.*, **178**, 1561 (1977).
5916. Prud'homme, R. E., *J. Polym. Sci. A2*, **15**, 1619 (1977).
5917. Prud'homme, R. E., Bourland, L., Natarajan, R. T., and Stein, R. S., *J. Polym. Sci. A2*, **12**, 1955 (1974).
5918. Prud'homme, R. E., and Marchessault, R. H., *Macromolecules*, **7**, 541 (1974).
5919. Prud'homme, R. E., and Stein, R. S., *Macromolecules*, **4**, 668 (1971).
5920. Prud'homme, R. E., and Stein, R. S., *J. Polym. Sci A2*, **11**, 1683 (1973).
5921. Prud'homme, R. E., and Stein, R. S., *J. Polym. Sci. A2*, **12**, 1805 (1974).
5922. Prud'homme, R. E., and Stein, R. S., *Europ. Polym. J.*, **13**, 365 (1977).
5923. Prud'homme, R. E., Yoon, D., and Stein, R. S., *J. Polym. Sci. A2*, **11**, 1047 (1973).
5924. Przybylski, M., Ringsdorf, H., and Ritter, H., *Makromol. Chem.*, Suppl. 1, 297 (1975).
5925. Ptitsin, O. B., *Vysokomol. Soedin*, **3**, 1673 (1961).
5926. Purdon, J. R., Jr, and Mate, R. D., *J. Polym. Sci. A1*, **6**, 243 (1968).
5927. Purvis, J., and Bower, D. I., *Polymer*, **15**, 645 (1974).
5928. Purvis, J., and Bower, D. I., *J. Polym. Sci. A2*, **14**, 1461 (1976).
5929. Purvis, J., Bower, D. I., and Ward, I. M., *Polymer*, **14**, 398 (1973).
5930. Puterman, M., Kolpak, F. J., Blackwell, J., and Lando, J. B., *J. Polym. Sci. A2*, **15**, 805 (1977).
5931. Pyżuk, W., and Krupkowski, T., *Makromol. Chem.*, **178**, 817 (1977).

5932. Quach, A., and Simha, R., *J. Appl. Phys.*, **42**, 4592 (1971).
5933. Quack, G., and Fetters, L. J., *Macromolecules*, **11**, 369 (1978).
5934. Quaegebeur, J. P., Lablache-Combier, A., and Chachaty, C., *Makromol. Chem.*, **178**, 1507 (1977).
5935. Quayle, D. V., *Brit. Polym. J.*, **1**, 15 (1969).
5936. Rabek, J. F., and Rånby, B., *J. Polym. Sci. A1*, **12**, 273 (1974).
5937. Rabek, J. F., and Rånby, B., *J. Polym. Sci. A1*, **14**, 1463 (1976).
5938. Rabek, J. F., Shur, Y. J., and Rånby, B., *J. Polym. Sci. A1*, **13**, 1285 (1975).
5939. Rabek, J. F., Shur, Y. J., and Rånby, B., in Singlet Oxygen: Reactions with Organic Compounds and Polymers (B. Rånby and J. F. Rabek, eds), Wiley, Chichester, 1978, p. 264.
5940. Rabel, W., and Ueberreiter, K., *Kolloid Z.*, **198**, 1 (1964).

814

5941. Rabieska, J., and Kovacs, A. J., *J. Appl. Phys.*, **32**, 2314 (1961).
5942. Rabold, G. P., *J. Polym. Sci. A1*, **7**, 1203 (1969).
5943. Rabolt, J. F., and Fanconi, B., *J. Polym. Sci. B*, **15**, 121 (1977).
5944. Rabolt, J. F., and Fanconi, B., *Macromolecules*, **11**, 740 (1978).
5945. Raczek, J., and Meyerhoff, G., *Makromol. Chem.*, **177**, 1199 (1976).
5946. Radhakrishna, S., and Murthy, M. R. K., *J. Polym. Sci. A2*, **15**, 987 (1977).
5947. Radhakrishna, S., and Murthy, M. R. K., *J. Polym. Sci. A2*, **15**, 1261 (1977).
5948. Radoc, D., and Gargilo, L., *J. Polym. Sci. A2*, **16**, 977 (1978).
5949. Raha, S., and Bowden, P. B., *Polymer*, **13**, 174 (1972).
5950. Rahlwes, D., and Kirste, R. G., *Makromol. Chem.*, **178**, 1793 (1977).
5951. Raine, H. C., Richards, R. B., and Ryder, H., *Trans. Faraday Soc.*, **41**, 56 (1945).
5952. Ram, A., and Miltz, J., *J. Appl. Polym. Sci.*, **12**, 2639 (1971).
5953. Ram, A., and Miltz, J., *Polym. Plast. Technol., Eng.*, **4**, 23 (1975).
5954. Ramachandran, G. N., Sasisekharan, V., and Ramakrishnan, C., *Biochem. Biophys. Acta*, **112**, 168 (1966).
5955. Ramasseul, R., Rassat, A., Rey, P., and Rinaudo, M., *Macromolecules*, **9**, 186 (1976)
5956. Ramey, K. C., *J. Phys. Chem.*, **70**, 2525 (1966).
5957. Ramey, K. C., *J. Polym. Sci. B*, **5**, 859 (1967).
5958. Ramey, K. C., and Field, N. D., *J. Polym. Sci. B*, **3**, 69 (1965).
5959. Ramey, K. C., Field, N. D., and Borchert, A. E., *J. Polym. Sci. A1*, **3**, 2885 (1965).
5960. Ramey, K. C., Hayes, M. W., and Altenau, A. G., *Macromolecules*, **6**, 795 (1973).
5961. Rami Reddy, C., and Kalpagam, V., *J. Polym. Sci. A2*, **14**, 579 (1976).
5962. Rami Reddy, C., and Kalpagam, V., *J. Polym. Sci. A2*, **14**, 749 (1976).
5963. Rånby, B., and Carstensen, P., Advances in Chemistry Series No. 66, American Chemical Society, Washington, D.C., 1967, p. 256.
5964. Rånby, B., Chan, K. S., and Brumberger, H., *J. Polym. Sci.*, **58**, 545 (1962).
5965. Rånby, B., Rabek, J. F., and Canbäck, G., *J. Macromol. Sci. A*, **12**, 587 (1978).
5966. Rånby, B., Woltersdorf, O. W., and Battista, O. A., *Svensk Papertid.*, **60**, 373 (1957).
5967. Rånby, B., and Yoshida, H., *J. Polym. Sci. C*, **12**, 263 (1966).
5968. Randall, J. C., *J. Polym. Sci. A2*, **12**, 703 (1974).
5969. Randall, J. C., *J. Polym. Sci. A2*, **13**, 889 (1975).
5970. Randall, J. C., *J. Polym. Sci. A2*, **13**, 901 (1975).
5971. Randall, J. C., *J. Polym. Sci. A2*, **13**, 1975 (1975).
5972. Randall, J. C., *J. Polym. Sci. A2*, **14**, 1693 (1976).
5973. Randall, J. C., *J. Polym. Sci. A2*, **14**, 2083 (1976).
5974. Randall, J. C., *J. Polym. Sci. A2*, **15**, 1451 (1977).
5975. Randall, J. C., *Macromolecules*, **11**, 592 (1978).
5976. Rapaport, D. C., *Macromolecules*, **7**, 64 (1974).
5977. Rashidov, D., Brestkin, Yu. V., and Ginzburg, B. M., *Vysokmol. Soedin. A*, **16**, 230 (1974).
5978. Ratajczak, R. D., and Jones, M. T., *J. Chem. Phys.*, **56**, 3898 (1972).
5979. Ravens, D. A. S., and Ward, I. M., *Trans. Faraday Soc.*, **57**, 150 (1961).
5980. Ravey, J. C., *J. Polym. Sci. C*, **61**, 129 (1977).
5981. Razumovskii, S. D., Kefeli, A. A., and Zaikov, G. E., *Europ. Polym. J.*, **7**, 275 (1971).
5982. Ray, G. J., Johnson, P. E., and Knox, J. R., *Macromolecules*, **10**, 773 (1977).
5983. Ray, N. H., *Trans. Faraday Soc.*, **48**, 809 (1952).
5984. Rayner, M. G., *Polymer*, **10**, 827 (1969).
5985. Read, B. E., SCI Monograph No. 17, Society for Chemistry in Industry, London, 1963.
5986. Read, B. E., and Hughes, D. A., *Polymer*, **13**, 495 (1972).
5987. Read, B. E., Hughes, D. A., Barnes, D. C., and Drury, F. W. M., *Polymer*, **13**, 485 (1972).
5988. Read, B. E., and Stein, R. S., *Macromolecules*, **1**, 116 (1968).
5989. Reading, F. P., Faucher, J. A., and Whitman, R. D., *J. Polym. Sci.*, **54**, S56 (1961).
5990. Reading, F. P., Faucher, J. A., and Whitman, R. D., *J. Polym. Sci.*, **57**, 483 (1962).
5991. Reading, F. P., and Walter, E. R., *J. Polym. Sci.*, **38**, 141 (1959).
5992. Reardon, J. P., and Zisman, W. A., *Macromolecules*, **7**, 920 (1974).
5993. Rebenfeld, L., Makarewicz, P. J., Weigmann, H. D., and Wilkes, G. L., *J. Macromol. Sci. C*, **15**, 279 (1976).
5994. Reddish, W., *J. Polym. Sci. C*, **14**, 123 (1966).
5995. Redlich, O., Jacobson, A. L., and McFadden, W. H., *J. Polym. Sci. A*, **1**, 393 (1963).

5996. Reed, C., *Brit. Polym. J.*, **6**, 1 (1974).
5997. Reed, D. H., Critchfield, F. E., and Elder, D. K., *Anal. Chem.*, **35**, 571 (1963).
5998. Reed, P. J., and Urwin, J. R., *Austral. J. Chem.*, **23**, 1743 (1970).
5999. Reed, R., and Barlow, J. R., *Polymer*, **13**, 226 (1972).
6000. Reed, T. F., *ACS Polym. Preprints*, **13**, 197 (1972).
6001. Reed, T. F., *Macromolecules*, **5**, 771 (1972).
6002. Reed, T. F., and Frederick, J. E., *Macromolecules*, **4**, 72 (1971).
6003. Rees, D. V., and Bassett, D. C., *J. Polym. Sci. A2*, **9**, 385 (1971).
6004. Reese, W., *J. Appl. Phys.*, **37**, 3959 (1966).
6005. Reese, W., *J. Macromol. Sci. A*, 3, 1257 (1969).
6006. Reese, W., and Tucker, J. E., *J. Chem. Phys.*, **43**, 105 (1965).
6007. Regen, S. L., *Macromolecules*, **8**, 689 (1975).
6008. Regen, S. L., *J. Amer. Chem. Soc.*, **97**, 3108 (1975).
6009. Rehage, G., and Koningsveld, R., *J. Polym. Sci. B*, **6**, 421 (1968).
6010. Reich, L., *J. Polym. Sci. B*, **2**, 621 (1964).
6011. Reich, L., *J. Appl. Polym. Sci.*, **9**, 3033 (1965).
6012. Reich, L., Lee, H. T., and Levi, D. W., *J. Polym. Sci. B*, **1**, 535 (1963).
6013. Reich, L., Lee, H. T., and Levi, D. W., *J. Appl. Polym. Sci.*, **9**, 351 (1965).
6014. Reich, L., and Levi, D. W., *Makromol. Chem.*, **66**, 102 (1963).
6015. Reid, R. F., and Soutar, I., *J. Polym. Sci. B*, **15**, 153 (1977).
6016. Reid, R. F., and Soutar, I., *J. Polym. Sci. A2*, **16**, 231 (1978).
6017. Reiff, T. R., and Yiengst, M. J., *J. Lab. Clin. Med.*, **53**, 291 (1959).
6018. Reilly, C. B., and Orchin, M., *Ind. Eng. Chem.*, **48**, 59 (1956).
6019. Reimschuessel, H. K., *J. Polym. Sci. A1*, **16**, 1229 (1978).
6020. Rempp, P., *J. Chem. Phys.*, **54**, 421 (1957).
6021. Reneker, D. H., *J. Polym. Sci.*, **59**, S39 (1962).
6022. Reneker, D. H., and Geil, P. H., *J. Appl. Phys.*, **31**, 1916 (1960).
6023. Renninger, A. L., and Uhlman, D. R., *J. Polym. Sci. A2*, **13**, 1481 (1975).
6024. Renninger, A. L., Wicks, G. G., and Uhlmann, D. R., *J. Polym. Sci. A2*, **13**, 1247 (1975).
6025. Renuncio, J. A. R., and Prausnitz, J. M., *Macromolecules*, **9**, 898 (1976).
6026. Reynolds, G. E. J., *J. Polym. Sci. C*, **18**, 3957 (1968).
6027. Reynolds, J., and Sternstein, S. S., *J. Chem. Phys.*, **41**, 47 (1964).
6028. Rhee, C. K., and Ferry, J. D., *J. Appl. Polym. Sci.*, **21**, 467 (1977).
6029. Rhee, C. K., and Ferry, J. D., *J. Appl. Polym. Sci.*, **21**, 733 (1977).
6030. Rhee, C. K., Ferry, J. D., and Fetters, L. J., *J. Appl. Polym. Sci.*, **21**, 783 (1977).
6031. Rheim, R. A., and Lawson, D., *Chem. Technol.*, **1971**, 122 (1971).
6032. Rhodes, M. B., *Appl. Opt.*, **13**, 2263 (1974).
6033. Rhodes, M. B., and Stein, R. S., *J. Polym. Sci. A2*, **7**, 1539 (1969).
6034. Riande, E., *J. Polym. Sci. A2*, **14**, 2231 (1976).
6035. Riande, E., *J. Polym. Sci. A2*, **15**, 1397 (1977).
6036. Richards, E. G., and Schachman, H. K., *J. Phys. Chem.*, **63**, 1578 (1959).
6037. Richards, R. R., and Rogowski, R. S., *J. Polym. Sci. A2*, **12**, 89 (1974).
6038. Richards, R. W., Macconnachie, A., and Allen, G., *Polymer*, **19**, 266 (1978).
6039. Richardson, I. D., and Ward, I. M., *J. Polym. Sci. A2*, **16**, 667 (1978).
6040. Richardson, M. J., *Proc. Roy. Soc. A*, **279**, 50 (1964).
6041. Richardson, M. J., *Trans. Faraday Soc.*, **61**, 1876 (1965).
6042. Richardson, M. J., *J. Polym. Sci. C*, **38**, 251 (1972).
6043. Richardson, M. J., Flory, P. J., and Jackson, J. B., *Polymer*, **4**, 221 (1963).
6044. Richardson, M. J., and Sacher, A., *J. Polym. Sci.*, **10**, 353 (1953).
6045. Richardson, M. J., and Savill, N. G., *Polymer*, **16**, 753 (1975).
6046. Richardson, M. J., and Savill, N. G., *Polymer*, **18**, 3 (1977).
6047. Rickert, S. E., and Baer, E., *J. Appl. Phys.*, **47**, 4304 (1976).
6048. Rickert, S. E., Baer, E., Wittmann, J. C., and Kovacs, A. J., *J. Polym. Sci.*, **16**, 895 (1978).
6049. Riddiford, C. L., *Macromolecules*, **11**, 427 (1978).
6050. Rider, J. G., and Watkinson, K. M., *Polymer*, **19**, 645 (1978).
6051. Rietsch, F., *Macromolecules*, **11**, 477 (1978).
6052. Rietsch, F., Daveloose, D., and Froelich, D., *Polymer*, **17**, 859 (1976).
6053. Rietveld, B. J., *J. Polym. Sci. A2*, **8**, 1837 (1970).
6054. Rietveld, B. J., *Brit. Polym. J.*, **6**, 181 (1974).

6055. Rietveld, B. J., Scholte, Th. G., and Pijpers, J. P. L., *Brit. Polym. J.*, **4**, 109 (1972).
6056. Riew, C. K., and Smith, R. W., *J. Polym. Sci. A1*, **9**, 2739 (1971).
6057. Rijke, A. M., and Mandelkern, L., *Macromolecules*, **4**, 594 (1971).
6058. Ring, W., Cantow, H. J., and Holtrup, W., *Europ. Polym. J.*, **2**, 151 (1966).
6059. Rippon, W. B., and Hiltner, W. A., *Macromolecules*, **6**, 282 (1973).
6060. Ritscher, T. A., and Elias, H. G., *Makromol. Chem.*, **30**, 48 (1959).
6061. Ritter, W., Möller, M., and Cantow, H. J., *Makromol. Chem.*, **179**, 823 (1978).
6062. Rheineck, A. E., Peterson, R. H., and Sastry, P., *J. Paint. Technol.*, **39**, 484 (1967).
6063. Rigbi, Z., *Polymer*, **19**, 1229 (1978).
6064. Rigby, S. J., and Dwe-Hughes, D., *Polymer*, **15**, 639 (1974).
6065. Robard, A., Patterson, D., and Delmas, G., *Macromolecules*, **10**, 706 (1977).
6066. Robard, A., and Patterson, D., *Macromolecules*, **10**, 1021 (1977).
6067. Roche, R. S., *J. Appl. Polym. Sci.*, **18**, 3555 (1974).
6068. Roberts, R. C., *Polymer*, **10**, 113 (1969).
6069. Roberts, K. R., Torkington, J. A., Gordon, M., and Ross-Murphy, S. B., *J. Polym. Sci. C*, **61**, 45 (1977).
6070. Robertson, R. E., *Macromolecules*, **2**, 250 (1969).
6071. Robertson, R. E., *J. Polym. Sci. A2*, **10**, 2437 (1972).
6072. Robinson, C., *Proc. Roy. Soc. A*, **204**, 339 (1950).
6073. Robinson, M. E. R., Bower, D. I., Allsopp, M. W., Willis, H. A., and Zichy, V., *Polymer*, **19**, 1225 (1978).
6074. Robinson, M. E. R., Bower, D. I., and Maddams, W. F., *Polymer*, **17**, 355 (1976).
6075. Robinson, M. E. R., Bower, D. I., and Maddams, W. F., *Polymer*, **19**, 773 (1978).
6076. Robinson, W. T., Cundiff, R. H., and Markunas, P. C., *Anal. Chem.*, **33**, 1030 (1961).
6077. Rodriguez, F., Kulakowski, R. A., and Clark, O. K., *Ind. Eng. Chem. Prod., Res. Develop.*, **5**, 121 (1966).
6078. Roe, R. J., *J. Polym. Sci. A2*, **8**, 1187 (1970).
6079. Roe, R. J., and Bair, H. F., *Macromolecules*, **3**, 454 (1970).
6080. Roe, R. J., and Gniewski, C., *Macromolecules*, **6**, 212 (1973).
6081. Roe, R. J., Gniewski, C., and Vadimsky, R. G., *J. Polym. Sci. A2*, **11**, 1673 (1973).
6082. Roe, R. J., and Tonelli, A. E., *Macromolecules*, **11**, 114 (1978).
6083. Rogalski, M., Rybakiewicz, K., and Malanowski, S., *Ber. Bunseges. Chem. Phys. Chem.*, **81**, 1070 (1977).
6084. Rogers, C. E., Stannett, V., and Szwarc, M., *J. Phys. Chem.*, **63**, 1406 (1959).
6085. Rogers, R. N., *Anal. Chem.*, **39**, 730 (1967).
6086. Roinishvili, E. Yu, Tavkhelidze, N. N., and Akopyan, V. B., *Vysokomol. Soedin, B*, **9**, 254 (1967).
6087. Rolfe, J., and Moore, S. E., *Appl. Opt.*, **9**, 63 (1970).
6088. Rolfson, F. B., and Coll, H., *Anal. Chem.*, **36**, 888 (1964).
6089. Rollins, M. L., Goyenes, W. R., and Brysson, R. J., *Appl. Polym. Symp.*, No. 16, 35 (1971).
6090. Romberger, A. B., Eastman, D. P., and Hunt, J. L., *J. Chem. Phys.*, **51**, 3723 (1969).
6091. Roots, J., and Nyström, B., *J. Polym. Sci. A2*, **16**, 695 (1978).
6092. Roots, J., and Nyström, B., *Europ. Polym. J.*, **14**, 773 (1978).
6093. Roovers, J. E. L., and Bywater, S., *Macromolecules*, **5**, 384 (1972).
6094. Roovers, J. E. L., and Bywater, S., *Macromolecules*, **7**, 443 (1974).
6095. Rosenberg, J. L., and Beckmann, C. O., *J. Colloid Sci.*, **3**, 483 (1948).
6096. Rosenthal, A., *Macromolecules*, **4**, 35 (1971).
6097. Rosenthal, A., *Macromolecules*, **5**, 310 (1972).
6098. Rosenthal, I., Frisone, G. J., and Coberg, J. K., *Anal. Chem.*, **32**, 1713 (1960).
6099. Rosenthal, J., and Yarmus, L., *Rev. Sci. Instrum.*, **37**, 381 (1966).
6100. Roshchupkin, V. P., and Kochervisnkii, V. V., *Vysokomol. Soedin. B*, **13**, 194 (1971).
6101. Rosie, D. M., and Grob, R. L., *Anal. Chem.*, **29**, 1263 (1957).
6102. Ross, J. H., and Casto, M. E., *J. Polym. Sci. C*, **21**, 143 (1968).
6103. Ross, W. D., and Jefferson, R. T., *J. Chromatogr. Sci.*, **8**, 386 (1970).
6104. Rossi, C., and Perico, A., *J. Chem. Phys.*, **53**, 1223 (1970).
6105. Ross-Murphy, S. B., *Polymer*, **19**, 497 (1978).
6106. Rosso, J. C., and Persoz, B., *J. Polym. Sci. C*, **16**, 4395 (1969).
6107. Rotschild, W. G., *Macromolecules*, **1**, 43 (1968).

6108. Rotschild, W. G., *Macromolecules*, **5**, 37 (1972).
6109. Powell, R. L., Kratochvil, J. P., and Kerker, M., *J. Colloid Interface Sci.*, **27**, 501 (1968).
6110. Rowland, T. J., and Labun, L. C., *Macromolecules*, **11**, 466 (1978).
6111. Roylance, D. K., and DeVries, K. L., *J. Polym. Sci. B*, **9**, 443 (1971).
6112. Rubčic, A., and Zerbi, G., *Macromolecules*, **7**, 755 (1974).
6113. Rubčic, A., and Zerbi, G., *Macromolecules*, **7**, 759 (1974).
6114. Rubin, R. J., and Mazur, J., *Macromolecules*, **10**, 139 (1977).
6115. Rubingh, D. N., and Yu, H., *Macromolecules*, **9**, 681 (1976).
6116. Rudin, A., *J. Polym. Sci. A1*, **9**, 2587 (1971).
6117. Rudin, A., and Burgin, D., *Polymer*, **16**, 291 (1975).
6118. Rudin, A., and Chang, R. J., *J. Appl. Polym. Sci.*, **22**, 781 (1978).
6119. Rudin, A., and Chee, K. K., *Macromolecules*, **6**, 613 (1973).
6120. Rudin, A., and Hoegy, L. W., *J. Polym. Sci. A1*, **10**, 217 (1972).
6121. Rudin, A., Hoegy, L. W., and Kirkjohnston, H., *J. Appl. Polym. Sci.*, **16**, 1281 (1972).
6122. Ruland, W., *Acta Crystal.*, **14**, 1180 (1961).
6123. Ruland, W., *Polymer*, **5**, 89 (1964).
6124. Ruland, W., *Faserforsch., Textiltechn.*, **15**, 533 (1964).
6125. Ruland, W., *Faserforsch., Textiltechn.*, **18**, 59 (1967).
6126. Runyon, J. R., Barnes, D. E., Rudi, J. F., and Tung, L. H., *J. Appl. Polym. Sci.*, **13**, 2359 (1969).
6127. Ruscher, C. H. J., *J. Polym. Sci. C*, **16**, 2923 (1967).
6128. Ruscher, C., and Schmolke, R., *Faserforsch. Textiltechn.*, **11**, 383 (1960).
6129. Rush, C. A., Crukshank, S. S., Schrock, J. J., and Rosenblatt, D. H., *Microchem. J.*, **10**, 522 (1966).
6130. Ruskin, A. M., and Parravano, G., *J. Appl. Polym. Sci.*, **8**, 565 (1964).
6131. Rusnock, J. A., and Hansen, D., *J. Polym. Sci. A1*, **3**, 647 (1965).
6132. Russel, E. W., *Nature*, **165**, 91 (1950).
6133. Russell, G. A., and Pascale, J. V., *J. Appl. Polym. Sci.*, **7**, 959 (1963).
6134. Rutherford, H., and Soutar, I., *J. Polym. Sci. A2*, **15**, 2213 (1977).
6135. Ryabova, L. G., Berestneva, Z. Ya., and Pravikova, N. A., *Vyskomol. Soedin.*, **7**, 1796 (1965).
6136. Ryska, M., Schüddemage, H. D. R, and Hummel, D. O., *Makromol. Chem.*, **126**, 32 (1969).

6137. Sacher, E., *J. Polym. Sci. A2*, **6**, 1813 (1968).
6138. Sacher, E., *J. Macromol. Chem. B*, **4**, 449 (1970).
6139. Sacher, E., *J. Polym. Sci. B*, **15**, 395 (1977).
6140. Sacher, E., *J. Appl. Polym. Sci.*, **22**, 2137 (1978).
6141. Sadler, D. M., and Keller, A. A., *Polymer*, **17**, 37 (1976).
6142. Sadler, D. M., and Keller, A. A., *Macromolecules*, **10**, 1128 (1977).
6143. Saeda, S., *J. Polym. Sci. A2*, **11**, 1465 (1973).
6144. Saegusa, T., Yatsu, T., Miyaji, S., and Fujii, J., *Polym. J.*, **1**, 7 (1970).
6145. Saeki, S., Kuwahara, N., and Kaneko, M., *Macromolecules*, **9**, 101 (1976).
6146. Saeki, S., Konno, S., Kuwahara, N., Nakata, M., and Kaneko, M., *Macromolecules*, **7**, 521 (1974).
6147. Saeki, S., Kuwahara, N., Konno, S., and Kaneko, M., *Macromolecules*, **6**, 246 (1973).
6148. Saeki, S., Kuwahara, N., Konno, S., and Kaneko, M., *Macromolecules*, **6**, 589 (1973).
6149. Saeki, S., Kuwahara, N., Nakata, M., and Kaneko, M., *Polymer*, **17**, 685 (1976).
6150. Safford, G. J., Danner, H. R., Boutin, H., and Berger, M., *J. Chem. Phys.*, **40**, 1426 (1964).
6151. Safford, G. J., and Simon, F. T., *J. Chem. Phys.*, **45**, 3787 (1966).
6152. Saika, A., Ohya-Nishiguchi, H., Satokawa, T., and Ohmori, A., *J. Polym. Sci. A1*, **15**, 1073 (1977).
6153. Saito, S., Sasabe, H., Nakajima, T., and Yada, K., *J. Polym. Sci.*, **6**, 1297 (1968).
6154. Saito, T., and Yamaguchi, K., *Polymer*, **15**, 219 (1974).
6155. Saiz, E., Mark, J. E., and Flory, P. J., *Macromolecules*, **10**, 967 (1977).
6156. Sakaguchi, F., Mandelkern, L., and Maxfield, J., *J. Polym. Sci. A2*, **14**, 2137 (1976).
6157. Sakaguchi, M., and Sohma, J., *J. Polym. Sci. A2*, **13**, 1233 (1975).
6158. Sakai, M., Fujimoto, T., and Nagasawa, M., *Macromolecules*, **5**, 786 (1972).
6159. Sakai, T., *J. Polym. Sci. A2*, **6**, 1535 (1968).

818

6160. Sakai, T., *Macromolecules*, **3**, 96 (1970).
6161. Sakakihara, H., Takahashi, Y., Tadekoro, H., Oguni, N., and Tani, H., *Macromolecules*, **6**, 205 (1973).
6162. Sakurada, I., Nakajima, A., and Fujiwara, H., *J. Polym. Sci.*, **35**, 497 (1959).
6163. Sakurada, I., Nakajima, A., and Shibatani, K., *Makromol. Chem.*, **87**, 103 (1965).
6164. Sakurada, K., Miyasaka, K., and Ishikawa, K., *J. Polym. Sci. A2*, **12**, 1587 (1974).
6165. Sakurai, K., Oota, T., Miyasaka, K., and Ishikawa, K., *J. Polym. Sci. A2*, **14**, 1527 (1976).
6166. Salee, G., *J. Appl. Polym. Sci.*, **15**, 1049 (1971).
6167. Salovey, R., Albarino, R. V., and Luongo, J. P., *Macromolecules*, **3**, 314 (1970).
6168. Salovey, R., Cortellucci, R., and Roaldi, A., *Polym. Eng. Sci.*, **14**, 120 (1974).
6169. Salovey, R., and Yager, W. A., *J. Polym. Sci. A*, **2**, 219 (1964).
6170. Samuels, R. J., *J. Polym. Sci. A*, **3**, 1741 (1965).
6171. Samuels, R. J., *J. Polym. Sci. C*, **13**, 37 (1966).
6172. Samuels, R. J., *J. Polym. Sci. A2*, **6**, 1101 (1968).
6173. Samuels, J. R., *J. Macromol. Sci. B*, **4**, 701 (1970).
6174. Samuels, R. J., *J. Polym. Sci. A2*, **9**, 2165 (1971).
6175. Samuels, R. J., *J. Polym. Sci. A2*, **12**, 1417 (1974).
6176. Samuels, R. J., *J. Polym. Sci. A2*, **13**, 1417 (1975).
6177. Samuels, R. J., and Yee, R. Y., *J. Polym. Sci. A2*, **10**, 385 (1972).
6178. Sanchez, I. C., and DiMarzio, E. A., *J. Chem. Phys.*, **55**, 893 (1971).
6179. Sanchez, I. C., and DiMarzio, E. A., *Macromolecules*, **4**, 677 (1971).
6180. Sanchez, I. C., and Lacombe, R. H., *J. Polym. Sci. B*, **15**, 71 (1977).
6181. Sanders, J. M., and Komoroski, R. A., *Macromolecules*, **10**, 1214 (1977).
6182. Sanders, J. W., *J. Polym. Sci.*, **11**, 447 (1953).
6183. Sands, J. D., and Turner, G. S., *Anal. Chem.*, **27**, 791 (1952).
6184. Santee, E. R., Jr, Malotky, L. O., and Morton, M., *Rubber Chem. Technol.*, **46**, 1156 (1973).
6185. Santee, E. R., Jr, Mochel, V. D., and Morton, M., *J. Polym. Sci. B*, **11**, 453 (1973).
6186. Sarazin, D., and Francois, J., *Polymer*, **19**, 695, 699 (1978).
6187. Sarazin, D., Le Moigne, J., and Francois, J., *J. Appl. Polym. Sci.*, **22**, 1377 (1978).
6188. Sasabe, H., *J. Polym. Sci. A2*, **11**, 2413 (1973).
6189. Sasabe, H., Saito, S., Asahina, M., and Kakutani, H., *J. Polym. Sci. A2*, **7**, 1405 (1969).
6190. Sasabe, M., and Moynihan, C. T., *J. Polym. Sci. A2*, **16**, 1447 (1978).
6191. Sato, K., Eirich, F. R., and Mark, J. E., *J. Polym. Sci. A2*, **14**, 619 (1976).
6192. Sato, T., Abe, M., and Otsu, T., *Makromol. Chem.*, **178**, 1061 (1977).
6193. Sato, T., Abe, M., and Otsu, T., *Makromol. Chem.*, **178**, 1259, 1267 (1977).
6194. Sato, T., Kita, S., and Otsu, T., *Makromol. Chem.*, **176**, 561 (1975).
6195. Sato, T., and Otsu, T., *Polymer*, **16**, 389 (1975).
6196. Sato, Y., and Yashiro, T., *J. Appl. Polym. Sci.*, **22**, 2141 (1978).
6197. Satoh, S., *J. Polym. Sci. A*, **2**, 5221 (1964).
6198. Satoh, S., Chûjô, R., Ozeki, T., and Nagai, E., *J. Polym. Sci.*, **62**, S101 (1962).
6199. Sauer, J. A., Morrow, D. R., and Richardson, G. C., *J. Appl. Phys.*, **36**, 3017 (1965).
6200. Sauer, J. A., and Pae, K. D., *J. Appl. Polym. Sci.*, **12**, 1921 (1968).
6201. Saunders, D., and Pecsok, R. L., *Anal. Chem.*, **40**, 44 (1968).
6202. Saunders, D. G., and Winnik, M. A., *Macromolecules*, **11**, 18 (1978).
6203. Savolainen, A., and Törmälä, P. J., *J. Polym. Sci. A2*, **12**, 1251 (1974).
6204. Schaefer, J., *Macromolecules*, **2**, 533 (1969).
6205. Schaefer, J., *Macromolecules*, **4**, 107 (1971).
6206. Schaefer, J., *Macromolecules*, **4**, 110, (1971).
6207. Schaefer, J., *Macromolecules*, **5**, 427 (1972).
6208. Schaefer, J., *Macromolecules*, **5**, 590 (1972).
6209. Schaefer, J., *Macromolecules*, **6**, 882 (1973).
6210. Schaefer, J., Bude, D. A., and Katnik, R. J., *Macromolecules*, **2**, 289 (1969).
6211. Schaefer, J., Chin, S. H., and Weissman, S. I., *Macromolecules*, **5**, 798 (1972).
6212. Schaefer, J., Katnik, R. J., and Kern, R. J., *J. Amer. Chem. Soc.*, **90**, 2476 (1968).
6213. Schaefer, J., Kern, R. J., and Katnik, R. J., *Macromolecules*, **1**, 107 (1968).
6214. Schaefer, J., and Natusch, D. F. S., *Macromolecules*, **5**, 416 (1972).
6215. Schard, M. P., and Russell, C. A., *J. Appl. Polym. Sci.*, **8**, 985, 997 (1964).
6216. Schaufele, R. F., *J. Opt. Soc. Amer.*, **57**, 105 (1967).

6217. Schaufele, R. F., and Koenig, J. L., *Macromolecules*, **3**, 597 (1970).
6218. Schausberger, A., and Pilz, I., *Makromol. Chem.*, **178**, 211 (1977).
6219. Scheffler, K., and Stegmann, H. B., *Ber. Bunsenges Phys. Chem.*, **67**, 864 (1963).
6220. Schelten, J., Wignal, G. D., and Ballard, D. G. H., *Polymer*, **5**, 682 (1974).
6221. Schelten, J., Wignall, G. D., Ballard, G. D. H., and Schmatz, W., *Colloid Polym. Sci.*, **252**, 749 (1974).
6222. Schelten, J., Wignall, G. D., Ballard, D. G. H., and Longman, G. W., *Polymer*, **18**, 1111 (1977).
6223. Schelten, J., Ballard, D. H. G., Wignall, G. D., Longman, G. W., and Schmatz, W., *Polymer*, **17**, 751 (1976).
6224. Schermann, W., and Zachmann, H. G., *Kolloid Z. Z. Polym.*, **241**, 921 (1970).
6225. Schiling, F. C., Cais, R. E., and Bovey, F. A., *Macromolecules*, **11**, 325 (1978).
6226. Schipmann, R. H., and Farber, E., *J. Polym. Sci.*, **B**, **1**, 65 (1963).
6227. Schlag, E. W., Schneider, S., and Chandler, D. W., *Chem. Phys. Lett.*, **11**, 474 (1971).
6228. Schlag, E. W., Selze, H. L., Schneider, S., and Larsen, J. G., *Rev. Sci. Instrum.*, **45**, 364 (1974).
6229. Schlag, E. W., Yao, S. J., and Weyssenhoff, H., *J. Chem. Phys.*, **50**, 732 (1969).
6230. Schmatz, W., Springer, T., Schelten, J., and Ibel, K., *J. Appl. Crystal.*, **7**, 96 (1974).
6231. Schmidt, M., and Burchard, W., *Macromolecules*, **11**, 460 (1978).
6232. Schmidt, M., Burchard, W., and Ford, N. C., *Macromolecules*, **11**, 452 (1978).
6233. Schmidt, P. G., *J. Polym. Sci. A1*, **1**, 1271 (1963).
6234. Schmieder, F., and Wolf, K., *Kolloid Z.*, **134**, 149 (1953).
6235. Schneider, B., Pivcova, H., and Doskocilova, D., *Macromolecules*, **5**, 120 (1972).
6236. Schneider, E. E., *J. Chem. Phys.*, **23**, 978 (1955).
6237. Schneider, I. A., Onu, A., and Aalenei, N., *Europ. Polym. J.*, **10**, 315 (1974).
6238. Schneider, I. A., Vasile, C., Onu, A., and Furnica, D., *Makromol. Chem.*, **117**, 41 (1968).
6239. Schneider, N. S., and Holmes, L. G., *J. Polym. Sci.*, **38**, 552 (1959).
6240. Schneider, N. S., Holmes, L. G., Mijal, C. F., and Loconti, J. D., *J. Polym. Sci.*, **37**, 551 (1959).
6241. Schneider, N. S., Loconti, J. D., and Holmes, L. G., *J. Appl. Polym. Sci.*, **3**, 251 (1960).
6242. Schneider, N. S., Loconti, J. D., and Holmes, L. G., *J. Appl. Polym. Sci.*, **5**, 354 (1961).
6243. Schneider, N. S., Traskos, R. T., and Hoffman, A. S., *J. Appl. Polym. Sci.*, **12**, 1567 (1968).
6244. Schnell, H., *Makromol. Chem.*, **2**, 172 (1948).
6245. Scholtan, W., *Makromol. Chem.*, **24**, 104 (1957).
6246. Scholte, T. G., *J. Polym. Sci. A2*, **6**, 91 (1968).
6247. Scholte, T. G., *J. Polym. Sci. A2*, **6**, 111 (1968).
6248. Scholte, T. G., *Ann. N.Y. Acad. Sci.*, **164**, 156 (1969).
6249. Scholte, T. G., *J. Polym. Sci. A2*, **8**, 841 (1970).
6250. Scholte, T. G., *J. Polym. Sci. A2*, **9**, 1533 (1971).
6251. Scholte, T. G., *J. Polym. Sci. A2*, **10**, 519 (1972).
6252. Scholz, R. G., Bednarczyk, J., and Yamauchi, T., *Anal. Chem.*, **38**, 331 (1966).
6253. Schön, K. G., and Schulz, G. V., *Z. Phys. Chem.*, **2**, 197 (1954).
6254. Schonhorn, H., *Macromolecules*, **1**, 145 (1968).
6255. Schonhorn, H., and Luongo, J. P., *Macromolecules*, **2**, 366 (1969).
6256. Schoon, T. G. F., and Kiyek, H., *Kolloid Z.*, **208**, 22 (1966).
6257. Schoon, T. G. F., and Kiyek, H., *Mikroskopie*, **21**, 3 (1966).
6258. Schoon, T. G. F., and Kretschmer, R., *Mikroskopie*, **21**, 7 (1966).
6259. Schorn, I. H., Frisch, J. L., and Kwei, T. K., *J. Appl. Phys.*, **37**, 4967 (1966).
6260. Schreiber, H. P., Tewari, Y. B., and Patterson, D., *J. Polym. Sci. A2*, **11**, 15 (1973).
6261. Schriever, J., and Leyte, J. C., *Polymer*, **18**, 1185 (1977).
6262. Schröder, J., *Rev. Sci. Instrum.*, **34**, 615 (1962).
6263. Schüddemage, H. D. R., and Hummel, D. O., *Adv. Mass Spectr.*, **4**, 857 (1967).
6264. Schué, F., Worsfold, D. J., and Bywater, S., *J. Polym. Sci. B*, **7**, 821 (1969).
6265. Schulken, R. M., Roy, R. E., and Cox, R. H., *J. Polym. Sci. C*, **6**, 1725 (1964).
6266. Schultz, J. M., *J. Polym. Sci. A2*, **7**, 821 (1969).
6267. Schultz, J. M., *J. Polym. Sci. A2*, **14**, 2291 (1976).
6268. Schultz, J. M., and Long, T. C., *J. Mater. Sci.*, **10**, 567 (1975).
6269. Schultz, J. M., Robinson, W. H., Pound, C. M., *J. Polym. Sci. A2*, **5**, 511 (1967).
6270. Schultz, J. M., and Scott, R. D., *J. Polym. Sci. A2*, **7**, 659 (1969).

820

6271. Schulz, G. V., Z. Phys. Chem. A, **176**, 317 (1936).
6272. Schulz, G. V., Z. Phys. Chem. B, **43**, 25 (1939).
6273. Schulz, G. V., Z. Phys. Chem. B, **52**, 1 (1942).
6274. Schulz, G. V., Z. Phys. Chem. A, **194**, 1 (1944).
6275. Schulz, G. V., Z. Phys. Chem. B, **47**, 489 (1963).
6276. Schulz, G. V., and Baumann, H., Makromol. Chem., **60**, 120 (1963).
6277. Schulz, G. V., and Baumann, H., Makromol. Chem., **114**, 122 (1968).
6278. Schulz, G. V., Baumann, H., and Darskus, R., J. Phys. Chem., **70**, 3647 (1966).
6279. Schulz, G. V., Berger, K. C., and Scholz, A. G. R., Ber. Bunseges Phys. Chem., **68**, 856 (1965).
6280. Schulz, G. V., and Blaschke, F., J. Prakt. Chem., **158**, 130 (1941).
6281. Schulz, G. V., and Cantow, H. J., Z. Elektrochem., **60**, 517 (1956).
6282. Schulz, G. V., Cantow, H. J., and Bodman, O., J. Polym. Sci., **10**, 73 (1953).
6283. Schulz, G. V., Cantow, H. J., and Meyerhoff, G., J. Polym. Sci., **10**, 79 (1952).
6284. Schulz, G. V., and Coll, H., Z. Elektrochem., **56**, 248 (1952).
6285. Schulz, G. V., Deussen, P., and Scholz, A. G. R., Makromol. Chem., **69**, 47 (1963).
6286. Schulz, G. V., Haug, A., and Kirste, R., Z. Phys. Chem., **38**, 1 (1963).
6287. Schulz, G. V., Inagaki, H., and Kirste, R., Z. Phys. Chem., **24**, 390 (1960).
6288. Schulz, G. V., and Kirste, R., Z. Phys. Chem., **30**, 171 (1969).
6289. Schulz, G. V., and Kirste, R., Z. Phys. Chem., **27**, 301 (1969).
6290. Schulz, G. V., and Kuhn, W. H., Makromol. Chem., **50**, 37 (1961).
6291. Schulz, G. V., and Lehner, M., Europ. Polym. J., **6**, 945 (1970).
6292. Schulz, R. C., and Jung, R. H., Makromol. Chem., **96**, 295 (1966).
6293. Schulz, R. C., and Schwaab, J., Makromol. Chem., **87**, 90 (1965).
6294. Schulze, P. E., and Wenzel, M., Angew. Chem. Intern. Ed., **1**, 580 (1962).
6295. Schurer, J. W., DeBoer, A., and Challa, G., Polymer, **16**, 201 (1975).
6296. Schurmans, H. J. L., J. Polym. Sci., **57**, 557 (1962).
6297. Schurz, J., Bayzer, H., and Stübchen, H., Makromol. Chem., **23**, 152 (1957).
6298. Schurz, J., and Immergut, E. H., J. Polym. Sci., **9**, 279 (1952).
6299. Schurz, J., and Muller, H. G., Polymer, **17**, 246 (1976).
6300. Schwarcz, A., J. Polym. Sci. A2, **12**, 1195 (1974).
6301. Schwartz, M., Biochim. Bophys. Acta, **22**, 175 (1956).
6302. Schwenker, R. F., Jr, and Zuccarello, R. K., J. Polym. Sci. C, **6**, 1 (1964).
6303. Scornaux, J., and Van Leemput, R., Macromol. Chem., **177**, 2721 (1976).
6304. Scott, R. G., J. Appl. Phys., **28**, 1089 (1957).
6305. Scott, R. G., J. Polym. Sci., **57**, 405 (1962).
6306. Scott, R. A., and Scheraga, H. A., J. Chem. Phys., **44**, 3054 (1966).
6307. Scott, R. L., J. Chem. Phys., **17**, 279 (1949).
6308. Scott, R. L., J. Chem. Phys., **17**, 268, 274 (1949).
6309. Scott, R. L., and van Konynenburg, P. H., Discuss. Faraday Soc., **49**, 87 (1970).
6310. Scott, R. L., and Magat, M., J. Polym. Sci., **4**, 555 (1949).
6311. Scott, R. P. W., and Kucera, P., Anal. Chem., **45**, 749 (1973).
6312. Screaton, R. M., and Seemann, R. W., J. Polym. Sci. C, **21**, 297 (1968).
6313. Seanor, D., J. Polym. Sci., C, **17**, 195 (1967).
6314. See, B., and Smith, T. G., J. Appl. Polym. Sci., **10**, 1625 (1966).
6315. Seeger, M., Barrall, E. M., II, and Shen, M., J. Polym. Sci. A1, **13**, 1541 (1975).
6316. Seeger, M., and Gritter, R. J., J. Polym. Sci. A1, **15**, 1393 (1977).
6317. Seeger, M., Gritter, R. J., Tibbitt, J. M., and Bell, A. T., J. Polym. Sci. A1, **15**, 1403 (1977).
6318. Seely, G. R., Macromolecules, **2**, 302 (1969).
6319. Sefton, M. V., and Merrill, E. W., J. Polym. Sci. A1, **14**, 1581 (1976).
6320. Sefton, M. V., and Merrill, E. W., J. Polym. Sci. A1, **14**, 1829 (1976).
6321. Segal, B. G., Kaplan, M., and Fraenkel, G. K., J. Chem. Phys., **43**, 4191 (1965).
6322. Segal, B. G., Reymond, A., and Fraenkel, G. K., J. Chem. Phys., **51**, 1336 (1969).
6323. Segal, L., J. Polym. Sci. C, **21**, 267 (1968).
6324. Segal, L., Creely, J. J., and Conrad, C. M., Rev. Sci. Instrum., **21**, 421 (1950).
6325. Segal, L., Creely, J. J., Martin, A. E., and Conrad, C. M., Text. Res. J., **29**, 786 (1959).
6326. Segre, A. L., Macromolecules, **1**, 93 (1968).
6327. Segre, A. L., Ciampell, F., and Dallásta, G., J. Polym. Sci. B, **4**, 633 (1966).
6328. Segre, A. L., Delfini, M., Conti, F., and Boicelli, A., Polymer, **16**, 338 (1975).

821

6329. Segre, A. L., Ferruti, P., Toja, E., and Danusso, F., *Macromolecules*, **2**, 35 (1969).
6330. Seguchi, T., and Namura, N., *J. Phys. Chem.*, **77**, 40 (1973).
6331. Sellen, D. B., *Polymer*, **11**, 374 (1970).
6332. Sellen, D. B., *Polymer*, **14**, 359 (1973).
6333. Sellen, D. B., *Polymer*, **16**, 169 (1975).
6334. Sellen, D. B., *Polymer*, **16**, 561 (1975).
6335. Sellen, D. B., *Polymer*, **16**, 773 (1975).
6336. Senti, F. R., Hellman, N. N., Ludwig, N. H., Babock, G. E., Tobin, R., Glass, C. A., and Lamberts, B. L., *J. Polym. Sci.*, **17**, 527 (1955).
6337. Seow, P. K., Gallot, Y., and Skoulios, A., *Makromol. Chem.*, **177**, 199 (1976).
6338. Serdzuk, I. N., and Grenader, A. K., *Makromol. Chem.*, **175**, 1881 (1974).
6339. Servotte, A., and De Bruille, R., *Makromol. Chem.*, **176**, 203 (1975).
6340. Šesták, J., *Thermochim. Acta*, **3**, 150 (1971).
6341. Seyler, R. J., and Kalbfleish, E., *Intern. Lab.*, March/April 1978, p. 40.
6342. Seymour, R. W, Allegrezza, A. E., Jr, and Cooper, S. L., *Macromolecules*, **6**, 896 (1973).
6343. Seymour, R. W., and Cooper, S. L., *Macromolecules*, **6**, 48 (1973).
6344. Seymour, R. W., Estes, G. M., and Cooper, S. L., *Macromolecules*, **3**, 579 (1970).
6345. Seymour, R. W., Overton, J., R., and Corley, L. S., *Macromolecules*, **8**, 331 (1975).
6346. Shaefgen, J. R., and Flory, P. J., *J. Amer. Chem. Soc.*, **70**, 2709 (1948).
6347. Sharma, R. K., and Mandelkern, L., *Macromolecules,*, **2**, 266 (1969).
6348. Sharma, R. K., and Mandelkern, L., *Macromolecules*, **3**, 758 (1970).
6349. Sheldon, R. P., and Lane, K., *Polymer*, **6**, 205 (1965).
6350. Shen, K. P., and Eirich, F. R., *J. Polym. Sci.*, **53**, 81 (1961).
6351. Shepherd, F., and Harwood, H. J., *J. Polym Sci.*, *B*, **10**, 799 (1972).
6352. Shepherd, F., and Harwood, H. J., *J. Polym. Sci. B*, **9**, 419 (1971).
6353. Sheppard, N., *Adv. Spectr.*, **1**, 288 (1959).
6354. Shetter, J. A., *J. Polym. Sci. B*, **1**, 209 (1963).
6355. Shibasaki, Y., *J. Polym. Sci. A1*, **5**, 21 (1967).
6356. Shick, M. J., and Harvey, E. M., *J. Polym. Sci. B*, **7**, 495 (1969).
6357. Shimada, S., Hori, Y., and Kashiwabara, H., *Polymer*, **18**, 25 (1977).
6358. Shimada, S., Hori, Y., and Kashiwabara, H., *Polymer*, **19**, 763 (1978).
6359. Shimada, S., and Kashiwabara, H., *Polym. J.*, **6**, 448 (1974).
6360. Shimada, S., Kashiwabara, H., and Sohma, J., *J. Polym. Sci. A2*, **8**, 1291 (1970).
6361. Shimada, S., Maeda, M., Hori, Y., and Kashiwabara, H., *Polymer*, **18**, 19 (1977).
6362. Shimanouchi, T., Asahina, M., and Enomoto, S., *J. Polym. Sci.*, **59**, 93 (1962).
6363. Shimanouchi, T., and Mizushima, S., *J. Chem. Phys.*, **23**, 707 (1955).
6364. Shimanouchi, T., and Tasumi, M., *Bull. Chem. Soc. Japan*, **34**, 359 (1961).
6365. Shimanouchi, T., Tasumi, M., and Abe, Y., *Makromol. Chem.*, **86**, 43 (1965).
6366. Shimazu, T., and Miyaoka, U., *J. Soc. Textile Cellulose Ind. Japan*, **14**, 557, 563, 637 (1958).
6367. Shimizu, K., Yano, O., and Wada, Y., *Polymer J.*, **5**, 107 (1973).
6368. Shimizu, K., Yano, O., and Wada, Y., *J. Polym. Sci. A2*, **13**, 1959 (1975).
6369. Shimizu, K., Yano, O., Wada, Y., and Kawamura, Y., *J. Polym. Sci. A2*, **11**, 1641 (1973).
6370. Shindo, H., Murakami, I., and Yamamura, H., *J. Polym. Sci. A1*, **7**, 297 (1969).
6371. Shindo, Y., Read, B. E., and Stein, R. S., *Makromol. Chem.*, **169**, 272 (1968).
6372. Shinouda, H. G., Hanna, A. A., and Kinawi, A., *J. Polym. Sci. A1*, **15**, 1991 (1977).
6373. Shipman, J. J., Folt, V. L., and Krimm, S., *Spectrochim. Acta*, **18**, 1603, (1962).
6374. Short, J. M., and Crystal, R. G., *Appl. Polym. Symp.*, No. 16, 137 (1971).
6375. Shrawagi, S. R., and Thomas, E. L., *J. Polym. Sci. A2*, **14**, 799 (1976).
6376. Shroff, R. N., *J. Appl. Polym. Sci.*, **9**, 1547 (1965).
6377. Shull, C. G., and Wollan, E. O., *Solid State Phys.*, **2**, 137 (1956).
6378. Shultz, A. R., *J. Polym. Sci. A2*, **10**, 983 (1972).
6379. Shultz, A. R., and Beach, B. M., *Macromolecules*, **7**, 902 (1974).
6380. Shultz, A. R., and Flory, P. J., *J. Amer. Chem. Soc.*, **74**, 4760 (1952).
6381. Shultz, A. R., and Flory, P. J., *J. Amer. Chem. Soc.*, **75**, 3888 (1953).
6382. Shultz, A. R., and Flory, P. J., *J. Amer. Chem. Soc.*, **75**, 5681 (1953).
6383. Shultz, A. R., and Gendron, B. M., *J. Appl. Polym. Sci.*, **16**, 461 (1972).
6384. Shultz, A. R., and Stockmayer, W. H., *Macromolecules*, **2**, 178 (1969).
6385. Shur, Y. J., and Rånby, B., *J. Appl. Polym. Sci.*, **20**, 3105, 2121 (1976).

6386. Shyluk, S., *J. Polym. Sci.*, **62**, 317 (1962).
6387. Siano, D. B., and Applequist, J., *Macromolecules*, **8**, 858 (1975).
6388. Sicotte, Y., and Rinfret, M., *Trans. Faraday Soc.*, **58**, 1090 (1961).
6389. Sidorova, E. E., and Baiteryakova, L. Kh., *Zh. Fiz. Khim.*, **39**, 2842 (1965).
6390. Sie, S. T., and Rijnders, G. W. A., *Anal. Chim. Acta*, **38**, 3 (1967).
6391. Siegfried, D. L., Manson, J. A., and Sperling, L. H., *J. Polym. Sci. A2*, **16**, 583 (1978).
6392. Siegel, B. M., Johnson, D. H., and Mark, J., *J. Polym. Sci.*, **5**, 111 (1950).
6393. Siegel, S., and Hedgpeth, H., *J. Chem. Phys.*, **46**, 3904 (1967).
6394. Sieglaff, C. L., and O'Leary, K. J., *Trans. Soc. Rheol.*, **14**, 49 (1970).
6395. Siesleer, H. W., *Makromol. Chem.*, **176**, 2451 (1975).
6396. Signer, R., *Liebigs Ann.*, **478**, 246 (1930).
6397. Sikora, A., *Makromol. Chem.*, **176**, 3501 (1975).
6398. Silberberg, A., *J. Chem. Phys.*, **48**, 2835 (1968).
6399. Silberberg, A., *Discuss. Faraday Soc.*, **49**, 162 (1970).
6400. Silias, R. J., Yates, J., and Thornton, V., *Anal. Chem.*, **31**, 529 (1959).
6401. Sillescu, H., *Makromol. Chem.*, **178**, 2759 (1977).
6402. Sillescu, H., and Brussau, R., *Chem. Phys. Lett.*, **5**, 525 (1970).
6403. Silverman, M. P., and Oyama, V. I., *Anal. Chem.*, **40**, 1833 (1968).
6404. Simha, R., *Macromolecules*, **10**, 1025 (1977).
6405. Simha, R., and Boyer, R. F., *J. Chem. Phys.*, **37**, 1003 (1962).
6406. Simha, R., and Wilson, P. S., *Macromolecules*, **6**, 908 (1973).
6407. Simon, F. T., and Rutherford, J. M., Jr, *J. Appl. Phys.*, **35**, 82 (1964).
6408. Simon, J., *J. Therm. Anal.*, **5**, 271 (1973).
6409. Simon, W., Clerc, J. T., and Dohner, R. E., *Microchem., J.*, **10**, 495 (1966).
6410. Simon, W., and Tomlinson, C., *Chimia*, **14**, 301 (1960).
6411. Simpson, R. B., Smith, J. S., and Irving, H. M. N. H., *Analyst*, **96**, 550 (1971).
6412. Singer, L. S., *J. Appl. Phys.*, **30**, 1463 (1959).
6413. Singh, S., and Hamielec, A. E., *J. Appl. Polym. Sci.*, **22**, 577 (1978).
6414. Siow, K. S., Dekmas, G., and Patterson, D., *Macromolecules*, **5**, 29 (1972).
6415. Siow, K. S., and Patterson, D., *Macromolecules*, **4**, 26 (1971).
6416. Sirianni, A. F., Wise, L. M., and McIntosh, R. L., *Can. J. Res. B*, **25**, 301 (1947).
6417. Sirianni, A. F., Worsfold, D. J., and Bywater, S., *Trans. Faraday Soc.*, **55**, 2124 (1959).
6418. Skirrow, G., and Young, K. R., *Polymer*, **15**, 771 (1974).
6419. Škvára, F., and Šesták, J., *Chem. Listy*, **68**, 225 (1974).
6420. Slagowski, E. L., Fetters, L. J., and McIntyre, D., *Macromolecules*, **7**, 394 (1974).
6421. Slagowski, E. L., Tsai, B., and McIntyre, D., *Macromolecules*, **9**, 687 (1976).
6422. Slichter, W. P., *J. Polym. Sci.*, **24**, 173 (1957).
6423. Slichter, W. P., *Makromol. Chem.*, **34**, 67 (1959).
6424. Slichter, W. P., *ACS Polym. Preprints*, **10**, 2 (1969).
6425. Slichter, W. P., and Davis, D. D., *J .Appl. Phys.*, **35**, 3103 (1964).
6426. Slichter, W. P., and Davis, D. D., *Macromolecules*, **1**, 47 (1968).
6427. Slichter, W. P., and McCall, D. W., *J. Polym. Sci.*, **25**, 230 (1957).
6428. Slomp, G., and Lindberg, J. G., *Anal. Chem.*, **39**, 60 (1967).
6429. Slow, K. S., Delmas, G., and Patterson, D., *Macromolecules*, **5**, 29 (1972).
6430. Small, H., *J. Colloid Interface Sci.*, **48**, 147 (1974).
6431. Small, H., *Chem. Technol.*, **7**, 196 (1977).
6432. Small, P. A., *Polymer*, **13**, 536 (1972).
6433. Small, R. L., *J. Appl. Chem. (London)*, **3**, 71 (1953).
6434. Smekov, F. M., Nepomnyashchii, A. I., Sanzharovskii, A. T., and Yakubovich, S. V., *Vysokomol. Soedin. A*, **13**, 2102 (1971).
6435. Smidsrød, O., and Guillet, J. E., *Macromolecules*, **2**, 272 (1969).
6436. Smith, C. W., and Dole, M., *J. Polym. Sci.*, **20**, 37 (1956).
6437. Smith, D., and Cromey, P. R., *J. Sci. Instrum.*, **2**, 523 (1968).
6438. Smith, D. A., *Rubber Chem. Technol.*, **37**, 937 (1964).
6439. Smith, D. M., and Wiggins, T. A., *Appl. Opt.*, **11**, 2680 (1972).
6440. Smith, E. G., *Polymer*, **17**, 761 (1976).
6441. Smith, E. G., and Robb, I. D., *Polymer*, **15**, 713 (1974).
6442. Smith, F. S., and Steward, R. D., *Polymer*, **15**, 283 (1974).
6443. Smith, H., *Trans. Faraday Soc.*, **52**, 402 (1956).
6444. Smith, J. A. S., *Discuss. Faraday Soc.*, **19**, 207 (1955).

6445. Smith, J. B., Manuel, A. J., and Ward, I. M., *Polymer*, **16**, 57 (1975).
6446. Smith, J. S., Irving, H. M. N. H., and Simpson, R. B., *Analyst*, **95**, 743 (1970).
6447. Smith, P., and Pennings, A. J., *Polymer*, **15**, 413 (1974).
6448. Smith, P., and Pennings, A. J., *Brit. Polym. J.*, **7**, 343 (1975).
6449. Smith, P., and Pennings, A. J., *J. Polym. Sci. B*, **14**, 29 (1976).
6450. Smith, P., and Pennings, A. J., *J. Polym. Sci. A2*, **15**, 523 (1977).
6451. Smith, R. P., and Mortensen, E. M., *J. Chem. Phys.*, **35**, 714 (1961).
6452. Smith, T. E., and Carpenter, D. K., *Macromolecules*, **1**, 204 (1968).
6453. Smith, W. B., May, J. A., and Kim, C. W., *J. Polym. Sci. A2*, **4**, 365 (1966).
6454. Smith, W. N., *J. Appl. Polym. Sci.*, **11**, 639 (1967).
6455. Smith, W. V., *J. Polym. Sci. A2*, **8**, 207 (1970).
6456. Smith, W. V., and Feldman, G. A., *J. Polym. Sci. A2*, **7**, 163 (1969).
6457. Smith, W. V., and Thiruvengeda, S., *Rubber Chem. Technol.*, **43**, 1439 (1970).
6458. Smuts, T. W., and van Niekerk, F. A., *J. Gas Chromatogr.*, **5**, 190 (1967).
6459. Snyder, F., *Anal. Biochem.*, **9**, 182 (1964).
6460. Snyder, F., *Anal. Biochem.*, **11**, 510 (1965).
6461. Snyder, R. G., *J. Chem. Phys.*, **47**, 1316 (1967).
6462. Snyder, R. G., *J. Mol. Spectr.*, **23**, 224 (1967).
6463. Snyder, R. G., *J. Mol. Spectr.*, **31**, 464 (1969).
6464. Snyder, R. G., *J. Mol. Spectr.*, **37**, 353 (1971).
6465. Snyder, R. G., Krause, S. J., and Scherer, J. R., *J. Polym. Sci. A2*, **16**, 1593 (1978).
6466. Snyder, R. G., and Poore, M. W., *Macromolecules*, **6**, 708 (1973).
6467. Snyder, L. R., and Saunders, D. L., *J. Chromatogr.*, **44**, 1 (1969).
6468. Snyder, R. G., and Schatschneider, J. H., *Spectr. Acta*, **20**, 853 (1964).
6469. Sobue, H., and Fukuhara, S., *J. Chem. Soc. Japan*, **58**, 946 (1955).
6470. Sobue, H., and Fukuhara, S., *J. Chem. Soc. Japan*, **60**, 86 (1957).
6471. Sochova, I. V., *Dokl. Akad. Nauk USSR*, **130**, 126 (1960).
6472. Sochova, I. V., and Trapeznikova, O. N., *Dokl. Akad. Nauk USSR*, **113**, 784 (1957).
6473. Soignet, D. M., Berni, R. J., and Benerito, R. R., *J. Appl. Polym. Sci.*, **20**, 2483 (1976).
6474. Šolc, K., *Macromolecules*, **3**, 665 (1970).
6475. Šolc, K., *J. Polym. Sci. A2*, **12**, 555 (1974).
6476. Šolc, K., *J. Polym. Sci. A2*, **12**, 1865 (1974).
6477. Šolc, K., *Macromolecules*, **8**, 819 (1975).
6478. Šolc, K., *Macromolecules*, **10**, 1101 (1977).
6479. Solomon, I., and Bloembergen, N., *J. Chem. Phys.*, **25**, 261 (1955).
6480. Solomon, O. F., and Ciuta, I., *J. Appl. Polym. Sci.*, **6**, 683 (1962).
6481. Somersall, A. C., Dan, E., and Guillet, J. E., *Macromolecules*, **7**, 233 (1974).
6482. Somersall, A. C., and Guillet, J. E., *Macromolecules*, **6**, 218 (1973).
6483. Sotobayashi, H., Amussen, F., and Chen, J. T., *Makromol. Chem.*, **178**, 3025 (1977).
6484. Southern, E., and Thomas, A. G., *J. Sci. Instrum.*, **39**, 645 (1962).
6485. Southern, J. H., Ballman, R. L., Burroughs, J. A., and Paul, D. R., *J. Polym. Sci. B*, **16**, 157 (1978).
6486. Southwart, D. W., *Polymer*, **17**, 147 (1976).
6487. Spatorico, A. L., and Coulter, B., *J. Chromatog.*, **79**, 121 (1973).
6488. Spatorico, A. L., and Coulter, B., *J. Polym. Sci. A1*, **11**, 1139 (1973).
6489. Speak, R., and Shepherd, I. W., *J. Polym. Sci. C*, **44**, 209 (1974).
6490. Speak, R., and Shepherd, I. W., *J. Polym. Sci. A2*, **13**, 997 (1975).
6491. Spegt, P., Terrise, J., Gilg, B., and Skoulios, A. E., *Makromol. Chem.*, **107**, 29 (1967).
6492. Spells, S. J., Shepherd, I. W., and Wright, C. J., *Polymer*, **18**, 905 (1977).
6493. Spencer, H. G., and Barrie, J. A., *J. Appl. Polym. Sci.*, **20**, 2557 (1976).
6494. Spencer, R. D., and Weber, G., *Ann. N.Y. Acad. Sci.*, **158**, 361 (1969).
6495. Spencer, R. S., *J. Polym. Sci.*, **3**, 606 (1948).
6496. Sperling, L. H., and Ferguson, K. B., *Macromolecules*, **8**, 691 (1975).
6497. Spěváček, J., *J. Polym. Sci. A2*, **16**, 523 (1978).
6498. Spěváček, J., *Polymer*, **19**, 1149 (1978).
6499. Spěváček, J., and Schneider, B., *Makromol. Chem.*, **175**, 2939 (1974).
6500. Spěváček, J., and Schneider, B., *Makromol. Chem.*, **176**, 729 (1975).
6501. Spěváček, J., and Schneider, B., *J. Polym. Sci. A2*, **14**, 1789 (1976).
6502. Spěváček, J., and Schneider, B., *Polymer*, **19**, 63 (1978).
6503. Spit, B. J., *Polymer*, **4**, 109 (1963).

6504. Springer, J., Ueberreiter, K., and Weinle, W., *Europ. Polym. J.*, **6**, 87 (1970).
6505. Srinivasan, B. N., Konoshita, M., Rabalais, J. W., and McGlynn, S. P., *J. Chem. Phys.*, **48**, 1924 (1968).
6506. Srinivasan, K. S. V., and Santappa, M., *Polymer*, **14**, 5 (1973).
6507. Stabin, J. V., and Immergut, E. H., *J. Polym. Sci.*, **14**, 209 (1954).
6508. Stabinger, H., and Kratky, O., *Makromol. Chem.*, **179**, 1655 (1978).
6509. Stacy, C. J., *J. Appl. Polym. Sci.*, **21**, 2231 (1977).
6510. Stacy, C. J., and Arnett, R. L., *J. Phys. Chem.*, **69**, 3109 (1965).
6511. Stafford, J. W., and Frey, H., *J. Polym. Sci. A2*, **11**, 2489 (1973).
6512. Stahl, E., *J. Chromatogr.*, **33**, 273 (1968).
6513. Stanford, G. F., *Analyst*, **90**, 266 (1965).
6514. Starkweather, H. W., Jr, *J. Polym. Sci. A2*, **11**, 587 (1973).
6515. Statton, W. O., *J. Polym. Sci.*, **22**, 385 (1956).
6516. Statton, W. O., *J. Polym. Sci.*, **58**, 205 (1962).
6517. Statton, W. O., *J. Appl. Polym. Sci.*, **7**, 803 (1963).
6518. Statton, W. O., *J. Polym. Sci.*, **58**, 205 (1964).
6519. Staub, H., and Perror, W., *Anal. Chem.*, **46**, 128 (1974).
6520. Staudinger, H., and Henel-Immendorfer, I., *Makromol. Chem.*, **1**, 185 (1944).
6521. Staverman, A. J., *Rec. Trav. Chim.*, **70**, 344 (1951).
6522. Staverman, A. J., *Rec. Trav. Chim.*, **71**, 623 (1952).
6523. Staverman, A. J., *Ind. Chim. Belge*, **18**, 235 (1953).
6524. Stavermann, A. J., Pals, D. T. F., and Kruissink, C. A., *J. Polym. Sci.*, **23**, 57 (1957).
6525. Stearne, J. M., and Urwin, J. R., *Makromol. Chem.*, **56**, 76 (1962).
6526. Steel, B. J., *J. Sci. Instrum.*, **42**, 751 (1965).
6527. Steere, R. C., *J. Appl. Phys.*, **37**, 3338 (1966).
6528. Steger, T. R., and Nielsen, L. E., *J. Polym. Sci. A2*, **16**, 613 (1978).
6529. Stegman, G. I. A., and Stoicheff, B. P., *Phys. Rev. Lett.*, **21**, 202 (1968).
6530. Stehling, F. C., *J. Polym. Sci. A*, **2**, 1815 (1964).
6531. Stehling, F. C., and Knox, J. R., *Macromolecules*, **8**, 595 (1975).
6532. Stehling, F. C., and Mandelkern, K., *Macromolecules*, **3**, 242 (1970).
6533. Steidl, J., and Pelzbauer, Z., *J. Polym. Sci. C*, **38**, 345 (1972).
6534. Stein, A. N., Gray, D. G., and Guillet, J. E., *Brit. Polym. J.*, **3**, 175 (1971).
6535. Stein, R. A., and Slawson, V., *Anal. Chem.*, **35**, 1008 (1963).
6536. Stein, R. S., *J. Polym. Sci.*, **31**, 327, 335 (1958).
6537. Stein, R. S., *J. Polym. Sci.*, **50**, 339 (1961).
6538. Stein, R. S., *J. Polym. Sci. C*, **15**, 185 (1966).
6539. Stein, R. S., *J. Polym. Sci. A2*, **7**, 1021 (1969).
6540. Stein, R. S., Erhardt, P., van Aartsen, J. J., Clough, S., and Rhodes, M., *J. Polym. Sci. C*, **13**, 1 (1966).
6541. Stein, R. S., Erhardt, P. F., and Chu., W., *J. Polym. Sci. A2*, **7**, 271 (1969).
6542. Stein, R. S., Holmes, F. E., and Tobolsky, A. V., *J. Polym. Sci.*, **14**, 443 (1954).
6543. Stein, R. S., and Keane, J. J., *J. Polym. Sci.*, **17**, 21 (1955).
6544. Stein, R. S., and Keane, J. J., *J. Polym. Sci.*, **20**, 327 (1956).
6545. Stein, R. S., and Norris, F. H., *J. Polym. Sci.*, **21**, 381 (1956).
6546. Stein, R. S., Norris, F. H., and Plaza, A., *J. Polym. Sci.*, **24**, 455 (1957).
6547. Stein, R. S., Onogi, S., and Keedy, D. A., *J. Polym. Sci.*, **57**, 801 (1962).
6548. Stein, R. S., and Picot, C., *J. Polym. Sci. A2*, **8**, 1955 (1970).
6549. Stein, R. S., and Picot, C., *Macromolecules*, **4**, 467 (1971).
6550. Stein, R. S., and Powers, J., *J. Polym. Sci.*, **56**, S9 (1962).
6551. Stein, R. S., and Rhodes, M. B., *J. Appl. Polym. Sci.*, **31**, 1873 (1960).
6552. Stein, R. S., and Stidham, S. N., *J. Polym. Sci. A2*, **4**, 89 (1966).
6553. Stein, R. S., and Sutherland, G. B. B. M., *J. Chem. Phys.*, **21**, 370 (1953).
6554. Stein, R. S., and Wilkes, G. L., *J. Polym. Sci. A2*, **7**, 1695 (1969).
6555. Stein, R. S., and Wilson, P. R., *J. Appl. Phys.*, **33**, 1914 (1962).
6556. Steingraber, O. J., and Berlman, I. B., *Rev. Sci. Instrum.*, **34**, 524 (1963).
6557. Stejskal, E. O., and Schaefer, J., *Macromolecules*, **7**, 14 (1974).
6558. Stejskal, E. O., and Schaefer, J., *Macromolecules*, **7**, 767 (1974).
6559. Stejskal, E. O., and Tanner, J. E., *J. Chem. Phys.*, **42**, 288 (1965).
6560. Stejskal, E. O., and Tanner, J. E., *Macromolecules*, **4**, 586 (1971).
6561. Stejskal, J., Beneš, M. J., Kratochvil, P., and Peška, J., *J. Polym. Sci. A2*, **11**, 1803 (1973).

6562. Stejskal, J., Beneš, M. J., Kratochvil, P., and Peška, J., *J. Polym. Sci. A2*, **12**, 1941 (1974).
6563. Stejskal, J., Janča, J., and Kratochvil, P., *Polym. J.*, **8**, 549 (1976).
6564. Stellman, J. M., and Woodward, A. E., *J. Polym. Sci. B*, **7**, 755 (1969).
6565. Stellman, J. M., and Woodward, A. E., *J. Polym. Sci. A2*, **9**, 59 (1971).
6566. Stellman, J. M., Woodward, A. E., and Stellman, S. D., *Macromolecules*, **6**, 330 (1973).
6567. Stellman, S. D., and Gans, P. J., *Macromolecules*, **5**, 516 (1972).
6568. Sten, R. A., and Slawson, V., *Anal. Chem.*, **35**, 1008 (1963).
6569. Stepko, R. F. T., and Waywell, D. R., *Makromol. Chem.*, **152**, 247 (1972).
6570. Stern, S. A., and DeMeringo, A. H., *J. Polym. Sci. A2*, **16**, 735 (1978).
6571. Stern, S. A., Sinclair, T. F., and Gareis, P. J., *Modern Plast.*, **42**, 154 (1964).
6572. Stetzler, R. S., and Smullin, C. F., *Anal. Chem.*, **34**, 194 (1962).
6573. Stevens, J. R., and Edwards, M. J., *J. Polym. Sci. C*, **30**, 297 (1970).
6574. Stevens, J. R., and Lichtenberger, P. C., *Phys. Rev. Lett.*, **29**, 166 (1972).
6575. Stevens, J. R., and Mao, A. C., *J. Appl. Phys.*, **41**, 4273 (1970).
6576. Stevens, J. R., and Rowe, R. M., *J. Appl. Phys.*, **44**, 4328 (1973).
6577. Still, J. E., and Cluley, H. J., *Chem. Ind.* (*London*), **1969**, 1777.
6578. Stockmayer, W. H., *Makromol. Chem.*, **35**, 54 (1960).
6579. Stockmayer, W. H., *J. Polym. Sci. C*, **14**, 29 (1966).
6580. Stockmayer, W. H., and Baur, M. E., *J. Amer. Chem. Soc.*, **86**, 3485 (1964).
6581. Stockmayer, W. H., and Burke, J. J., *Macromolecules*, **2**, 647 (1969).
6582. Stockmayer, W. H., and Fixman, M., *Ann. N.Y. Acad. Sci.*, **57**, 334 (1953).
6583. Stockmayer, W. H., and Fixman, M., *J. Polym. Sci. C*, **1**, 137 (1963).
6584. Stockmayer, W. H., Jones, A. A., and Treadwell, T. L., *Macromolecules*, **10**, 762 (1977).
6585. Stockmayer, W. H., Moore, L. D., Jr, Fixman, M., and Epstein, B. N., *J. Polym. Sci.*, **16**, 517 (1955).
6586. Stoeckli, H. F., *Helv. Chim. Acta*, **55**, 101 (1972).
6587. Stoelting, J., Karasz, F. E., and McKnight, W. J., *Polym. Eng. Sci.*, **10**, 133 (1970).
6588. Stoesser, P. R., and Gill, S. J., *Rev. Sci. Instrum.*, **38**, 422 (1967).
6589. Stoistis, R. F., Poehlein, G. W., and Vanderhoff, J. W., *J. Colloid Interface Sci.*, **57**, 337 (1976).
6590. Stölning, J., and Müller, F. H., *Kolloid Z.*, **238**, 459 (1970).
6591. Stone, J., Lynch, G., and Pontinen, R., *Appl. Opt.*, **5**, 653 (1966).
6592. Stone, N. W. B., and Williams, D., *Appl. Opt.*, **5**, 353 (1966).
6593. Straff, R., and Uhlmann, D. R., *J. Polym. Sci. A2*, **14**, 1087 (1976).
6594. Strassburger, J., Brauer, G. M., Tyron, M., and Forziati, A. F., *Anal. Chem.*, **32**, 454 (1960).
6595. Strazielle, C., and Benoit, H., *Macromolecules*, **8**, 203 (1975).
6596. Strazielle, C., and Czlonkowska-Kohutnicka, Z., *J. Appl. Polym. Sci.*, **22**, 1135 (1978).
6597. Strella, S., *J. Appl. Polym. Sci.*, **7**, 569 (1963).
6598. Strella, S., and Erhardt, P. F., *J. Appl. Polym. Sci.*, **13**, 1373 (1969).
6599. Strobl, G. R., and Eckel, R., *J. Polym. Sci. A2*, **14**, 913 (1976).
6600. Strobl, G. R., and Hagedorn, W., *J. Polym. Sci. A2*, **16**, 1181 (1978).
6601. Strobl, G. R., and Muller, N., *J. Polym. Sci. A2*, **11**, 1219 (1974).
6602. Stuart, H. A., and Peterlin, A., *J. Polym. Sci.*, **5**, 551 (1950).
6603. Stuart, J. M., and Smith, D. A., *J. Appl. Polym. Sci.*, **9**, 3195 (1965).
6604. Stubbs, W. H., Gore, C. R., and Marvel, C. S., *J. Polym. Sci. A1*, **4**, 1898 (1966).
6605. Sturm, E., *Thermochim. Acta*, **4**, 461 (1972).
6606. Su, C. S., and Patterson, D., *Macromolecules*, **10**, 708 (1977).
6607. Su, C. S., Patterson, D., and Schreiber, H. P., *J. Appl. Polym. Sci.*, **20**, 1025 (1976).
6608. Sudduth, R. D., and Rogers, C. E., *J. Polym. Sci. A2*, **12**, 1667 (1974).
6609. Suehiro, K., Tanizaki, H., and Takayanagi, M., *Polymer*, **17**, 1059 (1976).
6610. Sugamiya, K., Kuwahara, N., and Kaneko, M., *Macromolecules*, **7**, 66 (1974).
6611. Sugeta, H., *Polymer, J.*, **1**, 226 (1970).
6612. Sugeta, H., and Kajiura, T., *J. Polym. Sci. B*, **7**, 251 (1969).
6613. Sugeta, H., and Miyazawa, T., *Rept. Progr. Polym. Phys. Japan*, **9**, 177 (1966).
6614. Suh, K. W., and Clarke, D. H., *J. Polym. Sci. A1*, **5**, 1671 (1967).
6615. Suh, K. W., and Clarke, D. H., *J. Appl. Polym. Sci.*, **12**, 1775 (1968).
6616. Sultan, J. N., Laible, R. C., and McGarry, T. J., *Appl. Polym. Symp.*, No. 16, 137 (1971).
6617. Suman, P. T., and Werstler, D. D., *J. Polym. Sci. A1*, **13**, 1963 (1975).

6618. Sumi, M., Chokki, Y., Nakai, Y., Nakabayashi, M., and Kazawa, T., *Makromol. Chem.*, **78**, 146 (1964).
6619. Sumi, Y. O., Higashimura, T., and Okamura, S., *J. Polm. Sci. A1*, **4**, 923 (1966).
6620. Summerfield, G. C., *J. Chem. Phys.*, **43**, 1079 (1965).
6621. Summerfield, G. C., King, J. S., and Ullman, R., *Macromolecules*, **11**, 218 (1978).
6622. Summers, W. R., Tewari, Y. B., and Schreiber, H. P., *Macromolecules*, **5**, 12 (1972).
6623. Sund., E., Haanaes, E., Simdsrød, O., and Ugelstad, J., *J. Appl. Polym. Sci.*, **16**, 1869 (1972).
6624. Sundararajan, P. R., *Macromolecules*, **10**, 623 (1977).
6625. Sundararajan, P. R., *J. Polym. Sci. B*, **15**, 699 (1977).
6626. Sundararajan, P. R., *Macromolecules*, **11**, 256 (1978).
6627. Sundararajan, P. R., *J. Appl. Polym. Sci.*, **22**, 1391 (1978).
6628. Sundararajan, P. R., and Flory, P. J., *J. Amer. Chem. Soc.*, **96**, 5025 (1974).
6629. Sundelöf, L. O., *Arkiv Kemi*, **25**, 1 (1966).
6630. Sundelöf, L. O., *Arkiv Kemi*, **29**, 279 (1968).
6631. Sundelöf, L. O., *Chem. Scripta*, **6**, 55 (1974).
6632. Sundelöf, L. O., and Nyström, B., *J. Polym. Sci. B*, **15**, 377 (1977).
6633. Sutherland, G. B. B. M., and Willis, H. A., *Trans. Faraday Soc.*, **41**, 181 (1945).
6634. Suter, U. W., and Flory, P. J., *Macromolecules*, **8**, 765 (1975).
6635. Sutter, W., and Kuppel, A., *Makromol. Chem.*, **149**, 271 (1971).
6636. Suwandi, M. S., and Stern, S. A., *J. Polym. Sci. A2*, **11**, 663 (1973).
6637. Suzuki, E., *Spectrochim. Acta A*, **23**, 2303 (1967).
6638. Suzuki, H., *Macromolecules*, **3**, 373 (1970).
6639. Suzuki, H., Hiyoshi, T., and Inagaki, H., *J. Polym. Sci. C*, **61**, 291 (1977).
6640. Suzuki, H., and Leonis, C. G., *Brit. Polym. J.*, **5**, 485 (1973).
6641. Suzuki, H., Leonis, G. G., and Gordon, M., *Makromol. Chem.*, **172**, 227 (1973).
6642. Suzuki, J., *J. Appl. Polym. Sci.*, **20**, 2791 (1976).
6643. Suzuki, J., Iizuka, S., and Suzuki, S., *J. Appl. Polym. Sci.*, **22**, 2109 (1978).
6644. Suzuki, J., Nakagawa, H., and Ito, H., *J. Appl. Polym. Sci.*, **20**, 2791 (1976).
6645. Suzuki, T., Koeshiro, S., and Takegami, Y., *Polymer*, **14**, 549 (1973).
6646. Svenson, H., *Acta Chem. Scand.*, **3**, 1170 (1949).
6647. Svenson, H., *Acta Chem. Scand.*, **5**, 72 (1951).
6648. Swanson, C. L., Ernst, J. O., and Gugliemelli, L. A., *J. Appl. Polym. Sci.*, **18**, 1549 (1974).
6649. Swan, P. R., *J. Polym. Sci.*, **42**, 525 (1960).
6650. Swan, W. B., and Dux, J. P., *Anal. Chem.*, **33**, 654 (1961).
6651. Swartz, T. D., Bly, D. D., and Edwards, A. S., *J. Appl. Polym. Sci.*, **16**, 3353 (1972).
6652. Swell, J. H., *Modern Plast.*, **48**, 66 (1971).
6653. Swenson, H. A., Kaustinen, H. M., Bachhuber, J. J., and Carlson, J. A., *Macromolecules*, **2**, 142 (1969).
6654. Swingle, R. S., *Ind. Res.*, **14**, 40 (1972).
6655. Szewczyk, P., *Polymer*, **17**, 90 (1976).
6656. Szöcs, F., and Klimova, M., *J. Polym. Sci. A2*, **15**, 1983 (1977).
6657. Szöcs, F., and Placek, J., *J. Polym. Sci. A2*, **13**, 1789 (1975).
6658. Szu, S. C., and Hermans, J. J., *J. Polym. Sci A2*, **12**, 1743 (1974).
6659. Sybarndt, L. B., and Montoya, E. F., *Intern. Lab.*, July/Aug. 1977, p. 51.
6660. Tabata, T., Ito, Y., and Oshima, K., *Rept. Progr. Polym. Phys. Japan*, **11**, 489 (1968).
6661. Tabb, D. L., and Koenig, J. L., *J. Polym. Sci. A2*, **13**, 1159 (1975).
6662. Tabb, D. L., and Koenig, J. L., *Macromolecules*, **8**, 929 (1975).
6663. Tabb, D. L., Koenig, J. L., and Coleman, M. M., *J. Polym. Sci. A2*, **13**, 1145 (1975).
6664. Tachidbana, T., and Kambara, H., *Kolloid Z. Z. Polym.*, **219**, 40 (1967).
6665. Tadekoro, H., *Bull. Chem. Soc., Japan*, **28**, 559 (1955).
6666. Tadekoro, H., *Bull. Chem. Soc. Japan*, **32**, 1252, 1334 (1959).
6667. Tadekoro, H., *J. Chem. Phys.*, **33**, 1558 (1960).
6668. Tadekoro, H., *J. Polym. Sci C*, **15**, 1 (1966).
6669. Tadekoro, H., Chatani, Y., Kusanagi, H., and Yokoyama, M., *Macromolecules*, **3**, 441 (1970).
6670. Tadekoro, H., Chatani, Y., Yoshihara, T., Tahara, S., and Murahashi, S., *Makromol. Chem.*, **73**, 109 (1964).

6671. Tadekoro, K., Kitazawa, T., Mozakura, S., and Murahashi, S., *Bull. Chem. Soc. Japan*, **34**, 1209 (1961).

6672. Tadekoro, H., Kobayashi, M., Kawaguchi, Y., Kobaysahi, A., and Murahashi, S., *J. Chem. Phys.*, **38**, 703 (1963).

6673. Tadekoro, H., Kobayashi, M., Ukita, M., Yasufuku, K., and Murahashi, S., *J. Chem. Phys.*, **42**, 1432 (1965).

6674. Tadekoro, H., Kozai, K., Seiki, S., and Nitta, I., *J. Polym. Sci.*, **26**, 379 (1957).

6675. Tadekoro, H., Nagai, H., Seki, S., and Nitta, I., *Bull. Chem. Soc. Japan*, **34**, 1504 (1961).

6676. Tadekoro, H., Nishiyama, N., Nozakura, S., and Murahashi, S., *J. Polym. Sci.*, **36**, 553 (1959).

6677. Tadekoro, H., Nishiyama, Y., Nozakura, S., and Murahashi, S., *Bull. Chem. Soc. Japan*, **34**, 381 (1961).

6678. Tadekoro, H., Seiki, S., and Nitta, I., *J. Chem. Phys.*, **23**, 1351 (1955).

6679. Tadekoro, H., Seiki, A., and Nitta, I., *J. Polym. Sci.*, **22**, 563 (1956).

6680. Tedekoro, H., Seiki, S., Nitta, I., and Yamadera, R., *J. Polym. Sci.*, **28**, 244 (1958).

6681. Tadekoro, H., Tai, K., Yokoyama, M., and Kobayashi, M., *J. Polym. Sci. A2*, **11**, 825 (1973).

6682. Tadekoro, H., Ukita, M., Yasufuku, K., Murahashi, S., and Torii, T., *J. Chem. Phys.*, **42**, 1432 (1965).

6683. Taga, T., and Inagaki, H., *Angew. Makromol. Chem.*, **33**, 129 (1973).

6684. Tagami, Y., and Pecora, R., *J. Chem. Phys.*, **51**, 3293 (1969).

6685. Tagata, N., and Homma, T., *J. Chem. Soc. Japan*, **1972**, 1330 (1972).

6686. Tager, A. A., Andreeva, V. M., Vshivkov, S. A., and Tjukova, I. S., *J. Polym. Sci. C*, **61**, 283 (1977).

6687. Taham, M., *J. Paint Technol.*, **46**, 35 (1974).

6688. Taham, M., and Tighe, B. J., *J. Paint. Technol.*, **46**, 48 (1974).

6689. Tai, K., and Tadekoro, H., *Macromolecules*, **7**, 507 (1974).

6690. Tait, P. J. T., and Abushihada, A. M., *Polymer*, **18**, 810 (1977).

6691. Tait, P. J. T., and Livesey, P. J., *Polymer*, **11**, 359 (1970).

6692. Takagi, S., and Kimura, T., *Makromol. Chem.*, **179**, 557 (1978).

6693. Takahashi, T., Inamura, M., and Tsujimoto, I., *J. Polym. Sci. B*, **8**, 65 (1970).

6694. Takahashi, T., and Ogata, N., *J. Polym. Sci. B*, **9**, 895 (1971).

6695. Takahashi, T., Kohyama, M., and Tadekoro, H., *Macromolecules*, **9**, 870 (1976).

6696. Takahashi, Y., and Tadekoro, H., *Macromolecules*, **6**, 672 (1973).

6697. Takahashi, Y., Tadekoro, H., Hirano, T., Sato, A., and Tsuruta, T., *J. Polym. Sci. A2*, **13**, 285 (1975).

6698. Takahashi, Y., Sumita, I., and Tadekoro, H., *J. Polym. Sci. A2*, **11**, 2113 (1973).

6699. Takashima, K., Nakae, K., Shibata, M., and Yamakawa, H., *Macromolecules*, **7**, 641 (1974).

6700. Takazawa, A., Negishi, T., and Ishikawa, K., *J. Polym. Sci. A1*, **6**, 475 (1968).

6701. Takeda, S., *J. Appl. Phys.*, **47**, 5480 (1976).

6702. Takemura, T., *Progr. Polym. Sci. Japan*, **7**, 225 (1974).

6703. Takenaka, T., Shimura, Y., and Gotoh, R., *Kolloid Z.*, **237**, 193 (1970).

6704. Takeshita, T., Tsuji, K., and Seiki, T., *J. Polym. Sci. A1*, **10**, 2315 (1972).

6705. Takeuchi, T., Tsuge, S., and Okumoto, T., *J. Gas Chromatogr.*, **6**, 542 (1968).

6706. Tamura, K., Murakami, S., and Fujishiro, R., *Polymer*, **14**, 237 (1973).

6707. Tamura, S., and Gillham, J. K., *J. Appl. Polym. Sci.*, **22**, 1867 (1978).

6708. Tan, J. S., and Gasper, S. P., *J. Polym. Sci. A2*, **12**, 1785 (1974).

6709. Tan, Y. Y., and Challa, G., *Polymer*, **17**, 739 (1976).

6710. Tanabe, Y., Hirose, J., Okano, K., and Wada, Y., *Polymer J.*, **1**, 107 (1970).

6711. Tanabe, Y., and Kanetsuna, H., *J. Appl. Polym. Sci.*, **22**, 2619 (1978).

6712. Tanaka, A., Chang, E. P., Delf, B., Kimura, I., and Stein, R. S., *J. Polym. Sci. A2*, **11**, 1891 (1973).

6713. Tanaka, A., and Ishida, Y., *J. Polym. Sci. A2*, **12**, 335 (1974).

6714. Tanaka, A., and Ishida, Y., *J. Polym. Sci. A2*, **12**, 1283 (1974).

6715. Tanaka, A., and Ishida, Y., *J. Polym. Sci. A2*, **13**, 431 (1975).

6716. Tanaka, A., and Ishida, Y., *J. Polym. Sci. A2*, **13**, 437 (1975).

6717. Tanaka, A., Sawada, K., and Ishida, Y., *J. Polym. Sci. A2*, **12**, 2157 (1974).

6718. Tanaka, H., and Okajima, S., *J. Polym. Sci. B*, **15**, 349 (1977).

828

6719. Tanaka, N., *Polymer*, **19**, 770 (1978).
6720. Tanaka, S., Nakamura, A., and Morikawa, H., *Makromol. Chem.*, **85**, 164 (1965).
6721. Tanaka, T., Chatani, Y., and Tadekoro, H., *J. Polym. Sci. A2*, **12**, 515 (1974).
6722. Tanaka, T., Hocker, L., and Benedek, G. B., *J. Chem. Phys.*, **59**, 9 (1973).
6723. Tanaka, T., Kotaka, T., Ban, K., Hattori, M., and Inagaki, H., *Macromolecules*, **10**, 960 (1977).
6724. Tanaka, T., Kotaka, T., and Inagaki, H., *Macromolecules*, **7**, 311 (1974).
6725. Tanaka, T., Yokoyama, T., and Yamaguchi, Y., *J. Polym. Sci. A1*, **6**, 2137 (1968).
6726. Tanaka, Y., and Hatada, K., *J. Polym. Sci. A1*, **11**, 2057 (1973).
6727. Tanaka, Y., and Hatada, K., *J. Polym. Sci. B*, **11**, 569 (1973).
6728. Tanaka, Y., and Sato, H., *Polymer*, **17**, 113 (1976).
6729. Tanaka, Y., and Sato, H., *Polymer*, **17**, 413 (1976).
6730. Tanaka, Y., and Sato, H., *J. Polym. Sci. B*, **14**, 335 (1976).
6731. Tanaka, Y., Sato, H., Hatada, K., Terawaki, Y., and Okuda, H., *Makromol. Chem.*, **178**, 1823 (1977).
6732. Tanaka, Y., Sato, H., Ogawa, M., Hatada, K., and Terawaki, Y., *J. Polym. Sci. B*, **12**, 369 (1974).
6733. Tanaka, Y., Sato, H., Ogura, A., and Nagoya, I., *J. Polym. Sci. A1*, **14**, 73 (1976).
6734. Tanaka, Y., Sato, H., and Seimya, T., *Polym. J.*, **7**, 264 (1975).
6735. Tanaka, Y., Takeuchi, Y., Kobyashi, M., and Tadekoro, H., *J. Polym. Sci. A2*, **9**, 43 (1971).
6736. Tang, C. C. H., *J. Appl. Phys.*, **28**, 628 (1957).
6737. Tanner, J. E., *Macromolecules*, **4**, 748 (1971).
6738. Tanner, J. E., and Liu, K. J., *Makromol. Chem.*, **142**, 309 (1971).
6739. Tanner, J. E., Liu, K. J., and Anderson, J. E., *Macromolecules*, **4**, 586 (1971).
6740. Tanner, R. I., *J. Polym. Sci. A2*, **8**, 2067 (1970).
6741. Tao, S. J., and Green, J. H., *Proc. Phys. Soc. (London)*, **85**, 463 (1965).
6742. Tarasova, G. I., Pavlova, S. A., and Korshak, V. V., *Vyskomol. Soedin. A*, **15**, 929 (1973).
6743. Tashiro, K., Kobyashi, M., and Tadekoro, H., *Macromolecules*, **10**, 731 (1977).
6744. Tasumi, M., and Krimm, S., *J. Chem. Phys.*, **46**, 755 (1967).
6745. Tasumi, M., and Krimm, S., *J. Polym. Sci. A2*, **6**, 995 (1968).
6746. Tasumi, M., and Shimanouchi, T., *Spectrochim. Acta*, **17**, 731 (1961).
6747. Tasumi, M., and Shimanouchi, T., *J. Chem. Phys.*, **43**, 1245 (1965).
6748. Tasumi, M., and Shimanouchi, T., *Polym. J.*, **2**, 62 (1971).
6749. Tasumi, M., Shimanouchi, T., and Miyazawa, T., *J. Mol. Spectr.*, **9**, 261 (1962).
6750. Tasumi, M., Shimanouchi, T., and Miyazawa, T., *J. Mol. Spectr.*, **11**, 422 (1963
6751. Tautz, H., Glueck, M., Hartman, G., and Leuteritz, R., *Plaste Kaustch.*, **10**, 648 (1963)
6752. Tavel, P., and Schaller, J., *Makromol. Chem.*, **179**, 677 (1978).
6753. Tazuke, S., and Banba, F., *Macromolecules*, **9**, 451 (1976).
6754. Tazuke, S., and Matsuyama, Y., *Macromolecules*, **8**, 280 (1975).
6755. Tazuke, S., and Matsuyama, Y., *Macromolecules*, **10**, 215 (1977).
6756. Tazuke, S., Sato, K., and Banba, F., *Macromolecules*, **10**, 1224 (1977).
6757. Taylor, C. R., Greco, R., Kramer, O., and Ferry, J. D., *Trans. Soc. Rheol.*, **20**, 141 (1976).
6758. Taylor, D. L., *J. Polym. Sci. A1*, **2**, 614 (1964).
6759. Taylor, G. B. and Hall, M. B., *Anal. Chem.*, **23**, 947 (1951).
6760. Taylor, W. C., and Tung, L. H., *SPE Trans.*, **2**, 119 (1962).
6761. Teas, J. P., *J. Paint Technol.*, **40**, 19 (1968).
6762. Teer, D., and Dole, M., *J. Polym. Sci. A2*, **13**, 985 (1975).
6763. Temin, S. C., *J. Appl. Polym. Sci.*, **9**, 471 (1965).
6764. Teramoto, A., and Fujita, H., *Adv. Polym. Sci.*, **18**, 65 (1975).
6765. Terry, S. L., and Rodriguez, F., *J. Polym. Sci. C*, **21**, 191 (1968).
6766. Tewari, Y. B., and Schreiber, H. P., *Macromolecules*, **5**, 329 (1972).
6767. Theil, M. H., *J. Polym. Sci. A2*, **13**, 1097 (1975).
6768. Thomas, E. L., and Ast, D. G., *Polymer*, **15**, 37 (1974).
6769. Thomas, G. B., and Mark, J. E., *J. Polym. Sci. A2*, **12**, 2393 (1974).
6770. Thomas, J. M., Williams, J. O., Evans, W. C., and Griffiths, E., *Nature*, **211**, 181 (1966).
6771. Thomassin, J. M., Walckiers, E., Warin, R., and Teyssie, Ph., *J. Polym. Sci. B*, **11**, 229 (1973).

829

6772. Thomasson, C. N., and Cunningham, D. A., *J. Sci. Instrum.*, **41**, 308 (1964).
6773. Thompson, D. S., *J. Phys. Chem.*, **75**, 789 (1971).
6774. Thompson, D. S., *J. Chem. Phys.*, **54**, 1411 (1971).
6775. Thompson, G. H., Myers, M. N., and Giddings, J. C., *Separation Sci.*, **2**, 707 (1967).
6776. Thompson, G. H., Myers, M. N., and Giddings, J. C., *Anal. Chem.*, **41**, 1219 (1969).
6777. Thornley, P. W., and Shepherd, I. W., *J. Polym. Sci. A2*, **15**, 97 (1977).
6778. Thornley, P. W., and Shepherd, I. W., *J. Polym. Sci. A2*, **15**, 1339 (1977).
6779. Thosar, B. V., Kulkarni, V. G., Lagu, R. G., and Chandra, G., *J. Phys. Lett.*, **21**, 647 (1965).
6780. Thosar, B. V., Kulkarni, V. G., Lagu, R. G., and Chandra, G., *Phys. Lett. A*, **28**, 760 (1969).
6781. Thümmler, W., and Häntzsch, G., *Plaste Kautsch.*, **14**, 881 (1967).
6782. Thurmond, C. D., *J. Polym. Sci.*, **8**, 607 (1952).
6783. Thurmond, C. D., and Zimm, B. H., *J. Polym. Sci.*, **8**, 477 (1952).
6784. Till, P. H., Jr, *J. Polym. Sci.*, **24**, 301 (1957).
6785. Tincher, W. C., *J. Polym. Sci.*, **62**, 5148 (1962).
6786. Tincher, W. C., *Makromol. Chem.*, **85**, 20 (1965).
6787. Tincher, W. C., *Makromol. Chem.*, **85**, 34 (1965).
6788. Ting, R. Y., *J. Appl. Polym. Sci.*, **20**, 3017 (1976).
6789. Ting, R. Y., and Little, R. C., *J. Appl. Polym. Sci.*, **17**, 3345 (1973).
6790. Tirrell, M., Malone, M. F., *J. Polym. Sci. A2*, **15**, 1569 (1977).
6791. Tobiason, F. L., Cain, G. H., and Anderson, J. W., *J. Polym. Sci. A1*, **16**, 275 (1978).
6792. Tobin, M. C., *J. Chem. Phys.*, **23**, 891 (1955).
6793. Tobin, M. C., *J. Phys. Chem.*, **64**, 216 (1960).
6794. Tobin, M. C., *J. Polym. Sci. A2*, **12**, 399 (1974).
6795. Tobin, M. C., *J. Polym. Sci. A2*, **14**, 2253 (1976).
6796. Tobin, M. C., and Carrano, M. J., *J. Chem. Phys.*, **25**, 1044 (1956).
6797. Tobin, M. C., and Carrano, M., J., *J. Polym. Sci.*, **24**, 93 (1957).
6798. Tobolsky, A. V., and Canter, N. H., *J. Polym. Sci. C*, **14**, 21 (1966).
6799. Tobolsky, A. V., and Goebel, J. C., *Macromolecules*, **3**, 556 (1970).
6800. Toi, K., *J. Polym. Sci. A2*, **11**, 1829 (1973).
6801. Toi, K., Igarashi, K., and Tokuda, T., *J. Appl. Polym. Sci.*, **20**, 703 (1976).
6802. Tomimatsu, Y., Votello, L., and Fong, K., *J. Colloid Interface Sci.*, **27**, 573 (1968).
6803. Tomlinson, C., Chylewski, C., and Simon, W., *Tetrahedron*, **19**, 949 (1963).
6804. Tompa, A. S., *Anal. Chem.*, **44**, 628 (1972).
6805. Tompa, H., *Trans. Faraday Soc.*, **45**, 1142 (1949).
6806. Tompa, H., *J. Polym. Sci.*, **8**, 51 (1952).
6807. Tompa, H., and Bamford, C. H., *Trans. Faraday Soc.*, **46**, 310 (1950).
6808. Tonelli, A. E., *Macromolecules*, **4**, 653 (1971).
6809. Tonelli, A. E., *Polymer*, **15**, 194 (1974).
6810. Tonelli, A. E., *Macromolecules*, **7**, 632 (1974).
6811. Tonelli, A. E., *Macromolecules*, **9**, 547 (1976).
6812. Tonelli, A. E., *Polymer*, **17**, 695 (1976).
6813. Tonelli, A. E., *Macromolecules*, **10**, 153 (1977).
6814. Tonelli, A. E., *Macromolecules*, **11**, 634 (1978).
6815. Tonelli, A. E., and Helfand, E., *Macromolecules*, **7**, 59 (1974).
6816. Toniolo, C., and Bonora, G. M., *Makromol. Chem.*, **176**, 2547 (1975).
6817. Toporowski, P. M., and Roovers, J., *Macromolecules*, **11**, 365 (1978).
6818. Torikai, A., Takahashi, Y., and Kuri, Z. I., *J. Polym. Sci. A1*, **15**, 1519 (1977).
6819. Toriyama, K., and Iwasaki, M., *J. Phys. Chem.*, **73**, 2919 (1969).
6820. Törmälä, P., Lättilä, H., and Lindberg, J. J., *Polymer*, **14**, 481 (1973).
6821. Törmälä, P., and Tulikoura, J., *Polymer*, **15**, 248 (1974).
6822. Törmälä, P., and Weber, G., *Polymer*, **19**, 1026 (1978).
6823. Torza, S., *J. Polym. Sci.*, *A2*, **13**, 43 (1975).
6824. Toy, M. S., and Stringham, R. S., *J. Polym. Sci. B*, **14**, 717 (1976).
6825. Trap, H. J. L., and Hermans, J. J., *Rec. Trav. Chim.*, **73**, 167 (1954).
6826. Traskos, R. T., Schneider, N. S., and Hoffman, A. S., *J. Appl. Polym. Sci.*, **12**, 509 (1968).
6827. Trautman, R., *J. Phys. Chem.*, **60**, 1211 (1956).
6828. Traylor, P. A., *Anal. Chem.*, **33**, 1629 (1961).

6829. Treloar, L. R. G., *Polymer*, **17**, 143 (1976).
6830. Tremaine, P. R., and Gray, D. G., *J. Chem. Soc. Faraday Trans.*, **71**, 2170 (1975).
6831. Tremaine, P. R., and Gray, D. G., *Anal. Chem.*, **48**, 380 (1976).
6832. Trementozzi, Q. A., *J. Polym. Sci.*, **23**, 887 (1957).
6833. Trevino, S. F., *J. Chem. Phys.*, **45**, 757 (1966).
6834. Trevino, S. F., *J. Macromol. Sci. A*, **1**, 723 (1967).
6835. Trevino, S. F., and Boutin, H., *J. Chem. Phys.*, **45**, 2700 (1966).
6836. Tricot, M., Bleus, J. P., Riga, J. P. and Desreux, V., *Makromol. Chem.*, **175**, 913 (1974).
6837. Trott, G. F., *J. Appl. Polym. Sci.*, **14**, 2421 (1970).
6838. Troxell, T. C., and Scheraga, H. A., *Macromolecules*, **4**, 519, 528 (1971).
6839. Trudelle, Y., and Spach, G., *Polymer*, **16**, 16 (1975).
6840. Tsebrenko, M. V., Yudin, A. V., Ablazova, T. I., and Vinogradov, G. V., *Polymer*, **17**, 831 (1976).
6841. Tshudy, J. A., and von Frankenberg, C., *J. Polym. Sci. A2*, **11**, 2027 (1973).
6842. Tsuchida, E., Abe, K., and Honsma, M., *Macromolecules*, **9**, 112 (1976).
6843. Tsuchihashi, N., Hatano, M., and Sohma, J., *Makromol. Chem.*, **177**, 2739 (1976).
6844. Tsuge, S., Hiramitsu, S., Horibe, T., Yamaoka, M., and Takeuchi, T., *Macromolecules*, **8**, 721 (1975).
6845. Tsuge, S., Okumoto, T., and Takeuchi, T., *Makromol. Chem.*, **123**, 123 (1969).
6846. Tsuge, S., Okumoto, T., and Takeuchi, T., *Macromolecules*, **2**, 200 (1969).
6847. Tsuge, S., Okumoto, T., and Takeuchi, T., *Macromolecules*, **2**, 277 (1969).
6848. Tsuji, K., *J. Polym. Sci. B*, **11**, 351 (1973).
6849. Tsuji, K., *J. Polym. Sci. A1*, **11**, 467 (1973).
6850. Tsuji, K., *J. Polym. Sci. A1*, **11**, 1407 (1973).
6851. Tsuji, K., and Nagata, H., *J. Polym. Sci. A1*, **11**, 897 (1973).
6852. Tsuji, K., and Seiki, T., *J. Polym. Sci. B*, **7**, 839 (1969).
6853. Tsuji, K., and Seiki, T., *Polym. J.*, **1**, 133 (1970).
6854. Tsuji, K., and Seiki, T., *J. Polym. Sci. B*, **8**, 817 (1970).
6855. Tsuji, K., and Seiki, T., *Polymer*, **2**, 606 (1971).
6856. Tsuji, K., and Seiki, T., *J. Polym. Sci. A1*, **9**, 3063 (1971).
6857. Tsuji, K., and Seiki, T., *J. Polym. Sci. A1*, **10**, 123 (1972).
6858. Tsuji, K., and Takeshita, T., *J. Polym. Sci. B.*, **10**, 185 (1972).
6859. Tucker, J. E., and Reese, W., *J. Chem. Phys.*, **46**, 1388 (1967).
6860. Tung, L. H., *J. Polym. Sci.*, **20**, 495 (1956).
6861. Tung, L. H., *J. Polym. Sci.*, **61**, 449 (1962).
6862. Tung, L. H., *J. Appl. Polym. Sci.*, **10**, 1271 (1966).
6863. Tung, L. H., *J. Appl. Polym. Sci.*, **10**, 375, 1271, 1861 (1966).
6864. Tung, L. H., *J. Polym. Sci. A2*, **7**, 47 (1969).
6865. Tung, L. H., *J. Appl. Polym. Sci.*, **13**, 775 (1969).
6866. Tung, L. H., *J. Polym. Sci. A2*, **9**, 759 (1971).
6867. Tung, L. H., *J. Polym. Sci. A2*, **11**, 1249 (1973).
6868. Tung, L. H., Moore, J. C., and Knight, G. W., *J. Appl. Polym. Sci.*, **10**, 1261 (1966).
6869. Tung, L. H., and Runyon, J. R., *J. Appl. Polym. Sci.*, **13**, 2397 (1969).
6870. Turner-Jones, A., Aizlewood, J. M., and Beckett, D. R., *Makromol. Chem.*, **74**, 134 (1964).
6871. Turska, E., and Janeczek, H., *Polymer*, **19**, 81 (1978).
6872. Turska, E., and Jantas, R., *J. Polym. Sci. C*, **47**, 359 (1974).
6873. Tuzar, Z., and Kratochvil, P., *Macromolecules*, **10**, 1108 (1977).

6874. Ubbelohde, L., *Ind. Eng. Chem.*, **9**, 85 (1937).
6875. Ueberreiter, K., *Kolloid Z. Z. Polym.*, **216**, 217 (1967).
6876. Ueberreiter, K., and Asmussen, F., *J. Polym. Sci.*, **23**, 75 (1957).
6877. Ueberreiter, K., and Asmussen, F., *Makromol. Chem.*, **43**, 324 (1961).
6878. Ueberreiter, K., and Asmussen, F., *J. Polym. Sci.*, **57**, 187 (1962).
6879. Ueberreiter, K., and Götze, T., *Makromol. Chem.*, **29**, 61 (1959).
6880. Ueberreiter, K., and Kanig, G., *J. Colloid Sci.*, **7**, 569 (1952).
6881. Ueberreiter, K., and Nens, S., *Kolloid Z.*, **123**, 92 (1951).
6882. Ueberreiter, K., Orthmann, H. W., and Sorge, G., *Makromol. Chem.*, **8**, 21 (1952).
6883. Ueberreiter, K., and Rhode-Liebenau, U., *Makromol. Chem.*, **49**, 164 (1961).
6884. Ueberreiter, K., and Sotobayashi, H., *Makromol. Chem.*, **59**, 71 (1963).

6885. Ueda, M., *Polymer, J.*, **3**, 431 (1972).
6886. Ueda, S., and Kataoka, T., *J. Polym. Sci. A2*, **11**, 1975 (1973).
6887. Uemura, S., *J. Polym. Sci. A2*, **12**, 1177 (1974).
6888. Uemura, S., Stein, R. S., and MacKnight, W. J., *Macromolecules*, **4**, 490 (1971).
6889. Ueno, A., Osa, T., and Toda, F., *J. Polym. Sci. B*, **14**, 521 (1976).
6890. Ueno, A., Osa, T., and Toda, F., *Macromolecules*, **10**, 130 (1977).
6891. Ukita, M., *Bull. Chem. Soc., Japan*, **39**, 742 (1966).
6892. Ullman, R., *J. Polym. Sci. C*, **12**, 317 (1966).
6893. Ullman, R., *Macromolecules*, **2**, 27 (1969).
6894. Ullman, R., Summerfield, G. C., and King, J. S., *J. Polym. Sci. A2*, **16**, 1641 (1977).
6895. Unbehend, J., and Sarko, A., *J. Polym. Sci. A2*, **12**, 545 (1974).
6896. Urwin, J. R., Stearne, J. M., Jordan, D. O., and Mills, R. A., *Makromol. Chem.*, **72**, 53 (1964).
6897. Utiyama, H., Takenaka, K., Mizumori, M. and Fukuda, M., *Macromolecules*, **7**, 28 (1974).
6898. Utiyama, H., and Tsunashima, Y., *Appl. Opt.*, **9**, 1330 (1970).
6899. Utiyama, H., and Tsunashina, Y., *J. Chem. Phys.*, **56**, 1626 (1972).
6900. Utiyama, H., Tsunashina, Y., and Kurata, M., *J. Chem. Phys.*, **55**, 3133 (1971).
6901. Utiyama, H., Utsumi, S., Tsunashima, Y., and Kurata, M., *Macromolecules*, **11**, 506 (1978).
6902. Utracki, L. A., and Roovers, J. E. L., *Macromolecules*, **6**, 366, 373 (1973).
6903. Utracki, L. A., and Simha, R., *Macromolecules*, **1**, 505 (1968).
6904. Uy, W. C., and Graessley, W. W., *Macromolecules*, **4**, 458 (1971).

6905. Vala, M. T., Haebig, J., and Rice, S. A., *J. Chem. Phys.*, **43**, 886 (1965).
6906. Van, N. B., and Noel, C., *J. Polym. Sci. A1*, **14**, 1627 (1976).
6907. van Aartsen, J. J., and Stein, R. S., *J. Polym. Sci. A2*, **9**, 293 (1971).
6908. van Antwerpen, P., and van Krevelen, D. W., *J. Polym. Sci. A2*, **10**, 2409 (1972).
6909. van Antwerpen, P., and van Krevelen, D. W., *J. Polym. Sci. A2*, **10**, 2423 (1972).
6910. Van Beek, W. M., Odijk, T. J., Van der Touw, F. and Mandel, M., *J. Polym. Sci. A2*, **14**, 773 (1976).
6911. van Dam, J., *Rec. Trav. Chim.*, **83**, 129 (1964).
6912. Van Deemeter, J. J., Zuiderweg, F. J., and Klinkenberg, A., *Chem. Eng. Sci.*, **5**, 271 (1956).
6913. Van den Esker, M. W. J., Laven, J., Broeckman, A., and Vrij, A., *J. Polym. Sci. A2*, **14**, 1953 (1976).
6914. Van den Esker, M. W. J., and Vrij, A., *J. Polym. Sci. A2*, **14**, 1967 (1976).
6915. Van den Esker, M. W. J., and Vrij, A., *J. Polym. Sci. A2*, **14**, 1943 (1976).
6916. van den Hul, H. J., and Vanderhoff, J. W., *Brit. Polym. J.*, **2**, 121 (1970).
6917. Vanderhoff, J. W., *Brit. Polym. J.*, **2**, 161 (1971).
6918. Vanderschueren, J., *J. Polym. Sci. A2*, **10**, 543 (1972).
6919. Vanderschueren, J., *J. Polym. Sci. A2*, **15**, 873 (1977).
6920. Van Dijk, J. A. P. P., Henskens, W. C. M., and Smit, J. A. M., *J. Polym. Sci. A2*, **14**, 1485 (1976).
6921. Van Emmerik, P. T., and Smolders, C. A., *J. Polym. Sci. C*, **38**, 73 (1972).
6922. Van Kreveld, M. E., *J. Polym. Sci. A2*, **13**, 2253 (1975).
6923. Van Leemput, R., and Stein, R. S., *J. Polym. Sci. A2*, 4039 (1964).
6924. van Ligten, R. F., *J. Opt. Soc. Amer.*, **60**, 709 (1970).
6925. van Ligten, R. F., and Osterberg, H., *Nature*, **211**, 282 (1966).
6926. Van Lingen, R. L. M., *Z. Anal. Chem.*, **247**, 232 (1969).
6927. Van Schooten, J., and Evenhuis, K., *Polymer*, **6**, 561 (1965).
6928. Van Schooten, J., van Hoorn, and Boerma, J., *Polymer*, **2**, 161 (1961).
6929. Varoqui, R., Jacob, M., Freund, L., and Daune, M., *J. Chim. Phys.*, **59**, 161 (1962).
6930. Vasko, P. D., and Koenig, J. L., *Macromolecules*, **3**, 597 (1970).
6931. Vassallo, O. A., *Anal. Chem.*, **33**, 1823 (1961).
6932. Vaughan, M. F., *J. Polym. Sci.*, **33**, 417 (1958).
6933. Vaughan, M. F., *Chem. Ind. (London)*, **1958**, 555 (1958).
6934. Vaughan, M. F., and Francais, M. A., *J. Appl. Polym. Sci.*, **21**, 2409 (1977).
6935. Veksli, Z., and Miller, W. G., *Macromolecules*, **8**, 248 (1975).
6936. Veksli, Z., and Miller, W. G., *Macromolecules*, **10**, 686 (1977).

832

6937. Veksli, Z., and Miller, W. G., *Macromolecules*, **10**, 1245 (1977).
6938. Venyaminov, S. Yu., and Chirgadze, Yu. N., *Macromolecules*, **6**, 515 (1973).
6939. Verdier, J. C., and Guyot, A., *Makromol. Chem.*, **175**, 1543 (1974).
6940. Verdier, P. H., *Macromolecules*, **10**, 913 (1977).
6941. Verhoff, F. H., and Sylvester, N. D., *J. Polym. Sci. A1*, **6**, 243 (1968).
6942. Verma, G. P. S., and Peterlin, A., *J. Polym. Sci. B*, **7**, 587 (1969).
6943. Vettegren, V. I., and Noval, I. I., *J. Polym. Sci. A2*, **11**, 2135 (1977).
6944. Vink, H., *Arkiv Kemi*, **15**, 149 (1960).
6945. Vink, H., *Arkiv Kemi*, **19**, 15 (1962).
6946. Vink, H., *J. Polym. Sci. A2*, **4**, 830 (1966).
6947. Vink, H., *Europ. Polym. J.*, **7**, 1411 (1971).
6948. Vink, H., *Europ. Polym. J.*, **10**, 149 (1974).
6949. Vink, H., and Wikström, R., *Svensk Papperstid.*, **66**, 55 (1963).
6950. Vinogradov, G. V., Isayev, A. I., Mustafaev, D. A., and Podolsky, Y. Y., *J. Appl. Polym. Sci.*, **22**, 665 (1978).
6951. Vinogradov, G. V., Yarlykov, B. V., Tsebrenko, M. V., Yudin, A. V., and Ablazova, T. I., *Polymer*, **16**, 609 (1975).
6952. Visconti, S., Hien, N. V., Borch, J., and Marchessault, R. H., *J. Polym. Sci. A2*, **14**, 631 (1976).
6953. Visconti, S., and Marchessault, R. H., *Macromolecules*, **7**, 913 (1974).
6954. Visconti, S., Marchessault, R. H., *J. Polym. Sci. B*, **13**, 203 (1975).
6955. Vitale, G. G., and LeGrand, D. G., *Macromolecules*, **9**, 749 (1976).
6956. Vivilecchia, R. V., Cotter, R. L., Limpert, R. J., Thimot, N. Z., and Little, J. N., *J. Chromatogr.*, **99**, 407 (1974).
6957. Vladimiroff, T., *J. Chromatogr.*, **55**, 175 (1971).
6958. Vofsi, D., and Katchalsky, A., *J. Polym. Sci.*, **26**, 127 (1957).
6959. Voigt, J., *Kunststoffe*, **54**, 2 (1964).
6960. Vold, M. J., *Anal. Chem.*, **21**, 683 (1949).
6961. Volkenshtein, M. V., and Sharanov, Yu., A., *Vysokomol. Soedin.*, **3**, 1738 (1961).
6962. Volman, D. H., and Seed, J. R., *J. Amer. Chem. Soc.*, **86**, 5095 (1964).
6963. Vond, V., Vedam, K., and Stein, R., *J. Appl. Phys.*, **37**, 2551 (1966).
6964. von Falkai, B., *Makromol. Chem.*, **41**, 86 (1960).
6965. Vonk, C. G., *J. Polym. Sci. C.*, **38**, 429 (1972).
6966. Von Meerwall, E., Creel, R. B., Griffin, C. F., DiCato, E., Lin, F. T., and Lin, F. M., *J. Appl. Polym. Sci.*, **21**, 1489 (1977).
6967. von Zahn, U., *Rev. Sci. Instrum.*, **34**, 1 (1963).
6968. Voskotter, G., and Kosfeld, R., *Kolloid Z.*, **216**, 85 (1967).
6969. Vošnický, V., and Bohdanecký, M., *J. Polym. Sci. A2*, **15**, 757 (1977).
6970. Vrentas, J. S., *J. Polym. Sci. A2*, **15**, 441 (1977).
6971. Vrentas, J. S., *J. Appl. Polym. Sci.*, **21**, 1715 (1977).
6972. Vrentas, J. S., *J. Appl. Polym. Sci.*, **22**, 2325 (1978).
6973. Vrentas, J. S., and Duda, J. L., *Macromolecules*, **9**, 785 (1976).
6974. Vrentas, J. S., and Duda, J. L., *J. Polym. Sci. A2*, **15**, 403 (1977).
6975. Vrentas, J. S., and Duda, J. L., *J. Polym. Sci. A2*, **15**, 417 (1977).
6976. Vrentas, J. S., Duda, J. L., and Ni, Y. C., *J. Polym. Sci. A2*, **15**, 2039 (1977).
6977. Vrentas, J. S., Jarzebski, C. M., and Duda, J. L., *AIChE J.*, **21**, 894 (1975).
6978. Vrij, A., *J. Polym. Sci. A2*, **6**, 1919 (1968).
6979. Vrij, A., *J. Polym. Sci. A2*, **7**, 1627 (1969).
6980. Vrijbergen, R. R., Soeteman, A. A., and Smit, J. A. M., *J. Appl. Polym. Sci.*, **22**, 1267 (1978).
6981. Wada, A., Ford, N. C., and Larasz, F. E., *ACS Polym. Preprints*, **13**, 191 (1972).
6982. Wada, Y., and Yamamoto, K., *J. Phys. Soc. Japan*, **11**, 887 (1956).
6983. Wachter, A. H., and Simon, W., *Anal. Chem.*, **41**, 90 (1969).
6984. Wagner, H. L., and Hoeve, C. A. J., *J. Polym. Sci.*, **9**, 1763 (1971).
6985. Wagner, H. L., and Hoeve, C. A. J., *J. Polym. Sci. A1*, **11**, 1189 (1973).
6986. Wagner, H. L., and Hoeve, C. A. J., *J. Polym. Sci. A2*, **9**, 1763 (1971).
6987. Wagner, H. L., and Hoeve, C. A. J., *J. Polym. Sci. A2*, **11**, 1189 (1973).
6988. Wakelin, J. H., Virgin, H. S., and Crystal, E., *J. Appl. Phys.*, **30**, 1654 (1959).
6989. Wakelyn, N. T., and Young, P. R., *J. Appl. Polym. Sci.*, **10**, 1421 (1966).
6990. Wales, M., and Rehfeld, S. J., *J. Polym. Sci.*, **62**, 179 (1962).

6991. Wales, M., and Van Holde, K. E., *J. Polym. Sci.*, **14**, 81 (1954).
6992. Wales, M., and Rehfeld, S. J., *J. Polym. Sci.*, **62**, 179 (1962).
6993. Wall, F. T., and Miller, D. G., *J. Polym. Sci.*, **13**, 157 (1954).
6994. Wall, F. T., and White, R. A., *Macromolecules*, **7**, 849 (1974).
6995. Wallace, J. R., Kozak, P. J., and Noel, F., *SPE J.*, **26**, 43 (1970).
6996. Wallace, T. P., and Kratochvil, J. P., *J. Polym. Sci. C*, **25**, 89 (1968).
6997. Wallach, M. L., *ACS Polym. Preprints*, **6**, 53 (1965).
6998. Wallach, M. L., and Benoit, H., *J. Polym. Sci.*, **57**, 41 (1962).
6999. Waltz, J. E., and Taylor, G. B., *Anal. Chem.*, **19**, 448 (1947).
7000. Wang, F. W., and Zimm, B. H., *J. Polym. Sci. A2*, **12**, 1619, 1639 (1974).
7001. Wang, T. T., and Kwei, T. K., *Macromolecules*, **6**, 919 (1973).
7002. Wang, Y. C., and Morawetz, H., *Makromol. Chem.*, Suppl. 1, 283 (1975).
7003. Warburg, O., and Schocken, V., *Arch. Biochem.*, **21**, 363 (1949).
7004. Ward, I. M., *Nature*, **180**, 141 (1957).
7005. Ward, I. M., *Text. Res. J.*, **34**, 806 (1964).
7006. Ward, I. M., *Trans. Faraday Soc.*, **53**, 1406 (1957).
7007. Ward, I. M., and Wilding, M. A., *Polymer*, **18**, 327 (1977).
7008. Ward, I. M., and Williams, T., *J. Polym. Sci. A2*, **7**, 1585 (1969).
7009. Ward, T. C., and Books, J. T., *Macromolecules*, **7**, 207 (1974).
7010. Wardel, G. E., Douglass, D. C., and McBrierty, V. J., *Polymer*, **17**, 41 (1976).
7011. Ware, W. R., *J. Amer. Chem. Soc.*, **83**, 4374 (1961).
7012. Warfield, R. W., *SPE J.*, **15**, 625 (1959).
7013. Warfield, R. W., and Petree, M. C., *J. Polym. Sci.*, **55**, 497 (1961).
7014. Warfield, R. W., and Petree, M. C., *J. Polym. Sci. A*, **1**, 1701 (1963).
7015. Warfield, R. W., and Petree, M. C., *J. Polym. Sci. A1*, **4**, 532 (1966).
7016. Warfield, R. W., Petree, M. C., and Donovan, P., *SPE J.*, **15**, 1055 (1959).
7017. Warwicker, J. O., *J. Appl. Polym. Sci.*, **22**, 187 (1978).
7018. Wasai, G., Saegusa, T., and Furukawa, J., *Makromol Chem.*, **86**, 1 (1965).
7019. Wasiak, A., Peiffer, D., and Stein, R. S., *J. Polym. Sci. B*, **14**, 381 (1976).
7020. Waterman, B., and Jarzynski, J., *J. Acoust. Soc. Amer.*, **36**, 1485 (1964).
7021. Waterman, D. C., and Dole, M., *J. Phys. Chem.*, **74**, 1913 (1970).
7022. Waters, J. L., *ACS Polym. Preprints*, **6**, 1061 (1965).
7023. Waters, J. L., *J. Polym. Sci. A2*, **8**, 411 (1970).
7024. Watillon, A., and Dauchot, J., *J. Colloid Interface Sci.*, **27**, 507 (1968).
7025. Watson, E. S., O'Neill, M. J., Justin, J., and Brenner, N., *Anal. Chem.*, **36**, 1233 (1964).
7026. Weakley, T. J. R., Williams, R. J. P., and Wilson, J. D., *J. Chem. Soc.*, **1960**, 3963.
7027. Weeks, N. E., Mori, S., and Porter, R. S., *J. Polym. Sci. A2*, **13**, 2031 (1975).
7028. Weeks, N. E., and Porter, R. S., *J. Polym. Sci. A2*, **13**, 2049 (1975).
7029. Wegmann, D., Tomlinson, C., and Simon, W., *Microchem. J.*, **2**, 1069 (1962).
7030. Wegner, G., Fischer, E. W., and Munoz-Escalona, A., *Makromol. Chem.*, Suppl. 1, 521 (1975).
7031. Weill, G., and Herman, G., *J. Polym. Sci. A2*, **5**, 1293 (1967).
7032. Weiner, J. H., and Pear, M. R., *Macromolecules*, **10**, 317 (1977).
7033. Weir, N. A., *J. Polym. Sci. A1*, **16**, 13 (1978).
7034. Weiss, A. R., and Cohn-Ginsberg., *J. Polym. Sci. B*, **7**, 379 (1969).
7035. Weiss, A. R., and Cohn-Gimsberg, E., *J. Polym. Sci. A2*, **8**, 148 (1970).
7036. Weiss, G. H., and Yphantis, D. A., *J. Chem. Phys.*, **42**, 2117 (1965).
7037. Weisz, P. B., *Trans. Faraday Soc.*, **63**, 1801 (1967).
7038. Weisz, P. B., and Hicks, J. S., *Trans. Faraday Soc.*, **63**, 1807 (1967).
7039. Weisz, P. B., and Zollinger, H., *Trans. Faraday Soc.*, **63**, 1815 (1967).
7040. Weisz, P. B., and Zollinger, H., *Trans. Faraday Soc.*, **64**, 1693 (1968).
7041. Weitz, A., and Wunderlich, B., *J. Polym. Sci. A2*, **12**, 2473 (1974).
7042. Welch, G. J., *Polymer*, **15**, 429 (1974).
7043. Welch, G. J., *Polymer*, **16**, 69 (1975).
7044. Welling, W., *Sci. Tools*, **15**, 24 (1968).
7045. Wellinghoff, S. T., and Baer, E., *J. Appl. Polym. Sci.*, **22**, 2025 (1978).
7046. Wellinghoff, S. T., Koenig, J. L., and Baer, E., *J. Polym. Sci. A2*, **15**, 1913 (1977).
7047. Wellinghoff, S. T., Rybnikar, F., and Baer, E., *J. Macromol. Sci. B*, **10**, 1 (1974).
7048. Wendlandt, W. W., *Anal. Chim. Acta*, **27**, 309 (1962).

7049. Wendlandt, W. W., *Anal. Chim. Acta*, **49**, 187 (1970).
7050. Wendlandt, W. W., *Thermochim. Acta*, **12**, 109 (1975).
7051. Wenig, W., *J. Polym. Sci. A2*, **16**, 1635 (1978).
7052. Werner, F. P., McKnight, W. J., and Stein, R. S., *J. Polym. Sci. A2*, **15**, 2113 (1977).
7053. Wesslau, H., *Makromol. Chem.*, **20**, 111 (1956).
7054. Wesslau, H., *Makromol. Chem.*, **26**, 96, 102 (1958).
7055. Wesslén, B., Lenz, R. W., and Bovey, F. A., *Macromolecules*, **4**, 709 (1971).
7056. Wesslén, B., Lenz, R. W. MacKnight, W. J., and Karasz, F. E., *Macromolecules*, **4**, 24 (1971).
7057. Westerman, L., and Clark, J. C., *J. Polym. Sci. A2*, **11**, 559 (1973).
7058. Westlake, J. F., and Johnson, M., *J. Appl. Polym. Sci.*, **19**, 319 (1975).
7059. Wettermark, G., *Arkiv Kemi*, **18**, 1 (1961).
7060. Weyland, H. G., Hoftyzer, P. J., and Krevelen, D. W., *Polymer*, **11**, 79 (1970).
7061. Whipple, E. B., and Green, P. J., *Macromoleclues*, **6**, 38 (1973).
7062. White, J. L., Salladay, D. G., Quisenberry, D. Q., and Maclean, D. L., *J. Appl. Polym. Sci.*, **16**, 2811 (1972).
7063. White, J. R., *J. Polym. Sci. A2*, **11**, 2173 (1973).
7064. White, J. R., *J. Polym. Sci. A2*, **12**, 2375 (1974).
7065. White, J. R., *Polymer*, **16**, 157 (1975).
7066. White, J. U., Alport, N. L., Ward, W. M., and Gallawey, W. S., *Anal. Chem.*, **31**, 1267 (1957).
7067. White, J. V., and Liston, M. O., *J. Opt. Soc. Amer.*, **40**, 29 (1950).
7068. Whitehouse, B. A., *Macromolecules*, **4**, 463 (1971).
7069. Whitehouse, R. S., Counsell, P. J. C., and Lewis, G., *Polymer*, **17**, 699 (1976).
7070. Wichterle, K., *J. Polym. Sci. B*, **13**, 613 (1975).
7071. Wiedmann, H. G., and van Tets, A., *Z. Anal. Chem.*, **233**, 161 (1968).
7072. Wiedmann, H. G., and van Tets, A., *Thermochim. Acta*, **1**, 159 (1970).
7073. Wieland, T., and Determan, H., *Experientia*, **18**, 421 (1962).
7074. Wiff, D. R., and Gehatia, M., *J. Macromol. Sci. B*, **6**, 287 (1972).
7075. Wiff, D. R., and Gehatia, M., *J. Polym. Sci. C*, **43**, 219 (1973).
7076. Wiff, D. R., Gehatia, M., and Wereta, A., *J. Polym. Sci. A2*, **13**, 275 (1975).
7077. Wignall, G. D., Ballard, D. G. H., and Schelten, J., *Europ. Polym. J.*, **10**, 861 (1974).
7078. Wignall, G. D., Ballard, D. G. H., and Schelten, J., *J. Appl. Phys.*, **12**, 75 (1976).
7079. Wikjord, A. G., and Manley, R. S. J., *J. Macromol. Sci. B*, **2**, 501 (1968).
7080. Wild, L., and Guiliana, R., *J. Polym. Sci. A2*, **5**, 1087 (1967).
7081. Wild, L., Ranganath, R., and Barlow, A., *J. Appl. Polym. Sci.*, **21**, 3331 (1977).
7082. Wild, L., Ranganath, R., and Ryle, T., *ACS Polym. Preprints*, **12**, 266 (1971).
7083. Wild, L., Ranganath, R., and Ryle, T., *J. Polym. Sci. A2*, **9**, 2137 (1971).
7084. Wiley, R. H., *J. Polym. Sci.*, **2**, 10 (1947).
7085. Wiley, R. H., *Trans. N.Y. Acad. Sci.*, **32**, 688 (1970).
7086. Wiley, R. H., *Macromolecules*, **4**, 254 (1971).
7087. Wilhoit, R. C., and Dole, M., *J. Phys. Chem.*, **57**, 14 (1953).
7088. Wilke, K. C. B., Jones, J. K. N., Excell, B. J., and Semple, R. E., *Can. J. Chem.*, **35**, 795 (1957).
7089. Wilkes, C. E., *Macromolecules*, **4**, 443 (1971).
7090. Wilkes, C. E., Carman, C. J., and Harrington, R. A., *J. Polym. Sci. C*, **43**, 237 (1973).
7091. Wilkes, C. E., Folt, V. L., and Krimm, S., *Macromolecules*, **6**, 235 (1973).
7092. Wilkes, G. L., *J. Polym. Sci. A2*, **9**, 1531 (1971).
7093. Wilkes, G. L., and Stein, R. S., *J. Polym. Sci. A2*, **7**, 1525 (1969).
7094. Wilkes, G. L., Uemura, Y., and Stein, R. S., *J. Polym. Sci. A2*, **9**, 2151 (1971).
7095. Wilkinson, R. W., and Dole, M., *J. Polym. Sci.*, **58**, 1089 (1962).
7096. Willeboordse, F., and Meeker, R. L., *Anal. Chem.*, **38**, 854 (1966).
7097. Willems, J., *Discuss. Faraday Soc.*, **25**, 111 (1957).
7098. Willems, J., *Experientia*, **17**, 344 (1961).
7099. Willems, J., *Naturwiss.*, **5c**, 92 (1963).
7100. Willems, J., and Willems, I., *Experientia*, **13**, 465 (1957).
7101. Willenberg, B., and Sillescu, H., *Makromol. Chem.*, **178**, 2401 (1977).
7102. Williams, A. D., and Flory, P. J., *J. Amer. Chem. Soc.*, **91**, 3111 (1969).
7103. Williams, D. J., *Macromolecules*, **3**, 602 (1970).
7104. Williams, G., *Trans. Faraday Soc.*, **62**, 1321 (1966).

835

7105. Williams, G., and Watts, D. C., *Trans. Faraday Soc.*, **67**, 1971 (1973).
7106. Williams, J. L., and Peterlin, A., *J. Polym. Sci. A2*, **9**, 1483 (1971).
7107. Williams, J. W., *J. Polym. Sci.*, **12**, 351 (1954).
7108. Williams, J. W., Baldwin, R. L., Saunders, W. M., and Squire, P. G., *J. Amer. Chem. Soc.*, **74**, 1542 (1952).
7109. Williams, J. W., and Saunders, W. M., *J. Phys. Chem.*, **58**, 854 (1954).
7110. Williams, J. W., Saunders, W. M., and Cicirelli, J. S., *J. Phys. Chem.*, **58**, 774 (1954).
7111. Williams, M. L., Landel, R. F., and Ferry, J. D., *J. Amer. Chem. Soc.*, **77**, 3701 (1955).
7112. Williams, T., Blundel, D. J., Keller, A., and Ward, I. M., *J. Polym. Sci. A2*, **6**, 1613 (1968).
7113. Williamson, G. R., and Cervenka, A., *Europ. Polym. J.*, **10**, 295 (1974).
7114. Willis, H. A., Cudby, M. E. A., Chantry, G. W., Nicol, E. A., and Fleming, J. W., *Polymer*, **16**, 74 (1975).
7115. Willis, H. A., and Miller, R. G. J., *Spectr. Acta*, **14**, 119 (1959).
7116. Willmott, F. W., *J. Chromatogr. Sci.*, **7**, 101 (1969).
7117. Wilska, S., and Oy, V., *J. Paint. Technol.*, **43**, 65 (1971).
7118. Wilske, J., and Heusinger, H., *J. Polym. Sci. A1*, **7**, 955 (1969).
7119. Wilski, H., *Kunststoffe*, **54**, 10 (1964).
7120. Wilski, H., *Angew. Makromol. Chem.*, **6**, 101 (1969).
7121. Wilski, H., *Kolloid Z. Z. Polym.*, **238**, 426 (1970).
7122. Wilski, H., *Kolloid Z. Z. Polym.*, **248**, 867 (1971).
7123. Wilski, H., and Grewer, T., *J. Polym. Sci. C*, **6**, 33 (1964).
7124. Wilson, A. J. C., *Rev. Sci. Instrum.*, **20**, 831 (1949).
7125. Wilson, C. W., *J. Polym. Sci. A*, **1**, 1305 (1963).
7126. Wilson, C. W., and Pake, G. E., *J. Polym. Sci.*, **10**, 503 (1953).
7127. Wilson, C. W., and Pake, G. E., *J. Chem. Phys.*, **27**, 115 (1957).
7128. Wilson, C. W., and Santee, E. R., Jr, *J. Polym. Sci. C*, **8**, 97 (1965).
7129. Wilson, F. C., and Starkweather, H. W., Jr, *J. Polym. Sci. A2*, **11**, 919 (1973).
7130. Wilson, P. S., and Simha, R., *Macromolecules*, **5**, 903 (1973).
7131. Wilson, P. S., Lee, S., and Boyer, R. F., *Macromolecules*, **6**, 914 (1973).
7132. Wilson, T., *J. Amer. Chem. Soc.*, **88**, 2898 (1966).
7133. Winslow, F. H., Hellman, M. Y., Matreyek, W., and Salovey, R., *J. Polym. Sci. B*, **5**, 89 (1967).
7134. Winston, A., and Wichachewa, P., *Macromolecules*, **6**, 200 (1973).
7135. Wintle, H. J., *Polymer*, **15**, 425 (1974).
7136. Wippler, C., and Scheibling, G., *J. Chim. Phys.*, **51**, 201 (1954).
7137. Witholt, B., and Brand, L., *Rev. Sci. Instrum.*, **39**, 1271 (1968).
7138. Witschonke, G. R., *Anal. Chem.*, **24**, 350 (1952).
7139. Wittman, J. C., and Manley, R. S. J., *J. Polym. Sci. A2*, **15**, 1089 (1977).
7140. Wladimirow, W., *Photochem. Photobiol.*, **5**, 243 (1966).
7141. Wlochowicz, A., and Jeziorny, A., *J. Polym. Sci. A2*, **10**, 1407 (1972).
7142. Wolf, B. A., *Ber. Bunsenges. Phys. Chem.*, **75**, 924 (1971).
7143. Wolf, B. A., *J. Polym. Sci. A2*, 847 (1972).
7144. Wolf, B. A., *Makromol. Chem.*, **178**, 1869 (1977).
7145. Wolf, B. A., Bieringer, H. F., and Breitenbach, J. W., *Polymer*, **17**, 605 (1976).
7146. Wolf, B. A., and Blaum, G., *J. Polym. Sci.*, *A2*, **13**, 1115 (1975).
7147. Wolf, B. A., and Blaum, G., *J. Polym. Sci. C*, **61**, 251 (1977).
7148. Wolf, B. A., Breitenbach, J. W., and Rigler, J. K., *Angew Makromol. Chem.*, **34**, 177 (1973).
7149. Wolf, B. A., Breitenbach, J. W., and Sentfl, H., *J. Polym. Sci. C*, **31**, 345 (1970).
7150. Wolf, B. A., and Glaum, G., *Makromol. Chem.*, **177**, 1073 (1976).
7151. Wolf, B. A., and Jend, R., *Makromol. Chem.*, **178**, 1811 (1977).
7152. Wolf, B. A., and Molinari, R. J., *Makromol. Chem.*, **173**, 241 (1973).
7153. Wolf, S., and Mobus, B., *Anal. Chem.*, **186**, 194 (1962).
7154. Wolfram, L. E., Grasselli, J. G., and Koenig, J. L., *Appl. Spectra.*, **24**, 263 (1970).
7155. Wolfram, L. E., Grasselli, J. G., and Koenig, J. L., *ACS Polym. Preprints*, **14**, 640 (1973).
7156. Wolfsgruber, C., Zannoni, G., Rigamonti, E., and Zembelli, A., *Makromol. Chem.*, **176**, 2765 (1975).
7157. Wolinski, L. M., *Makromol. Chem.*, **176**, 2079 (1975).
7158. Wolpert, S. M., Weitz, A., and Wunderlich, B., *J. Polym. Sci. A2*, **9**, 1887 (1971).

836

7159. Wong, C. P., and Schrag, J. L., *Macromolecules*, **3**, 468 (1970).
7160. Wong, C. P., Schrag, J. L., and Ferry, D., *J. Polym. Sci. A2*, **8**, 991 (1970).
7161. Wong, K. C., Chen, F. C., and Choy, C. L., *Polymer*, **16**, 649 (1975).
7162. Wong, K. C., Chen, F. C., and Choy, C. L., *Polymer*, **16**, 858 (1975).
7163. Wong, P. K., *Polymer*, **19**, 785 (1978).
7164. Wood, J. A., and Bekkedahl, N., *J. Polym. Sci. B*, **5**, 169 (1967).
7165. Wood, L. A., *J. Polym. Sci.*, **28**, 319 (1958).
7166. Wood, L. J., and Philipps, G., *Nature*, **174**, 801 (1954).
7167. Woodbrey, J. C., *J. Polym. Sci. B*, **2**, 315 (1964).
7168. Woobrey, J. C., and Trementozzi, Q. A., *J. Polym. Sci. C*, **8**, 113 (1965).
7169. Wool, R. P., *J. Polym. Sci. A2*, **13**, 1795 (1975).
7170. Wool, R. P., *J. Polym. Sci. A2*, **14**, 1921 (1976).
7171. Wool, R. P., and Statton, W. O., *J. Polym. Sci. A2*, **12**, 1575 (1974).
7172. Wrasidlo, W., *Macromolecules*, **4**, 642 (1971).
7173. Wrasidlo, W., *J. Polym. Sci. A2*, **11**, 2143 (1973).
7174. Wrasidlo, W., and Stille, J. K., *Macromolecules*, **9**, 505 (1976).
7175. Wristers, J., *J. Polym. Sci. A2*, **11**, 1619 (1973).
7176. Wu, C. K., *J. Polym. Sci. A2*, **12**, 2493 (1974).
7177. Wu, S., *J. Polym. Sci. C*, **34**, 19 (1971).
7178. Wu, T. K., and Overnall, D. W., *Macromolecules*, **6**, 582 (1973).
7179. Wu, T. K., and Overnall, D. W., *Macromolecules*, **7**, 776 (1974).
7180. Wu, T. K., and Sheer, M. L., *Macromolecules*, **10**, 529 (1977).
7181. Wun, K. L., *Macromolecules*, **8**, 190 (1975).
7182. Wun, K. L., and Prins, W., *J. Polym. Sci. A2*, **12**, 533 (1974).
7183. Wunderlich, B., *J. Polym. Sci. A*, **1**, 1245 (1963).
7184. Wunderlich, B., *Polymer*, **5**, 611 (1964).
7185. Wunderlich, B., *J. Phys. Chem.*, **69**, 2078 (1965).
7186. Wunderlich, B., and Arakawa, T., *J. Polym. Sci. A*, **2**, 3697 (1964).
7187. Wunderlich, B., and Baur, H., *Adv. Polym. Sci.*, **7**, 151 (1970).
7188. Wunderlich, B., Bodily, D. M., and Kaplan, M. H., *J. Appl. Phys.*, **35**, 95 (1964).
7189. Wunderlich, B., and Bopp, R., *Makromol. Chem.*, **147**, 79 (1971).
7190. Wunderlich, B., and Cormier, C. M., *J. Phys. Chem.*, **70**, 1844 (1966).
7191. Wunderlich, B., and Cormier, C. M., *J. Polym. Sci. A2*, **5**, 987 (1967).
7192. Wunderlich, B., and Czornyj, G., *Macromolecules*, **10**, 906 (1977).
7193. Wunderlich, B., and Davidson, T., *J. Polym. Sci. A2*, **7**, 2043 (1969).
7194. Wunderlich, B., and Dole, M., *J. Polym. Sci.*, **24**, 201 (1957).
7195. Wunderlich, B., and Dole, M., *J. Polym. Sci.*, **32**, 125 (1958).
7196. Wunderlich, B., James, E. A., and Shu, S. W., *J. Polym. Sci. A*, **2**, 2759 (1964).
7197. Wunderlich, B., and Mehta, A., *J. Mater. Sci.*, **5**, 248 (1970).
7198. Wunderlich, B., and Mehta, A., *J. Polym. Sci. A2*, **12**, 255 (1974).
7199. Wunderlich, B., and Melillo, L., *Science*, **154**, 1329 (1966).
7200. Wunderlich, B., and Melillo, L., *Makromol. Chem.*, **118**, 250 (1968).
7201. Wunderlich, B., Melillo, L., Cormier, C. M., Davidson, T., and Synder, G., *J. Macromol. Sci. B*, **3**, 485 (1967).
7202. Wunderlich, B., and Sullivan, P., *J. Polym. Sci.*, **56**, 19 (1962).
7203. Wunderlich, B., and Sullivan, P., *J. Polym. Sci.*, **61**, 195 (1965).
7204. Wyard, S. J., *J. Sci. Instrum.*, **42**, 769 (1965).

7205. Yajnik, M., Witeczek, J., and Heller, W., *J. Polym. Sci. C*, **25**, 99 (1968).
7206. Yakubchik, A. I., *Rubber Chem. Technol.*, **32**, 284 (1959).
7207. Yamada, A., and Yanagita, M., *J. Polym. Sci. B*, **9**, 103 (1971).
7208. Yamada, M., Miyasaka, K., and Ishikawa, K., *J. Polym. Sci. A2*, **11**, 2393 (1973).
7209. Yamada, R., Hayashi, C., Onogi, S., and Horio, M., *J. Polym. Sci. C*, **5**, 123 (1964).
7210. Yamada, R., and Stein, R. S., *J. Polym. Sci. B*, **2**, 1131 (1964).
7211. Yamadera, R., *J. Polym. Sci.*, **50**, S4 (1961).
7212. Yamaguchi, K., *Makromol. Chem.*, **128**, 19 (1969).
7213. Yamaguchi, K., *Makromol. Chem.*, **132**, 143 (1970).
7214. Yamaguchi, K., Kojima, H., and Takahashi, A., *Intern. Chem. Eng.*, **5**, 169 (1965).
7215. Yamaguchi, K., and Saeda, S., *J. Polym. Sci. A2*, **7**, 1303 (1969).
7216. Yamakawa, H., *J. Chem. Phys.*, **34**, 1360 (1961).

837

7217. Yamakawa, H., *Macromolecules*, **10**, 692 (1977).
7218. Yamakawa, H., Aoki, A., and Tanaka, G., *J. Chem. Phys.*, **45**, 1938 (1966).
7219. Yamakawa, H., and Fuji, M., *Macromolecules*, **7**, 128 (1974).
7220. Yamakawa, H., and Fuji, M., *Macromolecules*, **7**, 649 (1974).
7221. Yamamoto, M., and White, J. L., *Macromolecules*, **5**, 58 (1972).
7222. Yamamoto, S., Nozawa, T., and Hatano, M., *Polymer*, **15**, 330 (1974).
7223. Yamamoto, Y., Tsuge, S., and Takeuchi, T., *Bull. Chem. Soc. Japan*, **44**, 1145 (1971).
7224. Yamamoto, Y., Tsuge, S., and Takeuchi, T., *Macromolecules*, **5**, 325 (1972).
7225. Yamaoka, K., and Charney, E., *Macromolecules*, **6**, 66 (1973).
7226. Yamashita, Y., Ito, K., Ishi, H., Hoshino, S., and Kai, M., *Macromolecules*, **1**, 529 (1969).
7227. Yamaura, K., Matsuzawa, S., and Go, Y., *Kolloid Z. Z. Polym.*, **240**, 820 (1970).
7228. Yamaura, K., Hoe, Y., Matsuzawa, S., and Go, Y., *Kolloid Z. Z. Polym.*, **243**, 7 (1971).
7229. Yamaura, K., Kinugasa, S., and Matsuzawa, S., *Kolloid Z. Z. Polym.*, **248**, 893 (1971).
7230. Yang, F. J. F., Myers, M. N., and Giddings, J. C., *Anal. Chem.*, **46**, 1924 (1974)
7231. Yang, F. J. F., Myers, M. N., and Giddings, J. C., *J. Colloid Interface Sci.*, **60**, 754 (1977).
7232. Yang, H. W. H., and Chien, J. C. W., *Macromolecules*, **11**, 759 (1978).
7233. Yanko, J. A., *J. Polym. Sci.*, **19**, 437 (1956).
7234. Yano, O., Kamoshida, T., Sekiyama, S., and Wada, Y., *J. Polym. Sci. A2*, **16**, 679 (1978).
7235. Yano, O., Saiki, K., Tarucha, S., and Wada, Y., *J. Polym. Sci. A2*, **15**, 43 (1977).
7236. Yano, O., and Wada, Y., *J. Polym. Sci. A2*, **12**, 665 (1974).
7237. Yano, S., *J. Polym. Sci. A2*, **8**, 1057 (1970).
7238. Yano, S., Rahalkar, R. R., Hunter, S. P., Wang, C. H., and Boyd, R. H., *J. Polym. Sci. A2*, **14**, 1877 (1976).
7239. Yano, S., Tadano, K., Aoki, K., and Koizumi, N., *J. Polym. Sci. A2*, **12**, 1875 (1974).
7240. Yanri, S. S., Bovey, F. A., and Lumry, R., *Nature*, **200**, 242 (1963).
7241. Yarborough, V. A., *Anal. Chem.*, **25**, 1914 (1953).
7242. Yariv, A., and Gordon, J. P., *Rev. Sci. Instrum.*, **32**, 462 (1961).
7243. Yasuda, H., *J. Appl. Polym. Sci.*, **19**, 2529 (1975).
7244. Yasuda, H., and Hirotsu, T., *J. Appl. Polym. Sci.*, **21**, 105 (1977).
7245. Yasuda, H., and Peterlin, A., *J. Appl. Polym. Sci.*, **17**, 433 (1973).
7246. Yasuda, H., and Peterlin, A., *J. Appl. Polym. Sci.*, **18**, 531 (1974).
7247. Yasuda, H., and Takayanagi, M., *Progr. Polym. Sci. Japan*, **7**, 245 (1964).
7248. Yasuda, H., and Tsai, J. T., *J. Appl. Polym. Sci.*, **18**, 805 (1974).
7249. Yasuda, M., Marsh, H. C., Brandt, S., and Reilley, C. N., *J. Polym. Sci. A1*, **15**, 991 (1977).
7250. Yasuda, S. K., *J. Chromatogr.*, **27**, 72 (1967).
7251. Yasuniwa, M., and Takemura, T., *Polymer*, **15**, 661 (1974).
7252. Yathindra, N., and Rao, V. S. R., *J. Polym. Sci. A2*, **8**, 2033 (1970).
7253. Yathindra, N., and Rao, V. S. R., *Biopolymers*, **9**, 783 (1970).
7254. Yau, W. W., *J. Polym. Sci. A2*, **7**, 483 (1969).
7255. Yau, W. W., Malone, C. P., and Fleming, S. W., *J. Polym. Sci. B*, **6**, 803 (1968).
7256. Yau, W. W., Malone, C. P., and Suchan, H. L., *ACS Div. Petrol. Chem. Preprints*, **15**, A63 (1970).
7257. Yau, W. W., and Stein, R. S., *J. Polym. Sci. A2*, **6**, 1 (1968).
7258. Yau, W. W., Stoklosa, H. J., and Bly, D. D., *J. Appl. Polym. Sci.*, **21**, 1911 (1977).
7259. Yau, W. W., Suchan, H. L., and Malone, C. P., *J. Polym. Sci. A2*, **6**, 1349 (1968).
7260. Yeh, G. S. Y., *Poly. Eng. Sci.*, **16**, 138, 145 (1976).
7261. Yeh, G. S. Y., Hoseman, R., Loboda-Čačković, J., and Čačković, H., *Polymer*, **17**, 309 (1976).
7262. Yen, L. Y., and Eichinger, B. E., *J. Polym. Sci. A2*, **16**, 117, 121 (1978).
7263. Yeh, Y., and Cummins, H. Z., *Appl. Phys. Lett.*, **4**, 176 (1964).
7264. Yin, T. P., *J. Phys. Chem.*, **73**, 2413 (1969).
7265. Yoka, K., Sakai, Y., and Ishii, A., *J. Polym. Sci., B*, **3**, 839 (1965).
7266. Yokota, K., Abe, A., Hosaka, S., Sakai, H., and Saito, H., *Macromolecules*, **11**, 95 (1978).
7267. Yokouchi, M., Sakakibara, Y., Chatani, Y., Tadekoro, H., Tanaka, T., and Yoda, K., *Macromolecules*, **9**, 267 (1976).
7268. Yokouchi, M., Tadekoro, H., and Chatani, Y., *Macromolecules*, **7**, 769 (1974).

838

7269. Yokouchi, M., Chatani, Y., Tadekoro, H., Teranishi, K., and Tani, H., *Polymer*, **14**, 267 (1973).
7270. Yokoyama, M., Ishihara, H., Iwamoto, R., and Tadekoro, H., *Macromolecules*, **2**, 184 (1969).
7271. Yokoyama, M., Tamamura, H., Atsumi, M., Yoshimura, M., Shirota, Y., and Mikawa, H., *Macromolecules*, **8**, 101 (1975).
7272. Yoon, D. Y., Chang, C., and Stein, R. S., *J. Polym. Sci. A2*, **12**, 2091 (1974).
7273. Yoon, D. Y., and Flory, P. J., *Polymer*, **16**, 645 (1975).
7274. Yoon, D. Y., and Flory, P. J., *Macromolecules*, **9**, 294, 299 (1976).
7275. Yoon, D. Y., and Flory, P. J., *Polymer*, **18**, 509 (1977).
7276. Yoon, D. Y., and Flory, P. J., *Macromolecules*, **10**, 562 (1977).
7277. Yoon, D. Y., Sundararajan, P. R., and Flory, P. J., *Macromolecules*, **8**, 776 (1975).
7278. Yoon, D. Y., Suter, U. W., Sundararajan, P. R., and Flory, P. J., *Macromolecules*, **8**, 784 (1975).
7279. Yoon, D. Y., and Stein, R. S., *J. Polym. Sci. A2*, **12**, 735 (1974).
7280. Yoon, D. Y., and Stein, R. S., *J. Polym. Sci. A2*, **12**, 763 (1974).
7281. Yoshida, H., and Rånby, B., *Acta Chem. Scand.*, **19**, 72 (1965).
7282. Yoshida, S., Sakiyama, M., and Seiki, S., *Polymer, J.*, **1**, 573 (1970).
7283. Yoshimoto, S., Akana, Y., Kimura, A., Hirata, H., Kusabayashi, S., and Kikawa, H., *Chem. Commun.*, **1969**, 987.
7284. Yoshino, T., Kikuchi, Y., and Komiyama, J., *J. Phys. Chem.*, **70**, 1059 (1966).
7285. Yoshino, T., and Komiyama, J., *J. Polym. Sci. B*, **3**, 311 (1965).
7286. Yoshino, T., and Shinomiya, M., *J. Polym. Sci. A1*, **3**, 2811 (1965).
7287. Yoshioka, H., Matsumoto, H., Uno, S., and Higashide, F., *J. Polym. Sci. A1*, **14**, 1331 (1976).
7288. Yoshizaki, T., and Yamakawa, H., *Macromolecules*, **10**, 359 (1977).
7289. Young, A. T., *Appl. Opt.*, **8**, 2431 (1969).
7290. Young, J. W., and Christien, G. D., *Anal. Chem.*, **45**, 1296 (1973).
7291. Young, M. A., and Pysh, E. S., *Macromolecules*, **6**, 790 (1973).
7292. Young, R. J., *Polymer*, **16**, 450 (1975).
7293. Yphantis, D. A., *J. Phys. Chem.*, **63**, 1742 (1959).
7294. Yphantis, D. A., *Ann. N.Y. Acad. Sci.*, **88**, 586 (1960).
7295. Yphantis, D. A., *Biochemistry*, **3**, 297 (1964).
7296. Yu, C. U., and Mark, J. E., *Macromolecules*, **6**, 751 (1973).
7297. Yu, C. U., and Mark, J. E., *Macromolecules*, **7**, 229 (1974).
7298. Yuen, H. K., and Kinsinger, J. B., *Macromolecules*, **7**, 329 (1974).
7299. Yuki, H., Ohta, K., Okamoto, Y., and Hatada., *J. Polym. Sci.*, **15**, 589 (1977).

7300. Zachmann, H. G., *Z. Naturforsch.*, **19a**, 1397 (1964).
7301. Zachmann, H. G., *Kolloid Z. Z. Polym.*, **216**, 180 (1967).
7302. Zachmann, H. G., *J. Polym. Sci. C*, **43**, 111 (1973).
7303. Zachmann, H. G., and Gölz, W., *J. Polym. Sci. C*, **42**, 693 (1973).
7304. Zachmann, H. G., and Stuart, H. A., *Makromol. Chem.*, **41**, 131 (1960).
7305. Zachmann, H. G., and Stuart, H. A., *Makromol. Chem.*, **44/46**, 522 (1961).
7306. Zafar, M. M., and Mahmood, R., *Makromol. Chem.*, **175**, 903 (1974).
7307. Zahn, H., Rathgeber, P., *Melliand Textilber*, **34**, 749 (1953).
7308. Zambelli, A., Bajo, G., and Rigamonti, E., *Makromol. Chem.*, **179**, 1249 (1978).
7309. Zambelli, A., Dorman, D. E., Brewster, A. I. R., and Bovey, F. A., *Macromolecules*, **6**, 925 (1973).
7310. Zambelli, A., and Gatti, G., *Macromolecules*, **11**, 485 (1978).
7311. Zambelli, A., Gatti, G., Sacchi, C., Crain, W. O., and Roberts, J. D., *Macromolecules*, **4**, 475 (1971).
7312. Zambelli, A., Giongo, M. G., and Natta, G., *Makromol. Chem.*, **112**, 183 (1968).
7313. Zambelli, A., Locatelli, P., Bajo, G., and Bovey, F. A., *Macromolecules*, **8**, 687 (1975).
7314. Zambelli, A., and Segre, A., *J. Polym. Sci. B*, **6**, 473 (1968).
7315. Zambelli, A., Segre, A., Farina, M., and Natta, G., *Makromol. Chem.*, **110**, 1 (1967).
7316. Zeigler, I., Freund, L., Benoit, H., and Kern, H., *Makromol. Chem.*, **37**, 217 (1960).
7317. Zeitler, V. A., and Brown, C. A., *Anal. Chem.*, **29**, 1904 (1957).
7318. Zerbi, G., Ciampelli, F., and Tramboni, V., *J. Polym. Sci. C*, **7**, 141 (1984).
7319. Zerbi, G., and Hendra, P. J., *J. Mol. Spectr.*, **27**, 17 (1968).

7320. Zerbi, G., and Sacchi, M., *Macromolecules*, **6**, 692 (1973).
7321. Zetta, L., and Gatti, G., *Macromolecules*, **5**, 535 (1972).
7322. Zharkov, V. V., *Zav. Lab.*, **32**, 436 (1966).
7323. Zhurkov., S. N., and Korsukov, V. E., *J. Polym. Sci. A2*, **12**, 385 (1974).
7324. Zhurkov, S. N., Vettegren, V. I., Korsukov, V. E., and Novak, I. I., *Fiz. Tverd. Tela*, **11**, 190 (1969).
7325. Zhurkov, S. N., Vettegren, V. I., Novak, I. I., and Kashincheva, K. N., *Dokl. Akad. Nauk USSR*, **176**, 623 (1967).
7326. Ziabicki, A., *Macromolecules*, **7**, 501 (1974).
7327. Ziegel, K. D., and Eirich, F. R., *J. Polym. Sci. A2*, **12**, 1127 (1974).
7328. Zielinski, M., and Kryszewski, M., *J. Electrostatics*, **3**, 69 (1977).
7329. Zielinski, M., and Kryszewski, M., *Phys. Status Solid. A*, **42**, 305 (1977).
7330. Zielinski, M., Swiderski, T., and Kryszewski, M., *Polymer*, **19**, 883 (1978).
7331. Zimm, B. H., *J. Chem. Phys.*, **14**, 164 (1946).
7332. Zimm, B. H., *J. Chem. Phys.*, **16**, 1093, 1099 (1948).
7333. Zimm, B. H., Crothers, D. M., *Proc. Nat. Acad. Sci. U.S.A.*, **48**, 905 (1962).
7334. Zimm, B. H., and Kilb, R. W., *J. Polym. Sci.*, **37**, 19 (1959).
7335. Zimm, B. H., and Myerson, I., *J. Amer. Chem. Soc.*, **68**, 911 (1946).
7336. Zimm, B. H., Stein, R. S., and Doty, P., *Polymer Bull.*, **1**, 90 (1945).
7337. Zimm, B. H., and Stockmayer, W. H., *J. Chem. Phys.*, **17**, 230 (1949).
7338. Zimm, B. H., and Stockmayer, W. H., *J. Chim. Phys.*, **17**, 1301 (1949).
7339. Zimmermann, H., and Kolbig, C., *Faserforsch. Textiltechn.*, **18**, 536 (1967).
7340. Zimmermann, H., and Tryonadt, A., *Faserforsch. Textiltechn.*, **18**, 487 (1967).
7341. Zoller, P., *Polymer*, **17**, 167 (1976).
7342. Zoller, P., *J. Appl. Polym. Sci.*, **22**, 633 (1978).
7343. Zoller, P., *J. Polym. Sci. A2*, **16**, 1261 (1978).
7344. Zott, H., and Heusinger, H., *Macromolecules*, **8**, 182 (1975).
7345. Zubov, Y. A., and Tsvankin, D. Y., *Vyskomol. Soedin.*, **7**, 1848 (1965).
7346. Zupančič, I., Lahajnar, G., Blinc, R., Reneker, D. H., and Peterlin, A., *J. Polym. Sci. A2*, **16**, 1399 (1978).
7347. Zweig, A., and Henderson, W. A., Jr, *J. Polym. Sci.*, **13**, 717 (1975).
7348. Zwijnenburg, A., and Pennings, A. J., *Kolloid Z. Z. Polym.*, **253**, 452 (1975).
7349. Zwijnenburg, A., and Pennings, A. J., *J. Polym. Sci. A2*, **14**, 339 (1976).
7350. Zymonds, J., Santee, E., Jr, and Harwood, H. J., *Macromolecules*, **6**, 129 (1973).

Index

X-ray diffraction analysis—*continued*
  Bragg equation, 490
  coherently scattering, 490
  Compton scattering, 490
  cross-section, 490
  crystal rotation method, 496
  Debye–Scherrer technique for, 494
  degree of crystallinity from X-ray diffraction, 507
  diffraction, 490
  diffraction pattern, 493
  diffuse X-ray diffraction effect, 490
  dual wavelength X-ray diffractometry, 497
  fibre-diffraction technique, 496
  fibre pattern, 496
  geometry of diffraction, 490
  goniometer, 501
  incoherently scattering, 490
  intensity region, 492
  lattice factor, 492
  Laue technique for, 494, 502
  linear absorption coefficient, 489
  mass absorption coefficient, 489
  Miller indexes, 490
  properties of X-rays, 488, 489
  reciprocal lattice, 491
  sample preparation for, 502

X-ray diffraction analysis—*continued*
  scattering, 490
  small-angle X-ray camera, 499
  small-angle X-ray diffraction (scattering) (SAXS), 497, 507
  sphere of reflection, 492
  structure factor, 492
  wide-angle X-ray diffraction (scattering) (WAXS), 493, 503
  X-ray cameras, 498
  X-ray diffractometer, 500
  X-ray spectrum, 488
  X-rays, 488

Young equation, 603
Young's modulus, 601

Z average molar masses, 58
Z average molecular weights, 58
Z average molecular weight determination, 113
Zeeman splitting for electron, 333
Zeeman splitting for proton, 302
Zimm equation, 194
Zimm method, 194
Zimm plot, 36, 196